LIVERPOOL
JOHN MOORES UNIVERSITY
AVRIL ROBARTS LRC
TITHEBARN STREET
LIVERPOOL L2 2ER
TEL. 0151 231 4022

Selected Papers on
Foundations of Linear Elastic Fracture Mechanics

SEM Classic Papers
Volume CP 1

SPIE Milestone Series
Volume MS 137

Selected Papers on Foundations of Linear Elastic Fracture Mechanics

R. J. Sanford, Editor
University of Maryland, College Park
Dept. of Mechanical Engineering

Brian J. Thompson
General Editor, SPIE Milestone Series

THE SOCIETY FOR EXPERIMENTAL MECHANICS
Bethel, Connecticut USA

SPIE OPTICAL ENGINEERING PRESS

SPIE—The International Society for Optical Engineering
Bellingham, Washington USA

Library of Congress Cataloging-in Publication Data

Selected papers on foundations of linear elastic fracture mechanics / R.J. Sanford, editor.
  p. cm. — (SEM classic papers ; v. CP 1) (SPIE milestone series ; v. MS 137)
  Includes bibliographical references and index.
  ISBN 0-8194-2620-2 (SPIE) ISBN 0-912053-55-0 (SEM) (alk. paper)
  1. Fracture mechanics. 2. Elastic solids. 3. Elastoplasticity. I. Sanford, R.J. II. Series. III. Series: SPIE milestone series : v. MS 137.
  TA409.S434 1997   97-20467
  620.1'126—dc21   CIP

The Society for Experimental Mechanics
ISBN 0-912053-55-0
SPIE—The International Society for Optical Engineering
ISBN 0-8194-2620-2

Copublished by

The Society for Experimental Mechanics (SEM)
7 School Street, Bethel, CT 06801 USA
Telephone 203 790 6373 (Eastern Time)• Fax 203 790 4472
http://www.sem.org/

SPIE—The International Society for Optical Engineering
P.O. Box 10, Bellingham, Washington 98227-0010 USA
Telephone 360 676 3290 (Pacific Time) • Fax 360 647 1445
http://www.spie.org/

This book is a compilation of outstanding papers selected from the world literature on optical and optoelectronic science, engineering, and technology. SPIE Optical Engineering Press and SEM acknowledge with appreciation the authors and publishers who have given their permission to reprint material included in this volume. An earnest effort has been made to contact the copyright holders of all papers reproduced herein.

Copyright © 1997 The Society for Experimental Mechanics (SEM) and The Society of Photo-Optical Instrumentation Engineers (SPIE)

Copying of SEM- and SPIE-copyrighted material in this book for internal or personal use, or the internal or personal use of specific clients, beyond the fair use provisions granted by the U.S. Copyright Law is authorized by SEM and SPIE subject to payment of copying fees. The Transactional Reporting Service base fee for this volume is $10.00 per SEM- or SPIE-copyrighted article and should be paid directly to the Copyright Clearance Center (CCC), 222 Rosewood Drive, Danvers MA 01923. Other copying for republication, resale, advertising, or promotion or any form of systematic or multiple reproduction of any SEM- or SPIE-copyrighted material in this book is prohibited except with permission in writing from the publisher. The CCC fee code for users of the Transactional Reporting Service is 0-8194-2620-2/97/$10.00.

Readers desiring to reproduce materials contained herein copyrighted by other than SEM or SPIE must contact the copyright holder for appropriate permissions.

Printed in the United States of America

Introduction to the Milestone Series

There is no substitute for reading the original papers on any subject even if that subject is mature enough to be critically written up in a textbook or a monograph. Reading a well-written book only serves as a further stimulus to drive the reader to seek the original publications. The problems are, which papers, and in what order?

As a serious student of a field, do you really have to search through all the material for yourself, and read the good with the not-so-good, the important with the not-so-important, and the milestone papers with the merely pedestrian offerings? The answer to all these questions is usually yes, unless the authors of the textbooks or monographs that you study have been very selective in their choices of references and bibliographic listings. Even in that all-too-rare circumstance, the reader is then faced with finding the original publications, many of which may be in obscure or not widely held journals.

From time to time and in many disparate fields, volumes appear that are collections of reprints that represent the milestone papers in the particular field. Some of these volumes have been produced for specific topics in optical science, but as yet no systematic set of volumes has been produced that covers connected areas of optical science and engineering.

The editors of each individual volume in the series have been chosen for their deep knowledge of the world literature in their fields; hence, the selection of reprints chosen for each book has been made with authority and care.

On behalf of SPIE, I thank the individual editors for their diligence, and we all hope that you, the reader, will find these volumes invaluable additions to your own working library.

Brian J. Thompson

Selected Papers on
Foundations of Linear Elastic Fracture Mechanics

Contents

xi **Preface**
R.J. Sanford

xiii **Introduction**
R.J. Sanford

Section One
Precursors to Modern Fracture Mechanics

3 **Stresses in a plate due to the presence of cracks and sharp corners**
C.E. Inglis *(Transactions of the Institution of Naval Architects* 1913)

18 **Bearing pressures and cracks**
H.M. Westergaard *(Journal of Applied Mechanics* 1939)

23 **The distribution of stress in the neighbourhood of a crack in an elastic solid**
I.N. Sneddon *(Proceedings of the Royal Society A* 1946)

55 **The distribution of stresses in the neighbourhood of a flat elliptical crack in an elastic solid**
A.E. Green, I.N. Sneddon *(Proceedings of the Cambridge Philosophical Society* 1950)

60 **The phenomena of rupture and flow in solids**
A.A. Griffith *(Philosophical Transactions of the Royal Society of London* 1921)

96 **The theory of rupture**
A.A. Griffith (in *Proceedings of the First International Conference for Applied Mechanics,* C.B. Biezeno and J.M. Burgers, editors, 1925)

105 **Direct measurements of the surface energies of crystals**
John J. Gilman *(Journal of Applied Physics* 1960)

116 **Fracture dynamics**
G.R. Irwin (in *Fracturing of Metals* 1948)

136 **Critical energy rate analysis of fracture strength**
G.R. Irwin, J.A. Kies *(Welding Journal, Research Supplement* 1954)

142 **Energy criteria of fracture**
E. Orowan *(Welding Journal, Research Supplement* 1955)

	146	**Experiments on brittle fracture of steel plates** D.K. Felbeck, E. Orowan *(Welding Journal, Research Supplement* 1955)
	152	**The propagation of cracks and the energy of elastic deformation** H.F. Bueckner *(Transactions of the ASME* 1958)
Section Two **The Modern Theory of Fracture Mechanics**	161	**On the stress distribution at the base of a stationary crack** M.L. Williams *(Journal of Applied Mechanics* 1957)
	167	**Analysis of stresses and strains near the end of a crack traversing a plate** G.R. Irwin *(Journal of Applied Mechanics* 1957)
	171	**Crack-extension force for a part-through crack in a plate** G.R. Irwin *(Journal of Applied Mechanics* 1962)
	175	**Fracture** G.R. Irwin (in *Handbuch der Physik*, S. Flügge, editor, 1958)
	215	**A continuum-mechanics view of crack propagation** G.R. Irwin, A.A. Wells *(Metallurgical Reviews* 1965)
	267	**Plastic zone near a crack and fracture toughness** G.R. Irwin (in *Mechanical and Metallurgical Behavior of Sheet Materials [Seventh Sagamore Conference]* 1960)
	280	**Yielding of steel sheets containing slits** D.S. Dugdale *(Journal of the Mechanics and Physics of Solids* 1960)
	285	**The mathematical theory of equilibrium cracks in brittle fracture** G.I. Barenblatt (in *Advances in Applied Mechanics*, H.L. Dryden and Th. von Kármán, editors, 1962)
	360	**On the Westergaard method of crack analysis** G.C. Sih *(International Journal of Fracture Mechanics* 1966)
	363	**On the modified Westergaard equations for certain plane crack problems** J. Eftis, H. Liebowitz *(International Journal of Fracture Mechanics* 1972)
	373	**A critical re-examination of the Westergaard method for solving opening-mode crack problems** R.J. Sanford *(Mechanics Research Communications* 1979)
	379	**Influence of non-singular stress terms and specimen geometry on small-scale yielding at crack tips in elastic-plastic materials** S.G. Larsson, A.J. Carlsson *(Journal of the Mechanics and Physics of Solids* 1973)
	393	**Limitations to the small scale yielding approximation for crack tip plasticity** J.R. Rice *(Journal of the Mechanics and Physics of Solids* 1974)
	403	**Biaxial load effects in fracture mechanics** H. Liebowitz, J.D. Lee, J. Eftis *(Engineering Fracture Mechanics* 1978)
	424	**A novel principle for the computation of stress intensity factors** H.F. Bueckner *(Zeitschrift für Angewandte Mathemetik und Mechanik* 1970)
	442	**Some remarks on elastic crack-tip stress fields** James R. Rice *(International Journal of Solids and Structures* 1972)
Section Three **Subcritical Crack Growth Analysis**	453	**Recent observations on fatigue failure in metals** W.A. Wood (in *Symposium on Basic Mechanisms of Fatigue* 1958)
	465	**A two stage process of fatigue crack propagation** P.J.E. Forsyth (in *Proceedings of the Crack Propagation Symposium* 1962)
	484	**Mechanisms and theories of fatigue** Campbell Laird (in *Fatigue and Microstructure* 1979)

539 **A rational analytic theory of fatigue**
P.C. Paris, M.P. Gomez, W.E. Anderson *(The Trend in Engineering* 1961)

545 **A critical analysis of crack propagation laws**
P. Paris, F. Erdogan *(Journal of Basic Engineering* 1963)

552 **Numerical analysis of crack propagation in cyclic-loaded structures**
R.G. Forman, V.E. Kearney, R.M. Engle *(Journal of Basic Engineering* 1967)

558 **Spectrum loading and crack growth**
O.E. Wheeler *(Journal of Basic Engineering* 1972)

564 **Fatigue crack closure under cyclic tension**
Wolf Elber *(Engineering Fracture Mechanics* 1970)

574 **The significance of fatigue crack closure**
Wolf Elber (in *Damage Tolerance in Aircraft Structures* 1971)

587 **Four lectures on fatigue crack growth**
J. Schijve *(Engineering Fracture Mechanics* 1979)
- 589 I. Fatigue crack growth and fracture mechanics
- 600 II. Fatigue cracks, plasticity effects and crack closure
- 613 III. Fatigue crack propagation, prediction and correlation
- 623 IV. Fatigue crack growth under variable-amplitude loading

639 **Author Index**
641 **Subject Index**

Preface

This collection of papers consists of those articles which provide fundamental contributions to the development of the theory of linear elastic fracture mechanics. The book is divided into three sections. The first section, "Precursors to Modern Fracture Mechanics," includes historical articles from the period 1913 to 1956 which influenced the thought of later researchers or predicted the development of a theory of brittle fracture long before its actual formulation.

The next section, "The Modern Theory of Fracture Mechanics," focuses on the fundamental papers which now constitute the mathematical theory of LEFM. The "birthdate" of modern fracture mechanics (if there is such a thing) is generally recognized as 1957, the year in which Irwin published his landmark paper in the *Journal of Applied Mechanics* on the state of stress surrounding a stationary crack. Starting from this date and ending in the early 1980s, the fundamental principles of what is now called linear elastic fracture mechanics were developed.

The final section of this volume, "Subcritical Crack Growth Analysis," is devoted to the development of the theory of subcritical crack propagation (primarily fatigue) within the framework of fracture mechanics theory. Although the time frame overlaps that of the middle section of this volume, the focus on the behavior of materials and structures at loads below the critical value for sudden fracture sets this section apart from the preceding one.

The 38 papers in this volume, encompassing over 630 pages, were selected with a special emphasis on those historic papers which are not generally available. Not included here are those papers devoted to the determination of the stress intensity factor; these papers are collected together in a companion volume in this series, *Selected Papers on Crack Tip Stress Fields* (SEM Classic Papers Vol. CP2; SPIE Milestone Series Vol. MS 138).

Also specifically excluded from this volume were those papers which developed the concepts needed for extension of fracture mechanics beyond the linear theory. The justification here is twofold. First, there needs to be a logical cutoff in coverage in order to limit the number of pages to the maximum permitted for one volume. In addition, the fundamental papers on the nonlinear theory of fracture mechanics logically form the core of a possible third volume in this series devoted to the theory of elasto-plastic fracture mechanics, which is currently in the conceptual stage.

The selection of papers to be included within the topic of this volume while confining the collection to a single volume has proven to be a formidable task. Fortunately, I have had the guidance and cooperation of distinguished leaders in the field of fracture mechanics. I wish to express my appreciation to Drs. Albert Kobayashi, J.C. Newman, Jr., and George R. Irwin for their review of my initial list of potential candidates for this volume and their insightful comments. I also wish to thank my colleagues and former students now at other institutions for their assistance (and that of their respective libraries) in locating some of the papers included in this collection. Finally, and most importantly, we, the researchers and readers in the field of fracture mechanics, express our thanks to the authors and publishers of the papers contained herein for their contributions to the body of knowledge and their permission to reprint their works.

R.J. Sanford
Professor Emeritus
University of Maryland
May 1997

Introduction

Section One: Precursors to Modern Fracture Mechanics

The first papers in this section focus on an understanding of the stress field surrounding crack tips. Inglis (p. 3) in 1913 solved exactly the problem of the state of stress in a sheet containing an elliptical hole and then considered the consequences of letting the local curvature along the major axis of the ellipse approach zero to represent a crack. (Editor's note: There is a typographic error in the equation for shear stress under biaxial load; see Griffith, p. 60, for the correct form.) This paper is generally regarded as the first attempt to look specifically at the stress field surrounding crack tips. The paper is not only significant from that perspective but, more interestingly, gives insights into fracture behavior that the author gleans from his mathematical results.

As a historical sidelight it might be noted that, had the designers and fabricators of the RMS Titanic (launched in 1911) been aware of the warning of Inglis, who was himself a naval architect, concerning the brittle fracture behavior of cold-punched rivet holes, the Titanic disaster might never had happened.

The next three papers also examine the stresses at crack tips. Westergaard (p. 18), using a new complex variable formulation, treated the two-dimensional problem, whereas Sneddon (p. 23) and later Green and Sneddon (p. 55) solved the flat crack problem in three dimensions. At the time these papers were published, they produced only minimal interest. Their true significance had to wait until the late 1950s when their results were incorporated into the concepts of fracture mechanics.

In 1921 Alan Griffith (p. 60) investigated the observation that "old" glass had a dramatically lower fracture strength than freshly drawn glass rods. He attempted to explain this discrepancy by postulating the existence of microscopic cracks formed probably due to moisture and proposed a weakest link theory to explain the fracture strength. In a footnote at the end of the paper Griffith explains that there is an error in the mathematics but that the error does not affect the fundamental conclusion of his proposed theory.

In his paper presented at the First International Congress for Applied Mechanics in Delft in 1924 (p. 96), Griffith explained that the expression for strain energy in the earlier paper incorrectly represents the Poisson ratio's contribution and briefly discusses the correct formulation. However, the Poisson ratio's influence for plane strain is again stated incorrectly. Sneddon (p. 23) provided the correct expression (without proof) in the Introduction to his paper. For the reader interested in the complete derivation of Griffith's equations, the 1967 paper by Sih and Liebowitz[1] provides the missing steps.

Griffith's theory of brittle fracture uses as its critical parameter the surface energy which, although a fundamental property of materials, is most difficult to measure directly. John Gilman's measurements (p. 105) in 1960 of the surface energy using the cleavage method are highly regarded within the glass and ceramics community.

Twenty-five years later, the arguments of Griffith were reexamined by Irwin (pp. 116 and 136) and Orowan (pp. 142 and 146) in terms of the energy of propagation of cracks. These pioneering fracture mechanics practitioners argued that the Griffith theory could, in principle, be extended to brittle-behaving ductile materials if the surface energy were replaced by a more appropriate parameter germane to these materials. The final paper in this section by Bueckner (p. 152) provides a rigorous mathematical foundation for the strain energy released as that parameter.

Section Two: Modern Theory of Fracture Mechanics

The year 1957 proved to be a pivotal year in the history of fracture mechanics. In that year both George Irwin and Max Williams published in the same journal (but different issues) detailed analyses of the stress distribution at the end of a crack. Williams (p. 161) used a classical formulation in real polar coordinates, whereas Irwin (p. 167) used for his analysis the complex variable method proposed by Westergaard (p. 18). (Editor's note: the Williams paper contains a large number of typographical errors, including Eqs. [4], [8], [9], [33], and [34]. The key parts of Williams's analysis are corrected in Ref. 2.)

These two papers and a third by Irwin (p. 171) for a part-through crack on the surface of a plate based on the three-dimensional results of Green and Sneddon (p. 55) form the basis for all of modern fracture mechanics theory by demonstrating the form invariance of the near-field equations for all geometries. This result allowed Irwin and others to shift the focus of attention from the details of the stress analysis to the magnitude of the stress field at the crack tip as represented by the parameter K. This parameter, called the stress intensity factor, became the focus of attention throughout the fracture mechanics community in the 1960s. In Irwin's part-through crack paper (p. 171) he also introduced the concepts of surface and plastic zone correction factors into the basic equations.

In the 1958 encyclopedia article in the Handbuch der Physik (in English), Irwin (p. 175) placed in perspective those mathematical concepts which led to the development of a fracture theory based on the crack-tip singularity concept. In a later (1965) review with Alan Wells (p. 215), the continuum mechanics view of fracture theory was reexamined but with more attention to the mechanisms of fracture than to its mathematical details.

At this point I should note that progress in the development of fracture mechanics theory could not have proceeded without the large number of specific solutions for the elasticity parameter *K*, generated mostly by NASA and the academic community. However, this volume is focused on the concepts leading to the development of the field of fracture mechanics into a scientific discipline in its own right, so I have purposefully excluded all of these landmark papers. Rather, a separate volume in this series entitled *Selected Papers on Crack Tip Stress Fields* (SEM Classic Papers Vol. CP2; SPIE Milestone Series Vol. MS 138) treats this topic in detail. This companion volume is not a handbook collection of *K* solutions but instead contains those papers which present new analytical, numerical, or experimental approaches for determining the stress intensity factor.

Fundamental to the extension of linear elastic fracture mechanics beyond fracture of truly brittle material to include materials of limited ductility is the concept of small scale yielding (SSY). The SSY approximation permits the inclusion of a small plastic zone at the crack tip which is fully enclosed by sufficient elastic material into the mathematical formulation of the near-field equations without resorting to an elasto-plastic analysis.

In the next three papers in this collection the notational crack tip approach of Irwin (p. 267) and the strip zone models of Dugdale (p. 280) and Barenblatt (p. 285) describe the concept. In its strictest interpretation, the small scale yielding approximation requires that the actual processes which govern crack extension occur within a small region, called the fracture process zone, of the plastic zone and that the plastic zone is itself small compared to the region around the crack tip, in which the near-field equations dominate the state of stress. In reality, it is difficult to adhere to these restrictions for materials and forces of practical interest, and the requirements of the SSY argument are often compromised to some extent. When these limits are stretched beyond practical bounds or the plastic zone extends to a boundary, the realm of a newer branch of fracture theory, elasto-plastic fracture mechanics, is employed.

Put in its simplest terms, the strength of the fracture mechanics approach to the analysis of brittle fracture is the assertion that a material's fracture resistance can be characterized by a single parameter, the fracture toughness K_c, regardless of the geometry (within certain restrictions), in a manner analogous to the characterization of yield behavior by the material's yield strength. This argument has its origins in the papers of Irwin and Williams in 1957 (pp. 167 and 161).

In the years immediately following these landmark papers, the Westergaard method of stress analysis was widely used to develop expressions for the stress intensity factors needed for the design of fracture toughness testing specimens. In a frequently cited paper (p. 360), George Sih reported an apparent error in Westergaard's formulation and proposed a "modified Westergaard formulation" which introduced an additional constant ($=A$) into the analysis. Westergaard was probably aware of this limitation in that he states explicitly in his paper that the analysis is valid only for "a limited but important class of problems."

Later, Eftis and Liebowitz (p. 363) provided a detailed mathematical history in support of Sih's observation and examined the consequences of incorporating this additional term into the near-field description of the state of stress, strain, and displacement. A major limitation of both of these criticisms of Westergaard's work are that they themselves are limited in scope. In 1978 I demonstrated that both the original Westergaard formulation and the modified Westergaard formulation of Sih (and Eftis and Liebowitz) are special cases of a more general representation of the two-dimensional crack problem in Westergaard notation. This analysis (p. 373) produced a description which is fully compatible with the classical complex variable formulation popularized by Muskhelishvili.[3]

The debate over the correct formulation of the Westergaard equations was of more than academic interest. The modified formulation of Sih predicts an additional term ($=2A$) in the near-field equations. This additional term corresponds to an arbitrary uniform stress parallel to the crack line. In fact, the existence of this additional term was generally known long before Sih's analysis. Williams's 1957 solution (p. 161) includes this term ($=c_{12}$,) but the author mistakenly argued that it must be zero for finite geometries. In contrast, Irwin[4] in 1958 in a discussion of the experiments of Wells and Post[5] demonstrated that this additional term ($=\sigma_{ox}$) was necessary to explain the tilt in the photoelastic fringe patterns at a crack tip. These observations beg the question: is a two-parameter theory of fracture necessary to account for variations in the σ_x stress component in different geometries?

The influence of the stress parallel to the crack line on fracture behavior has received limited but important scrutiny. Larsson and Carlsson (p. 379) in 1972, using the finite element method, demonstrated that different specimen geometries generate variations in plastic zone size and shape for the same nominal stress state (as characterized by the applied K level). In the following year J.R. Rice (p. 393) examined the impact of the Larsson and Carlsson work on the small scale yielding approximation of linear elastic fracture mechanics and introduced yet another symbol ($=T$ stress) to describe the stress parallel to the crack line. Rice observed that the influence of this stress on the crack opening displacement and the J-integral (an equivalent energy-based description of the fracture mechanics principle) were minimal and, accordingly, the one-parameter representation was sufficiently accurate for engineering applications.

Liebowitz, Lee, and Eftis (p. 403) hold an opposing view and provided experimental results to support their position. The debate over the significance of this additional term ($=c_{12}=\sigma_{ox}=2A=T$) in the near-field representation of the stress state is still unresolved. However, there is general agreement that it does influence the anti-plane constraint behavior of materials and, hence, the plane stress to plane strain transition behavior of fracture in metals.

The remaining two papers in this section develop, in vastly different styles, the concept of weight functions for determining the stress intensity factor. Bueckner (p. 424) and Rice (p. 442) presented mathematical treatments of the underlying theory and examples of the application of this important concept for determining the SIF through the use of fundamental solutions for the same geometry.

Section Three: Subcritical Crack Growth Analysis

It is generally recognized within the fracture mechanics community that the vast majority of brittle fractures observed in service are preceded by a period (possibly quite long) of subcritical crack growth. It is somewhat ironic that linear elastic fracture mechanics and its *elastic* stress field parameter K provide highly accurate modeling laws for this type of crack growth, even though the underlying mechanism of subcritical crack growth is a purely *plastic* phenomenon.

The process of fatigue crack growth goes through three distinct stages as the crack advances into the material. Starting with a smooth surface with no apparent cracks or defects, the first stage is that of crack initiation. The first three papers in this section describe the underlying mechanism which transitions a specimen from its defect-free state to one in which an identifiable crack perpendicular to the maximum applied tension is present.

The first two papers, by Wood (p. 453, 1958) and Forsyth (p. 465, 1962) present optical microscope evidence of the role of slipband-generated extrusions and intrusions in the initial formation of cracklike defects at the surface. Forsyth goes on to discuss the transition of these inclined cracks into macro-cracks perpendicular to the applied tension and shows the one-to-one correspondence between fracture striations and crack growth cycles. The more recent paper by Laird (p. 484, 1979) provides a rather complete description of the mechanistic behavior of fatigue crack growth.

Stage II crack growth consists of that portion of the crack growth between initiation and the final stage of rapid crack acceleration followed by gross fracture. It is in this stage that the crack growth per cycle of load can be accurately described by the fracture parameter K. The remaining papers in this collection highlight significant contributions to the development of the empirical relation between subcritical crack growth and linear elastic fracture mechanics.

This fundamental relation was first described by Paris and his co-workers at the University of Washington in 1961 (p. 539). When initially proposed, the use of the stress intensity factor range ΔK as the germane parameter to describe a plasticity-dominated event was not widely accepted. However, as the body of data and the wide range of specimen types which could be accommodated grew, the validity of the "Paris law" became apparent (Paris and Erdogan, p. 545).

The Paris law was proposed as a simple model for the stage II, linear portion of the fatigue crack growth rate versus ΔK data. As more data under varying conditions became available, a legend of "modified" Paris law equations were proposed. I will make no attempt to chronicle all of them here; however, there are just a few which stand out and are named for their proposers.

It had been experimentally observed that changing the mean stress produced a systematic shift in the data fatigue data curves. The Forman law (p. 552) incorporates an additional parameter into the Paris-type equation which compensates for the mean stress effect. Separately, the effect of single-cycle overload to retard the growth of a crack led Wheeler (p. 558) to propose a crack retardation model which takes the form of a multiplier (less than one) to be applied to the unretarded crack growth model. Upon careful examination, both of these effects have their origin in the localized plastic deformation field at the crack tip.

Elber, in two highly significant papers (pp. 564 and 574), described the principal conclusions of his Ph.D. research on the phenomenon of crack closure. He demonstrated that under some conditions the freshly generated crack surface closed upon itself during the unloading half-cycle and that, if only the open portion is considered, all of the load-dependent effects could be embodied in a single model using the "effective ΔK range" as the controlling parameter.

The final paper in this section (and volume) breaks no new ground in the theory of fatigue crack growth. Rather it is a systematic review of the field organized into four lectures by J. Schijve (p. 587). The paper places the various aspects of fatigue crack growth and modeling in perspective and contains ample references for further independent study. For the newcomer to fatigue crack growth, this series is the logical place to start the learning process.

References

1. Sih, G.C., and Liebowitz, H., "On the Griffith Energy Criterion for Brittle Fracture," Int'l J. Solids Structures **3**, 1-22 (1967).
2. Gross, B., Roberts, E., Jr., and Srawley, J.E., "Elastic Displacements for Various Edge-Cracked Plate Specimens," Int'l J. Fracture Mechanics **4**(3), 267-276 (1968).
3. Muskhelishvili, N.I., *Some Basic Problems of the Mathematical Theory of Elasticity*, Noordhoff Ltd., Netherlands (1953).
4. Irwin, G.R., "Discussion: The Dynamic Stress Distribution Surrounding a Running Crack—A Photoelastic Analysis," Proc. Soc. Experimental Stress Analysis **16**(1), 93-96 (1958).
5. Wells, A.A., and Post, D., "The Dynamic Stress Distribution Surrounding a Running Crack—A Photoelastic Analysis," Proc. Soc. Experimental Stress Analysis **16**(1), 69-92 (1958).

Section One
Precursors to Modern Fracture Mechanics

STRESSES IN A PLATE DUE TO THE PRESENCE OF CRACKS AND SHARP CORNERS.

By C. E. INGLIS, Esq., M.A., Fellow of King's College, Cambridge.

[Read at the Spring Meetings of the Fifty-fourth Session of the Institution of Naval Architects, March 14, 1913; Professor J. H. BILES, LL.D., D.Sc., Vice-President, in the Chair.]

PART I.

THE methods of investigation employed for this problem are mathematical rather than experimental, and this mathematical treatment is given in some detail in Part II. of this paper. Part I. is a summary of the more important results and conclusions, and in this part mathematics are kept as far as possible in the background.

The main work of the paper lies in the determination of the stresses around a hole in a plate, the hole being elliptic in form. The results are exact, and are consequently applicable to the extreme limits of form which an ellipse can assume. If the axes of the ellipse are equal, a circular hole is obtained; by making one axis very small the stresses due to the existence of a fine straight crack can be investigated.

The destructive influence of a crack is a matter of common knowledge, and is particularly pronounced in the case of brittle non-ductile materials. This influence is turned to useful account in the process of glass cutting. A fine scratch made on the surface produces such a local weakness to tension that a fracture along the line of the scratch can be brought about by applied forces which produce in the rest of the plate quite insignificant stresses.

In ductile materials some easing off of the local stresses at the end of a crack will be effected by plastic yield of the substance, but for the case of alternating loads the protection against breakdown gained thereby must be limited. A small load tending to open the crack will produce overstrain at its ends. On reversing the load the crack closes again, but not before it has set up some reversed stress at the ends of the crack. In this manner a small alternating load may produce in the material an alternating stress far in excess of its elastic range, and under these circumstances, if a crack has once fairly started, no amount of ductility will prevent it spreading through the substance.

In a paper read before the Sheffield Society of Engineers and Metallurgists in January, 1910, Professor B. Hopkinson made some general observations concerning the stresses around a crack. In this paper Professor Hopkinson did not attempt to compute the intensities of these stresses, but the desirability of doing so he strongly emphasised. The present paper is an endeavour to satisfy this requirement.

Fig. 1 (Plate XXVIII.) shows an elliptic hole in a plate, the major and minor axes being $2a$ and $2b$ respectively. The plate is subjected to a tensile stress of intensity R applied in the direction shown. If the hole is nowhere near the edge of the plate and the material is nowhere strained beyond its elastic limit, a tensile stress occurs at A of the value $R\left[1 + \frac{2a}{b}\right]$, and a compression stress at B of magnitude R. On exploring the plate in the direction A P, the tensile stress across A P, which has the value $R\left[1 + \frac{2a}{b}\right]$ at A, rapidly decreases, and in a short distance attains approximately its normal value R. Advancing along B Q the compression stress across B Q which starts with the value R at B soon changes to a small tensile stress, and this gradually dies out entirely. Stresses in the directions A P, B Q are likewise brought into existence. These radial stresses are zero at A and B. At considerable distances from the edge of the hole the former dies down again to zero, and the latter attains the value R.

The variations in the stresses along and across A B and B Q are shown by the curves of Fig. 2, which are accurately drawn for the case $a = 3b$. An inspection of these curves clearly indicates that the tension at A is by far the greatest stress. In this case it amounts to 7 R. The rapidity with which this stress decreases with the distance from the edge of the hole is also very noticeable. These effects become more marked as the ratio $\frac{a}{b}$ increases.

When $\frac{a}{b} = 100$, the tension at A is 201 times the mean tension.

,, $\frac{a}{b} = 1,000$, ,, ,, 2,001 ,, ,, ,,

The ellipse in this latter case would appear as a fine straight crack, and a very small pull applied to the plate across the crack would set up a tension at the ends sufficient to start a tear in the material. The increase in the length due to the tear exaggerates the stress yet further and the crack continues to spread in the manner characteristic of cracks.

So far we have taken the major axis of the ellipse to be perpendicular to the direction of the pull. If the tension is applied in the direction of the major axis, tensile

PLATE XXVIII

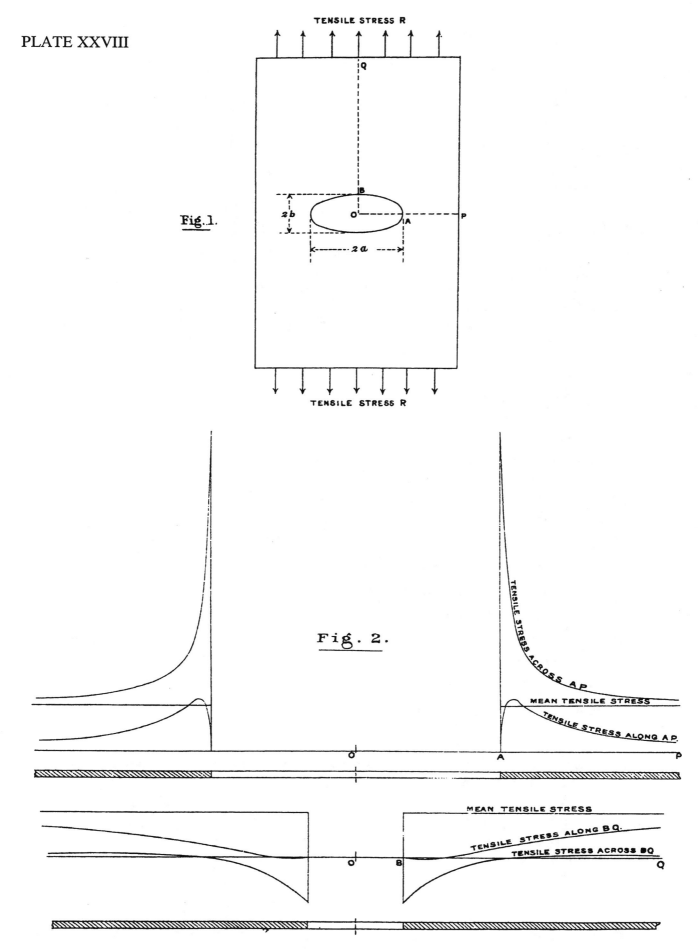

stress of amount $R\left[1+\frac{2b}{a}\right]$ is set up at B, accompanied by a compression stress of magnitude R at A. Hence a crack running in the direction of the pull does not produce great local stress, a conclusion which is almost self-evident.

If the major axis of the ellipse makes an angle θ with the direction of the pull, the tensile stress at the ends of this axis is—

$$R\left[\frac{a}{b} - \cos 2\theta \left(1 + \frac{a}{b}\right)\right]$$

For such a case, however, the greatest tension does not occur exactly at these ends, and the value given above may be considerably exceeded. An example of special importance illustrating this point is given in Fig. 3 (Plate XXVIII.). Here the axis of the ellipse is inclined at an angle of 45° to the direction of the pull.

At A the tensile stress is $R\frac{a}{b}$.

At P the tensile stress is—

$$R\left[1 + \frac{a}{b} + \frac{b}{a}\right]$$

Between A and P the tensile stress reaches a maximum value, and if $\frac{b}{a}$ is fairly small, a good approximation to this maximum value is—

$$R\frac{a}{2b}\left[1 + \frac{\sqrt{2a^2 + 2b^2}}{a - b}\right]$$

At Q there is compression stress of magnitude R.

For the case $a = 3b$—

The tensile stress at A is 3 R;

The tensile stress at P is $4\frac{1}{3}$ R;

The compression stress at Q is R.

The maximum tensile stress which occurs between A and P has the accurate value 4·64 R.

The approximation given above makes it 4·85.

For smaller values of $\frac{b}{a}$ this approximation is yet more accurate.

EXTENSION TO THE CASE OF CRACKS WHICH ARE NOT NECESSARILY ELLIPTIC IN FORM.

Consider an elliptic hole in the plate modified by cutting away the portion shown shaded in Fig. 4 (Plate XXVIII.).

From the point of view of stress, this is equivalent to applying along the boundaries P Q R and S V W distributions of stress, which in each case is an equilibrium system, and which dies out as the points of contact with the ellipse are approached. The application of such a system does not substantially affect the stress outside the actual region where the distribution is applied, and for the case shown the stresses at A and A' will not be altered to any appreciable extent, provided that the change in the boundary does not extend to these ends. In other words, the stresses at the ends of a cavity depend almost entirely upon the length of the cavity and the form of its ends.

If the ends of the cavity are approximately elliptic in form, it is legitimate, in calculating the stresses at these points, to replace the cavity by a complete ellipse having the same total length and end formation. If ρ is the radius of curvature at the ends of the major axis of an ellipse, $b = \sqrt{a\rho}$. This substitution for b gives the tensile stress at the ends of the ellipse in the form—

$$R\left[1 + 2\sqrt{\frac{a}{\rho}}\right]$$

This formula will accordingly apply to a cavity of any shape, the length of the cavity being $2a$ and the ends having a radius of curvature ρ; provided that the cavity near its ends merges smoothly into an ellipse.

Thus for the star-shaped hole represented in Fig. 5 (Plate XXVIII.), in which the ends merge into ellipses in the manner indicated, the tension at A will be $R\left[1 + 2\sqrt{\frac{a}{\rho}}\right]$, where ρ is the radius of curvature at the ends of the projecting arms. The compression stress at B will be R.

CASE OF A SQUARE HOLE WITH ROUNDED CORNERS.

To get the square hole with corners rounded to a radius ρ illustrated in Fig. 6, the cutting away process has to be more extended, and this may ease the stress at A somewhat below the value $R\left[1 + 2\sqrt{\frac{a}{\rho}}\right]$, but the error contained in this formula is in all probability quite small.

To this same degree of accuracy the tensile stress at A for the case illustrated in Fig. 7 is

$$R\sqrt{\frac{l}{\rho}}.$$

At P there is a tensile stress

$$R\left[1 + \sqrt{\frac{l}{\rho}} + \sqrt{\frac{\rho}{l}}\right]$$

At Q there is a compression stress of magnitude R. The greatest tensile stress along the rounded corner occurs at a point between A and P, and, if ρ is small compared with l, its approximate value is

$$\frac{R}{2}\sqrt{\frac{l}{\rho}}\left[1 + \frac{\sqrt{2l + 2\rho}}{\sqrt{e} - \sqrt{\rho}}\right].$$

PLATE XXVIII (continued)

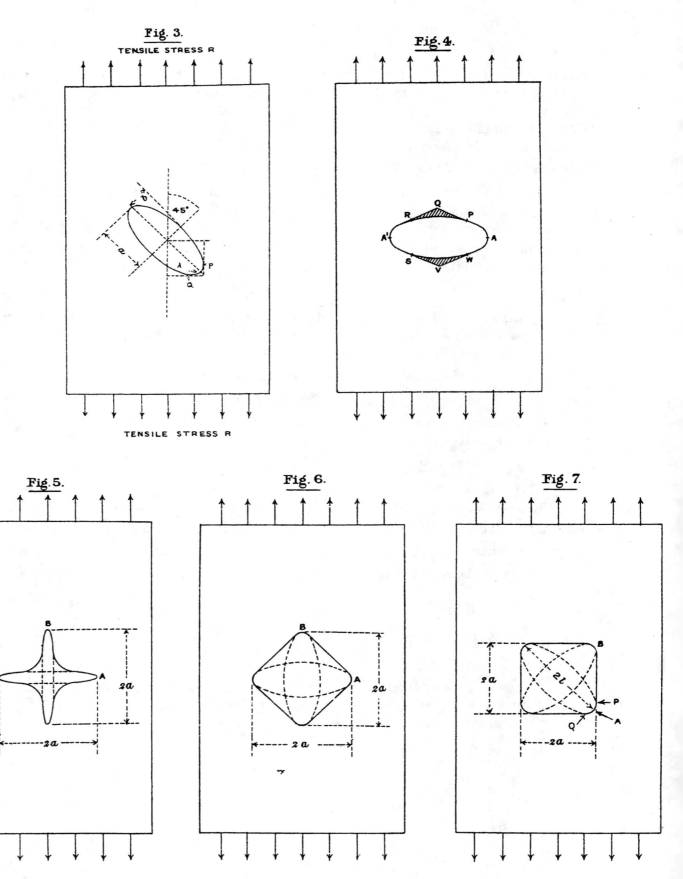

EXTENSION TO THE CASE OF A CRACK STARTING FROM THE EDGE OF A PLATE.

Consider once again the plate illustrated in Fig. 1 (Plate XXVIII.). If one half is isolated from the other, the notched plate shown in Fig. 8 is the result. This plate, in addition to a horizontal pull of intensity R, is subjected to normal actions distributed along B Q, B' Q', according to one of the curves shown in Fig. 2. If this latter distribution can be neutralised, the stresses in a plate with an elliptic notch in its top edge, subjected to longitudinal tension only, can be deduced.

The tensile stress at A is accordingly $R\left[1 + 2\sqrt{\frac{a}{\rho}}\right]$, less the tension due to the distribution of normal stress along the top edge. When the notch is narrow and deep, these actions along the top edge produce little or no effect at A. If the pressure at B were maintained all along B Q, the tension produced at B would only amount to R.

Actually this pressure dies out rapidly, and since the total action on B Q amounts to a zero force, the tensile stress set up at A will be much less than R; it may very likely vanish altogether. At any rate, we may conclude that these actions along the top edge are unimportant, so far as the stress at A is concerned, and for the case of a narrow, deep, elliptic notch, the tensile stress at the bottom of the notch lies between the limits—

$$R\left[1 + 2\sqrt{\frac{a}{\rho}}\right] \text{ and } R\left[2\sqrt{\frac{a}{\rho}}\right],$$

where a is the depth of the notch and ρ is the radius of curvature at its end. The former limit—

$$R\left[1 + 2\sqrt{\frac{a}{\rho}}\right]$$

is probably the closer approximation, but for the case of a deep sharp-ended notch, the difference between these limits is relatively unimportant.

EXTENSION TO THE CASE OF A NOTCH WHICH IS NOT NECESSARILY ELLIPTIC IN FORM.

Having deduced the tensile stress at the bottom of a narrow notch of elliptic form, the formula can be seen to have a wider application.

Thus, in Fig. 9 (Plate XXVIII.), the shaded portions can be removed without appreciably modifying the tension at A, which will continue to have the value—

$$R\left[1 + 2\sqrt{\frac{a}{\rho}}\right],$$

where a is the depth of the notch and ρ the radius of curvature at its end. The argument which leads to this conclusion is identical with that given in connection with Fig. 3, and calls for no repetition.

By similar reasoning the tensile stress at the point A for the cases illustrated in Fig. 10 (Plate XXVIII.) will be—

$$R\left[1 + 2\sqrt{\frac{a}{\rho}}\right]$$

For the rectangular notch represented in Fig. 11—

The tensile stress at A is

$$R\sqrt{\frac{l}{\rho}}.$$

The tensile stress at P is

$$R\left[1 + \sqrt{\frac{l}{\rho}} + \sqrt{\frac{\rho}{l}}\right].$$

The compression stress of Q is R.

The greatest tensile stress along the rounded corner occurs at some point between A and P, and, if ρ is small compared with l, a good approximation to its value is

$$\frac{R}{2}\sqrt{\frac{l}{\rho}}\left[1 + \frac{\sqrt{2l} + 2\rho}{\sqrt{l} - \sqrt{\rho}}\right].$$

The last case to be considered is a small crack or notch springing from the side of a hole in the manner shown in Fig. 12 (Plate XXVIII.).

In this case there is a double magnification of stress. The mean stress R is concentrated to the value $R\left[1 + 2\sqrt{\frac{a}{\rho}}\right]$ in the neighbourhood of the crack, and this, again, is magnified to the value

$$R\left[1 + \sqrt{\frac{a}{\rho}}\right]\left[1 + 2\sqrt{\frac{a'}{\rho'}}\right]$$

at the end of the crack, where a and ρ refer to the hole and $a'\rho'$ refer to the crack.

This example offers an explanation of the weakening of a plate which has been punched with holes. Around the edge of the holes fine cracks may be started. These cracks will have every inducement to extend, because the metal round the hole has been rendered hard and brittle by the punch. By the time the crack has extended through this hardened region, it has got such a hold on the plate that no amount of ductility will prevent the crack from advancing. The advisability of removing this hard and probably fractured rim round the edge of a punched hole is very apparent.

PLATE XXVIII (continued)

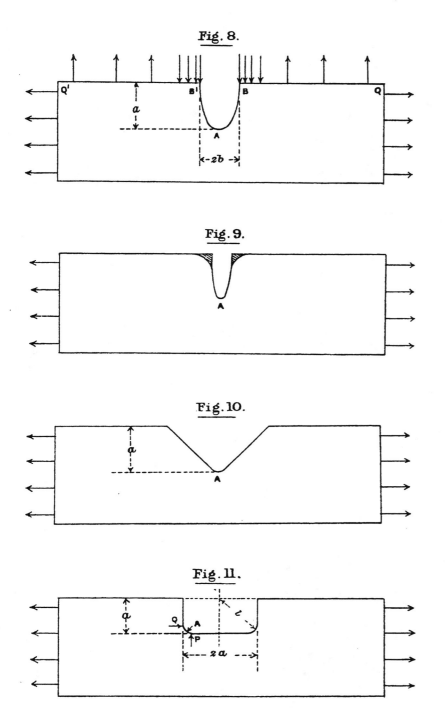

Fig. 8.

Fig. 9.

Fig. 10.

Fig. 11.

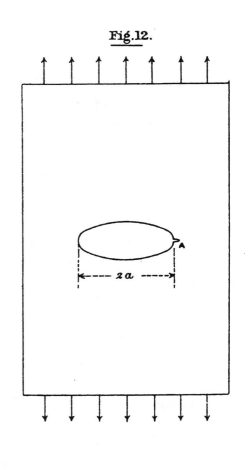

Fig. 12.

PART II.

DETERMINATION OF THE STRESSES IN A PLATE WHICH HAS AN ELLIPTIC HOLE.

Throughout the work curvilinear co-ordinates are employed.

Let α = constant and β = constant, define two systems of curves intersecting at right angles.

At any point let u_α, u_β denote the shifts in the direction of the normals to α and β.

Let $e_{\alpha\alpha}$, $e_{\beta\beta}$, $e_{\alpha\beta}$ denote the two stretches and the slide corresponding to these directions.

Then—

$$e_{\alpha\alpha} = h_1 \frac{\partial u_\alpha}{\partial \alpha} + h_1 h_2 u_\beta \frac{\partial}{\partial \beta}\left(\frac{1}{h_1}\right)\text{*}$$

$$e_{\beta\beta} = h_2 \frac{\partial u_\beta}{\partial \beta} + h_1 h_2 u_\alpha \frac{\partial}{\partial \alpha}\left(\frac{1}{h_2}\right)$$

$$e_{\alpha\beta} = \frac{h_1}{h_2}\frac{\partial}{\partial \alpha}(h_2 u_\beta) + \frac{h_2}{h_1}\frac{\partial}{\partial \beta}(h_1 u_\alpha)$$

where—

$$h_1^2 = \left(\frac{\partial \alpha}{\partial x}\right)^2 + \left(\frac{\partial \alpha}{\partial y}\right)^2$$

$$h_2^2 = \left(\frac{\partial \beta}{\partial x}\right)^2 + \left(\frac{\partial \beta}{\partial y}\right)^2$$

The dilation Δ is given by—

$$\Delta = h_1 h_2 \left[\frac{\partial}{\partial \alpha}\left(\frac{u_\alpha}{h_2}\right) + \frac{\partial}{\partial \beta}\left(\frac{u_\beta}{h_1}\right)\right]$$

The rotation ω is given by—

$$2\omega = h_1 h_2 \left[\frac{\partial}{\partial \alpha}\left(\frac{u_\beta}{h_2}\right) - \frac{\partial}{\partial \beta}\left(\frac{u_\alpha}{h_1}\right)\right]$$

For the particular problem under consideration the curvilinear co-ordinates α, β, are such that—

$$x = c \cosh \alpha \cos \beta$$
$$y = c \sinh \alpha \sin \beta.$$

α = constant is accordingly the ellipse $\dfrac{x^2}{c^2 \cosh^2 \alpha} + \dfrac{y^2}{c^2 \sinh^2 \alpha} = 1.$

β = constant is the hyperbola $\dfrac{x^2}{c^2 \cos^2 \beta} - \dfrac{y^2}{c^2 \sin^2 \beta} = 1.$

In this case $\qquad h_1^2 = h_2^2 = h^2 = \dfrac{2}{c^2(\cosh 2\alpha - \cos 2\beta)}.$

---
* See Love's "Elasticity."

General Stress Strain Equations.

These take the form—

$$(1 - \sigma)\frac{\delta \Delta}{\delta \alpha} - (1 - 2\sigma)\frac{\delta \omega}{\delta \beta} = 0 \Big\}$$
$$(1 - \sigma)\frac{\delta \Delta}{\delta \beta} + (1 - 2\sigma)\frac{\delta \omega}{\delta \alpha} = 0 \Big\}$$

Where σ is the value of Poisson's Ratio.

These two equations state that—

$$(1 - \sigma)\Delta + \iota(1 - 2\sigma)\omega \text{ is a function of } \alpha + \iota\beta$$

Take
$$(1 - \sigma)\Delta + \iota(1 - 2\sigma)\omega = \text{constant} \times \frac{e^{-n(\alpha + \iota\beta)}}{\sinh(\alpha + \iota\beta)}$$

So that
$$(1 - \sigma)\Delta = \text{const.} \frac{-e^{-(n-1)\alpha}\cos(n+1)\beta + e^{-(n+1)\alpha}\cos(n-1)\beta}{\cosh 2\alpha - \cos 2\beta}$$

$$(1 - 2\sigma)\omega = \text{const.} \frac{e^{-(n-1)\alpha}\sin(n+1)\beta - e^{-(n+1)\alpha}\sin(n-1)\beta}{\cosh 2\alpha - \cos 2\beta}$$

if
$$u = \frac{u_\alpha}{h}\alpha \text{ and } v = \frac{u}{h}\beta$$

$$\frac{\delta u}{\delta \alpha} + \frac{\delta v}{\delta \beta} = \frac{\Delta}{h^2} = \frac{a_n}{1 - \sigma}\left[e^{-(n-1)\alpha}\cos(n+1)\beta - e^{-(n+1)\alpha}\cos(n-1)\beta\right]$$

$$\frac{\delta u}{\delta \beta} - \frac{\delta v}{\delta \alpha} = -\frac{2\omega}{h^2} = \frac{2 a_n}{1 - 2\sigma}\left[e^{-(n-1)\alpha}\sin(n+1)\beta - e^{-(n+1)\alpha}\sin(n-1)\beta\right]$$

From these two partial differential equations u and v can be determined—

$$u = A_n\left[(n + p)e^{-(n-1)\alpha}\cos(n+1)\beta + (n - p)e^{-(n+1)\alpha}\cos(n-1)\beta\right] + \phi$$
$$v = A_n\left[(n - p)e^{-(n-1)\alpha}\sin(n+1)\beta + (n + p)e^{-(n+1)\alpha}\sin(n-1)\beta\right] + \psi$$

Where p stands for $3 - 4\sigma$, and ϕ and ψ are conjugate functions of α and β satisfying Laplace's equation.

Suitable values for ϕ and ψ are, const. $\times e^{-n\alpha}\cos n\beta$ and const. $\times e^{-n\alpha}\sin n\beta$.

From these general values of u and v values of $e_{\alpha\alpha}$, $e_{\beta\beta}$, $e_{\alpha\beta}$ can be deduced by means of the relations—

$$e_{\alpha\alpha} = h^2\frac{\delta u}{\delta \alpha} + \frac{u}{2}\frac{\delta h^2}{\delta \beta} - \frac{v}{2}\frac{\delta h^2}{\delta \beta}$$

$$e_{\beta\beta} = h^2\frac{\delta v}{\delta \beta} + \frac{v}{2}\frac{\delta h^2}{\delta \beta} - \frac{u}{2}\frac{\delta h^2}{\delta \alpha}$$

$$e_{\alpha\beta} = \frac{\delta}{\delta \alpha}(h^2 v) + \frac{\delta}{\delta \beta}(h^2 u)$$

The stress components can then be determined by the equations—

$$R_{\alpha\alpha} = \frac{E}{1+\sigma}\left[e_{\alpha\alpha} + \frac{\sigma}{1-2\sigma}\Delta\right]$$

$$R_{\beta\beta} = \frac{E}{1+\sigma}\left[e_{\beta\beta} + \frac{\sigma}{1-2\sigma}\Delta\right]$$

$$S_{\alpha\beta} = \frac{E}{2(1+\sigma)}\left[e_{\alpha\beta}\right]$$

These two sets of calculations are too lengthy to be given in detail; the results are as follows:—

$$R_{\alpha\alpha}(\cosh 2\alpha - \cos 2\beta)^2 = \begin{bmatrix} (n+1)e^{-(n-1)\alpha}\cos(n+3)\beta + (n-1)e^{-(n+1)\alpha}\cos(n-3)\beta \\ -\{4e^{-(n+1)\alpha} + (n+3)e^{-(n-3)\alpha}\}\cos(n+1)\beta \\ +\{4e^{-(n-1)\alpha} - (n-3)e^{-(n+3)\alpha}\}\cos(n-1)\beta \end{bmatrix} A_n$$
$$+ \begin{bmatrix} ne^{-(n+1)\alpha}\cos(n+3)\beta + (n+2)e^{-(n+1)\alpha}\cos(n-1)\beta \\ -\{(n+2)e^{-(n-1)\alpha} + ne^{-(n+3)\alpha}\}\cos(n+1)\beta \end{bmatrix} B_n$$

$$R_{\beta\beta}(\cosh 2\alpha - \cos 2\beta)^2 = \begin{bmatrix} -(n-3)e^{-(n-1)\alpha}\cos(n+3)\beta - (n+3)e^{-(n+1)\alpha}\cos(n-3)\beta \\ +\{(n-1)e^{-(n-3)\alpha} - 4e^{-(n+1)\alpha}\}\cos(n+1)\beta \\ +\{(n+1)e^{-(n+3)\alpha} + 4e^{-(n-1)\alpha}\}\cos(n-1)\beta \end{bmatrix} A_n$$
$$- \begin{bmatrix} ne^{-(n+1)\alpha}\cos(n+3)\beta + (n+2)e^{-(n+1)\alpha}\cos(n-1)\beta \\ -\{(n+2)e^{-(n-1)\alpha} + ne^{-(n+3)\alpha}\}\cos(n+1)\beta \end{bmatrix} B_n$$

$$S_{\alpha\beta}(\cosh 2\alpha - \cos 2\beta)^2 = \begin{bmatrix} (n-1)e^{-(n-1)\alpha}\sin(n+3)\beta + (n+1)e^{-(n+1)\alpha}\sin(n-3)\beta \\ -(n+1)e^{-(n-3)\alpha}\sin(n+1)\beta - (n-1)e^{-(n+3)\alpha}\sin(n-1)\beta \end{bmatrix} A_n$$
$$+ \begin{bmatrix} ne^{-(n+1)\alpha}\sin(n+3)\beta + (n+2)e^{-(n+1)\alpha}\sin(n-1)\beta \\ -\{(n+2)e^{-(n-1)\alpha} + ne^{-(n+3)\alpha}\}\sin(n+1)\beta \end{bmatrix} B_n$$

In these formulæ for $R_{\alpha\alpha}$, $R_{\beta\beta}$, $S_{\alpha\beta}$, n can be any integer positive or negative, and the general expression for these three stresses takes the form of an infinite series, involving a number of arbitrary constants, which have to be determined by the conditions existing at the inner and outer boundaries of the plate.

CASE OF A PLATE SUBJECTED TO A TENSILE STRESS R IN ALL DIRECTIONS, THE PLATE HAVING AN ELLIPTIC HOLE DEFINED BY $\alpha = \alpha_0$.

The boundary conditions for this case are—

$$R_{\alpha\alpha} = S_{\alpha\beta} = 0, \text{ when } \alpha = \alpha_0,$$
and $R_{\alpha\alpha} = R$, $R_{\beta\beta} = R$, $S_{\alpha\beta} = 0$, when α is large.

These conditions can be satisfied by making—

$$A_{-1} = -\frac{R}{8},$$

$$A_{+1} = -\frac{R}{8},$$

$$B_{-1} = \frac{R}{2} \cosh 2\alpha_0.$$

Adding together the three terms corresponding to A_{-1}, A_{+1}, and B_{-1}, the exact values of the stress for this case are—

$$R_{\alpha\alpha} = \frac{R \sinh 2\alpha [\cosh 2\alpha - \cosh 2\alpha_0]}{[\cosh 2\alpha - \cos 2\beta]^2}$$

$$R_{\beta\beta} = \frac{R \sinh 2\alpha [\cosh 2\alpha + \cosh 2\alpha_0 - 2\cos 2\beta]}{[\cosh 2\alpha - \cos 2\beta]^2}$$

$$S_{\alpha\beta} = \frac{R \sin 2\beta [\cosh 2\alpha - \cos 2\beta]}{[\cosh 2\alpha - \cos 2\beta]^2}$$

Along the edge of the elliptic hole—

$$R_{\beta\beta} = \frac{2 R \sinh 2\alpha_0}{\cosh 2\alpha_0 - \cos 2\beta}$$

CASE OF A PLATE SUBJECTED TO A TENSILE STRESS R IN THE DIRECTION $\beta = \frac{\pi}{2}$, THE PLATE HAVING AN ELLIPTIC HOLE DEFINED BY $\alpha = \alpha_0$.

The boundary conditions for this case are—

$$R_{\alpha\alpha} = S_{\alpha\beta} = 0, \text{ when } \alpha = \alpha_0,$$

and $R_{\alpha\alpha} = \frac{R}{2}(1 - \cos 2\beta)$, $R_{\beta\beta} = \frac{R}{2}(1 + \cos 2\beta)$, $S_{\alpha\beta} = -\frac{R}{2} \sin 2\beta$,

when α is large.

The conditions can be satisfied by making—

$$A_{-1} = -\frac{R}{16}; \quad B_{-1} = \frac{R}{4}\left[1 + \cosh 2\alpha_0\right]$$

$$A_{+1} = -\frac{R}{16} - \frac{R}{8} e^{2\alpha_0}; \quad B_{+1} = \frac{R}{8} e^{4\alpha_0};$$

$$B_{-3} = -\frac{R}{8}.$$

Adding together the five terms corresponding to A_{-1}, B_{-1}, A_{+1}, B_{+1}, B_{-3}, selected from the general formulæ stated above, the exact expression for $R_{\alpha\alpha}$, $R_{\beta\beta}$, $S_{\alpha\beta}$ can be determined. These expressions are complicated in appearance, and too lengthy to be written out here in full. The most important and interesting information they contain is the

EXPRESSION FOR THE TENSILE STRESS AT THE EDGE OF THE ELLIPTIC HOLE.

This latter takes the comparatively simple form—

$$R_{\beta\beta} = R \frac{\sinh 2\alpha_0 + e^{2\alpha_0} \cos 2\beta - 1}{\cosh 2\alpha - \cos 2\beta}$$

when $\alpha = \alpha_0$.

If the semi-major and minor axes of the ellipse are a and b respectively:—

At the end of the major axis ($\beta = 0$) the tensile stress is—

$$R \left[1 + \frac{2a}{b} \right]$$

At the end of the minor axis $\left(\beta = \frac{\pi}{2}\right)$ there is a compression stress of magnitude R.

CASE OF A PLATE SUBJECTED TO A SHEARING STRESS S, THE PLATE HAVING AN ELLIPTIC HOLE DEFINED BY $\alpha = \alpha_0$.

The boundary conditions for this case are—

$$R_{\alpha\alpha} = S_{\alpha\beta} = 0, \text{ when } \alpha = \alpha_0.$$

If the shearing stress is applied to planes at right angles to the area of the ellipse, the conditions which hold when α is large are—

$$R_{\alpha\alpha} = S \sin 2\beta$$
$$R_{\beta\beta} = -S \sin 2\beta$$
$$S_{\alpha\beta} = S \cos 2\beta$$

To deal with this case, the values of $R_{\alpha\alpha}$, $R_{\beta\beta}$, $S_{\alpha\beta}$, consequent on taking—

$$(1 - \sigma)\Delta + \iota(1 - 2\sigma)\omega = \iota \times \text{constant} \times \frac{e^{-n(\alpha + \iota\beta)}}{\sinh(\alpha + \iota\beta)}$$

have to be employed.

$R_{\alpha\alpha}$, $R_{\beta\beta}$ will be found to have exactly the same form as before, except that sines replace cosines. The new form of $S_{\alpha\beta}$ is obtained by reversing the sign of the original form and replacing sines everywhere with cosines.

The conditions stated above can be satisfied by making—

$$A_1 = \frac{S}{4} e^{2a_0}$$

$$B_{+1} = -\frac{S}{4} e^{4a_0}$$

$$B_{-3} = -\frac{S}{4}$$

Adding together the three corresponding terms, the value of $R_{\beta\beta}$ at the edge of the hole takes the form—

$$R_{\beta\beta} = -\frac{2 S \sin 2\beta \, e^{2a_0}}{\cosh 2a_0 - \cos 2\beta}$$

CASE OF A PLATE HAVING AN ELLIPTIC HOLE DEFINED BY $a = a_0$ SUBJECT TO A TENSILE STRESS R APPLIED IN A DIRECTION MAKING AN ANGLE ϕ WITH THE MAJOR AXIS OF THE ELLIPSE.

This case can be arrived at by a combination of cases previously considered. The expression for the tension along the inside edge of the cavity is—

$$R_{\beta\beta} = \frac{\sinh 2a_0 + \cos 2\phi - e^{2a_0} \cos 2(\phi - \beta)}{\cosh 2a_0 - \cos 2\beta}.$$

If $\phi = \frac{\pi}{4}$

$$R_{\beta\beta} = \frac{\sinh 2a_0 - e^{2a_0} \sin 2\beta}{\cosh 2a_0 - \cos 2\beta}$$

MAXIMUM VALUE OF $R_{\beta\beta}$ WHEN $\phi = \frac{\pi}{4}$.

Let m be the slope of the tangent at the point where $R_{\beta\beta}$ has its maximum value.

m is given by the expression $e^{2a_0} + \sqrt{1 + e^{-4a_0}}$.

The value of β for this point of maximum stress is given by—

$$\tan \beta = -\frac{1}{m} \tanh a_0,$$

and the value of the stress is—

$$- R \cdot \frac{\sin 2\beta}{\cos 2\beta} e^{2a_0}.$$

At the point where the slope of the tangent is $+1$

$$R_{\beta\beta} = R \left[1 + \tanh a_0 + \frac{1}{\tanh a_0} \right]$$

At the point where the slope of the tangent is -1

$$R_{\beta\beta} = -R.$$

Reprinted from *Journal of Applied Mechanics,* Vol. 6, pp. 49-53 (1939).

Bearing Pressures and Cracks

Bearing Pressures Through a Slightly Waved Surface or Through a Nearly Flat Part of a Cylinder, and Related Problems of Cracks

By H. M. WESTERGAARD,[1] CAMBRIDGE, MASS.

The task is undertaken of determining the bearing pressures, and the stresses and deformations created by them, in some cases that differ from those considered by Hertz[2] in his classical study of contact. Thus two solids are examined which, before loading, are in contact along a row of evenly spaced lines in a horizontal plane, as indicated in Fig. 1(*a*). Between these lines the surfaces have a separation defined by a nearly flat cosine wave. A uniform pressure on top of the upper solid creates contact over an area consisting of a row of strips, reduces the separation of the solids between the strips, as suggested in Fig. 1(*b*), and creates contact pressures distributed as indicated in Fig. 1(*c*), with vertical rises in the diagram of pressure at the edges of the strips. At a greater load the width of the strip becomes equal to the wave length, and the contact is complete. At still greater loads the stresses increase as if the two solids were one. The procedure by which this problem is solved is demonstrated first by showing its easy application to some well-known cases, especially Hertz's problem of circular cylinders in contact.[2]

Further applications are to a noncircular cylinder resting on a solid with a flat top, with an initial separation of the surfaces varying as the fourth power of the distance from the initial line of contact; to partial contact of two surfaces which are initially plane, except that one of them has a ridge or several parallel ridges; and to some related problems in which two parts of the same body are partially separated by the forming of one or more cracks.

Notation

x, y	= rectangular coordinates, y vertical
r, θ	= corresponding polar coordinates
z	= $x + iy = re^{i\theta}$ = complex variable
Z	= function of z, Equation [1], defining the stresses by Equations [4] to [6]
Z', $\bar{Z}$, $\bar{\bar{Z}}$	= derivative and first and second integral of Z, Equations [2]
σ_x, σ_y, τ_{xy}	= normal stresses and shearing stress in the directions of x and y
ξ, η	= displacements in the directions of x and y
η_0	= displacement η at $y = 0$
s	= initial separation of two surfaces
E, G, μ	= Young's modulus, modulus of elasticity in shear, and Poisson's ratio
F	= Airy's stress function
P	= force on slice parallel to the x,y-plane one unit thick, measurable in pounds per inch
p	= average pressure or tension, measurable in pounds per square inch
a, l	= horizontal distances on axis of x
c, c_1	= constants

Function of a Complex Variable Used as Stress Function

A stress function will be applied of a type which was introduced by Carothers[3] in 1920 and, evidently independently, by Nádai[4] in 1921. Both expressed the significant values in terms of harmonic functions, and both made use of the following fact: A harmonic function of x and y can be obtained as the real part ReZ or the imaginary part ImZ of an analytic function Z of the complex variable $z = x + iy$, with Z being written in the forms

$$Z = Z(z) = Z(x + iy) = \text{Re}Z + i\text{Im}Z \qquad [1]$$

In the present applications it is expedient, as done by MacGregor,[5] to use the function Z itself as stress function.

The further functions Z', $\bar{Z}$, and $\bar{\bar{Z}}$ are the derivative and first and second integrals of Z, so that

$$Z' = \frac{dZ}{dz}, \quad Z = \frac{d\bar{Z}}{dz}, \quad \bar{Z} = \frac{d\bar{\bar{Z}}}{dz} \qquad [2]$$

The properties of derivatives are noted

$$\frac{\partial \text{Re}Z}{\partial x} = \frac{\partial \text{Im}Z}{\partial y} = \text{Re}Z', \quad \frac{\partial \text{Im}Z}{\partial x} = -\frac{\partial \text{Re}Z}{\partial y} = \text{Im}Z' \qquad [3]$$

In a restricted but important group of cases the normal stresses and the shearing stress in the directions of x and y can be stated in the form

$$\sigma_x = \text{Re}Z - y\text{Im}Z' \qquad [4]$$

$$\sigma_y = \text{Re}Z + y\text{Im}Z' \qquad [5]$$

$$\tau_{xy} = -y\text{Re}Z' \qquad [6]$$

[1] Gordon McKay Professor of Civil Engineering and Dean of the Graduate School of Engineering, Harvard University. Mem. A.S.M.E.

Presented by title at the Joint Meeting of The Applied Mechanics and Hydraulic Divisions of The American Society of Mechanical Engineers, Ithaca, N. Y., June 25–26, 1937.

Discussion of this paper should be addressed to the Secretary, A.S.M.E., 29 West 39th Street, New York, N. Y., and will be accepted until August 10, 1939, for publication at a later date. Discussion received after the closing date will be returned.

Note: Statements and opinions advanced in papers are to be understood as individual expressions of their authors, and not those of the Society.

[2] Heinrich Hertz, Crelle's *Journal für die reine und angewandte Mathematik,* vol. 92, 1881, p. 156 (also in his Gesammelte Werke, vol. 1, 1895, p. 155). See, for example "Theory of Elasticity," by S. Timoshenko, McGraw-Hill Book Co., Inc., New York, N. Y., 1934, pp. 339–350.

[3] "Plane Strain: The Direct Determination of Stress," by S. D. Carothers, Proceedings of the Royal Society of London, series A, vol. 97, 1920, pp. 110–123, especially p. 119.

[4] "Über die Spannungsverteilung in einer durch eine Einzelkraft belasteten rechteckigen Platte," by A. Nádai, *Der Bauingenieur,* vol. 2, 1921, pp. 11–16, especially p. 12. Nádai applied the function to express curvatures and twists of elastic slabs. The curvatures and twists can be interpreted as stresses through Airy's stress function.

[5] "The Potential Function Method for the Solution of Two-Dimensional Stress Problems," by C. W. MacGregor, Trans. American Mathematical Society, vol. 38, no. 1, July, 1935, pp. 177–186.

By referring to Equations [3] it is observed that these stresses satisfy the two conditions of equilibrium of the form

$$\frac{\partial \sigma_x}{\partial x} + \frac{\partial \tau_{xy}}{\partial y} = 0 \quad \ldots \ldots \ldots \ldots [7]$$

The limitation of this type of solution appears in Equations [4] to [6], which require that

$$\sigma_x = \sigma_y \text{ and } \tau_{xy} = 0 \text{ at } y = 0 \ldots \ldots \ldots [8]$$

With deformation in the direction perpendicular to the x, y-plane prevented, the displacements ξ and η in the directions of x and y are defined by the formulas

$$2G\xi = (1 - 2\mu)\,\text{Re}\overline{Z} - y\,\text{Im}Z \ldots \ldots \ldots [9]$$

$$2G\eta = 2(1 - \mu)\,\text{Im}\overline{Z} - y\,\text{Re}Z \ldots \ldots \ldots [10]$$

For, it is found that these displacements define the stresses in Equations [4] to [6] through Hooke's law, which can be stated in the form

$$\sigma_x = 2G\left[\frac{\partial \xi}{\partial x} + \frac{\mu}{1-2\mu}\left(\frac{\partial \xi}{\partial x} + \frac{\partial \eta}{\partial y}\right)\right] \text{ and } \tau_{xy} = G\left(\frac{\partial \xi}{\partial y} + \frac{\partial \eta}{\partial x}\right) \quad \ldots [11]$$

A useful observation from Equation [10] is that the value of η at $y = 0$ is

$$\eta_0 = \frac{1-\mu}{G}\,\text{Im}\overline{Z} = \frac{2(1-\mu^2)}{E}\,\text{Im}\overline{Z} \ldots \ldots [12]$$

It is noted, furthermore, that the Airy function defining the stresses by the equations

$$\sigma_x = \frac{\partial^2 F}{\partial y^2}, \quad \sigma_y = \frac{\partial^2 F}{\partial x^2}, \text{ and } \tau_{xy} = -\frac{\partial^2 F}{\partial x \partial y} \ldots [13]$$

is

$$F = \text{Re}\overline{\overline{Z}} + y\,\text{Im}\overline{Z} \ldots \ldots \ldots [14]$$

In a slice parallel to the x, y-plane one unit thick the total vertical force transmitted between two points is the increase of the derivative

$$\frac{\partial F}{\partial x} = \text{Re}\overline{Z} + y\,\text{Im}Z \ldots \ldots \ldots [15]$$

between the points. Similarly, the total horizontal force transmitted between two points is the increase of

$$\partial F/\partial y = y\,\text{Re}Z \ldots \ldots \ldots [16]$$

between the points.

Introductory Application to Boussinesq's Problem

The semi-infinite solid $y \geq 0$, with y positive downward, is under consideration. The function

$$Z = P/(i\pi z) \ldots \ldots \ldots [17]$$

gives

$$\overline{Z} = \frac{P}{i\pi} \log(z/c_1) = \frac{P}{i\pi}\left[\log(r/c) + i\left(\theta - \frac{\pi}{2}\right)\right] \ldots [18]$$

that is

$$\text{Re}\overline{Z} = P[\theta - (\pi/2)]/\pi \ldots \ldots \ldots [19]$$

According to Equations [15] and [19], in a slice parallel to the xy-plane and one unit thick the total vertical force transmitted between $\theta = \pi$ and $\theta = 0$ is $-P$. It is concluded that Equation [17] represents the solution of Boussinesq's problem in two dimensions for a normal pressure P concentrated at $z = 0$. The familiar formulas for stresses and displacements are obtained readily by substituting from Equations [17] and [18] in Equations [4], [5], [6], [9], and [10].

Rows of Forces

Equation [17] suggests consideration of two modified functions

$$Z_1 = -i\frac{P}{l}\cot(\pi z/l) \text{ and } Z_2 = -i\frac{P}{l\sin(\pi z/l)} \ldots [20]$$

Near $z = 0$ both approach Z in Equation [17]. Further inspection shows that Z_1 represents a row of equal pressures P at $z = 0, \pm l, \pm 2l, \ldots$, and Z_2 represents a row of pressures P at $z = 0, \pm 2l, \pm 4l, \ldots$ and a row of pulls P at $z = \pm l, \pm 3l, \pm 5l, \ldots$ on the solid $y \geq 0$. When y becomes great, Z_1 converges toward $-P/l$, making the stresses in Equations [4] to [6] converge toward a uniform pressure P/l; while Z_2 converges toward zero, making the stresses converge toward zero, as they should under the self-balancing load.

Demonstration by Application to Hertz's Problem of Two Circular Cylinders in Contact

The solid $y \geq 0$ is considered again. As stress function is chosen

$$Z = -\frac{2P}{\pi a^2}\left[\sqrt{(a^2 - z^2)} + iz\right] \ldots \ldots [21]$$

or

$$Z = -\frac{2P}{\pi a^2}\left[\sqrt{a^2 - x^2 + y^2 - i\,2xy} + ix - y\right] \ldots [22]$$

At $y = 0$ the shearing stress $\tau_{xy} = 0$, and the normal stresses, according to Equations [4] and [5], are both equal to $\text{Re}Z$. Accordingly

$$\sigma_x = \sigma_y = 0 \text{ at } y = 0, x < -a \text{ or } x > a \ldots \ldots [23]$$

$$\sigma_x = \sigma_y = -(2P/\pi a^2)\sqrt{(a^2 - x^2)}$$
$$\text{at } y = 0, \, -a < x < a \ldots \ldots [24]$$

Equations [23] and [24] show that the diagram of pressures on the surface $y = 0$ can be drawn as a half-ellipse between $x = -a$ and $x = a$; outside there is no load. The total pressure on the slice one unit thick is P.

When z becomes numerically great, with y remaining positive, one may write

$$\sqrt{(a^2 - z^2)} = -iz\sqrt{[1 - (a^2/z^2)]} = -iz(1 - a^2/2z^2 - \ldots)$$
$$\ldots [25]$$

Therefore, Z in Equation [21] converges toward Z in Equation [17], which represents Boussinesq's problem.

In the interval $-a < x < a$ at $y = 0$ Equations [12] and [22] give

$$\frac{d\eta_0}{dx} = \frac{2(1-\mu^2)}{E}\,\text{Im}Z = -\frac{4(1-\mu^2)Px}{\pi E a^2} \ldots \ldots [26]$$

that is, along the axis of x there is produced a constant concave curvature

$$\frac{1}{R} = -\frac{d^2\eta_0}{dx^2} = \frac{4(1-\mu^2)P}{\pi E a^2} \ldots \ldots \ldots [27]$$

If instead of being initially flat along the axis of x the surface has an initial convex curvature equal to that in Equation [27], under the pressures defined by Equation [24] the surface will be flattened out and become plane in the interval $-a < x < a$; outside this interval it will be flattened out less.

It follows that if two parallel cylinders with radii R are pressed together by the load P per unit of length, the width $2a$ of the strip of contact will be defined by a in Equation [27], which agrees with Hertz's classical solution. With a known, the contact pressures are defined by Equation [24] and the stresses and displacements in the surrounding region by Equations [21], [4], [5], [6], [9], and [10].

BEARING PRESSURE THROUGH SLIGHTLY WAVED SURFACE

Equation [21] suggests investigation of the stress function

$$Z = -\frac{2p \cos(\pi z/l)}{\sin^2(\pi a/l)} \{\sqrt{[\sin^2(\pi a/l) - \sin^2(\pi z/l)]} + i \sin(\pi z/l)\} \quad \ldots [28]$$

as applying to the solid $y \geq 0$. It is assumed that $a < l/2$. By computing as in Equation [25], it is found that when y is positive and great compared with a, Equation [28] may be replaced by

$$Z = -ip \cot(\pi z/l) \ldots [29]$$

According to the comments on Equations [20], Z in Equation [29] represents a row of pressures pl with spacing l at $y = 0$, and a uniform pressure p at great values of y.

At the surface $y = 0$ one finds in the interval $-a < x < a$

$$\sigma_x = \sigma_y = \mathrm{Re}Z = -\frac{2p \cos(\pi x/l)}{\sin^2(\pi a/l)} \sqrt{[\sin^2(\pi a/l) - \sin^2(\pi x/l)]} \ldots [30]$$

and in the interval $a < x < l - a$

$$\sigma_x = \sigma_y = \mathrm{Re}Z = 0 \ldots [31]$$

The function Z is periodic, and the period is l. The values are repeated in the similar intervals. The strips $nl - a < x < nl + a$ are loaded by pressures $-\sigma_y$ defined numerically by Equation [30]; the remaining strips are unloaded.

Within the loaded strips of the surface Equation [28] gives

$$\mathrm{Im}Z = -\frac{p \sin(2\pi x/l)}{\sin^2(\pi a/l)} \ldots [32]$$

Over the whole surface $\mathrm{Im}Z$ is antisymmetrical with respect to the center lines $x = nl/2$ of the strips. By referring to Equation [12] it is then found that within the loaded strips the deflection of the surface can be stated as

$$\eta_0 = \frac{2(1-\mu^2)}{E} \mathrm{Im}\overline{Z} = \frac{(1-\mu^2)pl[\cos(2\pi x/l) - 1]}{\pi E \sin^2(\pi a/l)} \ldots [33]$$

with the integration constant being the same for all the loaded strips.

Assume now that instead of being initially flat the surface is slightly waved, having the equation

$$y_0 = \frac{c}{4}[1 - \cos(2\pi x/l)] \ldots [34]$$

with

$$c = \frac{4(1-\mu^2)pl}{\pi E \sin^2(\pi a/l)} \ldots [35]$$

Then under the pressures defined by Equation [30] the ordinates $y_0 + \eta_0$ of the deformed surface will be zero within the loaded strips. The loaded strips will be flattened out and be contained in a single plane. A further examination of $\mathrm{Im}Z$ as defined by Equation [28] shows that $y_0 + \eta_0$ will be positive between the loaded strips.

It is concluded that if another solid of the same material and shape is placed in contact with the one considered, so that the axis of x becomes an axis of symmetry, and if thereafter a uniform pressure p is produced at numerically large values of y, the contact pressures will be as defined by Equation [30]; the

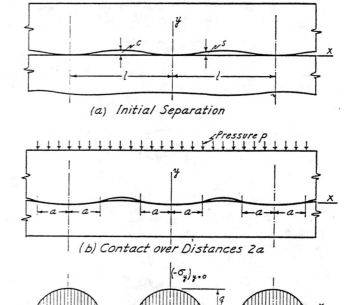

FIG. 1 BEARING PRESSURES THROUGH SLIGHTLY WAVED SURFACE

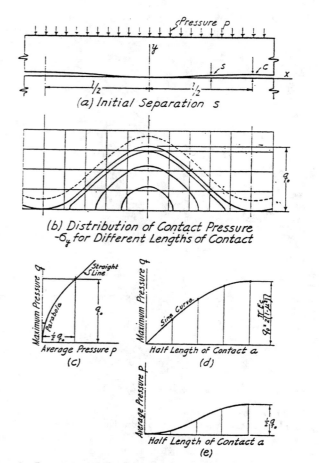

FIG. 2 DIAGRAMS OF BEARING PRESSURES THROUGH SLIGHTLY WAVED SURFACE

area of contact through which the pressures are transmitted will consist of the strips of width $2a$ defined by Equation [35]. Equations [34] and [35] are verified by Equation [27] when a is small compared with l.

It is noted that the initial separation of the two surfaces, before pressure is applied, is

$$s = (c/2)[1 - \cos(2\pi x/l)] = c \sin^2(\pi x/l), \quad s_{\max} = c \ldots [36]$$

The conclusions that were drawn continue to apply if the two nearly flat surfaces in contact have a different shape, as long as the initial separation is defined by Equations [36].

Fig. 1 illustrates this case. Fig. 2 shows some results obtained from Equations [30] and [35].

Noncircular Cylinder With Nearly Flat Bottom

The function

$$Z = -\frac{8P}{3\pi a^4}\left[(z^2 + a^2/2)\sqrt{(a^2 - z^2)} + iz^3\right] \ldots [37]$$

applied to the solid $y \geq 0$, is examined first for numerically great values of z. By writing

$$\sqrt{(a^2 - z^2)} = -iz\left(1 - \frac{a^2}{2z^2} - \frac{a^4}{8z^4} - \ldots\right) \ldots [38]$$

Z in Equation [37] is found to converge toward Z in Equation [17], which represents Boussinesq's problem. Again, at distances great compared with a the stresses are as in Boussinesq's problem, and the total load on the slice one unit wide is P. At $y = 0$ only the interval $-a < x < a$ is loaded; the pressures are $-\text{Re} Z$. At $x = 0$ the pressure is $8/3\pi$ times the average, that is, less than the average; the maximum pressure occurs at some distance from the center of the load. These pressures can be produced by contact of two solids. The required initial separation s is computed by considering the interval $-a < x < a$. One finds

$$s = -\frac{4(1-\mu^2)}{E}\,\text{Im}\,\overline{Z} = \frac{8(1-\mu^2)P}{3\pi E a^4}\,x^4 \ldots [39]$$

The lower solid may have a flat top while the upper solid is a noncircular cylinder shaped at the bottom according to a parabola of fourth degree.

Flat Surfaces With One or More Ridges

Fig. 3(a) shows two solids with surfaces that are initially plane except for a single ridge on one of the surfaces at $x = 0$. Under the pressure p contact is missing in the intervals $-a < x < 0$ and $0 < x < a$. The same situation may be created by driving a plug in between the two surfaces. The stress function

$$Z = -p\sqrt{(1 - a^2/z^2)} \ldots [40]$$

represents this case, with the provision that a uniform horizontal tension, for example, $\sigma_x = p$ may be superposed. Fig. 3(b) shows the distribution of the pressures of contact. The force P at the ridge is found by stating Z near $z = 0$ for $y > 0$ in the two forms

$$Z = -ipa/z = P/(i\pi z) \ldots [41]$$

which gives

$$P = \pi p a \ldots [42]$$

The value of a will depend not only on p but also on the height and sharpness of the ridge.

Fig. 3(c) shows the related problem of a number of equal parallel ridges with spacing l. The corresponding stress function is

$$Z = -p\sqrt{\left[1 - \frac{\sin^2(\pi a/l)}{\sin^2(\pi z/l)}\right]} \ldots [43]$$

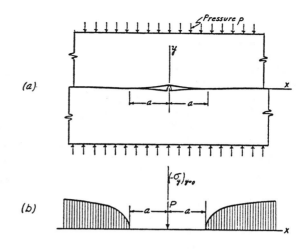

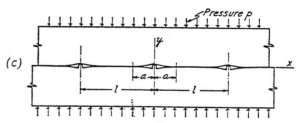

Fig. 3 Contact Between Initially Flat Surfaces With Ridges

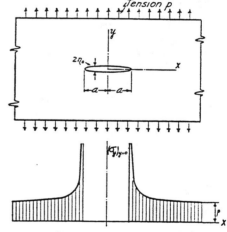

Fig. 4 Internal Crack

Internal Crack

Fig. 4 shows an internal crack which has opened from $z = -a$ to $z = a$ under the influence of an average tension p. The function

$$Z = p/\sqrt{[1 - (a^2/z^2)]} \ldots [44]$$

solves the problem. Z converges toward p when z becomes numerically great. At $y = 0$ one finds outside the crack the tension

$$\sigma_y = p/\sqrt{[1 - (a^2/x^2)]} \ldots [45]$$

and within the length of the crack the opening

$$2\eta_0 = \frac{4(1-\mu^2)}{E}\,\text{Im}\,\overline{Z} = \frac{4(1-\mu^2)p}{E}\sqrt{(a^2 - x^2)} \ldots [46]$$

which shows the shape of the crack to be elliptic. The concentration of stress and the infinite slope $d\eta_0/dx$ at $x = \pm a$ are subject to the usual interpretation applicable to singularities. A uniform horizontal compressive stress p may be superposed without disturbing the remaining features of the solution.

Equation [44] suggests examination of the function

$$Z = p \bigg/ \sqrt{\left[1 - \frac{\sin^2(\pi a/l)}{\sin^2(\pi z/l)}\right]} \quad \ldots \ldots \ldots \ldots [47]$$

At numerically great values of y this function converges toward p and defines a uniform tension p. At $y = 0$ the function accounts for a system of cracks, each of length $2a$, with centers at $x = 0, \pm l, \pm 2l, \ldots$.

The function

$$Z_1 = Z - p \ldots \ldots \ldots \ldots \ldots [48]$$

with Z as in Equation [44] or [47], accounts for a crack or a system of cracks at $y = 0$, created by a liquid pressure p in the cracks as the only load.

Crack Opened by Wedge

Fig. 5(a) shows a crack opened by a wedge exerting pressures P. The stress functions

$$Z_1 = \frac{P}{\pi(a+z)}\sqrt{\frac{a}{z}} \quad \text{and} \quad Z_2 = -\frac{P}{\pi(a+z)}\sqrt{\frac{z}{a}} \ldots [49][50]$$

represent two possible solutions, which require different loads at the outer boundary. Fig. 5(b) and (c), show the corresponding diagrams of stresses at $y = 0$. A change of the load on the outer boundary may bring about the change from Z_1 to Z_2, replacing the concentration of tension in Fig. 5(b) by the diagram of finite compressive stresses in Fig. 5(c). The form of the latter diagram near $x = 0$, with the vertical tangent at $x = 0$, should be considered as characteristic of brittle materials, such as concrete.[6]

---

[6] "Stresses at a Crack, Size of the Crack, and the Bending of Reinforced Concrete," by H. M. Westergaard, *Journal* American Concrete Inst., November-December, 1933, or, Proceedings, vol. 30, 1934, pp. 93–102. Contains an analysis of this feature of cracks.

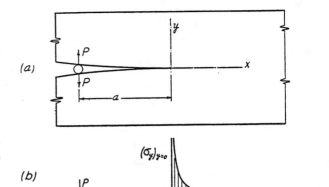

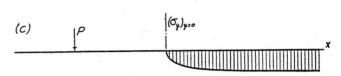

Fig. 5 Crack Opened by Wedge

An internal crack which has been opened between $z = -a$ and $z = a$ by a wedge exerting the pressure P at $z = 0$ is accounted for by the stress function

$$Z = Pa/[\pi z \sqrt{(z^2 - a^2)}] \ldots \ldots \ldots \ldots [51]$$

This function shows concentration of tension at $z = \pm a$, and vanishing stresses at great distances from the crack. If an external pressure is superposed, of the magnitude p defined by Equation [42], Z in Equation [51] will be replaced by Z in Equation [40], and the concentration of tension is replaced by moderate compressive stresses.

Concluding Comment

It is easy to add further examples. Those that have been shown indicate a type of problem to which the method that was used lends itself.

The distribution of stress in the neighbourhood of a crack in an elastic solid

By I. N. Sneddon, *Bryce Fellow of the University of Glasgow*

(*Communicated by N. F. Mott, F.R.S.—Received* 23 October 1945)

The distribution of stress produced in the interior of an elastic solid by the opening of an internal crack under the action of pressure applied to its surface is considered. The analysis is given for 'Griffith' cracks (§ 2) and for circular cracks (§ 3), it being assumed in the latter case that the applied pressure varies over the surface of the crack. For both types of crack the case in which the pressure is constant over the entire crack surface is considered in some detail, the stress components being tabulated and the distribution of stress shown graphically. The effect of a crack (of either type) on the stress produced in an elastic body by a uniform tensile stress is considered and the conditions for rupture deduced.

1. Introduction

1·1. A theory of rupture of non-ductile materials such as glass has been put forward by A. A. Griffith (1920, 1924) on the basis of the existence of a large number of cracks in the interior of the solid body. The fundamental concept of the Griffith theory is that the bounding surfaces of a solid possess a surface tension, just as those of a liquid do, and that when a crack spreads the decrease in the strain energy is balanced by an increase in the potential energy due to this surface tension. The calculation of the effect of the presence of a crack on the energy of an elastic body is based on Inglis's solution (Inglis 1913 a, b) of the two-dimensional equations of elastic equilibrium in the space bounded by two concentric ellipses, the crack being then taken to be an ellipse of zero eccentricity. Denoting the surface tension of the material of the solid body by T, the width of the crack by $2c$, and the Young's modulus of the material of the body by E, Griffith showed that, in the case of plane stress, the crack will spread when the stress P, applied normally to the direction of the crack, exceeds the critical value

$$P_c = \sqrt{\frac{2ET}{\pi c}}. \qquad (1\cdot1\cdot1)$$

The use of the equations of elastic equilibrium restricts the analysis to materials which obey Hooke's law fairly closely. This will not be true in general, as the high concentration of stress at the edges of the crack will induce a certain amount of plastic flow. Because of the smallness of the region to which the plastic flow is confined the distribution of stress at points remote from the edges of the crack will not be affected and the strain energy will not differ appreciably from that derived from the solution of the elastic equations. A formula differing from (1·1·1) by a factor 0·8 has been derived by Orowan (1934) from rather similar assumptions. In the case of plane strain the formula (1·1·1) is replaced by*

$$P_c = \sqrt{\frac{2ET}{\pi c(1-\sigma^2)}}. \tag{1·1·2}$$

1·2. Griffith's theory of rupture has recently been extended to three dimensions by Sack (1946). Observing that the length of internal cracks does not greatly exceed their width, Sack calculates the conditions of rupture for a solid containing a plane crack bounded by a circle—a 'penny-shaped' crack—when one of the principal stresses is acting normally to the plane of the crack. By treating the crack as an oblate spheroid whose elliptic section has zero eccentricity Sack establishes that rupture will occur when the tensile stress P normal to the crack exceeds the critical value

$$P_c = \sqrt{\frac{\pi ET}{2c(1-\sigma^2)}}, \tag{1·2·1}$$

where c is the radius of the crack, and, as before, T denotes the surface tension of the material of the body, and E its Young's modulus.

The three-dimensional model introduced by Sack thus gives a critical tensile stress differing from the Griffith value (1·1·1) by a factor $\pi/\{2(1-\sigma^2)^{\frac{1}{2}}\}$, and from (1·1·2) by a factor $\frac{1}{2}\pi$.

1·3. In the present paper main consideration will be given to the distribution of stress in the neighbourhood of a circular crack of the type considered by Sack. As is usual it is assumed that the crack is of such dimensions that its depth exceeds the radius of molecular action at all points not very near its edges. The mathematical theory of elasticity then gives the components of stress accurately at all points of the elastic body other than those in the immediate vicinity of the edges of the crack. If the crack is sufficiently large, the error incurred in the calculation of the strain energy is then negligible, for the stress near the edge of the crack is proportional to $r^{-\frac{1}{2}}$, so that the strain energy in any small sphere whose centre coincides with the edge is finite and negligible in comparison with the strain energy in the rest of the solid. The elastic body is assumed to consist of homogeneous isotropic material and to be so large that its dimensions are very much greater than the radius of the crack; the crack may then be considered to be situated in the interior of an infinite elastic medium. In the first instance the elastic body is considered to be deformed by an internal pressure acting across the surfaces of the crack, the effect of tensile stresses applied to a body with an internal crack free from stress being deduced later.

* This is given incorrectly in Griffith (1924) as

$$P_c = \sqrt{\left[\frac{2ET}{\pi c}(1-\sigma^2)\right]}.$$

To afford a comparison with the three-dimensional case the distribution of stress in the neighbourhood of a crack in two dimensions (plane strain) is first considered (§ 2); this corresponds to the two-dimensional Griffith model. The elastic body is supposed to be deformed by the opening of the crack under the action of a uniform hydrostatic pressure acting over its surface. The calculations are based on a solution of the elastic equations given by Westergaard (1939); analytical expressions for the components of stress and for the principal shearing stress are derived from Westergaard's stress function. The advantage of the analysis given here is not only that it is simpler than Inglis's but that Cartesian co-ordinates are employed throughout, thus facilitating the interpretation of the results. To give some idea of the distribution of stress in the body the intensity of the principal shearing stress is computed at a network of points in the two-dimensional co-ordinate plane and the system of 'isochromatic' lines (contours of equal principal shearing stress) plotted in the neighbourhood of the crack. The solution corresponding to the opening of an internal crack under the influence of a tensile stress applied to the surface of the body is then indicated and the Griffith criterion (1·1·2) derived.

In the case of a circular crack (§ 3) it is first of all assumed that the internal pressure is a function of the distance from the centre of the crack. Sack's calculations are based on Neuber's solution (Neuber 1934) of the equations of elastic equilibrium in oblate spheroidal co-ordinates. This method could no doubt be adapted to the case of a variable internal pressure but would probably be rather laborious. Even in the case of constant applied pressure considered by Sack the expressions for the components of stress do not lend themselves readily to computation; and, as in the case of Inglis's analysis, the choice of co-ordinate system makes the interpretation of the results somewhat difficult. In the analysis given below cylindrical polar co-ordinates are used. The solution of the equations of elastic equilibrium in these co-ordinates by the method of Hankel transforms, developed recently by Harding & Sneddon (1945), is employed to reduce the problem to that of the solution of a pair of dual integral equations. A relation giving the shape of the crack in terms of the pressure applied to its surface is derived from the known solution of this pair of equations. The converse problem, that of determining the distribution of internal pressure necessary to preserve a given shape of crack, is also considered.

The distribution of stress in the case in which the applied internal pressure remains constant over the entire surface of the crack is considered in some detail. It is first shown that in this instance the crack, assumed to be originally an infinitely thin crevice, has, after the application of the pressure, the shape of a very flat ellipsoid of revolution. The components of stress are calculated at various points in the interior of the elastic body, and the results given in a set of tables which enable the stress components at any point to be obtained by interpolation; the variation of the various stress components in certain planes parallel to the crack is also illustrated graphically. The principal shearing stress is tabulated in a similar fashion and the contours of equal principal shearing stress plotted to show the distribution of stress in the neighbourhood of the crack.

Finally, the effect of a circular crack, free from stress, on the distribution of stress in an elastic body subjected to a uniform tensile stress is considered, and Sack's criterion (1·2·1) derived on energy considerations.

1·4. Since an exact analysis of the stresses near the base of a crack using a model of the kind introduced by Griffith and Sack would be a necessary preliminary to the consideration of such theoretical problems as the determination of the velocity of propagation of cracks, it is perhaps of some interest to quote here expressions for the components of stress in the vicinity of the edge of the crack. In the case of a two-dimensional crack, if the origin of co-ordinates is chosen to be at the edge of the crack and take axes of co-ordinates (ξ, η) along the axis of the crack and perpendicular to that direction, then the shape of the crack is given approximately by the formula

$$\eta^2 = -b\xi. \tag{1·4·1}$$

Taking polar co-ordinates defined by the relations

$$\xi = \delta \cos \psi, \quad \eta = \delta \sin \psi,$$

then in the immediate vicinity of the crack the components of stress are given by the expressions

$$\sigma_x = p_0 \left(\frac{c}{2\delta}\right)^{\frac{1}{2}} (\tfrac{3}{4} \cos \tfrac{1}{2}\psi + \tfrac{1}{4} \cos \tfrac{5}{2}\psi), \tag{1·4·1}$$

$$\sigma_y = p_0 \left(\frac{c}{2\delta}\right)^{\frac{1}{2}} (\tfrac{5}{4} \cos \tfrac{1}{2}\psi - \tfrac{1}{4} \cos \tfrac{5}{2}\psi), \tag{1·4·2}$$

$$\tau_{xy} = p_0 \left(\frac{c}{8\delta}\right)^{\frac{1}{2}} \sin \psi \cos \tfrac{3}{2}\psi. \tag{1·4·3}$$

It follows immediately from these formulae that the principal shearing stress is given by the equation

$$\tau = p_0 \left(\frac{c}{8\delta}\right)^{\frac{1}{2}} \sin \psi. \tag{1·4·4}$$

The most striking feature of the analysis in the three-dimensional case is that the expressions for the components of stress in the neighbourhood of the crack differ from those of the two-dimensional case by a numerical factor only. The stresses are derived from (1·4·1–3) by replacing $\sigma_x, \sigma_y, \tau_{xy}$ on the left-hand sides by $\tfrac{1}{2}(\pi\sigma_r), \tfrac{1}{2}(\pi\sigma_2), \tfrac{1}{2}(\pi\tau_{2r})$, where δ and ψ have a similar interpretation in three dimensions. The hoop stress σ_θ has no analogue in two dimensions; it is given by the relation

$$\sigma_\theta = \frac{4\sigma p_0}{\pi} \left(\frac{c}{2\delta}\right)^{\frac{1}{2}} \cos \tfrac{1}{2}\psi. \tag{1·4·5}$$

2. The two-dimensional model

2·1. It has been shown by Westergaard (1939)* that the stress function

$$Z = p_0\left(\frac{z}{\sqrt{(z^2-c^2)}} - 1\right), \quad (z = x+iy), \qquad (2\cdot1\cdot1)$$

describes the distribution of stress in the interior of an infinite 'two-dimensional' elastic medium produced by the opening of an internal crack from $z = -c$ to $z = c$ (c being real) under the action of a uniform liquid pressure p_0 in the crack as the only load. If the value of the normal component of the displacement, u_y, be denoted by w when $y = 0$, then it can readily be shown that for $|x| < c$,

$$\left(\frac{x}{c}\right)^2 + \left(\frac{w}{\epsilon}\right)^2 = 1, \qquad (2\cdot1\cdot2)$$

where the value of the semi-minor axis ϵ is given by the relation

$$\epsilon = 2(1-\sigma^2)p_0 c/E. \qquad (2\cdot1\cdot3)$$

Equation $(2\cdot1\cdot2)$ shows that as a result of the elastic strain the shape of the internal crack is elliptic. Furthermore, equation $(2\cdot1\cdot1)$ shows that w is zero for $|x| > c$.

If one adopts the notation

$$z = re^{i\theta}, \quad z-c = r_1 e^{i\theta_1}, \quad z+c = r_2 e^{i\theta_2}, \qquad (2\cdot1\cdot4)$$

then the components of stress are given by the equations

$$\left.\begin{aligned}
\tfrac{1}{2}(\sigma_x+\sigma_y) &= \mathrm{Re}\,Z = p_0\left\{\frac{r}{r_1^{\frac{1}{2}} r_2^{\frac{1}{2}}}\cos(\theta - \tfrac{1}{2}\theta_1 - \tfrac{1}{2}\theta_2) - 1\right\}, \\
\tfrac{1}{2}(\sigma_y-\sigma_x) &= y\,\mathrm{Im}\,Z' = p_0 \frac{r\sin\theta}{c}\left(\frac{c^2}{r_1 r_2}\right)^{\frac{3}{2}} \sin\tfrac{3}{2}(\theta_1+\theta_2), \\
\tau_{xy} &= -y\,\mathrm{Re}\,Z' = p_0 \frac{r\sin\theta}{c}\left(\frac{c^2}{r_1 r_2}\right)^{\frac{3}{2}} \cos\tfrac{3}{2}(\theta_1+\theta_2).
\end{aligned}\right\} \qquad (2\cdot1\cdot5)$$

By means of these expressions the components of stress at any point in the medium can be calculated; the maximum shearing stress across any plane through the point (x,y) can then be obtained from the relation

$$\tau^2 = (\tfrac{1}{2}\sigma_y - \tfrac{1}{2}\sigma_x)^2 + \tau_{xy}^2. \qquad (2\cdot1\cdot6)$$

Substituting from $(2\cdot1\cdot5)$ into $(2\cdot1\cdot6)$ then, for the maximum shearing stress,

$$\tau = p_0 \frac{r\sin\theta}{c}\left(\frac{c^2}{r_1 r_2}\right)^{\frac{3}{2}}. \qquad (2\cdot1\cdot7)$$

* *Note added in proof.* The solution of the elastic equations due to Westergaard, and used in §2 of this paper, corresponds to a uniform distribution of pressure along the surfaces of the crack. By using a method similar to that of §3, but involving Fourier cosine transforms instead of Hankel transforms, it is possible to extend the analysis to cases in which the pressure along the crack is variable. This has been done and will be published shortly in *The Quarterly of Applied Mathematics* by Sneddon, I. N. & Elliott, H. A. under the title: 'The opening of a Griffith crack under internal pressure.'

It is easily verified from the equations (2·1·5) that when $y = 0$ and $-c < x < c$

$$\sigma_x = \sigma_y = -p_0, \quad \tau_{xy} = 0,$$

and that when $|x| > c$

$$\sigma_x = \sigma_y = p_0\left(\frac{x}{\sqrt{(c^2-x^2)}} - 1\right), \quad \tau_{xy} = 0.$$

Thus if $x-c$ is small and positive the normal component of stress $[\sigma_y]_{y=0}$ is given approximately by the expression

$$[\sigma_y]_{y=0} \doteqdot p_0 \sqrt{\frac{c}{2(x-c)}}. \tag{2·1·8}$$

To determine the distribution of stress in the immediate vicinity of the crack r_1 is taken to be a small quantity, δ say; to conform with the notation of § 1·4 take θ_1 to be ψ, but it should be noted that there are no restrictions on the magnitude of ψ. Since δ/c is assumed small then, from (2·1·4),

$$\begin{aligned} r &= c + \delta \cos \psi, & r_2 &= 2c + \delta \cos \psi, \\ \theta &= \delta \sin \psi / c, & \theta_2 &= \delta \sin \psi / 2c, \end{aligned} \tag{2·1·9}$$

to the first order of small quantities. Substituting from (2·1·9) into equations (2·1·5) then the expressions

$$\tfrac{1}{2}(\sigma_x + \sigma_y) = p_0\left[\left(\frac{c}{2\delta}\right)^{\frac{1}{2}} \cos \tfrac{1}{2}\psi \left\{1 - \frac{3\delta}{4c} + O(\delta^2/c^2)\right\} - 1\right], \tag{2·1·10}$$

$$\tfrac{1}{2}(\sigma_y - \sigma_x) = p_0\left(\frac{c}{8\delta}\right)^{\frac{1}{2}} \sin \psi \left\{\sin \tfrac{3}{2}\psi - \frac{3\delta}{4c} \sin \tfrac{1}{2}\psi + O(\delta^2/c^2)\right\}, \tag{2·1·11}$$

$$\tau_{xy} = p_0\left(\frac{c}{8\delta}\right)^{\frac{1}{2}} \sin \psi \left\{\cos \tfrac{3}{2}\psi - \frac{3\delta}{4c} \cos \tfrac{1}{2}\psi + O(\delta^2/c^2)\right\}, \tag{2·1·12}$$

are obtained for the determination of the components of stress near the edge of the crack. Similarly, one obtains from equation (2·1·7)—or directly from equations (2·1·11) and (2·1·12)—the expression

$$\tau = p_0\left(\frac{c}{8\delta}\right)^{\frac{1}{2}} \sin \psi \left\{1 - \frac{3\delta}{4c} \cos \psi + O(\delta^2/c^2)\right\} \tag{2·1·13}$$

for the principal shearing stress. If terms in δ/c of order greater than $-\tfrac{1}{2}$ are neglected in these expressions then one obtains the approximate formulae (1·4·1–3). Again, putting $x = c + \xi$, $w = \eta$, $b = 2\epsilon^2/c$ in equation (2·1·2) then the approximate relation

$$\eta^2 = -b\xi \tag{2·1·14}$$

is obtained on neglecting terms of the second order in ξ.

2·2. To give some idea of the distribution of stress in the neighbourhood of the crack, the maximum shearing stress τ was calculated for several values of x and y by means of equation (2·1·7). The results are embodied in table 1 and the variation of τ with x and y shown graphically in figure 1. In the case of plane strain (which is the case considered here) the condition of constant energy of distortion states (Nadai 1931, p. 184) that plastic flow is initiated when τ reaches the value k, where the constant k is related to s_0, the yield point in tension of the material, by the equation

$$3k^2 = s_0^2. \qquad (2 \cdot 2 \cdot 1)$$

The maximum shear theory gives precisely the same condition except that now

$$2k = s_0. \qquad (2 \cdot 2 \cdot 2)$$

Thus in both cases the inception of plastic flow can be predicted from the behaviour of the function τ defined by equation $(2 \cdot 1 \cdot 7)$.

Table 1. Variation of τ/p_0 with x and y in the two-dimensional problem

y \ x	0·0	0·2	0·4	0·6	0·8	1·0	1·2	1·4	1·6	1·8	2·0	3·0
0·2	0·189	0·199	0·236	0·327	0·546	0·785	0·405	0·179	0·094	0·058	0·037	0·009
0·4	0·320	0·332	0·372	0·444	0·534	0·543	0·400	0·248	0·153	0·099	0·068	0·017
0·6	0·378	0·386	0·408	0·439	0·456	0·428	0·346	0·252	0·176	0·124	0·089	0·024
0·8	0·381	0·384	0·390	0·395	0·386	0·354	0·298	0·235	0·178	0·134	0·101	0·031
1·0	0·354	0·354	0·352	0·345	0·329	0·299	0·262	0·210	0·171	0·135	0·106	0·036
1·2	0·315	0·313	0·308	0·298	0·281	0·256	0·225	0·192	0·159	0·130	0·106	0·039
1·4	0·275	0·273	0·267	0·257	0·242	0·222	0·198	0·172	0·147	0·124	0·103	0·042
1·6	0·238	0·236	0·231	0·222	0·209	0·193	0·174	0·154	0·134	0·115	0·099	0·044
1·8	0·206	0·205	0·200	0·192	0·182	0·169	0·154	0·138	0·123	0·107	0·093	0·044
2·0	0·179	0·178	0·174	0·168	0·159	0·149	0·137	0·124	0·112	0·099	0·087	0·044
3·0	0·095	0·094	0·093	0·091	0·088	0·084	0·080	0·076	0·071	0·066	0·061	0·039

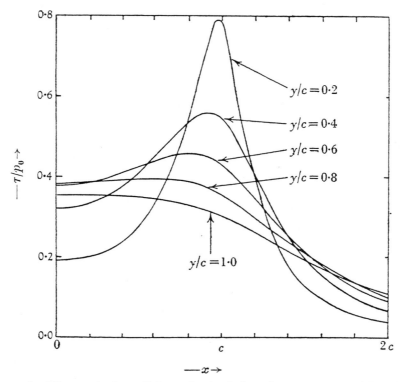

Figure 1. The variation of the principal shearing stress, τ, with x and y.

A convenient method of showing the variation of this function and of visualizing the distribution of stress in the interior of the medium consists of plotting the contours of equal principal shearing stress, i.e. constructing the family of curves $\tau = \alpha p_0$, or

$$c^2 y^4 = \alpha \{(x-c)^2 + y^2\}^{\frac{3}{2}} \{(x+c)^2 + y^2\}^{\frac{3}{2}}, \qquad (2\cdot2\cdot3)$$

where α is a parameter. These curves are the 'isochromatic lines' of photoelasticity. The family of curves was constructed by determining from table 1 and figure 1 the points in the vicinity of the crack at which the parameter α has the values 0·7, 0·5, 0·4, 0·3 and 0·2. The result is shown in figure 2. The shape of the isochromatic lines in the neighbourhood of the edge of the crack can readily be deduced from equation (2·1·13); in polar co-ordinates δ and ψ with origin at $x = c$ the equation of an isochromatic curve reduces to the simple form

$$\delta = \beta \sin^2 \psi, \qquad (2\cdot2\cdot4)$$

where $\beta = c/8\alpha^2$. Thus near the edge of the crack the isochromatics are curves of two loops situated symmetrically with regard to the axis of the crack.

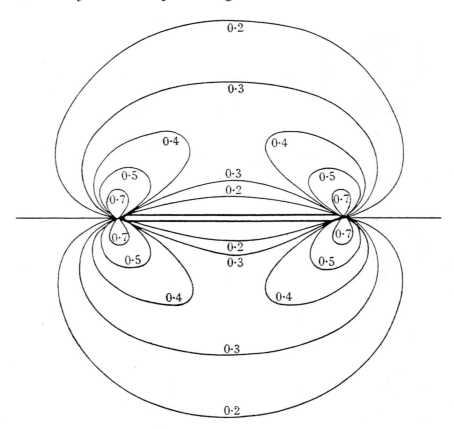

FIGURE 2. The isochromatic lines in the vicinity of a Griffith crack.

The fact that all the isochromatic lines pass through the points $(\pm c, 0)$ shows that at both of these points the principal shearing stress is infinite; thus even for small internal pressures p_0 plastic flow occurs at the corners of the crack to remove this infinite stress. There is, in fact, no purely *elastic* solution of the problem; if, however, the internal pressure is not too large the region of plastic flow will be small and will not appreciably affect the distribution of stress at points in the solid at a distance from the corners of the crack.

Although there seems to be no photoelectric determinations of the stress distribution in the vicinity of a crack there is a certain degree of similarity between the theoretical isochromatic lines of figure 2 and those in the vicinity of a tunnel as determined experimentally by Farquharson (Frocht 1941, p. 347). The shape of the tunnel differs appreciably from that of a crack of small transverse thickness, but the resemblance of the shape of the isochromatic lines in the neighbourhood of the lower straight edge of the tunnel to that of the lines of figure 2 is striking.

2·3. Now consider the energy of the crack. When the semi-minor axis of the crack is ϵ the internal pressure has the value
$$p = \frac{E}{2(1-\sigma^2)}\left(\frac{\epsilon}{c}\right).$$

Considering an element of length of the crack it is found that the work done in increasing the semi-minor axis to $\epsilon + d\epsilon$ is
$$2p\,dx\,dw = \frac{E}{1-\sigma^2}\left(\frac{\epsilon + \tfrac{1}{2}d\epsilon}{c}\right) dx \sqrt{\left(1-\frac{x^2}{c^2}\right)} d\epsilon,$$
so that the energy of the crack is
$$W = \frac{2E}{(1-\sigma^2)c}\int_0^c \sqrt{\left(1-\frac{x^2}{c^2}\right)} dx \int_0^\epsilon \epsilon\, d\epsilon$$
$$= \frac{\pi E \epsilon^2}{4(1-\sigma^2)} = \frac{\pi(1-\sigma^2)p_0^2 c^2}{E}, \qquad (2\cdot3\cdot1)$$
where p_0 refers to the final pressure.

The theory of the Griffith crack, considers the distribution of stress in the vicinity of a crack when the crack is opened under the influence of an average tension p_0 applied to the surface of the body; the surface of the crack is assumed to be free from stress. The solution of this problem is obtained by superposing on the solution (2·1·1) a second solution
$$Z = +p_0. \qquad (2\cdot3\cdot2)$$
This ensures that $\sigma_x = \sigma_y = p_0$ for large values of x and y, and that $\sigma_x = \tau_{xy} = 0$ across the surface of the crack. Equation (2·3·1) then shows that the presence of a crack of length $2c$ lowers the potential energy of the solid by an amount W. On the other hand, the crack has a surface energy
$$U = 4cT, \qquad (2\cdot3\cdot3)$$
where T is the surface tension of the material. Thus the total diminution of the potential energy due to the presence of the crack is
$$W - U = \pi(1-\sigma^2)p_0^2 c^2/E - 4cT.$$

The condition
$$\frac{\partial}{\partial c}(W-U) = 0$$
that the crack may extend then leads to the equation
$$c = \frac{2ET}{\pi p_0^2(1-\sigma^2)}.$$
If c is less than this value or if p_0 exceeds the critical value p_c given by the relation
$$p_c = \sqrt{\left\{\frac{2ET}{\pi c(1-\sigma^2)}\right\}}, \tag{2.3.4}$$
the crack will become unstable and spread. Equation (2·3·4) is the (corrected) Griffith criterion for rupture in the case of plane strain, equation (1·1·2) above.

3. The three-dimensional model

3·1. In the three-dimensional case assume that a crack is created in the interior of an infinite elastic medium and that it is 'penny-shaped', occupying the circle $r^2 = x^2 + y^2 = c^2$ in the plane $z = 0$. An examination of the equations of §2·1 shows immediately that in the case of the two-dimensional model the distribution of stress in the neighbourhood of a crack of length $2c$ lying along the axis $y = 0$, is identical to that in the semi-infinite elastic medium $y \geqslant 0$ when the conditions on the boundary $y = 0$ are prescribed by the relations

$$\begin{aligned}&\text{(i)} \quad \tau_{xy} = 0 \quad \text{for all values of } x,\\ &\text{(ii)} \quad \sigma_y = -p_0 \ (|x|<c), \quad u_y = 0 \ (|x|>c).\end{aligned} \tag{3.1.1}$$

In the three-dimensional case assume that the distribution of stress in the neighbourhood of the crack is the same as that produced in a semi-infinite elastic medium when certain conditions are prescribed on its bounding plane.

For a crack of the shape described above there is symmetry about the z-axis, so cylindrical co-ordinates r, θ, z may be employed; in this co-ordinate system the displacement vector assumes the form $(u_r, 0, u_z)$, and the stress in the interior of the medium is specified completely by the stress components $\sigma_r, \sigma_z, \sigma_\theta, \tau_{rz}$, the remaining components being identically zero.

By analogy with equations (3·1·1) it may now be assumed that the distribution of stress in the neighbourhood of the crack is the same as that produced in the interior of the semi-infinite elastic solid $z \geqslant 0$ by the boundary conditions

$$\begin{aligned}&\text{(i)} \quad \tau_{rz} = 0 \quad \text{for all values of } r,\\ &\text{(ii)} \quad \sigma_z = -p(r) \ (r<c), \quad u_z = 0 \ (r>c),\end{aligned} \tag{3.1.2}$$

on the plane $z = 0$. In the first instance assume that the applied internal pressure p is a function of r; the case of a constant internal pressure will be examined later.

The only restriction on the function $p(r)$ imposed by the subsequent analysis is that it is such that the integral
$$\int_0^c r p(r) J_0(\xi r) dr$$
exists for all real values of the parameter ξ.

The conditions (3·1·2) do not correspond to a hydrostatic pressure in the crack in all cases. In the analyses of Griffiths and Sack the crack is assumed to be of very small depth, so that the surface of the crack may be taken to be coincident with the z-axis; in this instance the pressure across the surface of the crack is $-\sigma_z$ and the conditions (3·1·2) are exact. Even in cases where the depth of the crack is taken into account the distribution of stress given by the conditions (3·1·2) will not differ much from that given by a hydrostatic pressure inside the crack—except possibly near the boundary of the crack. It will be seen later (§ 3·5) that in the case where the internal pressure is constant throughout the crack and the medium is incompressible the conditions (3·1·2) correspond exactly to a hydrostatic pressure.

With the axial symmetry the equations of elastic equilibrium reduce in the absence of body forces to (Love 1934, p. 90)

$$\frac{\partial \sigma_r}{\partial r} + \frac{\partial \tau_{rz}}{\partial z} + \frac{\sigma_r - \sigma_\theta}{r} = 0, \tag{3·1·3}$$

$$\frac{\partial \tau_{rz}}{\partial r} + \frac{\partial \sigma_z}{\partial z} + \frac{\tau_{rz}}{r} = 0. \tag{3·1·4}$$

It has been shown (Harding & Sneddon 1945) that these equations and the conditions of compatibility are satisfied by the forms

$$\sigma_r = \int_0^\infty \xi \{\lambda G''' + (\lambda + 2\mu) \xi^2 G'\} J_0(\xi r) d\xi - \frac{2(\lambda+\mu)}{r} \int_0^\infty \xi^2 G' J_1(\xi r) d\xi, \tag{3·1·5}$$

$$\sigma_z = \int_0^\infty \xi \{(\lambda + 2\mu) G''' - (3\lambda + 4\mu) \xi^2 G'\} J_0(\xi r) d\xi, \tag{3·1·6}$$

$$\sigma_\theta = \lambda \int_0^\infty \xi \{G''' - \xi^2 G'\} J_0(\xi r) d\xi + \frac{2(\lambda+\mu)}{r} \int_0^\infty \xi^2 G' J_1(\xi r) d\xi, \tag{3·1·7}$$

$$\tau_{rz} = \int_0^\infty \xi^2 \{\lambda G'' + (\lambda + 2\mu) \xi^2 G\} J_1(\xi r) d\xi, \tag{3·1·8}$$

and as noted above
$$\tau_{z\theta} = \tau_{r\theta} = 0. \tag{3·1·9}$$

In these expressions λ and μ are Lamé's elastic constants and the function $G(\xi, z)$ is a solution of the ordinary differential equation

$$\left(\frac{d^2}{dz^2} - \xi^2\right)^2 G = 0. \tag{3·1·10}$$

The constants of integration introduced into the solution of this fourth-order equation are determined from the imposed boundary conditions. The solutions given above are such that all the components of stress tend to zero as $r \to \infty$; to ensure further that all these components tend to zero as $z \to \infty$ assume a solution of equation (3·1·10) of the form

$$G = (A + Bz) e^{-\xi z}, \qquad (3·1·11)$$

where A and B are functions of the parameter ξ.

The corresponding expressions for the non-vanishing components of the displacement vector are

$$u_r = \frac{\lambda + \mu}{\mu} \int_0^\infty \xi^2 G' J_1(\xi r) \, d\xi, \qquad (3·1·12)$$

$$u_z = \int_0^\infty \xi \left\{ G'' - \frac{\lambda + 2\mu}{\mu} \xi^2 G \right\} J_0(\xi r) \, d\xi. \qquad (3·1·13)$$

To determine the values of the arbitrary functions A and B of equation (3·1·11) it is necessary to evaluate τ_{rz}, u_z, σ_z on the plane $z = 0$. By repeated differentiation of (3·1·11) and by use of the results

$$\lambda = \frac{E\sigma}{(1+\sigma)(1-2\sigma)}, \quad \mu = \frac{E}{2(1+\sigma)} \qquad (3·1·14)$$

expressing the Lamé constants in terms of the Young's modulus E and the Poisson ratio σ, then, on substituting $z = 0$ into equation (3·1·8),

$$[\tau_{rz}]_{z=0} = \frac{E}{(1+\sigma)(1-2\sigma)} \int_0^\infty \xi^3 (\xi A - 2\sigma B) J_1(\xi r) \, d\xi.$$

Inverting this result by means of the Hankel inversion theorem, then

$$\xi^2(\xi A - 2\sigma B) = \frac{(1+\sigma)(1-2\sigma)}{E} \int_0^\infty r[\tau_{rz}]_{z=0} J_1(\xi r) \, d\xi.$$

If the boundary conditions $[\tau_{rz}]_{z=0} = 0$ holds for all values of r it follows that

$$\xi A = 2\sigma B \qquad (3·1·15)$$

between the two constants of integration A and B. Substituting from equation (3·1·15) into equation (3·1·11), differentiating with respect to z and then putting $z = 0$ it is found from equations (3·1·6) and (3·1·13) that

$$[\sigma_z]_{z=0} = \frac{E}{(1+\sigma)(1-2\sigma)} \int_0^\infty \xi^3 B J_0(\xi r) \, d\xi, \qquad (3·1·16)$$

$$[u_z]_{z=0} = -\frac{2(1-\sigma)}{1-2\sigma} \int_0^\infty \xi^2 B J_0(\xi r) \, d\xi. \qquad (3·1·17)$$

Writing (3·1·11) in the form

$$G = \frac{B(c\xi)}{\xi} (2\sigma + \xi z) e^{-\xi z} \qquad (3·1·18)$$

and then putting $\eta = \xi c$ in equations (3·1·16) and (3·1·17), one obtains finally

$$[\sigma_z]_{z=0} = \frac{E}{c^4(1+\sigma)(1-2\sigma)} \int_0^\infty \eta f(\eta) J_0(\rho\eta) \, d\eta \qquad (3\cdot 1\cdot 19)$$

and

$$[u_z]_{z=0} = -\frac{2(1-\sigma)}{c^3(1-2\sigma)} \int_0^\infty f(\eta) J_0(\rho\eta) \, d\eta, \qquad (3\cdot 1\cdot 20)$$

where

$$f(\eta) = \eta^2 B(\eta) \qquad (3\cdot 1\cdot 21)$$

and $\rho = r/c$.

3·2. Inserting the boundary conditions

$$z = 0, \quad u_z = 0 \; (r > c), \quad \sigma_z = -p(r) \; (r < c)$$

into equations (3·1·19) and (3·1·20), the pair of integral equations

$$\left.\begin{aligned}\int_0^\infty \eta f(\eta) J_0(\rho\eta) \, d\eta &= g(\rho) \quad (0 < \rho < 1), \\ \int_0^\infty f(\eta) J_0(\rho\eta) \, d\eta &= 0 \quad (\rho > 1),\end{aligned}\right\} \qquad (3\cdot 2\cdot 1)$$

is obtained for the determination of the unknown function $f(\eta)$ and hence of the constant $B(c\xi)$ of equation (3·1·18); in the former of these two equations is written

$$g(\rho) = -\frac{(1+\sigma)(1-2\sigma)}{E} c^4 p(\rho c). \qquad (3\cdot 2\cdot 2)$$

Dual integral equations of the type (3·2·1) have been considered by Titchmarsh (1937, p. 334) and Busbridge (1938); the solution of the pair (3·2·1) can be derived easily from Busbridge's analysis in the form

$$f(\eta) = \frac{2}{\pi} \int_0^1 \mu \sin \mu\eta \, d\mu \int_0^1 \frac{\rho g(\rho\mu)}{\sqrt{(1-\rho^2)}} \, d\rho. \qquad (3\cdot 2\cdot 3)$$

Substituting from (3·2·3) into (3·1·20) and making use of the result (Watson 1944, p. 405)

$$\int_0^\infty \sin \mu\eta J_0(\rho\eta) \, d\eta = \begin{cases} 0 & (\rho > \mu), \\ (\mu^2 - \rho^2)^{-\frac{1}{2}} & (\rho < \mu), \end{cases}$$

it is found that when the applied pressure is $p(r)$ the value of the normal component of the surface displacement is given by the equation

$$[u_z]_{z=0} = \frac{4(1-\sigma^2)c}{\pi E} \int_\rho^1 \frac{\mu \, d\mu}{\sqrt{(\mu^2-\rho^2)}} \int_0^1 \frac{xp(x\mu c)}{\sqrt{(1-x^2)}} \, dx \quad (\rho < 1). \qquad (3\cdot 2\cdot 4)$$

If, for example, the applied pressure is given by the power series

$$p(r) = p_0 \sum_{n=0}^\infty \alpha_n \left(\frac{r}{c}\right)^n, \quad (0 < r < c), \qquad (3\cdot 2\cdot 5)$$

then substituting into equation (3·2·3) one obtains

$$f(\eta) = -\frac{(1+\sigma)(1-2\sigma)}{\sqrt{\pi} E} c^4 p_0 \sum_{n=0}^\infty \frac{\Gamma(1+\frac{1}{2}n)}{\Gamma(\frac{3}{2}+\frac{1}{2}n)} \alpha_n \int_0^1 \mu^{n+1} \sin \mu\eta \, d\mu, \qquad (3\cdot 2\cdot 6)$$

so that the normal component of the surface displacement now has the value

$$[u_z]_{z=0} = \frac{2}{\sqrt{\pi}}(1-\sigma^2)\frac{p_0}{E}\sum_{n=0}^{\infty}\frac{\Gamma(1+\tfrac{1}{2}n)}{\Gamma(\tfrac{3}{2}+\tfrac{1}{2}n)}\alpha_n I_n\left(\frac{r}{c}\right)^{n+1} \quad (0<r<c), \qquad (3\cdot2\cdot7)$$

where I_n denotes the integral

$$I_n = c\int_1^{c/r}\frac{\eta^{n+1}}{\sqrt{(\eta^2-1)}}\,d\eta. \qquad (3\cdot2\cdot8)$$

Integrating by parts it can readily be shown that I_n satisfies the recurrence relation

$$I_n = \frac{n}{n+1}I_{n-2} + \frac{1}{n+1}\left(\frac{c}{r}\right)^{n+1}(c^2-r^2)^{\frac{1}{2}}. \qquad (3\cdot2\cdot9)$$

By means of these equations the value of the normal component of the surface displacement can be calculated when the law of variation of the applied pressure is known.

The value of the radial component of the surface displacement is found from equation (3·1·12) to be

$$[u_r]_{z=0} = \frac{1}{c^3}\int_0^{\infty} f(\eta)\,J_1(\rho\eta)\,d\eta, \qquad (3\cdot2\cdot10)$$

whence by equations (3·2·2) and (3·2·3) and making use of the result (Watson 1944, p. 405) we obtain

$$\int_0^{\infty} J_1(\rho\eta)\sin\mu\eta\,d\eta = \begin{cases} \dfrac{\mu}{\rho}(\rho^2-\mu^2)^{-\frac{1}{2}} & (\mu<\rho), \\ 0 & (\mu>\rho), \end{cases} \qquad (3\cdot2\cdot11)$$

then, in the general case,

$$[u_r]_{z=0} = -\frac{2(1+\sigma)(1-2\sigma)c}{\pi E}\int_0^1 \mu^2\,d\mu \int_\mu^1 \frac{p(\rho\mu c)\,d\rho}{\sqrt{\{(1-\rho^2)(\rho^2-\mu^2)\}}}. \qquad (3\cdot2\cdot12)$$

It will be observed from this equation that the radial component of the surface displacement is zero when the medium is incompressible ($\sigma = \tfrac{1}{2}$).

3·3. If it be supposed that the applied pressure $p(r)$ is constant over a circular area of radius $a \leqslant c$, then

$$p(r) = \begin{cases} p_0 & (0<r<a), \\ 0 & (a<r<c). \end{cases} \qquad (3\cdot3\cdot1)$$

Substituting into equation (3·2·4) one obtains the expression

$$w = \frac{4p_0(1-\sigma^2)}{\pi E}(c^2-r^2)^{\frac{1}{2}}\{1-(1-a^2/c^2)^{\frac{1}{2}}\} \qquad (3\cdot3\cdot2)$$

for the normal component of the surface displacement, $w = [u_z]_{z=0}$. If $w = \epsilon$ when $r = 0$ then

$$\epsilon = \frac{4(1-\sigma^2)p_0 c}{\pi E}\{1-(1-a^2/c^2)^{\frac{1}{2}}\}, \qquad (3\cdot3\cdot3)$$

and equation (3·3·2) may be written in the form

$$\frac{r^2}{c^2}+\frac{w^2}{\epsilon^2} = 1, \tag{3·3·4}$$

showing that for all values of $a \leqslant c$, the crack resulting from the application of an internal pressure is ellipsoidal in shape provided the applied pressure is constant. For a point force applied at the centre of symmetry allow the radius a to tend to zero and the applied pressure p_0 to tend to infinity in such a way that the total force, $\pi a^2 p_0$, remains constant, P say; in this instance one obtains from equation (3·3·3)

$$\epsilon = \frac{2(1-\sigma^2)}{\pi^2}\frac{P}{cE}.$$

If the pressure p_0 acts over a circle of radius c, i.e. over the entire surface of the crack, and produces a displacement ϵ of the centre of the circle then

$$p_0 = \frac{\pi E \epsilon}{4(1-\sigma^2)c}, \tag{3·3·5}$$

from which it will be observed that p_0 is constant for constant values of the ratio ϵ/c. Now consider the work done in forming an ellipsoidal depression of circular section of radius c and of depth ϵ. When the depth of the depression is ϵ the pressure is given by equation (3·3·5); when the depth of the depression has increased to $\epsilon + d\epsilon$ the pressure will have increased to $\pi E(\epsilon+d\epsilon)/[4(1-\sigma^2)c]$, so that in bringing about a change $d\epsilon$ of the depth of the ellipsoidal depression the average pressure is

$$p_0 = \frac{\pi E(\epsilon + \tfrac{1}{2}d\epsilon)}{4(1-\sigma^2)c}.$$

Considering a ring element of surface intersecting the plane $\theta = 0$ at the points P, P' (see figure 3) it is found that the work done in making a small normal displacement dw is

$$p_0 2\pi r\, dr\, dw = \frac{\pi E(\epsilon + \tfrac{1}{2}d\epsilon)}{4(1-\sigma^2)c} 2\pi r\, dr\, d\epsilon (1-r^2/c^2)^{\tfrac{1}{2}}$$

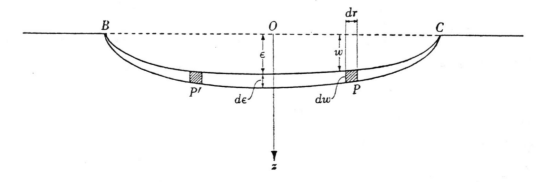

FIGURE 3

by equation (3·3·2). Thus the total work done in forming the ellipsoidal depression on the surface of a semi-infinite elastic solid is

$$W_1 = \frac{\pi^2 E}{2(1-\sigma^2)} \int_0^c r(1-r^2/c^2)^{\frac{1}{2}} dr \int_0^\epsilon \epsilon \, d\epsilon,$$

which gives
$$W_1 = \frac{\pi^2 E c \epsilon^2}{12(1-\sigma^2)}.$$

The energy of a crack formed by internal pressure is twice this energy; multiplying the last result by 2 and substituting for the value of ϵ in terms of the final pressure p_0 one obtains for the energy of a crack of radius c the formula

$$W = 2W_1 = \frac{8(1-\sigma^2)}{3E} p_0^2 c^3. \tag{3·3·6}$$

3·4. It was observed above (equations (3·3·3) and (3·3·4)) that as long as the internal applied pressure is constant the crack will be ellipsoidal in shape whatever the radius of the circle over which it is applied. It is natural then to try to determine whether cracks of a shape other than ellipsoidal are possible, and if they are to determine the distribution of internal pressure necessary to preserve their shape. This is equivalent to assuming that the value of $[u_z]_{z=0}$ is known when $r \leqslant c$, but that the function $p(r)$ is unknown. One might then regard equation (3·2·4) as an integral equation determining $p(r)$ when $[u_z]_{z=0}$ is known. It is, however, simpler to consider the problem of determining the distribution of stress in a semi-infinite elastic medium bounded by the plane $z = 0$ when the surface value of the normal component of the displacement vector is prescribed for all values of r. The analysis proceeds along similar lines except that now the boundary conditions

(i) $\tau_{zr} = 0$, for all values of r,

(ii) $u_z = \begin{cases} w(r) & (r < c), \\ 0 & (r > c), \end{cases}$

replace the set (3·1·2). Substituting the condition (ii) into equation (3·1·17) then

$$\int_0^\infty \xi^2 B J_0(\xi r) \, d\xi = -\begin{cases} \dfrac{(1-2\sigma)}{2(1-\sigma)} w(r) & (r < c), \\ 0 & (r > c). \end{cases}$$

Inverting this result by means of the Hankel inversion theorem

$$\xi B = -\frac{(1-2\sigma)}{2(1-\sigma)} \int_0^c r w(r) J_0(\xi r) \, dr. \tag{3·4·1}$$

Denoting the integral on the right-hand side by $\overline{w}(\xi)$ then, from equation (3·1·16), it is found that

$$[\sigma_z]_{z=0} = -\frac{E}{2(1-\sigma^2)} \int_0^\infty \xi^2 \overline{w}(\xi) J_0(\xi r) \, d\xi. \tag{3·4·2}$$

Equation (3·4·2) gives the value of the applied pressure on the circle $r \leqslant c$ which must be applied to maintain the prescribed surface displacement.

For example, if it is assumed that
$$w(r) = \epsilon(1 - r^2/c^2), \tag{3·4·3}$$
then from equation (3·4·1) it follows that
$$\overline{w}(\xi) = 2\epsilon J_2(c\xi)/\xi^2,$$
and so by equation (3·4·2) that the normal component of the surface stress is given by the expression
$$[\sigma_z]_{z=0} = -\frac{E\epsilon}{1-\sigma^2} \int_0^\infty J_2(c\xi) J_0(r\xi) \, d\xi.$$
Using the recurrence relation
$$J_2(c\xi) = J_0(c\xi) - \frac{2}{c} \frac{\partial}{\partial \xi} J_1(c\xi),$$
after an integration by parts it is found that
$$\int_0^\infty J_0(r\xi) J_2(c\xi) \, d\xi = \int_0^\infty J_0(r\xi) J_0(c\xi) \, d\xi - \frac{2r}{c} \int_0^\infty J_1(r\xi) J_1(c\xi) \, d\xi.$$
Now by Gübler's formula (Watson 1944, p. 410) the value of the first integral is found to be
$$\left(\frac{1}{c}\right) {}_2F_1\!\left[\tfrac{1}{2}, \tfrac{1}{2}; 1; \frac{r^2}{c^2}\right] = \frac{2}{\pi c} K(r/c) \quad (r < c),$$
in the usual notation for elliptic integrals (Jahnke-Emde 1938, p. 85). The second integral could also be evaluated by the Gübler formula, but the hypergeometric series involved in this instance is not readily summed. The integration can be carried out more easily by making use of Neumann's result
$$J_1(\xi r) J_1(\xi c) = \frac{1}{\pi} \int_0^\pi J_0(\xi R) \cos\theta \, d\theta, \quad R^2 = c^2 + r^2 - 2cr\cos\theta.$$
Multiplying both sides of this equation by $\exp\{-\xi z\}$ and integrating with respect to ξ from 0 to ∞, then
$$\int_0^\infty e^{-\xi z} J_1(c\xi) J_1(r\xi) \, d\xi = \frac{1}{\pi} \int_0^\pi \cos\theta \, d\theta \int_0^\infty J_0(\xi R) e^{-\xi z} \, d\xi = \frac{1}{\pi} \int_0^\pi \frac{\cos\theta \, d\theta}{\sqrt{(R^2 + z^2)}}.$$
Letting z tend to zero on both sides of the last equation one obtains the required result
$$\int_0^\infty J_1(\xi r) J_1(\xi c) \, d\xi = \frac{1}{\pi} \int_0^\pi \frac{\cos\theta \, d\theta}{R}.$$
The integral on the right-hand side can easily be evaluated in terms of elliptic integrals; with the usual notation for complete elliptic integrals, then
$$\frac{2r}{c} \int_0^\infty J_1(\xi r) J_1(\xi c) \, d\xi = \frac{2\{1 + (r/c)^2\}}{\pi(c+r)} K(k) - \frac{2(c+r)}{\pi c^2} E(k),$$
where
$$k^2 = 4rc/(r+c)^2.$$

Thus the displacement (3·4·3) is maintained by an applied internal pressure

$$p(r) = -[\sigma_z]_{z=0} = \frac{2E\epsilon}{\pi c(1-\sigma^2)} f(\rho), \tag{3·4·4}$$

where the function $f(\rho)$ is defined by the equation

$$f(\rho) = \frac{1+\rho^2}{1+\rho} K(k) - K(\rho) - (1+\rho) E(k) \tag{3·4·5}$$

and $\rho = r/c$, $k^2 = 4\rho/(1+\rho)^2$. The variation of the function $f(\rho)$ with ρ is shown graphically in figure 4. From this curve it follows that the applied internal force is *compressive* in the body of the crack, but that near the circular edge of the crack the applied force becomes *tensile* until round the edge it is both infinite and tensile. This is precisely the kind of variation which would be predicted on physical grounds for a crack of this shape.

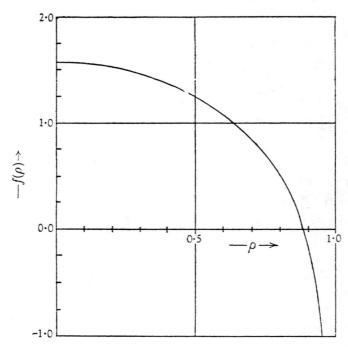

FIGURE 4. Variation of the function $f(\rho)$ with ρ.

3·5. Now return to the problem of determining the distribution of stress in the interior of the medium, in the case where the applied pressure $p(r)$ is a constant p_0 over the entire surface area $r \leqslant c$ of the crack. Putting $p(\rho c) = p_0$, then from equations (3·2·2) and (3·2·3),

$$f(\eta) = \frac{2p_0 c^4(1+\sigma)(1-2\sigma)}{\pi E} \frac{d}{d\eta}\left(\frac{\sin\eta}{\eta}\right). \tag{3·5·1}$$

Substituting this value for $f(\eta)$ into equation (3·1·19) it follows that, for the normal component of stress across the plane $z = 0$,

$$[\sigma_z]_{z=0} = \frac{2p_0}{\pi}\left[\rho \int_0^\infty \sin\eta\, J_1(\rho\eta)\, d\eta - \int_0^\infty \frac{\sin\eta}{\eta} J_0(\rho\eta)\, d\eta\right]. \tag{3·5·2}$$

By means of the results (Watson 1944, p. 405)

$$\int_0^\infty \frac{\sin\eta}{\eta} J_0(\rho\eta)\,d\eta = \begin{cases} \sin^{-1}(1/\rho) & (\rho \geq 1), \\ \tfrac{1}{2}\pi & (\rho \leq 1), \end{cases}$$

$$\int_0^\infty \rho \sin\eta\, J_1(\rho\eta)\,d\eta = \begin{cases} (\rho^2-1)^{-\frac{1}{2}} & (\rho > 1), \\ 0 & (\rho < 1), \end{cases}$$

it is verified that $\sigma_z = -p_0$ when $\rho < 1$, and for the value of the normal component of the surface stress when $\rho > 1$ the expression

$$[\sigma_z]_{z=0} = -\frac{2p_0}{\pi}\left[\sin^{-1}\frac{1}{\rho} - \frac{1}{\sqrt{(\rho^2-1)}}\right] \tag{3.5.3}$$

is obtained. It will be observed that when $\rho = 1$ the stress $[\sigma_z]_{z=0}$ is infinite and that when $\rho - 1$ is small and positive

$$[\sigma_z]_{z=0} \doteqdot \frac{p_0}{\pi}(\rho-1)^{-\frac{1}{2}}, \tag{3.5.4}$$

showing that there is a large *tensile* stress in the neighbourhood of the circle $\rho = 1$, $z = 0$. Furthermore, as $\rho \to \infty$,

$$[\sigma_z]_{z=0} \sim \frac{4p_0}{3\pi}\rho^{-3}, \tag{3.5.5}$$

the stress being still tensile. Since

$$\frac{d}{d\rho}[\sigma_z]_{z=0} = -\frac{2p_0}{\pi}\{\rho(\rho^2-1)^{\frac{3}{2}}\}^{-1}$$

does not change sign in the region $\rho > 1$, it follows that the normal component of the surface stress $[\sigma_z]_{z=0}$ never becomes compressive in that region. The similarity of the expression (3.5.4) to that for the corresponding stress in the two-dimensional analysis (equation (2.1.8)) is also of interest.

For the values of the other stress components on the plane $z = 0$ one obtains from equations (3.1.5)–(3.1.8) the expressions

$$[\sigma_r + \sigma_\theta]_{z=0} = (1+2\sigma)[\sigma_z]_{z=0},$$

$$[\sigma_r - \sigma_\theta]_{z=0} = \frac{2(1-\sigma)p_0}{\pi}\int_0^\infty \eta J_2(\rho\eta)\frac{d}{d\eta}\left(\frac{\sin\eta}{\eta}\right)d\eta$$

$$= \frac{2(1-2\sigma)p_0}{\pi}\left[\int_0^\infty J_2(\rho\eta)\cos\eta\,d\eta - \int_0^\infty J_2(\rho\eta)\frac{\sin\eta}{\eta}\,d\eta\right].$$

Evaluating the integrals in the square bracket it is found that the bracket vanishes when $\rho < 1$ and has the value

$$-(\rho^2-1)^{-\frac{1}{2}}$$

when $\rho > 1$. Thus if $\rho < 1$ and $z = 0$, then

$$\sigma_\theta = \sigma_r = -(\sigma + \tfrac{1}{2})p_0,$$

and, if $\rho > 1$, $z = 0$, $\quad \sigma_r = \dfrac{2p_0}{\pi}\left[(\rho^2-1)^{-\frac{1}{2}} - (\sigma+\tfrac{1}{2})\sin^{-1}\dfrac{1}{\rho}\right],$

$$\sigma_\theta = \dfrac{2p_0}{\pi}\left[2\sigma(\rho^2-1)^{-\frac{1}{2}} - (\sigma+\tfrac{1}{2})\sin^{-1}\dfrac{1}{\rho}\right].$$

It will be noted that in the case $\sigma = \tfrac{1}{2}$ (incompressible medium) $\sigma_r = \sigma_z = \sigma_\theta = -p_0$ ($\rho < 1$), so that in this instance the internal pressure is truly hydrostatic.

3·6. The evaluation of the stress components in the interior of the medium is similar to that involved in the solution of the Boussinesq problem for a cylinder (Sneddon 1946). Obtaining the value of $B(c\xi)$ from equations (2·1·21) and (3·5·1) and substituting from equation (3·1·18) into equations (3·1·12) and (3·1·13), then, for the non-vanishing components of the displacement vector,

$$u_r = \dfrac{2p_0 c}{\pi E}(1+\sigma)\int_0^\infty (1-2\sigma-\zeta\eta)\dfrac{d}{d\eta}\left(\dfrac{\sin\eta}{\eta}\right)e^{-\zeta\eta}J_1(\rho\eta)\,d\eta, \tag{3·6·1}$$

$$u_z = -\dfrac{4p_0 c(1-\sigma^2)}{\pi E}\int_0^\infty \left(1+\dfrac{\zeta\eta}{2(1-\sigma)}\right)\dfrac{d}{d\eta}\left(\dfrac{\sin\eta}{\eta}\right)e^{-\zeta\eta}J_0(\rho\eta)\,d\eta, \tag{3·6·2}$$

where $\rho = r/c$ and $\zeta = z/c$. Similarly, from equations (3·1·5)–(3·1·8) one obtains for the components of stress at a general point in the interior of the elastic solid

$$\sigma_z = \dfrac{2p_0}{\pi}\{C_1^0(\rho,\zeta) - S_0^0(\rho,\zeta) + \zeta C_2^0(\rho,\zeta) - \zeta S_1^0(\rho,\zeta)\}, \tag{3·6·3}$$

$$\tau_{zr} = \dfrac{2p_0\zeta}{\pi}\{C_2^1(\rho,\zeta) - S_1^1(\rho,\zeta)\}, \tag{3·6·4}$$

$$\sigma_r + \sigma_\theta + \sigma_z = \dfrac{4(1+\sigma)}{\pi}p_0\{C_1^0(\rho,\zeta) - S_0^0(\rho,\zeta)\}, \tag{3·6·5}$$

where C_n^m and S_n^m denote the integrals

$$C_n^m(\rho,\zeta) = \int_0^\infty \eta^{n-1}e^{-\zeta\eta}J_m(\rho\eta)\cos\eta\,d\eta, \quad S_n^m(\rho,\zeta) = \int_0^\infty \eta^{n-1}e^{-\zeta\eta}J_m(\rho\eta)\sin\eta\,d\eta. \tag{3·6·6}$$

The fourth relation for the complete determination of all the components of stress is found from equations (3·1·5) and (3·1·7) to be

$$\sigma_\theta - \sigma_r = 2(\lambda+\mu)\int_0^\infty \xi^3 G'\left\{\dfrac{2J_1(\xi r)}{\xi r} - J_0(\xi r)\right\}d\xi.$$

Putting $n = 1$, $z = \xi r$ in the recurrence formula

$$J_{n-1}(z) + J_{n+1}(z) = \dfrac{2n}{z}J_n(z), \tag{3·6·7}$$

then

$$\sigma_\theta - \sigma_r = 2(\lambda+\mu)\int_0^\infty \xi^3 G' J_2(\xi r)\,d\xi, \tag{3·6·8}$$

whence follows the relation

$$\sigma_\theta - \sigma_r = \dfrac{2p_0}{\pi}[(1-2\sigma)\{C_1^2(\rho,\zeta) - S_0^2(\rho,\zeta)\} - \zeta\{C_2^2(\rho,\zeta) - S_1^2(\rho,\zeta)\}]. \tag{3·6·9}$$

Along the axis of symmetry, $r = 0$, it is found that $\tau_{rz} = 0$, $\sigma_r - \sigma_\theta = 0$, so that the maximum shearing stress across any plane through a point on the axis of symmetry is given by the expression $|\tau|$, where

$$\tau = \tfrac{1}{2}(\sigma_z - \sigma_r) = -\frac{p_0}{2\pi}\left\{(1-2\sigma)\left(\int_0^\infty \cos\eta\, e^{-\zeta\eta}\, d\eta - \int_0^\infty \frac{\sin\eta}{\eta} e^{-\zeta\eta}\, d\eta\right) \right.$$
$$\left. + 3\zeta\left(\int_0^\infty \eta \cos\eta\, e^{-\zeta\eta}\, d\eta - \int_0^\infty \sin\eta\, e^{-\zeta\eta}\, d\eta\right)\right\}.$$

Evaluating the integrals

$$\tau = -\tfrac{1}{4}(1-2\sigma)p_0 + \frac{p_0}{2\pi}\left[(1-2\sigma)\left(\frac{\zeta}{1+\zeta^2} + \tan^{-1}\zeta\right) - \frac{6\zeta}{(1+\zeta^2)^2}\right]. \quad (3\cdot 6\cdot 10)$$

Differentiating this expression with respect to ζ then

$$\frac{d\tau}{d\zeta} = \frac{4p_0}{\pi}\frac{(5-\sigma)\zeta^2 - (1+\sigma)}{(1+\zeta^2)^3},$$

showing that $|\tau|$ increases from the value $\tfrac{1}{4}(1-2\sigma)p_0$ at $\zeta = 0$ to a maximum at $\zeta = \{(1+\sigma)/(5-\sigma)\}^{\frac{1}{2}}$ and then decreases steadily to zero at $\zeta = \infty$.

Writing, in the general case, $w = \zeta + i$, then

$$Z_n^m = C_n^m - iS_n^m = \int_0^\infty \eta^{n-1} e^{-w\eta} J_m(\rho\eta)\, d\eta,$$

and the integrals Z_n^m can be evaluated by means of the formulae

$$\int_0^\infty \eta^{n-1} e^{-w\eta} J_m(\rho\eta)\, d\eta = \begin{cases} \dfrac{(n-m+1)!}{(w^2+\rho^2)^{\frac{1}{2}n}} P_{n-1}^m\left(\dfrac{w}{\sqrt{(w^2+\rho^2)}}\right) & (m \leqslant n-1), \\ \dfrac{(n+m-1)!}{(w^2+\rho^2)^{\frac{1}{2}n}} P_{n-1}^{-m}\left(\dfrac{w}{\sqrt{(w^2+\rho^2)}}\right) & (m > n-1), \end{cases}$$

where P_n^m denotes the associated Legendre function.

In this way one obtains the results

$$Z_1^0 = (\rho^2 + w^2)^{-\frac{1}{2}}, \qquad Z_2^0 = w(\rho^2+w^2)^{-\frac{3}{2}},$$
$$Z_0^1 = \frac{1}{\rho}\{(\rho^2+w^2)^{\frac{1}{2}} - w\}, \qquad Z_1^1 = \frac{1}{\rho} - \frac{W}{\rho\sqrt{(w^2+\rho^2)}},$$
$$Z_2^1 = \rho(\rho^2+w^2)^{-\frac{3}{2}},$$

and by virtue of equation $(3\cdot 6\cdot 7)$ the further relation

$$Z_n^2 = \frac{2}{\rho}Z_{n-1}^1 - Z_n^0. \quad (3\cdot 6\cdot 11)$$

Adopting the notation

$$\left.\begin{aligned} r^2 &= 1+\zeta^2, & \zeta\tan\theta &= 1, \\ R^2 &= (\rho^2+\zeta^2-1)^2 + 4\zeta^2, & 2\zeta\cot\phi &= \rho^2 + \zeta^2 - 1, \end{aligned}\right\} \quad (3\cdot 6\cdot 12)$$

and equating real and imaginary parts one obtains the formulae

$$\left.\begin{aligned}
C_0^1 &= R^{-\frac{1}{2}}\cos\tfrac{1}{2}\phi, & S_1^0 &= R^{-\frac{1}{2}}\sin\tfrac{1}{2}\phi, \\
C_2^0 &= rR^{-\frac{3}{2}}\cos(\tfrac{3}{2}\phi-\theta), & S_2^0 &= rR^{-\frac{3}{2}}\sin(\tfrac{3}{2}\phi-\theta), \\
C_0^1 &= \frac{1}{\rho}(R^{\frac{1}{2}}\cos\tfrac{1}{2}\phi-\zeta), & S_0^1 &= \frac{1}{\rho}(1-R^{\frac{1}{2}}\sin\tfrac{1}{2}\phi), \\
C_1^1 &= \frac{1}{\rho}-\frac{r}{\rho}R^{-\frac{1}{2}}\cos(\theta-\tfrac{1}{2}\phi), & S_1^1 &= \frac{r}{\rho}R^{-\frac{1}{2}}\sin(\theta-\tfrac{1}{2}\phi), \\
C_2^1 &= \rho R^{-\frac{3}{2}}\cos\tfrac{3}{2}\phi, & S_2^1 &= \rho R^{-\frac{3}{2}}\sin\tfrac{3}{2}\phi,
\end{aligned}\right\} \quad (3\cdot6\cdot13)$$

and from $(3\cdot6\cdot12)$

$$\left.\begin{aligned}
C_1^2 &= \frac{2}{\rho}C_0^1 - C_1^0, \quad S_1^2 = \frac{2}{\rho}S_0^1 - S_1^0, \\
C_2^2 &= \frac{2}{\rho}C_1^1 - C_2^0.
\end{aligned}\right\} \quad (3\cdot6\cdot14)$$

It only remains to evaluate S_0^0 and S_0^2.

Integrating the expressions for Z_1^0, Z_0^1 with respect to w, then

$$\int_0^\infty \frac{1-e^{-wx}}{x} J_0(\rho x)\, dx = \log\frac{\surd(\rho^2+w^2)+w}{\rho}, \qquad (3\cdot6\cdot15)$$

$$\int_0^\infty \frac{1-e^{-wx}}{x^2} J_1(\rho x)\, dx = \frac{1}{2\rho}\left[w\surd(w^2+\rho^2)-w^2+\rho^2\log\frac{\surd(\rho^2+w^2)+w}{\rho}\right], \quad (3\cdot6\cdot16)$$

it being assumed in both cases that $\rho \neq 0$. With the notation of $(3\cdot6\cdot12)$ then from $(3\cdot6\cdot16)$ by equating imaginary parts

$$S_0^0 = \tan^{-1}\frac{R^{\frac{1}{2}}\sin\tfrac{1}{2}\phi + r\sin\theta}{R^{\frac{1}{2}}\cos\tfrac{1}{2}\phi + r\cos\theta} \quad (\rho\neq 0). \qquad (3\cdot6\cdot17)$$

Using the recurrence relation $(3\cdot6\cdot7)$ it is deduced, from equations $(3\cdot6\cdot15)$ and $(3\cdot6\cdot16)$, that if $\rho \neq 0$,

$$\int_0^\infty \frac{1-e^{-wx}}{x} J_2(\rho x)\, dx = \frac{1}{\rho^2}\{w\surd(w^2+\rho^2)-w^2\} = wZ_0^1/\rho.$$

Putting $w = \zeta+i$, $Z_0^1 = C_0^1 - iS_0^1$ and equating imaginary parts, then

$$S_0^2 = \frac{1}{\rho}(C_0^1 - \zeta S_0^1). \qquad (3\cdot6\cdot18)$$

3·7. Before proceeding to the determination of the distribution of stress in the medium at points remote from the crack consider the expressions for the stress components at points in the immediate vicinity of the periphery of the crack. In this way results are obtained analogous to those of §1·4; with a notation similar to that of that section the integrals of § 3·6 are computed when $\zeta = \delta \sin \psi / c$, and $\rho = 1 + \delta \cos \psi / c$. Then

$$r = 1, \qquad \phi = \psi - \tfrac{1}{2}\delta \sin \psi / c,$$
$$\theta = \tfrac{1}{2}\pi - \delta \sin \psi / c, \qquad R = 2\delta(c + \delta \cos \psi)/c^2. \tag{3.7.1}$$

Substituting from equations (3·7·1) into the formulae (3·6·13–18) and retaining only those terms which are of first or higher order in $(c/\delta)^{\frac{1}{2}}$, it follows that

$$S_1^1 = C_1^0 = -C_1^2 = \left(\frac{c}{2\delta}\right)^{\frac{1}{2}} \cos \tfrac{1}{2}\psi, \quad C_2^1 = \left(\frac{c}{2\delta}\right)^{\frac{3}{2}} \cos \tfrac{3}{2}\psi,$$

$$C_2^0 = -C_2^2 = \left(\frac{c}{2\delta}\right)^{\frac{3}{2}} \sin \tfrac{3}{2}\psi, \qquad S_1^0 = -S_1^2 = \left(\frac{c}{2\delta}\right)^{\frac{1}{2}} \sin \tfrac{1}{2}\psi,$$

the other integrals being negligible to this approximation. With these values for the integrals involved the following expressions are obtained from equations (3·6·3–5) and (3·6·9) for the stress components σ_r, σ_z and τ_{rz}:

$$\sigma_r = \frac{2p_0}{\pi}\left(\frac{c}{2\delta}\right)^{\frac{1}{2}} \{\tfrac{3}{4}\cos \tfrac{1}{2}\psi + \tfrac{1}{4}\cos \tfrac{5}{2}\psi\}, \tag{3.7.2}$$

$$\sigma_z = \frac{2p_0}{\pi}\left(\frac{c}{2\delta}\right)^{\frac{1}{2}} \{\tfrac{5}{4}\cos \tfrac{1}{2}\psi - \tfrac{1}{4}\cos \tfrac{5}{2}\psi\}, \tag{3.7.3}$$

$$\tau_{zr} = \frac{p_0}{\pi}\left(\frac{c}{2\delta}\right)^{\frac{1}{2}} \sin \psi \cos \tfrac{3}{2}\psi. \tag{3.7.4}$$

It will be observed that these formulae differ from the expressions for the components of stress in the two-dimensional case (equations (1·4·1–3)) only by the presence of a factor $2/\pi$. There is no parallel in the two-dimensional analysis to the hoop stress σ_θ; in the three-dimensional case its value near the edge of the crack is given by the equation

$$\sigma_\theta = \frac{4\sigma p_0}{\pi}\left(\frac{c}{2\delta}\right)^{\frac{1}{2}} \cos \tfrac{1}{2}\psi. \tag{3.7.5}$$

3·8. By means of the formulae of § 3·6 the components of stress at any point $r = \rho a$, $z = \zeta a$, in the interior of the elastic body can be calculated for any prescribed value of the Poisson ratio. The results of these calculations for the case where the Poisson ratio has the value 0·25 and where both ρ and ζ assume a sequence of values are given in tables 2–5. The variation of the various components of stress in certain planes is shown graphically in figures 5–8. The rapid decay of the stress components σ_r and σ_θ as z increases is illustrated by figures 5 and 6; the variation with z of the normal component of stress, σ_z, is much more gradual as is seen from figure 7. Perhaps the most striking feature of these results is the marked similarity between the variation of the shearing stress τ_{zr} and that of the principal shearing stress τ in the two-dimensional case (compare figures 1 and 8).

Table 2. The variation of σ_r/p_0 with ρ and ζ

ζ \ ρ	0·0	0·2	0·4	0·6	0·8	1·0	1·2	1·4	1·6	1·8	2·0
0·0	−0·750	−0·750	−0·750	−0·750	−0·750	∞	0·010	−0·055	−0·068	−0·069	−0·066
0·2	−0·446	−0·433	−0·386	−0·297	−0·180	−0·070	0·007	0·060	0·072	0·054	0·039
0·4	−0·214	−0·207	−0·160	−0·121	−0·087	−0·057	−0·041	−0·019	0·000	0·018	0·014
0·6	−0·075	−0·068	−0·060	−0·050	−0·039	−0·029	−0·019	−0·013	−0·007	0·000	0·007
0·8	−0·006	−0·001	0·007	0·013	0·022	0·014	0·006	0·001	0·000	−0·016	−0·026
1·0	0·017	0·015	0·012	0·010	0·005	−0·011	−0·022	−0·026	−0·028	−0·031	−0·027
1·2	0·031	0·029	0·026	0·023	0·022	0·019	0·015	0·009	−0·001	−0·007	−0·009
1·4	0·030	0·028	0·025	0·020	0·015	0·009	0·005	0·002	0·000	−0·001	−0·001
1·6	0·028	0·027	0·025	0·020	0·015	0·011	0·007	0·003	0·001	0·000	−0·001
1·8	0·024	0·023	0·022	0·019	0·015	0·010	0·007	0·004	0·001	0·001	0·000
2·0	0·021	0·020	0·018	0·016	0·013	0·010	0·006	0·005	0·003	0·002	0·000
3·0	0·009	0·009	0·008	0·008	0·007	0·006	0·005	0·004	0·003	0·002	0·002
4·0	0·004	0·004	0·004	0·004	0·003	0·003	0·003	0·003	0·002	0·002	0·002

Table 3. The variation of σ_θ/p_0 with ρ and ζ

ζ \ ρ	0·0	0·2	0·4	0·6	0·8	1·0	1·2	1·4	1·6	1·8	2·0
0·0	−0·750	−0·750	−0·750	−0·750	−0·750	∞	0·010	−0·055	−0·068	−0·069	−0·066
0·2	−0·446	−0·438	−0·412	−0·344	−0·242	−0·100	0·045	0·022	0·010	0·005	0·004
0·4	−0·214	−0·219	−0·178	−0·131	−0·067	−0·014	0·025	0·018	0·013	0·009	0·006
0·6	−0·075	−0·072	−0·051	−0·009	0·016	0·020	0·023	0·013	0·007	0·006	0·005
0·8	−0·006	−0·009	−0·017	−0·023	−0·019	−0·008	0·009	0·011	0·014	0·017	0·012
1·0	0·017	0·019	0·026	0·033	0·030	0·026	0·022	0·018	0·017	0·015	0·003
1·2	0·031	0·039	0·045	0·053	0·061	0·058	0·039	0·021	0·011	0·019	0·006
1·4	0·030	0·035	0·039	0·044	0·048	0·054	0·048	0·041	0·033	0·025	0·016
1·6	0·028	0·032	0·037	0·040	0·043	0·046	0·042	0·038	0·033	0·029	0·021
1·8	0·024	0·027	0·029	0·032	0·034	0·036	0·038	0·036	0·032	0·028	0·020
2·0	0·021	0·023	0·025	0·027	0·029	0·031	0·032	0·030	0·027	0·025	0·019
3·0	0·009	0·009	0·010	0·010	0·011	0·011	0·013	0·012	0·011	0·010	0·010
4·0	0·004	0·005	0·005	0·006	0·006	0·006	0·006	0·006	0·006	0·005	0·005

Table 4. The variation of σ_z/p_0 with ζ and ρ

ζ \ ρ	0·0	0·2	0·4	0·6	0·8	1·0	1·2	1·4	1·6	1·8	2·0
0·0	−1·000	−1·000	−1·000	−1·000	−1·000	∞	0·333	0·143	0·080	0·050	0·034
0·2	−0·987	−0·985	−0·975	−0·937	−0·729	0·232	0·358	0·166	0·089	0·055	0·035
0·4	−0·917	−0·912	−0·862	−0·742	−0·451	−0·006	0·189	0·150	0·095	0·060	0·040
0·6	−0·788	−0·768	−0·700	−0·560	−0·334	−0·081	0·068	0·097	0·080	0·053	0·041
0·8	−0·639	−0·617	−0·548	−0·429	−0·263	−0·108	0·002	0·048	0·055	0·048	0·038
1·0	−0·508	−0·482	−0·424	−0·333	−0·237	−0·113	−0·032	0·014	0·018	0·022	0·030
1·2	−0·386	−0·373	−0·329	−0·262	−0·185	−0·109	−0·049	−0·009	0·012	0·021	0·023
1·4	−0·302	−0·286	−0·256	−0·210	−0·154	−0·094	−0·055	−0·022	−0·001	0·010	0·015
1·6	−0·230	−0·224	−0·201	−0·168	−0·129	−0·090	−0·055	−0·027	−0·010	0·001	0·008
1·8	−0·180	−0·175	−0·159	−0·138	−0·108	−0·079	−0·054	−0·032	−0·010	−0·005	0·002
2·0	−0·142	−0·139	−0·127	−0·111	−0·091	−0·070	−0·050	−0·033	−0·019	−0·012	−0·003
3·0	−0·052	−0·051	−0·049	−0·045	−0·040	−0·035	−0·029	−0·024	−0·019	−0·014	−0·010
4·0	−0·025	−0·024	−0·023	−0·022	−0·021	−0·019	−0·017	−0·014	−0·013	−0·011	−0·009

TABLE 5. THE VARIATION OF τ_{zr}/p_0 WITH ρ AND ζ

ζ \ ρ	0·0	0·2	0·4	0·6	0·8	1·0	1·2	1·4	1·6	1·8	2·0
0·0	0·000	0·000	0·000	0·000	0·000	0·000	0·000	0·000	0·000	0·000	0·000
0·2	0·000	−0·020	−0·051	−0·121	−0·327	−0·420	0·066	0·052	0·030	0·017	0·011
0·4	0·000	−0·055	−0·126	−0·233	−0·354	−0·319	−0·097	0·003	0·021	0·019	0·014
0·6	0·000	−0·074	−0·155	−0·239	−0·292	−0·252	−0·133	−0·041	−0·006	0·018	0·007
0·8	0·000	−0·074	−0·145	−0·202	−0·225	−0·200	−0·131	−0·069	−0·029	−0·009	0·000
1·0	0·000	−0·061	−0·119	−0·159	−0·170	−0·156	−0·117	−0·075	−0·046	−0·021	−0·009
1·2	0·000	−0·050	−0·092	−0·123	−0·133	−0·124	−0·100	−0·076	−0·047	−0·028	−0·016
1·4	0·000	−0·040	−0·070	−0·092	−0·102	−0·097	−0·084	−0·065	−0·048	−0·032	−0·020
1·6	0·000	−0·029	−0·053	−0·070	−0·078	−0·077	−0·069	−0·057	−0·044	−0·032	−0·022
1·8	0·000	−0·021	−0·040	−0·053	−0·060	−0·061	−0·057	−0·049	−0·040	−0·031	−0·023
2·0	0·000	−0·016	−0·031	−0·041	−0·047	−0·049	−0·046	−0·042	−0·035	−0·029	−0·022
3·0	0·000	−0·005	−0·009	−0·012	−0·015	−0·017	−0·018	−0·018	−0·017	−0·016	−0·014
4·0	0·000	−0·002	−0·004	−0·005	−0·006	−0·007	−0·007	−0·008	−0·008	−0·008	−0·008

The principal stresses at any point are determined by the roots of the discriminating cubic

$$\begin{vmatrix} \sigma - \sigma_r & 0 & \tau_{zr} \\ 0 & \sigma - \sigma_\theta & 0 \\ \tau_{zr} & 0 & \sigma - \sigma_z \end{vmatrix} = 0, \qquad (3\cdot8\cdot1)$$

so that they are equal to

$$\sigma_\theta, \quad \tfrac{1}{2}(\sigma_r + \sigma_z) \pm \{(\tfrac{1}{2}\sigma_r - \tfrac{1}{2}\sigma_z)^2 + \tau_{rz}^2\}^{\frac{1}{2}}. \qquad (3\cdot8\cdot2)$$

With the values of the stress components given in tables 2–5 the components of principal stress may readily be deduced from the formulae (3·8·2). The numerical value of the principal shearing stress can then be determined from the fact that it is equal to one-half of the algebraic difference between the maximum and the minimum components of the principal stress. The results of such a calculation are embodied in table 6, and the variation of the principal shearing stress in planes parallel to the plane of the crack exhibited graphically in figure 9. In general outline this variation is similar to that of the principal shearing stress in the two-dimensional case.

TABLE 6. THE VARIATION OF THE PRINCIPAL SHEARING STRESS (DIVIDED BY p_0) WITH ρ AND ζ

ζ \ ρ	0·0	0·2	0·4	0·6	0·8	1·0	1·2	1·4	1·6	1·8	2·0
0·0	0·125	0·125	0·125	0·125	0·125	∞	0·172	0·099	0·074	0·059	0·050
0·2	0·270	0·277	0·299	0·342	0·427	0·446	0·188	0·085	0·051	0·033	0·021
0·4	0·351	0·356	0·373	0·388	0·398	0·320	0·151	0·085	0·047	0·033	0·023
0·6	0·357	0·358	0·356	0·349	0·327	0·254	0·140	0·069	0·044	0·032	0·025
0·8	0·317	0·316	0·313	0·299	0·266	0·201	0·131	0·073	0·040	0·032	0·032
1·0	0·263	0·256	0·248	0·234	0·209	0·164	0·117	0·075	0·044	0·031	0·009
1·2	0·209	0·207	0·200	0·188	0·168	0·139	0·105	0·077	0·048	0·030	0·017
1·4	0·169	0·162	0·157	0·147	0·123	0·110	0·089	0·066	0·048	0·032	0·022
1·6	0·129	0·129	0·125	0·117	0·107	0·092	0·076	0·059	0·045	0·032	0·023
1·8	0·102	0·100	0·099	0·095	0·086	0·076	0·064	0·052	0·040	0·031	0·023
2·0	0·082	0·081	0·079	0·075	0·070	0·063	0·054	0·046	0·037	0·030	0·022
3·0	0·030	0·030	0·030	0·029	0·028	0·026	0·025	0·023	0·020	0·018	0·016

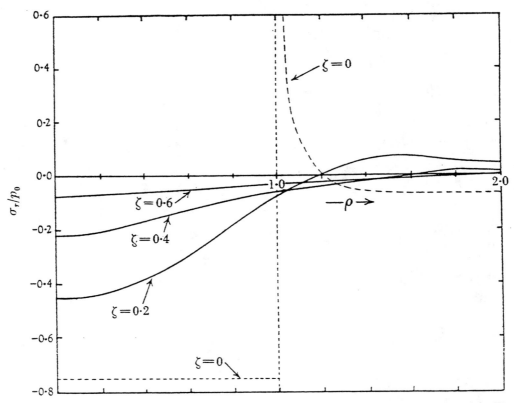

FIGURE 5. The variation of the radial component of stress, σ_r, with ρ and ζ. The broken curve shows the variation of σ_r in the plane of the crack ($z = 0$).

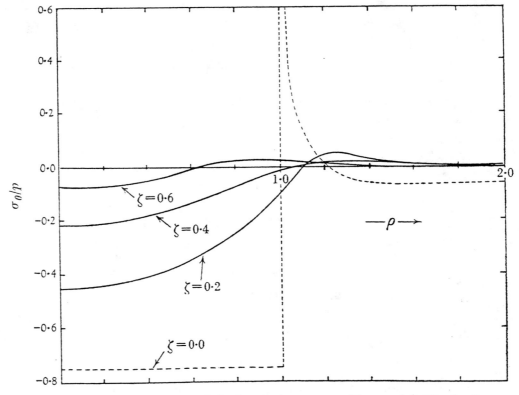

FIGURE 6. The variation of the hoop stress, σ_θ, with ρ and ζ. The broken curve shows the variation of σ_θ in the plane of the crack ($z = 0$).

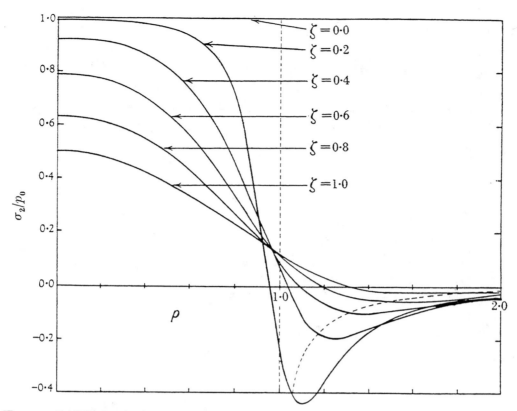

FIGURE 7. The variation of the normal component of stress, σ_z, with ρ and ζ. The broken curve shows the variation of σ_z in the plane of the crack ($z = 0$).

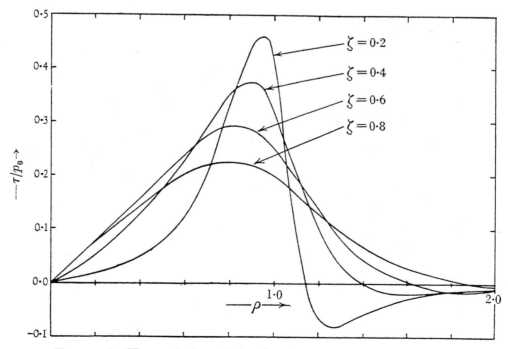

FIGURE 8. The variation of the shearing stress, τ_{zr}, with ρ and ζ.

The most convenient way of visualizing the distribution of stress in an elastic body is probably to draw the contours of equal principal shearing stress—the three-dimensional analogue of the isochromatic lines of photoelasticity. By means of table 6 and figure 9, points in the z-r plane at which the principal shearing stress

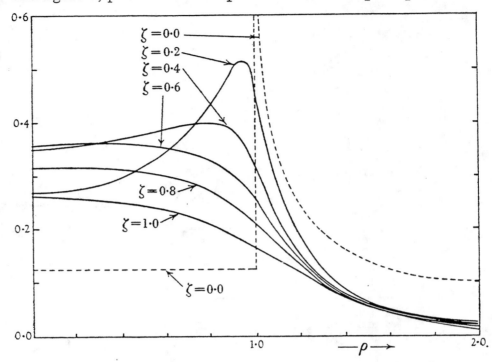

FIGURE 9. The variation of the principal shearing stress, τ, with ρ and ζ. The broken curve shows the variation of τ in the plane of the crack.

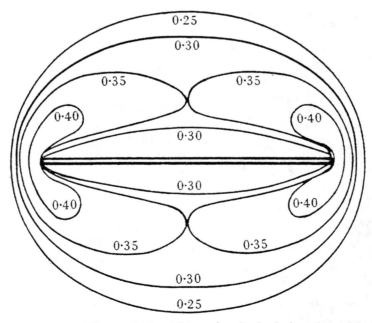

FIGURE 10. The contours of equal principal shearing stress in the vicinity of a circular crack.

reaches the values 0·25, 0·30, 0·35, 0·40 can readily be determined (in terms of p_0 as unit), and by drawing the curves through corresponding values the contours of (equal) principal shearing stress can be plotted. The result is shown in figure 10. It will be observed that except in the vicinity of the edges of the crack—and apart from numerical values—these contours of equal principal shearing stress are very like the isochromatic lines of the two-dimensional case (figure 2). The main difference between the two systems of curves lies in their behaviour near the edges of the crack. In the two-dimensional case the points $x = \pm c$ are nodes separating two loops of the curves (cf. the curves corresponding to the values $\tau/p_0 = 0·4, 0·5, 0·7$ in figure 2), but in the three-dimensional case the point $r = c$ is a simple point of the curve, the curve cutting the r-axis again at a point where $r > c$ (cf. curves with τ/p_0 in figure 10). It should, of course, be remembered that these 'contours' are in reality *surfaces* of equal principal shearing stress obtained by rotating figure 10 about the z-axis.

As in the two-dimensional case the principal shearing stress becomes infinite at the edges of the crack, indicating that a certain amount of plastic flow will occur and hence that there is in reality no purely elastic problem. Except in the neighbourhood of the crack, however, the elastic stresses are dominant and the energy relation (3·3·6) approximates very closely to the true result.

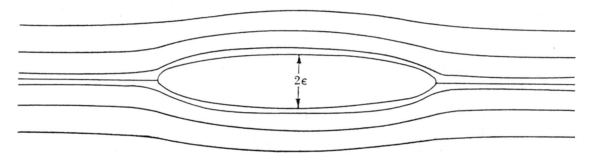

FIGURE 11. The normal displacement of planes parallel to the plane of an ellipsoidal crack of small depth. The depth of the crack, 2ϵ, has been greatly exaggerated in this diagram to show the variation of the normal displacement more clearly.

3·9. The distribution of stress in the vicinity of a crack can also be illustrated by calculating the normal component of the displacement vector for points in the elastic body and then drawing the normal displacement of planes which were originally parallel to the plane of the crack. Since the normal component of stress σ_z is infinitely large at the edge of the crack it might at first sight appear that the displacement u_z should also be infinite in that region; that this is not so can readily be seen from an examination of equation (3·6·2). It follows immediately from this equation that the normal component of the displacement vector is given by the formula

$$u_z = \epsilon \left[1 - \rho S_0^1 - \zeta S_0^0 - \frac{\zeta}{2(1-\sigma)} (\zeta S_1^0 + \rho S_1^1 - S_0^0) \right] \qquad (3·9·1)$$

with the notation of (3·6·6). By means of this formula the variation of the normal displacement along the planes $z/c = 0·05, 0·2, 0·4$ was calculated. The results are shown graphically in figure 11, which shows the position of the planes $z/c = 0·05, 0·2,$

0·4 after the crack has been opened out to a depth 2ϵ, by a uniform internal pressure; in this diagram the ratio ϵ/c has been very much exaggerated to show more clearly the variation of the normal displacement in planes near to the crack.

Approximate formulae for the determination of the normal component of the displacement vector in planes close to the crack can be derived from equation (3·9·1). If ρ is small, then
$$J_0(\rho\eta) = 1 - \tfrac{1}{4}\rho^2\eta^2, \quad J_1(\rho\eta) = \tfrac{1}{2}\rho\eta$$
may be written in the integrals for C_n^m and S_n^m; making this substitution, evaluating the integrals now involved and retaining terms of (3·9·1) of order two or less one obtains the approximate result
$$u_z = \epsilon\{f_1(\zeta) - \rho^2 f_2(\zeta)\} + O(\epsilon\rho^4), \tag{3·9·2}$$
where the functions $f_1(\zeta)$ and $f_2(\zeta)$ are defined by the relations
$$\left.\begin{aligned}f_1(\zeta) &= 1 - \frac{(1-2\sigma)}{2(1-\sigma)}\zeta\left(\frac{\pi}{2} - \tan^{-1}\zeta\right) - \frac{1}{2(1-\sigma)}\frac{\zeta^2}{1+\zeta^2},\\ f_2(\zeta) &= \left(\frac{1}{2} - \frac{3-\sigma}{2(1-\sigma)}\zeta^2\right)(1+\zeta^2)^{-3}.\end{aligned}\right\} \tag{3·9·3}$$

By means of equation (3·9·2) may be calculated the normal displacement at points just off the axis of symmetry $r = 0$.

Similarly, by taking ρ to be large the integrals involved by means of equations (3·6·13) can be evaluated; since as a first approximation one may take $R = \rho$, $\phi = 2\zeta/\rho^2$ in these equations. Thus for ρ large and $\zeta \ll \rho$ one has the approximate formula
$$u_z = \frac{\epsilon\zeta}{\rho}\left\{1 - \frac{\zeta}{2(1-\sigma)}\right\} + O\left(\frac{\epsilon}{\rho^2}\right). \tag{3·9·4}$$

The value of the displacement vector in the neighbourhood of the periphery of the crack $r = c$ can be determined by the methods of §3·7. With the notation of that section
$$u_z = \epsilon\sqrt{\left(\frac{2\delta}{c}\right)}\sin\tfrac{1}{2}\psi\left[1 - \frac{1+\cos\psi}{4(1-\sigma)}\right], \tag{3·9·5}$$
where $z = \delta\sin\psi$ and $r = c + \delta\cos\psi$.

The relation of the approximate formulae (3·9·2–5) to the exact result (3·9·1) is shown in figure 12. In this diagram the value of the displacement u_z for points along the plane $z/c = 0.05$ are shown; the full line gives the values of the displacement as calculated by equation (3·9·1) for $\sigma = 0.25$; the broken lines show the values given by the approximate formulae, for the same value of the Poisson ratio. It will be observed that the agreement is good within the range specified. It is to be expected that for smaller values of δ the expression (3·9·5) will give more accurate results.

3·10. The effect of a circular crack on the strength of a brittle solid subjected to a uniform tensile stress can be deduced immediately from the analysis of §§ 3·5 and 3·6. Assume that the radius c of the crack (which may be taken to be situated in the plane $z = 0$ with its centre at the origin of co-ordinates) is considerably smaller than the dimensions of the elastic body. Then replace the finite elastic body by an infinite elastic body with a crack in its interior which is subjected to the stress system

$$\sigma_z = p_0, \quad \sigma_r = \sigma_\theta = p_1 \tag{3.10.1}$$

when r is very large. Since the surface of the crack may be assumed to be free from external forces the further boundary conditions

$$\sigma_z = 0, \quad \tau_{zr} = 0 \tag{3.10.2}$$

follow when $z = 0$ and $r < c$.

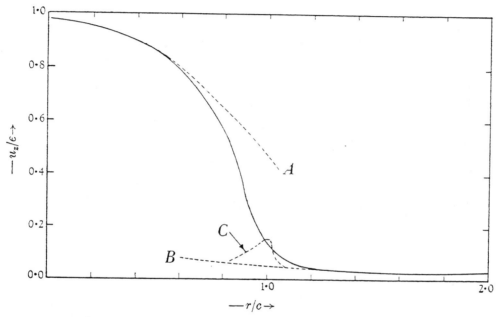

FIGURE 12. The variation of the normal displacement u_z in the plane $z = 0.05$ c.c. The full curve corresponds to the exact values calculated from equation (3.9.1). The broken curves correspond to the approximate formulae—A to (3.9.2), B to (3.9.4) and C to (3.9.5).

If the solution
$$\tau_{zr} = 0, \quad \sigma_z = p_0, \quad \sigma_r = \sigma_\theta = p_1 \tag{3.10.3}$$

be added to equations (3.6.3), (3.6.4), (3.6.5) and (3.6.9), the expressions so obtained for the components of stress satisfy the equations of equilibrium (3.1.3) and (3.1.4) and the boundary conditions (3.10.1) and (3.10.2). The distribution of stress in the interior of the solid can then be determined from the results of §3.7 by a very simple calculation, or from figures 5–8 by a simple change of scale along the axis corresponding to the stress component.

It follows immediately from equation (3.3.6) that if a brittle body is acted upon by uniform stress $\sigma_r = \sigma_\theta = p_1$, $\sigma_z = p_0$, the presence of a circular crack of radius c lowers the potential energy of the medium by an amount

$$W = \frac{8(1-\sigma^2)p_0^2 c^3}{3E} \tag{3.10.4}$$

which is independent of the value of p_1. The crack has also a surface energy

$$U = 2\pi c^2 T, \qquad (3\cdot10\cdot5)$$

where T denotes the surface tension of the material of the elastic body. The condition that the crack may spread is

$$\frac{\partial}{\partial c}(W - U) = 0$$

or

$$c = \frac{\pi E T}{2 p_0^2 (1-\sigma^2)},$$

so that the crack will become unstable and spread if p_0 exceeds the critical value

$$p_{cr} = \sqrt{\frac{\pi E T}{2c(1-\sigma^2)}}$$

This is the criterion derived by Sack (1946) (equation $(1\cdot2\cdot1)$ above).

I should like to take this opportunity of expressing my thanks to Professor N. F. Mott for acquainting me with this problem and for various suggestions received during many helpful discussions on the theory of cracks. I am also indebted to Mr R. Sack, of the H. H. Wills Physical Laboratory, the University of Bristol, for putting his unpublished work at my disposal, and to Miss June Tenwick for her assistance in checking the calculations of §3·8.

References

Busbridge, I. W. 1938 *Proc. Lond. Math. Soc.* (2), **44**, 115.
Frocht, M. M. 1941 *Photoelasticity*, p. 347. New York: McGraw Hill.
Griffith, A. A. 1920 *Phil. Trans.* A, **221**, 163.
Griffith, A. A. 1924 *Proc. Int. Congr. Appl. Mech. Delft*, p. 55.
Harding, J. W. & Sneddon, I. N. 1945 *Proc. Camb. Phil. Soc.* **41**, 16.
Inglis, C. E. 1913a *Engineering*, **95**, 415.
Inglis, C. E. 1913b *Trans. Instn Naval Archit.* **55**, 219.
Jahnke, E. & Emde, F. 1938 *Tables of Functions*. Leipzig: Teubner.
Love, A. E. H. 1934 *Mathematical theory of elasticity*, 4th ed. Cambridge.
Nadai, A. 1931 *Plasticity*. New York: McGraw Hill.
Neuber, H. 1934 *Z. angew. Math. Mech.* **14**, 203.
Orowan, E. 1934 *Zs. Kristallogr.* **89**, 327.
Sack, R. 1946 To be published shortly.
Sneddon, I. N. 1945 *Proc. Camb. Phil. Soc.* **42**, 29.
Titchmarsh, E. C. 1937 *An introduction to the theory of Fourier integrals*. Oxford Univ. Press.
Watson, G. N. 1944 *The theory of Bessel functions*, 2nd ed. Cambridge Univ. Press.
Westergaard, H. M. 1939 *J. Appl. Mech.* **6**, A, 49.

THE DISTRIBUTION OF STRESS IN THE NEIGHBOURHOOD OF A FLAT ELLIPTICAL CRACK IN AN ELASTIC SOLID

By A. E. GREEN and I. N. SNEDDON

Received 26 January 1949

1. *Introduction.* This paper is mainly concerned with the distribution of stress near a flat elliptical crack in a body of infinite extent under a uniform tension at infinity perpendicular to the plane of the crack. After an analytical solution of this problem was found the authors received a copy of a paper by Sadowsky and Sternberg (3) in which they solved the more general problem of the stress concentration around a tri-axial ellipsoidal cavity in an elastic body of infinite extent, the body at infinity being in a uniform state of stress whose principal axes are parallel to the axes of the cavity. The method of solution adopted by the present writers for the special case of the elliptical crack is somewhat different from the more general work of Sadowsky and Sternberg, and is, moreover, surprisingly simple, so it seems to be of value to present this solution here.

In addition, the problem of the indentation of a flat semi-infinite surface by a flat-ended elliptical cylindrical punch is also considered. This problem is, however, only partly solved.

2. *Fundamental equations.* If rectangular cartesian coordinates x, y, Z, with $z = x + iy$ the complex variable and $\bar{z} = x - iy$ its complex conjugate are used, it is known (1) that the elastic equations of equilibrium and the stress-strain relations for isotropic materials may be put in the form*

$$\left.\begin{array}{l} D = 4\dfrac{\partial F(z, \bar{z}, Z)}{\partial \bar{z}}, \quad w = \dfrac{\partial G(z, \bar{z}, Z)}{\partial Z}, \\[6pt] \Theta = -\dfrac{\mu'}{1-\eta}\dfrac{\partial^2}{\partial Z^2}(H - 2\eta G), \\[6pt] \Phi = 16\mu'\dfrac{\partial^2 F}{\partial \bar{z}^2}, \quad \Psi = 2\mu'\dfrac{\partial L}{\partial \bar{z}}, \\[6pt] \widehat{zz} = -\mu'\nabla_1^2 H = -\dfrac{\mu'}{1-\eta}\dfrac{\partial^2}{\partial Z^2}(\eta H - 2G), \end{array}\right\} \quad (2\cdot 1)$$

where u, v, w are the components of displacements and where

$$D = u + iv, \quad \Theta = \widehat{xx} + \widehat{yy}, \quad \Phi = \widehat{xx} - \widehat{yy} + 2i\widehat{xy}, \quad \Psi = \widehat{xz} + i\widehat{yz}, \quad (2\cdot 2)$$

$$\left.\begin{array}{l} 2(1-2\eta)\nabla^2 F = (1-2\eta)\nabla^2 G = -\Delta, \\[6pt] \Delta = \nabla_1^2(F + \bar{F}) + \dfrac{\partial^2 G}{\partial Z^2}, \quad \nabla_1^2 = 4\dfrac{\partial^2}{\partial z \partial \bar{z}}, \\[6pt] H = F + \bar{F} + G, \quad L = \dfrac{\partial}{\partial Z}(2F + G), \quad L + \bar{L} = 2\dfrac{\partial H}{\partial Z}. \end{array}\right\} \quad (2\cdot 3)$$

* The z in the notation for the stresses need not be confused with the complex variable.

A bar placed over a quantity denotes the complex conjugate of that quantity. Poisson's ratio η is related to Lamé's constants λ', μ' by $\eta = \frac{1}{2}\lambda'/(\lambda'+\mu')$.

For the present work, functions F, G, H which satisfy (2·3) may be expressed in the form

$$\left. \begin{array}{l} F = 2\omega + 4(1-\eta)\phi + 2Z\dfrac{\partial \phi}{\partial Z}, \\[6pt] G = 4\omega - 8(1-\eta)\phi + 4Z\dfrac{\partial \phi}{\partial Z}, \\[6pt] H = 8\omega \qquad\qquad\quad + 8Z\dfrac{\partial \phi}{\partial Z}, \end{array} \right\} \quad (2\cdot 4)$$

where ω, ϕ are real harmonic functions, i.e. they satisfy the equations

$$\nabla^2 \omega = 0, \quad \nabla^2 \phi = 0. \qquad (2\cdot 5)$$

3. *Statement of the problem for an elliptical crack.* Suppose that the crack occupies the ellipse

$$\frac{x^2}{a^2} + \frac{y^2}{b^2} = 1 \qquad (3\cdot 1)$$

in the plane $Z = 0$. A uniform tension p at infinity acts parallel to the Z-axis. The resulting distribution of stress consists of the system

$$\widehat{zz} = p, \quad \Theta = \Phi = \Psi = 0, \qquad (3\cdot 2)$$

together with a distribution of stress in the region $Z \geqslant 0$ which vanishes at infinity and which satisfies the conditions

$$\left. \begin{array}{ll} \text{(i)} \ \Psi = 0 & (Z = 0), \\[4pt] \text{(ii)} \ \widehat{zz} = -p & \left(\dfrac{x^2}{a^2} + \dfrac{y^2}{b^2} < 1,\ Z = 0\right), \\[6pt] \phantom{\text{(ii)}}\ w = 0 & \left(\dfrac{x^2}{a^2} + \dfrac{y^2}{b^2} > 1,\ Z = 0\right). \end{array} \right\} \quad (3\cdot 3)$$

From (2·1), (2·3) and (2·4) it is seen that the condition (i) in (3·3) can be satisfied by taking
$$\omega + \phi = 0$$
everywhere. The displacements and stresses may then be reduced to the forms

$$\left. \begin{array}{l} D = 8\dfrac{\partial}{\partial \bar z}\left\{(1-2\eta)\phi + Z\dfrac{\partial \phi}{\partial Z}\right\}, \\[8pt] w = -8(1-\eta)\dfrac{\partial \phi}{\partial Z} + 4Z\dfrac{\partial^2 \phi}{\partial Z^2}, \\[8pt] \Theta = -8\mu'\left\{(1+2\eta)\dfrac{\partial^2 \phi}{\partial Z^2} + Z\dfrac{\partial^3 \phi}{\partial Z^3}\right\}, \\[8pt] \Phi = 32\mu'\dfrac{\partial^2}{\partial \bar z^2}\left\{(1-2\eta)\phi + Z\dfrac{\partial \phi}{\partial Z}\right\}, \\[8pt] \widehat{zz} = -8\mu'\dfrac{\partial^2 \phi}{\partial Z^2} + 8\mu' Z\dfrac{\partial^3 \phi}{\partial Z^3}, \\[8pt] \Psi = 16\mu' Z\dfrac{\partial^3 \phi}{\partial \bar z\, \partial Z^2}. \end{array} \right\} \quad (3\cdot 4)$$

It will be noticed that the expressions (3·4) are valid for any problem in which condition (i) in (3·3) is satisfied, independently of (ii). On using (3·4) condition (ii) may now be put in the form

$$8\mu' \frac{\partial^2 \phi}{\partial Z^2} = p \quad \left(\frac{x^2}{a^2}+\frac{y^2}{b^2}<1,\ Z=0\right),$$
$$\frac{\partial \phi}{\partial Z} = 0 \quad \left(\frac{x^2}{a^2}+\frac{y^2}{b^2}>1,\ Z=0\right). \quad (3\cdot 5)$$

Since ϕ, and therefore $\partial\phi/\partial Z$, are harmonic functions, $\partial\phi/\partial Z$ is equivalent to the velocity potential due to a flat elliptical disk moving with uniform velocity $p/8\mu'$ perpendicular to its plane through an infinite incompressible fluid which is at rest at infinity*. The solution of this problem is well known (see (2)) and will be given in the next section in terms of ellipsoidal coordinates.

4. *Ellipsoidal coordinates.* The coordinates (x, y, Z) of any point may be expressed in terms of a triply orthogonal system (λ, μ, ν) in the form

$$a^2(a^2-b^2)x^2 = (a^2+\lambda)(a^2+\mu)(a^2+\nu),$$
$$b^2(b^2-a^2)y^2 = (b^2+\lambda)(b^2+\mu)(b^2+\nu),$$
$$a^2 b^2 Z^2 = \lambda\mu\nu, \quad (4\cdot 1)$$

where
$$\infty > \lambda \geq 0 \geq \mu \geq -b^2 \geq \nu \geq -a^2. \quad (4\cdot 2)$$

In the plane $Z = 0$ the inside of the ellipse $x^2/a^2 + y^2/b^2 = 1$ is given by $\lambda = 0$, and the outside by $\mu = 0$.

The following partial derivatives are needed, namely,

$$\frac{\partial \lambda}{\partial x} = \frac{x}{2(a^2+\lambda)h_1^2},\quad \frac{\partial \lambda}{\partial y} = \frac{y}{2(b^2+\lambda)h_1^2},\quad \frac{\partial \lambda}{\partial Z} = \frac{Z}{2\lambda h_1^2}, \quad (4\cdot 3)$$

where
$$4h_1^2 Q(\lambda) = (\lambda-\mu)(\lambda-\nu),\quad Q(\lambda) = \lambda(a^2+\lambda)(b^2+\lambda). \quad (4\cdot 4)$$

Alternative expressions for x, y, Z in terms of elliptic functions may also be obtained if necessary (see (4)).

On recalling the hydrodynamical analogy of the previous section, $\partial\phi/\partial Z$ may be put in the form

$$\frac{\partial \phi}{\partial Z} = AZ \int_\lambda^\infty \frac{ds}{s\sqrt{\{Q(s)\}}}, \quad (4\cdot 5)$$

where A is a constant, and hence

$$\phi = \frac{A}{2}\int_\lambda^\infty \left\{\frac{x^2}{a^2+s}+\frac{y^2}{b^2+s}+\frac{Z^2}{s}-1\right\}\frac{ds}{\sqrt{\{Q(s)\}}}. \quad (4\cdot 6)$$

The function ϕ in (4·6) is known to be harmonic, since, apart from a multiplying constant, it represents the gravitational potential at an external point of a uniform elliptical plate, and it can at once be verified that $\partial\phi/\partial Z$ is given by (4·5), since

$$\frac{x^2}{a^2+\lambda}+\frac{y^2}{b^2+\lambda}+\frac{Z^2}{\lambda}-1 = 0. \quad (4\cdot 7)$$

Except for the determination of the constant A the solution of the problem is now complete. Displacements and stresses in cartesian coordinates (x, y, Z) may be obtained

* S. G. Michlin (*App. Math. Mech.* (*Prikl. Mat. Mekh.*), 10 (1946), 304) has also reduced stress problems of the type considered in this paper to an equivalent problem in potential theory.

from (3·4) and (4·6) by differentiations. For numerical calculations it is convenient to express some of the integrals in terms of Jacobian elliptic functions by writing

$$\lambda = a^2 \operatorname{cn}^2 u / \operatorname{sn}^2 u = a^2 (\operatorname{sn}^{-2} u - 1), \qquad (4·8)$$

where $\operatorname{sn} u$ is the Jacobian elliptic function which has real and imaginary periods $4K$, $2iK'$ respectively, corresponding to the modulus k and complementary modulus k', where

$$ka = (a^2 - b^2)^{\frac{1}{2}}, \quad ak' = b. \qquad (4·9)$$

The variable u takes all real values between 0 and K.

The integral in (4·5) diverges when $\lambda = 0$, so it is convenient to express this integral in the alternative form

$$\frac{\partial \phi}{\partial Z} = AZ \left\{ \frac{2}{\sqrt{\{Q(\lambda)\}}} - \int_\lambda^\infty \frac{2s + a^2 + b^2}{(a^2+s)(b^2+s)\sqrt{\{Q(s)\}}} ds \right\}$$

$$= \frac{2AZ}{ab^2} \left\{ \frac{\operatorname{sn} u \operatorname{dn} u}{\operatorname{cn} u} - E(u) \right\}, \qquad (4·10)$$

where

$$E(u) = \int_0^u \operatorname{dn}^2 t \, dt, \qquad (4·11)$$

in order to evaluate $\partial \phi / \partial Z$ when $\lambda = 0$. Differentiating $\partial \phi / \partial Z$ with respect to Z with the help of (4·3) gives

$$\frac{\partial^2 \phi}{\partial Z^2} = A \left\{ \frac{2\lambda^{\frac{1}{2}}[\lambda(a^2 b^2 - \mu\nu) - a^2 b^2(\mu+\nu) - (a^2+b^2)\mu\nu]}{a^2 b^2 (\lambda - \mu)(\lambda - \nu)(a^2+\lambda)^{\frac{1}{2}}(b^2+\lambda)^{\frac{1}{2}}} - \int_\lambda^\infty \frac{2s+a^2+b^2}{(a^2+s)(b^2+s)\sqrt{\{Q(s)\}}} ds \right\}$$

$$= A \left\{ \frac{2\lambda^{\frac{1}{2}}[\lambda(a^2 b^2 - \mu\nu) - a^2 b^2(\mu+\nu) - (a^2+b^2)\mu\nu]}{a^2 b^2 (\lambda - \mu)(\lambda - \nu)(a^2+\lambda)^{\frac{1}{2}}(b^2+\lambda)^{\frac{1}{2}}} - \frac{2}{ab^2}\left[E(u) - \frac{\operatorname{sn} u \operatorname{cn} u}{\operatorname{dn} u}\right] \right\}. \qquad (4·12)$$

When $\mu = 0$, i.e. $Z = 0$, outside the ellipse, it can be verified from (4·5) or (4·10) that $\partial \phi / \partial Z = 0$. When $\lambda = 0$, i.e. $Z = 0$, inside the ellipse, $u = K$, and, from (4·12),

$$\frac{\partial^2 \phi}{\partial Z^2} = -\frac{2A}{ab^2} E(K) = \frac{p}{8\mu'}. \qquad (4·13)$$

Hence the constant A is given by

$$A = -\frac{ab^2 p}{16\mu' E(K)}. \qquad (4·14)$$

Also, when $\lambda = 0$,

$$w = -8(1-\eta)\frac{\partial \phi}{\partial Z} = -\frac{16(1-\eta) A \mu^{\frac{1}{2}} \nu^{\frac{1}{2}}}{a^2 b^2} = -\frac{16(1-\eta) A}{ab}\left\{1 - \frac{x^2}{a^2} - \frac{y^2}{b^2}\right\}^{\frac{1}{2}}, \qquad (4·15)$$

which is the value of the normal displacement over all points of the elliptical crack.

The solution of the problem has now been taken to a point at which the displacements and stresses at any point of the medium may be found by straightforward calculations.

5. *The elliptical punch problem.* Consider a semi-infinite medium $Z \geq 0$ indented by a flat-ended elliptical cylindrical punch. It is assumed that there is zero shear stress on the plane boundary $Z = 0$ so that the general formulae (3·4) will still be applicable. The remaining conditions to be satisfied on the plane boundary are

$$\left. \begin{array}{ll} \text{(i)} \ w = -8(1-\eta)\dfrac{\partial \phi}{\partial Z} = \epsilon & (\lambda = 0), \\[6pt] \text{(ii)} \ \dfrac{\partial^2 \phi}{\partial Z^2} = 0 & (\mu = 0), \end{array} \right\} \qquad (5·1)$$

where ϵ is a small constant. Condition (ii) ensures that there is zero normal stress on the boundary outside the punch.

Since $\partial\phi/\partial Z$ is harmonic it is seen that $\partial\phi/\partial Z$ represents the velocity potential of the perfect fluid flow through an elliptical aperture in a thin rigid boundary. The solution of this problem (2) is

$$\frac{\partial\phi}{\partial Z} = -\frac{\epsilon a}{16(1-\eta)K}\int_\lambda^\infty \frac{ds}{\sqrt{\{Q(s)\}}} = -\frac{\epsilon}{8(1-\eta)}\frac{u}{K}, \qquad (5\cdot 2)$$

where u is given by (4·8). It can be verified that this satisfies the conditions (5·1). In order to complete the solution of the problem it is necessary to find a harmonic function ϕ such that $\partial\phi/\partial Z$ is given by (5·2). So far this function has not been found. It is, however, possible to obtain all the components of displacement and stress in (3·4) except D and Φ. In particular, the value of $\widehat{zz}$ on the boundary $Z=0$ under the elliptical punch is

$$\widehat{zz} = -8\mu'\frac{\partial^2\phi}{\partial Z^2} = -\frac{\mu'\epsilon}{(1-\eta)bK}\left(1-\frac{x^2}{a^2}-\frac{y^2}{b^2}\right)^{-\frac{1}{2}} \quad (\lambda=0). \qquad (5\cdot 3)$$

On integrating this over the ellipse the total pressure P exerted by the punch is found to be

$$P = -\iint \widehat{zz}\,dx\,dy = \frac{2\pi\mu'a\epsilon}{(1-\eta)K}. \qquad (5\cdot 4)$$

REFERENCES

(1) GREEN, A. E. *Proc. Roy. Soc.* A, 195 (1949), 533.
(2) LAMB, H. *Hydrodynamics*, 6th. edit. (Cambridge, 1932), pp. 151–2.
(3) SADOWSKY, M. A. and STERNBERG, E. *J. Appl. Mech.* 16 (1949), 149.
(4) WHITTAKER, E. T. and WATSON, G. N. *A course of modern analysis* (Cambridge, 1927), Ch. 23.

KING'S COLLEGE
NEWCASTLE-ON-TYNE

THE UNIVERSITY
GLASGOW

VI. *The Phenomena of Rupture and Flow in Solids.*

By A. A. Griffith, *M. Eng. (of the Royal Aircraft Establishment).*

Communicated by G. I. Taylor, *F.R.S.*

Received February 11,—Read February 26, 1920.

Contents.

1. Introduction
2. A Theoretical Criterion of Rupture
3. Application of the Theory to a Cracked Plate
4. Experimental Verification of the Theory
5. Deductions from the Foregoing Results
6. The Strength of Thin Fibres
7. Molecular Theory of Strength Phenomena
8. Extended Application of the Molecular Orientation Theory
9. Practical Limitations of the Elastic Theory
10. Methods of Increasing the Strength of Materials
11. Application of the Theory to Liquids
12. Summary of Conclusions

1. *Introduction.*

In the course of an investigation of the effect of surface scratches on the mechanical strength of solids, some general conclusions were reached which appear to have a direct bearing on the problem of rupture, from an engineering standpoint, and also on the larger question of the nature of intermolecular cohesion.

The original object of the work, which was carried out at the Royal Aircraft Establishment, was the discovery of the effect of surface treatment—such as, for instance, filing, grinding or polishing—on the strength of metallic machine parts subjected to alternating or repeated loads. In the case of steel, and some other metals in common use, the results of fatigue tests indicated that the range of alternating stress which could be permanently sustained by the material was smaller than the range within which it was sensibly elastic, after being subjected to a great number of reversals. Hence it was inferred that the safe range of loading of a part, having a scratched or grooved surface of a given type, should be capable of estimation with the help of one of the two hypotheses of rupture commonly used for solids which are elastic to fracture. According to these hypotheses rupture may be expected if (a) the maximum tensile stress, (b) the maximum extension, exceeds a certain critical value. Moreover, as the behaviour of the materials under consideration, within the safe range of alternating stress, shows very little departure from Hooke's law, it was thought that the necessary stress and strain calculations could be performed by means of the mathematical theory of elasticity.

The stresses and strains due to typical scratches were calculated with the help of the mathematical work of Prof. C. E. Inglis,* and the soap-film method of stress estimation developed by Mr. G. I. Taylor in collaboration with the present author.† The general conclusions were that the scratches ordinarily met with could increase the maximum stresses and strains from two to six times, according to their shape and the nature of the stresses, and that these maximum stresses and strains were to all intents and purposes independent of the absolute size of the scratches. Thus, on the maximum tension hypothesis, the weakening of, say, a shaft 1 inch in diameter, due to a scratch one ten-thousandth of an inch deep, should be almost exactly the same as that due to a groove of the same shape one-hundredth of an inch deep.

These conclusions are, of course, in direct conflict with the results of alternating stress tests. So far as the author is aware, the greatest weakening due to surface treatment, recorded in published work, is that given by J. B. Kommers,‡ who found that polished specimens showed an increased resistance over turned specimens of 45 to 50 per cent. The great majority of published results indicate a diminution in strength of less than 20 per cent. Moreover, it is certain that reducing the size of the scratches increases the strength.

To explain these discrepancies, but one alternative seemed open. Either the ordinary hypotheses of rupture could be at fault to the extent of 200 or 300 per cent., or the methods used to compute the stresses in the scratches were defective in a like degree.

The latter possibility was tested by direct experiment. A specimen of soft iron wire, about 0·028-inch diameter and 100 inches long, which had a remarkably definite elastic limit, was selected. This was scratched spirally (i.e., the scratches made an angle of about 45 degrees with the axis) with carborundum cloth and oil. It was then normalised to remove initial stresses and subjected to a tensile load. Under these conditions the effect of the spiral scratches was to impart a twist to the wire, the twisting couple arising entirely from the stress-system due to the scratches. It was found that if the load exceeded a certain critical value, a part of the twist, amounting

* "Stresses in a Plate due to the Presence of Cracks and Sharp Corners," 'Proc. Inst. Naval Architects,' March 14, 1913.

† 'Proc. Inst. Mech. Eng.' December 14, 1917, pp. 755–809.

‡ 'Intern. Assoc. for Testing Materials,' 1912, vol. 4A and vol. 4B.

in some cases to 15 per cent., remained after the removal of the load. It was inferred that at this critical load the maximum stresses in the scratches reached the elastic limit of the material. This load was about one-quarter to one-third of that which caused the wire to yield as a whole, so that the scratches increased the maximum stress three or four times. The readings were quite definite even in the case of scratches produced by No. 0 cloth, which were found by micrographic examination to be but 10^{-4}-inch deep. Control experiments with longitudinal and circumferential scratches gave twists only 2 or 3 per cent. of those found with spiral scratches, and there was no permanent twist.

This substantial confirmation of the estimated values of the stresses, even in very fine scratches, shows that the ordinary hypotheses of rupture, as usually interpreted, are inapplicable to the present phenomena. Apart altogether from the numerical discrepancy, the observed difference in fatigue strength as between small and large scratches presents a fundamental difficulty.

2. *A Theoretical Criterion of Rupture.*

In view of the inadequacy of the ordinary hypotheses, the problem of the rupture of elastic solids has been attacked from a new standpoint. According to the well-known "theorem of minimum energy," the equilibrium state of an elastic solid body, deformed by specified surface forces, is such that the potential energy of the whole system* is a minimum. The new criterion of rupture is obtained by adding to this theorem the statement that the equilibrium position, if equilibrium is possible, must be one in which rupture of the solid has occurred, if the system can pass from the unbroken to the broken condition by a process involving a continuous decrease in potential energy.

In order, however, to apply this extended theorem to the problem of finding the breaking loads of real solids, it is necessary to take account of the increase in potential energy which occurs in the formation of new surfaces in the interior of such solids. It is known that, in the formation of a crack in a body composed of molecules which attract one another, work must be done against the cohesive forces of the molecules on either side of the crack.† This work appears as potential surface energy, and if the width of the crack is greater than the very small distance called the "radius of molecular action," the energy per unit area is a constant of the material, namely, its surface tension.

In general, the surfaces of a small newly formed crack cannot be at a distance apart greater than the radius of molecular action. It follows that the extended theorem of minimum energy cannot be applied unless the law connecting surface energy with distance of separation is known.

* POYNTING and THOMSON, 'Properties of Matter,' ch. xv.
† The potential energy of the applied surface forces is, of course, included in the "potential energy of the system."

There is, however, an important exception to this statement. If the body is such that a crack forms part of its surface in the unstrained state, it is not to be expected that the spreading of the crack, under a load sufficient to cause rupture, will result in any large change in the shape of its extremities. If, further, the crack is of such a size that its width is greater than the radius of molecular action at all points except very near its ends, it may be inferred that the increase of surface energy, due to the spreading of the crack, will be given with sufficient accuracy by the product of the increment of surface into the surface tension of the material.

The molecular attractions across such a crack must be small except very near its ends; it may therefore be said that the application of the mathematical theory of elasticity on the basis that the crack is assumed to be a traction-free surface, must give the stresses correctly at all points of the body, with the exception of those near the ends of the crack. In a sufficiently large crack the error in the strain energy so calculated must be negligible. Subject to the validity of the other assumptions involved, the strength of smaller cracks calculated on this basis must evidently be too low.

The calculation of the potential energy is facilitated by the use of a general theorem which may be stated thus: In an elastic solid body deformed by specified forces applied at its surface, the sum of the potential energy of the applied forces and the strain energy of the body is diminished or unaltered by the introduction of a crack whose surfaces are traction-free.

This theorem may be proved* as follows: It may be supposed, for the present purpose, that the crack is formed by the sudden annihilation of the tractions acting on its surface. At the instant following this operation, the strains, and therefore the potential energy under consideration, have their original values; but, in general, the new state is not one of equilibrium. If it is not a state of equilibrium, then, by the theorem of minimum energy, the potential energy is reduced by the attainment of equilibrium; if it is a state of equilibrium the energy does not change. Hence the theorem is proved.

Up to this point the theory is quite general, no assumption having been introduced regarding the isotropy or homogeneity of the substance, or the linearity of its stress-strain relations. It is necessary, of course, for the strains to be elastic. Further progress in detail, however, can only be made by introducing Hooke's law.

If a body having linear stress-strain relations be deformed from the unstrained state to equilibrium by given (constant) surface forces, the potential energy of the latter is diminished by an amount equal to twice the strain energy.† It follows that the net reduction in potential energy is equal to the strain energy, and hence the total decrease in potential energy due to the formation of a crack is equal to the increase in strain energy less the increase in surface energy. The theorem proved above shows that the former quantity must be positive.

* The proof is due to Mr. C. Wigley, late of the Royal Aircraft Establishment.
† A. E. H. Love, 'Mathematical Theory of Elasticity,' 2nd ed., p. 170.

3. *Application of the Theory to a Cracked Plate.*

The necessary analysis may be performed in the case of a flat homogeneous isotropic plate of uniform thickness, containing a straight crack which passes normally through it, the plate being subjected to stresses applied in its plane at its outer edge.

If the plate is thin, its state is one of "plane stress," and in this case it may, without additional complexity, be subjected to any uniform stress normal to its surface, in addition to the edge tractions. If it is not thin, it may still be dealt with provided it is subjected to normal surface stresses so adjusted as to make the normal displacement zero. Here the plate is in a state of "plane strain." The equations to the two states are of the same form,* differing only in the value of the constants; they will therefore be taken together.

The strain energy may be found, with sufficient accuracy, in the general case where the edge-tractions are arbitrary; it is necessary in the present application, however, for the resulting stress-system to be symmetrical about the crack, as otherwise it is not obvious that the latter will remain straight as it spreads. The only stress distribution which will be considered, therefore, is that in which the principal stresses in the plane of the plate, at points far from the crack, are respectively parallel and perpendicular to the crack, and are the same at all such points. This is equivalent to saying that, in the absence of the crack, the plate would have been subjected to uniform principal stresses in and perpendicular to its plane. It is also necessary, on physical grounds, for the stress perpendicular to the crack and in the plane of the plate to be a tension, otherwise the surfaces of the crack are forced together instead of being separated, and they cannot remain free from traction.

In calculating the strain energy of the plate use will be made of the solution obtained by Prof. INGLIS for the stresses in a cracked plate, to which reference has already been made. The notation of Prof. INGLIS's paper will be employed. In that notation α, β, are elliptic co-ordinates defined by the family of confocal ellipses; $\alpha = $ const. and the orthogonal family of hyperbolæ $\beta = $ const. The crack is represented by the limiting ellipse or focal line $\alpha = 0$. The axis of x coincides with the major axes, and the axis of y with the minor axes of the ellipses. The cartesian co-ordinates x, y, are connected with the elliptic co-ordinates α, β, by the relation

$$x + iy = c \cosh (\alpha + i\beta).$$

$R_{\alpha\alpha}, u_\alpha$, are the tensile stress and displacement respectively along the normal to
  $\alpha = $ const.

$R_{\beta\beta}, u_\beta$, are the corresponding quantities in the case of the normal to $\beta = $ const.

$S_{\alpha\beta}$ is the shear stress in the directions of these normals.

c is the half-length of the focal line.

---

* A. E. H. LOVE, 'Mathematical Theory of Elasticity,' 2nd ed., p. 205.

h is the modulus of transformation, $\sqrt{\dfrac{2}{c^2\,(\cosh 2\alpha - \cos 2\beta)}}$.

μ is the modulus of rigidity of the material.

E is YOUNG's modulus.

σ is POISSON's ratio.

$p = 3 - 4\sigma$ in the case of plane strain, and

$\dfrac{3-\sigma}{1+\sigma}$ in the case of plane stress.

The state of uniform stress existing at points far from the crack (*i.e.* where α is large) will be specified by the three principal tensions P, Q and R. P is normal to the plate, and in the case of plane stress it is the same everywhere. Q and R are parallel respectively to the axes of x and y, and R is positive.

The strain energy of the plate is a quadratic function of P, Q and R, and hence, in accordance with the theorem proved above, the increase of strain energy due to the crack must be a positive quadratic function of P, Q and R. The general form of this function may be found by evaluating a sufficient number of particular cases.

The following particular cases are sufficient :—

I.—Q = R (and P = 0 in the case of plane stress).

Boundary of crack given by $\alpha = \alpha_0$.

The stresses are

$$R_{\alpha\alpha} = R\,\frac{\sinh 2\alpha\,(\cosh 2\alpha - \cosh 2\alpha_0)}{(\cosh 2\alpha - \cos 2\beta)^2}, \quad\quad\quad\quad (1)$$

$$R_{\beta\beta} = R\,\frac{\sinh 2\alpha\,(\cosh 2\alpha + \cosh 2\alpha_0 - 2\cos 2\beta)}{(\cosh 2\alpha - \cos 2\beta)^2}, \quad\quad (2)$$

$$S_{\alpha\beta} = R\,\frac{\sin 2\beta\,(\cosh 2\alpha - \cosh 2\alpha_0)}{(\cosh 2\alpha - \cos 2\beta)^2}, \quad\quad\quad\quad (3)$$

while the displacements are given by

$$\left. \begin{aligned} \frac{u_\alpha}{h} &= \frac{c^2 R}{8\mu}\{(p-1)\cosh 2\alpha - (p+1)\cos 2\beta + 2\cosh 2\alpha_0\} \\ \frac{u_\beta}{h} &= 0 \end{aligned} \right\} \quad (4)$$

The strain energy of the material within the ellipse α, per unit thickness of plate is

$$\tfrac{1}{2}\int_0^{2\pi} \frac{u_\alpha}{h}\cdot R_{\alpha\alpha}\cdot d\beta + \tfrac{1}{2}\int_0^{2\pi} \frac{u_\beta}{h}\cdot S_{\alpha\beta}\cdot d\beta. \quad\quad\quad\quad (5)$$

On substituting and integrating, it is found that, as α becomes large, the strain energy tends towards the value

$$\frac{\pi c^2 R^2}{8\mu}\{\tfrac{1}{2}(p-1)e^{2\alpha} + (3-p)\cosh 2\alpha_0\}. \quad\quad (6)$$

Hence W, the increase of strain energy due to the cavity α_0, is given by

$$W = \frac{\pi c^2 R^2}{8\mu}(3-p)\cosh 2\alpha_0 \quad\quad (7)$$

or, on proceeding to the limit, $\alpha_0 = 0$,

$$W = \frac{(3-p)\pi c^2 R^2}{8\mu} \quad\quad (8)$$

for a very narrow crack of length $2c$.

II.—$R = 0$ ($= P$ in the case of plane stress) $\alpha_0 = 0$.

Here the stresses are entirely unaltered by the crack, at every point of the plate except the two points $x = \pm c$, $y = 0$, where $R\alpha\alpha = -Q$. It follows that $W = 0$.

III.—$Q = R = 0$, $\alpha_0 = 0$.

Here, again, the stresses are unaltered, and $W = 0$.

The only positive quadratic function of P, Q and R which is compatible with these three particular cases is that given by equation (8); this is therefore the general form of W, and rupture is determined entirely by the stress R, perpendicular to the crack.

A point of some interest, with regard to equation (8), may be noticed in passing. Since W cannot be negative it follows that, in real substances, where μ is positive, $3 - p$ must be positive. Hence σ cannot be negative in real isotropic solids.

The potential energy of the surface of the crack, per unit thickness of the plate is

$$U = 4cT \quad\quad (9)$$

where T is the surface tension of the material.

Hence the total diminution of the potential energy of the system, due to the presence of the crack, is

$$W - U = \frac{(3-p)\pi c^2 R^2}{8\mu} - 4cT. \quad\quad (10)$$

The condition that the crack may extend is

$$\frac{\partial}{\partial c}(W - U) = 0,$$

or

$$(3-p)\pi c R^2 = 16\mu T, \quad\quad (11)$$

so that the breaking stress is

$$R = 2\sqrt{\frac{\mu T}{\pi \sigma c}}, \quad \ldots \ldots \ldots \ldots (12)$$

in the case of plane strain, and

$$R = \sqrt{\frac{2ET}{\pi \sigma c}}. \quad \ldots \ldots \ldots \ldots (13)$$

in the case of plane stress.

Formula (13) has been verified experimentally. In connection with the experiments, interest attaches not only to the magnitude of R, but also to the value of the maximum tension in the material, which occurs at the extremities of the crack. This stress may be estimated if the radius of curvature of the boundary of the crack, at the points in question, can be found.

Expression (2) gives the maximum tension as

$$2R\sqrt{\frac{c}{\rho}} \quad \ldots \ldots \ldots \ldots (14)$$

in case I. above, ρ being the radius of curvature at the corners of the elliptic crack. Prof. INGLIS shows that this expression may also be used, with little error, for cracks which are elliptic only near their ends. The foregoing expressions for the stresses are obtained, however, on the assumption that the displacements are everywhere so small that their squares may be neglected. At the corner of a very sharp crack, it cannot be assumed, without proof, that the change in ρ leaves formula (14) substantially unaffected.

In the case under consideration the displacements at the surface of the crack, due to a small tension dR at distant points, are given by

$$\left. \begin{array}{l} \dfrac{u_\alpha}{h} = \dfrac{c^2 dR}{E} (\cosh 2\alpha_0 - \cos 2\beta) \\ \dfrac{u_\beta}{h} = 0 \end{array} \right\} \quad \ldots \ldots \ldots (15)$$

Whence, by resolution, the displacements parallel respectively to the major and minor axes are

$$\left. \begin{array}{l} u_x = \dfrac{2dR}{E} c \sinh \alpha_0 \cos \beta \\ u_y = \dfrac{2dR}{E} c \cosh \alpha_0 \sin \beta \end{array} \right\}, \quad \ldots \ldots \ldots (16)$$

which may be written

$$\left. \begin{array}{l} u_x = \dfrac{2dR}{E} x \tanh \alpha_0 \\ u_y = \dfrac{2dR}{E} y \coth \alpha_0 \end{array} \right\} \quad \ldots \ldots \ldots (17)$$

Equations (17) show that the effect of the small stress $d\mathrm{R}$ on the elliptic cavity is to deform it into another ellipse. If a and b are the major and minor semi-axes of the ellipse, when the plate is subjected to a stress R, then, by (17),

$$\left. \begin{array}{l} \dfrac{da}{d\mathrm{R}} = \dfrac{2b}{\mathrm{E}} \\[1em] \dfrac{db}{d\mathrm{R}} = \dfrac{2a}{\mathrm{E}} \end{array} \right\} \qquad \ldots \ldots \ldots \ldots \quad (18)$$

on making use of the relation $b = a \tanh \alpha_0$.

The solution of these simultaneous differential equations is

$$\left. \begin{array}{l} a = a_0 \cosh \dfrac{2\mathrm{R}}{\mathrm{E}} + b_0 \sinh \dfrac{2\mathrm{R}}{\mathrm{E}} \\[1em] b = a_0 \sinh \dfrac{2\mathrm{R}}{\mathrm{E}} + b_0 \cosh \dfrac{2\mathrm{R}}{\mathrm{E}} \end{array} \right\} \qquad \ldots \ldots \ldots \quad (19)$$

where a_0 and b_0 are the values of a and b in the unstrained state.

With the help of equations (19) it is possible to find the maximum stress, F, due to an applied stress, R, taking account of the change in the shape of the cavity. From (2)

$$\frac{d\mathrm{F}}{d\mathrm{R}} = 2\frac{a}{b} \qquad \ldots \ldots \ldots \ldots \quad (20)$$

whence

$$\mathrm{F} = 2\int_0^{\mathrm{R}} \frac{a_0 \cosh \dfrac{2\mathrm{R}}{\mathrm{E}} + b_0 \sinh \dfrac{2\mathrm{R}}{\mathrm{E}} \, d\mathrm{R}}{a_0 \sinh \dfrac{2\mathrm{R}}{\mathrm{E}} + b_0 \cosh \dfrac{2\mathrm{R}}{\mathrm{E}}}$$

$$= \mathrm{E} \log \left(\cosh \frac{2\mathrm{R}}{\mathrm{E}} + \frac{a_0}{b_0} \sinh \frac{2\mathrm{R}}{\mathrm{E}} \right), \ldots \ldots \ldots \quad (21)$$

and in the case of a narrow crack which is elliptic only near its ends, $\dfrac{a_0}{b_0}$ may, as in (14), be replaced by $\sqrt{\dfrac{c}{\rho}}$.

In the general case, where Q is not equal to R, the quantity $\mathrm{R} - \mathrm{Q}$ must be added to the value of F given by (21).

Formulæ (19) and (21) are not, of course, exactly true. The application of integration to equations (18) and (20) involves the assumption that the strains are so small that they can be superposed. If the strains are finite, this involves an error in the stresses depending on the square of the strains. In the case of ordinary solids, it is improbable that this assumption can alter the calculated stress by as much as 1 per cent.

4. *Experimental Verification of the Theory.*

In order to test formula (13), it was necessary to select an isotropic material which obeyed Hooke's law somewhat closely at all stresses, and whose surface tension at ordinary temperatures could be estimated. For these reasons glass was preferred to the metals in common use. A comparatively hard English glass,* having the following properties, was employed:—

Composition—SiO_2, 69·2 per cent.; K_2O, 12·0 per cent.; Na_2O, 0·9 per cent.; Al_2O_3, 11·8 per cent.; CaO, 4·5 per cent.; MnO, 0·9 per cent.

Specific gravity—2·40.

Young's modulus—9·01 × 10^6 lbs. per sq. inch.

Poisson's ratio—0·251.

Tensile strength—24,900 lbs. per sq. inch.

The three last-named quantities were determined by the usual tension and torsion tests on round rods or fibres about 0·04-inch diameter and 3 inches long between the gauge points. The fibres had enlarged spherical ends which were fixed into holders with sealing wax. A slight load was applied while the wax was still soft, to ensure freedom from bending. The possible error of the extension measurements was about ± 0·3 per cent., and Hooke's law was obeyed to this order of accuracy. No "elastic after-working" was observed with this glass, though more accurate measurements would doubtless have indicated its existence.

The problem of estimating the surface tension of glass, in the solid state, evidently requires special consideration. Direct determinations appeared to be impracticable, and ultimately an indirect method was decided on, in which the surface tension was found at a number of high temperatures and the value at ordinary temperatures deduced by extrapolation.

On the accepted theory of matter, intermolecular forces in solids and liquids consist mainly of two parts, namely, an attraction which increases rapidly as the distance between the molecules diminishes, balanced by a repulsion (the intrinsic pressure), which is due to the thermal vibrations of the molecules. It is reasonable to assume that the attraction, at constant volume, is sensibly independent of the temperature; this amounts merely to supposing that the attraction exerted by a molecule does not depend on its state of motion. On this view, the temperature variation, at constant volume, of the intermolecular forces is determined entirely by the change in thermal energy. Hence, it may be inferred, on the accepted theory of surface tension,† that the surface tension of a material, at constant volume, is equal to a constant diminished by a quantity proportional to the thermal energy of the substance. In the case of solids, nearly the same result should hold at constant pressure, as the temperature-volume change is small.

---

* Supplied in the form of test-tubes by Messrs. J. J. Griffin, Kingsway, London.
† Poynting and Thomson, 'Properties of Matter,' ch. xv.

The specific heat of glass is greater at high than at low temperatures, but the temperature coefficient is not large. Hence its surface tension may be expected to be nearly a linear function of the temperature, and extrapolation should be fairly reliable. This was found to be the case with the glass selected for the present experiments.

In the neighbourhood of 1100° C. the surface tension was found by QUINCKE's drop method. At lower temperatures this method was not satisfactory, on account of the large viscosity of the liquid glass; but between 730° C. and 900° C. the method described below was found to be practicable. Fibres of glass, about 2 inches long and from 0·002-inch to 0·01-inch diameter, with enlarged spherical ends, were prepared. These were supported horizontally in stout wire hooks and suitable weights were hung on their mid-points. The enlarged ends prevented any sagging except that due to extension of the fibres. The whole was placed in an electric resistance furnace maintained at the desired temperature. Under these conditions viscous stretching of the fibre occurred until the suspended weight was just balanced by the vertical components of the tension in the fibre. The latter was entirely due, in the steady state, to the surface tension of the glass, whose value could therefore be calculated from the observed sag of the fibre. In the experiments the angle of sag was observed through a window in the furnace by means of a telescope with a rotating cross wire. If w is the suspended weight, d the diameter of the fibre, T the surface tension, and θ the angle at the point of suspension between the two halves of the fibre, then, evidently,

$$\pi \cdot d \cdot T \cdot \sin \tfrac{1}{2}\theta = w.$$

For this method of determining the surface tension to be valid, it is evidently necessary that the angle of sag shall reach a steady value before the development of local contractions, arising from the instability of liquid cylinders, becomes appreciable. That this requirement is satisfied is shown by the following experimental results. After heating for two hours at about 750° C. the angle of sag of a particular fibre was 18°·25. Two hours later it had increased by less than 0°·1. The temperature was then raised momentarily to 940° C., and quickly reduced again to 750° C. The angle was then found to be 20°·2. After two hours further heating at 750° C. the angle had *decreased* to 18°·4, agreeing within permissible limits of error with the former value. That is to say, substantially the same limiting angle of sag was reached whether the initial angle was above or below that limit.

Above 900° C. it was found that the viscosity was insufficient to enable an observation to be made before the fibre commenced to break up into globules. Below 730° C., on the other hand, observations made on fibres of different diameters were inconsistent, the apparent surface tension being higher for the larger fibres. The obvious meaning of this result is that below 730° C. the glass used was not a perfect viscous liquid and hence the method was inapplicable. The transition from the viscous liquid state was quite gradual. The maximum tension (apart from surface tension) which could be permanently sustained, was zero at 730° C., 1·3 lbs. per sq. inch at 657° C., and 24 lbs.

per sq. inch at 540° C. At lower temperatures the rates of increase, both of this "solid stress" and the viscosity, were enormously greater. At 540° C. a fibre took about 70 hours to reach the steady state.

Table I. below gives the values of the surface tension obtained from these experiments. That for the temperature 1110° C. is the mean of five determinations by the drop method. The remaining figures were obtained from the sag of fibres.

TABLE I.—Surface Tension of Glass.

Temperature.	Surface Tension.
° C.	lb. per inch.
1110	0·00230
905	0·00239
896	0·00250
852	0·00249
833	0·00254
820	0·00249
801	0·00257
760	0·00255
745	0·00251
15	0·0031*

So far as they go, these figures confirm the deduction that the surface tension of glass is approximately a linear function of temperature. Moreover, as the actual variation is not great, the error involved in assuming such a law and extrapolating to 15° C. is doubtless fairly small. The value so obtained, 0·0031 lb. per inch, will be used in the present application.

Rigorously, expressions (13) and (21) above are true only for small cracks in large flat plates. In view, however, of the difficulties attendant on annealing and loading large flat glass plates, it was decided to perform the breaking tests on thin round tubes and spherical bulbs. These were cracked and then annealed and broken by internal pressure. The calculation cannot be exact for such bodies, but the error may obviously be reduced by increasing the ratio of the diameter of the bulb or tube to the length of the crack. It will be seen from the results of the tests that the variation of this ratio from two to ten caused little, if any, change in the bursting strength, and hence it may be inferred that the error in question is negligible for the present purpose.

The cracks were formed either with a glass-cutter's diamond, or by scratching with a hard steel edge and tapping gently. The subsequent annealing was performed by heating to 450° C. in a resistance furnace, maintaining that temperature for about one hour, and then allowing the whole to cool slowly. The question of the best annealing temperature required careful consideration, as it was evidently necessary to relieve the

* By extrapolation.

initial stresses due to cracking as much as possible, while at the same time keeping the temperature so low that appreciable deformation of the crack did not occur. It was found that the bursting strength increased with the annealing temperature up to about 400° C., while between 400° C. and 500° C. very little further change was perceptible. From this it was inferred that relief of the initial stresses was sufficient for the purpose in view at a temperature of 450° C.

The principal stresses at rupture, Q and R, were calculated from the observed bursting pressure by means of the usual expressions for the stresses in thin hollow spheres and circular cylinders, the thickness of the glass near the crack being measured after bursting. In the case of the tubes the cracks were parallel to the generators, and provision was made for varying Q by the application of end loads.

In dealing with the longer cracks, leakage was prevented by covering the crack on the inside with celluloid jelly, the tube being burst before the jelly hardened. In the case of the smaller cracks leakage was imperceptible and this precaution was unnecessary.

The time of loading to rupture varied from 30 seconds to five minutes. No evidence was observed of any variation of bursting pressure with time of loading.

The results of the bursting tests are set down in Tables II. and III. below. $2c$ is the length of the crack, Q and R are the calculated principal stresses respectively

TABLE II.—Bursting Strength of Cracked Spherical Bulbs.

$2c$	D	Q	R	$R\sqrt{c}$
inch.	inch.	lbs. per sq. inch.	lbs. per sq. inch.	
0·15	1·49	864	864	237
0·27	1·53	623	623	228
0·54	1·60	482	482	251
0·89	2·00	366	366	244

TABLE III.—Bursting Strength of Cracked Circular Tubes.

$2c$	D	Q	R	$R\sqrt{c}$
inch.	inch.	lbs. per sq. inch.	lbs. per sq. inch.	
0·25	0·59	−621	678	240
0·32	0·71	−176	590	232
0·38	0·74	− 31	526	229
0·28	0·61	55	655	245
0·26	0·62	202	674	243
0·30	0·61	308	616	238

parallel and perpendicular to the crack, and D is the diameter of the bulb or tube. The thickness of the bulbs was about 0·01 inch and the tubes 0·02 inch.

The average value of $R\sqrt{c}$ is 239, and the maximum 251.

According to the theory, fracture should not depend on Q, and $R\sqrt{c}$ should have, at fracture, the constant value

$$\sqrt{\frac{2ET}{\pi\sigma}}.$$

In the case of the glass used for these experiments, $E = 9\cdot01 \times 10^6$ lbs. per sq. inch, $T = 0\cdot0031$ lbs. per inch, and $\sigma = 0\cdot251$, so that the above quantity is equal to 266.

These conclusions are sufficiently well borne out by the experimental results, save that the maximum recorded value of $R\sqrt{c}$ is 6 per cent., and the average 10 per cent., below the theoretical value. It must be regarded as improbable that the error in the estimated surface tension is large enough to account for this difference, as this view would render necessary a somewhat unlikely deviation from the linear law.

A more probable explanation is to be obtained from an estimate of the maximum stress in the cracks. An upper limit to the magnitude of the radius of curvature at the ends of the cracks was obtained by inspection of the interference colours shown there. Near the ends a faint brownish tint was observed, and this gradually died out, as the end was approached, until finally nothing at all was visible. It was inferred that the width of the cracks at the ends was not greater than one-quarter of the shortest wave length of visible light, or about 4×10^{-6} inch. Hence ρ could not be greater than 2×10^{-6} inch.

Taking as an example the last bulb in Table II. and substituting in formula (21), it is found that

$$\frac{a_0}{b_0} = \sqrt{\frac{c}{\rho}} \geq 478$$

$$\frac{2R}{E} = 8\cdot13 \times 10^{-5},$$

whence the maximum stress $F \geq 344{,}000$ lbs. per sq. inch. The value given by the first order expression

$$F = 2R\sqrt{\frac{c}{\rho}}$$

is 350,000 lbs. per sq. inch.

A possible explanation of the discrepancy between theory and experiment is now evident. In the tension tests, the verification of Hooke's law could only be carried to the breaking stress, 24,900 lbs. per sq. inch. There is no evidence whatever that the law is still applicable at stresses more than ten times as great. It is much more probable that there is a marked reduction in modulus at such stresses. But a decrease in modulus at any point of a body deformed by given surface tractions involves an increase in strain energy, and therefore in the foregoing experiments a decrease in strength. This is in agreement with the observations.

5. *Deductions from the Foregoing Results.*

The estimate of maximum stress obtained above appears to lead to a peremptory disproof of the hypothesis that the maximum tension in this glass, at rupture, is always equal to the breaking stress in ordinary tensile tests. If Hooke's law was obeyed to rupture, and the squares of the strains were negligible, the maximum tension in the above cracked tube could not have been less than 344,000 lbs. per sq. inch; but, in the tensile tests, Hooke's law was obeyed up to the breaking stress, the squares of the strains were negligible, and the maximum stress was only 24,900 lbs. per sq. inch. Hence the stresses could not have been the same in the two cases. Moreover, the order of the results obtained suggests (though this is not rigorously proved, as the assumptions have not been checked at stresses above 24,900 lbs. per sq. inch) that the actual strength may be more than ten times that given by the hypothesis.

Similar conclusions may be drawn regarding the "maximum extension," "maximum stress-difference" and "maximum shear strain" hypotheses which have been proposed from time to time for estimating the strength of brittle solids.

These conclusions suggest inquiries of the greatest interest. If the strength of this glass, as ordinarily interpreted, is not constant, on what does it depend? What is the greatest possible strength, and can this strength be made available for technical purposes by appropriate treatment of the material? Further, is the strength of other materials governed by similar considerations?

Some indication of the probable maximum strength of this glass may be obtained from the bursting tests already described. There is no reason for supposing that, in those tests, the radii of curvature at the corners of the cracks were as great as 2×10^{-6} inch. It is much more likely that they were of the same order as the molecular dimensions. Considering, as before, the last bulb in Table II., and putting $\rho = 2 \times 10^{-8}$ inch in formula (21), it is found that the maximum stress, F, is about 3×10^6 lbs. per sq. inch. Elastic theory cannot, of course, be expected to apply with much accuracy to cases where the dimensions are molecular, on account of the replacement of summation by integration, and the probable diminution of modulus at very high stresses must involve a further error. Taking these circumstances into consideration, however, it may still be said that the probable maximum strength of the glass used in the foregoing experiments is of the order 10^6 lbs. per sq. inch.

It is of interest to enquire at this stage whether there is any reason for ascribing similar maximum strengths to other materials. On the molecular theory of matter the tensile strength of an isotropic solid or liquid is of the same order as, though less than, its "intrinsic pressure," and may therefore be estimated either from a knowledge of the total heat required to vaporise the substance or by means of Van der Waal's equation.* It may be noted that these methods of estimating the stress indicate that

---

* Poynting and Thomson, 'Properties of Matter,' ch. xv.

it should be, approximately at least, a constant of the material. TRAUBE* gives the following as the intrinsic pressures of a number of metals, at ordinary temperatures:—

TABLE IV.—Intrinsic Pressures of Metals (TRAUBE).

Metal.	Intrinsic Pressure.
	lbs. per sq. inch.
Nickel	$4\cdot71 \times 10^6$
Iron	$4\cdot70 \times 10^6$
Copper	$3\cdot42 \times 10^6$
Silver	$2\cdot34 \times 10^6$
Antimony	$1\cdot74 \times 10^6$
Zinc	$1\cdot58 \times 10^6$
Tin	$1\cdot06 \times 10^6$
Lead	$0\cdot75 \times 10^6$

These are of the same order as the direct estimate obtained above for glass, but they are from 20 to 100 times the strengths found in ordinary tensile and other mechanical tests.

In the case of liquids, the discrepancy between intrinsic pressure and observed tensile strength is much greater. According to VAN DER WAAL's equation, water has an intrinsic pressure of about 160,000 lbs. per sq. inch, whereas its tensile strength is found to be about 70 lbs. per sq. inch. It has been suggested that this divergence may be due to impurities, such as dissolved air, but DIXON and JOLY† have shown that dissolved air has no measurable effect on the tensile strength of water.

Thus the matters under discussion appear to be of general incidence, in that the strengths usually observed are but a small fraction of the strengths indicated by the molecular theory.

Some further discrepancies between theory and experiment may now be noticed. In the theory it is assumed that rupture occurs in a tensile test at the stress corresponding with the maximum resultant pull which can be exerted between the molecules of the material. On this basis the applied stress must have a maximum value at rupture, and hence, if intermolecular force is a continuous function of molecular spacing, the stress-strain diagram must have zero slope at that point. This, of course, is never observed in tensile tests of brittle materials; in no case has any evidence been obtained of the existence of such a maximum anywhere near the breaking stress.

Again, the observed differences in strength as between static and alternating stress tests are at first sight inexplicable from the standpoint of the molecular theory, if the

* 'Zeitschr. für Anorganische Chemie,' 1903, vol. xxxiv., p. 413.
† 'Phil. Trans.,' B, 1895, p. 568.

breaking load is regarded as the sum of the intermolecular attractions. According to the theory, large changes in the latter can only occur as a result of large changes in the thermal energy of the substance, such as would be immediately evident in alternating stress tests, if they took place.

Lastly, as indicated above, the strain energy at rupture of an elastic solid or liquid should on the molecular theory be of the same order as its heat of vaporisation. Hence rupture should be accompanied by phenomena, such as a large rise of temperature, indicative of the dissipation of an amount of energy of this order. It is well known that tensile tests of brittle materials show no such phenomena.

If, as is usually supposed, the materials concerned are substantially isotropic, there is but one hypothesis which is capable of reconciling all these apparently contradictory results. The theoretical deduction—that rupture of an isotropic elastic material always occurs at a certain maximum tension—is doubtless correct; but in ordinary tensile and other tests designed to secure uniform stress, the stress is actually far from uniform so that the average stress at rupture is much below the true strength of the material.

Now it may be shown, with the help of elastic theory, that the stress must be substantially uniform, in such tests, unless the material of the test-pieces is heterogeneous or discontinuous. It is known that all substances are in fact discontinuous, in that they are composed of molecules of finite size, and it may be asked whether this type of discontinuity is sufficient to account for the observed phenomena.

With the help of formula (13) above, this question may be answered in the negative. Formula (13) shows that a thin plate of glass, having in it the weakest possible crack of length $2c$ inch, will break at a tension, normal to the crack, of not less than $266/\sqrt{c}$ lbs. per sq. inch. This result, however, is subject to certain errors, and experiment shows that the true breaking stress is about $240/\sqrt{c}$ lbs. per sq. inch. But such a crack is the most extreme type, either of discontinuity or heterogeneity, which can exist in the material. Hence it is impossible to account for the observed strength, 24,900 lbs. per sq. inch, of the simple tension test specimens, unless they contain discontinuities at least $2 \times \left(\frac{240}{24,900}\right)^2$ inch, or say, 2×10^{-4} inch wide. This is of the order 10^{-4} times the molecular spacing.

The general conclusion may be drawn that the weakness of isotropic solids, as ordinarily met with, is due to the presence of discontinuities, or flaws, as they may be more correctly called, whose ruling dimensions are large compared with molecular distances. The effective strength of technical materials might be increased 10 or 20 times at least if these flaws could be eliminated.

It is easy to see why the presence of such small flaws can leave the strength of cracked plates, such as those of the foregoing experiments, practically unaffected. The most extreme case of weakening is that where there is a flaw very near the end of the crack and collinear with it. Here the result is merely to increase the effective length of the crack by less than 10^{-3} inch. This involves a weakening of less than 0·1 per cent.

6. *The Strength of Thin Fibres.*

Consideration of the consequences of the foregoing general deduction indicated that very small solids of given form, *e.g.*, wires or fibres, might be expected to be stronger than large ones, as there must in such cases be some additional restriction on the size of the flaws. In the limit, in fact, a fibre consisting of a single line of molecules must possess the theoretical molecular tensile strength. In this connection it is, of course, well known that fine wires are stronger than thick ones, but the present view suggests that in sufficiently fine wires the effect should be enormously greater than is observed in ordinary cases.

This conclusion has been verified experimentally for the glass used in the previous tests, strengths of the same order as the theoretical tenacity having been observed. Incidentally, information of interest has been obtained, somewhat unexpectedly, concerning the genesis of the flaws, and it has been found to be possible to prepare quite thick fibres in an unstable condition in which they have the theoretical strength.

Fibres of glass, about 2 inches long and of various diameters, were prepared. One end of a fibre was attached to a stout wire hanging on one arm of a balance, and the other end to a fixed point, the medium of attachment being sealing wax. A slight tension was applied while the wax was still soft, in order to eliminate bending of the fibre at the points of attachment. The other arm of the balance carried a beaker into which water was introduced from a pipette or burette. The weight of water necessary to break the fibre was observed, and the diameter of the latter at the fracture was found by means of a high-power measuring microscope. Hence the tensile strength was obtained.

At first the results were extremely irregular, though the general tendency of the strength to increase with diminishing diameter was clear. It was found that the irregularities were due to the dependence of the strength on the following factors:—

(1) The maximum temperature of the glass.—To secure the best results it was found necessary to heat the glass bead to about 1400° C. to 1500° C. before drawing the fibre.

(2) The temperature during drawing.—If the glass became too cool before drawing was complete, a weak fibre was obtained. This temperature could not be very closely defined, but it is perhaps the same as the limiting temperature of the viscous liquid phase, namely, 730° C. This effect made the drawing of very fine fibres a matter of some difficulty, as the cooling was so rapid.

(3) The presence of impurities and foreign bodies.

(4) The age of the fibre.—For a few seconds after preparation, the strength of a properly treated fibre, whatever its diameter, was found to be extremely high. Tensile strengths ranging from 220,000 to 900,000 lbs. per sq. inch were observed in fibres up to about 0·02 inch diameter. These strengths were estimated by measuring the radii to which it was necessary to bend the fibres in order to break them. They are therefore probably somewhat higher than the actual tenacities. The glass appeared to be

almost perfectly elastic up to these high stresses. The strength diminished, however, as time went on, until after the lapse of a few hours it reached a steady value whose magnitude depended on the diameter of the fibre.

Similar phenomena have been observed with other kinds of glass, and also with fused silica.

The relation between diameter and strength in the steady state was investigated in the following manner. Fibres of diameters ranging from 0.13×10^{-3} inch to 4.2×10^{-3} inch, and 6 inches long, were prepared by heating the glass to about 1400° C. to 1500° C. in an oxygen and coal-gas flame and drawing the fibre by hand as quickly as possible. The fibres were then put aside for about 40 hours, so that they might reach the steady state. The test specimens were prepared by breaking these fibres in tension several times until pieces about 0·5-inch long remained; these were then tested by the balance method already described. The object of this procedure was the elimination of weak places due to minute foreign bodies, local impurities and other causes.

Table V below gives the results of these tests. Diameters are in thousandths of an inch, and breaking stresses in lbs. per sq. inch.

TABLE V.—Strength of Glass Fibres.

Diameter.	Breaking Stress.	Diameter.	Breaking Stress.
0·001 inch.	lbs. per sq. inch.	0·001 inch.	lbs. per sq. inch.
40·00	24,900*	0·95	117,000
4·20	42,300	0·75	134,000
2·78	50,800	0·70	164,000
2·25	64,100	0·60	185,000
2·00	79,600	0·56	154,000
1·85	88,500	0·50	195,000
1·75	82,600	0·38	232,000
1·40	85,200	0·26	332,000
1·32	99,500	0·165	498,000
1·15	88,700	0·130	491,000

It will be seen that the results are still somewhat irregular. No doubt more precise treatment of the fibres would lead to some improvement in this respect, but such refinement is scarcely necessary at the present stage.

The limiting tensile strength of a fibre of the smallest possible (molecular) diameter may be obtained approximately from the figures in Table V. by plotting reciprocals of the tensile strength and extrapolating to zero diameter. This maximum strength is found to be about 1.6×10^6 lbs. per sq. inch, which agrees sufficiently well with the rough estimate previously obtained from the cracked plate experiments.

* From the tensile tests previously described.

In 1858, KARMARSCH* found that the tensile strength of metal wires could be represented within a few per cent. by an expression of the type

$$F = A + \frac{B}{d} \quad \ldots \ldots \ldots \ldots (22)$$

where d is the diameter and A and B are constants. In this connection it is of interest to notice that the figures in Table V. are given within the limits of experimental error by the formula

$$F = 22,400 \frac{4.4 + d}{0.06 + d} \quad \ldots \ldots \ldots (23)$$

where F is in lbs. per sq. inch and d is in thousandths of an inch. Within the range of diameters available to KARMARSCH, this expression differs little from

$$F = 22,400 + \frac{98,600}{d}, \quad \ldots \ldots \ldots (24)$$

which is of the form given by KARMARSCH. Moreover, the values of B found by him for the weaker metals, *e.g.*, silver and gold, in the annealed state, are of the same order as that given by formula (24) for glass.

To a certain extent this correspondence suggests that the mechanism of rupture, as distinct from plastic flow, in metals, is essentially similar to that in brittle amorphous solids such as glass.

The remarkable properties of the unstable strong state referred to on p. 180 above are exhibited most readily in the case of clear fused silica. If a portion of a rod about 5 mm. diameter be made as hot as possible in an oxyhydrogen flame, and then drawn down to, say, 1 mm. or less and allowed to cool, the drawn-down portion may be bent to a radius of 4 or 5 mm. without breaking, and if then released will spring back almost exactly to its initial form. If instead of being released it is held in the bent form it will break spontaneously after a time which usually varies from a few seconds to a few minutes, according to the degree of flexure. To secure the best results the drawing should be performed somewhat slowly.

When fracture occurs it is accompanied by phenomena altogether different from those associated with the fracture of the normal substance. The report is much louder and deeper than the sharp crack which accompanies the rupture of an ordinary silica or glass rod, and the specimen is invariably shattered into a number of pieces, parts being frequently reduced to powder. This shattering is not confined to the highly stressed drawn-out fibre; it usually occurs also in the unchanged parts of the original thick rod and sometimes in the free ends, which are not subjected to the bending moment. The experiment is most striking, for it appears at first sight that the 5 mm. rod has been broken by a couple applied through the fibre, which may be only 0·5 mm. in

---

* 'Mittheilungen des gew. Ver. für Hannover,' 1858, pp. 138–155.

diameter. As a matter of fact, however, the shattering is probably merely one of the means of dissipating the strain energy of the strong fibre, which at fracture is perhaps 10,000 times that of silica in the ordinary weak state. An elastic wave is doubtless propagated from the original fracture, and the stresses due to this wave shatter the rod.

Confirmation of this view is obtainable if the fibre is broken by twisting instead of by bending. The thick part of the rod is in this case found to contain a number of spiral cracks, at an angle of about 45° to the axis, showing that the material has broken in tension, but the cracks run in both right- and left-handed spirals, so that the surface of the rod is divided up into little squares. This shows that the cracking must be due to an alternating stress, such as would result from the propagation of a torsional wave along the rod.

Another phenomenon which has been observed in these fibres is that fracture at any point appears to cause a sudden large reduction in the strength of the remaining pieces. Thus, in one case a glass fibre was found to break in bending at an estimated stress of 220,000 lbs. per sq. inch. One of the pieces, on being tested immediately afterwards, broke at about 67,000 lbs. per sq. inch.

7. *Molecular Theory of Strength Phenomena.*

From the engineering standpoint the chief interest of the foregoing work centres round the suggestion that enormous improvement is possible in the properties of structural materials. Of secondary, but still considerable, importance is the demonstration that the methods of strength estimation in common use may lead in some cases to serious error.

Questions relating to methods of securing the indicated increase in tenacity, or of eliminating the uncertainty in strength calculations, can scarcely be answered without some more or less definite knowledge of the way in which the properties of molecules enter into the phenomena under consideration. In this connection it is of interest to enquire whether any indication can be obtained of the nature of the properties which are requisite for an explanation of the observed facts.

For this purpose it is convenient to start with molecules of the classical type, whose properties may be defined as (a) a central attraction between each pair of molecules which decreases rapidly as their central distance increases, and which depends only on that distance and the nature of the molecules; (b) translational and possibly rotational vibrations whose energy is the thermal energy of the substance. In the unstrained state, the kinetic reactions due to (b) balance the central attractions (a).

In a body composed of such molecules, the flaws which have been shown to exist in real substances might consist of actual cracks. But experiment shows that under certain conditions the strength of glass diminishes with lapse of time. On the present hypothesis this would require the potential energy of the system to increase spontaneously by the amount of the surface energy of the cracks. This view must therefore be regarded as untenable.

Again, the observed weakening might conceivably occur if at any instant the vibrations of a large number (at least 10^8) of near molecules synchronised and were in phase, provided the energy of these molecules was approximately that corresponding with the temperature of ebullition of the substance. Except in the case of a material very near its boiling point, the probability of such an occurrence must be so small as to be quite negligible. Hence this hypothesis also must be discarded.

The foregoing discussion seems to suggest that the assumed type of molecule is too simple to permit of the construction of an adequate theory. An increase in generality may be obtained by supposing that the attraction between a pair of molecules depends not only on their distance apart, but also on their relative orientation. The properties of crystals seem definitely to require the molecules of anisotropic materials to be of this type, but those of isotropic substances have usually been assumed to be of the simpler kind. In view of the author, however, molecular attraction must be a function of orientation even in substances, such as glass, metals and water, which are usually referred to as "isotropic."

Consider a solid made up of a number of such molecules, initially oriented at random. Doubtless the mechanical properties of the substance, while it is in this amorphous condition, will differ little from those of a substance composed of molecules of the simpler type, having an attraction of appropriate strength. If this is so, the tensile strength of the material must be that corresponding with its average intrinsic pressure.

In general, however, this initial condition cannot be one of minimum potential energy.

It is clear that under suitable conditions the tendency to attain stable equilibrium can cause the molecules to rotate and set themselves in chains or sheets, with their maxima of attraction in line. The formation of sheets will commence at a great number of places throughout the solid, *i.e.*, wherever the initial random arrangement is sufficiently favourable. Evidently it is possible for the number of such "sufficiently favourable" arrangements to be enormously less than the total number of molecules, so that the ultimate result will be the formation of a number of units or groups, each containing a large number of molecules oriented according to some definite law. The relative arrangement of the units will, of course, be haphazard.

Now, in each unit there will, in general, be a direction which is, approximately at least, that of the minimum attractions of the majority of the molecules in the unit. Hence if the ratio of the maximum to the minimum attractions is sufficiently great, each unit can constitute a "flaw," and there appears to be no reason why the units should not be as large as the flaws have been shown to be in the case of glass. Thus, in order to explain the spontaneous weakening of glass, it is only necessary to suppose that the thermal agitation at about 1400° C. is sufficient to bring about the initial random formation.

It will be remembered that in the case of the freshly drawn fibres the reduction in tenacity required several hours for completion, so that the time taken was large compared with the time of cooling. Expressed in terms of molecular motion, this means that the molecules resist rotation very much more than they resist translation. This is in keeping with the conclusions of DEBYE,* who found that, on the basis of the quantum theory, the phenomena associated with the specific heat of solids could be explained only if the thermal vibrations of the molecules were regarded as practically irrotational. The same thing is shown more roughly, without introducing the quantum theory, by the law of DULONG and PETIT, which requires that each molecule shall have only three degrees of freedom.

The theory here put forward makes the spontaneous weakening a consequence of the attainment of a molecular configuration of stable equilibrium; it therefore suggests that the weakening should be accompanied, in general, by a change in the dimensions of the solid. This has been verified by direct observation with a high-power microscope; in the course of half an hour a spontaneously weakening glass fibre increased in length by about 0·1 per cent., while the length of a silica fibre decreased by about 0·03 per cent.

On account of the random arrangement of the molecular groups, this spontaneous change in unstrained volume must set up internal stresses, which may be sufficiently large to start cracks along the directions of least strength. In this connection it may be mentioned that irregularly shaped pieces of glass, of which some parts had been put into the strong unstable state by heating, have sometimes been observed to break spontaneously about an hour after cooling was practically complete.

It was remarked on p. 184 that cracks could not form spontaneously in a substance composed of molecules having spherical fields of force, as the process would involve an increase in potential energy. This is no longer true when the attraction is a function of orientation, as the surface energy of the cracks may be more than counterbalanced by the decrease in potential energy accompanying the molecular rearrangement.

For this reason, it is impossible to deduce the ratio of the maximum to the minimum molecular attractions from the ratio of the maximum and minimum strengths of the material, as it is possible that the spontaneous weakening is always accompanied by the formation of minute cracks, of the same size as the molecular groups.

It is probable that, in many cases, the most stable orientation of the molecules at a free surface is that in which their maxima of attraction lie along the surface. Such an orientation would in turn lead to a similar tendency on the part of the next layer of molecules, and so on, the tendency diminishing with increasing distance from the surface. There would therefore be a surface layer having the special property that in it the "flaws" ran parallel to the surface.

Hence this layer would be of exceptional strength in the direction of the surface. This suggests a reason for the experimental fact that the breaking load of wires and

* 'Ann. der Physik,' 39 (1912), p. 789.

fibres consists mainly of two parts, one proportional to the area, and the other to the perimeter of the cross-section. The process of drawing, too, might predispose the molecules to take up positions with their maxima of attraction parallel to the surface.

If a perfectly clean glass plate be covered with gelatine and set aside, the gelatine gradually contracts, and as it does so it tears from the glass surface thin flakes up to about 0·06-inch diameter and shaped like oyster shells.* This tendency to flake at the surface is also observed when glass is broken by bending. This was particularly well shown in the specially prepared fibres used for the experiments described in the present paper. In almost all cases of flexural fracture the crack curled round on approaching the compression side, till it was nearly parallel to the surface. On two occasions the fracture divided before changing direction, the two branches going opposite ways along the fibre and a flake of length several times the diameter of the fibre was detached.

Surface flaking is also observed when some kinds of steel are subjected to repeated stress. Here the flakes are usually very small.

All these facts are evidently in complete agreement with the " surface layer " theory and, indeed, it is difficult to account for them on any other basis.

8. *Extended Application of the Molecular Orientation Theory.*

On the basis of the present theory, the physical properties of materials must be intimately related to the geometrical properties of the molecular sheet-formation. In order that a substance may exhibit the characteristic properties of crystals, it is clearly necessary for the sheets of molecules to be plane. In this case the crystals are, of course, the molecule groups or " units " referred to above. In " amorphous " materials, on the other hand, the sheets are probably curved.†

In materials of the former type, there must exist planes on which, if they are subjected to a sufficiently large shearing stress, the portions on either side of the planes can undergo a mutual sliding through a distance equal to any integral multiple of the molecular spacing, without fundamentally affecting the structure of the crystal. It is well known that the phenomenon of yield in crystals, and especially in metals, is of this nature. The planes in question are, of course, the well-known " gliding planes," and it is further possible that they may be identified also with the surfaces of least attraction. The stress at which gliding occurs in a single crystal must be determined in the following manner. The molecules of a crystal are normally in a configuration of stable equilibrium, and if two parts of the crystal slide on a gliding plane through one molecular space the resulting configuration is also stable. Between these two positions there must, in general, be one of higher potential energy, in which the equilibrium is unstable, and the shearing stress is determined by the condition that the rate at which work is done, in

* Lord RAYLEIGH, 'Engineering,' 1917, vol. 103, p. 111, and H. E. HEAD, 'Engineering,' 1917, vol. 103, p. 138.
† See QUINCKE, 'Ann. der Physik,' (4), 46, 1915, p. 1025.

sliding from the stable to the unstable state, must be equal to the greatest rate of increase in potential energy which occurs during the passage between the two states. This rate will depend on the shape of the molecular fields of force, and may in particular cases be zero. Liquid crystals are doubtless of this type. The average shear stress, during yield, of a random aggregation of a large number of crystals, is doubtless greater than that of a single crystal, as the angle between the gliding planes and the maximum shear stress must vary from crystal to crystal and can be zero in only a few of them.

As the mutually gliding portions of a crystal pass from the stable to the unstable state, the molecular cohesion between them (normal to the gliding plane) must, in general, become less. In particular instances it may diminish to zero before the position of unstable equilibrium is reached. In these cases, shearing fracture along the gliding planes will occur, unless the material is subjected to a sufficiently high "hydrostatic" pressure, in addition to the shearing stress. Thus, a crystalline substance may be either ductile or brittle, according to nature of the applied stress, or it may be ductile at some temperatures and brittle at others, under the same kind of stress as has been actually observed by BENGOUGH and HANSON in the case of tensile tests of copper. This rupture in shear explains the characteristic fracture of short columns of brittle crystalline material under axial compression. The theory indicates that such fracture can always be prevented and yield set up by applying sufficient lateral pressure in addition to the longitudinal load; this is in agreement with experiments on rocks such as marble and sandstone.* Conversely, a ductile substance might be made brittle if it were possible to apply to it a sufficiently large hydrostatic tension.

In the case of an alloy of, say, two metals A and B, suppose, as an example, that the sequence of molecules on either side of a gliding plane is

$$.A.B.B.A.B.B$$
$$B.B.A.B.B.A.$$

Let sliding occur (through one molecular space) to an adjoining position of stable equilibrium, or, say, to the configuration

$$.A.B.B.A.B.B. \leftarrow$$
$$.B.B.A.B.B.A \rightarrow$$

Evidently, the structure in the neighbourhood of the gliding plane is in this case no longer the same as in the original crystal formation. It is therefore likely that the new state is one of higher potential energy, whence it is reasonable to suppose that the maximum rate of increase in potential, in sliding, is greater than it would have been had the potential of the two states been the same. Thus an alloy may be expected to have a higher yield-point than its most ductile constituent. This is in accordance with experience. For example, it is known that quenching from a high temperature

* T. V. KARMAN, 'Zeitschr. Ver. Deutsch. Ing.,' 55, 1911.

hardens tool steel by preventing the separation of "ferrite," or iron containing no carbon.

In a single crystal the molecules are presumably in an equilibrium configuration of maximum stability. In this event, the equilibrium of molecules at or near intercrystalline boundaries, in a body composed of a large number of crystals, must, in general, be less stable than that of the molecules in the interior of the crystals. In fact, where the orientation of the component crystals is haphazard, the stability of the boundary molecules may be expected to range from the maximum of normal crystallisation down to zero, i.e., neutral equilibrium. If such a body be subjected to a shear stress, some of the molecules in or near neutral equilibrium must, in general, become unstable, and these will tend to rotate to new positions of equilibrium. This rotation, however, will be strongly resisted, as has been seen, by forces doubtless of a viscous nature, and its amount will accordingly depend on the time during which the stress is applied. If, therefore, the strain is observed it will be found to increase slowly as time goes on, but at a constantly decreasing rate, as the molecules concerned approach equilibrium. If now the load is removed, these molecules must rotate in order to regain their original positions of equilibrium, and this process in turn will be retarded by viscous forces. Hence a small part of the observed strain will remain after the removal of the load, and this will gradually disappear as time goes on. These properties, known as "elastic after-working," are, of course, well known to belong to crystalline materials. Moreover, the theory shows that they should not be possessed by single crystals, and this has been demonstrated experimentally.*

There is a special type of gliding or yield which may occur at stresses below the normal yield point. Consider a pair of adjacent crystals, separated by a plane boundary. If these crystals are thought of as sliding relatively to each other, it will be seen that only in a finite number of the positions so taken up can the two be in stable equilibrium. Between each pair of such positions there must in general be one of unstable equilibrium. Suppose that, while near such an unstable position, the two crystals are embedded in a number of others. Under these conditions the boundary molecules of the two crystals will be pulled over in the direction of one or other of the two adjoining stable positions, and they will strain the solid in the process. If now the body is subjected to a shearing stress tending to cause relative displacement of the two crystals towards the other stable position, then at a certain value of this stress the molecules on either side of the boundary will be wrenched away, will pass through the position of instability, and will then take up a new position bearing the same relation to the second stable position as their original state did to the first. This new condition will, of course, persist after the removal of the load, as the original state cannot be regained without passing through unstable equilibrium, i.e., a condition of maximum potential energy. To cause the crystals to pass through this condition it would be necessary to apply a load of opposite sign, and in this way the process might be repeated indefinitely. In a body composed

* H. v. WARTENBERG, 'Deutsch. Phys. Gesell., Verh.,' 20, pp. 113–122, August 30, 1918.

of a large number of crystals there must be many arrangements of this type, in which adjacent crystals can execute inelastic oscillations about positions of unstable equilibrium, under alternating shear stresses below the ordinary yield stress. The consequent observed phenomena would correspond exactly with those known to be manifested in metals, under the name "elastic hysteresis."*

Experimentally, elastic hysteresis is distinguished from elastic after-working by the circumstance that it is completed very much more quickly. This is just what would be expected theoretically, on the view that molecular translation occurs much more readily than rotation.

It has been remarked that when a single crystal of a pure substance is caused to yield, its structure is fundamentally unaltered. This cannot hold, however, in the case of an aggregate of a large number of crystals arranged at random, or a crystal embedded in amorphous material. True, the material in the interior of each crystal can retain its original properties, but near the crystalline boundaries the structure must be violently distorted. As a result, it may be expected that the number of the molecules of inferior stability will be largely increased. Elastic after-working in metals should therefore be increased by overstraining or "cold-working." This, again, agrees with experience.

The foregoing considerations lend support to the view that each crystal of a severely cold-worked piece of metal is surrounded by an amorphous layer of appreciable thickness. If such a piece of metal undergoes a shear strain greater than that which can initiate yield in the normal crystalline substance, the average stress which is set up must be above the normal yield stress, for the part due to the amorphous layers must be the elastic stress corresponding with the strain, and this, by hypothesis, is greater than the yield stress. This part, moreover, will increase with the strain. It follows that yield in cold-worked metal should be less sharply defined, and should occur at a higher shear stress than in the normal crystalline variety. That this is actually the case is, of course, well known.

In the case of very large strains an important part of the shear stress must be taken by the amorphous boundary layers, and as a result the maximum tensile stress may reach a value sufficient to cause rupture of some favourably disposed crystals across their planes of least strength. This is, perhaps, the actual mode of rupture in ductile materials. On this view, the "ductility" of a metal depends simply on the relation between the tensile strength of the "flaws" and the normal yield stress. A substance whose ductility is small may still be "malleable," as hammering need not give rise to large tensile stresses.

The formation of non-crystalline material at the intercrystalline boundaries, when a piece of metal is over-strained, appears to provide an explanation of the sudden drop in stress which occurs immediately after the initiation of yield† in ductile metals.

* GUEST and LEA, "Torsional Hysteresis of Mild Steel," 'Roy. Soc. Proc.,' A, June, 1917.
† ROBERTSON and COOK, 'Roy. Soc. Proc.,' A, vol. 88, 1913, pp. 462–471.

Remembering that the surface tension of a substance is the work done in forming unit area of new surface, it will be seen that the tension of any surface of a crystal must depend on the angle it makes with the crystal axis. Thus the surface tension parallel to the planes of least strength must be less than that in any other direction. Consequently, in a body composed of a number of crystals there must exist a mutual surface tension at each intercrystalline boundary. Now, the theory of surface tension shows that the magnitude of such a mutual tension is greatly diminished by making the transition between the two bodies more gradual. Hence the formation of the amorphous boundary layer involves a reduction in the surface energy of the crystals, and this is shown in the experiments by a drop in the stress. If this account of the phenomenon is complete, the drop in stress must be determined by the condition that the loss of strain energy equals the reduction in surface energy. The mechanism of the process appears to be that the breaking up of the boundary, which must accompany yield, is resisted by the surface tension, and yielding therefore requires a higher stress for its initiation than for its maintenance.

According to this view, the loss of strain energy should be inversely proportional to the linear dimensions of the crystals. Hence the results of different experiments should show considerable variation in the magnitude of the drop in stress. This is actually the case; a single series of experiments on mild steel, by ROBERTSON and COOK, gave drops varying from 17 per cent. to 36 per cent., while in other experiments as little as 7 per cent. has been observed.

In the above series of experiments the average loss of strain energy was about 12 inch-lbs. per cubic inch. Assuming, for simplicity, that the crystals were cubes, of, say, $0 \cdot 001$-inch side (which is a fair value for well-treated mild steel), the area of the intercrystal surface was 3000 sq. inches per cubic inch. These figures give the average intercrystal surface tension as $0 \cdot 004$ lbs. per inch. This is certainly of the right order of magnitude.

Many of the phenomena discussed above will be more complicated, in practice, if the coefficient of expansion of the crystals is not the same in all directions. In such an event, internal stresses will be set up in cooling, on account of the random arrangement of the crystals, and these stresses must be taken into consideration in applying the theory.

There remains for consideration the problem of the fracture of metals under alternating stress. It is known that fatigue failure occurs as the result of cracking after repeated slipping on gliding planes, and the theory has been advanced* that this cracking is due to repeated to and fro sliding and consequent attrition and removal of material from the gliding planes. This theory presents some difficulties, in that it does not explain how the attrition can occur, or the method of disposing of the debris.

A theory which is free from these objections may be constructed if it is supposed that a change in volume occurs on the passage of the metal from the crystalline to the amorphous state. This assumption is, of course, known to be valid for many substances

* EWING and HUMFREY, 'Phil. Trans.,' A, 1902, p. 200.

at their melting points, but at lower temperatures there seems to be no definite information available.

This assumption being granted, suppose that a piece of material which contracts on decrystallising is being subjected to a stress cycle just sufficient to cause repeated slipping in the most favourably disposed crystals. As a result, the material at the boundaries of these crystals will become amorphous, and the quantity of amorphous material will increase continuously as long as the repeated slipping goes on. But, by hypothesis, the unstrained volume of the amorphous phase is less than the space it filled when in the crystalline state. Hence all the material in the immediate neighbourhood will be subjected to a tensile stress, and as soon as this exceeds a certain critical value a crack will form. It has been observed above that the application of a sufficiently large hydrostatic tension may be expected to make a ductile substance brittle. Hence the crack may occur either in tension or in shear, according to the properties of the material and the nature of the applied stress. Further alternations of stress will cause this crack to spread until complete rupture occurs. This theory makes the limiting safe range of stress equal to that which just fails to maintain repeated sliding in the most favourably disposed crystals.

It may be asked why such cracking does not take place in a static test where the quantity of amorphous material, once yield has fairly started, is presumably much greater. The answer to this is two-fold. In the first place, if the material becomes amorphous round all, or nearly all, the crystals of a piece of metal, it is evident that it will contract as a whole and no great tensile stress will be set up. In the case where only a few crystals yield, the tension arises from the rigidity of the unchanged surrounding metal.

In the second place, even if some crystals do crack, the cracks will not, in general, tend to spread through the ductile cores of the neighbouring crystals, unless the applied load is alternating, on account of the equalisation of stress due to yield.

The safe limit of alternating stress will usually be less than the apparent stress necessary to initiate yield in a static test, on account of initial stresses, including those due to unequal contraction of the crystals.

The theory indicates that the cracking of the first crystal marks a critical point in the history of the piece. At any earlier stage the effects of the previous loading may be removed by heat treatment, or possibly by a rest interval, but once a crack has formed this cannot be done. True, the tension may be relieved and the ruptured crystal may even be compressed somewhat, but this cannot, in general, close the crack, as cracking is not a "reversible" operation. An exception may occur if the top temperature of the heat-treatment is sufficient to bring the molecules on either side of the crack within mutual range by thermal agitation, but it is unlikely that this can happen save in the case of very small cracks.

If this theory is correct, it appears at first sight that the phenomenon of fatigue failure must be confined to substances which contract on decrystallising. This, however is not necessarily so. If, for instance, a small thickness of material at the interface between two crystals were to increase in volume, it could not be said without proof that tensile stresses would not be set up thereby, in addition to compressions. In some cases, in fact, it is obvious that there must be tensions. Thus, if the outer layer of a sphere increases in volume, the matter inside must be subjected to a tensile stress.

The effect of overstrain on the density of metals is at present under investigation at the Royal Aircraft Establishment. The work is not yet sufficiently complete for detailed publication, but it may be mentioned here that the expected change in density has been found, and that the results already obtained are such as to leave little room for doubt that this change is in fact the cause of fatigue failure in metals. Thus, in overstraining mild steel by means of a pure shearing stress, a decrease in average density of as much as $0 \cdot 25$ per cent. has been observed.

Some progress has also been made in the direction of estimating the internal stresses set up as a result of the change in density, and it has been found that an average change of the magnitude mentioned above could give rise to a hydrostatic tensile stress in the cores of the crystals, of the order of 30,000 lbs. per sq. inch.

Dealing now with materials whose molecular sheet-formations are curved, it is at once evident that all yield, or slide, phenomena must be absent, as possible gliding planes do not exist. Thus, this case, though geometrically more complicated, is practically much simpler than that in which the sheets are plane. The theoretical properties of materials having the curved type of formation appear to correspond exactly with those known to belong to brittle "amorphous" substances. Exactly as in the case of crystalline materials, elastic after-working is explained by the inferor stability of molecules near the boundaries of the units of molecular configuration, but elastic hysteresis should not occur. If adequate precautions are taken to avoid secondary tensile stresses, fracture of short columns in compression should occur at stresses of an altogether higher order than in the case of crystals. In this connection it may be remarked that the compressive strength of fused silica is about 25 times as great as its ordinary tensile strength.

It appears from the foregoing discussion that the molecular orientation theory is capable of giving a satisfactory general account of many phenomena relating to the mechanical properties of solids, though closer investigation will perhaps show that the agreement is in some cases superficial only. Such questions as the effects of unequal cooling, foreign inclusions and local impurities, and the behaviour of mixtures of different crystals, have not been dealt with; it is thought that these are matters of detail whose discussion cannot usefully precede the establishment of the general principles on which they depend.

9. *Practical Limitations of the Elastic Theory.*

It is now possible to indicate the directions in which the ordinary mathematical theory of elasticity may be expected to fail when applied to real solids.

It is a fundamental assumption of the mathematical theory that it is legitimate to replace summation of the molecular forces by integration. In general this can only be true if the smallest material dimension, involved in the calculations, is large compared with the unit of structure of the substance. In crystalline metals the crystals appear, from the foregoing investigation, to be anisotropic and they must therefore be regarded as the units of structure. Hence the theory of isotropic homogeneous solids may break down if applied to metals in cases where the smallest linear dimension involved is not many times the length of a crystal.

Similar considerations apply to solids such as glass, save that here the units of structure are probably curved.

The most important practical case of failure is that of a re-entrant angle or groove. Here the theory may break down if the radius of curvature of the re-entering corner is but a small number of crystals long. An extreme instance is that of a surface scratch, where the radius of curvature may be but a fraction of the length of the crystals.

In the case of brittle materials the general nature of the effect of scratches on strength may be inferred from the theoretical criterion of rupture enunciated in section 2 above. Whether the material be isotropic or anisotropic, homogeneous or heterogeneous, it is necessary on dimensional grounds that the strain energy shall depend on a higher power of the depth of the scratch than the surface energy. It follows that small scratches must reduce the strength less than large ones of the same shape. Hence, where the tenacity of the material, under " uniform " stress, is determined by the presence of " flaws," it must always be possible to find a certain depth of scratch whose breaking stress is equal to that of the flaws. Evidently such a scratch can have no influence on the strength of the piece. Deeper scratches must have some weakening effect, which must increase with the depth, until in the limit the strength of very large grooves may be found by means of the elastic theory and the appropriate empirical hypothesis of rupture.

In the case of ductile metals, the effect of scratches is important only under alternating or repeated stresses. On the theory advanced in the preceding section, fatigue failure under such stresses is determined by phenomena which occur at the intercrystalline boundaries. Hence the strength of a scratched piece is fixed, not by the maximum stress range in the corner of the scratch, but by the stress range at a certain distance below the surface. This distance cannot be less than the width of one crystal, and it may be greater. Elastic theory suggests that the stress due to a scratch falls off very rapidly with increasing distance from the re-entrant corner, so that the relatively small effect of scratches in fatigue tests is readily explained.

Possibly many published results bearing on this matter depend more on initial skin stresses than on sharp corner effects.

10. *Methods of Increasing the Strength of Materials.*

The most obvious means of making the theoretical molecular tenacity available for technical purposes is to break up the molecular sheet-formation and so eliminate the "flaws." In the case of crystalline material this has the further advantage of eliminating yield and probably also fatigue failure.

In materials which normally have curved sheets, the molecular fields of force are presumably asymmetrical, and the process indicated above would lead of necessity to a random arrangement, which might be unstable. It has been seen that in glass and fused silica it is actually unstable, except in the case of the finest fibres.

As regards crystalline materials, however, in which the fields of force must have some sort of symmetry, there seems to be no reason why there should not be possible a very fine grained stable configuration, which could be derived from the ordinary crystalline form by appropriate rotations of certain molecules to new positions of stable equilibrium, in such a way as to break up the gliding planes. The grain of such a structure need be but a few molecules long, and its strength would approximate to the theoretical value corresponding with the heat of vaporisation.

There is some evidence that mild steel which has been put into the amorphous condition by over-strain tends, under certain conditions, to take up a stable fine-grained formation of this kind, in preference to resuming its original coarse crystalline configuration, in that a temperature of 0° C. appears to prevent recovery from tensile over-strain.*

These considerations suggest that if a piece of metal were rendered completely amorphous by cold-working, and then suitably heat-treated, its molecules might take up the stable strong configuration already described. The theory indicates, however, that over-straining tends to set up tensile stresses in the unchanged parts of the crystals which may start cracks long before decrystallisation is complete. Such cracking could be prevented if the over-straining were carried out under a sufficiently great hydrostatic pressure, and this line of research seems to be well worth following up. It might, of course, be found that the requisite pressure was so enormous as to render the method unworkable, but if the theory is sound there seems to be no other reason why definite results should not be obtained.

The problem may be attacked in another way. As has been seen, the theory suggests that the drop in stress at the initiation of yield is due to the surface energy of the inter-crystal boundaries. Thus the yield point may be raised by "refining" the metal, *i.e.*, so heat-treating it as to reduce the size of the crystals. The limit of refinement is, doubtless, reached when each "crystal" contains but a single molecule and the material is then in the strong stable state already described.

Refining is also of great value in connection with resistance to fatigue failure. Suppose, in accordance with the foregoing theory of fatigue, that one crystal has been fractured,

* Coker, 'Phys. Rev.,' 15, August, 1902.

then the general criterion of rupture shows that the crack cannot spread unless the material is subjected to a certain minimum stress, which is greater the smaller the crack. Thus, reducing the size of the crystals increases the stress necessary to cause the initial crack to extend. There is therefore a critical size of crystal for which the stress-range necessary to spread the crack is equal to that necessary to start it. Until the refining has reached that stage it can have no effect on the magnitude of the safe stress range, but from that point on the range must increase progressively with refinement until the limit is reached, as before, when each "crystal" contains but one molecule.

It therefore appears that refining is one avenue of approach towards the ideal state of maximum strength. Strangely enough, another line of argument suggests that the reverse of refinement might be effective in securing the desired result, in certain special cases. If a wire is required to withstand a simple tension, it seems that the best arrangement is that in which the strongest directions of all the molecules are parallel to the axis of the wire. This is equivalent to making the wire out of a single crystal. The theoretical tenacity would not be obtained, however, if the gliding planes made with the axis angles other than 0° C. and 90° C., as yield would occur.

If, in passing from the normal crystalline to the strong fine-grained state, the necessary orientation of the molecules were performed in accordance with some regular plan, the resulting configuration would possess some kind of symmetry, and the material might therefore exhibit crystalline properties. In cases where a substance exists in nature in several different crystalline forms, of which one is much stronger than the others, it may be that the strong modification is of the fine-grained type here considered. Thus, diamond may be a fine-grained modification of carbon. If this view is correct, it suggests that the transformation of carbon into diamond requires, firstly, the existence of conditions of temperature and pressure under which diamond has less potential energy than carbon; and, secondly, the provision of means for causing relative rotation of the molecules. In the attempts which have so far been made in this direction, attention seems to have been concentrated on satisfying the former requirement, the possible existence of the latter one having been overlooked. The most obvious way of satisfying it, if the mechanical difficulties could be overcome, would appear to be the application of suitable shearing stresses in addition to the hydrostatic pressure.

11. *Application of the Theory to Liquids.*

A detailed discussion of the properties of liquids, in the light of the present theory, would scarcely fall within the scope of this paper. One prediction which has been made, however, and which has been verified experimentally, affords such a remarkable confirmation of the general theory that it is felt that no apology is necessary for introducing it here. Consider a solid composed of molecules whose attraction is a function of orientation, the molecules being arranged in groups, in accordance with the theory outlined in the preceding pages. If the temperature of this body be supposed to be increasing, it will be seen that at some temperature the kinetic reactions due to the thermal vibrations must overcome the minimum attractions of the molecules in each group. It is clear, therefore, that at this temperature the substance will be unable to withstand shearing stresses. At the same time it cannot vaporise, as the molecules must still be held together in chains by their maximum attractions. In other words, the transformation which has been discussed is simply the liquefaction of the solid.

This view of the phenomenon of melting indicates that the molecules of liquids are in general arranged in groups or chains, of a length comparable with that of the structure ascribed to solids in the preceding work, or, say, 10^4 molecules.

If, therefore, a liquid be contained in a solid boundary which it wets, the ends of these chains may be expected to attach themselves to the solid; and if at any point the distance between the bounding walls is less than the length of the chains, some of the latter will attach themselves to both walls and hinder the free flow of the liquid and the relative movement, if any, of the boundaries. At such a point the liquid will act as a solid under any stress which is insufficient to break the chains.

This has been verified experimentally. The apparatus consisted of a polished steel ball 1 inch in diameter, and a block of hard tool steel containing a circular hole about 4 inches long. The hole was carefully ground, after hardening, to a diameter about $0 \cdot 0001$ inch greater, at its smallest part, than the diameter of the ball. When both were dry the ball passed freely through the hole. If, however, they were wetted with a liquid, considerable pressure was necessary to force the ball through. This resistance possessed the characteristic "stickiness" of solid friction, and was exactly the kind of resistance which would have been expected in forcing the dry ball through a hole which was too small for it.

To show that the resistance was a true "solid stress" and not due merely to viscosity, the apparatus was on one occasion left for a week, with the weight of the ball supported by the stress in the liquid (paraffin oil). The hole was vertical, so that there was no normal pressure between its surface and the surface of the ball. During this period no motion whatever could be detected.

It is essential to the success of these experiments that the ball and hole should be thoroughly wetted by the liquid. For this reason the liquids used have been chiefly paraffin oil and lubricating oils, but on one occasion the effect was obtained with water.

The present theory suggests a reason for the very low tensile strength of liquids. If a liquid is composed of a random aggregation of chains of molecules, it may reasonably be expected to contain regions of dimensions comparable with, but smaller than, the length of the chains, across which no chains run. Rupture of the liquid will evidently occur by the enlargement of these cavities. Now the tension, R, necessary to enlarge a spherical cavity of diameter, D, in a liquid of surface tension, T, is given by

$$R = 4T/D.$$

In the case of water, the tensile strength, R, is about 70 lbs. per sq. inch at ordinary temperatures, while T is about $0 \cdot 00042$ lbs. per inch. Hence the cavities, if spherical, must be at least $0 \cdot 000024$ inch in diameter. This is of the order indicated by the theory.

The foregoing conclusions are of especial interest in their relation to the theory of ROSENHAIN,* on which many of the properties of metals, and particularly "season cracking" under prolonged stress, are explained by supposing that the crystals are cemented together by very thin layers of amorphous material having the properties of an extremely viscous undercooled liquid. The experiments described above show that fluidity is not a property which can be ascribed *a priori* to such films. Hence if the view of ROSENHAIN and ARCHBUTT were to be definitely established, it would be necessary to regard it, not as a theory of season cracking in terms of the known properties of materials, but as a deduction of the properties of the intercrystalline layers from the phenomena of season cracking. Looked at in this way, it would be of extreme interest, for it would show that the molecular arrangement of the intercrystalline layers could not be of the coarse-grained type characteristic of the normal states of solids and liquids.

It is clear that the foregoing theory of liquids is not free from objection, and that in some respects it appears to be less satisfactory than existing theories. The most obvious objection is that it seems to be incompatible with accepted determination of the molecular weight of liquids. Since, however, these experiments are based ultimately on kinetic considerations, the author believes that this difficulty will not in fact arise unless the requisite bonds between the molecules of each group are found to be sufficiently strong to cause appreciable modification of the average molecular kinetic energy.

12. *Summary of Conclusions.*

(1) The ordinary hypothesis of rupture cannot be employed to predict the safe range of alternating stress which can be applied to metal having a scratched surface. The safe range of an unscratched test piece appears to be slightly less than the yield range, but if the surface is scratched the safe range may be several times the range which causes yield in the corners of the scratches.

(2) The "theorem of minimum potential energy" may be extended so as to be capable of predicting the breaking loads of elastic solids, if account is taken of the increase of surface energy which occurs during the formation of cracks.

(3) The breaking load of a thin plate of glass having in it a sufficiently long straight crack normal to the applied stress, is inversely proportional to the square root of the length of the crack. The maximum tensile stress in the corners of the crack is more than ten times as great as the tensile strength of the material, as measured in an ordinary test.

(4) The foregoing observation is in agreement with the known fact that the observed strength of materials is less than one-tenth of the strength deduced indirectly from physical data, on the assumption that the materials are isotropic. The observed

* W. ROSENHAIN and D. EWEN, "Intercrystalline Cohesion in Metals," 'J. Inst. Metals,' vol. 8 (1912); and W. ROSENHAIN and S. L. ARCHBUTT, "On the Intercrystalline Fracture of Metals under Prolonged Application of Stress (Preliminary Paper)," 'Roy. Soc. Proc.,' A, vol. 96 (1919).

strength is, in fact, no greater than it would be, according to the theory, if the test pieces contained cracks several thousand molecules long.

(5) It has been found possible to prepare rods and fibres of glass and fused quartz which have a tenacity of about one million pounds per square inch (approximately the theoretical strength) when tested in the ordinary way. The strength so observed diminishes spontaneously, however, to a lower steady value, which it reaches a few hours after the fibre has been prepared. This steady value depends on the diameter of the fibre. In the case of large rods it is the same as the ordinary tenacity, whereas in the finest fibres the strength diminishes but little from its initial high value. The relation between diameter and strength is of practically the same form for glass fibres as for metal wires.

(6) If it is assumed that intermolecular attraction is a function of the relative orientation of the attracting molecules, it is possible to construct a theory of all the phenomena mentioned in (3), (4) and (5) above. In the case of crystalline substances the theory also appears to explain yield and shearing fracture; elastic hysteresis; elastic afterworking; the fracture in tension of ductile materials and the flow of brittle materials under combined shearing stress and hydrostatic pressure; the drop in stress which occurs on the initiation of yield in ductile substances; fatigue failure under alternating stress; and the relatively slight effect of surface scratches on fatigue strength. In the case of non-crystalline materials the theory explains elastic afterworking and the great strength of short columns in compression.

(7) The theory shows that the application of the mathematical theory of homogeneous elastic solids to real substances may lead to error, unless the smallest material dimension involved, *e.g.*, the radius of curvature at the corner in the case of a scratch, is not many times the length of a crystal.

(8) It should be possible to raise the yield point of a crystalline substance by "refining" it, until at the ultimate limit of refinement the yield stress should be of the same order as the theoretical strength. It should also be possible similarly to increase the tenacity. Up to a certain stage the fatigue range should be unaffected by refining, but thereafter it should increase in the same degree as resistance to static stress.

(9) The theory requires that a thin film of liquid enclosed between solid boundaries which it wets should act as a solid. Experimental confirmation of this has been obtained.

In conclusion, the author desires to place on record his indebtedness to many past and present members of the staff of the Royal Aircraft Establishment for their valuable criticism and assistance, and also to Prof. C. F. JENKIN, at whose request the work on scratches was commenced.

[*Note.*—It has been found that the method of calculating the strain energy of a cracked plate, which is used in Section 3 of this paper, requires correction. The correction affects the numerical values of all quantities calculated from equations (6), (7), (8), (10), (11), (12) and (13), but not their order of magnitude. The main argument of the paper is therefore not impaired, since it deals only with the order of magnitude of the results involved, but some reconsideration of the experimental verification of the theory is necessary.]

The Theory of Rupture

by A. A. GRIFFITH, Farnborough.

In the past, attempts to construct theories of the rupture of solids have usually arisen out of the mathematical theory of elasticity, and have been directed mainly towards the object of making the results of that theory useful in estimating the mechanical strength of solid bodies. These theories were all based on empirical assumptions, such as, for instance, the hypothesis that rupture will occur if a specific tensile stress is exceeded. [1])

The development of the soap-film and photo-elastic methods rendered it possible to apply the theory of elasticity to bodies of complex shape which could not previously be studied; indeed, these developments were largely directed towards the practical object of estimating the mechanical strength of such bodies with the help of one of the current theories of rupture. So soon, however, as attempts were made to verify experimentally the strenghts so estimated, very serious disagreements were disclosed, which could not be removed by any plausible adjustment of the classical hypotheses of rupture.

It was in consequence of these discrepancies, which rendered the new methods of stress estimation almost valueless for their main purpose, that a new theory of rupture, the elements of which are to be described, was developed.

The fundamental conception of the new theory is this. Just as in a liquid, so in a solid the bounding surfaces possess a surface tension which implies the existence of a corresponding amount of potential energy. If owing to the action of a stress a crack is formed, or a pre-existing crack is caused to extend, therefore, a quantity of energy proportional to the area of the new surface must be added, and the condition that this shall be possible is that such addition of energy shall take place without any increase in the total potential energy of the system. This means that the increase of potential energy due to the surface tension of the crack must be balanced by the decrease in the potential of the strain energy and the applied forces.

With the help of this fundamental conception, we may proceed at once to estimate, from thermal data, the approximate magnitude of the strain energy necessary to break any given solid.

The surface tension of any substance is zero at the critical temperature and pressure. It follows at once that the tensile strength is zero under these conditions. Hence the strain energy necessary for rupture under uniform tensile stress at a lower temperature must be equal to the work done against cohesion in raising the substance to the critical temperature and pressure. If we choose a substance which does not undergo chemical dissociation, this quantity of work is nearly equal to the total heat necessary to attain the critical temperature. This, again, is of the same order as the total heat of vaporisation at intermediate temperatures. In general, then, it may be said that the strain energy for rupture is of the same order as the total heat of vaporisation. We note further that, according to this method of attacking the problem, the rupture stress is comparable with the intrinsic pressure of the solid.

This latter relation discloses at once a great disparity between the theoretical rupture stresses of common metals and the strengths observed in ordinary tensile tests. According to TRAUBE,[2]) the intrinsic pressures of metals vary from 3,340 kg./sq. mm. for nickel and 3,330 kg./sq. mm. for iron to 530 kg./sq. mm. for lead. In tensile tests nothing approaching these figures is ever obtained. Even if we exclude ductile metals on the ground that plastic flow, whose nature is not well understood, may give rise to a type of rupture different from the simple tensile type considered here, we still find that there is a very great defect of strength even in the non-ductile metals. Thus, hard steel containing 0.9 % of carbon has a tenacity of only 118 kg./sq. mm., while the strongest alloy steel known has a tenacity not greater than 310 kg./sq. mm.

Taking the most favourable case, then, the observed rupture stress is less than 10 %, and the energy for rupture less than 1 %, of the theoretical values. Now, by the principle of the conservation of energy, the work done in forming a crack cannot actually be less than the value calculated from thermal data, because that amount of work must make its appearance in the form of the surface energy of the crack. Therefore since the average strain energy in the tensile test is small compared with the theoretical energy, it must be concluded that, at the moment of rupture, the energy of the test piece is not uniformly distributed, that is to say, at the points where the cracks originate there must be very severe concentrations of some form of energy.

Now it appears that there are but three possible ways in which the required energy concentrations can occur. These are as follows:
(1) If the material is heterogeneous there may exist a mutual surface tension, of sufficient magnitude, between different constituents.
(2) The material may contain severe initial stresses.
(3) The material may contain small cracks formed during manufacture or in the course of subsequent treatment.

At the present little is known of the mutual surface tension of solids; it is not yet possible, therefore, to make much progress in the discussion of the first-named of these hypotheses. The intercrystalline fracture of iron at the temperature of liquid air, is, however, suggestive of the possibility that this type of weakening does actually occur. With regard to the second, little can be done without considerable knowledge of the nature and distribution of the supposed initial stresses. We proceed, therefore, to consider in more detail the third possible cause of the defect of strength.

For this purpose, it is necessary to find the changes in the stresses and in the strain energy of the solid, due to the introduction of the cracks. In the present state of the theory of elasticity, this necessitates the restriction of the investigation to two-dimensional problems and to materials which obey HOOKE's law fairly closely. It will, however, be legitimate to utilise the results so obtained in making general inferences regarding three-dimensional systems. It may be remarked, moreover, that in view of recent developments it may soon be possible to extend the rigorous theory to certain three-dimensional cases.

For our present purpose, the fundamental two-dimensional elastic problem is the problem of finding the stresses in a plate pierced by a small crack whose surface has the form of an elliptic cylinder. The axis of the cylinder is perpendicular to the plate and one of the axes of the ellipse is very small compared with the other. The plate is loaded by tractions applied at the outer edge of the plate in directions parallel to its surface. The general case of this problem was solved by INGLIS [3]) in 1913, by the method of finding solutions of the equations of elasticity in the elliptic coordinates α, β obtained by the conformal transformation:

$$x + iy = c \cosh(\alpha + i\beta).$$

The first application which suggests itself is to find the theoretical breaking load of a plate of given material containing a crack of given length (the length of the crack is the length of the major axis of the narrow ellipse which forms its edge). In order to do this we must form the expression for the sum of the potential energy of the applied edge tractions, the strain energy of the plate and the surface energy of the crack. The condition that the crack may extend, that is, that rupture may occur, is then obtained by expressing the condition that the total energy is unchanged by small variations in the length of the crack.

A solution of this problem was given in a paper read in 1920 [1]) but in the solution there given the calculation of the strain energy was erroneous, in that the expressions used for the stresses gave values at infinity differing from the postulated uniform stress at infinity by an amount which, though infinitesimal, yet made a finite contribution to the energy when integrated round the infinite boundary. This difficulty has been overcome by slightly modifying the expressions for the stresses, so as to make this contribution to the energy vanish, and the corrected condition for the rupture of a thin cracked plate, under a uniform pull applied in its plane at its outer edge, is:

$$R = \sqrt{\frac{2ET}{\pi c}},$$

where R is the component of the applied edge traction normal to the direction of the crack, $2c$ is the length of the crack, and E and T are respectively the YOUNG's modulus and surface tension of the material. (This formula was obtained by the method used in the paper already mentioned — see pp. 167—170 of that paper, — the only difference being that the corrected value of W — see eq. 8, p. 169, — namely, $\frac{(p+1)\pi c^2 R^2}{8\mu}$, was used instead of $\frac{(3-p)\pi c^2 R^2}{8\mu}$, the value given in the paper). The component of edge traction S parallel to the crack has no influence on the rupture stress. Fig. 1 gives a diagrammatic representation of the crack and of the stress system.

The maximum stress, F, at the corners of the crack is:

$$F = 2R\sqrt{\frac{c}{\rho}},$$

where ρ is the radius of curvature of the corners of the crack, provided ρ is small compared with c. This formula is derived from INGLIS's expression:

$$F = R\left(1 + 2\frac{a}{b}\right),$$

for an elliptic hole of semi-axes a and b. Thus the actual rupture stress of the material is constant if ρ is constant.

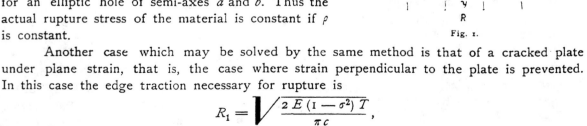

Fig. 1.

Another case which may be solved by the same method is that of a cracked plate under plane strain, that is, the case where strain perpendicular to the plate is prevented. In this case the edge traction necessary for rupture is

$$R_1 = \sqrt{\frac{2E(1-\sigma^2)T}{\pi c}},$$

where σ is the value of POISSON's ratio.

If we now calculate the strain energies of the material at the point of rupture, in the above cases, we find:

$$\frac{F^2}{2E} = \frac{F_1^2}{2E(1-\sigma^2)} = \frac{4T}{\pi \rho},$$

or the same amount of energy is required for rupture in each case. This is in accordance with the general theory.

Experimental verifications of the above formulae will be discussed later, but let us first consider the theoretical laws of rupture of the general type of solid which owes its weakness to cracks, namely, a solid containing a large number of small cracks oriented at random. As before, we are limited to the two-dimensional case and, since it will be necessary to employ INGLIS's solution for the stresses in a plate containing a single small crack, we must further postulate that the cracks are so far apart that the maximum stress due to any crack is not seriously affected by the presence of its neighbours. We may infer, however, from DE ST.-VENANT's equipollence principle, that the error due to this cause is unlikely to be large.

The problem before us is to find the laws of rupture of our solid under various kinds of applied edge traction, e.g., simple tension, simple compression (crushing), and tension in one direction combined with compression in the orthogonal direction, etc. The theory cannot, of course give the absolute strengths unless the dimensions of the cracks are known, but, as we shall see, the ratios are determinate, so that all the strengths can be found if one of them is known.

In accordance with the theory already discussed, we may infer that the general condition for rupture will be the attainment of a specific tensile stress at the edge of one of the cracks. It is therefore merely a question of finding in which of the cracks the greatest tensile stress occurs, and the magnitude of that stress in terms of the applied tractions.

Using INGLIS's results, therefore, we form the expression for the stress at all points of the elliptic edge of a small crack in a large plate, to whose outer edges are applied principal tractions Q and P, respectively making angles θ and $\frac{\pi}{2} - \theta$ with the major axis of the crack. These principal stresses may be resolved into two direct stresses and a shear stress, applied in directions parallel and perpendicular to the direction θ, whence the complete stress system may be found by addition of the appropriate solutions given by INGLIS. Using INGLIS's notation, we find that the required stress is:

$$R_{\beta\beta} = \frac{(P+Q)\sinh 2\alpha_0 + (P-Q)(e^{2\alpha_0}\cos 2\beta - 1)\cos 2\theta + (P-Q)e^{2\alpha_0}\sin 2\beta \sin 2\theta}{\cosh 2\alpha_0 - \cos 2\beta}$$

wherein α_0 is the parameter of the ellipse corresponding with the edge of the crack, so that α_0 is nearly equal to zero. β is the coordinate specifying the positions of points on the ellipse. The direction of $R_{\beta\beta}$ is parallel to the edge of the crack. Since we are interested only in tensile stresses, let us postulate that tensile stresses are positive and that P is algebraically greater than Q. An explanatory diagram is given in fig. 2 (for $R_{\alpha\alpha}$ read: $R_{\beta\beta}$).

By differentiation, we find the values of θ and β for which $R_{\beta\beta}$ is a maximum. We find that, in general, $R_{\beta\beta}$ is a maximum at two pairs of points on each crack. If $\theta = 0$ or $\frac{\pi}{2}$, these points are at the ends of the major and minor axes respectively, but for all other values of θ both pairs are very near the ends of the major axis. In general, one pair of maximum stresses is tensile and the other compressive. By expressing the conditions that the greatest tensile stress so found is constant, we arrive at the required laws of rupture. These are as follows:

(1) If $3P + Q$ is positive, the condition for rupture is:
$$P = K,$$
where K is a constant depending only on the properties of the material and the dimensions of the cracks.

(2) If $3P + Q$ is negative, rupture is determined by the equation:
$$(P - Q)^2 + 8K(P + Q) = 0,$$
where K is the same constant as in condition (1) above.

In the first case $\theta = 0$, so that the surface of rupture is perpendicular to the direction of P. In the second case we have the equation:
$$\cos 2\theta = -\frac{1}{2}(P - Q)/(P + Q),$$
so that the fracture is oblique.

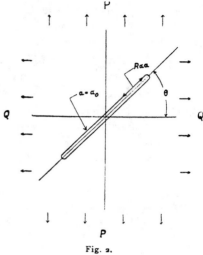

Fig. 2.

Now, notwithstanding the restriction of the rigorous application of the above formulae to two-dimensional cases, it is of interest to compare them with results obtained in practice in the testing of brittle solids. In view of the limitations of the theory a certain degree of numerical discrepancy will be tolerable provided that the general nature of the predictions is verified.

The usual mechanical tests are tensile, crushing (compressive), and torsional (shearing) tests. Taking these three cases and applying the formulae we find that the shearing and tensile strengths should be equal and that the fracture should in each case be perpendicular to the tensile stress, while the crushing strength should be eight times as large, the fracture being in this case oblique.

Comparing these predictions with the results of tests on brittle solids, we find that the tensile and torsional strengths are approximately equal in the cases of hard cast iron and carbon steel, and that the fractures run in the predicted directions, while in the crushing tests the fracture is oblique as required by the theory; the crushing strength of the cast iron is about 7 times, and of the steel about $5\frac{1}{2}$ times, the tensile strength. The accurate determination of the crushing strength, however, is not easy, particularly for the steel, owing to the difficulty of obtaining axial loading. In such tests of stone and like materials as are available the crushing strength is from 7 to 11 times the tensile strength, and the crushing fracture is oblique. I have not been able to find data for the torsional strength of stone.

In view of this remarkable agreement, we may conclude that there is strong evidence that the rupture of the above-mentioned solids is governed by the existence of a multitude of small internal cracks. The prediction of the essential difference between tensile and crushing failure, a difference which resisted all efforts at explanation on the basis of the classical hypotheses of rupture, is a particularly successful result of the new theory, and we may look forward with interest to the time when it will be possible to develop the three-dimensional form of the equations.

A remarkable and somewhat unexpected consequence of the theory is that the presence of internal cracks can give rise to tensile stresses large enough for fracture, even when the components of applied traction are both compressive, provided that they are unequal.

Some experiments will now be described which were performed in the course of the development of the theory.

First, it was sought to verify the theoretical formula for the strength of a plate containing a single crack. Various considerations, such as theoretical limitations and the necessity for estimating the surface tension of the material in the solid state, led to the selection of glass as a suitable material for the purpose. A fairly hard glass containing much potash and alumina was chosen. By the usual mechanical tests, its YOUNG's modulus was found to be 6,340 kg./sq. mm., and its tensile strength 18.3 kg./sq. mm.

The surface tension was determined at a series of temperatures at which the glass behaved as a viscous liquid; its value at ordinary temperatures was obtained by extrapolation on the assumption of linear variation with temperature. The value so found was 546 dynes/cm.

The experiments proper were performed on cracked spherical bulbs and also on round tubes; these were broken by internal hydraulic pressure *) Rigorously, the theoretical formula can only be applied to such bulbs and tubes if their diameters are very great, but preliminary experiments disclosed no serious error due to this cause, when using the dimensions which were found to be convenient.

The next question requiring consideration was this. Did the operation of forming the cracks leave initial stresses in the glass of sufficient magnitude to alter the breaking pressure seriously? If so, it was evidently necessary to remove the initial stresses by annealing. Annealing, however, had the disadvantage that it was liable to cause an increase in the radius of curvature at the corners of the crack, by viscous flow under surface tension. It was considered that the most satisfactory way of settling this question was to perform experiments on tubes subjected to various annealing processes, and also on unannealed tubes.

According to the theory, the condition for rupture is of the form:

$$R = m c^{-\frac{1}{2}},$$

where m is a constant of the material. The theoretical value of m for the material used is 0.471, c being in mm. and R in kg./sq. mm. The experimental verification was performed by comparing this value of m with the values obtained from the tests.

In the earlier experiments the cracked glass was annealed for half an hour at various temperatures up to 500° C. It was found that the value of m increased with the annealing temperature, at first rapidly and then more slowly until a fairly constant figure of 0.85 was obtained for temperatures between 400° C. and 500° C. Tests on unannealed tubes, on the other hand, gave values ranging from 0.30 to 0.39.

It appears probable from these figures that the unannealed tubes contained appreciable initial stresses, which, moreover, varied considerably from one tube to another. The high values obtained from the annealed tubes, on the other hand, can only be explained by supposing that considerable enlargement of the radius at the corners occurred during the half hour's heating at about 450° C. It was therefore decided to try the expedient of greatly reducing the heating period, as it was thought that by this means it would be possible to relieve the initial stresses sufficiently without giving time for appreciable flow to occur at the corners of the crack. This expedient was completely successful: it was found that by shutting off the heating current as soon as the temperature reached 450° C., and then allowing

---

*) The cracks were formed either with a glass-cutter's diamond, or by scratching with a hard steel edge and tapping gently.

the tubes to cool in the furnace, values of m were obtained which consistently fell between 0.44 and 0.46, with a mean value of 0.453.

The theoretical deduction that rupture is independent of the component of applied stress parallel to the crack was also verified by applying end loads to the tubes.

This good agreement between prediction and experiment is a very satisfactory achievement of the new theory, since it seems that this is the first occasion on which the breaking load of an elastic solid body has been successfully calculated from the form of the body and the physical constants of the material.

It is of interest to calculate at this point the order of magnitude of the tensile stress which must have been set up, in these experiments, at the corners of the crack. From the interference colours shown by the cracks, it was deduced that the radius of curvature of the corners could not have been greater than 5×10^{-5} mm., whence by the formula previously given the stress at fracture could not have been less than 130 kg./sq.mm. But there is no reason why the radius should have been so large. It is much more likely that it was of the order of the molecular dimensions. Allowing for the possibility that glass has a somewhat complex molecule, let us put $\rho = 5 \times 10^{-7}$ mm. Then we find that the stress must have been of the order of 1,300 kg./sq.mm. This is of the same order as the values previously quoted for the intrinsic pressures of metals, so that the theoretical deduction that the intrinsic pressure is a measure of the true rupture stress thus receives experimental confirmation.

Further experiments on the same glass have given some insight into the cause of the defect of strength characteristic of the material in its ordinary state, in which, as already mentioned, its strength is only 18.3 kg./sq. mm.

The glass was heated to a temperature somewhat below the temperature at which boiling commenced and was then drawn down to thin rods. Immediately after preparation, these rods were found to possess strengths as high as 630 kg./sq. mm. or about half the probable value of the theoretical strength. This strength diminished spontaneously with time, however, until after the lapse of a few hours a steady value was reached which depended on the diameter of the rod. Rods of 1 mm. dia. or over had a strength equal to the strength of the glass as ordinarily determined, while the permanent strength of the finest fibres which could be made (about 1/300th. mm. dia.) was about 350 kg./sq. mm., or more than a quarter of the theoretical strength. By extrapolation, it was inferred that the strength of fibres of infinitesimal diameter would be about 1,130 kg./sq. mm., which is the same for all practical purposes as the estimated breaking stress in the cracked tube experiments.

We may infer that the strong state of this glass is unstable, and that there occurs in it some spontaneous change, probably a species of phase change, which causes a loss of strength by one of the three possible means already mentioned. This is confirmed by the observation that the weakening is accompanied by a slight change in the length of the rod (of the order of 0.1 % increase), and also by the observation of BRIDGMAN that glass capillaries could withstand an external pressure of 24,000 atmospheres without apparent change, but sometimes broke several weeks or even months after the release of the pressure.

It is of interest to estimate the order of magnitude of the smallest cracks which could cause the observed defect of strength, though there is, of course, no evidence at present that the sources of weakness are actually cracks in the case of glass.

We have the relation:
$$18.3 = 0.471 \, c^{-\frac{1}{2}},$$
whence we find that $2c$ is roughly 1.5×10^{-3} mm.

The behaviour of vitreous silica which has been drawn down in the oxyhydrogen flame differs from that of glass. If it contains impurities it exhibits the same spontaneous weakening, though in a less degree, but if pure it does not. Rods of the pure substance may be prepared having a tensile strength of over 370 kg./sq.mm. and this strength can be maintained indefinitely. Occasionally as much as twice this strength is obtained. The question at once arises as to what is the cause of the defect of strength of the silica as ordinarily met with, if no spontaneous weakening occurs. The answer is that the weakness is due almost entirely to minute cracks in the surface, caused by various abrasive actions to which the material has been accidentally subjected after manufacture.

If a strong silica rod be lightly rubbed with any other solid body, it immediately loses its great strength, owing to the formation of these surface cracks. Even atmospheric dust particles have an appreciable weakening effect. The following is an instructive experiment illustrating this effect of abrasion. If two strong silica rods be touched lightly together and one of them be then bent, it breaks very easily if the point of contact is on the side subjected to tension, while if it is on the compression side the rod appears to be as strong as ever.

The theory of this surface cracking by abrasion has been worked out, in its two-dimensional form, with the help of INGLIS's results together with HERTZ's theory of bodies in contact, [5] but it is too long to be given here. The general conclusion is that, if two elastic solids are pressed together so as to touch over a small area, then surface cracks will be formed in both of them if the direction of the force which holds them together deviates appreciably from the normal to the surface of contact.

Strong silica may be protected from this surface abrasion to some extent by coating it with a lubricating liquid, so that a rod with an oily surface may be handled freely without risk of damage.

The rupture strain energy of the strongest silica rods is so great that if it were all converted into heat it would be enough to make the rod red-hot. It is of interest to enquire, therefore, what becomes of this energy when fracture occurs. It is not converted into heat; indeed, measurements have shown that the defect of elasticity of the strong material is so inappreciably small that such an operation would take several minutes at least. What actually happens is that the energy is used up almost entirely in forming new surfaces, that is, surfaces of fracture, in the material. The disintegration of the strong drawn-down part is sometimes so complete that recognisable portions of this part are difficult to find. The thick ends also are usually broken, owing to the propagation of an elastic wave from the original fracture.

In concluding, I wish to refer to some of the difficulties which impede the further development of the theory of rupture. On attempting to pass from isotropic (amorphous) solids to brittle crystals, we at once meet the difficulty that the surface tension is not a constant but is a function of the position of the surface in the space lattice, so that the theoretical strength is different for different faces. The anisotropic nature of the elasticity is a further obstacle.

In the case of plastic crystals, we are further hindered by the fact that rupture is almost invariably preceded by plastic flow, whose nature is still the subject of hot controversy. In fact, it may be said that, at the present day, there is no theory of the rupture of such solids, insomuch that no attempt has been made to cope with the basic problem, namely, the discovery of the source of the surface energy of the cracks.

References:

[1] A. E. H. LOVE, Mathematical Theory of Elasticity, 3rd ed., Ch. 4, pp. 119—121.

[2] TRAUBE, Zeitschr. für anorganische Chemie, 1903, vol. XXXIV, p. 413.

[3] C. E. INGLIS, Stresses in a Plate due to the Presence of Cracks and Sharp Corners, Trans. Inst. Naval Architects, London, vol. LV, pp. 219—230, 1913.

[4] A. A. GRIFFITH, The Phenomena of Rupture and Flow in Solids, Phil. Trans. Roy. Soc. London, A, vol. 221, pp. 163—198, 1921.

[5] A. E. H. LOVE, loc. cit., Ch. VIII, pp. 191—196.

Note of the editors: A German review of Dr. GRIFFITH's theory of rupture has been given by K. WOLF in the Zeitschrift für angewandte Mathematik und Mechanik (Zur Bruchtheorie von A. GRIFFITH, l. c. Bd. 3, pp. 107—112, 1923); further a paper by A. SMEKAL (Technische Festigkeit und molekulare Festigkeit, Die Naturwissenschaften, Bd. 10, pp. 799—803, 1922) may be mentioned.

Dr. GRIFFITH's paper, mentioned above under reference, [4] also contains some considerations on the molecular theory of strength phenomena.

The calculation of the cohesional forces, etc. of crystallised solids from the structure and the electrical forces of the atomic lattice, has in the last few years been made the subject of a series of papers, amongst which those of M. BORN (Göttingen) and his collaborators may be mentioned.

Direct Measurements of the Surface Energies of Crystals

JOHN J. GILMAN
*General Electric Research Laboratory, Schenectady, New York**
(Received July 25, 1960)

By means of quantitative cleavage experiments, the surface energies of several simple crystals have been measured at $-196°C$. The crystals and their cleavage planes are: LiF(100), MgO(100), CaF$_2$(111), BaF$_2$(111), CaCO$_3$(10$\bar{1}$0), Si(111), and Zn(0001). Measured values of their respective surface energies (ergs/cm^2) are: 340, 1200, 450, 280, 230, 1240, and 105. The measured values for LiF and MgO are in good agreement with simple ionic lattice theory. Values for the other crystals seem consistent with their binding energies.

Under irreversible conditions an effective surface energy is measured. This quantity increases rapidly with increasing temperature for the metallic crystals, Zn and Fe(3% Si). The increase correlates with increasing plastic flow in these crystals. In contrast, the effective surface energy of LiF and MgO is only moderately dependent on temperature.

A small amount of cadmium (0.1 at.%) markedly increases the cleavage surface energy of zinc.

INTRODUCTION

IN the experiments reported here surface energy measurements were made by the most direct of all methods. Although the surface energy of a solid is one of its most fundamental properties, there is a dearth of experimental values for it. This is because many difficulties are associated with measuring it; especially when indirect methods are used. Therefore, it seemed to be worthwhile to apply cleavage techniques that have been developed in other studies to this purpose. Since the surface energy can be defined as the work that

* Now at Brown University, Providence, Rhode Island.

is required to separate a crystal into two parts along a plane, cleavage is a particularly direct way of measuring it. The method consists of measuring the work that is required to cause cleavage per unit area of cleaved surface. If the cleavage is carried out in a reversible fashion, the work of cleavage is equal to the intrinsic surface energy of the crystal. If the cleavage is not carried out in a reversible fashion, the work of cleavage is increased and one speaks of an *effective* cleavage energy.

The cleavage method for measuring surface energies has been previously applied to the case of mica crystals by Obreimov.[1] Derjaguin et al.,[2] and Bailey[3]; but it has not been applied to simpler crystals. Some particularly simple crystals (LiF, MgO, CaCO$_3$, Zn, CaF$_2$, BaF$_2$, Si) have been used in the present study so the results can be readily compared with theoretical calculations.

Other methods that have been used for surface energy measurements are: (a) measurements of energy absorption during impact cleavage by Kusnetsov and Teterin[4]; (b) comparisons of the heats of solution of large and small particles of crystal, most recently by Benson, Schreiber, and van Zeggeren[5]; and (c) measurement of the force required to balance the surface tensions of metal wires near their melting points, by Udin, Shaler and Wulff.[6] Method (a) suffers from difficulties associated with achieving reversibility and precision under impact conditions. Both methods (a) and (b) have not yet been applied under conditions where irreversible plastic flow did not accompany the production of new surfaces. Therefore, they have yielded values too high by a factor of about two. Method (c) can only be used at high temperatures, and it does not yield values for individual crystallographic faces.

For historical reviews of the subject, the reader is referred to the books by Partington[7] and Kusnetsov.[8]

EXPERIMENTAL TECHNIQUE

The method used for measuring the work of cleavage is the same principle as the one used by Obreimov[1] for mica. The geometry has been modified somewhat, however, so it can be used for a variety of crystals. A schematic view of the experimental arrangement is shown in Fig. 1. The specimens had the form of rectangular blocks that were partially bisected by cracks. In preliminary experiments, the forces F were applied by slowly pushing wedges into the cracks. However, this method gave poor reproducibility (even when the wedges were carefully made and lubricated) so it was abandoned. The method that was finally adopted for

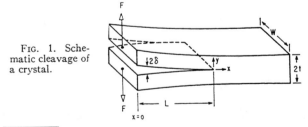

Fig. 1. Schematic cleavage of a crystal.

[1] J. W. Obreimov, Proc. Roy. Soc. (London) **A127**, 290 (1930).
[2] B. V. Derjaguin, N. A. Krotova, V. V. Karassev, and Y. M. Kirillova, Second Int. Cong. Surf. Act. **III**, 417 (1957).
[3] A. I. Bailey, Second Int. Cong. Surf. Act. **III**, 406 (1957).
[4] V. D. Kusnetsov and P. P. Teterin in *Surface Energy of Solids* (Her Majesty's Stationery Office, London, 1957), pp. 39–42.
[5] G. C. Benson, H. P. Schreiber, and F. van Zeggeren, Can. J. Chem. **34**, 1553 (1956).
[6] H. Udin, A. J. Shaler, and J. Wulff, Trans. AIME **185**, 186 (1949).
[7] J. R. Partington, *An Advanced Treatise on Physical Chemistry* (Longmans, Green and Company, London, 1952), Vol. 3, p. 242.
[8] V. D. Kuznetsov, *Surface Energy of Solids*, translation from the Russian (Her Majesty's Stationery Office, London, 1957).

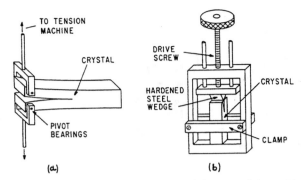

Fig. 2. Methods for applying force to specimens and for making partially cleaved crystals.

applying the forces was to attach a small yoke to each "arm" of a specimen by means of pivot pins. This is shown schematically in Fig. 2(a). The length L of each lever arm then becomes the distance between the crack tip and the pivot point.

Crystals were initially cracked part way along their lengths by means of the jig shown in Fig. 2(b). It consists of a screw-driven hardened steel wedge for starting cracks, and rubber-padded clamps for applying compressive forces to the crystal near the middle of its length. The wedge was ground to a 30° chisel edge with convex curvature so it made initial contact with a crystal at only one point. Compressive stresses produced by the clamps stopped the cracks from running the entire lengths of crystals. The entire jig was immersed in liquid nitrogen for starting cracks at a low temperature.

Measured forces were applied to the partially cracked specimens by means of an Instron tensile-testing machine. The cross-head of the machine moved at constant speed so as to increase the applied force at a rate of about 8 g/sec. When the cracks began to increase in length the applied forces decreased suddenly because of the increased deflections of ends of the crystals. These deflections relaxed the stiff elastic spring of the loading machine. The forces at which this happened were recorded as critical forces for crack propagation.

The dimensions of the crystals were measured by means of a standard micrometer. The crack lengths were measured approximately before each test and more accurately afterwards by microscopic examination of the cleaved surfaces. When a cleavage crack stops moving and later starts to move again, the small cleavage steps that it leaves behind suffer a discontinuous change in direction. These microscopic markings indicate the position of the stopped crack tip.

Several different kinds of crystals were studied: LiF, MgO, CaF$_2$, BaF$_2$, CaCO$_3$, Zn, Si, and Fe-Si. The LiF, CaF$_2$, and BaF$_2$ crystals were purchased from the Harshaw Chemical Company, MgO from the Norton Company, and the CaCO$_3$ crystals were natural Iceland Spar crystals purchased from Ward's Natural Science Establishment. The silicon crystals were high-purity semiconductor grade crystals provided by the Semi-

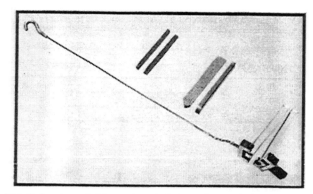

FIG. 3. Specimen holder and typical cleaved crystals. From left to right are shown a silicon, a zinc, and a lithium fluoride crystal.

conductor Products Department of the General Electric Company. The Fe-Si crystals were grown at the General Electric Research Laboratory by Dr. C. Dunn, and the zinc crystals were prepared by the author from 99.999% pure zinc purchased from the New Jersey Zinc Company. This particular collection of crystals was selected primarily on the basis of availability in large enough sizes to allow the present experimental methods to be applied, but an effort was made to include representatives of the ionic, metallic, and covalent types.

Specimens were initially prepared by cleavage or abrasive cutting followed by deep chemical polishing (at least $\frac{1}{4}$ mm removal from all surfaces) to remove the damaged surface layers. At this stage they were rectangular bars 2–6 cm long and with w and $2t$ (Fig. 1) in the range 2–6 mm. They were oriented with their cleavage planes parallel to one set of side surfaces. Cracks 1–3 cm long were then made in them with the jig of Fig. 2(b). Some of the actual specimens are shown in Fig. 3 together with the jig that was used to apply forces to them.

CLEAVAGE MECHANICS

In order to convert the measured critical forces for crack propagation into fracture surface energies one must analyze the mechanics of the system shown in Fig. 1. This has been done using isotropic elasticity theory and following the method of Benbow and Roesler.[9] Except for the selection of reasonable elastic modulus values for the various crystals, their elastic anisotropy was not taken into account. This seems justified because the experimental results show an amount of scatter that exceeds the error that might be expected from using the simple analysis instead of something more elaborate.

Mechanically, a partially cracked crystal is nearly equivalent to two built-in cantilever beams of length L, height t, width w, and moment of inertia $I = wt^3/12$. If a force F is applied at $x=0$, then the bending moment in each beam is $M(x) = Fx$, and the strain energy U due

to this bending moment (in each beam) is

$$U = \frac{1}{2EI} \int_0^L M^2(x)dx = \frac{F^2 L^3}{6EI}.$$

Then by Castigliano's theorem, the deflection δ due to bending of each beam at the point of application of the force is

$$\delta_0 = \frac{\partial U}{\partial F}\bigg|_{x=0} = \frac{FL^3}{3EI}.$$

According to elasticity theory,[10] the deflection curve of the top cantilever in Fig. 1 will be

$$\delta(\text{along } y=t/2) = \frac{Fx^3}{6EI} - \frac{FL^2 x}{2EI} + \frac{FL^3}{3EI} + \frac{Ft^2}{8GI}(L-x),$$

where the first three terms give the deflection caused by bending, and the last term gives the deflection caused by shearing; E and G are the Young's modulus and the shear modulus, respectively. The deflection caused by shearing depends on the boundary condition at the built-in end. At most, at the free end ($x=0$) it will be $\delta_1 = Ft^2 L/8GI$; or, since $G \simeq 3E/8$, $\delta_1 \simeq (t/L)^2 \delta_0$, and hence will be small in most cases. The strain energy Y caused by the shearing stress will be

$$Y = (F^2 L t^2 / 16 GI) \simeq (t/L)^2 U,$$

and hence also small in most cases. (In the most extreme case of the present work, that of zinc with its small shear modulus, this correction to the measured surface energy amounts to less than 5%.) The boundary condition for this maximum effect of shearing is that at the point $x=L$, $y=t/2$, a vertical element of the cross section is held fixed. A photoelastic study of the stress distribution near a cleavage crack, made by Guernsey and Gilman,[11] indicated that there is not much disturbance of the neutral bending planes on either side of the crack. Therefore, the boundary condition mentioned above is probably not suitable. The actual situation seems to be closer to the condition that at $x=L$, $y=t/2$, a horizontal element of the cross section remains fixed. Then there is no deflection caused by shear.[10] Since the effect of shear is small at most, and perhaps completely negligible, it will be neglected in what follows, and the deflection curve will be taken to be

$$\delta = (F/6EI)(x^3 - 3L^2 x + 2L^3).$$

For a moving crack, there is kinetic energy associated with the sidewise deflections of the cantilevers. The mass of an elemental volume in one of the cantilevers is $\rho w t dx$, where ρ is the density, and the velocity of the element is $(d\delta/dt)$, so its kinetic energy is $dT = \frac{1}{2}$

---

[9] J. J. Benbow and F. C. Roesler, Proc. Phys. Soc. 70, 201 (1957).
[10] S. Timoshenko, *Theory of Elasticity* (McGraw-Hill Book Company, Inc., New York, 1934), p. 33.
[11] R. Guernsey and J. Gilman, Proc. Soc. Exper. Stress Anal. (1960).

TABLE I. Elastic moduli used in this study.

Crystal	Cleavage plane	Parallel to cleavage plane		Perpendicular to cleavage plane	
		$1/E_{\parallel}$	$E_{\parallel}$ (10^{11} d/cm²)	$1/E_{\perp}$	$E_{\perp}$ (10^{11} d/cm²)
LiF[a]	(100)	S_{11}	10.2	S_{11}	10.2
MgO[b]	(100)	S_{11}	26.1	S_{11}	26.1
Fe(3% Si)[c]	(100)	S_{11}	13.2	S_{11}	13.2
CaF$_2$[d]	(111)	$S_{11} - \frac{1}{2}[(S_{11}-S_{12}) - S_{44}/2]$	8.62	$S_{11} - \frac{2}{3}[(S_{11}-S_{12}) - S_{44}/2]$	11.4
BaF$_2$[e]	(111)	$S_{11} - \frac{1}{2}[(S_{11}-S_{12}) - S_{44}/2]$	6.53	$S_{11} - \frac{2}{3}[(S_{11}-S_{12}) - S_{44}/2]$	6.52
Si[f]	(111)	$S_{11} - \frac{1}{2}[(S_{11}-S_{12}) - S_{44}/2]$	16.8	$S_{11} - \frac{2}{3}[(S_{11}-S_{12}) - S_{44}/2]$	18.6
Zn[g]	(0001)	S_{11}	13.5	S_{33}	4.13
CaCO$_3$[d]	(001)	h	14.0	h	9.45

[a] C. V. Briscoe and C. F. Squire, Phys. Rev. **106**, 1175 (1957).
[b] M. A. Durand, Phys. Rev. **50**, 449 (1936).
[c] E. Schmid and W. Boas, *Plasticity of Crystals* (F. A. Hughes & Company, Ltd., London, 1950).
[d] H. B. Huntington, *Solid State Physics* (Academic Press, Inc., New York, 1958), Vol. 7.
[e] L. Bergman, Z. Naturforsch. **12a**, 299 (1957).
[f] M. E. Fine, J. Appl. Phys. **26**, 862 (1955).
[g] G. A. Alers and J. R. Neighbours, Bull. Am. Phys. Soc. **2**, 121 (1957).
[h] In the trigonal system, $1/E = (1-n^2)^2 S_{11} + n^4 S_{33} + n^2(1-n^2)(S_{44}+2S_{13}) + 2mn(3l^2-m^2)S_{14}$, where l, m, n are direction cosines of the direction of E. For calcite in the perpendicular direction: $l=0$, $m=0.701$, $n=0.713$; in the parallel direction: $l=0$, $m=0.897$, $n=0.442$.

$\rho wt dx (d\delta/dt)^2$. Then the kinetic energy of the whole cantilever is

$$T = \frac{\rho w t}{2} \int_0^L \left(\frac{d\delta}{dt}\right)^2 dx$$

but

$$\frac{d\delta}{dt} = \frac{\partial \delta}{\partial L}\left(\frac{dL}{dt}\right) = \left(\frac{\partial \delta}{\partial L}\right) v_c,$$

where v_c is the crack velocity. Upon substitution and integration, one finds that

$$T = 12 (L/t)^2 (v_c/v_s)^2 U,$$

where v_s is the velocity of sound $(E/\rho)^{\frac{1}{2}}$. From this it can be seen that for the usual specimen dimensions ($L \simeq 10t$), the kinetic energy is small compared with the strain energy unless the crack velocity is greater than about $v_s/100$.

The energy of the new surfaces that are created when the crack moves forward a distance dx can be found by forming an energy balance for the whole system. This requires that the motion is reversible; a condition that obtains only for slow motion of the crack. This latter condition was obtained in the present work by forming the energy balance at the critical point when crack motion just begins. The work dW done when the crack length increases by dL, must then equal the increase in strain energy dU plus the energy of the newly created surfaces dS if the process is reversible. That is,

$$dW = dU + dS.$$

The incremental work is $Fd\delta_0 = (F^2L^2/EI)dL$; and $dU = (F^2L^2/2EI)dL = dW/2$; and, if γ is the specific surface energy per unit area, $dS = \gamma w dL$. Then, the above condition becomes

$$S = 3U$$

or

$$\gamma = 6F^2L^2/Ew^2t^3. \quad (1)$$

This is the equation that was used to interpret the measured critical force values in this study. The value of γ that it yields is a minimum when the cracking is reversible, and equals the intrinsic surface energy γ_s of the crystal. When the cracking is accompanied by irreversible processes, the value of γ given by Eq. (1) will take some larger value γ_p which includes the energy absorbed irreversibly plus the intrinsic surface energy.

The value of E that was used throughout is the Young's modulus in the direction of propagation of the crack. It is calculated from the elastic constants. For the cubic crystals with {100} cleavage (LiF, MgO, Fe-Si) it is simply $1/S_{11}$. For the cubic crystals with {111} cleavage (CaF$_2$, Si) it is $[\frac{1}{2}(S_{11}+S_{12}+S_{44}/2)]^{-1}$. For hexagonal zinc, it is $1/S_{11}$. For calcite (CaCO$_3$) it is $[0.65 S_{11} + 0.04 S_{33} + 0.16(S_{44}+2S_{13}) - 0.64 S_{14}]^{-1}$. Wherever values of the elastic constants at 78°K were available (LiF, MgO, Si, and Zn) they were used; otherwise it was necessary to use room temperature values. The various values are listed in Table I.

RESULTS

The experimental results will be described for each crystal species in turn.

a. Lithium Fluoride

The most extensive work was done on LiF crystals because they were more readily available and easily cleaved than the other crystals. At room temperature, plastic flow occurs near the tips of slowly moving cracks in them. Therefore, in order to measure their intrinsic surface energies, it was necessary to carry out the measurements at the boiling point of liquid nitrogen (−196°C) where plastic flow is almost completely suppressed. Proof that flow is suppressed was obtained by etching the cleaved surfaces to look for dislocations, and by examining the patterns of cleavage steps on the same surfaces. It was found that cracks could be stopped and started again in crystals at −196°C causing only barely perceptible changes in the directions and

Table II. Cleavage surface energies measured for LiF crystals.

Specimen number	Test conditions	Dimensions (cm) w	t	L	F^* (g)	γ (ergs/cm^2)	Author's choice
LF-2	As-grown crystal No. 3— cracked in N$_2$(l) and immediately tested in N$_2$(l)	0.262	0.322	2.85	164	530	
6		0.242	0.322	2.30	175	472	
4		0.244	0.338	2.05	210	450	
8		0.242	0.310	2.35	158	450	
9		0.264	0.330	2.35	172	370	
5		0.279	0.325	2.50	166	370	
3		0.274	0.328	2.15	186	360	
10		0.269	0.337	2.15	182	350	
7		0.294	0.322	2.10	195	330	340
LF-C	As-grown crystal No. 3— cracked in N$_2$(l) and then warmed to room temp. in N$_2$(g) and cooled again in N$_2$(l) for testing	0.274	0.315	3.0	193	800	
B		0.292	0.317	2.9	185	600	
A		0.290	0.327	1.9	290	580	
G		0.498	0.330	1.6	580	550	
D		0.277	0.319	2.8	174	530	
E		0.508	0.366	2.6	412	510	
F		0.485	0.368	2.5	390	460	
LF-R6	As-grown crystal No. 1— cracked in N$_2$(l) and then warmed to room temp. in N$_2$(g) for testing	0.122	0.100	2.4	205	830	
R3		0.284	0.333	2.3	270	730	
R4		0.292	0.332	3.0	188	580	
R7		0.422	0.345	1.1	785	580	
R5		0.269	0.332	2.9	178	570	
R2		0.259	0.332	2.8	150	390	
R1		0.259	0.342	1.6	220	260	
LF-BK5	Crystals compressed ~2.5% in compression at room temp and then cracked and tested in N$_2$(l)	0.290	0.322	2.2	350	1140	
BK2		0.224	0.345	2.5	250	1080	
BK1		0.233	0.350	1.9	335	980	
BK4		0.233	0.322	2.1	255	860	
LF-NA1	Crystals given neutron dose of 32×10^{14} nvt and then cracked and tested in N$_2$(l)	0.307	0.160	1.4	107	330	
NA2		0.312	0.157	0.9	160	310	

numbers of cleavage steps. Dislocation etch pits were observed, but these were confined to thin layers on the cleavage surfaces (less than 10^{-4} cm thick). Estimates of the amount of energy that could have been absorbed by forming and moving the dislocations indicate that it is negligible in comparison with the intrinsic surface energy. The estimates were based on the plastic work done by each dislocation as it moved into a crystal under a stress approximately equal to the flow stress. The amount of plastic flow that occurs depends, of course, on the hardness of a crystal. In the present work the crystals were moderately hard (room temperature bending yield stress equal to 890 g/mm^2).

At the top of Table II, the measured values for nine specimens that were cleaved from the same large crystal are given. These specimens were all partially cracked while they were immersed in liquid nitrogen, and then were quickly transferred to a liquid nitrogen bath that was mounted on the testing machine. It may be seen that the measured surface energy values spread over the range of 330–530 ergs/cm^2. Part of this spread is caused by experimental error, but much of it can be correlated with the density of cleavage steps on the cracked surfaces. This ranged from 7–66 step-lines/cm. The steps were introduced by errors in the alignment of the initial crack with respect to the cleavage plane, or by twist-type subgrain boundaries. Even without correcting for the effect of cleavage steps, the majority of the results lie in the range of 330–400 ergs/cm^2; and, since the *lowest* values measured by this method should approach the reversible value most closely, it is believed that 340 ergs/cm^2 represents the intrinsic surface energy of the {100} planes of LiF.

The second set of data in Table II shows the effect of warming up a crystal between the time it was initially cracked and the time of testing. Plastic relaxation occurred near the crack tips while they were warm, and this raised the subsequently measured surface energy values. The effect is measurable but not large, indicating that small amounts of plastic relaxation would not change the first set of data discussed above.

A surprising result is presented by the third set of data. It was expected that large amounts of energy would be absorbed by slowly moving cracks at room temperature in LiF because these crystals are quite plastic at this temperature. However, this is not the case. The measured values have a considerable spread, but even the highest one is only about two and one-half times the intrinsic surface energy of LiF. Such low cleavage surface energy reflects the brittleness of LiF at room temperature even though it exhibits plasticity in slow bend tests.

Table III. Cleavage surface energies measured for MgO crystals.

Specimen number	Test conditions	Dimensions (cm)			F^* (g)	γ (erg/cm²)	Author's choice
		w	t	L			
MG-2	As-grown crystal cracked	0.360	0.224	2.4	381	1370	
3	in N₂(1) and immediately	0.363	0.319	1.8	876	1270	
X2	tested in N₂(1)	0.386	0.250	2.3	1089	1140	1200
MG-Y1	As-grown crystal cracked in	0.363	0.238	1.8	863	3000	
Y3	air at 25°C and tested in	0.238	0.266	2.6	413	2380	
Y2	air at 25°C	0.238	0.266	2.3	404	1790	
MG-X1	As-grown cracked in N₂(1) and tested in air at 25°C	0.386	0.250	2.3	590	1720	
MG-4	As-grown cracked in air at	0.323	0.238	2.7	375	1610	
5	25°C and tested in N₂(1)	0.231	0.211	1.3	410	1260	

The fourth set of data in Table II shows how prestrain affects cleavage in liquid nitrogen. It increases the energy absorption by a factor of about three by markedly increasing the density of cleavage steps on the fractured surfaces.

Finally, it can be seen from the last set of data in the table that substantial radiation damage has but little effect on the surface energy. In other words, the surface energy of LiF is not very structure sensitive. As would be expected, the surface energy depends simply on the atomic binding strength per unit area of surface, and if a small fraction of the atomic bonds is damaged by radiation, this has only a small effect on it.

b. Magnesium Oxide (Periclase)

The results given in Table III for MgO follow the same pattern as for LiF, but the numbers are larger. The specimens in this case were cleaved out of large irregular chunks that were purchased from the Norton Company. The chunks were colorless, but not more than about 99.9% pure. Specimens were obtained by first sawing rectangular blocks from the chunks with the flat sides parallel to {100} planes. Then, by successively bisecting the blocks by means of cleavage, several slender rods could be obtained from each block. From tests of these rods it is concluded that the surface energy of the {100} planes of MgO is about 1200 ergs/cm². As in the case of LiF, although MgO exhibits plasticity at room temperature, its fracture surface energy is quite low. This is consistent with its brittle behavior at room temperature.

c. Calcium Fluoride (Fluorite)

Large crystals of CaF_2 were sawed into slabs with the large faces of the slabs parallel to (111) planes. The large faces were then optically polished with aluminum oxide powder. Next the slabs were sawed into rectangular rods, and the sawed faces were polished. Finally the rods were partially cleaved and tested. Their top and bottom faces were cleaved {111} surfaces while their side surfaces were polished {112} surfaces, and the direction of crack propagation was ⟨110⟩.

The cleavage surfaces of the CaF_2 crystals showed very few cleavage steps, and no evidence at all of plastic relaxation at the tips of stopped cracks. This is consistent with the relatively small amount of scatter in the data of Table IV. The surface energy of the {111} planes of CaF_2 is about 450 ergs/cm².

d. Barium Fluoride

Barium fluoride specimens were prepared in the same way as calcium fluoride. The formation of cleavage cracks was quite reversible in this case because plastic flow and cleavage steps were nearly absent. In fact, the most perfect crystal surfaces that the author has ever seen resulted. This made it difficult to determine the initial lengths of cracks and caused scatter in the results (Table V). The average of the measured values is taken to be the surface energy of the {111} plane of BaF_2; 280 ergs/cm².

Table IV. Cleavage surface energies measured for CaF₂ crystals.

Specimen number	Test conditions	Dimensions (cm)			F^* (g)	γ (erg/cm²)	Author's choice
		w	t	L			
CA-2a	As-grown crystal, cracked in	0.246	0.465	2.1	313	490	
4a	N₂(1) and immediately	0.236	0.454	1.5	405	470	
3	tested in N₂(1)	0.233	0.452	1.8	336	440	
4b		0.236	0.223	1.3	154	440	
2b		0.246	0.229	1.4	152	420	450
CA-1	As-grown cracked and tested	0.241	0.229	1.3	154	390	
2	in air at 25°C	0.243	0.452	2.3	222	320	

TABLE V. Cleavage surface energies measured for BaF$_2$ crystals.

Specimen number	Test conditions	Dimensions (cm)			F^* (g)	γ (ergs/cm^2)	Author's choice
		w	t	L			
BA-1	As-grown crystal, cracked in liquid N$_2$ and immed. tested in liquid N$_2$	0.188	0.257	1.68	86	304	
2		0.188	0.257	1.52	78	207	
3		0.188	0.258	1.60	79	230	
5		0.185	0.254	1.35	108	333	
6		0.183	0.250	3.31	43	335	
							280

e. Calcium Carbonate (Calcite)

Slabs about $\frac{3}{16}$-in. thick were cleaved from natural Iceland Spar rhombs. This can be readily accomplished if a rhomb is cut with two parallel faces that are perpendicular to the remaining four cleavage faces. Such a block of crystal can be successively bisected parallel to one set of cleavage faces if a convex chisel is driven into the center of one of the ground faces (excessive crushing occurs if one tries to drive a chisel into an edge).

The slabs (with cleavage planes as their largest faces) were cemented to glass slabs with glycol pthalate, while being careful to avoid thermal stresses by heating and cooling the whole assembly uniformly. After the slabs had been sawed into rods, the side faces of each rod were optically polished with carborundum powder.

After many attempts it was found that longitudinal cracks would not run straight through mechanically polished specimens. Therefore, several crystals were given additional chemical polishing treatments. They were repeatedly dipped into conc HCl for 5-10 sec followed by an alcohol rinse and drying in an air blast. This treatment was successful in eliminating the defects that had previously caused premature cracking.

Table VI shows the data for calcite crystals that were partially cracked in liquid nitrogen and then tested while they were still immersed in liquid nitrogen. Their cleavage surfaces were quite flat and only a few new cleavage steps appeared at the places where the cracks stopped in them. It is concluded that the surface energy of $\{10\bar{1}0\}$ planes in calcite is about 230 ergs/cm^2.

f. Silicon

Of all the crystals, silicon was the most difficult to test because longitudinal cracks in it have a strong tendency to turn and run out the side of a specimen on one of the transverse $\{111\}$ planes. Out of 12 specimens that were completely prepared for testing, only two yielded good results. Rectangular specimen rods were cut out of a cylindrical crystal of silicon that had been grown by the floating-zone technique. One set of side faces were cut parallel to a (111) plane and the lengths of the bars were parallel to $\langle 112 \rangle$ directions. A narrow slot was cut into one end of each bar parallel to the oriented (111) plane as a means for starting a crack. Small holes for the pivot points were ultrasonically drilled on either side of each slot. The specimens were then chemically polished in "white polish" (HF+HNO$_3$). Next the base of each slot was scratched by gently drawing a SiC-coated thread over it. The purpose of this was to facilitate the initiation of cleavage. Finally, each specimen was initially cleaved while immersed in liquid nitrogen, and then immediately tested in liquid nitrogen.

The results of the two successful tests on silicon in Table VII agree with each other quite well so it is concluded that the surface energy of the $\{111\}$ planes of silicon is about 1240 ergs/cm^2.

g. Zinc

High purity zinc crystals were grown by the author in the form of 6-mm square rods with the (0001) basal plane lying parallel to one set of side faces. Specimens

TABLE VI. Cleavage surface energies measured for CaCO$_3$ crystals.

Specimen number	Test conditions	Dimensions (cm)			F^* (g)	γ (ergs/cm^2)	Author's choice
		w	t	L			
CA-1	Natural Iceland spar cracked in N$_2$(l) and immediately tested in N$_2$(l)	0.282	0.386	1.30	431	280	
2		0.261	0.320	1.42	254	240	
3		0.259	0.312	1.12	297	220	
							230

TABLE VII. Cleavage surface energies measured for silicon crystals.

Specimen number	Test conditions	Dimensions (cm)			F^* (g)	γ (ergs/cm^2)	Author's choice
		w	t	L			
S-10	Semiconductor grade silicon, cracked and tested in N$_2$(l)	0.220	0.284	0.85	749	1250	
4		0.193	0.310	1.01	622	1230	
							1240

TABLE VIII. Cleavage surface energies measured for Zn and for $(Zn+\frac{1}{10}\%\ Cd)$ crystals.

Specimen number	Test conditions	Dimensions (cm) w	t	L	F^* (g)	γ (ergs/cm²)	Author's choice
Zn-3a	High purity zinc, cracked in	0.574	0.283	0.6	890	143	
1	N₂(l) and immediately	0.574	0.284	1.0	450	115	
2a	tested in N₂(l)	0.563	0.276	0.9	410	89	105
ZN-3b	High purity zinc, cracked in	0.574	0.283	1.1	710	347	
4B	N₂(l) and tested in acetone at (−78°C)	0.594	0.292	1.3	635	324	

		Test temperature (°C)					
ZC-4a	Zinc +0.1% cadmium in N₂(l) and then immediately tested at various temperatures	−196	0.568	0.279	0.9	1615	1150
1		−196	0.541	0.264	2.4	417	780
4b		−78	0.568	0.279	1.1	1920	2700
4c		−25	0.568	0.279	1.2	2360	4570
2a		+25	0.525	0.260	2.2	1180	5930
4d		+25	0.568	0.279	1.2	2530	5620

about 4 cm long were cut from these rods, and one end of each specimen was cut to a chisel shape as may be seen in Fig. 3. The purpose of this was to provide a place where cleavage could start at a point and would therefore tend to spread along a single cleavage plane. This is especially important for highly anisotropic crystals like zinc where the energies of cleavage steps are very high. After the chisel ends had been formed, the crystals were heavily chemically polished to remove the damaged surface layers. Finally, the zinc crystals were partially cracked, and then tested while they were immersed in liquid nitrogen. Table VIII gives the results.

The high purity zinc crystals were tested at −78°C in addition to the −196°C tests. At the higher temperature, the cleavage surface energy is considerably greater. At temperatures above −78°C the crystals were so soft that they bent excessively before enough force could be applied to them to cause cleavage.

Zinc crystals containing cadmium (0.1%) were also tested. They had considerably higher (an order of magnitude) yield strengths than the pure zinc crystals, so they could be tested at all temperatures from −196 to +25°C. A plot of the variation of the cleavage surface energy for zinc with temperature is shown in Fig. 4. It is not surprising that the cleavage surface energy increases rapidly with increasing temperature in the ductile-brittle transition range of zinc. However, it was surprising to find that cadmium increases the surface energy of zinc even at low temperatures. This seems to mean that the polarizable cadmium atoms form strong "bridges" between the (0001) cleavage planes of zinc.

As the cleavage surface energy of zinc increases with increasing temperature, characteristic changes in the appearances of the cleaved surfaces are observed. These have been interpreted as indications of plastic glide that occurs in directions transverse to the basal planes.[12] It is this transverse glide that absorbs increased amounts of energy during cleavage at higher temperatures.

---

[12] J. J. Gilman, J. Appl. Phys. **27**, 1262 (1956).

h. Iron-Silicon (3% Si)

The iron-silicon crystals were grown by the strain-anneal method in strip form with {100} planes parallel to the faces and edges of the strip. Specimens were sawed out of the strips and then heavily chemically polished. One end of each specimen was notched to facilitate crack initiation. Because of the large forces that had to be applied to these crystals in order to cleave them, they could not be held by pivot bearings. Instead, small holes were drilled near their notched ends, and hard steel pins were inserted in the holes, and used to apply the cleavage forces.

At −78°C the force required to make iron-silicon cleave is so large that specimens prefer to break on a transverse cleavage plane instead of giving longitudinal cleavage. The force at which this occurs corresponds to

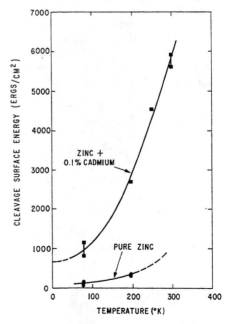

FIG. 4. Effect of temperature on the cleavage of zinc.

TABLE IX. Cleavage surface energies measured for Fe+3% Si crystals.

Specimen number	Test conditions	Dimensions (cm)			F^* (g)	γ (erg/cm²)
		w	t	L		
SF-I	As-grown crystal, cracked and tested in H₂(l) (−250°C)	0.079	0.304	1.30	204	1360
SF-E	As-grown, cracked and tested in N₂(l) (−196°C)	0.078	0.312	1.04	1040	27 700
A		0.116	0.429	1.80	1410	26 400
C		0.081	0.310	1.36	726	23 000
D		0.086	0.312	1.23	863	23 000
B		0.081	0.302	0.90	1071	22 500
SF-G	As-grown, cracked and tested in acetone at −78°C	0.080	0.308	1.2	>2630	>240 000
SF-L	As-grown, prestrained 3% at 25°C, then cracked and tested in N₂(l)	0.084	0.313	1.4	976	40 000
J		0.084	0.312	1.5	472	10 000

a cleavage surface energy in excess of 240 000 ergs/cm² at −78°C (Table IX). At −196° the surface energy is still quite high but is at least an order of magnitude less than it is at −78°C. By the time −259°C (liquid hydrogen) has been reached the surface energy is down by another order of magnitude to about 1360 ergs/cm². This is about what would be expected for the intrinsic surface energy of iron, but only one test result was obtained so no conclusion about its significance can be reached.

The bottom set of data in Table VII show that, just as in the case of LiF, prestrain has little effect on the surface energy.

Much of the energy that is absorbed by cracks in iron-silicon crystals at −196°C is a result of twinning. Twinned regions are found near and on the edges of cracks.

COMPARISON WITH THEORY

An estimate of the work that is required to separate two internal surfaces of a crystal can be obtained by approximating the attractive stress between the two surfaces by a sine function. The stress is taken to be zero when the surfaces have their normal separation distance d_0. As the distance between the surfaces increases by an amount y greater than normal, the stress rises to a maximum value σ_0 and then drops to zero again when y exceeds the range a of the attractive forces. This can be expressed as

$$\sigma = \sigma_0 \sin[(\pi y)/a] \quad 0 < y < a.$$

The strain between the surfaces is y/d_0 and for small strains the stress is given by Hooke's law; $\sigma = E(y/d_0)$. Therefore, since the sine is equal to its argument for small y, the value of the maximum stress can be evaluated; it is $Ea/\pi y$. The work that is done in moving the surfaces from their initial positions, $y=0$, to positions where they are no longer attracted by each other, $y=a$, is

$$\int_0^a \sigma dy = \frac{Ea}{\pi d_0} \int_0^a \sin\frac{\pi y}{a} dy.$$

Two surfaces are created by this amount of work, so the specific surface energy is

$$\gamma = (E/d_0)(a/\pi)^2.$$

Surface energy values estimated from the above elementary theory are listed in Table X. The d_0 values that were used in making the calculations are simply distances between the crystallographic cleavage planes. In the case of LiF and MgO, for example, d_0 equals one-half the cubic unit cell size. The a values were taken to equal the atomic radii of atoms lying in the cleavage planes. The elementary theory gives values that are uniformly high but of the same order of magnitude as the experimental values.

The simplest atomistic theory of surface energies is that of Born and Stern[13] which considers the electrostatic attractions between ions in rock-salt-type crystals, and hard-shell repulsions. The resulting surface energies for {100} and {110} surfaces are

$$\gamma_{100} = 0.0145[(Ze)^2/r^3] \quad \text{(cgs units)}$$

$$\gamma_{110} = 0.0394[(Ze)^2/r^3] \quad \text{(cgs units)},$$

where e is the electronic charge, Z is the valence, and r is the interionic distance. Table X shows that the agreement between values given by this theory and the present experimental results is fair. Better agreement is obtained with the theory of Glauberman[14] who used the same equation for the interionic potential as Born and Stern, but allowed some contraction of the surface layers to occur perpendicular to the surface plane. According to him, the cleavage surface energy should be

$$\gamma_{100} = 0.0124[(Ze)^2/r^3],$$

which is in remarkably good agreement with the experimental values.

Similar calculations, using somewhat different potential energy equations have been made by Lennard-Jones

---

[13] M. Born and O. Stern, Sitzber. preuss. Akad. Wiss. Physik. math. Kl. 901 (1919).
[14] A. E. Glauberman, Zhur. Fiz. Khim. 23, 124 (1949).

Table X. Comparison of experimental and theoretical surface energy values (ergs/cm²).

Crystal	NaCl	LiF	MgO	CaF$_2$	BaF$_2$	CaCO$_3$	Si	Zn	Fe(3% Si)
Cleavage plane	(100)	(100)	(100)	(111)	(111)	(10$\bar{1}$0)	(111)	(0001)	(100)
Experimental									
This study	...	340	1200	450	280	230	1240	105	1360(?)
Benson, Schreiber, Van Zeggeren [a]	276	...							
Hutchinson, Manchester [a]	366	...							
Jura, Garland [a]	...	...	1040						
Lipsett, Johnson, Mass [a]	381	...	...						
Kusnetsov, Teterin [b]	300	...	...	146	...	78			
Liquid at melting point (extrapolated to 0°K)	190	350							
Atomic theory									
Born, Stern	150	420	1440						
Glauberman	130	360	1230						
Lennard-Jones, Taylor	96	...	1360						
Dent	77	...							
Biemuller	87	170							
Shuttleworth	155	...							
Van Zeggeren, Benson	188	...							
van der Hoff, Benson	...	560							
Klobe	...	...	...	<1019					
Ormont (from subl. energy)	...	...	...	...	...	...	1232		
Estimated from elastic modulus	310	370	1300	540	350	380	890	185	1400

[a] Heat of solution method.
[b] From impact cleavage.

and Taylor,[15] and by Dent.[16] Biemuller,[17] Verwey,[18] Shuttleworth,[19] and Van Zeggeren and Benson[20] used the complete Born-Mayer potential equation, thereby attempting to include the effects of ionic polarization on the surface energy. The results for LiF do not agree with the experimental values as well as do the simpler theories. A quantum-mechanical approach was used by van der Hoff and Benson,[21] but it gave a value for LiF that is too large. The results of these various atomic theories are summarized in Table X. Also included are an estimate for CaF$_2$ by Klobe[22] and for Si by Ormont.[23] Data for NaCl are included in the table because more work has been done for this crystal than for any other.

Although there is considerable spread in the various values given in Table X, they change in a consistent way from one crystal to another. This, together with the good agreement between the simple ionic theory and the experimental values, indicates that the present experimental method is suitable and reliable. The experimental value determined by Jura and Garland[24] for MgO smokes using the heat of solution method agrees quite well with the present work. For NaCl, the measurements by Lipsett, Johnson and Mass,[25] by Hutchinson and Manchester,[26] and by Benson, Schreiber, and Van Zeggeren[5] all are high in comparison with the theoretical values. This probably resulted because they used powders that were prepared by filing or crushing. Such damaged powders would be expected to contain a variety of defects that would contribute to an increasing heat of solution with decreasing particle size.

Comparison may also be made with estimates based on measurements by Jaeger[27] of the surface tensions of the alkali halides just above their melting points. Correction for the effect of temperature gives the two values listed in Table X for the surface energies of NaCl and LiF at 0°K.

Since the present measurements were carried out with the specimens immersed in liquid nitrogen rather than in vacuum, the effect of the environment on the measurements should be considered. It is known that the environment can markedly influence the cleavage of mica,[2] but this is associated with the fact that the cleavage surfaces are charged and discharging occurs between them through the environment. Perhaps the best evidence that the environment had no major effect on the present measurements is provided by the consistencies of the results and their good agreement with theory and other experiments. Also, since the molecular

---

[15] J. E. Lennard-Jones and P. A. Taylor, Proc. Roy. Soc. (London) **A109**, 476 (1925).
[16] B. E. Dent, Phil. Mag. **8**, 530 (1929).
[17] J. Biemuller, Z. Physik **38**, 759 (1936).
[18] E. J. W. Verwey, Rec. trav. chim. **65**, 521 (1946).
[19] R. Shuttleworth, Proc. Phys. Soc. (London) **A62**, 167 (1949).
[20] F. van Zeggeren and G. C. Benson, J. Chem. Phys. **26**, 1077 (1957).
[21] M. E. van der Hoff and G. C. Benson, J. Chem. Phys. **22**, 475 (1954).
[22] Reported by M. Born and M. Goeppert-Mayer in Handbuch der Physik **24-2**, 764 (1933).
[23] B. F. Ormont, Doklady Akad. Nauk. SSSR **106**, 687 (1956).
[24] G. Jura and C. W. Garland, J. Am. Chem. Soc. **74**, 6033 (1952).
[25] S. G. Lipsett, F. M. G. Johnson, and O. Mass, J. Am. Chem. Soc. **49**, 1940 (1927).
[26] E. Hutchinson and K. E. Manchester, Rev. Sci. Instr. **26**, 364 (1955).
[27] Jaeger, Z. anorg. u. allgem. Chem. **1010**, 1 (1917).

diameter of N_2 is about 2.2 A, it would not be able to absorb on the fresh cleavage surfaces in the immediate vicinity of the crack tip. It would only be able to adsorb some distance behind the crack tip where the surfaces were separated by more than 2.2 A. Thus the heat of adsorption would probably be lost as heat to the crystal and would not reduce the measured surface energy. Furthermore, at least for the ionic crystals the free energy change upon adsorption of N_2 is small. Measurements by Orr[28] for N_2 on KCl give a value of about 1 kcal/mol of surface at $-196°$C. Since the adsorption is due to van der Waals forces, it should be proportional to the polarizability of the substrate. This would reduce the free energy to about 0.2 kcal/mol for N_2 on LiF which is only about 5% of the total surface energy of 3.9 kcal/mol. The free energy of N_2 adsorption would be only about 2% of the total surface energy in the case of MgO.

FUTURE IMPROVEMENTS

The *precision* of the cleavage method is relatively low because of experimental scatter. The primary cause of this seems to be variations in the cleavage step densities amongst specimens. These variations could be reduced by (a) using more nearly perfect crystals with fewer subgrain boundaries in them; (b) cleaving at lower temperatures to further reduce plastic flow; (c) improving the technique used to start cracks so they are started more nearly parallel to a cleavage plane.

The *accuracy* of surface energy measurement could be improved by studying the elastic deflections in the cleavage specimens in detail. This can be done experimentally by means of the optical interference fringes that can be formed between the surfaces of a crack. Anisotropic elastic theory can be used to predict the deflections more accurately than was done in this study. Although the gaseous atmosphere does not seem to play an important role in these experiments, this should be investigated systematically.

Many pure crystals are so soft (even at low temperatures) that the cleavage method cannot be applied to them, but adding impurities to many of them would cause enough hardening to allow the method to be applied. By making measurements as a function of purity one could probably find the surface energy of the pure crystal by extrapolation.

SUMMARY

By means of quantitative cleavage, the surface energies of LiF, MgO, CaF_2, BaF_2, $CaCO_3$, Zn, and Si crystals have been measured at $-196°$C. The measured values for LiF and MgO are in good agreement with simple ionic theory.

It has been found that the cleavage surface energy of LiF is only moderately dependent upon increasing temperature and upon plastic prestrain. It is quite independent of radiation damage. The cleavage surface energy of MgO is also relatively independent of temperature.

Zinc crystals that contain a small amount of cadmium (0.1%) are considerably more difficult to cleave than pure zinc crystals. The cleavage surface energy of zinc increases rapidly with increasing temperature, and the increase is accompanied by increasing nonbasal glide.

Iron (+3% Si) crystals absorb large amounts of energy through twinning when they are cleaved at $-196°$C, but they become completely brittle at $-250°$C. For temperatures of $-78°$C and greater these crystals cannot be cleaved under static conditions.

It is believed that the cleavage technique as described has been established as a reliable method for directly measuring crystal surface energies. Some suggestions for future improvement of the method are made.

ACKNOWLEDGMENTS

Considerable patience and skill were required to perform the experiments described here. These were provided by Mr. V. J. DeCarlo. Mr. R. Grubel of the Semiconductor Products Department of the General Electric Company kindly furnished a large silicon crystal. Dr. C. G. Dunn provided iron-silicon crystals. Mr. L. Nesbitt gave advice about performing tests in liquid hydrogen. Helpful comments on the manuscript were given by Dr. J. R. Low, Dr. G. W. Sears, Dr. G. Ehrlich and Dr. R. L. Fullman of the General Electric Research Laboratory. Dr. B. W. Roberts kindly checked the elastic moduli of calcite parallel and perpendicular to the cleavage plane by the pulse-echo technique.

---

[28] W. J. C. Orr, Proc. Roy. Soc. (London) **A173**, 349 (1939).

FRACTURE DYNAMICS

By George Irwin

BEFORE the war, the Charpy and Izod impact tests were in wide use to evaluate qualities called toughness, temper brittleness, and notch sensitivity. About 10 years ago pioneering tensile fracture studies at high straining speeds were made at several laboratories. Mann (1)[1] and Clark (2) showed experimental evidence for a rapid change in deformation work at high speeds of deformation of tensile impact specimens. Subsequently von Karman (3) developed a theory for the propagation of a plastic wave in a wire when moved rapidly at one end. Duwez (4, 5) provided experimental demonstrations of this theory. The von Karman theory predicted a critical pulling velocity for quick failure with highly localized deformation which was a function of the dynamic stress-strain relationship for the material of the wire. Ways of applying this type of analysis to studies of fracturing of plates are of interest but have yet to be developed.

During the war Zener and Hollomon (6) carried forward applications of the rate theory concept for associating temperature and strain rate effects, thus adding substantially to the appreciation of temperature effects in plastic flow and rupture of metals. Under the impetus of the welded ship brittleness problem numerous advances in the field of fracture dynamics were accomplished in a comparatively short time. The fracture velocity studies at David Taylor Model Basin (7) and at the University of California showed that cleavage fractures travel at velocities which are of the order of half the transverse sound velocity. Similar results have been shown for fast fractures of glass. At Bristol University, England, N. F. Mott (8) suggested a theoretical analysis of fracturing velocity which will be explained later. In different applications fracture dynamics presents a variety of interesting aspects. A complete recitation of these would be more confusing than helpful here. In this lecture the author will try to add some clarity to the dynamics of rapid fracturing in ductile metals. It is felt, however, that the features discussed are of central importance in many specialized applications. The

---

[1] The figures appearing in parentheses pertain to the references appended to this paper.

The author, George Irwin, is associate superintendent, Mechanics Division, Naval Research Laboratory, Washington, D. C.

illustrations employed are mainly from Navy research studies.

Fig. 1 illustrates some features of fractures at slow and at impact machine speeds. The steel used was a semi-killed low carbon steel of the composition generally used for ship plate in this country. The transition temperature for half-brittle fracture was about 65 °F both with slow speed and with impact measurements. In Fig. 1a, for slow loading, considerable deformation may be noted, which preceded onset of brittle fracture in the low temperature specimen. The total work to complete these fractures dropped only from 50 ft-lbs. at 80 °F to 36 ft-lbs. at 0 °F. In Fig. 1b, for impact loading, the work measured by the machine dropped from 54 ft-lbs. at 212 °F to 3 ft-lbs. at 0 °F., a typical impact test result. The impact specimens were double rather than single notched. However, the flat brittle fracture with nearly zero deformation of the 0 °F sample is typical also of brittle low temperature impacts on standard single V-notched Charpy specimens.

A rough calculation shows the ductile fractures to be running through at about 20 feet per second in the impact tests. The cleavage fractures were of course very fast in both slow and impact testing.

The regions of cleavage fracture differ in appearance. At slow speed, cleavage fracture usually occurs in bands or large spots. In the impact fractures a scattering of small ductile fracture-areas through the central region where cleavage fracture first occurs persists long after the fracture has become mainly brittle. A considerable strain aging effect occurs in slow fractures of this particular steel and one can see the opportunity for it in the extensive deformation near the fractures. Gensamer and others at Pennsylvania State College (9) recently reported that large effects upon impact work values of ship plate mild steels were obtained by prestraining by amounts of the order of 10%. Unless one has deliberately prestrained the material by amounts of this magnitude, strain aging will of course be absent from the impact results and this feature alone can cause important differences in results obtained by slow and impact methods.

Turn now to some details seen in sections magnified about 50-fold. Figs. 2, 3, 4, and 5 show various fracture contours. These were prepared from $3/8$-inch single notch, slow bend specimens of low carbon steel; they were partly cracked then sectioned normally to the axis of the notch, at midthickness, where the fracture has its deepest penetration. A. E. Ruark and George Hornbeck made extensive examinations at NRL (10) of fracture paths and crack head

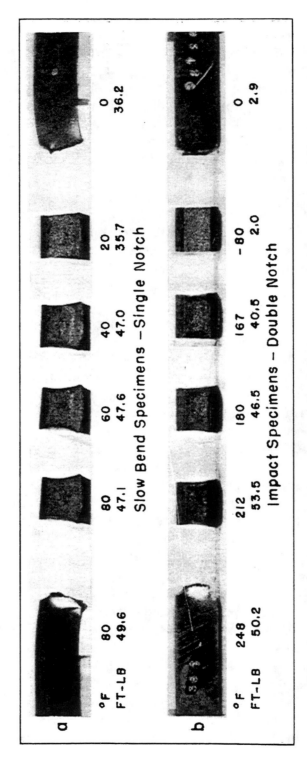

Fig. 1—Comparison of Deformation and Fracture Appearance in Slow Bend and Impact Fractures of Notched 3/8-Inch Square Bars Cut From a 3/4-Inch Mild Steel Ship Plate.

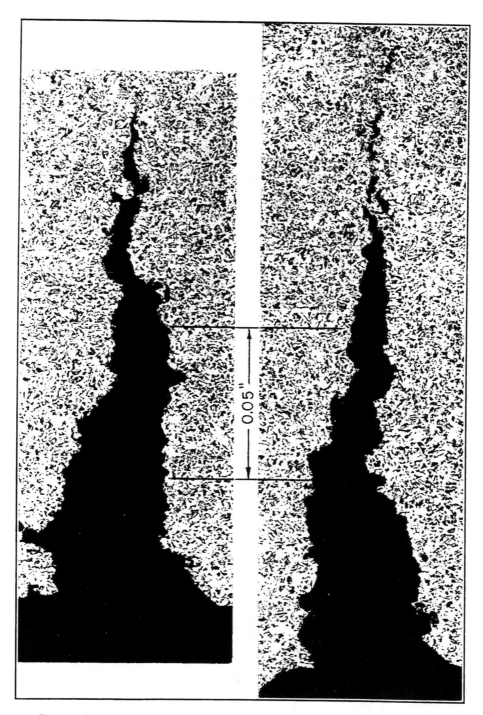

Fig. 2a—Fracture Contours in Mild Steel Ship Plate Showing Cracks Ending in Fast Fracture Segments of Increasing Length.

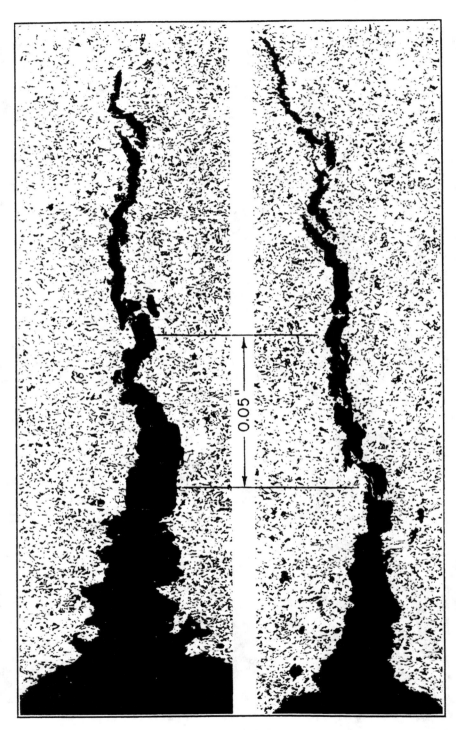

Fig. 2b—Fracture Contours in Mild Steel Ship Plate Showing Cracks Ending in Fast Fracture Segments of Increasing Length.

contours on samples of this character. These specimens were refinished recently by W. J. Ferguson at NRL. In some cases, before the partial cracking was completed, a short segment of fast fracture occurred with a drop in load on the machine. The per cent drop in load was 0, 30, 41, and 63% for the series shown in Figs. 2a and 2b. The deformation which accompanied fracturing reduced as the amount of fast fracturing increased in this series. The notch appears at the bottom.

Fig. 3 is of the same steel. In this case we have a ductile one, R109F, at 30°F (zero load drop) and, at the same temperature, one with a fast fracture segment which dropped the load 85%, R109E. The fracture in similar material shown in Fig. 4 ended with a fast fracture segment with 59% load drop. The lower view is a repolishing of the same specimen as shown in the upper view after removal of 0.01 inch of the specimen surface. Fig. 5 shows (above) a ductile fracture contour and (below) the fracture contour when a very short fast fracture segment occurred.

In introducing his theory of fracture strength, Griffith (11) compared the work required to extend a crack with the release of stored elastic energy which accompanies crack extension. If we imagine a very thin straight crack, through a plate or sheet so loaded that it has a uniform biaxial tensile stress, F, in regions remote from the crack, then Griffith's relations may be written as:

$$\frac{de}{dA} = \frac{d}{dx}\left(\pi \frac{F^2 x^2}{E}\right) \qquad \text{Equation I}$$

$$\frac{dW}{dA} = \frac{d}{dx}(4Tx) \qquad \text{Equation II}$$

In these expressions de is the release of strain energy and dW is the work done, each as associated with the increment of fracture area, dA. Griffith thought of the work, dW, as expended against surface tension, hence the symbol T. The length of the crack parallel to the plate width was given to be 2x, and the total work done in enlarging the crack to length 2x, proportional to surface area, is written as 4Tx. E is Young's modulus. When de/dA becomes slightly larger than dW/dA, the crack develops rapidly, driven by release of strain energy. Theoretically, this instability to fracturing on strain energy will occur in any case for a long enough crack since de/dA is proportional to x. Practically, for ductile metals, dW must include work done in plastic deformation. In this interpretation the work

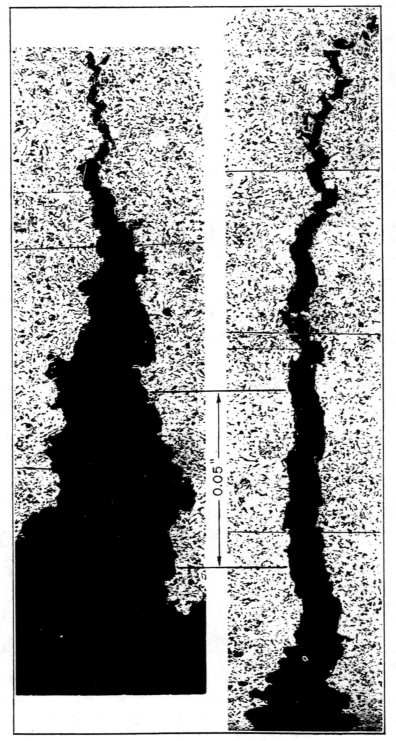

Fig. 3—Fracture Crack Contours in Mild Steel Ship Plate. R109F, above, fractured slowly. R109E, below, fractured fast except for the initial stages.

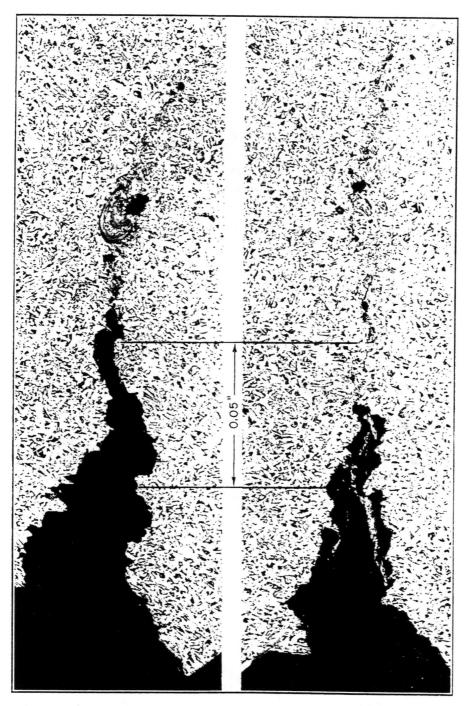

Fig. 4—Fracture Contours in Mild Steel Ship Plate. Approximately the right half of the crack formation is a fast fracture. Two views at different levels of the same crack.

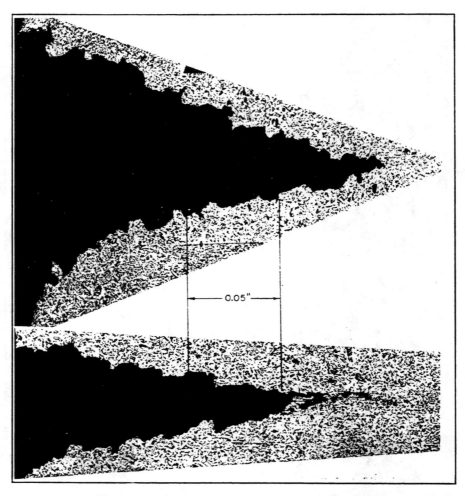

Fig. 5—Fracture Contours in Mild Steel Ship Plate. A short fast-fracture segment occurred in the case of G149AA, lower view, none in the case of G509AA, upper view.

done against surface tension is generally not significant. The term equivalent to $4Tx$ may be approximately represented by two terms, one proportional to area of fracture, and one proportional to volume of metal affected by plastic flow; these change in relative magnitude with changes in crack contour. In cases which are of practical experimental importance one must make extensive changes to obtain valid quantitative formulae for fracturing of ductile metals as a function of cracking distance. However, the idea of locating the point of fracturing instability by equating de/dA to dW/dA remains basic and useful. References (12) and (13) provide additional discussions of the Griffith theory of fracture strength.

For the case of cleavage fracturing Mott assumed he could consider T as approximately a constant in Griffith's expressions. To obtain a velocity estimate for this case Mott added a term to represent the rate of change of total kinetic energy of moving plate material:

$$\frac{d(K.E.)}{dA} = \frac{d}{dx}\left\{\tfrac{1}{2}\rho V^2 x^2 \int \left(\frac{du}{dx}\right)^2 \frac{ds}{x^2}\right\} \quad \text{Equation III}$$

Where K.E. is total kinetic energy; ρ is density of the material; V is the fracturing velocity, dx/dt; and $\frac{du}{dt}$ (equals $V\frac{du}{dx}$) is the displacement velocity of a small element, ds, of moving plate material. The integral, $\int \left(\frac{du}{dx}\right)^2 \frac{ds}{x^2}$, may be seen to be independent of x if the shape of the deformation remains similar. A rough estimate of the value of this integral suggests it is approximately $(F/E)^2$. Thus

$$\frac{d(K.E.)}{dA} = \frac{d}{dx}\left\{\tfrac{1}{2}\rho V^2 x^2 k \left(\frac{F}{E}\right)^2\right\} \quad \text{Equation IV}$$

where k is near unity.

By putting $\frac{de}{dA}$ minus dW/dA equal to $d(K.E.)/dA$ one obtains

$$V^2 = \frac{2\pi E}{k\rho} - \frac{4T}{\rho k (F/E)^2 x} \quad \text{Equation V}$$

Thus since the second term on the right becomes as small as we please with extension of the crack, the running velocity of brittle fracturing is of the order of magnitude of transverse sound velocity.

This brief mathematical outline is in reasonable agreement with the measurements of fast fracture velocities which I mentioned earlier. Presumably, by adding different specific descriptions of the movement of plate material during fracturing, various quantitative theories of fracture dynamics could be developed.

We know, of course, that ductile fracturing at velocities much greater than impact testing machine speeds of say 20 feet per second occurs in the case of some steel fractures. For example in armor perforation when rough fractures accompany the throwing off of pieces of the armor plate to the rear, these fractures may be either crystalline or ductile in appearance but are always fast. For point-of-view purposes it is interesting to ask the following question: Do we have fast uncontrolled fractures because of onset of cleavage fracture or do we have onset of cleavage as a result of a steadily increasing velocity of ductile fracturing?

At the Naval Research Laboratory T. W. George has been making observations of fracturing of metal foils. These studies arose from our desire to improvise a two-dimensional small-scale laboratory model of fracturing for point-of-view-type self-instruction. Fig. 6 shows the simple procedure used. The foils are centrally notched with a razor blade and loaded in tension with a dead

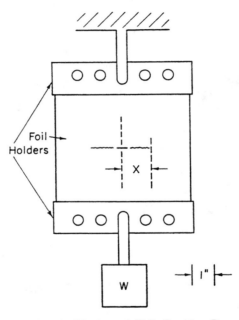

Fig. 6—Diagram of Foil Cracking Procedure Showing the Central Notch, the Creeping Extension of the Fracture From This Notch, and the Dead Weight Loading Method.

weight. When the load is properly chosen the initial cracking progresses slowly and can be profitably observed or measured with a microscope. Generally the fractures begin to move rapidly after several centimeters of creeping and then finish with a snapping apart and release of stored elastic energy. Figs. 7 and 8 show some of the results obtained for fracturing velocities. The loads used were of the order of several kilograms per centimeter of specimen width.

Recent experiments at the University of North Carolina (14) furnish an opportunity to try out the general instability criterion:

$$\frac{de}{dA} = \frac{dW}{dA}$$

on low carbon steel slow speed bend tests.

In order to obtain for this comparison conditions which were as uniform as possible along the breadth of the crack, test samples for bend testing were notched on all four sides so that the crack progressed through the piece without the usual lag at the sides of the bar. A staining technique was developed so that after some exten-

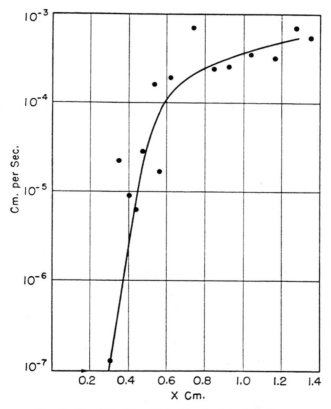

Fig. 7—Slow Cracking Velocities in 0.01-Inch Lead Foil. Arrow shows half length of the central razor-blade cut from which the cracking developed.

sion of fracturing the test bar could be unloaded, removed from the machine, stained, loaded in the machine again and the cycle repeated. Four bands of fracturing distinguished by staining could be obtained when the fracture remained ductile. From the machine records both deformation work and stored elastic energy were computed for the depth of each interval of fracturing.

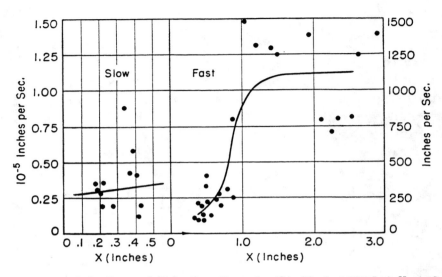

Fig. 8—Left, Slow, and Right, Fast, Fracturing Velocities in 0.0001-Inch Hard 2S Aluminum Foil. The fast velocities were estimated from the films of moving pictures taken at about 3000 frames per second.

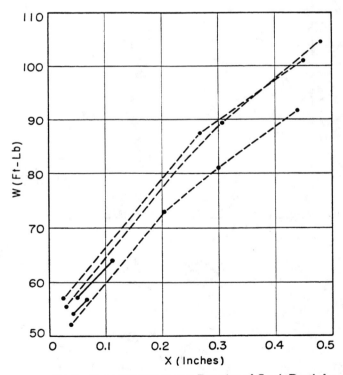

Fig. 9—Fracturing Work as a Function of Crack Depth for ¾-Inch Square Bars of Mild Steel Ship Plate Cut Parallel to the Rolling Direction. Tests were at 100°F.

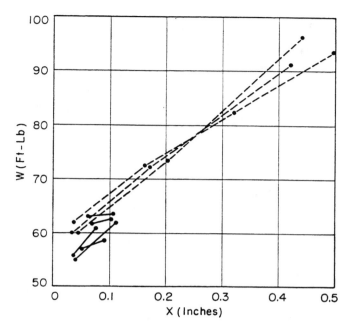

Fig. 10—Same as Fig. 9 but at 120°F.

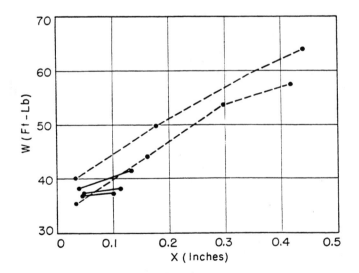

Fig. 11—Same as Figs. 9 and 10 but Samples Cut Transverse to the Rolling Direction and Tested at 90°F.

The illustrations, Figs. 9, 10, and 11, show typical preliminary results from studies of a ¾-inch mild steel ship plate, with this technique. That portion of the energy expended in fracturing to a depth, x, which was recoverable by removal of the load, was found from the machine unloading curves and this "stored elastic energy" was subtracted from total energy in computing values of W. Points connected with dashed lines are for specimens which ruptured with slow ductile fracturing throughout the test. Points connected with solid lines are for specimens whose fractures changed to the fast cleavage type at the depth, x, indicated by the upper end of the solid

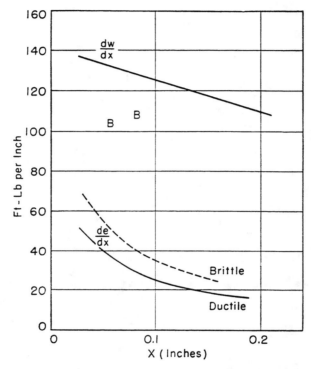

Fig. 12—Comparisons of the Fracturing Work Rate, dW/dx, With the Rate of Release of Stored Elastic Energy, de/dx, for the Tests of Fig. 9.

line. It is seen that the slope, dW/dx, averages less for the solid than for the dashed lines and in some instances at least these results suggest that a relatively small amount of work per unit area preceded development of fast cleavage fracturing.

Figs. 12, 13 and 14 show the first comparisons one might make to find some connection between fast fracturing and the ratios dW/dA and de/dA. The line marked dW/dx is based on data like those in the dashed curves of the preceding figures. The solid and dashed lines marked de/dx were obtained from reference 14 and represent respectively samples which broke entirely by slow fracture and samples in which the rupturing changed to fast cleavage fracture early in the test. The fact that the relationship between stored elastic energy and crack depth differs measurably for slow fracturing from that which was to develop into the cleavage type is interesting but does not in itself explain how this development came about. The slopes of the solid lines of Figs. 9, 10 and 11 are represented by the letters "B" on Figs. 12, 13 and 14. It is clear that dW/dx values small enough to be nearly unstable relative to release of stored elastic energy are shown in some cases by this technique. According to Griffith, where fast fracturing on released strain energy occurred, dW/dx values less than de/dx must have developed.

The manner of arrival at instability for fast fracturing requires careful attention. In these particular tests de/dx is a decreasing function of x. Thus the dW/dx curve can come to an intersection with the de/dx curve early in the test only by dropping rapidly from initially larger values. Such a development implies that the plastic flow per unit change of x becomes less. Considering the crack contours shown in Figs. 2, 3, 4 and 5, one sees that the usual result of cracking accompanied by less plastic working is a sharper pointed crack contour.

The development of fast fracturing, as suggested by our experiments, is as follows: As the specimen bends the notched region reaches its limit of localized plastic flow. The crack opens in the central portion of a relatively large region under the notch. As the crack deepens it acts as a sharper notch so that the region subject to stress relief by plastic flow is steadily and considerably reduced. In many cases the reduction of dW/dA accompanying this development leads at once to fast fracturing. In the notched bending of mild steel under the conditions used in the tests of reference 14 the fracturing was still stable at the point of completion of the crack along the entire length of the starting notch. The samples were unloaded, stained and reloaded to begin the second band of fracturing. Within a distance of less than one-tenth inch additional crack depth all of the fast fractures which were to occur did so. Evidently, in the case of these fast fractures, narrowing of the zone of plastic yielding and reduction of dW/dA, as shown schematically in Fig. 16, continued after reloading until instability for fast fracturing occurred.

The fracture creeping effect which is so pronounced in fracture of foils, the sharpening of the crackhead contour, and the release of stored elastic energy as the point of instability is reached appear to

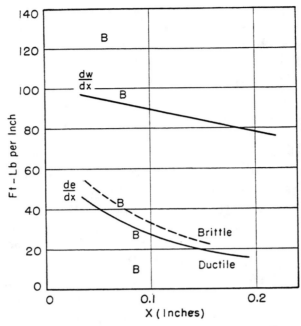

Fig. 13—Comparisons Like Fig. 12, but for the Tests of Fig. 10.

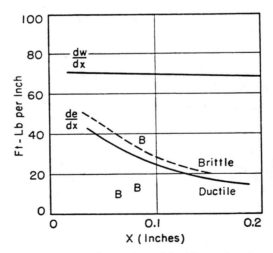

Fig. 14—Comparisons Like Fig. 12 but for the Tests of Fig. 11.

be the major features in the development of fast fracturing. The definite boundary usually observed separating ductile from cleavage portions of an initially slow fracture of mild steel suggests that the advance of the crack as velocities increase enough for cleavage to predominate is generally quite small.

The pioneering work of Mann, Clark, and von Karman led to the concept of a limiting velocity of elongation beyond which localized deformation and reduced total work to fracture occurred. It is also true in the case of progressive fracturing of ductile metals that high velocities of failure are likely to be accompanied by relatively small amounts of total plastic deformation. However, the

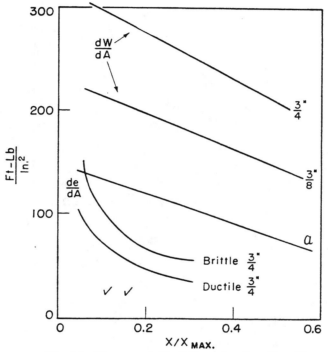

Fig. 15—Diagram of Work Rates for Two Test Sizes and of the Work Rate Component, a, from the Equation $dW/dA = a + \beta b$, Where b Is Specimen Size.

importance of onset of fracturing by release of stored elastic energy furnishes a different viewpoint in the latter case. This paper has endeavored to furnish a detailed description of the approach of fracturing to such an instability point.

In tracing this development comparisons of the rate of work done in advancing the crack, dW/dx, with the rate of release of stored elastic energy, de/dx, were employed. The experimental observations introduced suggested that, in the case of initially slow fractures of mild steel, dW/dx becomes sufficiently small for instability by gradual development of smaller volumes undergoing plastic flow at the head of the advancing crack. The transition to fast fracturing includes a creeping advance of the crack as may be seen in exaggerated form in the fracturing of centrally notched foils. The value of dW/dx may and in some cases demonstrably does attain

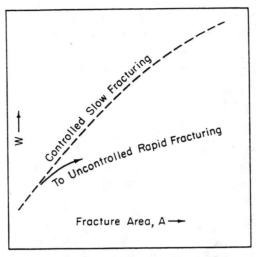

Fig. 16—Schematic Diagram Showing Change in Work Rate as Fracturing Approaches Instability.

critical smallness prior to onset of cleavage fracturing. It is suggested therefore that cleavage fracture of mild steel is associated with rapidity of fracturing rather than that cleavage fracture begins simultaneously with occurrence of the instability point.

References

1. H. C. Mann, "Tension Static and Dynamic Tests," *Proceedings*, American Society for Testing Materials, Part III, 1935, p. 323.
2. D. S. Clark and G. Datwyler, "Tension Impact Testing," *Mechanical Engineering*, Vol. 60, July 1938, p. 559.
3. T. von Karman, "The Propagation of Plastic Deformation in Solids," NDRC Report A-29 (OSRD No. 365), 1942.
4. P. E. Duwez, "Preliminary Experiments on the Propagation of Plastic Deformation", NDRC Report A-33 (OSRD No. 380), 1942.
5. T. von Karman and P. E. Duwez, "The Propagation of Plastic Deformation in Solids," *Proceedings*, Sixth International Congress for Applied Mechanics, Paris, 1946
6. C. Zener and J. H. Hollomon, "Effects of Strain Rate Upon Plastic Flow of Steel," *Journal of Applied Physics*, Vol. 15, 1944, p. 22.
7. G. Hudson and M. Greenfield, "Speed of Propagation of Brittle Cracks in Steel," *Journal of Applied Physics*, Vol. 18, 1947, p. 405.
8. Lecture by N. F. Mott at the Strength of Solids Symposium, Bristol University, Bristol, England, to be published in *Proceedings*, Physical Society (England) in spring of 1948.

9. M. Gensamer, E. P. Klier, T. A. Prater, F. C. Wagner, J. O. Mack and V. L. Fisher, "Progress Report on Correlation of Laboratory Tests With Full Scale Ship Plate Fracture Tests," National Research Council Report, Serial No. SSC-9, March 19, 1947. Navy Contract NObs-34231.
10. A. E. Ruark, I. R. Kramer, P. E. Shearin, and others, "Ductility and Fracture Resistance of Ship Plate," NRL Report No. 0-2796, Nov. 1946.
11. A. A. Griffith, "The Phenomenon of Rupture and Flow in Solids," *Philosophical Transactions*, Royal Society of London, A-221, 1920, p. 163-198.
12. E. Orowan, "Notch Brittleness and the Strength of Metals," *Proceedings*, Society of Engineers and Shipbuilders (Scotland), December 18, 1945, Paper No. 1063.
13. C. Zener and J. H. Hollomon, "Plastic Flow and Rupture of Metals," TRANSACTIONS, American Society for Metals, Vol. 33, 1943, p. 163.
14. P. E. Shearin, R. M. Trimble, M. M. Rogers, F. T. Rogers, Jr., J. W. Straley and W. A. Page, "Crack Propagation, Cone Penetration, and Speed of Propagation of Brittle Fracture Studies," University of North Carolina Progress Report No. 38, June 30, 1947. Navy Contract N6ori-227.

Critical Energy Rate Analysis of Fracture Strength

▶ *Mechanical concepts basic to an understanding of fracturing control possibilities and their applicability to large welded structures*

by G. R. Irwin and J. A. Kies

IN A RECENT article in this journal[1] the writers discussed how engineering design and materials property controls for the onset of unstable fracturing might be established. The discussion given there was incidental to a more general discussion of fracturing and fracture dynamics. This paper gives a review of the simple mechanical concepts basic to an understanding of these unstable fracture control possibilities and discusses their applicability to large welded structures.

The approach of fracture extension to the unstable or self-fracturing condition can be visualized in terms of the information presented in Fig. 1. The test piece represented in the upper right of this figure is a flat plate with an extending central crack of length, X. For such a test the strain energy release rate, dE/dX, rises steadily with crack length as shown in the lower left of the figure. The self-fracturing structure must include the testing machine and specimen gripping devices. However, if these components are very stiff compared to the test specimen, they will contribute very little strain energy during unstable fracturing and the unstable fracturing process may be thought of as one in which the separation, l, of the grips does not change. Thus the load-extension curve as shown in the upper left of Fig. 1 must become a straight down drop of the load, F, during onset of fracture instability. An instability of this kind can occur with the load-extension curve bending continuously over into a straight down drop as illustrated by Curve 1. In tests by the writers and their associates, fractures extending in thin sheets of ductile metals have shown this type of approach to instability. When the test piece is a plate of mild steel of average ship steel quality and $1/2$ in. or more in thickness, onset of self-fracturing may be quite abrupt with a discontinuity in slope of the load-extension curve as shown by the second case illustrated in the schematic curves of Fig. 1. Ample motivation for the abrupt change from predominately ductile to semibrittle cleavage fracturing may be found in the snapping of tough sections which always accompanies ductile fracturing of mild steel coupled with the sensitivity of that material to impact.

If extensive sections of cleavage fracturing of mild steel plate began and moved along with work rate values, dW/dX much less than the release rate of strain energy, dE/dX, then such instability events would be to a large extent capricious and unpredictable. Such measurements as are known indicate that cleavage fractures of mild steel run with velocities less than half of the transverse sound velocity. Under conditions of validity similar to those for the Hertz theory of impact, the kinetic energy of directed particle motion associated with the changing pattern of strain in the test plate is $(1/2)(C/C_2)^2$ times the strain energy release times a numerical factor of the order of unity. (Here C is the crack velocity and C_2 is the transverse sound velocity.) Both of these energies are proportional to the same power of the crack length.

If the kinetic energy is represented by the symbol K, then

$$F \frac{dl}{dX} + \frac{dE}{dX} = \frac{dW}{dX} + \frac{dK}{dX} \qquad (1)$$

For a small crack in a large plate both dE/dX and dK/dX are proportional to the length of the crack. The approximate proportionality of K and dK/dX to the square of (C/C_2) becomes ever more exact for representing their dependence upon crack speed as C approaches zero. For the conditions of unstable fracturing as discussed above dl/dX is zero. One concludes that the strain energy release rate and the fracturing work rate must be equal at onset of instability and that they are unlikely to differ widely in magnitude as fracturing continues.

The pattern and appearance of the fracture which occurs in a service or a laboratory case of unstable cleavage fracturing tells the extent to which one may rely in that instance on the near equality of dE/dX and effective dW/dX values as fracturing continues. If dW/dX becomes and continues substantially less

---

G. R. Irwin and J. A. Kies are with the Naval Research Laboratory, Washington, D. C.

than dE/dX, the crack velocity increases, the fracture appearance changes and repeated forking or shattering occurs. On the other hand if extensive lengths of cleavage fracturing are observed with little change of fracture appearance and with much less extensive branching than in shatter, then one concludes that a near equality of dE/dX and dW/dX existed all along except close to such points of forking or branching as did occur. In regions of cleavage fracturing such that dE/dX and dW/dX are most nearly equal, as at the start or halting of a segment of cleavage fracturing, calculation or experimental knowledge of dE/dX furnishes our best information of the effective fracturing work rate, dW/dX.

Referring again to Fig. 1, if, when the crack length is X, and the load is F, the testing machine is reversed then F and l decrease while the crack length remains fixed. We may represent at least the upper portions of the unloading curve by the relation

$$dF = M\,dl \qquad (2)$$

where M is the spring constant of the specimen corresponding to the crack length X. The spring constant is a decreasing function of X. For the case of a large structure, M would be the structural modulus to be determined in complicated cases by experiment.

It is useful to note that, for l constant during crack extension, the ratio of F to M should remain approximately unchanged. This is obvious for the case of a brittle crack. Experiments with cracks deepening in notched bars indicate that the constancy of F/M is also not significantly altered by extensive plastic deformation confined to the vicinity of the crack. For semibrittle cleavage fracturing of mild steel the assumption that F/M is constant when l is fixed involves negligible risk of error.

For the conditions of the starting and stopping of unstable fracturing we have

$$\frac{dW}{dX} = \frac{dE}{dX} = \frac{d}{dX}\left(-\tfrac{1}{2}\frac{F^2}{M}\right) \qquad (3)$$

The strain energy, E, is here represented as a negative quantity so that the release rate, dE/dX, will be positive. Since, for F/M constant,

$$\frac{dF}{dX} = -FM\,\frac{d}{dX}\left(\frac{1}{M}\right) \qquad (4)$$

then

$$\frac{dE}{dX} = \tfrac{1}{2}F^2\,\frac{d}{dX}\left(\frac{1}{M}\right) \qquad (5)$$

Having outlined the essential features with the simplifying assumption of fixed grips one may next observe that the stiffness of the loading fixtures is actually of secondary importance. If the grips move during an increment of fracture extension, dX, due to the springiness of the loading fixtures then the drop of the load is less than would occur under fixed grip conditions. The total change of strain energy in the test piece may be thought of as having two parts. One part is the loss of strain energy considered as due to the crack extension, dX, with grips fixed. The second part is the gain in strain energy considered as if produced by elastic extension of the test piece with the crack length fixed. Since it is recoverable spring energy, this second part is separate from and does not directly affect the energy release increment which may expend itself in inelastic work associated with crack extension. The effect of motion of the grips is that such motion tends to maintain the load on the specimen subsequent to the instability point. Thus, whether or not the experiment is done under fixed grip conditions, it is proper to compute the strain energy release rate available for unstable self-fracturing as if the grips were fixed.

The above relations are useful in connection with laboratory fracture tests conducted for the purpose of determining effective dW/dX values in terms of the strain energy release rate at onset of unstable fracturing. Experimentally $1/M$ can be plotted as a function of X by loading trials using specimens precracked by any

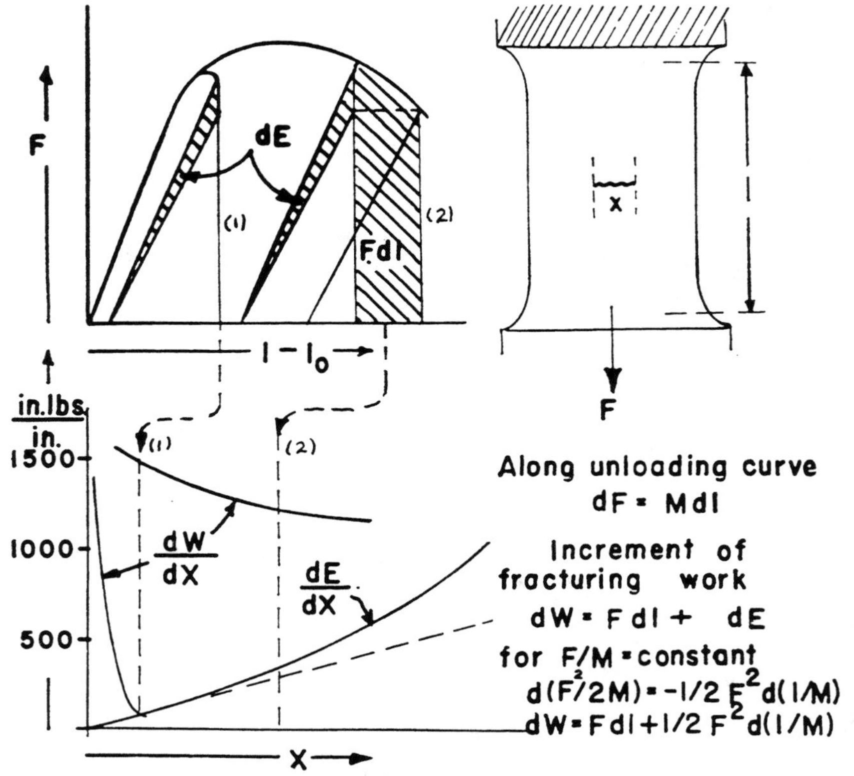

Fig. 1 Schematic diagrams showing continuous and discontinuous types of onset of fracturing instability

convenient means to various crack lengths, X. Other trials can then be made with centrally precracked specimens under condition such that when the crack extends to critical length and then runs on strain energy release across the test piece, the load, F, and the value of X at the start of the instability are observed. Equation 5 can then be used to obtain the effective work rate value governing onset of instability. This procedure[2] has been quite practical at the writers' laboratory in its application to flat plates and plate structures of transparent plastics and to thin sheets of strong light alloys. The results obtained appear to properly represent fracture toughness as affected by temperature, plate thickness and significant variations in fabrication. The measured effective work rates are less for moving than for static starting cracks.

Fracturing work rate values for onset of unstable fracturing in $3/4$- and 1-in. mild steel plates may be obtained by the above procedure. However, the testing machine capacity required is beyond the capacity of machines available to the writers. In this paper we shall rely on certain tests using notched bars in slow bending and on supplementary information from centrally notched plate tests done by other groups.

A good indication of the magnitude of the fracture work rate for semibrittle cleavage fracturing of that material is furnished by notched-bend tests conducted at the University of North Carolina during 1947 and 1948.[3] Three-quarter inch thick plates of two materials were used in these experiments. One was a semikilled steel probably similar to the poorer quality plates tested in investigations sponsored by the Ship Structures Committee. The other material was of similar composition but fine grained and completely killed with aluminum. The former was code labeled "Green" and the Al-killed plate was code labeled "Blue." The tests showed a marked fracture toughness superiority of the Blue over the Green plate material.

The notched-bar fracture tests were conducted in slow bending. The specimens were of square cross section, $3/4$ x $3/4$ in. outside the notches. Instead of a single notch on the tension side, each of the four sides of the bar at the central test section was notched to a depth of 15% with an Izod cutter. The purpose of this procedure was to reduce the customary lag of the fracture crack at the specimen sides so that the leading edge of the crack was nearly straight across the specimen at all crack depths.

Tests were conducted at a series of temperatures. Fracture work rates for slow ductile fracturing were obtained in accordance with the equation

$$\frac{dW}{dA} = R\frac{d\theta}{dA} + \frac{dE}{dA} \qquad (6)$$

where R is the torque applied to the specimen, θ is the over-all specimen-bend angle, and $Rd\theta$ is the increment of work done by externally applied forces on the test specimen during the fracture extension, dA. Here again the positive sign on dE means a release of elastic energy from the specimen. A series of fracture area increments, dA, were applied and were marked by staining, and measured after complete fracture of the test section. The unloading program after each increment of fracturing permitted measurement of specimen strain energy. The strain energy in the balance of the system was a known function of the applied load. Thus by plotting strain energy against fracture depth it was possible to determine dE/dA experimentally. The assumption that the ratio of the bending moment, R, to the specimen spring constant was an invariant of crack depth was not used. However, recent studies of the data[3] by the writers indicate the assumption would have been valid through the whole range of crack depths excluding the extreme portions, the portion where the crack was just forming and the final portion where closure of the back notch brought in new effects of uncertain magnitude.

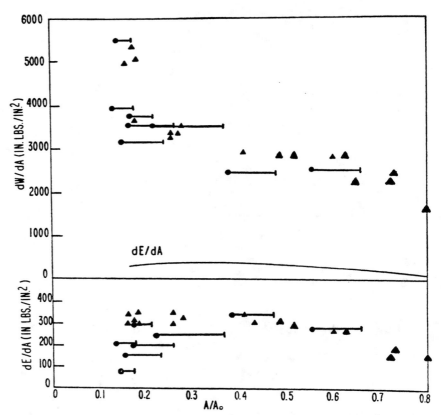

Fig. 2 Ductile fracture work rates (dW/dA) and strain energy release rates (dE/dA) for progressive fracturing in bending of $3/4$-in. mild steel bars notched on all four sides. Short verticle bars in lower graph are equivalent to measurements of effective work rates for onset of semibrittle cleavage fracturing. Data are from Shearin, Trimble and Rogers[3]

Figure 2 shows the comparison of ductile fracture work rate and dE/dA values for these tests. It is at once apparent that the advantage of having to apply smaller forces in notched bending in order to fracture a section of the plate is obtained at a considerable sacrifice. Instead of having available a steady increase of dE/dA values against which the fracture toughness of the material might be measured, this test provided only values in the range of 200 to 350 psi. Unless a material possessed an effective work rate fracture toughness as small as this range of values, then no fracture instability occurred. In a comparison of materials by such a test the most obvious result found, with a series of decreasing temperatures, is the temperature at which the effective work rate for unstable fracturing becomes low enough to show fracture instability. The existence of a progressively increasing strain-aging embrittlement as the fracture deepens makes the crack depth at onset of fracture instability somewhat uncertain. The data of Fig. 2 is for the Green plate. The abscissa used is the ratio of fracture area, as revealed by the staining technique, to the total area of the section severed when the fracture is completed.

Results from tests of several specimens at each of three temperatures, 76, 90 and 100° F are superimposed on Fig. 2. At the highest temperature the last fracture segment was cleavage in only two of seven specimens. At the lowest temperature all fractures were completed with cleavage. The plotted points are placed at the middle of the fracture band which was used to determine each point. Where the fracture band terminated with onset of cleavage this position is marked by a short vertical bar joined to the "middle of the band" point by a horizontal line. Even though the ductile fracture work rate values are depressed in these tests by side notching they are clearly larger by a factor of about ten than the effective work rate values characteristic of semibrittle cleavage fracturing which are the points joined to horizontal lines in the lower graph of the figure. For this particular slow-bend test the Green plate might be said to have a transition temperature of 85° F while the Blue plate had a transition temperature of 25° F. There was, in addition to this temperature difference, an average dW/dA advantage for the Blue plate of about 50 in. lbs/in.^{2} There is no indication in these tests that effective work rate values characteristic of a particular form of fracturing—for example, slow ductile as compared to semibrittle cleavage—undergo large changes as the temperature is reduced. These work rate values are, however, known to be influenced by plate thickness, orientations of the fracture relative to rolling direction and damaging types of heat treatment.

Reference may now be made to the results of several investigations in which centrally notched plates of mild steel were used. One well-known series of such tests was performed at the University of Illinois[4] using plate widths ranging from 12 to 72 in. The initial slot, with its jeweler's hacksaw-cut ends, in each case had a length of one-fourth of the plate width. Values of dE/dA at onset of fracture instability were estimated from the published data and showed a steady increase with the test plate width. A quantity with the dimensions of stress, $(1/B)(dE/dA)$ where B is plate width, may be compared for the various sizes as an indication of the effect of lateral dimensions upon the fracture strength. Such a comparison shows a normal size effect, that is, a steady decrease of moderate amount with increase of B. However, this decrease of $(1/B)(dE/dA)$ with increase of B is much less than would be required for (dE/dA) to remain constant with increasing B. Intuitively one would expect the effective cleavage fracture work rate to approach a constant value with lengthening of the starting crack. It is probable that careful study of experimental work already done in various tests of large steel plates and plate structures would reveal some firm evidence on this point. Currently one can only say that an increase of cleavage fracture work rate with lengthening of the crack beyond a starting size of five times the plate thicknesses appears most unlikely. All of our model experiments in which work rates for "saw cut" and natural cracks were compared have shown larger fracture work rates in the case of the "saw cut" starting cracks. Thus large values of dE/dA at instability which one can obtain from the wide plate tests of ship steels do not represent amounts of fracture toughness which can be counted on to resist cleavage fracture propagation continuing from natural cracks.

By confining attention to 12-in. wide plate tests of the University of Illinois[4] and at Swarthmore[5] one can select cases in which a thumbnail of ductile fracture preceding onset of cleavage indicated that the operative (dE/dA) driving energy was neither far below nor far above the critical value for onset of cleavage fracturing. Such instances of fracture instability of this nature which the writers could find indicated a cleavage fracturing work rate of about 200 in. lbs/in.^{2} This cleavage fracturing work rate appeared to be insensitive to temperature in the range of room temperature to −40° F.

The concept of a cleavage fracturing work rate value for plates of mild steel, as discussed above, is not independent of events which precede onset of cleavage. In the extreme case of onset of cleavage at the root of a notch from one or more ductile fracture origins of inconspicuous size, the computed dE/dA value at onset of instability is usually quite large, for example 800 in. lbs/in.^{2} in the 72-in. wide plates at the University of Illinois. Three other cases may be recognized. These are (a) cleavage fracturing starting from a "thumbnail" of ductile fracture at the root of a notch or from a section of ductile fracture of any length in which general plastic deformation was suppressed by side notching, (b) the onset of cleavage for a slowly extending crack under fatigue or repeated load conditions and (c) a continuation of cleavage fracturing when a cleavage crack is moved quickly into a plate subjected to tension.[6] It seems probable that measurements made under conditions corresponding to (a) and (b) will give similar cleavage fracture work rate values. From experience with other materials it might be anticipated that condition (c) would result in moderately smaller cleavage fracture work rate values. Condition (c) corresponds in service failures to sudden fracturing of a flawed region.

A fatigue crack ordinarily advances by sudden short jumps. The instability criterion applies to the start of each jump and for this case[7] dE/dA has been calculated to be as low as 35 in. lbs/in.[2] for a very small region in the immediate vicinity of the fatigue crack head. Intermittent stopping presumably occurs because the fatigue damage is localized and dW/dA is much more than 35 in. lbs/in.[2] beyond this region.

When the crack length at instability is less than one-third of the plate width and the equivalent length of the specimen, allowing for springiness in the grip and machine, is more than twice the plate width a simple method can be used for estimating values of dE/dA. The procedure shown here is in agreement with the equation used by Griffith and is strictly accurate for an infinite plate. For illustration consider the result of one of the Swarthmore tests. The test was made with 12-in. wide plates of $3/4$-in. steel having a central notch 3 in. wide. The nominal stress, σ, was 32,000 psi. We assume the strain energy relieved by the central slot plus its cracked extensions is just the strain energy which would be contained in an ellipse having the separation of the crack tips as its minor axis, twice this length as its major axis, and located in a position where the stress is the nominal stress. Thus, if X is the crack length,

$$\frac{dE}{dA} = \frac{1}{t}\frac{dE}{dX} = \frac{1}{t}\frac{d}{dX}\frac{\pi X^2 t}{2}\frac{\sigma^2}{2Y} \qquad (7)$$

where Y is Young's Modulus and t is the plate thickness. For $X = 3.5$ in.

$$\frac{dE}{dA} \cong 200 \text{ psi} \qquad (8)$$

Irwin and Kies published a more exact method for making this calculation.[1] This working formula has been simplified[2] and is

$$\frac{dE}{dA} = \frac{\pi F^2}{BYt^2}\frac{y(2+y^4)}{(2-y^2-y^4)^2} \quad \text{where } y = \frac{x}{B} \qquad (9)$$

B = width, t = thickness and Y = Young's Modulus. F is the load, decreasing as y increases, on the fixed grips when the relative crack length is y. This formula is independent of specimen length and applies to the onset of instability. Certain dynamic effects are length dependent but are not discussed here. It should be kept in mind that the above formulas apply strictly to simple plates under tension. In the case of major structures the simple formula of eq 3 would be used and the structural modulus M and dM/dX would not necessarily be calculated but might be determined experimentally under service loading conditions. If the measured structural modulus turned out to be less than the calculated one then an unexpectedly high energy release rate would be possible.

Both for its practical aspects and as a means for obtaining a long starting crack while at the same time avoiding excessive plastic deformation, the extension of a crack to critical length by fatigue or repeated loading deserves attention.

Indication of what may be done along these lines is furnished by the Cornell University ship plate steel fatigue tests.[8] These tests do not, actually, exhibit results in which the unstable fracture length in the sense of complete fracturing was attained. If we assume that each test, in which well-developed cracks appeared, was carried to the point where the speed of extension of the crack, say per 1000 cycles, indicated to the test operator serious danger of a quick break, then the dE/dA values for these crack lengths represent limiting values of practical interest. The tests of simplest geometry were with $3/4$-in. plates of 17 in. width containing a 4.25-in. diam central hole. Cracks extending 1 to 2 in. to each side of the hole were obtained in 20,000 to 100,000 cycles using a nominal stress of 22,500 psi and in nearly one million cycles for a nominal stress of 15,000 psi. More than half the test time was required to get any kind of a crack started. Estimates of dE/dA were made by several rough approximation methods. All of these computed values were in the range of 150 to 400 in. lbs/in.[2] The ability of ship steel plate to resist extension of fairly long cracks with effective dE/dA values in the range of 150 to 400 in. lbs/in.[2] suggests that cleavage fracturing of mild steel plates requires dE/dA values which are far from vanishingly small. Apparently, for $3/4$-in. mild steel plates, more than 100 in. lbs/in.[2] of fracture must ordinarily be supplied from the strain energy field.

Another notable fact which carries a similar message is that mild steel cleavage fractures have been observed to stop in plates subjected to tensile stress in the range of 10,000 to 15,000 psi.[6] The expected mechanics of having a fracture change from an unstable to a stable condition relative to self fracturing on released strain energy does not differ greatly from the reverse transition as represented in the experiments discussed above. Although information available to the writers does not permit calculation in this paper of effective dW/dA values from fracture arrest data, the observations discussed in Reference 6 are not inconsistent with work rate values of 100 in. lbs/in.[2] or more.

Estimates along the lines discussed above using a dW/dA of 200 in. lbs/in.[2] indicate that if the nominal stress in a structure of mild steel plates never exceeds 30,000 in. lbs/in.[2] then a crack forming and moving out into a plate of this structure should not go unstable until it has developed to a length of 4 in. The critical length for 15,000 in. lbs/in.[2] would be four times as large. Cleavage fracture origins as large as this are not often found because the starting crack length can be produced in large welded structures in so many ways other than by creep or by growth in fatigue. Most of these have to do with poor weldments and the residual stresses associated with welds. Efforts to find out more about the seriousness of various kinds of flaws are in progress at several places. In order that the information from these studies may be helpful in more than a qualitative sense, it is desirable that the experiments be designed so as to show the strain energy release rate considerations applicable either to the start or to the arrest of unstable fast fracturing. The same remark applies to studies of transition temperatures. In the latter case so much has been done with standard impact tests that an effort might be made to correlate results of these standard impact tests with some test capable of furnishing effective dW/dA values at crack lengths appropriate to specific applications. A centrally notched plate in fatigue is an example of one kind of measurement which might be employed for this purpose.

Several ways suggest themselves of using dW/dA values in connection with estimating danger of brittle fracture in large structures of mild steel plate. For example where a welded connection between two plates runs parallel to the direction of greatest tension, cracks to either side perpendicular to the weld and extending through the heat-affected zone might be considered to indicate a probable starting size. Allowing something for unnoticed extension of such a crack during periods of stress and temperature variation and assuming this happens in poor quality ship plate, a calculation can be made based on, say, a 3-in. long starting crack and an effective dW/dA of 150 in. lbs/in.^{2} For this situation it should not be far from correct to take

$$150 \text{ psi} = \frac{\pi \text{ (3 in.) } (\sigma^2)}{2 \times 30 \times 10^6 \text{ psi}}$$

Thus we would find σ equals about 30,000 psi as a condition for fracture instability. It has been suggested that brittle fractures of welded ships might be eliminated if no stresses large enough to cause yielding were permitted in the structure. The suggestion appears to be quantitatively appropriate for the starting crack situation just described.

If the weld parallel to the tension direction is joined by a crosswise weld coming in from one side, the situation is somewhat more alarming because we do not know how much of the crosswise weld participates virtually or actually as part of the starting crack. It may be a few inches or as much as a foot. If the latter is the case then a nominal stress of less than half of the material yield strength is a dangerous condition.

Unless some limitation can be placed upon the starting crack sizes which have fair probability of occurrence the stress limitations for safety against unstable fracturing become quite unreasonable. A logical conclusion is that inspection procedures might furnish upper bounds on the lengths of the starting cracks which need to be assumed as potentially present in the structure during its load time history. Granting that we have or can find out the critical strain energy release rates for various cracks and flaws then one can estimate their seriousness relative to one another and relative to the structure containing them. The inspectors can then be told what flaws are not tolerable and the design drawings can be checked in a realistic way for estimating danger of unstable fracturing. The checking procedure would be to suppose cracks of the maximum nonobservable length to exist in the worst location of each region of high stresses. One would then calculate dE/dA for each and compare with whatever appropriate effective dW/dA values are known.

The proposals listed above appear to be direct applications of certain information about fracturing which has emerged in recent years. It is realized that practical construction difficulties associated with large welded structures make a situation far from ideal for theoretical considerations, that inspection of some areas may be inconvenient, and that the workmen never cease from producing unpredictable flaws. However, it is believed the introduction into engineering design of rational quantitative procedures for control of unstable fracturing will, nevertheless, have many beneficial effects.

References

1. Irwin, G. R., and Kies, J. A., "Fracturing and Fracture Dynamics," THE WELDING JOURNAL, 31 (2), Research Suppl., 95-s to 100-s (1952).
2. Kies, J. A., "A Method for Evaluating the Shatter Resistance of Aircraft Canopy Materials," NRL Memorandum Report 237 (December 1953).
3. Shearin, P. E., Trimble, R. M., Rogers, Marguerite M., Scherrer, V. E., Progress Report 39 to Naval Research Laboratory, Dept. of Physics University of North Carolina (June 30, 1948).
4. Wilson, W. M., Hechtman, R. A., and Bruckner, W. H., "Cleavage Fractures of Ship Plates," University of Illinois Engineering Experiment Sta. Bull. Series No. 388, Vol. 48 (March 1951).
5. Carpenter, S. T., Roop, W. P., Barr, W. P., Kasten, N., and Zell, A., "Progress Report on Twelve Inch Flat Plate Tests," Swarthmore College Serial No. SSG-21 (Apr. 15, 1949).
6. Feely, F. J., Jr., Hrtko, D., Kleppe, S. R., and Northrup, M. S., "Report on Brittle Fracture Studies," Standard Oil Development Co., Rept. E. E.4M 53 (Oct. 7, 1953).
7. Wilson, W. M., and Burke, J. L., "Rate of Propagation of Fatigue Cracks in 12 inch by 3/4 inch Steel Plates with Severe Geometrical Stress Raisers," University of Illinois Engineering Experiment Station Bull. Series No. 371, Vol. 45 (Sept. 29, 1947).
8. Hollister, S. C., Garcia, J., and Cuykendall, T. R., "Final Report on Fatigue Test of Ship Welds," (Cornell University Serial SSC-14 (Feb. 4, 1948).

ENERGY CRITERIA OF FRACTURE

Modifications of the Griffith theory are presented to cover the case for a rapidly running crack and for starting up a stationary crack

BY E. OROWAN

SUMMARY. It is shown that, for fully brittle materials, the Griffith equation represents a necessary and sufficient condition of tensile fracture. The present writer's modification of the Griffith equation is a necessary and sufficient condition of fast crack propagation in low-carbon steels; in order to initiate and accelerate a cleavage crack under static load, however, an additional condition (e.g., one demanding a certain initial plastic deformation) must be fulfilled.

The Griffith energy principle from which the Griffith equation is derived cannot be applied to ductile fracture, except when the plastic deformation is confined to a thin layer of material at the surface of fracture.

The Griffith Energy Principle

In the course of the last few years, it has become clear that the Griffith equation for the tensile strength of a brittle solid cannot be applied in its original form to brittle fracture in normally ductile steels. X-ray back reflection photographs show[1] that a thin layer at the surface of apparently quite brittle fractures of low-carbon steels contains significant plastic distortion; the plastic work p in this layer amounts to roughly 2×10^6 ergs/cm^2 if the fracture has occurred not too far below room temperature. Compared with this value, the surface energy (a few times 10^3 ergs/cm^2) is negligible; consequently, if an expression of the Griffith type can be used at all in this case, the surface energy (representing the work for creating unit area of the surface of fracture) has to be replaced by the plastic surface work p. Thus, the crack propagation condition would be[2]

$$\sigma \approx \sqrt{Ep/c} \quad (1)$$

The presence of considerable plastic distortion at the surface of fracture raises the question under what conditions the Griffith principle of virtual work can be applied to fractures accompanied by plastic deformation. This principle can be stated in the following manner: Let dW be the free energy required for increasing the length* of a crack from c to $c + dc$, and $-dU$ the elastic energy released simultaneously in the specimen if this is held between rigidly fixed grips so that the external forces cannot do work. The critical length of the crack above which it can propagate spontaneously is then determined by the condition

$$dW = -dU \quad (2)$$

It is easily seen that the assumption of rigidly fixed grips is not essential; the same result is obtained if the crack propagation is assumed to occur under constant load. Let $M(c)$ be the elastic compliance, i.e., the reciprocal spring constant, of a specimen containing a crack of length c; thus,

$$x = MF \quad (3)$$

where F is the tensile force acting upon the specimen and x its elastic elongation. The elastic energy of a specimen containing a crack of fixed length c is

$$U = \int_{x=0}^{x=MF} F \cdot dx = \frac{M(c)F^2}{2}; \quad (4)$$

on the other hand, the differential dU of the elastic energy when both F and c change simultaneously is

$$dU = (F^2/2)dM + MF \cdot dF \quad (5)$$

$dM = (dM/dc) \cdot dc$ being the increment of the elastic compliance due to the increase of the crack length.

If the crack length increases while the specimen is held between rigidly fixed grips ($x = MF = $ const.),

$$dx = MdF + FdM = 0 \quad (6)$$

substitution of $MdF = -FdM$ in eq 5 gives

$$(dU)_x = -(F^2dM/2) \quad (7)$$

On the other hand, if the crack propagates while the load is kept constant ($dF = 0$), eq 5 gives

$$(dU)_F = F^2dM/2 \quad (8)$$

At the same time, the force F does the work

$$dL = F \cdot dx = F^2dM \quad (9)$$

since, at constant F, $dx = FdM$.

Eqs 8 and 9 show that, if the crack propagates at constant load, half of the external work is stored as additional elastic energy of the specimen, and the other half is available for increasing the free energy of the crack walls. If the length of the crack exceeds the critical value at which eq 2 is just satisfied, the work of the applied force is more than sufficient to provide the increment of its free energy; the balance creates kinetic energy and accelerates the rate of crack propagation.

If, on the other hand, the crack propagates between fixed grips, the elastic energy of the specimen decreases according to eq 7, and its decrement is available for increasing the free energy of the crack and the kinetic energy. Comparison of eqs 7, 8, and 9 shows that the energy available for crack propagation at fixed load is the same as at fixed grips; in the former case, $-dU$ in eq 2 has to be replaced by $dL - (dU)_F$ which is numerically equal to $-(dU)_x$ for a given increment dc of the crack length.

In the present paper, two questions will be treated that have been widely discussed in connection with the brittle fracture of structural and ship steel, and on which a wide divergence of opinions has arisen. They are:

(a) Does the Griffith equation

$$\sigma \approx \sqrt{E\alpha/c} \quad (\alpha = \text{surface energy}) \quad (10)$$

represent a necessary and sufficient condition of completely brittle fracture? And is the present writer's eq 1 a necessary condition of brittle fracture in low carbon steels?

(b) Under what conditions can the Griffith principle, eq 2, be applied to fractures involving plastic deformation?

The Griffith Equation as a Necessary and Sufficient Condition of Completely Brittle Fracture

The Griffith equation (10) expresses the condition that the elastic energy released (or the work of the applied forces) during crack propagation can

---

E. Orowan is George Westinghouse Professor of Mechanical Engineering; Head of Materials Division, Department of Mechanical Engineering, Massachusetts Institute of Technology.

* As in the original work of Griffith, only two-dimensional cases (cracks in plate specimens) will be considered here for simplicity. The general results can be easily extended to three-dimensional cases.

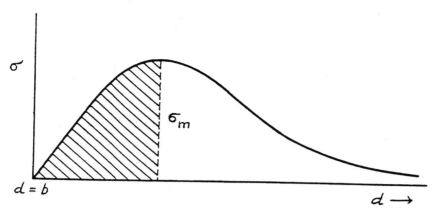

Figure 1

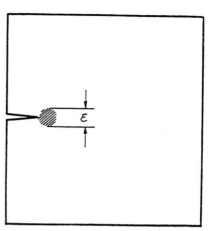

Figure 2

provide the additional surface energy of the increasing crack wall area. It is, therefore, a necessary condition of crack propagation in a completely brittle specimen under tension. If it is not satisfied, propagation of the crack with the accompanying increase of its (free) surface energy would violate the first or the second law of thermodynamics. In particular, thermal fluctuations (disruption of atomic bonds at the tip of the crack by thermal activation) cannot propagate the crack if the Griffith equation is not satisfied, because any such process would result in the creation of free energy from thermal energy without heat flowing from one reservoir to another of a lower temperature. Of course, thermal fluctuations of free energy do occur; however, they cannot lead to any significant crack propagation because the greatest energy fluctuation that may arise with any probability amounts to a few electron volts which is sufficient only for the disruption of a few individual atomic bonds at the tip of the crack. After the fluctuation has passed away, the disrupted bonds join up again.

From the fact that the Griffith equation is a necessary condition of completely brittle fracture, it does not follow that it is also a sufficient condition. However, it can be proved that once the condition is satisfied, crack propagation is not merely possible but is bound to follow. This can be shown by proving that, if the applied stress has the value given by the Griffith equation, the stress concentration at the tip of the crack reaches the value of the molecular cohesion (theoretical strength) at which fracture is bound to take place.

The molecular cohesion of a brittle material can be estimated in the following well known way. When a rod of unit cross-sectional area breaks with a smooth surface of fracture perpendicular to the axis of the rod, two new surfaces of unit area are created; the work required for this is 2α (α = surface energy). This work is done against the intermolecular attractive forces as the two fragments are pulled apart. Figure 1 shows the variation of the molecular forces between the two fragments, per unit of cross-sectional area, as a function of the distance d between the adjacent layers of molecules in the two fragments between which the fracture occurs. The force is zero when $d = b =$ the molecular spacing in the absence of stress; it rises to a maximum σ_m which is the molecular cohesion, and then falls to zero with increasing separation of the fragments. The area below the curve is the work of fracture per unit of the cross-sectional area; i.e., it is equal to 2α. At the maximum of the curve in Fig. 1, the amount of energy represented by the shaded area below the curve must be present between all neighboring pairs of molecular (or atomic) planes perpendicular to the tension; it is identical with the elastic energy stored in the material between two adjacent atomic planes. If, for an order-of-magnitude estimate, Hooke's law is assumed to be valid up to the theoretical maximum σ_m of the stress, the density of elastic energy between two atomic planes of unit area, spaced at b, is $b \cdot \sigma_m^2/2E$. If it is assumed that the shaded area is about one-half of the total area below the curve and therefore approximately equal to α, the relationship

$$b \cdot \frac{\sigma_m^2}{2E} = \alpha \quad (11)$$

gives the order of magnitude of the molecular strength as

$$\sigma_m \cong \sqrt{2E\alpha/b} \quad (12)$$

The next question is: what is the value of the applied tensile stress at which the critical value σ_m is reached at the tip of the crack? The stress concentration factor of a surface crack of depth c and (relatively small) root radius ρ is[3]

$$q = 2\sqrt{c/\rho} \: ; \quad (13)$$

this relationship shows that the maximum stress would be infinitely high for any finite value of σ and c in an elastic continuum containing a perfectly sharp crack, and therefore the tensile strength would be zero. The reason why brittle solids have a finite strength lies in the atomic structure of matter. Figure 1 shows that Hooke's law breaks down when the increment of the atomic spacing becomes comparable in magnitude with the atomic spacing itself: near the tip of the crack the stress versus strain curve levels out, and the situation can be regarded roughly as if a certain region at the tip, comparable in linear dimensions with the interatomic spacing, would be under the constant stress σ_m instead of obeying Hooke's law.

This case of the law of elasticity ceasing to be valid in a region at the tip of the crack has been treated by L. Föppl[4] and, in particular, by Neuber.[5] Neuber proved the following theorem: let there be a region of linear dimensions ϵ at the tip of the crack (Fig. 2), so that the specimen is Hookean elastic outside this region, whereas the stress in the shaded region in Fig. 2 is approximately constant, having the value existing at its boundary. The ratio of the stress in the constant-stress region to the tensile stress applied to the specimen is then approximately equal to the stress concentration factor of a crack of the same length and of root radius $\epsilon/2$ in a purely Hookean elastic material. (The quantity ϵ is assumed to be small compared with the length c of the crack which itself must be small compared with the dimensions of the specimen.)

In the present case, the diameter of the region in which Hooke's law breaks down and the stress levels out is obviously of the order of magnitude of the interatomic spacing b; if it is assumed to be approximately $2b$, Neuber's theorem indicates that the effective stress concentration factor is that of a crack of tip radius b in a purely Hookean specimen. According to eq 13, this is

$$q = 2\sqrt{c/b} \quad (13a)$$

Thus, the value σ of the applied tensile stress at which the molecular strength σ_m is reached at the tip of the crack is given by

$$\sigma_m = \sigma \cdot 2\sqrt{c/b}. \quad (14)$$

If σ_m is replaced from eq. 12, the tensile strength σ is obtained as

$$\sigma \approx \sqrt{E\alpha/2c} \quad (15)$$

which, within the accuracy of the estimate, is identical with the Griffith eq 10.

This derivation of the Griffith equation directly from the stress concentration factor of the crack shows that, when the applied tensile stress has the value given by the equation, the stress at the tip of the crack reaches the highest value that can be withstood by the interatomic forces in the material; any further straining is bound to produce crack propagation. In other words, the Griffith equation represents not only a necessary but also a sufficient condition of fracture in a completely brittle specimen.

Can the Griffith Principle Be Applied to Ductile Fracture?

In recent years the view has been expressed that the Griffith energy principle eq 2 may be applied to all types of fracture, not only to essentially brittle ones. In what follows, it should be pointed out that this is not so: the principle can only be applied if plastic deformation is either absent or confined to a thin layer at the crack walls so that the bulk of the specimen is still elastic.

Figure 3 indicates the manner of crack propagation in a purely elastic material: owing to elastic strain release around the crack, its walls are pulled apart, and its length increases. Figure 4, on the other hand, shows one of the simplest types of ductile fracture,[6] such as is observed in aluminum single crystals or (polycrystalline) plates of ductile metals in tension. The crack (which in this case is a channel of square cross section perpendicular to the plane of Fig. 4) is propagated by slip in the planes $AB + CD$ and $EB + CF$ as indicated by arrows; in the course of this process the cross section of the crack increases until fracture is complete.

The fundamental difference between the propagation of the brittle crack shown in Fig. 3 and the ductile mechanism of Fig. 4 is that the former relies essentially on the elasticity of the material, while the latter could work in the same way even if the elastic moduli were infinitely high. The Griffith eq 10 shows directly that the tensile strength of a brittle material would rise to infinity with an infinite increase of the value of Young's modulus: in such a material, the crack could not open up because there would be no elastic strains to release. On the other hand, the slip mechanism shown in Fig. 4 is quite independent of the elastic moduli.

It is immediately obvious that the force required for propagating the crack in Fig. 4 cannot be derived from the Griffith principle eq (2). Its value is simply

$$F = Y \cdot A \quad (16)$$

where Y is the yield stress of the material in tension and A the projection of the areas AB plus CD on the plane perpendicular to the direction of the tension. If F satisfies eq 16, the plastic deformation that opens up the crack can progress, and the crack propagates. The elastic moduli do not appear in eq 16; they could be infinitely high without any consequence to the propagation of the crack. On the other hand, infinitely high elastic moduli would make the right-hand side of eq 2 vanish: the tensile strength obtained by any application of the Griffith energy principle would rise to infinity with the elastic moduli.

The conclusion is, then, that the Griffith energy principle can only be applied to fully or substantially brittle fractures; ductile fractures are quite outside its scope.

In arguing the applicability of the elastic energy release principle to ductile fractures, occasionally the point has been made that, if the specimen is long enough, the elastic energy stored in it should be sufficient to produce rapid crack propagation even if the energy absorption of the crack is as high as it is in typically ductile fractures. The answer to this is that a *fast* fracture is not necessarily a *brittle* fracture (i.e., a fracture involving very low energy absorption). Any ductile fracture can be made to run fast, at least from a certain stage onwards, if the specimen is connected in series with a large spring (or, what is the same, if the specimen is long enough). It can be shown that the condition for a ductile fracture to become a fast fracture is not eq 2 but equality of the *second* derivatives of W and U.*

The Writer's Crack Propagation Condition for Brittle Fracture in Normally Ductile Steels

As mentioned in the first Section, the present writer has suggested that brittle fracture in ductile steels may obey the crack propagation condition

$$\sigma \approx \sqrt{Ep/c} \quad (1)$$

which results if, in the Griffith eq 10, the surface energy α is replaced by the surface plastic work p. It can be obtained by starting from the Griffith principle of elastic energy release eq 2 and equating the free energy required for producing unit area of the crack wall to p instead of α.

The first question is: Can the Griffith energy principle be applied to a fracture process that involves plastic deformation? It was seen in Section 2 that the Griffith equation can be derived from the elastic stress concentration factor of the crack; however, can this be done if plastic deformation takes place and redistributes stresses at the tip of the crack? The Neuber principle, mentioned in Section 2, shows that the

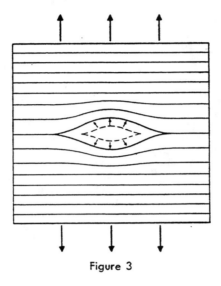

Figure 3

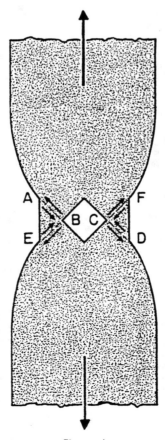

Figure 4

---

* To be published elsewhere.

stress concentration factor can be calculated approximately on the basis of the classical theory of elasticity if the plastically deformed region is small compared with the length of the crack. In this case it can be treated in the manner explained in connection with Fig. 2: the stress concentration factor is the same as that of a crack in a purely elastic body with a tip radius equal to half of the diameter of the plastically deformed region. In fact, this case is basically the same as that of the completely brittle material in which, in order to take into account the atomic structure of matter, the Neuber theorem had to be applied to the region at the tip of the crack in which the stress distribution flattens out owing to the presence of the maximum in the force-displacement curve Fig. 1. The only difference is that in the Griffith case the diameter of the non-Hookean region is of the order of the interatomic spacings, while in the brittle fracture of steel it is about twice the thickness t of the plastically deformed layer at the surface of the crack. According to the Inglis eq 13, the stress concentration factor is then

$$q = 2\sqrt{c/t} \qquad (17)$$

X-ray measurements indicate[1] that t is of the order of 0.2 to 0.4 mm in low-carbon steels broken not too far above or below room temperature.

In the Griffith theory, the tensile strength of the specimen was obtained by dividing the molecular cohesion by the stress concentration factor. What is the quantity corresponding to the molecular cohesion in the brittle fracture of steels? The clue is given by the important observation[7] that in steels the crack does not propagate continuously: before it has broken through a grain boundary, unconnected small cracks arise in grains ahead of the tip of the main crack. This shows that the brittle strength of steel cannot have the order of magnitude of the theoretical strength (molecular cohesion); in fact, it must be quite low if independent fracture processes can start ahead of the main crack at points where the stress cannot be much above the yield stress. This may be due to the presence of numerous invisible cracks scattered in the material; or to the well-known fact that plastic deformation can produce high microscopic internal stresses and subsequent crack formation. It seems that the cleavage strength of the material at the tip of the crack is not, or not much, higher than the ordinary brittle strength of steel obtained experimentally as the stress at which brittle fracture occurs. Since the cleavage strength of steel depends on the plastic strain which is difficult to estimate in the small region around the tip of a crack, only a rough idea at its magnitude can

be obtained; it is probably somewhere between 100,000 and 200,000 psi for a low-carbon steel. For an applied tensile stress σ of, say, 20,000 psi, therefore, a stress concentration factor between 5 and 10 would be needed in order that the cleavage strength σ_m may be reached at the tip of the crack. If the thickness t of the cold worked layer in eq 16 is taken as $1/100$ in., the necessary crack length, given by eq 17,

$$c = \frac{1}{4} t \cdot q^2 = \frac{1}{4} t \left(\frac{\sigma_m}{\sigma}\right)^2 \qquad (18)$$

is between $1/16$ and $1/4$ in.; for a tensile stress of 10,000 psi the stress concentration factor is four times higher, and the necessary crack length is between $1/4$ and 1 in. These orders of magnitude appear quite reasonable in the light of experimental observations: brittle fracture does not occur around room temperature unless a notch or crack of this order of magnitude is present.

The last question to be discussed is whether eq 1 represents a sufficient as well as a necessary condition of crack propagation. At this point a significant difference appears between the fracture, say, of glass and of low-carbon steel. The rise of stress at the tip of a crack in glass is not limited by plastic deformation; in steel, however, the stress at the crack tip cannot exceed the yield stress multiplied by a plastic constraint factor which probably has a value between 2 and 3.[1] If, therefore, the cleavage strength is higher than two or three times the yield stress Y in tension, the tensile stress at the tip of the crack cannot reach the fracture level, no matter how high the stress concentration factor of the given crack in a purely elastic material would be. An additional point of great importance is that the yield stress of steel increases with the rate of deformation more rapidly than the yield stresses of most metals; between the usual rates of "static" tests and the fastest rates at which measurements could be carried out it seems to increase by a factor approaching 3. It seems that, in typical cases of brittle fracture in low-carbon steels, the velocity increase of the yield stress is the salient feature of the phenomenon of brittle fracture. Although cleavage fracture can arise at slow deformation rates, it then requires so much plastic deformation for producing the necessary plastic constraint and strain hardening[8] that the resulting cleavage fracture is anything but brittle; its energy absorption may be almost equal to that of a ductile fracture. Typical brittle fracture in a low-carbon steel, therefore, can occur usually only after the crack propagation has reached a sufficiently high velocity; in laboratory experiments, the fracture is almost always initiated

by some ductile (fibrous) cracking, accompanied by considerable local plastic deformation.

It can be said, therefore, that a characteristic feature of brittle fracture in ductile steels is the enormous decrease of the crack propagation work with increasing velocity of the crack. The crack propagation condition eq 1 may well be fulfilled for a rapidly running crack with its low value of p but not for a stationary crack, the propagation of which may require, per unit of crack length, an energy of a higher order of magnitude. In such cases, cleavage fracture must be initiated in laboratory experiments by large deformations producing strong plastic constraint and strain hardening, and usually some fibrous cracking; the plastic deformation may have to extend across the entire specimen, so that the yield load has to be reached before cleavage cracking can start. After a cleavage crack has arisen, it may accelerate rapidly provided that the condition eq 1 is satisfied, so that there is sufficient elastic energy released during the crack propagation to increase the kinetic energy around the running crack. Consequently, it may be assumed that eq 1 represents a necessary and sufficient condition for the fast and, therefore, brittle propagation of a cleavage crack; in order to initiate a cleavage crack and accelerate it by static loading, however, an additional condition may have to be satisfied, demanding a certain amount of plastic deformation for producing plastic constraint and strain hardening.

It should be remarked that many service fractures seem to start without significant plastic deformation in spite of static loading. An interesting suggestion for explaining this has been put forward by Wells;[8] another possibility will be discussed in a subsequent paper.

This paper represents an expanded version of remarks that were stimulated by the work done under Office of Naval Research Contract No. N5ori-07870, and contributed to the Conference on Brittle Fracture Mechanics held at the Massachusetts Institute of Technology on October 15 and 16, 1953, under the auspices of the Committee on Ship Structural Design, advisory to the Ship Structure Committee, National Research Council.

References

1. Orowan, E., *Trans. Inst. Engrs. Shipbuild. in Scotland*, 165 (1945).
2. Orowan, E., *Proc. Symp., Fatigue and Fracture of Metals*, Massachusetts Institute of Technology, John Wiley & Sons, New York, p. 139 (1950).
3. Inglis, C. E., *Trans. Inst. Naval Arch.*, 55, No. 1, 219 (1913).
4. Föppl, L., *Ing.-Arch.*, 7, 229 (1936).
5. Neuber, H., *Kerbspannungslehre*, Julius Springer, Berlin, p. 142 ff (1937).
6. Orowan, E., "Fracture and Strength of Solids," in *Rept. on Progress in Physics*, Physical Society of London, 12, 185 (1949).
7. Tipper, C. F., *Rept. of Conference on Brittle Fracture in Steel Plates*, Cambridge, England, British Iron and Steel Research Assn., p. 24 (1945).
8. Wells, A. A., *Welding Research (London)*, 7, 34r (1953).

EXPERIMENTS ON BRITTLE FRACTURE OF STEEL PLATES

High velocity dependence of the yield stress found to be the decisive factor in the brittle fracture of low-carbon steels

BY D. K. FELBECK and E. OROWAN

SUMMARY. Tensile tests at room temperature on ship-plate specimens provided with brittle edge cracks of various lengths are described. The fracture stress was found to be inversely proportional to the square root of the initial crack length. The initial brittle crack did not propagate directly as such but first changed into a very short fibrous crack, with considerable plastic deformation around its tip, and this changed back into a brittle crack.

The conclusion is drawn that the decisive factor in the brittle fracture of low-carbon steels is the high velocity dependence of the yield stress. If a cleavage crack propagates slowly, it requires under the conditions of the experiments large local plastic deformations for producing the necessary triaxiality of tension; it becomes a really brittle crack only when it runs so fast that the high velocity of any plastic deformation at its tip raises the yield stress there nearly or fully to the level of the cleavage stress, so that a considerable triaxiality of tension and the corresponding intense plastic deformations are no longer necessary.

1. Extension of the Griffith Theory

According to the Griffith theory of brittle fracture,[1,2] the tensile strength of a brittle body containing a surface crack of length* c is

$$\sigma = \sqrt{\frac{2\alpha E}{\pi c}} \quad (1)$$

if the body is a plate thin compared with the length of the crack, and

D. K. Felbeck is Assistant Professor of Mechanical Engineering and E. Orowan is George Westinghouse Professor of Mechanical Engineering; head of Materials Division, Department of Mechanical Engineering, Massachusetts Institute of Technology, Cambridge, Mass.

This investigation was carried out in the Materials Division of the Department of Mechanical Engineering, Massachusetts Institute of Technology, for the Office of Naval Research under Contract No. N5ori-07870, D.I.C. 6949.

* The original calculations of Griffith refer to the two-dimensional case (e.g., a plate containing a crack). The "length" of the crack is then measured on the face of the plate; in the case of an edge crack, it is identical with its depth measured from the edge. The results of the calculation are approximately valid for any surface crack if the radius of curvature of the surface is large compared with the dimensions of the crack and if the width of the crack (measured along the surface) is large compared with its depth. In this case, the depth of the crack corresponds to the crack length in the two-dimensional case and is traditionally called its "length."

$$\sigma = \sqrt{\frac{2\alpha E}{c(1-\nu^2)}} \quad (2)$$

if it is thick compared with c. In these equations, E is Young's modulus, α the specific surface energy of the surface of fracture, and ν Poisson's ratio.

These equations cannot be applied to brittle fracture in normally ductile steels, for the following reason.[3] In the Griffith theory, the surface energy α represents the work required for enlarging the crack, per unit area of additional surface of fracture. In a completely brittle material, fracture is not accompanied by any plastic deformation; in a low-carbon steel, however, a thin layer adjacent to the surfaces of fracture is plastically distorted, even though the fracture may appear quite brittle: this is recognized from X-ray back reflection photographs.[4] The cause of the distortion is easy to understand: the cleavage planes in neighboring grains do not intersect, in general, at the grain boundary, so that a cleavage in one grain cannot progress smoothly into a neighboring grain. As a rule, cleavage starts independently in several adjacent grains, and the process of separation is then completed by tearing involving plastic deformation at the grain boundaries. In an annealed or hot-rolled steel the X-ray diffraction spots from the individual grains are normally sharp; the intensity of the plastic distortion at the surface of fracture, therefore, can be estimated from the diffuseness of the spots in back-reflection photographs. The effective thickness of the cold-worked layer can be obtained from the rate at which the diffraction spots in successive X-ray photographs become sharper as more and more material is removed by etching from the surface of fracture (Fig. 1 a, b, c). Previous estimates of this kind[4] have led to the order of magnitude of 10^6 ergs/cm² for the plastic distortion work in the brittle fracture surface of a ship steel broken at room temperature; the steel from which the photographs shown in Fig. 1 (a to c), were taken gave a probable value of about $2\cdot10^6$ ergs/cm².

In general, the Griffith theory can be applied only if no plastic deformation occurs during the fracture process. However, if the deformation is confined to a layer at the surface of fracture the thickness of which is small compared with the length of the crack, the plastic work is proportional to the area of the surface of fracture and its value per unit area can be added to the surface energy, since their sum is the total work required for enlarging the area of the crack wall by unit amount. According to the X-ray investigations just mentioned, the effective thickness of the cold-worked layer is of the order of 0.3 or 0.4 mm. which is far above the usual lengths of Griffith cracks in completely brittle materials like glass. However, brittle fracture in low-carbon steel at or around room temperature cannot start without the presence of a crack or a notch far deeper than the thickness of the plastically deformed surface layer; it seems justified, therefore, to assume that the plastic surface work can be taken into account by adding its value p per unit area to the surface energy in the Griffith equation. In this way, eq 1 becomes[16]

$$\sigma \cong \sqrt{\frac{E(\alpha + p)}{c}} \quad (3)$$

if the factor $\sqrt{2/\pi}$ is omitted. If this equation is satisfied, the elastic energy released during the crack propagation is just sufficient to cover the work $p + \alpha$ required for enlarging the walls of the crack by unit area.

Since, as just mentioned, the plastic surface work p in low-carbon steel at room temperature is probably of the order of 10^6 ergs/cm², it is about 1000 times greater than the surface energy

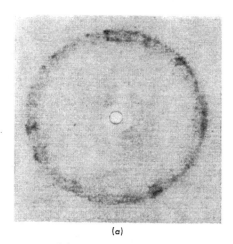 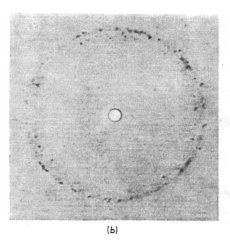 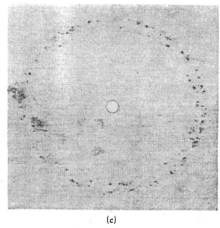

Fig. 1 (a) X-ray back reflection photograph from surface of brittle fracture of ship plate. (b) Same as (a), after removal of a layer of 0.25 mm thickness by etching. (c) Same as (a), after removal of a layer of 0.58 mm thickness by etching

which, for hard metals, is between 1000 and 2000 ergs/cm². Consequently, α can be neglected beside p, and eq 3 written as

$$\sigma \cong \sqrt{\frac{Ep}{c}} \qquad (4)$$

The initial purpose of the present investigation was to test experimentally the crack propagation condition eq 4. The most satisfactory and complete way of doing this would be to measure p independently by the X-ray method mentioned and to compare the value of σ given by eq 4 with measurements of the tensile strength of plates containing atomically sharp cracks of varying lengths c. However, the X-ray measurement of p could not be carried out with an accuracy going beyond an order-of-magnitude estimate, for the following reason. In order to establish a relationship between the diffuseness of the X-ray diffraction spots and the amount of plastic strain, a comparison chart of X-ray photographs of the same material for a series of known plastic strains has to be made. For the present purpose, strains of the order of 1 or 2% are of interest, since X-ray back reflection photographs of the surface of brittle fracture indicate plastic strains of such magnitude. However, plastic strains below the magnitude of the Lüders strain (that present in the Lüders bands before they have spread over the entire specimen) cannot be produced in macroscopic volumes, so that the required region of the comparison chart would be missing. Furthermore, the distortion in the surface of fracture is very unevenly distributed; some fragments, almost completely torn out of the surface, are severely distorted, while the grains underneath may be only slightly deformed. Owing to this circumstance, the type of spottiness seen in the X-ray photographs of the fracture surface (cf. Fig. 1, a) is markedly different from that obtained with a specimen distorted in tension or compression (see Fig. 2), and the comparison of the two photographs is difficult: the surface of the fracture reveals a mixture of widely different plastic strains.

After realizing this difficulty, it was decided to restrict the investigation to the measurement of fracture stresses of steel plates provided with cracks of different lengths, and the comparison of the observed relationship between σ and c with that in eq 4. If the measured σ-c curve is sufficiently well approximated by eq 4, it can be used for obtaining a value for the quantity p.

It is well known that no notch or saw-cut, however sharp, can start a cleavage crack in a low-carbon steel around room temperature without extensive plastic deformation at its root. However, when the present investigation was started, it seemed that this was probably due to the fact that the root of a machined notch did not have the atomic sharpness of a cleavage crack. The plan of the experimental work, therefore, was based on the assumption that a cleavage crack, put into the specimen at a low temperature, would propagate under tension at room temperature* without more initial plastic deformation than that in a thin surface layer containing the plastic work p per unit of area (see previous discussion).

As will be seen below, one of the main results of the work was that this assumption was not fulfilled: even an atomically sharp cleavage crack could not propagate in the ship plate specimens under static tension without large plastic deformation at its root and a temporary change of the crack from the cleavage to a ductile type. Simultaneously with this work, results agreeing with it were obtained by Robertson[5].

2. Production of Specimens Containing Sharp Cracks of Given Length

The experiments were carried out on plates from the tanker *Ponaganset* which broke in two in Boston Harbor on Dec. 9, 1947. All specimens discussed in this paper were cut from a ³/₄-in. plate through which the crack ran when the ship broke up; its position in the hull can be identified by means of the report on the *Ponaganset* failure,[6] where it was designated by the letters "PAD."

The final size of the plate tensile specimens was about 4 in. width by 12 in. length. The width represented the upper limit that could be handled with the largest available testing machine, a Southwark-Emery hydraulic machine of 300,000 lb capacity. The specimens were provided with an initial brittle crack at the middle of one edge in the following way. First, a specimen with a side flap according to Fig. 3 was machined, and the flap provided with a notch for inserting a splitting wedge, as shown in the figure. A brittle crack was then produced by hammering the

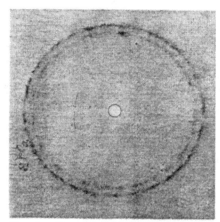

Fig. 2 X-ray back reflection photograph of annealed low-carbon steel strained 2.5% in tension

---

* Provided, of course, that the transition temperature of the material was above room temperature.

Fig. 3 Plate specimen with side flap, showing wedge inserted in saw cut

wedge into the notch after the specimen was cooled down by immersion in liquid nitrogen. The length of the crack could be controlled to a certain extent by progressive splitting with a succession of moderate blows with a hammer on the wedge. In addition, any desired length could be achieved accurately when the flap and a certain margin containing the excess length of the crack were trimmed off the plate along a line shown dotted in Fig. 3. The ends of the crack could be recognized clearly by the absence of frost in a narrow zone along the crack. Examination of the fractured specimen showed that the crack front was not straight; in the middle of the plate it was ahead of the ends of the crack visible on the surfaces of the plate.

The presence of a notch on only one side of the plate introduces a certain eccentricity. Although the effect of this can be taken into account approximately,[7] a number of experiments was carried out in which the bending moment due to the asymmetry was compensated in the way shown in Fig. 4. Two strips of 2 in. width, each provided with a crack as described above, were tacked together by welding near the ends with the cracks turned toward each other. Any bending due to the eccentricity of the individual strips would force them together and thus would be counteracted by the pressure between them. The experiments described in the following section have shown that the difference between the results obtained with the asymmetrical and the symmetrical (twin) specimens was comparatively small.

3. Experiments

The load was applied to the specimens through wedge grips with roughened inner surfaces; the specimen was gripped so that its free length was about 7 in. with the crack half way between the grips, and loaded at a rate of about 1000 lb/sec. In Table 1 the measured tensile strengths are given for all specimens except those which showed obvious disqualifying irregularities of fracture appearance. In most specimens represented in Table 1, the fracture went right across the plate; in a minority, the crack stopped before the separation was complete. The corresponding data for the symmetrical specimens, cut from the same plate (PAD) as those in Table 1, are given separately in Table 2.

The relationship between breaking stress and initial crack length is shown in Fig. 5, representing the data of Table 1. The curve represents eq 4, with the value of p that minimizes the sum of the squares of the ordinate difference between the measured points and the curve; this value is $p = 4.9 \times 10^6$ ergs/cm². Thus the experimentally determined specific plastic energy is somewhat higher than the value derived from X-ray observations (see Section 1). Figure 6 shows the corresponding relationship for the symmetrical specimens (see Fig. 4). The value of p derived from this plot is 3.4×10^6 ergs/cm².

The question is whether this discrepancy is a trivial consequence of the inherent inaccuracies of measurement and evaluation, or the expression of a real difference between the plastic surface work estimated from X-ray photographs and the constant p obtained by fitting eq 4 to the data given in Tables 1 and 2. The latter possibility will be discussed in the following section.

4. Conclusions

The idea of the experiments described in the preceding section was to provide the specimen before the tensile test with a brittle crack of atomic sharpness and measure the tensile stress required for starting it to propagate. This, it was hoped, would give a direct experimental test for the crack propagation condition, eq 4.

The plots of the observed fracture stresses against the lengths of the initial cracks, shown in Figs. 5 and 6, seem at first glance to provide impressive support for eq 4: not only do the measured points lie well on curves represent-

Table 1—*Ponaganset* Plate PAD: Asymmetrical Specimens with One Edge Crack
(Room temperature tests: plate thickness: ³/₄ in.; specimen length: 12 in.)

No.	Initial width, cm	Initial crack length, cm	Fracture load, lb	Avg stress psi	kg/cm²
3	10.4	1.7	111,000	34,300	2420
4	10.0	0.9	145,000	46,800	3300
5	10.0	1.3	134,000	43,100	3040
6	10.0	2.6	87,000	28,000	1970
8	9.9	2.4	84,000	27,400	1930
9	10.0	1.0	137,500	44,400	3130
10	10.3	0.5	191,000	59,900	4220

Table 2—*Ponaganset* Plate PAD: Symmetrical Twin Specimens
(Room temperature tests: plate thickness, ³/₄ in.; specimen length, 12 in.)

No.	Initial width, cm	½ initial crack length, cm	Fracture load, lb	Avg stress psi	kg/cm²
D-1	10.1	0.68	126,000	40,300	2840
D-2	10.1	1.2	106,000	33,800	2380
D-3	10.0	1.85	92,000	29,600	2090
D-4	10.1	2.0	91,000	29,100	2050
D-5	10.1	0.55	154,500	49,400	3480

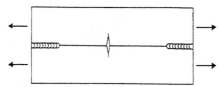

Fig. 4 Symmetrical twin specimen

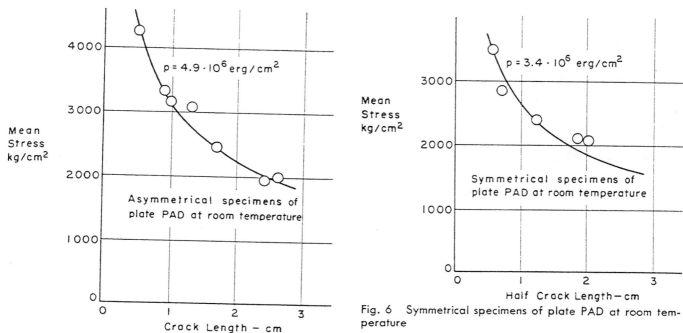

Fig. 5 Asymmetrical specimens of plate PAD at room temperature

Fig. 6 Symmetrical specimens of plate PAD at room temperature

ing eq 4, but the value of p derived from them agrees in the order of magnitude with the X-ray estimate. Yet an important feature observable on the fractured specimens casts doubt not only on the significance of this agreement but also on the foundations of the classical Mesnager-Ludwik triaxial tension theory of notch brittleness. It was easy to recognize at the first visual examination of the fragments that, contrary to the expectation that underlay the program of the investigation, the initial brittle crack never continued to propagate as a brittle crack when the fracture stress was reached. First, considerable plastic deformation took place at the tip of the crack which started to propagate as a fibrous crack. After a very short run, this changed again into a brittle crack which ran across the plate at high velocity (except in the cases discussed hereafter in which the brittle crack stopped due to load relaxation and had to be restarted by raising the load). This can be seen in Fig. 7b where 1 is the initial brittle crack, 2 the fibrous crack and 3 the brittle crack developed from the fibrous one in the course of the test. The contraction of the thickness of the plate by the plastic deformation around the tip of the initiating crack can be recognized in the photograph. Figure 8 shows the plastic deformation at the tip of the initiating crack as it appears in view upon the face of the plate. Figure 7c shows the surfaces of fracture in a specimen containing a shorter initiating crack.

Occasionally, when the initial crack was long and therefore the fracture stress low, the second brittle crack did not run through the plate but stopped in it. In such cases, it could be restarted by increasing the load which, owing to the inability of the (hydraulic) testing machine to maintain the load during the rapid extension of the specimen, had dropped drastically (e.g., to one-half of the initial fracture load) by the time the crack had stopped. However, as in the first instance, the brittle crack did not propagate directly when the load was raised. Again extensive plastic deformation leading to the formation of a narrow zone of fibrous or shear fracture took place at the tip of the crack, and then the ductile crack changed to a brittle one which ran at high velocity. In a few cases, the crack stopped for a second time owing to the fall of the load, and the whole process of restarting, with the conversion into a ductile and then again into a brittle crack, could be repeated again. Figure 7a shows the surface of fracture of a specimen in which the crack stopped twice before fracture was complete.

Why is it that a brittle crack produced in the plate before loading cannot continue to propagate as a brittle crack when the load reaches the necessary

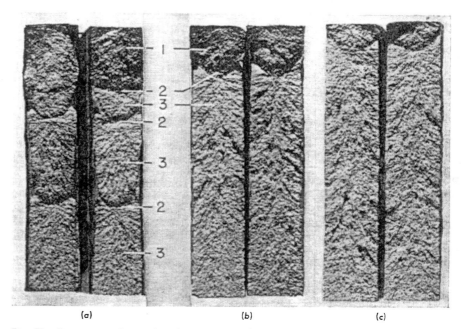

Fig. 7 Fracture surfaces showing transition from ductile to brittle crack propagation

Fig. 8 Appearance of the crack at the face of the plate

value? What are the factors that compel it to change into a ductile crack with copious local plastic deformation before it can change back into a brittle crack? And, in view of this change, do the tensile tests described above have the meaning initially ascribed to them, of giving corresponding values of crack length and crack propagating stress suitable for supporting eq 4?

The only simple explanation for the inability of an atomically sharp brittle crack to continue its propagation unchanged when a tensile stress is applied is to attribute it to the absence of velocity when the propagation is resumed under a slowly applied stress. This almost unavoidable conclusion requires a drastic revision of the classical theory of notch brittleness. For the last 50 years, the brittle fracture of ductile steels was attributed, after Mesnager[8] and Ludwik,[9] to the triaxial tension arising when plastic deformation starts at the tip of a notch or a crack. Let Y in Fig. 9 be the ordinary uniaxial (true) stress-strain curve of the material in tension; fracture of the ductile (fibrous or shear) type occurs at the point F. Curve B represents the brittle strength (brittle fracture stress); since it lies above the yield stress-strain curve OF,

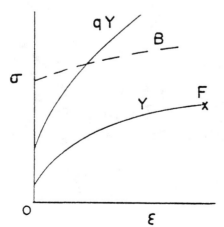

Fig. 9 To the triaxial tension theory of brittle fracture

no brittle fracture can occur in the ordinary tensile test. However, brittle fracture is possible if the maximum tensile stress required for yielding is raised above the values given by the curve OF. This can be done by:

1. lowering the temperature: ferrous materials become brittle at a sufficiently low temperature;
2. increasing the rate of straining: ferrous materials have an abnormally high velocity-dependence of the yield stress-strain curve;
3. superposing a hydrostatic tension, e.g., by plastic constraint in a specimen containing a crack or a notch.

The classical theory of notch brittleness saw its main cause in the triaxial tension produced by notch constraint. The way in which this acts can be described very simply, without any reference to triaxiality, in the following manner. In a specimen containing a notch or a crack, plastic deformation starts in the region of stress concentration at the tip of the notch. However, this region is embedded in less highly stressed surroundings that have not reached the point of yielding at the same time; consequently, the region of stress concentration cannot yield freely before the maximum tensile stress in it has increased to the magnitude required for overcoming both the constraint of the surroundings and its own resistance to plastic deformation. It can be shown that this effect can raise the maximum tensile stress reached at yielding up to about three times the value of the uniaxial tensile yield stress if the notch is very deep and sharp. In the presence of a notch, therefore, the maximum principal tensile stress plotted against the plastic strain will be represented by a curve like that denoted by qY in Fig. 9, and this may intersect the curve B of the brittle strength before the plastic deformation could reach the value necessary for ductile fracture to occur. In this case, the specimen undergoes brittle fracture.

As mentioned above, the Mesnager-Ludwik theory attributed brittle fracture in commonly ductile steels primarily to the development of a triaxial tension during plastic yielding at the tip of a notch or a crack. Such a triaxiality requires the occurrence of plastic deformation; without this, only a small effect due to elastic constraint can be present. How much plastic deformation is necessary for the development of full plastic constraint has never been calculated; it was implicitly assumed that deformation extending over a very small (perhaps microscopically small) region around the tip of the crack would be sufficient. The observations described above indicate strongly that this expectation was mistaken: quite extensive deformation around the tip of the crack is required, either because the region of highest tensile stress has to reach a certain critical size before cleavage can occur, or because the effect of plastic constraint in raising the tensile stress has to be complemented by strain hardening.[10] Once this is assumed, the explanation of the phenomenon shown in Fig. 7 presents no difficulty. If a fracture appears quite brittle, it cannot involve plastic deformation sufficient for the development of considerable plastic constraint and strain hardening: brittle fracture must then be due to the raising of the yield stress to the level of the brittle strength by the velocity effect. If a crack travels fast, any plastic deformation that may occur at its tip involves an extremely high strain rate and so requires a strongly increased yield stress. With most metals, even the highest velocities cannot increase the yield stress more than 20 or 30%; according to several investigators,[11,12] however, the yield stress (possibly only the upper yield point) of low-carbon steels increases by a factor of 2 or 3 at high rates of deformation. This would be quite sufficient to replace fully the highest possible plastic constraint effect. But if the brittleness is due mainly to a velocity effect, the crack put initially into the specimen cannot start propagating in the typical brittle manner under static load because it has no velocity. Consequently, local plastic deformation sets in, and plastic constraint develops until the triaxiality of tension is high enough to change the initial fibrous crack propagation into a brittle one. Once this has taken place, the crack accelerates rapidly and further plastic deformation becomes unnecessary because the velocity effect takes over from the plastic constraint effect the task of raising the maximum tensile stress to the level of the brittle fracture stress.

Figure 7a shows clearly that, at not too low temperatures, the velocity effect alone may not be sufficient for raising the yield tension to the value of the brittle strength. At the free surface of a plate, no triaxiality can exist: consequently, some plastic deformation with ductile (shear) fracture must occur here if the velocity effect alone is insufficient, and the well-known phenomenon of the "shear lip" arises. The plastic deformation produces a slight but sharp constriction running in the form of a shallow and narrow rounded groove along the line of fracture on the surfaces of the plate. This is the same phenomenon as the necking of a tensile specimen, and it is likewise accompanied by the development of transverse tensions in the interior which, added to the velocity effect, lead to brittle fracture everywhere except at the shear lip where, owing to proximity of the free surface, the degree of triaxiality becomes too low. Thus, the width of the shear lips gives a measure of how much the velocity effect falls short of being able to produce brittle fracture alone. This is impressively seen in Fig. 7a. When the brittle fracture starts at the fibrous "nail" 2, it soon attains high velocity, and the shear lip is quite narrow. However, as it progresses, the load drops because the testing machine cannot follow the fast expansion of the specimen, and the velocity of the crack decreases. With this, more and more of the velocity-effect upon the yield stress has to be replaced by a plastic constraint effect, and the width of the shear lip increases until finally the two shear lips join up to a parabolic arc. At this point, the crack stops, possibly after it has produced a very narrow margin of a ductile crack, a second fibrous "nail."

This picture of brittle fracture changes the classical concept in which triaxiality was the fundamental effect. It seems now that triaxiality alone cannot produce really brittle fracture except at low temperatures because, before it can lead to cleavage, it requires considerable plastic deformation. Whenever a fracture is truly brittle (i.e., of extremely low energy consumption), this must be due mainly to the velocity effect, unless the temperature is so low that the relatively small elastic constraint effect is sufficient. Triaxiality may be important in starting off brittle crack propagation, as seen in the above experiments; it is doubtful, however, whether it is indispensable for this purpose. Many service fractures do not reveal any visible trace of plastic deformation at their starting point. This effect could be reproduced in laboratory experiments by using plate specimens containing a weld.[13, 14] It seems, therefore, that heating or cooling under stress creates a condition in which cleavage can start under static load without preceding large deformations.

The observation that the initiating cleavage crack never spread under the applied stress but always changed first into a ductile crack before reverting to the cleavage type raises the question whether the use of the crack propagation condition eq 4, as it was done in Section 3, can still be justified. It is easy to see that the incisive change in the theoretical picture of notch brittleness to which the above experiments have led has clarified the significance of eq 4. The conditions of *brittle* cleavage now include a high velocity of the crack, and this can be reached only if the energy to be fed into it (in the form of the surface work p) can be obtained from the elastic energy released during its propagation. Otherwise, very high velocities of load application would be necessary. If the tensile stress σ is not below the value demanded by eq 4, the work needed for enlarging the crack can be obtained from the released elastic energy. Equation 4, therefore, is the condition for the high velocity of crack propagation required for brittle cleavage to be reached under static or nearly static loading. From this, it does not follow that the fracture stresses observed in the present experiments must have been those given by eq 4: they must have been higher than this value. In the present experiments, brittle fracture could not occur unless:

(a) the plastic deformation needed for starting cleavage could take place at the tip of the initial crack; and

(b) the condition for crack acceleration eq 4 was fulfilled.

It can be shown[15] that the Griffith-type equation (eq 4) is equivalent to a stress concentration condition. If it is satisfied, the applied tension is capable of producing plastic deformation at the tip of the crack in a region the radius of which is equal to the thickness of the cold-worked layer in which the plastic surface work p is done. The thickness of this layer, according to the X-ray evidence mentioned above, is of the order of 0.3 or 0.5 mm; since the tensile stress needed for producing yielding in a region of this size is lower than that required for yielding in the much larger area that has to be deformed before the crack can start to propagate, the stress required for satisfying condition (a) is higher than that demanded by (b). Consequently, the fracture stress is the starting, not the accelerating, stress.

That the observed fracture stresses are approximately inversely proportional to the square root of the crack length is not surprising. According to a theorem due to Neuber,[7, 15] the mean tensile stress σ needed to produce plastic yielding at the level of the (constrained) yield stress σ_m in a region of radius r at the tip of a crack of length c is

$$\sigma \approx \sigma_m \cdot 1/2 \sqrt{r/c} \qquad (5)$$

If, therefore, the size of the plastically deformed region required for starting cleavage is fairly independent of the crack length, the fracture stress should be inversely proportional to $\sqrt{c}$.

References

1. Griffith, A. A., *Phil. Trans. Roy. Soc.*, A221, 163 (1920).
2. Griffith, A. A., *First Internal. Congr. Appl. Mech.*, Delft, p. 55 (1924).
3. Orowan, E., *Repts. on Progress in Physics*, 12, 185 (1949).
4. Orowan, E., *Trans. Inst. Eng. and Shipbuild.*, Scotland, 89 (Paper No. 1063), 165 (1945).
5. Robertson, T. S., *Engineering*, 172, 445 (Oct. 5. 1951).
6. McCutcheon, E. M., Pittiglio, C. L., and Raring, R. H., THE WELDING JOURNAL, 29 (4), Research Suppl., 184-s to 194-s (1950).
7. Neuber, H., *Kerbspannungslehre*, Berlin (1937).
8. Mesnager, A., Internat. Assn. for Testing Materials, Brussels Congress 1906, Rapport non-offic., No. A 6f, pp. 1–16.
9. Ludwik, P., and Scheu, R., *Stahl u. Eisen*, 43, 999 (1923).
10. Wells, A. A., *Welding Research (London)*, 6, 68r (August 1952).
11. Clark, D. S., and Dätwyler, G., *Proc. ASTM*, 38, 98 (1938).
12. Taylor, G. I., *J. Inst. Civil Eng.*, 26, 486 (October 1946).
13. Weck, R., *Welding Research (London)*, 6, 70r (August 1952).
14. Greene, T. W., THE WELDING JOURNAL, 14 (5), Research Suppl., 193-s to 204-s (1949).
15. Orowan, E., "Energy Criteria of Fracture," THE WELDING JOURNAL. In press.
16. Orowan, E., in *Fatigue and Fracture of Metals* (Symposium held at MIT June 1950), Cambridge, Mass., New York and London, p. 136 (1952).

The Propagation of Cracks and the Energy of Elastic Deformation

By H. F. BUECKNER,[1] SCHENECTADY, N. Y.

The Griffith-model of brittle fracture of elastic solids and the model by Irwin and Orowan for the brittle fracture of elastic-plastic solids predict the propagation of cracks on the basis of energy supplied by the work of externally impressed forces and by the change of strain energy. Previous discussions have neglected the three-dimensional viewpoint and the body forces. Since the Irwin method of fracture-strength analysis has become of increased interest, especially with respect to rotor fracture, a general analysis of energy supply is presented. The analysis uses Clapeyron's theorem and Betti's reciprocal theorem. One of the results is that the energy supplied for crack extension equals the strain energy of the difference of the two stress fields before and after crack extension.

Introduction

THE following considerations deal with the change of strain energy and with the work of impressed forces during crack extension. It is assumed that the strains are small and that stresses and strains are correlated by Hooke's law. The following models of crack propagation are assumed:

The Griffith-Model. It has been demonstrated by Griffith (1)[2] that the onset of crack propagation in brittle material like glass can be explained by considering an elastic body with two forms of potential energy. These are the strain energy and the surface energy, the latter one stored with constant density H along the surface of the body. As an example we consider, Fig. 1, a plate of uniform thickness h and of rectangular shape with two opposite sides subject to a constant tension σ. Let there be a crack of length $2c$ (blacked area) parallel to the loaded sides. Let us assume that the crack extends to length $2c'$ while the two loaded sides are kept by fixed grips. The external forces will do no work. The extension creates new surfaces with $4(c' - c)h$ as gain in area. This requires the surface energy to increase by $4(c' - c)hH$. The strain energy U decreases to an amount U', and

$$U - U' \geq 4(c' - c)h \cdot H \qquad [1]$$

is a necessary condition in order to make the crack extension possible. The difference between the two sides of Inequality [1] will appear as kinetic energy at a real crack propagation. In the limit $c' \to c$ a certain strain-energy release rate

$$U^* = \lim_{c' \to c} \frac{U - U'}{2(c' - c)h} \qquad [2]$$

is obtained. Condition [1] implies

---

[1] Engineering Mathematician, Large Steam Turbine and Generator Department, General Electric Company.
[2] Numbers in parentheses refer to the Bibliography at the end of the paper.
Contributed by the Power Division and presented at the Annual Meeting, New York, N. Y., December 1–6, 1957, of THE AMERICAN SOCIETY OF MECHANICAL ENGINEERS.
NOTE: Statements and opinions advanced in papers are to be understood as individual expressions of their authors and not those of the Society. Manuscript received at ASME Headquarters, September 5, 1957. Paper No. 57—A-189.

$$U^* \geq 2H \qquad [3]$$

Assuming a virtual crack extension for any pair of parameters σ, c Griffith found

$$U^* = \pi c \sigma^2 / E \qquad E = \text{Young's modulus} \qquad [4]$$

which holds for a state of plane stress and for c small against the dimensions of the plate. For constant c and sufficiently small values of σ the release rate U^* is below $2H$. There will be no crack propagation. For a critical value of σ, namely

$$\sigma^* = (2HE/\pi c)^{1/2} \qquad [5]$$

the equality sign in [3] is reached. At this value crack propagation can start and, indeed, this was observed by Griffith.

The Model of Irwin and Orowan for the Brittle Fracture of Mild Steel. The experiment of Fig. 1 can be repeated with mild steel rather than glass as material of the plate. There is experimental evidence to support the existence of a critical stress σ^* and even the type of the law [5], (9). However, H has to be substituted by another quantity, say $^1/_2 G_c$, which is much larger than H, in some cases by a factor of 1000. Clearly the surface energy plays no role whatsoever. Due to Orowan (2, 3) the phenomenon can be explained by plastic deformation of a certain surface layer along the crack. In the realm of elasticity the stresses become infinitely large at the root of the crack. In reality a zone 1 of contained plastic deformation will appear at the root (see shaded area of Fig. 2). As the crack propagates zone 1 will be unloaded but some permanent plastic deformation will be left. Another zone 2 of plastic deformation will be set up, thereafter partially relieved, and so forth. Thus a surface layer of permanent plastic deformation is generated along the crack. This process is dis-

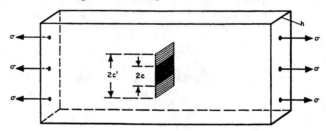

Fig. 1

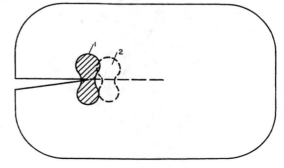

Fig. 2

sipative as mechanical energy is transformed into heat. If the layer is thin and of constant thickness then one may calculate the dissipated energy in proportion to the gain in crack surface. This makes it plausible why Formula [5] would stay valid with a dissipation density G_c substituted for $2H$. In this way the following model for the brittle fracture of mild steel is obtained:

The rupturing body is considered to be elastic; energy dissipation goes with the creation of new crack surfaces, and it is proportional to the area gained times a certain density factor $1/2 G_c$.

An extensive application of this model as well as theoretical results about the release rate of strain energy are due to Irwin (4 to 8). The model is simple, and it cannot be expected that it will explain all phenomena which occur in fractures. Even so it will be useful in order to separate one cause of crack propagation from others. Regardless of what the area of applicability of the model will turn out to be, it is useful to analyze the changes of elastic strain energy for virtual crack extension. This is the objective of the following investigation. Dynamical effects will not be considered. The results are applicable to the onset of crack propagation.

Release of Energy in General

With respect to Fig. 3 we consider an elastic body V with a crack inside. The two crack surfaces are C_1, C_2. The other part of the surface of the body is denoted by $O = O_1 + O_2$. Here O_1 may be subject to tractions T (T is a vector) while the displacements are prescribed along O_2. We also admit a distribution of body forces given by a vector field X. With respect to a cartesian co-ordinate system (x_1, x_2, x_3) we introduce the displacement vector $u = (u_1, u_2, u_3)$ and the following components of the strain tensor

$$e_{ik} = \frac{1}{2}\left(\frac{\partial u_i}{\partial x_k} + \frac{\partial u_k}{\partial x_i}\right) \quad\quad [6]$$

The components of the stress tensor may be denoted by σ_{ik}. It is assumed that C_1, C_2 are separated by the displacements and that they are free from tractions.

Let us consider a virtual crack extension which adds a new surface C_1' to C_1 and a new surface C_2' to C_2 (dotted lines in Fig. 3). With body forces and surface conditions unchanged, a new displacement vector u' together with new strains e_{ik}' and stresses σ_{ik}' will characterize the state that goes with the extended crack. Again it is assumed that the crack surfaces are separated and free from tractions.

In what follows the state of stresses σ_{ik} and strains e_{ik} will be called the first state; its strain energy will be denoted by U. The state of the stresses σ_{ik}' and strains e_{ik}' will be referred to as the second state with U' standing for its strain energy.

We are now going to set up the virtual work of all impressed forces due to crack extension. The virtual work of the externally impressed forces is

$$W_e = \iint_{O_1} T(u' - u)do + \iiint_V X(u' - u)dV$$

do = surface element, dV = volume element ..... [7]

The virtual work of the internal forces is

$$W_i = U - U' \quad\quad\quad [8]$$

The total virtual work is $W = W_e + W_i$, and the condition of crack propagation becomes

$$W = W_e + W_i > G_c \cdot |C_1'|; \quad |C_1'| = \text{area of } C_1' \quad [9]$$

If Equation [9] holds for at least one virtual crack extension then the crack C_1, C_2 cannot be stable and crack extension is bound to follow.

The following considerations are not restricted to infinitely small crack extensions which are attributed to the principle of virtual work. The strain energies U, U' are explicitly

$$U = \frac{1}{2} \cdot \iiint_V \sum_{i,k} \sigma_{ik} e_{ik} \cdot dV;$$

$$U' = \frac{1}{2} \cdot \iiint_V \sum_{i,k=1}^{3} \sigma_{ik}' e_{ik}' \cdot dV \quad\quad [10]$$

It is also useful to introduce the mixed energy

$$U_m = \frac{1}{2} \cdot \iiint_V \sum_{i,k} \sigma_{ik}' \cdot e_{ik} \cdot dV$$

$$= \frac{1}{2} \cdot \iiint_V \sum_{i,k} \sigma_{ik} \cdot e_{ik}' \cdot dV \quad\quad [11]$$

The two integrals in Equation [11] are equal to one another as a consequence of Hooke's law. From Equations [10] and [11] it follows

$$-W_i = U' - U$$

$$= \frac{1}{2} \cdot \iiint \sum_{i,k}(\sigma_{ik}' + \sigma_{ik})(e_{ik}' - e_{ik})dV \quad\quad [12]$$

In addition to the first and second states of stress and strain two more states will be introduced. These are the sum-state with the stresses $\sigma_{ik}' + \sigma_{ik}$ and strains $e_{ik}' + e_{ik}$ and the difference-state with the stresses $\sigma_{ik}' - \sigma_{ik}$ and strains $e_{ik}' - e_{ik}$. The displacement vectors are $v_s = u + u'$ and $v_d = u' - u$, respectively.

From now on the first state will be reinterpreted as a state of the body V with the extended crack. The stresses of the first state result in certain tractions T^* along C_1', C_2' (see Fig. 3). Vice versa, the first state is determined by the condition on O_2, by the distribution of the tractions T, T^* and by the body forces X, applied to the body V with the extended crack. It is to be noted that the vector T^* preserves its length but jumps into the opposite direction as we go from a point of C_1' to the opposite point of C_2'. The displacement vector u remains unchanged.

With the first state reinterpreted, the sum-state and the difference-state find a corresponding interpretation. The sum-state is the response to the body forces $2X$, to the surface tractions $2T$ on O_1, zero-tractions on C_1, C_2, and tractions T^* along C_1', C_2'; along O_2 the displacements are prescribed. The difference-state is the response to no body forces, no tractions on

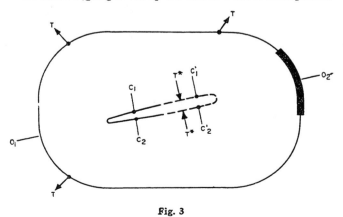

Fig. 3

O_1, C_1, C_2, tractions $-T^*$ along C_1', C_2', and no displacements along O_2.

The integral in Equation [12] is the mixed energy of the sum-state and of the difference state. According to Betti's theorem (10) the mixed energy equals one-half of the work done by the external forces (impressed and reactionary) of one state through the displacements of the other, no matter from which state the forces are taken. We decide to take the forces from the sum-state and the displacements from the difference-state. There is no contribution from the reactionary forces of the sum-state along O_2, since they go with zero-displacements of the difference-state; the externally impressed forces of the sum-state give rise to the work

$$U' - U = \iiint_V X v_d \cdot dV$$
$$+ \iint_{O_1} T \cdot v_d \cdot do + \frac{1}{2} \cdot \iint_{C_1'+C_2'} T^* \cdot v_d \cdot do \quad \ldots \ldots [13]$$

whence in combination with Equations [7] and [8]

$$W = W_e + W_i = -\frac{1}{2} \cdot \iint_{C_1'+C_2'} T^* \cdot v_d \cdot do \quad \ldots \ldots [14]$$

On the other hand the integral expression on the right-hand side equals the strain energy U_d of the difference-state

$$U_d = \frac{1}{2} \cdot \iiint_V \sum_{i,k} (\sigma_{ik}' - \sigma_{ik})(e_{ik}' - e_{ik}) dV =$$
$$-\frac{1}{2} \cdot \iint_{C_1'+C_2'} v_d \cdot T^* do \quad \ldots \ldots [15]$$

This follows from Clapeyron's theorem (10), according to which the strain-energy equals one half of the work done by the externally impressed forces through the displacements. The energy U_d is always non-negative. In general the same is not true for either W_e or W_i. Relations [14] and [15] are useful for a practical discussion of the Condition [9].

Since the strain energy of the difference-state is all that counts for crack propagation, it does not matter if a certain reference state of stress and strain is subtracted from both the first and the second state. As reference state one may take the state of the body V *without any crack* subject to the constraints and externally impressed forces of the first state. Subtraction of this special reference state from the first and second state will lead us to the modified first and the modified second state.

The reference state gives rise to tractions T_0 along C_1, C_1', C_2, C_2'. Each of the modified states is the response to these conditions: No body forces, no tractions along O_1, no displacements along O_2, and traction $-T_0$ along its crack surfaces. This makes it clear that the crack extension depends on T_0 only. Once T_0 is known along $C_1 + C_1'$, $C_2 + C_2'$ one may forget everything about the body forces X and the tractions T.

As an example we consider the state of plane strain of a rotor with a borehole, as shown in Fig. 4. Let ω be the angular velocity, a the inner and b the outer radius. With respect to polar co-ordinates r, ϕ a radial crack C_1, C_2 along a radial interval $a \leq r \leq c$ and a crack extension C_1', C_2' along the interval $c \leq r \leq c'$ of the same direction is assumed. No surface tractions and no constraints are prescribed.

As is well known the tangential stress of the uncracked rotor is

$$\sigma_\phi = \frac{3+\nu}{8} \rho \omega^2 \cdot \left(b^2 + a^2 + \frac{a^2 b^2}{r^2} - \frac{1+3\nu}{3+\nu} \cdot r^2 \right) = F(r)$$

ν = Poisson's ratio
ρ = mass density $\quad \ldots \ldots \ldots \ldots [16]$

The shear stress $\tau_{r\phi}$ vanishes.

The two modified states are determined by the following impressed forces: No body forces; no tractions along the surfaces $r = a$, $r = b$; normal tractions along the cracks as given by $\sigma_\phi = -F(r)$.

Therefore the difference state and its strain energy are entirely determined by the function $F(r)$ along the interval $a \leq r \leq c'$. Especially for very short cracks, that is $c' - a \ll a$ it is the value $F(a)$ which together with c, c' determines U_d with sufficiently good approximation.

Crack Extension Under Constant Load and Under Fixed Grips

We consider the special case where O_2 vanishes; $O_1 = O$ represents the full surface of the body V without cracks. This situation will be referred to as the case of constant load. The integral of Equation [12] will now be transformed by means of Betti's theorem in the other way; i.e., we will calculate the work done by the forces of the difference-state through the displacements of the sum-state. This results in

$$U' - U = \iint_{C_1'+C_2'} -\frac{1}{2} T^* \cdot v_s \cdot do \quad \ldots \ldots [17]$$

But

$$\iint_{C_1'+C_2'} T^* \cdot u \cdot do = 0 \quad \text{and} \quad \iint_{C_1'+C_2'} T^* \cdot v_d \cdot do$$
$$= \iint_{C_1'+C_2'} T^* \cdot v_s \cdot do \quad \ldots \ldots [18]$$

as u is the same and as T^* differs in direction at opposite points of C_1', C_2'. By comparison of Equations [13], [14], [17], and [18] we find

$$W_e = 2U_d \quad W_i = U - U' = -U_d \quad \ldots \ldots [19]$$

This means that the strain energy *increases* as the crack extends. But the virtual work of the externally impressed forces equals twice the increase of strain energy, and the surplus of energy is available for crack propagation. An example of crack propagation under constant load is the rotor of Fig. 4. Another

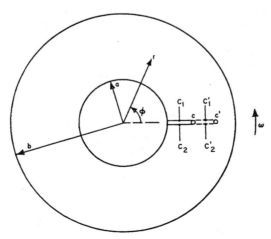

Fig. 4

facet of crack extension under constant load is expressed by the relation

$$\iiint_V \sum_{i,k} \sigma_{ik}(e_{ik}' - e_{ik})dV = 0 \quad \ldots\ldots\ldots [20]$$

The difference-state and the first state σ_{ik}, e_{ik} have no mixed energy; they are orthogonal. Relation [20] can be proved by means of Betti's theorem, if the forces are taken from the difference-state and the displacements from the first state. The proof uses Equation [18] and may be left to the reader. A special case of Equation [20] is

$$\iiint_V \sum_{i,k} \sigma_{ik}^{(0)} \cdot (e_{ik}' - e_{ik}^{(0)})dV = 0 \quad \ldots\ldots [21]$$

where $\sigma_{ik}^{(0)}$, $e_{ik}^{(0)}$ characterizes the state of the body without any crack. Hence from Equation [12] and from [21]

$$\begin{cases} U' - U_0 = \frac{1}{2} \cdot \iiint_V \sum_{i,k} (\sigma_{ik}' + \sigma_{ik}^{(0)})(e_{ik}' - e_{ik}^{(0)})dV \\ \qquad = \frac{1}{2} \iiint_V \sum_{i,k} (\sigma_{ik}' - \sigma_{ik}^{(0)})(e_{ik}' - e_{ik}^{(0)})dV \\ U_0 = \frac{1}{2} \iiint_V \sum_{i,k} \sigma_{ik}^{(0)} e_{ik}^{(0)} dV \end{cases} \quad \ldots\ldots [22]$$

Therefore the difference $U' - U = U_d$ of Equation [19] can be computed as the difference of the strain energies of the modified states $(\sigma_{ik}' - \sigma_{ik}^{(0)})$ and $(\sigma_{ik} - \sigma_{ik}^{(0)})$. This applies especially to the rotor of Fig. 4 which was studied in section 2 by means of the modified states.

Sometimes crack extension is considered under the condition of fixed grips. This means that the first state is defined by the general configuration of Fig. 3. One takes the displacements which this state induces along O_i, and it is required that the second state have the same displacements along O_1 and in addition the prescribed displacements along O_2. In this case the vector v_d of the difference-state vanishes along O_1 and O_2. Again Relations [14] and [15] are valid, but W_e represents the virtual work of the body forces only. If there are *no* body forces, then

$$W = U_d = U - U' \quad \ldots\ldots\ldots [23]$$

The strain energy *decreases* as the crack extends. The difference in the strain energies is available for crack propagation. The analog of Equation [20] turns out to be

$$\iiint_V \sum_{i,k} \sigma_{ik}'(e_{ik}' - e_{ik})dV = 0 \quad \ldots\ldots\ldots [24]$$

which is a direct consequence of Equations [9], [15], and [23]. Relation [24] says that the mixed energy of the difference-state and of the second state vanishes if no body forces exist.

It is useful to compare the difference-states of crack propagation under constant load and under fixed grips with one another. In both cases we have the same tractions $-T^*$ along C_1', C_2'; there are no tractions along C_1, C_2. In the case of constant load the tractions along O vanish; in the case of fixed grips the displacements along O vanish.

According to the so-called minimum principle of stresses the energy U_d in the case of fixed grips is smaller than the energy of any other stress field where the stresses satisfy the equilibrium conditions and the boundary conditions on the impressed forces. Therefore the strain energy of the difference field of fixed grips is *smaller* than the strain energy of the difference field of constant load.

Yet the difference between the two cases seems to become insignificant for infinitely small crack extensions as can be seen from the one-dimensional model of a spring subject to a tensile force F. The reciprocal spring constant M may characterize a crack. M increases as the crack extends. With $u = MF$ denoting the elongation, the strain energy of the spring is

$$U = uF/2 = MF^2/2 = u^2/2M \quad \ldots\ldots [25]$$

If M increases up to M' as the crack extends one finds

$$U_d = (M' - M)F^2/2 \quad \ldots\ldots\ldots [26]$$

for the case of constant load and

$$U_d = (M' - M)\frac{M}{M'} F^2/2 \quad \ldots\ldots\ldots [27]$$

for the case of fixed grips. The energy [27] is smaller than the energy [26]. However, the limit

$$\lim_{M' \to M} \frac{U_d}{M' - M} = F^2/2 \quad \ldots\ldots\ldots [28]$$

is the same for both cases.

It is to be mentioned that the difference between the two cases has been pointed out by Orowan (2 and 3) in connection with the spring model. He also found the results, Equations [19] and [23], for this special case. The identity of the limits, Equation [28], was observed by Irwin and also conveyed to this writer.

Another class of cases will be discussed in the next section. However, for a study of crack propagation in the large the emphasis should be on the case of constant load.

Irwin's Formula

We consider a state of plane strain of a cylindrical body V, of which Fig. 5 shows a cross-section V^*. Let σ_x, σ_y, τ_{xy} denote the stresses and u, v the displacements in x and y-direction respectively.

The following simplifying assumptions will be made:

1 The cross-section is symmetric with respect to the x-axis.

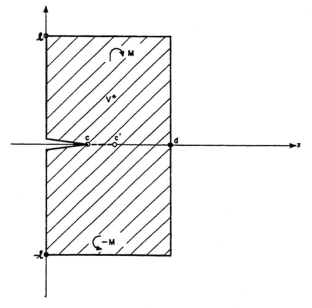

Fig. 5

2 The distribution of load is symmetric with respect to the x-axis.

3 The crack runs along part of the x-axis.

Consequently there will be no shear stress along the x-axis. Fig. 5 shows an example. We have a bar of length $2l$ and of width d. The crack shall run from $x = 0$ to $x = c$; its extension will run from $x = c$ to $x = c'$. The bar is bent by two moments M and $-M$, as indicated in Fig. 5.

Let us consider the case of the crack $0 \leq x \leq c$. In general the stress σ_y for $x > c, y = 0$ follows the law

$$\sigma_y = \sigma_y(x, c) = \left[\frac{EG}{(1-\nu^2)2\pi}\right]^{1/2} \cdot (x-c)^{-1/2} \cdot F_1(x, c); \quad F_1(c, c) = 1 \quad [29]$$

where G is a certain constant (introduced by Irwin) and where $F_1(x, c)$ is a continuous function of both x and c. Also the displacement v along the upper side of the crack behaves according to

$$v = v(x, c) = \frac{4(1-\nu^2)}{E}\left[\frac{E \cdot G}{(1-\nu^2)2\pi}\right]^{1/2} (c-x)^{1/2} \cdot F_2(x, c);$$

$$F_2(c, c) = 1 \quad [30]$$

where F_2 is a continuous function of x and c. Applying Equation [17] for the case of crack extension under constant load we find

$$U(c') - U(c) = \int_c^{c'} \sigma_y(x, c) v(x, c') dx \quad [31]$$

where $U(c)$ represents the strain energy per unit thickness of the state with crack length c. With Equations [29] and [30] taken into account we obtain from Equation [31]

$$U(c') - U(c) = \frac{2G}{\pi} \int_c^{c'} (x-c)^{-1/2}(c'-x)^{1/2} \cdot F_1(x, c) F_2(x, c') dx \quad [32]$$

Introducing $t = (x - c)/(c' - c)$ we obtain

$$\frac{U(c') - U(c)}{c' - c} = \frac{2G}{\pi} \int_0^1 \left(\frac{1-t}{t}\right)^{1/2} \cdot F_1[c + (c'-c)t; c]$$

$$F_2[c + (c'-c)t; c'] dt \quad [33]$$

and

$$\frac{dU}{dc} = \frac{2G}{\pi} \int_0^1 \left(\frac{1-t}{t}\right)^{1/2} \cdot dt = G \quad [34]$$

This is Irwin's formula. It shows that the energy-release rate depends on the parameter G of the first state only. The same reasoning can be applied to the case of crack extension under fixed grips. Again G is obtained as the energy-release rate. Hence for infinitesimal crack extension there is no difference between the energy-release rates at constant load and under fixed grips. The Formula [34] together with Equations [29] and [30] is a useful tool for the study and prediction of crack propagation. The criterion, Equation [9], of crack propagation takes the very simple form

$$G \geq G_c \quad [35]$$

Conclusions

The energy available for crack extension comes from two sources: The work of the externally impressed forces, and the change of the strain energy. The combined energy release from both sources is always positive; it equals the strain energy of the difference of the stress fields before and after crack extension. The difference field stays the same if the fields before and after crack extension are modified by subtracting from both of them the field of the uncracked specimen. The problem of crack extension is thus reduced to a study of the modified fields. These have no body forces; they show surface tractions at the crack surfaces only; these tractions are induced by the field of the uncracked specimen. The effects of different systems of impressed forces (e.g., one system with body forces, another with surface tractions) acting on the same body, can be compared with one another on the basis of the surface tractions, which the modified states display on their crack surfaces.

In addition to these general results, the following special statements can be made. When a crack extends under constant load, the strain energy decreases. However, the work of the externally impressed forces supplies twice the increment of strain energy and the surplus is available for crack extension. When a crack extends under fixed grips without body forces, there is no work of externally impressed forces. The strain energy decreases, and the decrement is available for crack extension. The energy supply for crack extension under constant load is larger than the supply under the condition of fixed grips. However, the energy release rates (energy per unit area of crack extension) are the same at the onset of crack propagation, at least for certain cases of plane strain.

Bibliography

1 "The Phenomenon of Rupture and Flow in Solids," by A. A. Griffith, *Philosophical Transactions of the Royal Society of London*, series A, vol. 221, 1929, pp. 163–198.

2 "Fundamentals of Brittle Behavior of Metals," by E. Orowan, Fatigue and Fracture of Metals (MIT Symposium, June, 1950), John Wiley & Sons, Inc., New York, N. Y.

3 "Energy Criteria of Fracture," by E. Orowan, *Welding Journal*, Research Supplement, March, 1955.

4 "Fracture Dynamics," by G. R. Irwin, Fracturing of Metals, ASM, 1948, p. 152.

5 "Fracturing and Fracture Dynamics," by G. R. Irwin and J. A. Kies, *Welding Journal*, Research Supplement vol. 31, February, 1952, pp. 95s–100s.

6 "Critical Energy Rate Analysis of Fracture Strength of Large Welded Structures," by G. R. Irwin and J. A. Kies, *Welding Journal*, Research Supplement, vol. 33, April, 1954, pp. 193s–198s.

7 "Onset of Fast Crack Propagation in High Strength Sheet and Aluminum Alloys," by G. R. Irwin, Proceedings of Sagamore Conference on High Strength Materials, Syracuse University, vol. 2, 1955, pp. 289–305.

8 "Analysis of Stresses and Strains Near the End of a Crack Traversing a Plate," by G. R. Irwin, *Journal of Applied Mechanics*, TRANS. ASME, vol. 79, 1957, pp. 361–364.

9 "The Relation of Microstructure to Brittle Fracture," by J. R. Low, Jr., ASM Symposium on Behavior of Materials at Low Temperature, American Society for Metals, Cleveland, Ohio, 1953, p. 163.

10 "Mathematical Theory of Elasticity," by I. S. Sokolnikoff, McGraw-Hill Book Company, New York, N. Y., second edition, 1956, pp. 86 and 391.

Discussion

G. R. Irwin.[3] This paper is a welcome contribution for several reasons. The effect of motion of the loading forces during crack extension, although explained in references (3) and (6), has not been well understood. The present paper provides a thorough review of crack-extension energy and strain energy on a 3-dimensional basis, and shows explicitly the applicability of classic theorems by Betti and by Clapeyron. The fact that the paper

---

[3] U. S. Naval Research Laboratory, Washington, D. C.

shows that the presence of body forces introduces no new type of difficulty into the crack-extension-force determination is of special value.

However, the difficulty of the analysis of stresses near a crack in a rotating cylinder or in a notched-bend specimen should not be underestimated. For example, the stresses in a rotating cylinder as shown by Fig. 4 of the paper are well known when no crack is present. As stated by the author, to obtain the stress system appropriate to a cylinder containing a radial crack, one may add a new stress system which cancels the normal stresses along the crack surfaces. The stresses which must be canceled, $F(r)$, are indeed known. However, it appears unlikely that any simple procedure exists for constructing a stress system which produces pressures equal to $-F(r)$ along the crack and at the same time satisfies free-surface conditions at the inner and outer radii of the cylinder. Of course, for purposes of fracture-mechanics analysis only the stresses near the leading edge of the crack are required. The comments of the author as to the actual difficulty of this stress calculation would be appreciated.

Author's Closure

The calculation of the intensity factor G for cracked specimen with a free boundary is indeed a difficult problem. For the notched bar in bending (Fig. 5) such calculations were carried out by means of an integral equation which relates the displacement v to the stress σ_y along the notch; more precisely an integral operator applied to v produces the stress σ_y. In order to match a prescribed stress distribution along the notch, a linear combination of the responses to four different v-functions was used; this procedure furnished the G-value with engineering accuracy. It appears from this experience that the method of integral equations is a useful tool for other problems as well.

Section Two
The Modern Theory of Fracture Mechanics

On the Stress Distribution at the Base of a Stationary Crack[1]

By M. L. WILLIAMS,[2] PASADENA, CALIF.

In an earlier paper it was suggested that a knowledge of the elastic-stress variation in the neighborhood of an angular corner of an infinite plate would perhaps be of value in analyzing the stress distribution at the base of a V-notch. As a part of a more general study, the specific case of a zero-angle notch, or crack, was carried out to supplement results obtained by other investigators. This paper includes remarks upon the antisymmetric, as well as symmetric, stress distribution, and the circumferential distribution of distortion strain-energy density. For the case of a symmetrical loading about the crack, it is shown that the energy density is not a maximum along the direction of the crack but is one third higher at an angle $\pm \cos^{-1}(1/3)$; i.e., approximately ± 70 deg. It is shown that at the base of the crack in the direction of its prolongation, the principal stresses are equal, thus tending toward a state of (two-dimensional) hydrostatic tension. As the distance from the point of the crack increases, the distortion strain energy increases, suggesting the possibility of yielding ahead of the crack as well as ± 70 deg to the sides. The maximum principal tension stress occurs on ± 60 deg rays. For the antisymmetrical stress distribution the distortion strain energy is a relative maximum along the crack and 60 per cent lower ± 85 deg to the sides.

MANY previous investigators have studied the elastic-stress distributions around cracks with some of the earliest contributions being from Inglis (1)[3] who studied an internal crack using elliptical bounding surfaces, Griffith (2) who set up an energy criterion for crack instability, Westergaard who initially treated the crack problem in the same way to be exploited in this paper (3), as well as by the complex variable technique in a later paper (4). The stress distribution at the base of cracks also has been examined photoelastically with the first attempt to measure isochromatic-fringe patterns in this specific application apparently being accomplished by Hollister (5). Recently Post (6) has published some interesting results of his photoelastic observations for the case of an edge crack.

It is the purpose of this paper to supplement the results of these investigators in certain respects which it is hoped will aid in the further understanding of the elastic-stress distribution at the base of a stationary crack.

---
[1] This investigation was sponsored in part by the National Advisory Committee for Aeronautics, under Contract NAw-6431.
[2] Associate Professor, California Institute of Technology.
[3] Numbers in parentheses refer to the Bibliography at the end of the paper.

Contributed by the Applied Mechanics Division for presentation at the Annual Meeting, New York, N. Y., November 25-30, 1956, of THE AMERICAN SOCIETY OF MECHANICAL ENGINEERS.

Discussion of this paper should be addressed to the Secretary, ASME, 29 West 39th Street, New York, N. Y., and will be accepted until one month after final publication of the paper itself in the JOURNAL OF APPLIED MECHANICS.

NOTE: Statements and opinions advanced in papers are to be understood as individual expressions of their authors and not those of the Society. Manuscript received by ASME Applied Mechanics Division, March 21, 1956. Paper No. 56—A–16.

In a previous investigation (7) the plane-stress distribution near the vertex of an infinite sector of included angle, α, was considered for various combinations of boundary conditions. For the particular purpose of this paper it is desired to consider the case where the two radial edges of the plate are unloaded and the included angle approaches 2π. It has been shown that stress functions, i.e., solutions of $\nabla^4 \chi(r, \theta) = 0$, of the form

$$\chi(r, \theta, \lambda) \equiv r^{\lambda+1} F(\theta; \lambda)$$
$$= r^{\lambda+1}[c_1 \sin(\lambda+1)\theta + c_2 \cos(\lambda+1)\theta$$
$$+ c_3 \sin(\lambda-1)\theta + c_4 \cos(\lambda-1)\theta] \quad [1]$$

will satisfy the conditions of stress-free edges along $\theta = 0$ and $\theta = \alpha$ if the λ are chosen as the positive roots of

$$\sin(\lambda \alpha) = \pm \lambda \sin \alpha \quad [2]$$

For the case where $\alpha = 2\pi$, corresponding to the case of a crack with flank angle $\omega = \alpha - 2\pi = 0$, the eigen equation takes the particularly simple form $\sin(2\pi\lambda) = 0$, thus requiring $\lambda = n/2$, $n = 1, 2, 3, \ldots$ and yielding the stress function

$$\chi(r, \theta; n/2) = r^{(n/2)+1}$$
$$\left[c_1 \sin\left(\frac{n}{2}+1\right)\theta + c_2 \cos\left(\frac{n}{2}+1\right)\theta\right.$$
$$\left. + c_3 \sin\left(\frac{n}{2}-1\right)\theta + c_4 \cos\left(\frac{n}{2}-1\right)\theta\right] \quad [3]$$

From the general definition of the stress function

$$\sigma_r = \frac{1}{r^2}\frac{\partial^2 \chi}{\partial \theta^2} + \frac{1}{r}\frac{\partial \chi}{\partial r} = r^{\lambda-1}[F''(\theta) + (\lambda+1)F(\theta)] \quad [4]$$

$$\sigma_\theta = \frac{\partial^2 \chi}{\partial r^2} = r^{\lambda-1}[\lambda(\lambda+1)F(\theta)] \quad [5]$$

$$\tau_{r\theta} = -\frac{1}{r}\frac{\partial^2 \chi}{\partial r \partial \theta} + \frac{1}{r^2}\frac{\partial \chi}{\partial \theta} = r^{\lambda-1}[-\lambda F'(\theta)] \quad [6]$$

one proceeds to require for this case that σ_θ and $\tau_{r\theta}$ vanish on $\theta = 0$ and $\theta = \alpha = 2\pi$. By reference to Equations [3–6] this implies that

$$F(0, n/2) = F'(0, n/2) = F(2\pi, n/2) = F'(2\pi, n/2) = 0$$

Ordinarily, that is, if $\alpha < 2\pi$, the four homogeneous boundary conditions permit three of the four c_i constants to be determined in terms of the fourth. In this case, however, all four boundary conditions can be satisfied by

$$\chi(r, \theta) = r^{(n/2)+1}$$
$$\left\{c_3\left[\sin\left(\frac{n}{2}-1\right)\theta - \frac{n-2}{n+2}\sin\left(\frac{n}{2}+1\right)\theta\right]\right.$$
$$\left. + c_4\left[\cos\left(\frac{n}{2}-1\right)\theta - \cos\left(\frac{n}{2}+1\right)\theta\right]\right\} \quad [7]$$

where the first term is equivalent to that used by Westergaard (3).

Turning now to a more convenient alternate form of expressing Equation [7] in terms of the bisector angle $\psi = \theta - \pi$, the stress

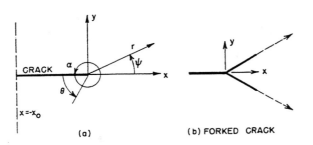

Fig. 1 Geometry

function $\chi(r, \psi)$ can be split into its even $\chi_e(r, \psi)$ and odd $\chi_0(r, \psi)$ parts with respect to ψ (Fig. 1)

$$\chi_e(r, \psi) = \sum_{n=1,2,3,\ldots} \left\{ (-1)^{n-1} a_{2n-1} r^{n+\frac{1}{2}} \right.$$
$$\left[-\cos\left(n - \frac{3}{2}\right)\psi + \frac{2n-3}{2n+1}\cos\left(n + \frac{1}{2}\right)\psi \right]$$
$$\left. + (-1)^n a_{2n} r^{n+1} \left[-\cos\left(n - \frac{2}{2}\right)\psi + \cos\left(n + \frac{2}{2}\right)\psi \right] \right\}. \quad [8]$$

$$\chi_0(r, \psi) = \sum_{n=1,2,3,\ldots} \left\{ (-1)^{n-1} b_{2n-1} r^{n+\frac{1}{2}} \right.$$
$$\left[\sin\left(n - \frac{3}{2}\right)\psi - \sin\left(n + \frac{1}{2}\right)\psi \right]$$
$$+ (-1)^n b_{2n} r^{n+1} \left[-\sin\left(n - \frac{2}{2}\right)\psi \right.$$
$$\left.\left. + \frac{2n-3}{2n+1} \sin\left(n + \frac{2}{2}\right)\psi \right] \right\} \ldots [9]$$

It should be observed that even though the field equation and the boundary conditions along the radial edges are satisfied, the constants a_i and b_i are undetermined. Their values of course depend upon the loading conditions; more specifically, either upon the boundary conditions at infinity in the case of an infinite sector, or upon those at some fixed radius when the plate has finite dimensions. For the latter practical case, which includes the problem under consideration, all the higher eigen functions in general will be present in order to determine a solution in the large.

Upon writing out the first few terms

$$\chi(r, \psi) = r^{3/2} \left[a_1 \left(-\cos\frac{\psi}{2} - \frac{1}{3}\cos\frac{3\psi}{2} \right) \right.$$
$$\left. + b_1 \left(-\sin\frac{\psi}{2} - \sin\frac{3\psi}{2} \right) \right]$$
$$+ a_2 r^2 [1 - \cos 2\psi] + 0(r^{5/2}) + \ldots \ldots [10]$$

from which the associated stresses may be found from Equations [4–6] as

$$\sigma_r(r, \psi) = \frac{1}{4r^{1/2}} \left\{ a_1 \left[-5\cos\frac{\psi}{2} + \cos\frac{3\psi}{2} \right] \right.$$
$$\left. + b_1 \left[-5\sin\frac{\psi}{2} + 3\sin\frac{3\psi}{2} \right] \right\}$$
$$+ 4a_2 \cos^2 \psi + 0(r^{1/2}) + \ldots \ldots [11]$$

$$\sigma_\psi(r, \psi) = \frac{1}{4r^{1/2}} \left\{ a_1 \left[-3\cos\frac{\psi}{2} - \cos\frac{3\psi}{2} \right] \right.$$
$$\left. + b_1 \left[-3\sin\frac{\psi}{2} - 3\sin\frac{3\psi}{2} \right] \right\}$$
$$+ 4a_2 \sin^2 \psi + 0(r^{1/2}) + \ldots \ldots [12]$$

$$\tau_{r\psi}(r, \psi) = \frac{1}{4r^{1/2}} \left\{ a_1 \left[-\sin\frac{\psi}{2} - \sin\frac{3\psi}{2} \right] \right.$$
$$\left. + b_1 \left[\cos\frac{\psi}{2} + 3\cos\frac{3\psi}{2} \right] \right\}$$
$$- 2a_2 \sin 2\psi + 0(r^{1/2}) + \ldots \ldots [13]$$

Before proceeding further, it is convenient to identify the stress state which is multiplied by the constant a_2, viz.

$$a_2 r^2 (1 - \cos 2\psi) = 2a_2 r^2 \sin^2 \psi = 2a_2 y^2$$

In the Cartesian co-ordinates analog of Equations [4–6], the stresses are

$$\sigma_x = 4a_2, \quad \sigma_y = 0, \quad \text{and } \tau_{xy} = 0$$

For most cases of interest, including the usual tension and bending specimens, σ_x along the edge $x = -x_0$ (see Fig. 1) is zero; hence for these cases $\sigma_x = 4a_2 = 0$ and thus Equations [11–13] with respect to the radial variation are all of the form $r^{-1/2} + 0(r^{1/2})$, and the local stress variations in the vicinity of the base of the crack, $r \to 0$, are proportional, up to vanishing terms in r, to the contribution of the first term. If the boundary $x = -x_0$ were loaded, however, say by a uniform pressure, this contribution would have to be superimposed.

It is convenient at this point to compute two quantities useful in photoelastic analysis; specifically, the sum of the normal stresses which are proportional to the isopachic lines, and the difference of the principal stresses which are proportional to the isochromatic lines.

Thus

$$\sigma_r + \sigma_\psi = \frac{1}{4r^{1/2}} \left\{ a_1 \left[-8\cos\frac{\psi}{2} \right] \right.$$
$$\left. + b_1 \left[-8\sin\frac{\psi}{2} \right] \right\} + \ldots = \sigma_1 + \sigma_2 \ldots [14]$$

$$\sigma_1 - \sigma_2 = [(\sigma_r - \sigma_\psi)^2 + 4\tau_{r\psi}^2]^{1/2}$$
$$= \frac{\pm 1}{4r^{1/2}} \left\{ 16 a_1^2 \sin^2 \psi + 64 b_1^2 \left(1 - \frac{3}{4} \sin^2 \psi \right) \right\}^{1/2} + \ldots$$
$$\ldots [15]$$

The direction of principal stress is found from the condition that the shear stress, τ_{ns}, on a plane whose normal is inclined at an angle α to the radius vector vanishes

$$\tau_{ns} = \tau_{r\psi} \cos 2\alpha - (1/2)(\sigma_r - \sigma_\psi) \sin 2\alpha = 0$$

$$0 = -r^{-1/2} \left\{ \left[\sin\frac{\psi}{2} \cos^2\frac{\psi}{2} \cos 2\alpha \right. \right.$$
$$\left. - \cos\frac{\psi}{2} \sin^2\frac{\psi}{2} \sin 2\alpha \right] a_1 + \left[\cos\frac{\psi}{2} \left(2 - 3\cos^2\frac{\psi}{2} \right) \cos 2\alpha \right.$$
$$\left.\left. + \sin\frac{\psi}{2} \left(2 - 3\sin^2\frac{\psi}{2} \right) \sin 2\alpha \right] b_1 \right\} + \ldots \ldots [16]$$

The angle, β, of a stress trajectory with the x-axis, Fig. 1, is then $\beta = \psi + \alpha$. Also the total strain-energy density

$$W = \frac{1}{2E}[\sigma_r{}^2 + \sigma_\psi{}^2 - 2\nu\sigma_r\sigma_\psi + 2(1+\nu)\tau_{r\psi}{}^2] \ldots [17]$$

and the strain energy due to distortion alone, i.e., the total strain energy less that due to change in volume

$$W_d = \frac{1}{12G}(\sigma_1 - \sigma_2)^2 + (\sigma_2 - \sigma_3)^2 + (\sigma_3 - \sigma_1)^2]$$

$$= \frac{1}{12G}(3\tau_{\text{oct}})^2 \ldots [18]$$

where the expression is given in terms of principal stresses ($\sigma_3 = 0$ in this case), and the definition of the octahedral shearing stress has been used, may be written, respectively, as

$$32ErW = a_1{}^2 \left[(34 - 30\nu)\cos^2 \frac{\psi}{2} + 2(1+\nu)\sin^2 \frac{\psi}{2} \right.$$
$$\left. + 2(1+\nu) - 4(1+\nu)\cos 2\psi \right] + b_1{}^2 \left[(34 - 30\nu) \sin^2 \frac{\psi}{2} \right.$$
$$\left. + 2(1+\nu)\cos^2 \frac{\psi}{2} + 18(1+\nu) + 12(1+\nu)\cos 2\psi \right]$$
$$+ 2a_1b_1 \left[(32 - 22\nu) \sin \frac{\psi}{2} \cos \frac{\psi}{2} \right.$$
$$\left. - 8(1+\nu)\sin 2\psi \right] + \ldots \ldots [19]$$

$$6GrW_d = a_1{}^2 \left(1 + 3\sin^2 \frac{\psi}{2}\right) \cos^2 \frac{\psi}{2}$$
$$+ 2b_1{}^2 \left[3 + \left(1 + 3\cos \frac{\psi}{2}\right)\left(1 - 3\cos \frac{\psi}{2}\right)\sin^2 \frac{\psi}{2} \right]$$
$$+ 2a_1b_1 \sin \psi + \ldots \ldots [20]$$

In the foregoing expressions it is seen that the term multiplied by the coefficient a_1b_1 represents a coupling between the symmetric and antisymmetric variations with respect to ψ.

Finally, the displacements (see reference 7) are found to be

$$2\mu U_r(r, \psi) = r^{1/2} \left\{ a_1 \left[\left(-\frac{5}{2} + 4\sigma\right)\cos \frac{\psi}{2} + \frac{1}{2} \cos \frac{3\psi}{2} \right] \right.$$
$$\left. + b_1 \left[\left(-\frac{5}{2} + 4\sigma\right)\sin \frac{\psi}{2} + \frac{3}{2}\sin \frac{3\psi}{2} \right] \right\} + \ldots \ldots [21]$$

$$2\mu U_\psi(r, \psi) = r^{1/2} \left\{ a_1 \left[\left(\frac{7}{2} - 4\sigma\right)\sin \frac{\psi}{2} - \frac{1}{2} \sin \frac{3\psi}{2} \right] \right.$$
$$\left. + b_1 \left[-\left(\frac{7}{2} - 4\sigma\right)\cos \frac{\psi}{2} + \frac{3}{2} \cos \frac{3\psi}{2} \right] \right\} + \ldots \ldots [22]$$

where

$$\sigma \equiv \nu/(1+\nu)$$

Symmetrical Stress Distribution

For the sake of simplicity it is convenient to analyze the two types of solutions separately; the symmetric solutions, i.e., $b_i = 0$, are perhaps the more common, occurring, for example, in the case of an edge crack in a thin plate subjected to bending or tension.

Specializing the stresses and the other quantities to this case gives

$$\sigma_r(r, \psi) = \frac{1}{4r^{1/2}} \left[-5 \cos \frac{\psi}{2} + \cos \frac{3\psi}{2} \right] a_1 + \ldots \ldots [23]$$

$$\sigma_\psi(r, \psi) = \frac{1}{4r^{1/2}} \left[-3 \cos \frac{\psi}{2} - \cos \frac{3\psi}{2} \right] a_1 + \ldots \ldots [24]$$

$$\tau_{r\psi}(r, \psi) = \frac{1}{4r^{1/2}} \left[-\sin \frac{\psi}{2} - \sin \frac{3\psi}{2} \right] a_1 + \ldots \ldots [25]$$

$$2\mu U_r(r, \psi) = r^{1/2} \left[\left(-\frac{5}{2} + 4\sigma\right) \cos \frac{\psi}{2} \right.$$
$$\left. + \frac{1}{2} \cos \frac{3\psi}{2} \right] a_1 + \ldots \ldots [26]$$

$$2\mu U_\psi(r, \psi) = r^{1/2} \left[\left(\frac{7}{2} - 4\sigma\right) \sin \frac{\psi}{2} \right.$$
$$\left. - \frac{1}{2} \sin \frac{3\psi}{2} \right] a_1 + \ldots \ldots [27]$$

$$\sigma_1 + \sigma_2 = \frac{1}{4r^{1/2}} \left[-8 \cos \frac{\psi}{2} \right] a_1 + \ldots \ldots [28]$$

$$\sigma_1 - \sigma_2 = \frac{1}{4r^{1/2}} [4a_1 \sin \psi] + \ldots \ldots [29]$$

$$\beta_{n-1} = (3\psi/4) \pm (\pi/4); \text{ also isotropic locus at } \psi = 0 \ldots [30]$$

$$W = \frac{a_1{}^2}{Er} \cos^2 \frac{\psi}{2} \left[1 - \nu + (1+\nu)\sin^2 \frac{\psi}{2} \right] + \ldots \ldots [31]$$

$$W_d = \frac{a_1{}^2}{Er} \cos^2 \frac{\psi}{2} \left[\frac{1+\nu}{3} + (1+\nu)\sin^2 \frac{\psi}{2} \right] + \ldots \ldots [32]$$

It is observed that for this elastic behavior in an isotropic homogeneous material, the stress system represented by the first term possesses the characteristic square-root singularity found by Inglis (1), Westergaard (4), and others; also, the shapes of the isopachics and isochromatic fringes are in agreement with photoelastic data obtained by Post (6) and corroborated independently at GALCIT. In addition, however, there are certain other interesting features which, because of the relative simplicity of the previous expressions, become readily apparent.

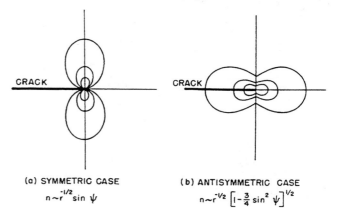

Fig. 2 Isochromatic-Fringe Patterns

The first of these relates to the observation that the shear stress is zero along the line of propagation of the crack ($\psi = 0$). The σ_r and σ_ψ stresses must therefore be principal stresses as expected, but moreover from Equation [29] the stresses are equal

$$\sigma_1(0) \to \sigma_2(0) = (-a_1)r^{-1/2} + \ldots$$

In other words, at the base of the crack there exists a strong tend-

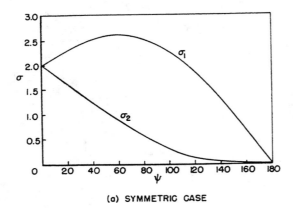

(a) SYMMETRIC CASE

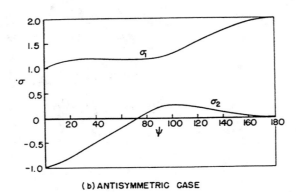

(b) ANTISYMMETRIC CASE

Fig. 3 Principal-Stress Variations

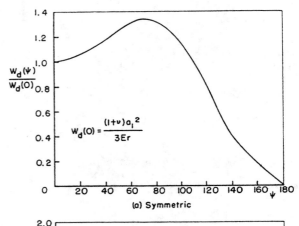

(a) Symmetric

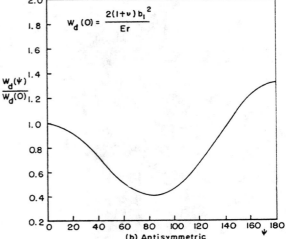

(b) Antisymmetric

Fig. 4 Strain-Energy-Density Distributions

ency toward a state of (two-dimensional) hydrostatic tension which consequently may permit the elastic analysis to apply closer to the point of the crack that was hitherto supposed, notwithstanding the square-root stress singularity.

In order to investigate this point a bit more fully, particularly for the higher terms in r, i.e., further away from the crack point, more terms can be considered in Equations [23] and [24] to obtain, for $\psi = 0$

$$\sigma_r(r, 0) = \sum_{n = 1, 2, 3 \ldots} (-1)^n (2n - 1) a_{2n-1} r^{n - \frac{3}{2}} \quad \ldots [33]$$

$$\sigma_\psi(r, 0) = \sum_{n = 1, 2, 3 \ldots} \left\{ (-1)^n (2n - 1) a_{2n-1} r^{n - \frac{3}{2}} \right.$$
$$\left. + (-1)^{n+1} 2 a_{2n} (2n - 1) r^{n-1} \right\} \ldots [34]$$

from which it may be concluded that, if the constant loading term $a_2 = 0$, the principal stresses are equal up to the linear term in r. From the previous relation the difference between the stresses can then be written

$$\sigma_r - \sigma_\psi = \sum_{n = 1, 2, 3 \ldots} (-1)^n 2 a_{2n} (2n - 1) r^{n-1} \ldots [35]$$

Therefore the principal stresses will become progressively unequal and the tendency toward hydrostatic tension reduced as the distance from the crack increases. As a matter of interest the stress trajectories for the lowest eigen solution, $n = 1$, are shown in Fig. 5 and they are observed to be of the interlocking type which explains the apparent contraction that $\beta = \pm \pi/4$ for $\psi = 0$ from Equation [30]. The equality of the principal stresses leads to a locus of isotropic points.

Another interesting characteristic of the solution is shown in Fig. 4 which shows the variation of distortion strain energy as a function of angle for a fixed radius. Because of the aforementioned hydrostatic tendency, the maximum energy of distortion does not occur along the line of crack direction, but rather at

$$\psi^* = \pm \cos^{-1}(1/3) \approx \pm 70 \text{ deg}$$

where it is one third higher.

The previous results also give the principal stress as

$$\sigma_{1,2} = -\frac{a_1}{r^{1/2}} \cos \frac{\psi}{2} \left[1 \mp \sin \frac{\psi}{2} \right]$$

for which the maximum value of $(3\sqrt{3}/4)(-a_1) r^{-1/2}$ occurs at $\overline{\psi} = \cos^{-1}(1/2) = \pi/3$ at which angle the stress trajectories are parallel to the x, y-co-ordinates; i.e., $\beta = 0, \pi/2$. The maximum shear stress $\tau_{\max} = 1/2 (\sigma_1 - \sigma_2)$ oriented at $\beta = \pi/8, 5\pi/8, \psi = \pi/2$ and equals exactly one half the hydrostatic tension value to which the normal stresses are subjected along $\psi = 0$ at the same radial distance. Some of the foregoing properties are summarized in Fig. 6.

Finally, the radial and circumferential displacements on the free edges from Equations [23] and [24] are found to be zero and $\pm 4(1 + \nu)^{-1} a_1 r^{1/2}$, thus leading to the expected fact that the two faces will have closed together and overlapped under a compression loading, an impossible situation by itself requiring $a_1 = 0$.

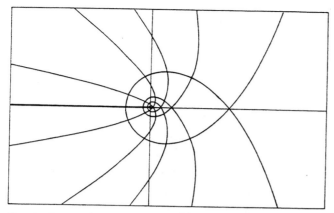

FIG. 5 STRESS TRAJECTORIES FOR LOWEST EVEN EIGENFUNCTION $\beta = (3\psi/4) \pm (\pi/4)$

In this case one is led to the contact problem treated by Westergaard (4) wherein there is no stress singularity at the point of the crack.

ANTISYMMETRICAL STRESS DISTRIBUTION

When the stress distribution is antisymmetric, which is a case that does not seem previously to have been treated explicitly, such as may exist about a crack parallel to the neutral axis of a beam in pure bending, there results

$$\sigma_r(r, \psi) = \frac{1}{4r^{1/2}} \left[-5 \sin^2 \frac{\psi}{2} + 3 \sin \frac{3\psi}{2} \right] b_1 + \ldots \quad [36]$$

$$\sigma_\psi(r, \psi) = \frac{1}{4r^{1/2}} \left[-3 \sin \frac{\psi}{2} - 3 \sin \frac{3\psi}{2} \right] b_1 + \ldots \quad [37]$$

$$\tau_{r\psi}(r, \psi) = \frac{1}{4r^{1/2}} \left[\cos \frac{\psi}{2} + 3 \cos \frac{3\psi}{2} \right] b_1 + \ldots \quad [38]$$

$$2\mu U_r(r, \psi) = r^{1/2} \left[(-3 + 4\sigma) \sin \frac{\psi}{2} \right.$$
$$\left. + \sin \frac{3\psi}{2} \right] b_1 + \ldots \quad [39]$$

$$2\mu V_\psi(r, \psi) = r^{1/2} \left[-\left(\frac{7}{2} - 4\sigma\right) \cos \frac{\psi}{2} \right.$$
$$\left. + \frac{3}{2} \cos \frac{3\psi}{2} \right] b_1 + \ldots \quad [40]$$

$$\sigma_1 + \sigma_2 = \frac{1}{4r^{1/2}} \left[-8 \sin \frac{\psi}{2} \right] b_1 + \ldots \quad [41]$$

$$\sigma_1 - \sigma_2 = \frac{1}{4r^{1/2}} \left[8 \left(1 - \frac{3}{4} \sin^2 \psi \right)^{1/2} \right] b_1 + \ldots \quad [42]$$

$$W = \frac{b_1^2}{Er} \left[(1 - \nu) \sin^2 \frac{\psi}{2} + (1 + \nu) \right.$$
$$\left. \left(1 - \frac{3}{4} \sin^2 \psi \right) \right] + \ldots \quad [43]$$

$$W_d = \frac{b_1^2}{Er} \left[\frac{1 + \nu}{3} \sin^2 \frac{\psi}{2} + (1 + \nu) \right.$$
$$\left. \left(1 - \frac{3}{4} \sin^2 \psi \right) \right] + \ldots \quad [44]$$

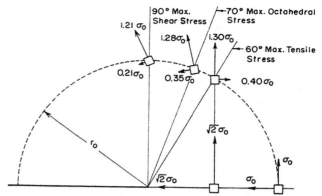

FIG. 6 CRITICAL PRINCIPAL STRESSES FOR THE LOWEST EVEN EIGENFUNCTION $\sigma_0 = -a_1 r_0^{-1/2}$

Again in this case the singular stress system varies as $r^{-1/2}$; the distortion-strain-energy-density variation is shown in Fig. 4. It is interesting to observe that in this case the energy drops on either side from a relative maximum in the direction of the crack to a minimum of 40 per cent of its $\psi = 0$ value at $\psi^* = \cos^{-1}(1/9)$.

It would be interesting to experiment photoelastically with the antisymmetric-loading condition, which is not too simple a matter, in an analysis similar to that of Post for the symmetrical case.

CONCLUSION

While it is not the intent to extend the range of validity of the elastic analysis by delving into the complicated phenomenon of failure or fracture mechanics, it does seem pertinent to remark upon some possible implications of the results obtained from elasticity theory.

First of all it seems that even in the presence of a partial (two-dimensional) hydrostatic-stress field there would be a reduced tendency for yielding at the base of the crack. Then, as the magnitude of the individual stresses increases as the inverse half power of the radius, the stress should become quite high with a tendency toward a cleavage failure. At the same time the elastic analysis indicates there should be a large amount of distortion off to the sides of the crack and presumably some yielding should take place in these areas which would tend toward a ductile failure. The character of failure actually occurring in a given specimen would depend upon the material. In this connection it is pertinent to mention some recent experimental evidence of Forsyth and Stubbington (8) called to the author's attention by S. R. Valluri, which tends to substantiate the remarks concerning an area of yielding off the crack direction, presumably ±70 deg for an exactly symmetrical loading.

In addition, because the stress becomes nonhydrostatic as the distance from the point of the crack increases, by Equation [35], there may be reason therefore to suspect a yielded region ahead of the crack also, although not as severe.

As a second remark, concerning the direction of cracking or forking, it is recognized that slow-moving cracks generally propagate more or less straight.[4] Because some cracks do fork,

---

[4] In an interesting piece of work Yoffe (9) has discussed the change of direction of a running crack due to its velocity and has found that if the material is such that a crack propagates in a direction normal to the maximum tensile stress (which, incidentally, is not σ_ψ as she apparently has assumed; indeed aside from the principal stresses, $\sigma_r \geq \sigma_\psi$), there is a critical velocity of about 0.6 times the velocity of shear waves in the material above which the crack tends to become curved. It may be of value to supplement her work by testing her hypothesis with respect to the maximum principal tensile stress; it is suspected, however, that the results will be qualitatively the same because, as she has remarked and as is shown in Fig. 6, the stress field is relatively uniform over a wide area in front of the crack.

however, it may be worth noting that any arbitrary crack will be subjected simultaneously to both symmetric and antisymmetric loading. In this connection it is observed that for the antisymmetric contributions the relative maximum and minimum energies, for example, tend to negate the effects which occur for a symmetrical loading. One might therefore conclude that for randomly oriented cracks there would be no preferred direction that the crack might take upon moving, with macroscopic structure and rate-of-energy release being controlling factors. On the other hand, it may be possible to relate the maximum principal tensile stresses occurring at ±60 deg, in the neighborhood of the maximum distortion, to the angle crack-forking phenomenon shown, for example, by Post (6) or in Fig. 1(b), although again for metals as opposed to plastics the effect of crystal orientation and slip planes probably would be quite strong.

BIBLIOGRAPHY

1 "Stresses in a Plate Due to the Presence of Cracks and Sharp Corners," by C. E. Inglis, Transactions of the Institution of Naval Architects, London, England, vol. 60, 1913, p. 219.

2 "The Phenomena of Rupture and Flow in Solids," by A. A. Griffith, Philosophical Transactions of the Royal Society of London, England, vol. 221, 1921, pp. 163–198.

3 "Stresses at a Crack, Size of the Crack and the Bending of Reinforced Concrete," by H. M. Westergaard, Proceedings of the American Concrete Institute, vol. 30, 1934, pp. 93–102.

4 "Bearing Pressures and Cracks," by H. M. Westergaard, JOURNAL OF APPLIED MECHANICS, Trans. ASME, vol. 61, 1939, pp. A-49-53.

5 "Experimental Study of Stresses at a Crack in a Compression Member," by S. C. Hollister, Proceedings of the American Concrete Institute, vol. 30, 1934, pp. 361–365.

6 "Photoelastic Stress Analysis for an Edge Crack in a Tensile Field," by D. Post, Proceedings of the Society for Experimental Stress Analysis, vol. 12, no. 1, 1954, p. 99.

7 "Stress Singularities Resulting From Various Boundary Conditions in Angular Corners of Plates in Extension," by M. L. Williams, JOURNAL OF APPLIED MECHANICS, Trans. ASME, vol. 74, 1952, p. 526.

8 "The Slip Band Extrusion Effect Observed in Some Aluminum Alloys Subjected to Cyclic Stresses," by P. J. E. Forsyth and C. A. Stubbington, Royal Aircraft Establishment, MET Report 78, January, 1954.

9 "The Moving Griffith Crack," by E. H. Yoffe, *Philosophical Magazine*, vol. 42, part 2, 1951, pp. 739–750.

Analysis of Stresses and Strains Near the End of a Crack Traversing a Plate

By G. R. IRWIN,[1] WASHINGTON, D. C.

A substantial fraction of the mysteries associated with crack extension might be eliminated if the description of fracture experiments could include some reasonable estimate of the stress conditions near the leading edge of a crack particularly at points of onset of rapid fracture and at points of fracture arrest. It is pointed out that for somewhat brittle tensile fractures in situations such that a generalized plane-stress or a plane-strain analysis is appropriate, the influence of the test configuration, loads, and crack length upon the stresses near an end of the crack may be expressed in terms of two parameters. One of these is an adjustable uniform stress parallel to the direction of a crack extension. It is shown that the other parameter, called the stress-intensity factor, is proportional to the square root of the force tending to cause crack extension. Both factors have a clear interpretation and field of usefulness in investigations of brittle-fracture mechanics.

INTRODUCTION

DURING and subsequent to the recent World War, investigations of fracturing have shared in the general growth of applied-mechanics research. Among the fracture failures responsible for interest in this field were those of welded ships, gas-transmission lines, large oil-storage tanks, and pressurized cabin planes. The propagation of a brittle crack across one or more plates in which the average tensile stress was thought to be safely below the yield strength is a prominent feature of these examples.

As a result of these investigations there was a revival of interest in the Griffith theory of fracture strength (1).[2] It was pointed out independently by Orowan (2) and by the author (3) that a modified Griffith theory is helpful in understanding the development of a rapid fracture which is sustained with energy from the surrounding stress field. Expositions of this idea have been given (3, 4, 5) using such terms as fracture work rate and strain-energy release rate.

The basic idea of the modified Griffith theory is that, at onset of unstable fast fracturing, one can equate the fracture work per unit crack extension to the rate of disappearance of strain energy from the surrounding elastically strained material. The term, disappearance of strain energy, refers to the loss of strain energy which would occur if the system were isolated from receiving energy, for example, from movement of the forces applying tension to the material. For convenience this is referred to here as the fixed-grip strain-energy release rate. Since the strain-energy disappearance rate at any moment depends on the load magnitudes rather than on movement of the points of load application, use of the fixed-grip strain-energy release-rate concept is not limited to fixed-grip experiments.

It is the purpose of this paper to describe the relation of these terms to the elastic stresses and strains near the leading edge of a somewhat brittle crack. For purposes of this paper "somewhat brittle" means that a region of large plastic deformations may exist close to the crack but does not extend away from the crack by more than a small fraction of the crack length.

Previous investigations (3–7) have established a viewpoint with respect to the mechanics of fracturing which may be summarized in part as follows:

The fixed-grip strain-energy release rate has the same role as an influence controlling time rate of crack extension as the longitudinal load has in controlling time rate of plastic extension of a tensile bar. In the latter case the force per unit area tending to cause plastic extension is the longitudinal stress. In the former case a motivating force per unit thickness can be defined quite generally in terms of the rate of conversion of strain energy to thermal energy during crack extension. This generalized force is the rate of decrease of the fixed-grip strain energy with crack extension on a unit-thickness basis. Also this energy rate can be regarded as composed of two terms: (a) The strain-energy loss rate associated with nonrecoverable displacements of the points of load application (assumed zero in this discussion); and (b) the strain-energy loss rate associated with extension of the fracture accompanied only by plastic strains local to the crack surfaces. The second of these two terms, herein called $\mathcal{G}$, appears to be the force component most directly related to crack extension and the one with the most practical usefulness.

Determination of characteristic values of $\mathcal{G}$ for onset or arrest of rapid fracturing and the applications of such measurements to "fail-safe" design procedures have been discussed elsewhere (4, 5, 8, 9). It will be shown here that the tensile stresses near the crack tip and normal to the plane of the crack are determined by the force tendency $\mathcal{G}$. The discussion is arranged so as to develop relationships useful in the analysis of fracture experiments whether the purpose of the work is to determine characteristic $\mathcal{G}$-values or simply to determine the stress field near the leading edge of the crack.

The material of this paper is, at one point, related to Sneddon's analysis of stresses near an embedded crack having the shape of a flat circular disk (10). Otherwise, for simplicity and bearing in mind the service fracture failures referred to in the foregoing, discussion is restricted to a straight crack in a plate. It is assumed the plate thickness is small enough compared to the crack length so that generalized plane stress constitutes a useful two-dimensional viewpoint. In addition it is assumed the crack is moving, as brittle cracks generally do move, along a path normal to the direction of greatest tension, so that the component of shear stress resolved on the line of expected extension of the crack is zero.

---

[1] Superintendent, Mechanics Division, U. S. Naval Research Laboratory.

[2] Numbers in parentheses refer to the Bibliography at the end of the paper.

Presented at the Applied Mechanics Division Summer Conference, Berkeley, Calif., June 13–15, 1957, of THE AMERICAN SOCIETY OF MECHANICAL ENGINEERS.

Discussion of this paper should be addressed to the Secretary, ASME, 29 West 39th Street, New York, N. Y., and will be accepted until October 10, 1957, for publication at a later date. Discussion received after the closing date will be returned.

NOTE: Statements and opinions advanced in papers are to be understood as individual expressions of their authors and not those of the Society. Manuscript received by ASME Applied Mechanics Division, February 19, 1956. Paper No. 57—APM-22.

Representative Stress Fields Associated with Cracks

A paper by Westergaard (11) gave a convenient semi-inverse method for solving a certain class of plane-strain or plane-stress problems. Let $\overline{\overline{Z}}$, $\overline{Z}$, Z, and Z' represent successive derivatives with respect to z of a function $\overline{\overline{Z}}(z)$, where z is $(x + iy)$. Assume that the Airy stress function may be represented by

$$F = \operatorname{Re}\overline{\overline{Z}} + y \operatorname{Im}\overline{Z} \quad \ldots \ldots \ldots \ldots [1]$$

then

$$\sigma_x = \frac{\partial^2 F}{\partial y^2} = \operatorname{Re} Z - y \operatorname{Im} Z' \quad \ldots \ldots \ldots \ldots [2]$$

$$\sigma_y = \frac{\partial^2 F}{\partial x^2} = \operatorname{Re} Z + y \operatorname{Im} Z' \quad \ldots \ldots \ldots \ldots [3]$$

$$\tau_{xy} = -\frac{\partial^2 F}{\partial x \partial y} = -y \operatorname{Re} Z' \quad \ldots \ldots \ldots \ldots [4]$$

By choices of the function $Z(z)$, Westergaard showed solutions for stress distribution as influenced by bearing pressures or cracks in a variety of situations. The class of problems which can be solved in this way is limited to those such that τ_{xy} is zero along the x-axis.

In particular, if a large plate contains a single crack on the x-axis whose length is small compared to the plate dimensions or a colinear series of such cracks, and if the applied loads are such that τ_{xy} is zero along the x-axis, then the stress distribution is readily constructed with the aid of Westergaard's semi-inverse procedures.

Two examples of such problems were given by Westergaard (11) as follows:

1 A central straight crack of length $2a$ along the x-axis in an infinite plate with a biaxial field of tension σ at large distances from the crack

$$Z(z) = \frac{\sigma}{[1 - (a/z)^2]^{1/2}} \quad \ldots \ldots \ldots \ldots [5]$$

2 A series of equally spaced straight cracks of length $2a$, on the x-axis in an infinite plate with biaxial stress σ, as before, and with the distance between the crack centers, l

$$Z(z) = \frac{\sigma}{\left[1 - \left(\dfrac{\sin \pi a/l}{\sin \pi z/l}\right)^2\right]^{1/2}} \quad \ldots \ldots \ldots \ldots [6]$$

Three additional examples obtainable with the semi-inverse procedure suggested by Westergaard are as follows:

3 Single crack along the x-axis extending from $-a$ to a with a wedge action applied to produce a pair of "splitting forces" of magnitude P located at $x = b$ (see Fig. 1)

$$Z(z) = \frac{Pa}{\pi(z - b)z} \left[\frac{1 - (b/a)^2}{1 - (a/z)^2}\right]^{1/2} \quad \ldots \ldots \ldots \ldots [7]$$

4 The situation of example 3 with an additional pair of forces of magnitude P at $x = -b$

$$Z(z) = \frac{2Pa}{\pi(z^2 - b^2)} \left[\frac{1 - (b/a)^2}{1 - (a/z)^2}\right]^{1/2} \quad \ldots \ldots \ldots \ldots [8]$$

5 Example 3 repeated along the x-axis at intervals l, and with the wedge action centered so that b is zero

$$Z(z) = \frac{P \sin \dfrac{\pi a}{l}}{l \left(\sin \dfrac{\pi z}{l}\right)^2} \left[1 - \left(\frac{\sin \dfrac{\pi a}{l}}{\sin \dfrac{\pi z}{l}}\right)^2\right]^{-1/2} \quad \ldots \ldots \ldots \ldots [9]$$

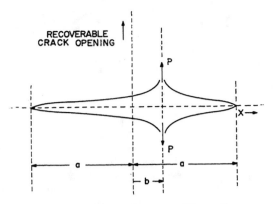

Fig. 1 Opening of a Crack by Wedge Forces

In all of these problems a uniform compression, $-\sigma_{ox}$, may be added to the value of σ_x given by Equation [2]. Since linearized elasticity relations are assumed to apply, one may obtain the Z-function for combined tension and wedge action by adding the appropriate Z-function for tension to the appropriate Z-function for a pair of wedge forces.

As an extending crack moves across a plate of finite width the crack may attain sufficient length so that the tensile forces acting to cause crack extension are not sufficiently accurate when obtained using infinite plate relations such as those of Equations [5] and [7]. The major adjustment required is that of the total load across the x-axis from the end of the crack to the side of the plate. A convenient way to make this adjustment, if the crack is centered, is to use expressions for Z such as in examples 2 and 5. The side boundaries of the plate would then be taken to occur at $x = -l/2$ and $x = +l/2$. In the stress distributions resulting from examples 2 and 5 the shearing stress τ_{xy} is zero along the side boundaries. However, the side boundaries are represented as possessing a distribution of x-direction loads which should be absent. Depending upon the objectives of the stress analysis this defect may be outweighed in importance by the convenience of having an approximate solution of the problem in compact form.

Suppose, next, that the situation to be studied is a crack extending across a finite-width plate from one of the plate side boundaries. Let the intersection of the crack with the side boundary be the origin of co-ordinates and let the line of crack extension be the positive portion of the x-axis, the end of the crack being at $x = a$. It will be assumed that weights or blocks have been set against the side boundaries so as to prevent or greatly reduce the tendency of these boundaries to move in the negative x-direction as the crack extends. In this event Z-functions similar to those of examples 2 and 5 may again be employed as a convenient means for obtaining a compact approximation to the stress distribution. In this situation the side boundaries of the plate would be assumed to be at $x = 0$ and at $x = l/2$.

In any of the foregoing examples the only stress acting at the edges of the crack is the optional added stress in the x-direction $-\sigma_{ox}$. An uncertainty as to proper choice of σ_{ox} exists for the example discussed previously of a crack extending from one side of a finite-width plate. In addition, if the crack moves rapidly, determination of the stress distribution away from the crack will require a dynamic-stress analysis.

Stress Environment of the End of the Crack

However, the stress distribution near the end of the crack can be expressed (a) independently of uncertainties of both magnitude of applied loads and of the dynamic unloading influences, and (b) in such a way that records from several strain gages placed near the

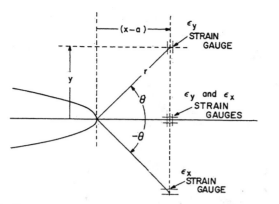

FIG. 2 RELATION OF r AND θ TO y AND $(x - a)$ AND EXAMPLES OF LOCATIONS FOR STRAIN GAGES

end of the crack serve to determine the "crack-tip stress distribution."

Consider for all of the five examples the substitution of variables

$$z = a + re^{i\theta}$$

where

$$r^2 = (x - a)^2 + y^2 \text{ and } \tan \theta = y/(x - a)$$

as shown in Fig. 2.

If one assumes quantities such as r/a and $r/(a - b)$ may be neglected in comparison to unity, one finds in each case

$$\sigma_y = \left(\frac{E\mathcal{G}}{\pi}\right)^{1/2} \frac{\cos \theta/2}{\sqrt{(2r)}} \left(1 + \sin \frac{\theta}{2} \sin \frac{3\theta}{2}\right) \ldots \ldots [10]$$

and

$$\sigma_x = \left(\frac{E\mathcal{G}}{\pi}\right)^{1/2} \frac{\cos \frac{\theta}{2}}{\sqrt{(2r)}} \left(1 - \sin \frac{\theta}{2} \sin \frac{3\theta}{2}\right) - \sigma_{ox} \ldots [11]$$

where E is Young's modulus. $\mathcal{G}$ is independent of r and of θ and will be discussed in following sections of this paper.

For a crack traversing a plate, the thickness of which is considerably smaller than the crack length, a generalized plane-stress viewpoint is appropriate and σ_z is zero. However, for comparison with results obtained by Sneddon (10) one may consider for the moment the set of three extensional stresses which would pertain to a plane-strain analysis. Sneddon studied the stress distribution predicted by linear elastic theory in the vicinity of a "penny-shaped" crack embedded in a much larger solid material and subjected to tension perpendicular to the plane of the crack. For the extensional stresses in the close neighborhood of the crack outer boundary, Sneddon gave expressions identical to Equations [10] and [11] with regard to the functional relationship of σ_x and σ_y to r and θ. A third extensional stress directed parallel to the outer boundary of the penny-shaped crack was given by Sneddon with the remark that no counterpart to this third extensional stress existed in a two-dimensional analysis of stresses near a crack. However, the remark applies only to the two-dimensional analysis assuming generalized plane stress. For the two-dimensional analysis assuming plane strain the third extensional stress, which is Poisson's ratio times the sum of σ_x and σ_y, as in Equations [10] and [11], is the counterpart to Sneddon's third extensional stress component. Thus for any small region around the outer boundary of Sneddon's penny-shaped crack, the stresses, strains, and displacements correspond to a situation which is locally one of plane strain. The preceding comment becomes intuitively obvious when one considers that, in Sneddon's example, all particle displacements lie in planes which contain the axis of symmetry. These planes would approximate to a set of parallel planes within any region whose dimensions are very small compared to distance from the region to the axis of symmetry.

FORCE TENDING TO CAUSE CRACK EXTENSION

As the crack extends, an energy transfer from mechanical or strain energy into other forms occurs in the vicinity of the crack. The process is such that transfer of strain energy to heat dominates.

$\mathcal{G}$ is the magnitude of this energy exchange associated with unit extension of the crack and may be regarded as the force tending to cause crack extension. This may be seen as follows:

The linear elasticity relations resulting from Equations [5] through [11] correspond to a parabolic shape for the crack opening near the crack tip. In Fig. 3 the origin of x, y-co-ordinates has been shifted so that the crack opening, shown by the dashed line,

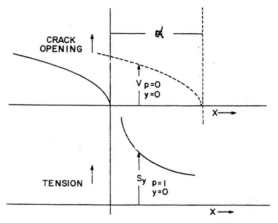

FIG. 3 LINEAR-ELASTIC-THEORY CRACK OPENINGS AND STRESSES NEAR END OF A CRACK

extends to $x = \alpha$. It is assumed α is very small compared to the length of the crack. If y-direction tensions given by

$$S_y(p) = p \left(\frac{E\mathcal{G}}{\pi}\right)^{1/2} \frac{1}{\sqrt{(2x)}} \ldots \ldots \ldots [12]$$

are exerted on the edges of the crack from $x = 0$ to $x = \alpha$, and p is increased from zero to 1, the crack is closed up so that the crack opening appears to end at the origin as shown by the full line. The factor p may be regarded as a proportional loading parameter. To the same approximation as Equation [10], the crack opening from $x = 0$ to $x = \alpha$ at any time during the closure operation is given by

$$v(p) = (1 - p) \frac{2}{E} \left(\frac{E\mathcal{G}}{\pi}\right)^{1/2} \sqrt{[2(\alpha - x)]} \ldots \ldots [13]$$

Since the degree of closure is a linear function of S_y the work done by the closing forces as p is varied from zero to 1 is given by

$$\int_0^\alpha S_y(1)v(0)dx = \frac{2\mathcal{G}}{\pi} \int_0^\alpha \left(\frac{\alpha - x}{x}\right)^{1/2} dx = \alpha\mathcal{G} \ldots [14]$$

Thus $\alpha\mathcal{G}$ is the "fixed grip" loss of energy from the strain-energy field as the crack extends by the amount α and the generalized force interpretation of $\mathcal{G}$ is apparent.

For mathematical simplicity the foregoing calculation was

based upon the linear elasticity stresses and crack-opening displacements in the immediate vicinity of the crack tip. One should not assume, however, that local stress relaxation and crack-opening distortion by plastic flow necessarily change the rate of loss of strain energy with crack extension from that indicated in the foregoing by an appreciable amount. The procedure leading to Equation [14] is equivalent to finding the derivative with respect to crack length of the total strain energy under fixed-grip conditions. The contribution to this calculation from a small circular region enclosing the crack tip is relatively small. In situations such as those of Equations [5] through [9], the fraction of $\mathcal{G}$ contributed from this region is, in fact, only about a third of the ratio of the outer radius of the region to the half length a of the crack. Thus if plastic strains near a crack affect the stress field only within distances from the crack, small in relation to the crack length, then the influence of these plastic strains on the calculation of $\mathcal{G}$ is correspondingly small.

Remarks on Measurement Methods

Consider the situation suggested earlier of a crack moving in the x-direction across a large plate. As the crack moves under and beyond a strain gage placed close to its path for ϵ_y measurement, Fig. 2, the gage output is expected to rise and then fall to a small value. An uncertainty in interpretation of the gage record in terms of stress will exist if σ_{ox} is uncertain. If it can be assumed that σ_{ox} is zero then, using Equations [10] and [11]

$$E\epsilon_y = \sigma_y - \nu\sigma_x = \left(\frac{E\mathcal{G}}{\pi}\right)^{1/2} \frac{\cos\theta/2}{\sqrt{(2r)}} \left[(1-\nu) + (1+\nu)\sin\frac{\theta}{2}\sin\frac{3\theta}{2}\right] \quad [15]$$

where ν is Poisson's ratio. By putting $r = y \csc\theta$ and differentiating with respect to θ with y constant, one finds ϵ_y should be greatest when the gage position relative to the end of the crack is at $\theta = 70$ deg. This result is quite insensitive to the assumed value of ν (unpublished calculations by L. McFadden and J. H. Hancock based upon Equation [9]).

A better situation for analysis purposes exists if both ϵ_y and ϵ_x are measured. In this event one has

$$\sigma_y = \frac{E}{1-\nu^2}(\epsilon_y + \nu\epsilon_x) \quad\quad [16]$$

Differentiating Expression [10] for σ_y with respect to θ holding y constant, one finds the maximum will occur when the measurement position is at 73.4 deg. As a crack moves under and beyond the position of measurement of ϵ_y and ϵ_x the quantity $(\epsilon_y + \nu\epsilon_x)$ plotted against time should have a maximum at that angle. Thus with θ, r, and $(\epsilon_y + \nu\epsilon_x)$ known for a particular location of the crack, the stress-intensity factor $(E\mathcal{G}/\pi)^{1/2}$, and the crack extension force $\mathcal{G}$ existing at the moment of that crack location can be calculated.

The stresses near the crack tip predicted by linear elasticity theory are calculable from Equations [10] and [11] except for knowledge of the intensity factor $(E\mathcal{G}/\pi)^{1/2}$, which appears in the expressions both for σ_x and for σ_y and the additive uniform stress factor $-\sigma_{ox}$, which appears in the expression for σ_x. Any arrangement of strain gages which permits determination of these two factors serves to determine the crack-tip stress distribution. The region in which the stresses are thus represented is an annular region which excludes any large distortions close to the crack but which extends outward only a small fraction of the crack length.

Conclusions

The stress field near the end of a somewhat brittle tensile fracture, in situations of generalized plane stress or of plane strain, can be approximated by a two-parameter set of equations. The most significant of these parameters, the intensity factor, is $(E\mathcal{G}/\pi)^{1/2}$ for plane stress where $\mathcal{G}$ is the force tending to cause crack extension.[3] When the experimental situation permits use of strain gages at distances from the crack tip, small compared to the crack length, values of $\mathcal{G}$ and σ_{ox} may be evaluated conveniently by measuring local strain at selected positions.

Bibliography

1 "The Phenomenon of Rupture and Flow in Solids," by A. A. Griffith, Philosophical Transactions of the Royal Society of London, series A, vol. 221, 1920, pp. 163–198.
2 "Fundamentals of Brittle Behavior in Metals," by E. Orowan, Fatigue and Fracture of Metals (MIT Symposium, June, 1950), John Wiley & Sons, Inc., New York, N. Y., 1952, p. 154.
3 "Fracture Dynamics," by G. R. Irwin, Fracturing of Metals, ASM, 1948, p. 152.
4 "Fracturing and Fracture Dynamics," by G. R. Irwin and J. A. Kies, Welding Journal, vol. 31, Research Supplement, February, 1952, pp. 95s–100s.
5 "Critical Energy Rate Analysis of Fracture Strength of Large Welded Structures," by G. R. Irwin and J. A. Kies, Welding Journal, vol. 33, Research Supplement, April, 1954, pp. 193s–198s.
6 "Strain Energy Release Rates for Fractures Caused by Wedge Action," by A. A. Wells, NRL Report No. 4705, March, 1956.
7 "Onset of Fast Crack Propagation in High Strength Steel and Aluminum Alloys," by G. R. Irwin, Proceedings of 1955 Sagamore Conference on Strength Limitations of Metals, Syracuse University, N. Y., March, 1956, pp. 289–305.
8 "The Relation of Crack Extension Force to Onset of Fast Fracturing," by T. W. George, G. R. Irwin, and J. A. Kies, I.A.S. Meeting, January 23–26, 1956.
9 "Energy Release Rates During Fracturing of Perforated Plates," by M. W. Brossman and J. A. Kies, NRL Memorandum Report No. 370, November, 1954.
10 "The Distribution of Stress in the Neighborhood of a Crack in an Elastic Solid," by I. N. Sneddon, Proceedings of the Physical Society of London, vol. 187, 1946, p. 229.
11 "Bearing Pressures and Cracks," by H. M. Westergaard, Trans. ASME, vol. 61, 1939, pp. A-49–A-53.

---

[3] For plane strain, one substitutes $E/(1-\nu^2)$ for E in the expression for the stress-intensity factor. No change in the magnitude of the stress-intensity factor occurs because $\mathcal{G}$, for plane strain, is $(1-\nu^2)$ times $\mathcal{G}$ for plane stress.

Crack-Extension Force for a Part-Through Crack in a Plate

G. R. IRWIN
Superintendent,
Mechanics Division,
U. S. Naval Research Laboratory,
Washington, D. C.

The crack stress-field parameter $\mathcal{K}$ and crack-extension force $\mathcal{G}$ at boundary points of a flat elliptical crack may be derived from knowledge that normal tension produces an ellipsoidal crack opening. Rough correction procedures can be employed to adapt this result for application to a part-through crack in a plate subjected to tension. Experimental measurements suggest this adapted result has a useful range of accuracy.

STARTING cracks which initiate crack-propagation failures usually develop from a local flaw of some kind which was considered unimportant or was not found by inspection. Sometimes the defect overlooked is an actual crack. At other times it is an arc strike, or a patch of material severely weakened by grain coarsening, or an excessively sharp root radius from which a crack may develop in service. These flaws as well as the highest tensile stress tend to occur at one border of the component cross section. In the examination of fracture failures one rarely finds that the initial starting crack extended completely through the thickness or that the starting-crack development, say by fatigue, occurred along a line extending completely through a cross section of a component. A part-through crack, as illustrated in Fig. 1, is more typical of the starting-crack shape customarily found.

In order to assist the use of laboratory measurements of crack toughness to estimate the danger of crack propagation under various conditions of tensile load and crack size it is desirable to estimate by calculation the stress-intensity factor $\mathcal{K}$ and force $\mathcal{G}$ descriptive of stress environment at the crack border. For a crack of the kind shown in Fig. 1 this is not an easy task even when the boundary is assumed to be representable as half an ellipse, a shape easily expressed in mathematical terms. The only three-dimensional crack problem for which stresses near the crack have been discussed in terms of values of $\mathcal{G}$ (or $\mathcal{K}$) is the crack with a circular boundary, sometimes referred to as a "penny-shaped" crack [1].[1]

For a real crack similar to that illustrated in Fig. 1, the solution can be thought of as composed of (A) the solution to the mathematical problem of the flat elliptical crack in an infinite solid plus (B) consideration of three "real crack" factors as follows:

1 The correction for inserting a free surface normal to the crack through the major axis of the ellipse.

2 The correction for plastic strains near the crack boundary.

3 The correction for inserting a free surface representing the side of the plate opposite from the crack opening.

With the assistance of previous studies of the flat elliptical crack problem [2] it is possible to establish (A). This portion is discussed first. The discussion of the adjustments (B) which then follows is approximate only but is nevertheless desirable in order to clarify engineering use of the mathematical work.

[1] Numbers in brackets designate References at end of paper.
Contributed by the Applied Mechanics Division and presented at the Winter Annual Meeting, New York, N. Y., November 25–30, 1962, of THE AMERICAN SOCIETY OF MECHANICAL ENGINEERS.
Discussion of this paper should be addressed to the Editorial Department, ASME, United Engineering Center, 345 East 47th Street, New York 17, N. Y., and will be accepted until January 10, 1963. Discussion received after the closing date will be returned. Manuscript received by ASME Applied Mechanics Division, June 16, 1961. Paper No. 62—WA-13.

The Basic Mathematical Model

Studies of the flat elliptical crack problem by Green and Sneddon [2] provide principally the information that tension normal to the crack produces an ellipsoidal crack-opening shape. Assume the crack lies in the x,z-plane with its major dimension $2c$ along the z-axis and its minor dimension $2a$ along the x-axis in such a way that border points x_1, z_1 of the crack correspond to

$$\frac{x_1^2}{a^2} + \frac{z_1^2}{c^2} = 1 \qquad (1)$$

If the crack opening displacements in the y-direction are represented by η then the Green-Sneddon result may be expressed as

$$\eta = \eta_0 \left(1 - \frac{x^2}{a^2} - \frac{z^2}{c^2}\right)^{1/2} \qquad (2)$$

where η_0 is half the total separation of the walls of the crack at the origin.

The desired expressions for $\mathcal{G}$ and $\mathcal{K}$ will be derived from equation (2) rather than from stress equations for two reasons: (a) The stress relations given by the Green-Sneddon paper are not in a form convenient for the calculation of σ_y near the border of the crack; (b) the author wishes to establish the point that the general shape of the crack opening provides sufficient information to determine $\mathcal{G}$ and $\mathcal{K}$.

The procedure to be followed consists first in a discussion of the variation around the border of the crack of the elastic-opening displacements of the crack. It will be noted that deviations of the stress state from plane strain become negligible in the limit of small separations from the crack border. Plane-strain relations for $\mathcal{G}$ and $\mathcal{K}$ associated with crack-opening displacement then permit the desired calculations.

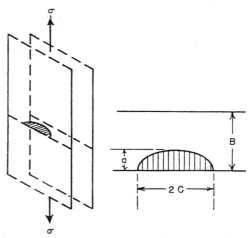

Fig. 1 Part-through crack in a plate showing dimensions a and c of crack and plate thickness, B

The position variables x_1 and z_1 which lie on the crack borders may be represented in parametric form by

$$x_1 = a \sin \phi \quad (3)$$
$$z_1 = c \cos \phi \quad (4)$$

A change $d\phi$ then corresponds to a segment ds of crack border given by

$$ds = (a^2 \cos^2 \phi + c^2 \sin^2 \phi)^{1/2} d\phi \quad (5)$$

At a small normal separation r (inward from the crack border) straightforward algebraic steps from equation (2) lead to the expression

$$\eta^2 = \eta_0^2 \frac{2r}{ac} (a^2 \cos^2 \phi + c^2 \sin^2 \phi)^{1/2} \quad (6)$$

The variation of displacement η at fixed distance r from the crack border as a function of ϕ will be considered in terms of the variation with ϕ of

$$\rho = \frac{\eta^2}{2r} \quad (7)$$

Thus

$$\rho = \frac{\eta_0^2}{ac} (a^2 \cos^2 \phi + c^2 \sin^2 \phi)^{1/2} \quad (8)$$

Differentiating with respect to ϕ and using equation (5) one finds

$$\frac{d\rho}{ds} = \left(\frac{\eta_0^2}{ac}\right) \frac{(c^2 - a^2) \sin \phi \cos \phi}{a^2 \cos^2 \phi + c^2 \sin^2 \phi} \quad (9)$$

At the value of ϕ where $d\rho/ds$ is largest the fractional change of ρ across the segment ds is

$$\frac{d\rho}{\rho} = \frac{(c^2 - a^2)(a^2 + c^2)^{1/2} ds}{2\sqrt{2} \, a^2 c^2} \quad (10)$$

Putting $c = 2a$ for specific illustration, the value of $d\rho/\rho$ becomes

$$\frac{d\rho}{\rho} = 0.6 \frac{ds}{a} \quad (11)$$

The quantity ρ can be interpreted as the root radius of the elastic crack opening along the crack border. As a reference point approaches the crack border, ρ becomes the principal length factor associated with the crack. If we know a solution for stresses and strains valid along a crack-border length δs large enough so that the dimension ρ is negligible in comparison, then such a solution can be used to supply an expression for η for substitution into equation (6).

In crack stress-field analysis problems ρ is always equal to $(\sigma/E)^2$ times a dimension comparable to the crack size. Bearing in mind that linear elasticity analysis regards σ/E as an infinitesimal in comparison to unity, we can, for example, choose[2]

$$\delta s = \left(\frac{E}{\sigma}\right) \rho \quad (12)$$

or, say,

$$\delta s = \left(\frac{\sigma}{E}\right) a \quad (13)$$

Such a crack-border-length segment is obviously short compared to crack size and long compared to ρ. Substituting δs from equation (13) for ds in equation (11) gives

$$\frac{d\rho}{\rho} = 0.6 \left(\frac{\sigma}{E}\right) \quad (14)$$

which can be regarded as meaning ρ changes only by an in-

[2] E is Young's modulus, ν is Poisson's ratio.

finitesimal fraction across the border length segment δs. The changes of displacements and stresses near the crack border must possess a similar degree of constancy relative to a coordinate parallel to the crack border. Since the problem is symmetrical about the plane containing the crack, a plane-strain stress field fits the conditions of the problem in local crack-border regions comparable to dimensions to δs.

The three crack-border stress fields for which the displacements are independent of the co-ordinate parallel to the border are discussed in reference [3]. The first of these, opening-mode plane strain, is appropriate here and provides the relations

$$\eta = \frac{2(1-\nu^2)}{E}(2r)^{1/2} \mathcal{K} \quad (15)$$

$$\mathcal{G} = \frac{\pi}{E}(1-\nu^2)\mathcal{K}^2 \quad (16)$$

In reference [3] the $\mathcal{G}$ and $\mathcal{K}$ for opening-mode plane strain are written $\mathcal{G}_\mathrm{I}$ and $\mathcal{K}_\mathrm{I}$. The subscript is to distinguish these quantities from the $\mathcal{G}$ and $\mathcal{K}$-values for the other two modes. Here the subscript can be temporarily omitted without risk of ambiguity. Inserting η^2 from equation (15) into equation (6) one finds

$$\mathcal{K}^2 = \frac{1}{4}\left(\frac{E}{1-\nu^2}\right)^2 \left(\frac{\eta_0^2}{ac}\right)(a^2 \cos^2 \phi + c^2 \sin^2 \phi)^{1/2} \quad (17)$$

and thus

$$\mathcal{G} = \frac{\pi}{4}\left(\frac{E}{1-\nu^2}\right)\left(\frac{\eta_0^2}{ac}\right)(a^2 \cos^2 \phi + c^2 \sin^2 \phi)^{1/2}. \quad (18)$$

One can, at this point, either use a relationship of η_0 to σ, a, and c from reference [2] or find this relation by computing the strain-energy change for a small expansion of the crack boundary. The latter is not difficult. If the ellipse is expanded by adding fa to a and fc to c where f is very small, the normal outward displacement, r, of any point on the elliptical boundary is

$$r = \frac{acf}{(a^2 \cos^2 \phi + c^2 \sin^2 \phi)^{1/2}} \quad (19)$$

Computation of the strain-energy change dU from

$$dU = 4 \int_0^{\pi/2} \mathcal{G} r \frac{ds}{d\phi} d\phi \quad (20)$$

leads to

$$dU = \frac{\pi E c f \eta_0^2 \Phi}{(1-\nu^2)} \quad (21)$$

where Φ is the elliptic integral

$$\Phi = \int_0^{\pi/2} \left[\sin^2 \phi + \left(\frac{a}{c}\right)^2 \cos^2 \phi\right]^{1/2} d\phi \quad (22)$$

We also know the strain-energy change is one half of σ times the change in crack volume. From this

$$dU = 2\pi \sigma a c f \eta_0 \quad (23)$$

The value of η_0 found by equating the two relations for dU, equations (21) and (23), may be substituted into (18) with the result that

$$\mathcal{G} = \frac{\pi(1-\nu^2)\sigma^2}{E\Phi^2}\left(\frac{a}{c}\right)(a^2 \cos^2 \phi + c^2 \sin^2 \phi)^{1/2} \quad (24)$$

By inspection of equation (24) one observes that $\mathcal{G}$ is greatest where the crack boundary intersects the minor axis. Thus with increase of tension on a flat elliptical crack, the crack-extension should, barring anisotropy, tend to produce a circular crack-boundary shape.

The function Φ increases with (a/c). When a is infinitesimal compared to c, Φ is unity. When the crack shape is circular Φ is $\pi/2$. Other values of Φ are easily found from published tables of elliptic integrals. Taking $\phi = \pi/2$ it is readily seen that the values of unity and $\pi/2$ for Φ correspond, respectively, to the Griffith equation

$$\mathcal{G} = \frac{\pi \sigma^2 a}{E}(1-\nu^2) \quad (25)$$

and to the energy-release rate for the penny-shaped crack [1]

$$\mathcal{G} = \frac{4\sigma^2 a}{\pi E}(1-\nu^2) \quad (26)$$

Approximate Estimates for Real Crack

Consider next the three corrections to this result which should be considered in constructing an estimate of $\mathcal{G}$ pertaining to a real part-through crack at one surface of a finite-thickness plate. No exact treatment of correction (2) is attempted. Although detailed numerical computations pertaining to corrections (1) and (3) might be somehow constructed, these would be overly elaborate considering that correction (2) can only be a rough approximation. In the subsequent discussion only estimates of a simple and approximate nature are employed. Some exercise of arbitrary choice will be noted in these approximations. Uncertainties remain in the final expression, equation (32), due to this. However, it is believed the treatment will serve to illuminate the nature of the "real crack" problem and that the end result has a useful degree of validity.

From general considerations it is clear that introduction of the two free surfaces allows larger crack-opening displacements. Corrections (1) and (3) therefore increase the crack-extension force. For reasons which will be discussed the influence of corrections (1) and (3) is assumed to be a 20 per cent increase of $\mathcal{G}$. Thus, for $\phi = \pi/2$,

$$\mathcal{G}_\mathrm{I} = \frac{1.2\pi(1-\nu^2)\sigma^2 a}{E\Phi^2} \quad (27)$$

When a is small both in relation to c and to the plate thickness, the factor 1.2 in the foregoing equation is known to represent correction (1) from the "crack at a free-surface boundary" solution discussed in reference [4]. Part-through cracks lie in a critical range of special interest when $a/c < 1.0$ and when $a/B \leq 0.5$, where B is the plate thickness as shown in Fig. 1. Within this range the small influence of the free surface opposite from the crack opening in causing increase of $\mathcal{G}$ should compensate somewhat for the overestimate of $\mathcal{G}$ by maintaining the 20 per cent correction as a/c increases. Thus it is tentatively assumed that equation (27) furnishes a useful estimate of the maximum $\mathcal{G}$ for (a/c) less than 1 and (a/B) less than 0.5. The fact that this estimate is not strictly limited to $\phi = \pi/2$ is discussed later.

The force-concept $\mathcal{G}$ employed in this paper pertains to a simple mathematical model of fracture consisting of a flat internal free surface, representing the crack, surrounded by a linear elastic-stress field. To avoid complexity and ambiguity in the force concept and yet make reasonable allowance for stress relaxation by plastic yielding near the crack edge, it is necessary to assume the influence of this local yielding on the surrounding stress field can be represented as equivalent to an additional spread of the internal free surface beyond the real crack size. It has been suggested in other papers [5, 6] that a suitable allowance for a situation of plane stress, where yielding is dominated by relaxation of the stress normal to a free surface, is an additional crack extension r_YS given by

$$r_\mathrm{YS} = \frac{\mathcal{K}^2}{2\sigma^2_\mathrm{YS}} \quad (28)$$

The term σ_YS represents the uniaxial tensile yield strength. When yielding is limited by elastic constraint so that crack extension occurs in the transverse-tensile (plane-strain) mode a smaller value would be expected. The additional or equivalent crack extension suggested in reference [6] is

$$r_\mathrm{I} = \frac{\mathcal{K}_\mathrm{I}^2}{4\sqrt{2}\,\sigma_\mathrm{YS}^2} \quad (29)$$

The subscript I implies the crack-edge stress field is governed by conditions which produce a situation of plane strain. With reference to this it can be noted that the lateral plate dimensions presumably are large compared to c. Thus plastic strains would not appreciably relax the stress σ_z close to the crack border and near $\phi = \pi/2$. Writing equation (27) with $\mathcal{K}_\mathrm{I}$ rather than $\mathcal{G}_\mathrm{I}$ terminology

$$\mathcal{K}_\mathrm{I}^2 \Phi^2 = 1.2\sigma^2 a \quad (30)$$

Adding the amount given by equation (29) to a one obtains

$$\mathcal{K}_\mathrm{I}^2 \Phi^2 = 1.2\sigma^2 \left(a + \frac{\mathcal{K}_\mathrm{I}^2}{4\sqrt{2}\,\sigma_\mathrm{YS}^2}\right) \quad (31)$$

and thus

$$\mathcal{K}_\mathrm{I}^2 = \frac{1.2\sigma^2 a}{\Phi^2 - 0.212\left(\dfrac{\sigma}{\sigma_\mathrm{YS}}\right)^2} \quad (32)$$

The value of Φ depends only upon the ratio, a/c. There is no simple way of determining a plastic zone correction to the ratio a/c. Since the major axis lies on a free surface, one would guess the plastic zone at the extremities of the major axis would be larger than that at $\phi = \pi/2$, but this would depend upon the a/c ratio. Rather than attempt to estimate how much the effective a/c ratio should be shifted to larger or smaller size to allow for local plastic strain near the crack border, it will be assumed, for simplicity, that Φ can be calculated using the a/c ratio for the actual crack.

As a reference point moves on the crack border away from $\phi = \pi/2$ in either direction, variations of $\mathcal{K}_\mathrm{I}^2$ in accordance with equation (24) might be expected. For $c = 2a$ the variation thus expected is less than 10 per cent across the central half of the z-direction span of the crack. This variation would be reduced by the tendency of the plastic-zone size to increase as the reference point approaches the intersection of the crack boundary with a free surface. Thus our estimate of $\mathcal{K}_\mathrm{I}^2$ along a substantial portion of the crack boundary near $\phi = \pi/2$ should have nearly a constant value. It is suggested that this value can be calculated with useful accuracy from equation (32).

A résumé of the way the three corrections are taken into account by equation (32) is as follows: Correction (1) is assumed to increase $\mathcal{K}_\mathrm{I}^2$ by 20 per cent. This correction factor is known to be valid when a/c is small. As a/c increases, the factor of increase representing correction (1) should decrease. However, increase of a/c usually corresponds to a closer approach of the crack boundary to the second free surface. Thus the moderate overestimate implied by maintaining the 20 per cent increase factor as the ratio a/c increases can be applied toward correction (3). Even a rough estimate of correction (3) would require extensive numerical computations. Since this discussion consistently avoids effects which cannot be handled in a simple way, it appears best to limit the magnitude of the second free-surface influence by limiting the a/B-values. In consideration of the fact that some allowance for correction (3) was implicit in the 20 per cent increase factor for correction (1), it is believed a range of a/B values extending to 0.5 would be permissible. Correction (2) was done by increasing the minor axis, a, an amount consistent with previous studies of plane-strain crack extension. The plastic-strain and stress-field conditions influencing the effective size of the major axis c are complex. For simplicity it was assumed the a/c ratio of the enlarged effective crack size would be the same as that for the actual crack.

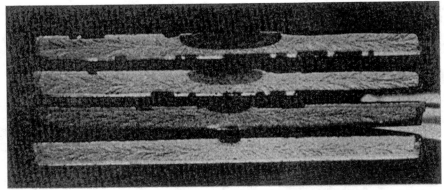

Fig. 2 Fracture appearance of tensile specimens which contained part-through cracks. Initial crack surface was darkened by furnace gases during heat-treatment which followed insertion of crack.

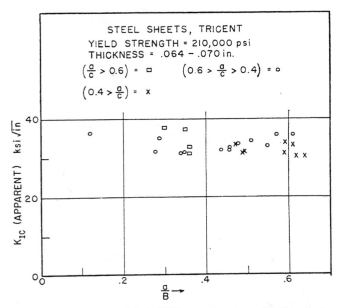

Fig. 3 Apparent values of $K_{Ic}(\sqrt{\pi}\mathcal{K}_{Ic})$ are shown as a function of relative crack depth for part-through cracks of different shapes as indicated by a/c ranges shown

Experimental Illustration

The tensile strengths of steel sheets containing various size part-through cracks have been measured by Srawley. The results appear in references [7] and [8]. After heat-treatment to produce a yield stress of 210,000 psi, the initial cracks were inserted by application of a corrosive agent to a limited region of the tensile side of the sheet while stressed in bending. Fig. 2 shows typical fracture surfaces from this work. The initial cracks were darkened by the furnace gases during a final baking for 16 hr at 400 F after cracking. From the tensile load at fracture and from measurements of the initial crack, values of σ, a, and c were obtained for use with equation (32).

References [7] and [8] give values K_{Ic} where

$$K_{Ic} = \sqrt{\pi}\mathcal{K}_{Ic} \tag{33}$$

Subscript c indicates that the computed crack edge-stress field parameter pertains to the stress field for onset of fast crack extension. There is some doubt as to this interpretation. The results shown in Fig. 3 are therefore given as "apparent" K_{Ic}-values.

The uncertainty stems from the fact that, although ink was used in the cracks to determine the amount of slow crack extension, the ink staining gave ambiguous indications and these were not usable for the purpose of establishing values of a and c based upon actual size of the crack at onset of rapid fracture. Customarily slow crack extensions marked by ink staining are comparable in magnitude to the size of the crack edge-plastic strain zone, less than 0.01 in. for the tests reported by Srawley. That some slow crack extension occurred and was assisted by the environmental effect of the ink can be inferred from a limited number of no-ink tests. These gave K_{Ic} results about 15 per cent above the average level of the results shown in Fig. 3.

It will be noted from the foregoing discussion that the uncertainties of the theoretical $\mathcal{K}_I$ estimate are scarcely greater than the uncertainties of experimental verification when one of the situations of practical interest, rocket vessel-wall material, is represented in a typical sheet thickness. For purposes of experimental verification of equation (32), $\frac{1}{4}$-in-thick plates and proportionately larger part-through cracks would have been preferable. Nevertheless, the results shown in Fig. 3 are encouraging with regard to the accuracy of equation (32). Within experimental scatter the relative depth and shape of the crack appear from these data to have no significant effect on the computed critical value of K_{Ic}.

References

1 I. N. Sneddon, "The Distribution of Stress in the Neighborhood of a Crack in an Elastic Solid," *Proceedings of the Physical Society of London*, vol. 187, 1946, p. 229.

2 A. E. Green and I. N. Sneddon, "The Distribution of Stress in the Neighborhood of a Flat Elliptical Crack in an Elastic Solid," *Proceedings of Cambridge Philosophical Society*, vol. 46, 1950, pp. 159–164.

3 G. R. Irwin, "Fracture Mechanics," Structural Mechanics, Pergamon Press, London, England, 1960, pp. 560–574.

4 G. R. Irwin, "Crack-Extension Force for a Crack at a Free Surface Boundary," NRL Report 5120, April, 1958.

5 "Fracture Testing of High-Strength Sheet Materials," First Report of a Special ASTM Committee, *ASTM Bulletin*, January, 1960, p. 38.

6 G. R. Irwin, "Plastic Zone Near a Crack and Fracture Toughness," 7th Sagamore Ordnance Materials Research Conference, August, 1960.

7 J. E. Srawley, "The Slow Growth and Rapid Propagation of Cracks," NRL Report No. 5617, May, 1961.

8 "The Slow Growth and Rapid Propagation of Cracks," Second Report of a Special ASTM Committee, *ASTM Materials Research and Standards*, vol. 1, no. 5, May, 1961.

## Fracture.

By

George R. Irwin.

With 10 Figures.

1. Introduction. In 1776 Coulomb [*1*] expressed the view that fracture of a solid would occur if the maximum shear strain at some point surpassed a critical value characterizing the mechanical strength of the material. Although the suggestion is of no more than historic interest it may be noted as the oldest of a number of empirical "critical stress" or "critical strain" relations [*2*].

In 1920 A. A. Griffith [*3*] proposed an explanation of fracture strength of glass based on the idea that glass contains crack-like flaws. Griffith reasoned that the largest crack-like flaw would become self-propagating when the rate of release of strain energy became greater than the rate of increase of surface energy of the extending crack. Subsequently, during the interval to 1938, a statistical concept of mechanical strength was introduced. A widely known exposition of this viewpoint, sometimes referred to as a "worst flaw" theory is due to Weibull [*4*]. Weibull showed that fracture strength might be expected to depend upon the extensiveness of the region of greatest stress as well as upon the stress magnitude.

It will be seen that the strain energy release rate idea basic to Griffith's proposal, and the statistical viewpoint inherent in the "worst-flaw" idea are essential for development of a basis for understanding fracturing and fracture strength of real materials. The additional concepts which are needed are best shown through illustrative examples and discussion.

I. Tensile strength of liquids.

2. Growth of a cavity to unstable size. A cavity in a large volume of liquid subjected to a hydrostatic tension p will expand or collapse depending upon whether p is greater or less than the value

$$p = \frac{2\gamma}{r} - p_v$$

where γ is the surface tension, r is the radius of a spherical cavity, and p_v is the vapor pressure. A rough estimate of the limiting tensile strength of a pure liquid may be obtained by neglecting p_v and by making the size of the cavity correspond to the specific volume per single molecule of liquid. Thus for carbon tetrachloride (CCl_4) at room temperature

$$p = \frac{2 \times 27 \text{ ergs/cm}^2}{7.3 \times 10^{-8} \text{ cm}} = 740 \times 10^6 \frac{\text{ergs}}{\text{cm}^3}.$$

Experimental values of tensile strength for CCl_4 of nearly half this magnitude have been measured by Briggs[1]. In contrast to this near agreement, similar

---
[1] L. J. Briggs: J. Chem. Phys. **19**, 970 (1951).

estimation procedures applied to pure solids lead to calculated strengths which are two or three orders of magnitude greater than the strengths found by actual measurement. Evidently the tensile strength of pure liquids is more easily related to fundamental theoretical considerations than is the case for solids. For this reason the tensile strength of pure liquids will be considered in sufficient detail to bring out features which pertain to fracturing generally.

The procedure used in this discussion extends somewhat an analysis of this problem by FISHER[1].

Assume at first that the liquid has no impurities, that it adheres perfectly to the walls of its container, and that the negative pressure is uniform throughout. One may imagine the liquid is in a rigid cylindrical container fitted with a piston. Suppose, then, that the effect of piston motion in exerting an initial tension or negative pressure p_0 has been to increase the liquid volume from V to $(V+w)$. If the modulus of compression is k_0

$$p_0 = \frac{k_0}{V} w. \tag{2.1}$$

We suppose next that the tension is relaxed by growth of one spherical cavity whose volume v is smaller than w. Then the pressure is

$$p = p_0(1 - v/w) \tag{2.2}$$

and the strain energy in the liquid is

$$U = \tfrac{1}{2} p(w - v). \tag{2.3}$$

In addition to affecting U the growth of the cavity increases the energy in the liquid by the surface energy term γA, where A is the surface area. As the cavity grows in volume with the walls and piston position fixed

$$\frac{d}{dv}(U + \gamma A) = -p + \frac{2\gamma}{r}. \tag{2.4}$$

The loss in strain energy is just offset by the increase in surface energy when

$$p = \frac{2\gamma}{r}. \tag{2.5}$$

When the test volume is large compared to the value

$$\omega = \frac{16\pi}{3}\left(\frac{8\gamma}{3p_0}\right)^3 \frac{k_0}{p_0} \tag{2.6}$$

Eq. (2.5) is satisfied at two values of v, an unstable equilibrium cavity size v_1 much smaller than w and a stable cavity size v_2 nearly as large as w. At the condition such that the cavity size is v_1, the loss in strain energy due to relaxation of pressure is $p_0 v_1(1 - \tfrac{1}{2}u)$ where $u = v_1/w$. The gain in surface energy is $\tfrac{3}{2} p_0 v_1(1 - u)$. Thus there is a net energy increase of

$$\Delta H_1 = \tfrac{1}{2} p_0 v_1(1 - 2u). \tag{2.7}$$

When the hydrostatic tension is large, say 3×10^8 ergs/cm^3, the magnitude of expression (2.6) is about 10^{-17} cm^3. In a volume of liquid equal to or less than ω, unstable cavity growth would be impossible without the aid of strain energy from neighboring volumes. In a volume of liquid of, say, 10ω sudden growth of a cavity of size v_1 would relax the pressure by only 1% even without

---
[1] J. FISHER: J. Appl. Phys. **19**, 1062 (1948).

motion of particles at the outer boundaries of that volume. Thus we may consider the whole test volume of the liquid to consist of a large number of cavity growth sites, each surrounded by a liquid volume of ample size, say, 10ω, so that the development of an unstable cavity might, conceptually, occur within any growth site independent of neighboring elements. At each cavity growth site thermal energy fluctuations are, from time to time, momentarily converted by various degrees into surface energy through cavity dilations. Since a considerable general movement of molecules must occur during the opening of a cavity to critical size, such an event is unlikely to occur in a time shorter than

$$\tau = \frac{(10\omega)^{\frac{1}{3}}}{c} \tag{2.8}$$

where c is the velocity of sound in the liquid. Thus we are concerned with thermal energy fluctuations integrated over time periods roughly equal to τ and over volumes roughly equal to 10ω. It will be assumed that when a thermal energy fluctuation meeting these conditions occurs then the probability of cavity growth to unstable size can without appreciable error be taken to be unity. It may be seen later that taking this probability to be 10^{-3} would increase the computed strength by only 4%.

3. The probability of unstable cavity growth.

The local thermal energy fluctuations constitute a statistical population. It is a property of this population that the a priori probability of any given magnitude of integrated energy fluctuation for the time period, τ, and the volume, 10ω, is not dependent upon the sizes of τ and of 10ω. For the temperature range of interest the a priori probability of occurrence of ΔH_1 is given by the expression, $\exp(-\Delta H_1/kT)$.

If the liquid test volume contains N molecules, there are, at most, N growth sites for cavities. From one viewpoint these must be counted as operating independently even though the regions, 10ω, surrounding adjacent sites would overlap. In any case it would appear safe to assume a number of independent cavity growth sites, N_1, such that

$$N > N_1 > \frac{V}{10\omega}. \tag{3.1}$$

Similarly the restriction of one's consideration to t/τ time elements, the testing time being t, would overlook the fact that a larger number of positions in time might act as the mid-points of the intervals τ. The frequency of fluctuations of magnitude kT or more must be about kT/h for consistency with HEISENBERG'S uncertainty principle. We will therefore assume a there are $v\,dt$ significant time sites in the interval dt where

$$\frac{kT}{h} > v > \frac{1}{\tau}. \tag{3.2}$$

Thus the probability for occurrence of cavity growth to unstable size in the interval of testing time dt becomes

$$f(t)\,dt = N_1 v\,dt \exp\left(-\frac{\Delta H_1}{kT}\right) \tag{3.3}$$

where ΔH_1 with the aid of (2.5) and (2.7) is given by

$$\Delta H_1 = \frac{16\pi}{3} \frac{\gamma^3}{p_0^2} \frac{1}{(1-u)}. \tag{3.4}$$

The tension p_0 is regarded as a function of time t obtainable from knowledge of the test procedure under consideration. In steps leading to expression (3.4) advantage was taken of the fact that u is very small to make a substitution of $(1-u)(1-2u)$ for $(1-u)^3$.

Define, next, a quantity F such that

$$F(t) = \int_0^t f(t)\, dt. \tag{3.5}$$

Then by the usual methods of extreme value statistics, the expected average time to fracture is given by

$$\bar{t} = \int_0^\infty t f(t)\, dt \exp[-F(t)]. \tag{3.6}$$

If (dp_0/dt) is greater than zero one may replace $f(t)\,dt$ by $g(p_0)\,dp_0$ so that with

$$g(p_0)\,dp_0 = N_1 \nu \frac{dt}{dp_0}\, dp_0 \exp\left[-\frac{\Delta H_1}{kT}\right] \tag{3.7}$$

and

$$G(p_0) = \int_0^{p_0} g(p_0)\, dp_0 \tag{3.8}$$

one obtains the average tension for fracture of the liquid $\bar{p}_0$ by the equation

$$(\bar{p}_0) = \int_0^\infty p_0 g(p_0)\, dp_0 \exp[-G(p_0)]. \tag{3.9}$$

The coefficient of p_0 in the integrand of (3.9) differs appreciably from zero over so small a range that the average value of p_0 will not differ significantly from the value at which $g(p_0)\exp(-G)$ attains maximum size. Thus the expected average strength, $(\bar{p}_0)$, is closely approximated by the value of p_0 which satisfies the relation

$$\frac{dg}{dp_0} - g^2 = 0. \tag{3.10}$$

Neglecting variation of ΔH_1 with p_0 through the negligible volume ratio u,

$$\frac{dg}{dp_0} = g\left(\frac{dN_1}{N_1 dp_0} + \frac{d\nu}{\nu\, dp_0} + \frac{2\Delta H_1}{p_0 kT}\right). \tag{3.11}$$

Thus

$$g(p_0) = \frac{dN_1}{N_1 dp_0} + \frac{d\nu}{\nu\, dp_0} + \frac{2\Delta H_1}{p_0 kT}. \tag{3.12}$$

For purposes of specific calculation it will be assumed that one may take

$$t = \beta p_0. \tag{3.13}$$

Where β is constant, as the relation of p_0 to time in experiments such as those of BRIGGS. The magnitude of the right side of (3.12) has so little influence upon the calculation that one may drop the terms involving N_1 and ν. In fact it may be shown that

$$0 < \frac{dN_1/N_1}{dp_0/p_0} + \frac{d\nu/\nu}{dp_0/p_0} < \frac{1}{20}\frac{\Delta H_1}{kT}. \tag{3.14}$$

Using (3.7), (3.13) and (3.12) one obtains

$$\log(N_1 \nu t) - \frac{\Delta H_1}{kT} = \log\left(2\frac{\Delta H_1}{kT}\right) \tag{3.15}$$

where t is the duration of the test. Taking t as one second, N_1 as the number of molecules in one cubic centimeter, and ν as (kT/h), the strength should be such that ΔH_1 is about 80 times kT or about 2 electron volts.

4. Pure liquid tensile strength measurements. In the experiments which have given the highest values of liquid fracture strength a strong capillary glass tube filled with a pure liquid is rotated about a transverse axis[1]. The speed of rotation is rapidly increased until the liquid column breaks. Calculation of the negative pressure at the axis of spin is based upon knowledge of the rotational speed at the instant the column separates. The assumption made above that p_0 increases linearly with time may be applied to this experiment without significant error. However, since the tension is not uniform but decreases as a quadratic function of distance from the axis of spin some consideration should be given to estimating the equivalence of the liquid volume used in this experiment with a liquid volume which would have the same strength in a uniform tension experiment. This estimate can be made with sufficient precision by a further application of extreme value statistics. The details need not be given here. If the length of the liquid column between points of zero tension lies between 5 and 10 cm and the diameter of the liquid column is 0.6 to 0.8 mm, the experiment may be taken as equivalent to applying uniform tension to a volume of 2.5×10^{-3} cm³. Taking the time of rise of the tension as 5 sec as a further adaptation to the BRIGGS type of experiment, calculations were made of theoretical strengths for a number of pure liquids. The extremes allowed by the inequalities (3.1) and (3.2) may be represented by

$$\log\left(\frac{kT}{h} \frac{75 \times 10^{20}}{V_m} \text{cm}^3 \text{sec}\right) = \frac{\Delta H_1}{kT} + \log\left(\frac{2\Delta H_1}{kT}\right) \tag{4.1}$$

and

$$\log\left(\frac{c}{(10\omega)^{\frac{1}{3}}} \times 6.25 \text{ cm}^3 \text{sec}\right) = \frac{\Delta H_1}{kT} + \log\frac{2\Delta H}{kT} \tag{4.2}$$

where V_m is the molar volume, c is the velocity of sound, and 10ω is given by

$$10\omega = 10\left(\frac{8}{3}\right)^3 \frac{k_0}{p_0^2} \Delta H_1. \tag{4.3}$$

Values of p_0 were obtained from equation (3.4) after ΔH_1 had been obtained from (4.1) and (4.2). The results are shown in Table 1 along with the highest strength values observed by BRIGGS for these same liquids. Any non-wetting areas of container wall, non-wetting impurities, or gas bubbles of about 10^{-5} cm size would cause the observed strengths to be considerably less than those given in the table. Experimental uncertainties of this kind are discussed by DONOGHUE, VOLLRATH, and GERGOUY[2]. The high strength of 150 atmospheres for pure benzene reported by BRIGGS was attained by DONAGHUE, VOLLRATH, and GERGOUY in only two out of a large number of trials. It is apparent that the multiplicity of factors which normally limit the tensile strength of a liquid to 1% or less of its theoretical value, calculated as above, are difficult to control experimentally. Thus the theoretical strength calculations for pure liquids are of doubtful practical utility. Nevertheless, the degree of completeness permitted in the theoretical considerations make the pure liquid tensile strength analysis of importance.

In the foregoing analysis it was observed that removal of flaws from the liquid reduced the fracture starting situation to one of cavity growth governed

---

[1] L. J. BRIGGS: J. Chem. Phys. **19**, 970 (1951).
[2] J. J. DONAGHUE, R. E. VOLLRATH and E. GERGOUY: J. Chem. Phys. **19** (1951).

Table 1. *Theoretical and measured tensile strengths of liquids*[1].

Liquid	Molecular weight	Temp. range °C	Surface tension ergs/cm²	Tensile Strength, p_0 (bars)		
				from (4.2)	from (4.1)	measured
H_2O	18	$+5$ to $+20$	74	1528	1640	275[2]
Acetic acid CH_3COOH	60	$+17$ to $+30$	27.6	352	385	288[3]
Aniline $C_6H_5NH_2$ . .	93	-2 to $+5$	43	685	744	300[3]
Benzene C_6H_6	78	$+15$ to $+20$	28.9	377	412	150[3]
Carbon Tet. CCl_4 . .	154	-15 to $+20$	26.8	336	368	275[3]
Chloroform $CHCl_3$ . .	119	-15 to $+10$	27.2	344	377	315[3]

by thermal fluctuations. The cavities thus generated from time to time constituted the strength controlling flaw population. Elimination of flaws of the impurity and non-wetting kinds from the theoretical analysis did not eliminate influence of test volume and testing time. Although changes by a factor of 10^7 in volume or in testing time would have been necessary in order to change the calculated strength by 10% nevertheless no estimate of strength could have been made without inserting specific values for the test volume and the testing time.

The tendency for conversion from potential or strain energy to thermal energy was restricted by an activation energy requirement ΔH_1 which was proportional to the inverse square of the applied tension. The size of ΔH_1 and its relation to the tension was obtained from the conditions describing the point of onset of unstable cavity growth. Relation (2.5) which expressed these conditions is a simple example of a strain energy release rate consideration which is of basic importance in every kind of macroscopic fracturing.

II. Stress and force relations in fracture.

5. Stresses near a flat internal free surface and the crack-extension-force.
When a fracture develops in a solid from minute origins, these origins are embedded in a larger field of inelastic or plastic strains. Knowledge of the complex patterns of distortion which characterize such early stages of fracturing is for the most part qualitative. As a definite crack develops and extends it is a characteristic of the fracturing process that the growth of crack length tends to overtake expansion of the field of plastic strains. Eventually, with continued crack extension, plastic strains are limited to a distance from the crack which is small relative to crack size and may be neglected in the elastic strain field analysis. Obviously no single descriptive model of fracturing could be applied to the entire range of the phenomenon. The condition of fracturing which closely resembles a thin internal cut is conceptually possible in any solid material[4]; its simplicities assist theoretical analysis and it is basic to development of a fracture strength concept of practical usefulness. The discussion of fracture in solids will therefore begin with extension of a well developed crack and with a "macro" rather than "micro" point of view.

A. A. GRIFFITH [3] presented a theory of fracture strength of a solid in terms of its surface energy. For this theory one assumes cracks exist or form in the solid as it is subjected to tensile forces. The theory then states that whenever elevation of the tensile forces produces a strain-energy release rate with crack extension which is larger than the rate of energy gain by formation of new free

---
[1] Table prepared by H. L. SMITH.
[2] L. J. BRIGGS: J. Appl. Phys. **21**, 721 (1950).
[3] L. J. BRIGGS: J. Chem. Phys. **19**, 970 (1951).
[4] This statement is discussed in Sect. 15.

surface area, rapid crack extension occurs and the solid breaks. Presumably the balance of energy rates for fracture strength determination would refer to the pre-existent crack with largest dimensions normal to the applied tensile load. It must also be assumed that the pattern of bonding forces characteristic of the solid internally are not easily re-established when an internal free surface closes so that any crack extensions are irreversible.

It was suggested by IRWIN [6] that the Griffith theory could be made generally applicable by substitution of energy spent in localized plastic strain for surface tension as a measure of resistance to crack extension. A similar proposal restricted to brittle fracture was advanced by OROWAN [7]. Direct applicability of the modified Griffith theory of fracture strength to the various large scale fracture propagation problems was shown by KIES and IRWIN[1]. Recognition of the role of GRIFFITH's strain energy release rate as the force conjugate to time rate of crack extension and as a measure of elevation of stresses near the end of a crack[2] contributed further to the body of theory now available for analysis of the fracturing process.

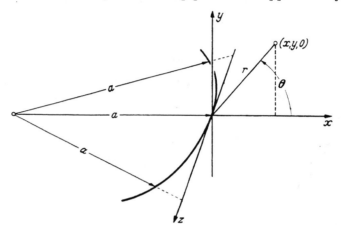

Fig. 1. Rectangular coordinates, x, y, z and polar coordinates r, ϑ, at the edge of a penny-shaped interior crack.

A paper by SNEDDON[3] gave a linear elastic theory analysis of the stresses and strains near a "penny-shaped" crack in an infinite solid[4]. The analysis assumed uniform tension at large distances from the crack. Superposition of uniform extensional stresses parallel to the plane of the crack would not alter the character of the results. SNEDDON used his analysis to obtain a value for the strain energy release rate appropriate to a Griffith theory [3] for predicting an unstable crack size. At one point in his paper SNEDDON gave approximate equations for the stress components in a region very close to the outer edge of the circular disc-shaped crack.

Consider a rectangular coordinate system with its y axis parallel to the axis of symmetry of SNEDDON's penny-shaped crack and with the x and z axes in the plane containing the crack. A uniform tension, σ, assumed to be applied at infinity causes a symmetrical pattern of displacement discontinuities which constitute the crack opening.

Assume next that the origin of coordinates is located at any point on the outer circular edge of the crack with the z axis tangent to the circle (see Fig. 1).

For points in the x, y plane and at distances from the origin which are small compared to the radius of the crack, a, the appropriate stress relations are those approached in the limit as r approaches zero and may be obtained from SNEDDON's paper. These are most simply written in terms of the coordinates r, ϑ shown

---

[1] G. R. IRWIN and J. A. KIES: Welding J. Res. Suppl. **31**, 95 (1952).
[2] G. R. IRWIN: IXth Int. Congr. of Appl. Mech., paper No. 101—II, Brussels, 1956.
[3] I. N. SNEDDON: Proc. Phys. Soc. Lond. **187**, 229 (1946).
[4] See also R. A. SACK: Proc. Phys. Soc. **58**, 729 (1946).

in Fig. 1 and may be expressed as follows

$$\sigma_x = K \frac{\cos \vartheta/2}{\sqrt{2r}} \left\{1 - \sin \frac{\vartheta}{2} \sin \frac{3\vartheta}{2}\right\}, \tag{5.1}$$

$$\sigma_y = K \frac{\cos \vartheta/2}{\sqrt{2r}} \left\{1 + \sin \frac{\vartheta}{2} \sin \frac{3\vartheta}{2}\right\}, \tag{5.2}$$

$$\sigma_z = \nu(\sigma_x + \sigma_y) = 2\nu K \frac{\cos \vartheta/2}{\sqrt{2r}}, \tag{5.3}$$

$$\tau_{xy} = K \frac{\cos \vartheta/2}{\sqrt{2r}} \sin \frac{\vartheta}{2} \cos \frac{3\vartheta}{2}, \tag{5.4}$$

$$\tau_{yz} = \tau_{xz} = 0 \tag{5.5}$$

where ν is Poisson's ratio.

Comparison with Sneddon's paper shows that the factor K in these equations is $\frac{2}{\pi}(\sigma^2 a)^{\frac{1}{2}}$.

To the same approximation as Eqs. (5.1) through (5.5) the y-direction displacements at the crack surface for small values of r/a are given by

$$v = \frac{(1-\nu)}{\mu} K \sqrt{2r} \tag{5.6}$$

where μ is the modulus of rigidity and r extends in the negative x direction.

Consider the edge region of a crack in plane strain corresponding to Eqs. (5.1) through (5.5) and assume the crack edge to be straight rather than a section of a circle. It is mathematically possible to close a short segment of length α to the left of the origin by superposition of tensile forces per unit area on the crack surfaces given by

$$S_y(p) = p K \frac{1}{\sqrt{2(\alpha - r)}}. \tag{5.7}$$

The factor, p, may be regarded as a proportional loading parameter. When p has been increased from 0 to 1 the crack will be closed along the segment α.

During the closure operation the y-direction displacements at the crack surfaces within the closing segment are given by

$$v(p) = (1-p) \frac{(1-\nu)K}{\mu} \sqrt{2r}. \tag{5.8}$$

Since the displacements, Eq. (5.8), are a linear function of the applied forces, Eq. (5.7), it is clear that the work to close up a length α at the edge of a crack is given by

$$\int_0^\alpha v(0) S_y(1) dr = \frac{(1-\nu)}{\mu} K^2 \int_0^\alpha \left(\frac{r}{\alpha - r}\right)^{\frac{1}{2}} dr = \frac{(1-\nu)\pi}{2\mu} K^2. \tag{5.9}$$

The calculation is on the basis of unit thickness of material in the z-direction. If the system is isolated from receiving energy, say by a fixed-grip situation, Eq. (5.9) represents the amount of energy which disappears from the strain energy field when the crack extends a small distance α. It is, therefore, equal to $\alpha \mathscr{G}$ where $\mathscr{G}$ is the strain energy release rate applicable to a Griffith theory analysis.

Eq. (5.9) was presented as pertaining to a crack whose edge was straight and coincided with the z axis. However, since α can be taken very small, since the calculation is per unit of length in the z direction, and since the stress equations might apply to any point on the circular boundary of the penny-shaped crack, one may insert for K Sneddon's value

$$K = \frac{2}{\pi}(\sigma^2 a)^{\frac{1}{2}}. \qquad (5.10)$$

One then obtains

$$\mathscr{G} = \frac{(1-\nu)\pi}{2\mu} K^2 = \frac{2(1-\nu)\sigma^2 a}{\pi\mu} \qquad (5.11)$$

which verifies the value for $\mathscr{G}$ obtained by Sneddon for this specific problem.

In agreement with precedent in other discussions of dislocation mechanics, $\mathscr{G}$ is the force per unit of edge length tending to cause extension of the crack.

One might question the applicability of the above discussion to a real crack, since linear elasticity relations were assumed valid to the edge of the crack, and, for cracks in real materials, substantial plastic distortion is found near the crack surfaces even when the fracture appears to be quite brittle. In this connection it is clear from physical considerations and may also be shown mathematically that the calculation of Eq. (5.9) is the equivalent of an overall integral of the strain energy loss rate due to crack extension. If Eqs. (5.1) through (5.5) are used to form an expression for strain energy density, ψ, the contribution to $\mathscr{G}$ from a region of radius r_1 around the edge of the crack is

$$\int_0^{r_1}\int_{-\pi}^{\pi} \frac{\partial \psi}{\partial a} r \, dr \, d\vartheta = \left(\frac{5-8\nu}{8-8\nu}\right) r_1 \frac{\partial \mathscr{G}}{\partial a}. \qquad (5.12)$$

As a first approximation it may be assumed that the preceding integral represents the order of magnitude of error due to neglect of plastic strains when r_1 includes the major portion of the plastic strain zone.

For a central crack of length $2a$ in a large plate, or for the interior penny-shaped crack subjected to uniform tensile stress at a large distance from the crack, the value of $\mathscr{G}$ is proportional to a, so that

$$r_1 \frac{\partial \mathscr{G}}{\partial a} = \left(\frac{r_1}{a}\right) \mathscr{G}. \qquad (5.13)$$

For cracks similarly located but opening under the action of wedge forces applied at the crack centers, $\mathscr{G}$ is inversely proportional to a, so that

$$r_1 \frac{\partial \mathscr{G}}{\partial a} = -\left(\frac{r_1}{a}\right) \mathscr{G}. \qquad (5.14)$$

When $\mathscr{G}$ is passing through a maximum or a minimum value as a function of crack extension the integral equivalent to Eq. (5.12) is zero. Under the assumed conditions of this analysis the plastic strains are confined to distances from the crack surfaces small in comparison to the crack size. Thus stress relaxation by plastic flow near the crack surfaces may be estimated as having an influence upon the calculation of $\mathscr{G}$ which varies with $\partial \mathscr{G}/\partial a$ but is relatively small.

For a crack traversing a plate under conditions such that r_1 can be considered to be a few plate thickness in magnitude without significant loss of accuracy, a generalized plane stress analysis can be used. The relations for stresses near the crack tip are as in Eqs. (5.1), (5.2), and (5.4). However, because of the increase in the displacement magnitudes when σ_z is put equal to zero, the value of $\mathscr{G}$ for a crack in plane stress is $\frac{1}{1-\nu^2}$ times the value of $\mathscr{G}$ for this crack subjected to similar σ_y values in a situation of plane strain. Thus for plane stress

$$K = \left[\frac{E\mathscr{G}}{\pi}\right]^{\frac{1}{2}} \tag{5.15}$$

where E is YOUNG's modulus. With this value of K, and remembering that a uniform stress can be added to the value of σ_x as well as σ_z, Eqs. (5.1), (5.2), and (5.4) provide the stress relations for the crack tip region of a separational crack dislocation in plane stress.

With simplifying approximations similar to those in the preceding discussion it is possible to write general expressions for elastic stresses and displacements near the edge of a crack. One assumes the internal cut, representing the crack, has an edge boundary which along any short segment can be replaced by a straight line, and that the locus of the crack near such a segment of its edge can be represented approximately as a plane. The portion of crack edge unter consideration may then be represented as on Fig. 1. The system of elastic stresses and displacements near the crack edge separates naturally into the sum of three systems. These refer to the three sets of displacement discontinuities which relax the three stresses σ_y, τ_{xy}, τ_{yz}, to constant or zero values at the locus of the crack.

The stress system related to σ_y is a special case of the set of problems solved by WESTERGAARD [5] starting from the assumed Airy stress function

$$F = \operatorname{Re}\bar{\bar{Z}} + y \operatorname{Im}\bar{Z}. \tag{5.16}$$

In WESTERGAARD's notation $\bar{Z}, Z$, and Z' are successive derivatives of a function, $\bar{\bar{Z}}(\zeta)$ where ζ is $(x+iy)$ or, when polar coordinates are more convenient, ζ is $re^{i\vartheta}$.

For the stresses, one has

$$\sigma_x = \frac{\partial^2 F}{\partial y^2} = \operatorname{Re} Z - y \operatorname{Im} Z', \tag{5.17}$$

$$\sigma_y = \frac{\partial^2 F}{\partial x^2} = \operatorname{Re} Z + y \operatorname{Im} Z', \tag{5.18}$$

$$\sigma_z = \nu(\sigma_x + \sigma_y) = 2\nu \operatorname{Re} Z, \tag{5.19}$$

$$\tau_{xy} = -\frac{\partial^2 F}{\partial x \partial y} = -y \operatorname{Re} Z', \tag{5.20}$$

$$\tau_{yz} = \tau_{xz} = 0. \tag{5.21}$$

By taking

$$\bar{Z} = K_1 \sqrt{2\zeta} \tag{5.22}$$

and $re^{i\vartheta}$ for ζ, one obtains Eqs. (5.1) through (5.4). For the displacements, u, v, w in the x, y, z directions one has

$$u = \frac{K_1 \sqrt{2r}}{2\mu} \cos\frac{\vartheta}{2} \left\{1 - 2\nu + \sin^2\frac{\vartheta}{2}\right\}, \tag{5.23}$$

$$v = \frac{K_1 \sqrt{2r}}{2\mu} \sin\frac{\vartheta}{2} \left\{2(1-\nu) - \cos^2\frac{\vartheta}{2}\right\}, \tag{5.24}$$

$$w = 0.$$

As in Eq. (5.11)

$$K_1^2 = \frac{2\mu\mathscr{G}_1}{(1-\nu)} \tag{5.25}$$

where $\mathscr{G}_1$ is the crack-extension-force for the opening mode of crack extension.

A similar procedure to that given above solves the problem for the stress system related to τ_{xy}. For this one assumes

$$F = - y \operatorname{Re} \bar{Z}. \tag{5.26}$$

By taking

$$\bar{Z} = K_2 \sqrt{2\zeta} \tag{5.27}$$

one obtains

$$\sigma_x = - \frac{K_2}{\sqrt{2r}} \sin \frac{\vartheta}{2} \left\{ 2 + \cos \frac{\vartheta}{2} \cos \frac{3\vartheta}{2} \right\}, \tag{5.28}$$

$$\sigma_y = \frac{K_2}{\sqrt{2r}} \sin \frac{\vartheta}{2} \cos \frac{\vartheta}{2} \cos \frac{3\vartheta}{2}, \tag{5.29}$$

$$\sigma_z = - 2\nu \frac{K_2}{\sqrt{2r}} \sin \frac{\vartheta}{2}, \tag{5.30}$$

$$\tau_{xy} = \frac{K_2}{\sqrt{2r}} \cos \frac{\vartheta}{2} \left\{ 1 - \sin \frac{\vartheta}{2} \sin \frac{3\vartheta}{2} \right\}, \tag{5.31}$$

$$\tau_{yz} = \tau_{xz} = 0. \tag{5.32}$$

For the displacements

$$u = \frac{K_2 \sqrt{2r}}{2\mu} \sin \frac{\vartheta}{2} \left\{ 2(1-\nu) + \cos^2 \frac{\vartheta}{2} \right\}, \tag{5.33}$$

$$v = - \frac{K_2 \sqrt{2r}}{2\mu} \cos \frac{\vartheta}{2} \left\{ (1-2\nu) - \sin^2 \frac{\vartheta}{2} \right\}, \tag{5.34}$$

$$w = 0.$$

By a calculation similar to that of Eq. (5.9)

$$K_2^2 = \frac{2\mu \mathscr{G}_2}{\pi (1-\nu)} \tag{5.35}$$

where $\mathscr{G}_2$ is the crack-extension-force for the first sliding or shear mode of crack extension.

For the stresses related to τ_{yz} one may assume

$$u = v = 0, \tag{5.36}$$

$$\mu w = K_3 \operatorname{Im} \left(\sqrt{2\zeta} \right) = K_3 \sqrt{2r} \sin \frac{\vartheta}{2}. \tag{5.37}$$

Thus

$$\sigma_x = \sigma_y = \sigma_z = \tau_{xy} = 0, \tag{5.38}$$

$$\tau_{xz} = - \frac{K_3}{\sqrt{2r}} \sin \frac{\vartheta}{2}, \tag{5.39}$$

$$\tau_{yz} = \frac{K_3}{\sqrt{2r}} \cos \frac{\vartheta}{2}. \tag{5.40}$$

By procedures similar to those used above

$$K_3^2 = \frac{2\mu \mathscr{G}_3}{\pi} \tag{5.41}$$

where $\mathscr{G}_3$ is the crack-extension-force for the second sliding or shear mode of crack extension.

Any constant terms representing superimposed uniform stress fields can be added to the stress components of the above stress systems. Otherwise linear combinations of the above three stress groups represent the stress fields possible at the edge of an internal cut representing a crack.

The three quantities $\mathscr{G}_1$, $\mathscr{G}_2$, and $\mathscr{G}_3$ are rates of transfer with crack extension of energy from the surrounding elastic strain field into other forms. In the case of metals the released strain energy is converted primarily to heat through local plastic deformation. In the case of all solids so far investigated the most rapid crack extension rates are much less than the velocity of sound. Thus the speed of the process is primarily limited by the reaction rates characteristic of local inelastic deformations rather than by inertia considerations.

For fracturing of a solid whose strength properties from point to point are homogeneous, the average time rate of crack extension is an increasing function of the crack-extension-force. The detailed relationship of crack extension velocity to crack-extension-forces is a complex rate theory problem. Knowledge of this relationship is currently a matter of experimental observation as is true also of the analogous problem for plastic yielding.

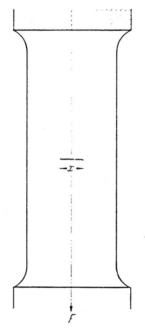

Fig. 2. Spreading central crack in simple tension.

The sliding processes associated with $\mathscr{G}_2$ and $\mathscr{G}_3$ if not accompanied by the first mode to provide actual separation of the crack surfaces, are not crack-like deformations in the usual interpretation. Solids can be severed, for example, in compression tests of ceramics, by a localized sliding action. However, because the conditions of restricted plastic deformation are poorly met in shear fracturing of metals and because the attention of investigations has been primarily on fractures of the opening mode type, discussion of the sliding modes of fracture will be quite limited. The symbol $\mathscr{G}$ with no subscript will refer to the crack-extension-force for the opening mode.

6. Calculation and measurement of the crack-extension-force. Consider next the energy changes which occur when segments of a solid material are separated by fracture. Assume the solid material under consideration is a plate in simple tension with a centrally located crack as in Fig. 2. Such a situation can be developed experimentally in various ways, for example, one may saw or cut a narrow slot perpendicular to the direction of the tension and apply wedging actions so that natural cracks are produced outward at either end of the slot. By increasing the tension, a load F can be found such that lengthening of the central crack occurs. In materials of moderate ductility the experimental situation can be adjusted so that the average time rate of crack extension increases from slow to fast in a manner which, at least from trend analysis viewpoint, can be considered as continuous. Consider an increment δx of crack extension during a period in which the process is slow enough so that kinetic energy may be omitted from the energy considerations. During this increment we assume the force F extended the plate by amount δl. A certain part l_e of the extension of the plate is recoverable by unloading. The balance which is not recoverable by unloading will be referred to as l_p. Then the total extension from the beginning of the experiment is

and
$$\left. \begin{array}{l} l = l_e + l_p \\ \delta l = \delta l_e + \delta l_p. \end{array} \right\} \tag{6.1}$$

For the purpose of this elementary illustration it may be assumed
$$F = M l_e \qquad (6.2)$$
where the spring constant, M, is a decreasing function of the length of the crack. Corresponding to this linear relation the strain energy U in the plate is
$$U = \tfrac{1}{2} F l_e \qquad (6.3)$$
and
$$\delta U = F \, \delta l_e - \tfrac{1}{2} F^2 \, \delta\!\left(\tfrac{1}{M}\right). \qquad (6.4)$$

It is useful to define a quantity δW by the relation,
$$F \, \delta l = \delta U + \delta W. \qquad (6.5)$$

Using (6.1) and (6.4) one observes that
$$\delta W = F \, \delta l_p + \tfrac{1}{2} F^2 \, \delta\!\left(\tfrac{1}{M}\right). \qquad (6.6)$$

Since the fracturing process with its accompanying plastic deformations is unaffected if, at any instant, the grips shown in Fig. 2 are considered to be fixed, one may write
$$\frac{dW}{dx} \delta x = -\delta U_f = F \, \delta l_p + \tfrac{1}{2} F^2 \frac{d}{dx}\!\left(\tfrac{1}{M}\right) \delta x \qquad (6.7)$$

where $(-\delta U_f)$ is the loss of strain energy for fixed grip conditions. Assuming a situation in which crack extension is accomplished by negligible plastic extensions δl_p, it is clear that the coefficient of δx in the last term of Eq. (6.7) is the crack-extension-force. Thus, on a unit thickness basis
$$\mathscr{G} = \tfrac{1}{2} F^2 \frac{d}{dx}\!\left(\tfrac{1}{M}\right). \qquad (6.8)$$

Consider a tensile test such as that of Fig. 2 but with no crack present. For this situation the last term of Eq. (6.7) is zero and the force acting to produce plastic extension, δl_p, is the longitudinal force, F. On a unit area basis one would say the force driving the deformation process is the longitudinal stress. A special relationship between time rate of plastic extension and longitudinal stress is usually found. For most materials the time rate is in the creep range and considered negligible for structural purposes so long as the stress is below the yield strength of the material. When, as the stress is increased, an abrupt increase occurs in the time rate of deformation, the material is said to have a sharp or well defined yield strength.

It is useful to define the fracture strength of a solid material in a way analogous to that of the yield strength. Since the change from slow to fast fracturing of a developing crack is usually abrupt, fracture strength may be described as the critical value, $\mathscr{G}_c$, of the crack-extension-force, $\mathscr{G}$, necessary for onset of rapid crack extension.

Eq. (6.8) suggests several practical procedures for measurement of $\mathscr{G}$. The elastic deflection of specimens containing cuts, representing cracks, of various length may be measured. From these measurements M as a function of crack length must be found experimentally with sufficient precision to permit computation of the slope of the curve representing $(1/M)$ as a function of crack length. With this information at hand one needs only to observe the applied force and the crack length for onset of rapid fracturing. Eq. (6.8) then permits calculation of $\mathscr{G}_c$. The method has been applied with reasonable success to tests of bars of steel and aluminum alloys broken in bending. However, a high degree of precision is required in the spring constant measurements for good accuracy.

Historically the second experimental method for measurement of the crack-extension-force was that used by WELLS[1] for cracks extending rapidly in steel plates. The method assumes the crack is in rapid motion and that the conversion from strain energy to heat is confined to a region close to the path of the crack. By means of thermocouples pressed into small holes near the path of the crack WELLS was able to measure the rise and fall of local temperature after rapid fracturing had occured. Assuming the path of the crack to be the source of the observed heat wave, WELLS was able to calculate the energy converted to heat per unit of fracture area. In contrast with the method based upon Eq. (6.8) which is applicable only to stationary cracks, the thermocouple method is applicable only to rapidly moving cracks. Satisfactory agreement of results from the thermocouple method with those expected from theoretical considerations was found.

For both of the experimental procedures discussed above a knowledge of the stress distribution is unnecessary. However, because use of the thermocouple method is limited to special materials and because of inconveniences inherent in both procedures, an approximate theoretical stress distribution is normally employed. An example of such a procedure using a photoelasticity technique is given in Sect. 10.

The relation of the stresses near a flat crack to the applied loads may be determined theoretically in several ways. Either semi-inverse procedures such as those of NEUBER[2] and WESTERGAARD [5] or integral equation procedures such as those of SNEDDON[3] and MUSKHELISHVILI[4] may be used. The Westergaard procedure was described in Sect. 5 of this article. The integral equation methods are potentially more powerful particularly when the desired answers require use of computing machines but will not be discussed in this article. NEUBER's methods of exact stress analysis have been widely used as a means of determining the largest stress at the root of a notch. For purposes of elastic theory analysis, a flat crack can be considered to be a notch, external or internal, having zero flank angle and an edge of nearly zero radius of curvature. Inspection of NEUBER's equations for the maximum stress at the notch σ_m shows that the product $\sigma_m \sqrt{q}$, where q is the radius of curvature of the notch, approaches a non-zero finite limit as q approaches zero. This product is related to the stress intensity factor K of this article by the equation

$$K = \lim (\tfrac{1}{2} \sigma_m \sqrt{q}) \quad \text{as} \quad q \to 0. \tag{6.9}$$

In his consideration of the influence of plastic yielding at a notch NEUBER suggested the effective stress concentration might be related to the average stress from the root of the notch across a "plastic particle" distance, ε. From Eq. (5.1) with ϑ equal to zero

$$\sigma_y = \frac{K}{\sqrt{2r}}.$$

The average stress from $r = 0$ to $r = \varepsilon$ is given by

$$\bar{\sigma} = \frac{1}{\varepsilon} \int_0^\varepsilon \sigma_y \, dr = K \sqrt{\frac{2}{\varepsilon}}. \tag{6.10}$$

---

[1] A. A. WELLS; Welding Res. 7, No. 2, 34-r (1953).
[2] H. NEUBER: Kerbspannungslehre: Grundlagen für genaue Spannungsrechnung. Berlin: Springer 1937.
[3] I. N. SNEDDON: Proc. Phys. Soc. Lond. 187, 229 (1946).
[4] N. I. MUSKHELISHVILI: Some Basic Problems of the Mathematical Theory of Elasticity. Groningen, Holland: P. Noordhoff 1953.

Fracture experiments do not determine ε and $\bar{\sigma}$ separately. However, by arbitrarily choosing $\bar{\sigma}$ to equal, say, the ultimate tensile strength found by standard tensile bars, a characteristic ε for onset of rapid crack extension can be determined. This procedure has been employed for prediction of applied loads which will cause crack propagation in steel and aluminum alloys[1]. The equivalence of a characteristic value of the product $\bar{\sigma}\sqrt{\varepsilon}$ to a characteristic value of K is obvious from Eq. (6.10).

Knowledge of the stresses near a flat crack in tension for a variety of situations is needed not only to assist fracture strength measurements but also to permit estimates of the crack-extension-force generally in the different laboratory or practical situations in which knowledge of this force tendency assists understanding of observed events. For two-dimensional stress fields associated with a straight crack the semi-inverse procedure suggested by WESTERGAARD has provided solutions for a number of problems. Three of these which have applications of general value are stated next in terms of the Westergaard stress function, Z.

Series of co-linear two-dimensional cracks. Consider the following stress function

$$Z = \sigma \prod_{i=1}^{N} \left\{1 - \frac{a_i^2}{(\zeta - b_i)^2}\right\}^{-\frac{1}{2}} \tag{6.11}$$

where b_i is an increasing series of real numbers, the values of a_i are positive, ζ is $(x+iy)$, and

$$b_i + a_i + a_{i+1} < b_{i+1}.$$

At infinity both σ_y and σ_x approach σ. Along the x axis there are N regions such that

$$|x - b_i| < a_i$$

where free surface boundary conditions are met. At the right end of the j-th crack, the stress intensity factor is

$$K_j = \sigma \sqrt{a_j} \prod_{i \neq j}^{N} \left\{1 - \frac{a_i^2}{(a_j + b_j - b_i)^2}\right\}^{-\frac{1}{2}}. \tag{6.12}$$

Periodically repeated two-dimensional crack. The limit of the stress function of Eq. (6.11) as N approaches infinity for equal length and equally spaced cracks is a stress function suggested by WESTERGAARD [5],

$$Z = \sigma \left\{1 - \left(\frac{\sin(\pi a/2A)}{\sin(\pi \zeta/2A)}\right)^2\right\}^{-\frac{1}{2}}. \tag{6.13}$$

Internal free surfaces representing cracks occur along the x axis whenever

$$\left(\sin \frac{\pi x}{2A}\right)^2 < \left(\sin \frac{\pi a}{2A}\right)^2.$$

The length of each crack is $2a$ and the period of repetition is $2A$. The stress intensity, K, near a crack end is

$$K = \sigma \left(\frac{2A}{\pi} \tan \frac{\pi a}{2A}\right)^{\frac{1}{2}}. \tag{6.14}$$

Eq. (6.14) is a useful approximation to the stress intensity for plates of width $2A$, containing either a central crack of length $2a$, or two colinear cracks of length a, extending inward from the side boundaries.

---

[1] P. KUHN: Stockholm Colloquium on Fatigue, Int. Union of Theor. and Appl. Mech., pp. 131—140. Berlin: Springer 1956.

Localized pressure in a two-dimensional straight crack. Consider the stress function

$$Z(\zeta) = \frac{P\sqrt{a^2 - b^2}}{\pi\zeta(\zeta - b)\sqrt{1 - (a/\zeta)^2}}. \tag{6.15}$$

The stress field corresponding to this function represents a crack of length $2a$, in which splitting or wedge forces of strength $\pm P$ per unit of length in the z direction act at $x = b$. The stress intensity factor for stresses surrounding $x = a$ is

$$K = \frac{P}{\pi\sqrt{a}}\sqrt{\frac{a+b}{a-b}}. \tag{6.16}$$

Since the Z and K of component stress fields are additive, values of Z and K appropriate to any prescribed pressure relation,

$$P = p(b)\,db$$

can be obtained from appropriate integrals of Eqs. (6.15) and (6.16).

The values of $\mathscr{G}_c$ given in Table 2 were for the most part obtained using a plate in tension with a central cut or slot as a starting crack. A stress distribution corresponding to the stress function of Eq. (6.13) was assumed. The corresponding $\mathscr{G}$ value is

$$\mathscr{G} = \frac{\pi K^2}{E} = \frac{2A\sigma^2}{E}\tan\frac{\pi a}{2A}. \tag{6.17}$$

Table 2. *Typical values of the crack-extension-force $\mathscr{G}_c$ necessary for onset of unstable fast fracturing*[1].

Material	Plate or sheet thickness cm	Method	Temperature deg. Cent.	$\mathscr{G}_c$ 10^5 ergs/cm²
Steel, ship plate cleavage	1.9	a	−20	175
Steel, ship plate ductile	1.9	d	+25	3500
Aluminum alloy 24 ST 4	2.5	b	+25	525
Aluminum alloy 24 ST 6	0.1	a	+25	1050[2]
Aluminum alloy 75 ST 6	0.1	a	+25	525[2]
Aluminum alloy 78 ST 6	0.1	a	+25	260[2]
Polymethylmethacrylate, as cast plates	0.3 to 1.2	a	+25	8.7
Polyesters, plates	0.3 to 1.2	a	+25	2.0
Polymethyl-alpha chloracrylate plates	0.3 to 0.6	a	+25	4.9
Vulcanized natural rubber	0.09	c	+25	120
Glass, lantern slide covers, moist	0.05	a	+25	0.07
Glass, lantern slide, in 2% RH air	0.05	a	+25	0.14
Cellulose acetate	0.06	a	+25	25
Cellulose acetate	0.005	a	+25	50

a) Central crack type tension specimens in which the unstable crack length was about one quarter of the plate width.
b) Slow bend tests of notched bars.
c) Tear tests by RIVLIN: J. Polymer Sci. **10**, 291—318 (1953). Various compositions give values of $\mathscr{G}_c$ from 10^6 to 2×10^6 ergs/cm².
d) Deduced from side-notched tear test fracture work rate results with the assumption that unstable fast ductile fracture would occur without significant decrease of work rate beyond the slow fracture value listed.

[1] Table prepared by J. A. KIES and H. L. SMITH.
[2] These values are increased by a factor of 2 when restraints are used to prevent buckling across the span of the initial crack.

III. Forming and spreading of cracks.

7. Illustrative models of fracture development. The processes by means of which small openings form, grow, and join to produce a visible crack are primarily understood only in qualitative terms. ZENER [6] and HOLLOMON [6] discussed several mechanisms for micro-crack formation in terms of the arrangement and movement of crystalline dislocations. They pointed out that the intensified stress field near the leading edge of a slip band contains regions of large tensile stress. Opportunities for openings to develop in this zone of increased tensions may be expected when continued slip is delayed, for example by arrest of a slip band at a grain boundary or inclusion.

If a uniform shear stress equal to the relaxed resolved shear stress at the slip plane is added, the equations for stresses pertaining to sliding modes of separation as discussed in Sect. 5 are applicable. From Eqs. (5.28) through (5.31) one finds the greatest tension is nearly four times the maximum shear stress at a ϑ value of $-\pi/2$. ZENER noted that large tensile stresses with a high ratio of tension to shear might also develop at grain corners due to relaxation of stress in the grain boundary. His discussion emphasized the close association of micro-crack development with plastic strain.

A similar viewpoint was developed in somewhat greater detail by PETCH[1] and by STROH[2]. Evidence was presented by PETCH[3] that fracture stress of polycrystalline pure iron has a linear relation to the inverse-square-root of grain size over a wide temperature range. The similarity of this result to the relation between yield stress and grain size suggested that both crack formation and slip propagation were related to attainment of certain critical stress intensities near the boundary of a blocked array of dislocations. The appearance of experimental evidence for a relation which can be justified in terms of a theoretical model in this instance, as in others to which reference will be made, provides a helpful view of one aspect.

The growth of a crack to macroscopic size results from a series of events ranging to submicroscopic scale. Proper conditions for formation of holes or cracks at smallest scale must exist as prerequisites for coalescence of these in groups to form fracture elements of next larger size. However, the events at large as well as small scale must be considered. A description of this whole complex series of processes which is both concise and general in application is not at hand. It is believed the broad features of fracture development can be seen in terms of the selected illustrative models which will presently be discussed.

Very small fracture origins in metals and plastics can often be identified on a fracture surface. Microscopic studies of these invariably show an appearance which suggests a rounded cavity rather than a thin crack. Although openings of more crack-like nature such as segments of cleavage may be present both in earlier and later stages of fracture development there are rather general grounds for considering that fracture development also includes openings which more nearly resemble holes than cracks.

Consider a small pennyshaped crack of radius r embedded in a polycrystalline rod. If the rod is subjected to a tensile stress of value S at its ends and if the plane of the crack is normal to the direction of the tensile stress, the enlargement of the volume of the crack by elastic strains is very nearly.

$$V_1 = \frac{4\pi r^3 S}{3E}. \tag{7.1}$$

---

[1] N. J. PETCH: J. Iron Steel Inst. **173**, 25 (1953).
[2] A. N. STROH: Proc. Roy. Soc. Lond., Ser A **223**, 404 (1954).
[3] N. J. PETCH: Phil. Mag., VIII. Ser. **1**, 186 (1956).

A factor of approximately 1.15 is neglected in this equation. If the area of the crack is extended by adding a row of vacancies all around the outer edge, the volume of vacancies added is

$$V_2 = 2\pi r \lambda^2 \tag{7.2}$$

where λ is the average atomic separation distance.

The new crack volume enlargement due to the tension S now exceeds V_1 by the amount

$$\lambda \frac{\partial V_1}{\partial r} = 4\pi r^2 \frac{S}{E} \lambda. \tag{7.3}$$

The ratio of this volume increment to V_2 is $2rS/\lambda E$. If one takes E/S to be about 10^3 as would be true for strong metals, the above ratio approaches unity when r is about 10^{-5} cm. The volume enlargement of a crack-like opening during a tensile strength test is, therefore, comparable to the volume of the added vacancies and somewhat insensitive as to their influence upon cavity shape when either the opening or the stress is extremely small.

Examination of polycrystalline metals usually discloses a wide variety of flaws and defects whose linear dimensions are very much larger than 10^{-5} cm. Normally, cracks started by separations of these larger magnitudes would be expected to extend by plastic deformation processes localized in edge regions of largest stress. The corresponding crack volume enlargement then results primarily from contractions in the elastic strain field and the shape of the opening becomes more crack-like. As the crack size increases, flaw containing regions approached by the leading edge are subjected to an environment of increasing tensile stress. The extension of the crack to encompass regions of this nature when they separate near a leading edge of the crack becomes the process contributing most to extension velocity as the dominant crack becomes large in comparison with the individual flaws. Two easily reproduced model type experiments are of value in bringing out these and other features characteristic of growth of cracks.

One model of this nature may be constructed by forming a raft of small equal sized bubbles[1]. Experimentally one finds that the opposite sides of the bubble raft will cling tenaciously to glass rods and that attempts to fracture the bubble raft can be performed by separating the glass rods. When the raft is essentially a single crystal and contains none or very few dislocations, all attempts to produce a fracture result in separation by local width reduction with no obvious crack. This action resembles the pinching or sliding apart action of a coarse crystalline cadmium rod pulled slowly in tension. However, if a polycrystalline raft of bubbles is formed, such a raft can be made to fracture by development and extension of interior openings. These interior openings always occur at the crystalline boundaries. This behavior appears to be due to the mobility of the bubble raft dislocations which prevents transcrystalline separation, and to the tendencies for dislocation movements to center in the neighborhood of crystalline boundaries, thus facilitating enlargement of grain boundary openings by addition of vacancies. It has been generally accepted that bubble raft models reproduce metal behavior at elevated rather than at low temperatures. The tendency of fracture to follow grain boundaries at elevated temperatures is illustrated by the polycrystalline bubble raft in an interesting fashion.

Consider next another experimental model of a separation process. This model consists of a 0.002″ thick zinc foil pulled in simple tension by a dead

---

[1] W. L. BRAGG: Proc. Roy. Soc. Lond., Ser. A **190**, 474 (1947).

weight[1]. If a central crack is created in such a foil by a razor blade cut and the subsequent extension of this crack observed, several significant characteristics of fracturing may be noted. A continuing steady general creep deformation of the sheet results in a continuing movement of the weight. Extensive local plastic

Fig. 3. Enlarged view of the end of a crack extending slowly in zinc foil. Fracture origins beyond end of crack are about to join together forming a quick extension of the main crack.

deformations near the ends of the crack result in thickness reductions which are unequal in magnitude in small neighboring regions. In part because of local thinning of the sheet and in part because of unequal strengths of different grains, holes appear near the ends of the extending crack. In the zinc foil used by McLean and George for these experiments, the grain size was approximately equal to the original sheet thickness. The number of such holes was counted and might be said to be either one to three per original grain diameter, or one to three per original sheet thickness. A preference for hole formation at grain corners would account for the observed spacing of "pin hole" fracture origins. Fig. 3 shows the appearance of a group of holes just prior to a joining up process which will result in a relatively fast advance of the crack.

When results of this experiment are studied by methods of trend analysis, the general crack extension behavior observed may be represented as shown

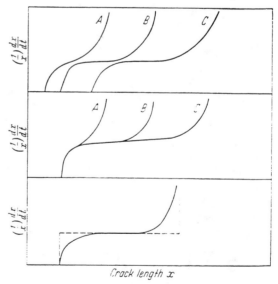

Fig. 4. Crack growth rate curves as a function of crack length for specimen sizes A, B, and C proportional respectively to $\frac{1}{2}$, 1, and 2.

on Fig. 4. In this schematic figure, crack length is indicated by x and time by t. In the upper diagram, the fractional time rate of extension is shown as a function of x for a size effect experiment. The arrangement of this experiment is such that the crack extension in a medium size sheet is compared to that in a half size and a double size sheet. In each case, the starting crack length and

---

[1] E. A. McLean, and T. W. George: Phys. Rev. **94**, 761 (1954) (abstract).

the load are proportional to the width of the test sheet. The velocity dx/dt changes from a slow "creeping" extension rate to a very fast rate during the course of the experiment, typically several hours. Only the approach to unstable fast fracturing is shown on the diagram. Midway between the starting crack length and the crack length for fast fracturing the fractional time rate of crack extension is about the same for each test size. The crack length for onset of unstable fast fracturing is not proportional to the width of test sheet. If this fast fracturing crack length is thought of as representing the end of the experiment, the ratio of the final to the initial crack length is, in this set of size effect experiment, a decreasing function of the size of the test sheet.

The middle diagram of Fig. 4 shows the result when the same starting crack size is used for each size of test. Again, the load is adjusted so that the nominal stress is a constant value. In this case the ratio of final to initial crack length increases with the size of the test sheet.

If the slow fractional extension rate period is represented by a constant average value as shown by the dashed lines on the lower diagram of Fig. 4, then the total time to fracture may be represented as follows:

$$\frac{1}{x}\frac{dx}{dt} = \frac{1}{t_0}, \qquad (7.4)$$

$$t_f = t_0 \log \frac{x_2}{x_1}. \qquad (7.5)$$

In these equations, t_0 is a constant, t_f is the total time to fracture, x_2 is the final crack length, and x_1 is the initial crack length.

One observes that, for the size effect experiment in which the starting crack was proportional to the test sheet size, t_f decreases with increasing test sheet size. This is a normal size effect in terms of time to fracture failure. However, the second set of tests discussed above in which the starting crack was the same for each sheet size possesses the opposite tendency in terms of time to failure as affected by sheet size. When these same tests are performed with no initial starting crack, the time to final fracture decreases with increasing sheet size.

8. Growth of damage and onset of rapid separation. For some purposes a fatigue fracture experiment may also be regarded as an illustrative model of a fracture developed during a single load application. When the nominal stress is held well below yield stress levels as in a fatigue test inelastic strains are detectable during the initial cycles only by internal friction measurements. However, some regions are weaker than others, some shifting of internal load distribution occurs, and configurations for load relaxation by localized plastic slip eventually form. It then becomes possible to develop cracks which grow to sufficient size for self-fracturing without the development of measurable general yielding. Because of the special conditions of long times, strain reversals, and local temperature fluctuation the balance of factors contributing to crack formation during fatigue differ from the case of fracturing under single load application, as discussed by FREUDENTHAL[1]. However, the suppression of general yielding assists observation of cracks during early stages of their growth and studies of development and growth of cracks during fatigue are of definite general interest.

Careful microscopic observations have been made by THOMPSON[2] and his associates of the appearance of minute fatigue cracks in polycrystalline pure

---
[1] A. M. FREUDENTHAL: "Fatigue", this volume.
[2] N. THOMPSON, N. WADSWORTH and N. LOUAT: Phil. Mag. **1**, 113—126 (1956).

copper. The first detectable cracks appeared to form directly from slip bands and their lengthening was accompanied by additional local deformation. As was shown in reference [11] the macroscopic extension process of large cracks consists in the formation, spreading, and joining to the main crack of new crack-like elements. Microscopic studies of fatigue fracture origins reveal an equally complex pattern of events. These are not discussed in further detail since the crack forming actions are somewhat specialized to surface cracks and to fatigue. However, one may note that new fracture origins, whether located in the elevated stress zone of an advancing large crack or developed within a zone of localized deformation during fatgiue, are consistently associated with substantial amounts of plastic strain.

Observable stages of crack lengthening are practically confined to the final half of the fatigue life. At the end of the fatigue life rapid fracturing occurs during one load cycle. $\mathscr{G}_c$ values have been calculated by KIES and IRWIN[1] from estimates of the crack size and load at instability using illustrations accompanying papers on fatigue of steel and aluminium alloys. The values obtained in this way compare will with $\mathscr{G}_c$ values measured on similar material not subjected to fatigue damage.

Large amounts of experimental information exist relating the creep rate of metal rods in tension to the tensile stress. Such studies are simplified by the fact that the force motivating creep, the longitudinal stress, can be held to a constant value as the deformation proceeds. It would be desirable at this point to discuss similar information relating time rate of crack extension to the crack-extension-force. However, the force definition appropriate when the crack length is large compared to the accompanying zone of plastic strains may not be appropriate for a small crack embedded in a region in which there is general plastic yielding. Furthermore, continual readjustment of the loads to hold the crack-extension-force constant while conceptually possible is experimentally difficult and little data based upon such tests is available[2]. Nondestructive measurement of the spreading of cracks within thick solids is also experimentally difficult and again the experimental information at hand is meager.

The first of the above difficulties may be resolved somewhat arbitrarily by computing the crack-extension-force as if no plastic strains were present throughout the entire range of crack lengths. This procedure has the virtue of uniformity and one is less at fault in ignoring influence of plastic strains than in ignoring influence of crack opening size. The second and third difficulties restrict this discussion to what can be inferred from illustrative models as discussed in Sect. 7 coupled with overall consistency considerations.

Such observations as have been reported indicate that the zinc foil results can be considered a good model of slow extension of cracks in glassy and polycrystalline materials. These results can be explained if one assumes that the development of fracture origins in each flawed region of a test specimen follows the same trend as do the fractional time rates of extension shown in Fig. 4. Thus, if a number of microscopic separations are growing in a competitive way with fractional or specific time rates of extension which tend to increase, then in terms of averages, the ratio of the length of the largest to that of the smallest cracks will also increase.

Consider two similarly stressed regions of the same material whose sizes are in the ratio, say, of one to two. Assume that each contains a normal distribution

---

[1] Unpublished notes.
[2] For an isolated exception see J. J. BENBOW and F. C. ROESLER: Proc. Phys. Soc. B **70**, 201 (1957).

of flaws. The larger region may be thought of as slightly weaker than the smaller one because of its greater chance of containing a more serious flaw. With growth of separations under the influence of tensile stress and time, the strength disadvantage of the larger region relative to the smaller region may be expected to increase steadily.

Amplification of an initially small "flaw probability" type of size effect by growth of fracture origins is a prominent characteristic of fatigue tests. A similarly large size effect is the normal expectancy of creep-rupture tests. The end point of slow fracture extension in fatigue, creep-rupture, and tensile tests is the onset of a fast fracture using locally released strain energy for propagation.

Judging by the curves of Fig. 4 the time rate of crack extension may be represented as a linear increasing function of $\mathscr{G}$ during stages of crack growth well removed from onset of fast fracturing. However, toward the last of the growth period the time rate of crack extension increases quite rapidly with increase of crack-extension-force. The general features of the relationship between time rate and force for crack extension are thus similar to the relationship which has been found to apply for creep rates as a function of stress.

In the case of strong metals, glass, and hard plastics, the onset of final self-fracturing is often observed to be abrupt and sudden. The reason for this is two-fold. In the first place, the resistance of the test specimen to fracture extension decreases in regions of largest stress or largest flaws due to growth of fracture origins. In the second place, the average fracture extension rate prior to fast fracture is composed of discontinuous segments of relatively fast fracture extension. The crack extension process waits for an accumulation of damage in material near the end of the crack sufficient to lower the resistance to crack extension within range of the existing crack-extension-force. Then the crack extends quickly through this damaged region. Finally, one such quick extension provides a sufficient increase of force so that the separation process no longer has to wait for time and stress to produce additional fracture strength damage. In extreme cases, the first visible crack extension process starts the final self-fracturing separation.

Characteristically, high temperature grain boundary fractures under creep-rupture conditions occur with little or no local reduction of area of the test specimen at the location of the fracture. The separation is of sudden brittle kind starting from one or more surface cracks of relatively small depth. Clearly, a general loss of strength is of at least comparable importance to lengthening of the starting crack in causing onset of rapid fracture in such instances.

In many situations of practical importance, there are residual stresses within fabricated metallic alloys, which are not removed by removal of the external loads. In such instances, crack growth, assisted by hydrogen embrittlement, corrosion, thermal fatigue, or by other causes, continues in unloaded structural members and may produce a substantial loss of fracture strength over a long period of time.

9. **Conditions for fracture in terms of stress.** The law of constant resolved shear stress represents rather well the influence of orientation upon yielding of ductile single crystals. NABARRO[1] has pointed out that this result corresponds to a characteristic critical force per unit dislocation line length necessary for expansion of dislocation loops in the acting slip planes. A similar inference would pertain to the passing of obstacles on the slip planes by crowded arrays or "pile-ups" of dislocations. Grain boundaries, interstitial atoms, and undissolved

---

[1] F. R. N. NABARRO: Phil. Mag., Ser. VII **12**, 213 (1951).

constituents add to the complexity of analysis of yielding of real materials. Whereas the force on a segment of a single dislocation loop depends only upon the resolved shear stress and not upon the loop size, the force on a dislocation "pile-up" is dependent upon spacing and extent of these arrays. Of the factors influencing force on a dislocation array only the average stress on the yielding region is directly observable. Thus, although the idea of relating forces on dislocations to their time rates of movement is of fundamental importance in dislocation mechanics, results of yield strength measurements are always given in terms of average stress.

Somewhat the reverse of this occurs in the application of a dislocation-type force concept to crack extension. As the crack becomes larger, restriction of

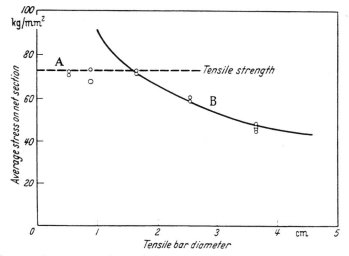

Fig. 5. Influence of size upon fracture stress for notched round bars of 7075-T6 Aluminum alloy. Notch reduced section area by one-half and had root radius of 0.25 mm. Curve B corresponds to a $\mathscr{G}_c$ value of 177×10^5 ergs/cm². Data from WEISS, reported by LUBAHN.

plastic strains on a relative scale to the zone of overstress near the crack edges leads finally to a clear physical situation well adapted for application of a crack-extension-force concept. Furthermore, fracture test data expressed in terms of the characteristic value of crack-extension-force for onset of rapid crack extension are directly useful in engineering strength calculations.

On the other hand, applications of the force idea to extension of very small cracks, while of assistance to qualitative understanding of their development and growth, are of little practical value. When a solid is tested to rupture under conditions such that the unstable crack size is not observed or under conditions such that a force calculation based upon applied loads and crack configuration is not feasible, then the conditions for fracturing can only be stated in terms of the average stresses and strains descriptive of the region within which the fracture instability developed.

Consider for example the test results discussed by LUBAHN[1] shown on Fig. 5. A somewhat brittle aluminum alloy was subjected to tests at various sizes in the form of round bars sharply side notched to a depth which reduced the net section-area by one half. The results shown are typical of those from a wide variety of fracture tests in which a notch or saw cut is used to assist the starting

---

[1] J. D. LUBAHN: Proc. of 1955 Sagamore Conference on Strength Limitations of Metals, Syracuse University, p. 159, 1956.

of a crack. When the test piece is sufficiently small, the zone of plastic deformation extends across the net section. The average stress on this section at the time of fracture then cannot exceed the yield stress by more than the moderate amounts due to work hardening and the influence of biaxiality of stresses upon the maximum shear stress. Thus line A on Fig. 5 is only about ten percent above the yield stress and coincides, in this case, with the ultimate tensile strength as measured on unnotched bars of the same material.

When the test piece size is large enough so that the plastic strains accompanying onset of rapid fracture are confined to the zone near the root of the sharp notch, the net section stress is less and reduces toward zero as the test piece size is increased. For points on curve B (Fig. 5) the product of the square root of the specimen size times the average net section stress has a constant value. Curve B corresponds to development and propagation of a crack from the vicinity of the notch root when the stress environment of that region reaches a critical magnitude and the result can be stated in terms of a characteristic crack extension force.

When a specimen with no notch is broken, as in the case of the most commonly used tests, the situation more or less resembles that of Fig. 5, curve A, depending upon the size and ductility of the specimen. Although the specimen has no intentional notch, as the load is increased, a starting crack of some type eventually must form. If at onset of fast extension of the starting crack the entire net section is yielding, then the average tensile stress for such a fracture, like that for the smallest test specimens of Fig. 5, can be estimated from the known resistance to plastic deformation of the material.

When the material is quite brittle, plastic strains accompanying development of the starting crack may be restricted to the close neighborhood of largest flaws to such an extent that the entire strain prior to fracture appears to be elastic and recoverable by unloading. All degrees of this condition occur in various crystalline and amorphous solids. Cleavage surfaces apparently devoid of plastic strains can be produced in some materials; in mica primarily because of the weakness of bonding across the cleavage plane, and in diamond primarily because of the absence of dislocation mobility. On the other hand, evidence of non-recoverable strain is found near fracture origins in glass and in all metals in which brittleness is brought about by lowering of temperature.

When the object of measurement is the determination of stress conditions for development of a small starting crack, the pertinent stresses are those averaged across the entire region within which crack development occurs. The results of such tests are often stated in terms of the average tensile stress normal to the plane of the section eventually severed by the fracture. This is usually the direction of greatest tension. Due to developed or inherent directional weakness the fracture may not have this orientation. For example, single crystal fractures under brittle conditions occur preferentially upon cleavage planes, usually the set of planes of widest spacing. In such cases the average tension resolved normal to the cleavage plane provides a fracture stress component which is essentially constant for fracture of the crystal in various orientations.

Cleavage of single crystals was discussed by BARRETT[1] who illustrated constancy of the critical normal tensile stress with results on bismuth. Results with zinc crystals in the temperature range $-253°$ C to $-80°$ C and with bismuth in the range $-80°$ C to $20°$ C substantiate the relative insensitivity of the critical normal tensile stress to temperature changes. According to BARRETT cleavage

---

[1] C. S. BARRETT: Structure of Metals, p. 317–320. New York: McGraw-Hill 1943.

in most crystals is facilitated by cooling to liquid air temperature and by driving a sharp blade into the crystal along the cleavage plane.

The constancy of critical normal tensile stress corresponds, in terms of crack-extension-force, to a characteristic value for $\mathscr{G}_1$ as defined in Sect. 5. The influences of temperature, strain rate, and biaxiality are common to many solids and are discussed in Sect. 15.

Fracture stress conditions for aluminum and magnesium alloys and steel were discussed by DORN [6] who concluded the results corresponded more closely to a critical maximum shear stress law than to invariance of the greatest tensile stress. A less ductile material, cast iron follows a critical maximum shear stress law only when fracture occurs by the sliding modes. Similar conclusions were reached by PARKER[1].

The general behavior pattern of materials relative to conditions for fracture in terms of average stresses is quite clear. For test specimens in which development of the starting crack is accompanied by general plastic yielding one would expect to predict the average stresses from known conditions for plastic yielding. In extremes of brittleness such that a condition of general yielding is not attained prior to propagation of the starting crack the tensile stress normal to the plane of separation dominates. When fracture occurs in compression tests of materials such as cast iron and ceramics, a shear fracture normally develops from sliding on a plane of greatest shear stress and results of tests are predictable in terms of a critical maximum shear stress.

For a wide variety of materials including single crystals experiments have established that both the yield stress and the fracture stress are dependent upon the specimen size. The size dependency is more noticeable for fracture stress than for yield stress. Reasons for this will be discussed more fully in Sects. 12 and 13.

When the object of measurement is the determination of stress conditions for extension of a well developed and somewhat brittle crack, the stress conditions in the region of the advancing edge of the crack may be given in terms of crack-extension-force as was brought out in Sect. 5. The description of fracture strength in terms of a $\mathscr{G}_c$ value while attractive because of its close association with the crack extension process is nevertheless a considerable oversimplification. Essentially it is a description of the average stress conditions for growth and joining of new fracture elements modified in a way appropriate for the region enclosing the advancing edge of a crack. Representation of fracture strength in terms of average stress is convenient and appropriate when no well formed crack can be observed prior to rupture. Primarily one needs to bear in mind (1) that fracture stress is not easily distinguished from plastic flow strength except for materials so brittle that yielding is not observed, (2) that fracture stress is substantially reduced by increase of specimen size, and (3) that both fracture stress and characteristic crack-extension-force are approximate descriptive concepts dependent in a complex way upon time rates of response to forces at a much smaller scale.

IV. Stress field, velocity, and division of a running crack.

10. Stresses near a running crack. A relatively complete photoelastic investigation of stresses near a rapidly moving crack was made by WELLS and POST [14]. The transparent material used was Columbia resin (CR-39) in the form of flat plates 3.2 mm thick and 12.7 cm wide. The length of plate in the direction

---

[1] E. R. PARKER: Brittle Behavior of Engineering Materials, p. 66. New York: John Wiley & Sons 1957.

of applied tension was 38 cm. The load, applied manually in a small fraction of a second, caused extension to occur from a pre-cracked notch at one of the plate side boundaries. Four spark light sources fired in sequence furnished four views of the stress field as shown in Fig. 6. The isochromatic loops close to the end of

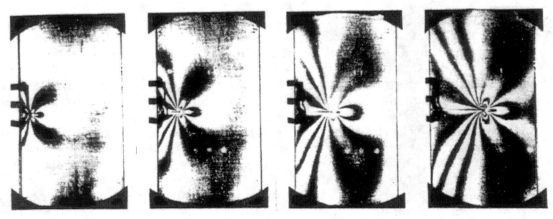

Fig. 6. Stress field near end of running crack from WELLS and POST [*14*]. Flash photographs show isochromatic fringes for four positions of the same crack traversing plate of Columbia resin.

the crack agree in appearance with calculated theoretical isochromatics based upon Eqs. (5.1), (5.2), (5.4), and an added uniform stress, $-\sigma_{0x}$, parallel to the crack. Close to the crack tip each isochromatic loop leans toward the direction

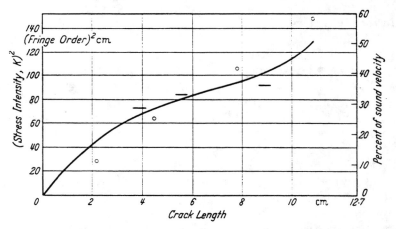

Fig. 7. Circles show average values of K^2 from isochromatics using photographs by WELLS and POST [*14*]. The curve is from Eqs. (6.14) and (10.4) with L proportional to time. Horizontal bars show average velocity measurements by WELLS and POST relative to a velocity of sound of 1.53 km/sec. One fringe order is 0.0092 kg/mm².

of crack motion by an amount which increases with σ_{0x} and has a size which increases with K. By measuring the angle, ϑ_m, from the x-axis (the extended path of the crack) to the position of greatest separation of the fringe from the crack tip and by measuring this greatest separation distance, r_m, it was possible, to obtain values of the two stress field parameters, K and σ_{0x}, for each of the WELLS-POST photographs[1]. The equations employed were

$$K = (2\tau_m)\sqrt{2r_m}\,f(\vartheta_m) \tag{10.1}$$

---

[1] See discussion by G. R. IRWIN published with reference [*14*].

and
$$\sigma_{0x} = (2\tau_m) \sec \frac{3\vartheta_m}{2} \left\{1 + \left(\frac{3}{2}\tan\vartheta_m\right)^2\right\}^{-\frac{1}{2}} \tag{10.2}$$
where
$$f(\vartheta_m) = \operatorname{cosec}\vartheta_m \left(1 + \frac{2}{3}\tan\frac{3\vartheta_m}{2}\cot\vartheta_m\right)\left\{1 + \left(\frac{2}{3}\cot\vartheta_m\right)^2\right\}^{-\frac{1}{2}}. \tag{10.3}$$

The results for K^2 are shown on Fig. 7. Each isochromatic loop corresponds to a constant maximum shearing stress, τ_m, in the plane of the plate. In the calculations the value of τ_m was taken equal to the fringe order and was not converted to stress units. The average values of K^2 obtained were nearly in simple proportionality to the crack length. The lateral compression factor, σ_{0x}, tended to be least in the central portion of the plate and was about half of the initial longitudinal stress, σ_0.

To obtain a theoretical calculation for purposes of comparison, use was made of a modified static stress analysis. Eq. (6.14) was employed with the value of A set equal to the plate width and σ adjusted for dynamic unloading due to rapid motion of the crack. To make the dynamic unloading adjustment, a study was made of the influence of the crack length upon the spring constant of a long plate. The result was

$$\frac{\sigma_0}{\sigma} = 1 + \left(\frac{8A}{\pi L}\right) \log\left(\sec\frac{\pi a}{2A}\right) \tag{10.4}$$

where L is the plate length and σ_0 is the initial tensile stress applied to the plate.

For the curve shown on Fig. 7 it was assumed the applicable value of L increased in proportion to time from the start of rapid crack extension and had the value $1.5\,A$ when the end of the crack reached the midpoint of the plate width. The values of crack-extension-force predicted in this way are within the scatter range of the observations. An influence of bending in the plane of the plate not considered in the calculations would tend to increase the observed values for the longest crack lengths.

It is evident that, since the crack velocity is substantially below the velocity of elastic disturbances, the stresses in the end region of the running crack must resemble the stress field characteristic of the end region of a stationary crack. This was verified by WELLS and POST [14] by appropriate comparisons with the isochromatic fringes near the ends of stationary cracks. Additional support for this conclusion is provided by the mathematical studies of YOFFE[1].

11. Texture, velocity, and division of fractures. Examination of the stress equations for the opening mode crack dislocation, Eqs. (5.1) to (5.4), shows that the locus of greatest tensile stress does not lie in the x, z plane which contains the crack. Table 3 gives the relative values of the largest tensile stress, σ_1, and of the largest shear stress, τ_m, for various angles, ϑ. For simplicity it is assumed the components of added constant stress are zero. The ϑ value of largest tensile stress at a fixed small distance, r, from the edge of the crack then becomes $\pm\pi/3$.

Previous discussion has brought out that plastic strains assist the development of the smallest elements of separation. Thus, in the elevated stress zone

Table 3. *Relative values of largest tensile stress and largest shear stress near the edge of an opening mode crack for $\sigma_{0x} = 0$ and $\nu = 0.3$.*

ϑ	$\sigma_1\left(\frac{\sqrt{2r}}{K}\right)$	$\tau_m\left(\frac{\sqrt{2r}}{K}\right)$
0°	1.00	0.200
30°	1.18	0.306
60°	1.30	0.433
90°	1.21	0.500
120°	0.93	0.433

---
[1] E. A. YOFFE: Phil. Mag. **42**, 739 (1951).

surrounding the crack edge, flaws develop into components of fracture most rapidly at positions near a ϑ value of $\pm\pi/3$ where substantial shearing stress as well as maximum tensile stress exists. As new fracture elements form away from the idealized locus plane of the main crack and attain sufficient size, sliding modes of separation develop joining them with the main crack and with each other thus advancing the composite edge of the main crack. The magnitude of the roughness of the fracture surface depends in part upon ductility factors which handicap new openings in attaining a crack-like shape and in part upon the numerosity of serious flaws which determine the largest r values at which the stress elevation is effective. Due to the nature of the opening mode stress field, fractures can avoid roughness only in exceptional situations, for example the cleavage of brittle single crystals.

At the position $r, \pi/3$ the direction of σ_1 is normal to the plane of the main crack. However, the character of new openings forming in this region depends upon ductility, directional weakness, and the nature of the local flaws as well as upon the stress field. In low carbon steel at room temperature a flat fracture may be primarily composed of small elements of sliding mode separation as discussed by PARKER[1]. These show a preference for planes of greatest shear stress although the locus of the main crack is a plane normal to the greatest tension of the general stress field. In the same material at low temperature, opening mode separations by cleavage dominate. The fracture surface then contains many small elements of separation in planes nearly parallel to the plane of the main crack.

In materials for which resistance to plastic flow increases with time rate of strain the velocity of crack extension has a marked influence upon the texture of the fracture surface near onset of rapid crack extension. As crack velocity increases, the development of outlying flaws is at first suppressed and the fracture surface becomes more smooth. However as will be clear from later discussion, the velocity increase lags behind increase of crack-extension-force. Thus the effectiveness of the stress field in development of flaws away from the crack plane soon dominates and increased roughening occurs.

The stress field situation for the sliding modes of fracture is not analogous to that for opening mode. For such fractures the role of new elements ahead of the main crack in the extension process has not been observed. Judging by position and direction of the largest shear stress, formation of new elements would be expected to occur most rapidly in the plane of expected extension of the crack.

As an aid to understanding the fracture strength of a material careful study of the fracture surface is always rewarding. Guiding principles to assist such studies are given in references [10] and [11].

Under the influence of a steadily increasing crack-extension-force, the velocity of a running crack approaches a limiting magnitude. SMITH and KIES[2] investigated the effect of the tensile stress at crack initiation upon the velocity of cracks traversing plates of cellulose acetate. It was found the velocity increased with the initial tension until the crack began to divide. Trials at successively greater magnitudes of initial tension produced, first, the occurrence of division at shorter crack lengths, then sub-division of the branches, and finally a condition which one might term shattering.

---

[1] E. R. PARKER: Brittle Behavior of Structural Materials, p. 67—70. New York: John Wiley & Sons 1957.

[2] H. L. SMITH, J. A. KIES and G. R. IRWIN: Phys. Rev. **83**, 872 (1951) (abstract).

At tensile loads sufficient to cause crack division the velocity was estimated in terms of movement of the ends of the most advanced cracks. This velocity appeared to be independent of the initial tensile stress. At tensile loads just less than that necessary for crack division the velocity across more than half of the plate width was equal within experimental error to the limiting velocity found after development of shattering.

It will be assumed negligible error is incurred in referring to the limiting crack velocity in a material either as the velocity of the locus of most advanced cracking or as the velocity of a single crack when the crack-extension-force is near the magnitude required for crack division. Table 4 gives a summary of limiting velocities on this basis.

Table 4. *Limiting velocities of running cracks*[1].

Material	Values of $0.5\,c_2$ km/sec	Experimental velocities km/sec
Glass (soda-lime-silica)	1.57	1.54
Glass (silica)	1.85	2.19
Cellulose acetate	0.34	0.30
Steel (0.18 C-annealed)	1.58	1.50
Columbia resin (CR-39)	0.47	0.55

$$\mu = \varrho\, c_2^2 = \frac{E}{2(1+\nu)}; \quad \mu = \text{shear modulus}; \quad \varrho = \text{density}.$$

Note: Glass data — H. RAWSON, Soc. of Glass Tech. **36**, 173 (1952). Cellulose acetate data — H. L. SMITH and W. J. FERGUSON, Naval Research Laboratory Progress Report, April 1950. Steel data — OSRD Report No. 6452, Jan. 1946. Columbia resin data — WELLS and POST [14].

Returning to the WELLS-POST experiment, on Fig. 7 the horizontal bars indicate average observed values of crack velocity as a function of crack length. One may note that as the crack extends the increase of velocity becomes much less than the increase with crack length of the measured crack-extension-force. A similar remark applies to comparison of measured velocities with the theoretical estimates of crack-extension-force although the margin of difference in this case is not so great.

The fact that a limiting velocity is at least nearly reached prior to branching of a crack suggests a calculation of limiting velocity may be allowable within the general frame work of comments on this topic by MOTT[2]. Exploratory calculations on this basis were made[3] which suggested the limiting velocity should be about half the velocity of elastic shear waves, c_2. Both the theoretical estimates and limiting velocity measurements are segments of incomplete work. The measurements at hand correspond closely to the value of one-half c_2 which suggests an inertial type velocity limitation whereas the appearance features of the experiments suggest the velocity is limited by onset of crack division.

Along the borders of long single cracks resulting from fast fractures of polymethylmetacrylate plates SMITH and KIES observed numerous small pairs of cracks suggestive of unsuccessful efforts to divide the main crack. There was an increased roughening of the texture of the fracture surface with distance from the region of onset of fast fracturing. This behavior, which is typical of most solids,

---
[1] Table prepared by H. L. SMITH.
[2] N. F. MOTT: Engineering **165**, 16 (1948).
[3] Correspondence between A. A. WELLS and G. R. IRWIN.

is apparently due to the expansion of the field of elevated stresses and, correspondingly, the increase of the crack-extension-force.

It was noted, however, that a doubling of the force, while sufficient to produce occasional small pairs of cracks, did not cause crack division. For plates of polymethylmetacrylate, cellulose acetate, and Columbia resin approximately a ten fold increase of crack-extension-force above the characteristic value, $\mathscr{G}_c$, was necessary to cause crack division.

V. Effects of size upon fracturing.

12. Analysis of strength by extreme value statistics. It was shown in Division I that, even for a pure liquid, an influence of size upon tensile strength is predicted. In glasses and metals, processes such as slip, twinning, cleavage, and growth of holes or micro-cracks which, in the aggregate, compose plastic flow and fracture depend in various ways upon pre-existing flaws or strength inequalities. The existence of strength inequalities implies existence of size effects. Thus, it would be self contradictory to assume that plastic deformation and separation processes occur and at the same time that no dimensional effects exist. Influences of size are fundamental to the nature of the deformation and separation processes. It is only the magnitude and character of the size effects which are in question. Larger size effects for fracture than for plastic flow are anticipated because fracturing tends to accentuate pre-existent strength inequalities to a greater degree than plastic flow. In fact, in a fracture test, the largest flaw developed by accentuation of strength inequalities is the crack, extension of which severs the test piece.

If one considers classes of fracture in which the volume element whose failure controls the measured strength is small compared to the total volume tested, statistical methods sometimes called "extreme value statistics" are useful. Fractures generally described as brittle and fractures of slender filaments and wires are of this class.

The fact that fracture origins grow at various rates under the influence of applied load and temperature still permits application of a statistical theory of fracture strength based on a flaw probability function. One merely assumes that the flaw probability function refers, not to the flaws in the material prior to the test, but to those which exist just prior to onset of unstable fast fracturing. The significant flaws in each small volume element are a function of the load-time history of that region. To a fair approximation, one may think of the effects of the load-time history as merely altering the values of the parameters in the flaw probability distribution function. These parameters, then, may be different for the same material under different conditions of testing.

For the statistical analysis it will be sufficient to use concepts developed by WEIBULL [4] with some modifications to show the influence of flaw growth. Assume, at first, the test volume is in uniform uniaxial tension. The influence of each weakening defect is assumed independent of other defects and such that it would by itself limit the supportable load on the test bar to a tensile stress S. Let P_S be the probability that a fixed small volume ω will contain an "S-flaw" in the range S to $(S+dS)$. The probability of no S-flaw of this kind in a large volume V is $(1-P_S)^{V/\omega}$. Assume the range of S values sub-divided and numbered so that

$$\delta S_K = S_{K+1} - S_K > 0. \qquad (12.1)$$

The probability for failure in the range, δS_K, for the whole test volume is the probability that V contains no flaw as serious as S_K diminished by the probability that V contains no flaw as serious as S_{K+1}. This difference is

$$(1 - G_{K-1})^{V/\omega} - (1 - G_K)^{V/\omega} \tag{12.2}$$

where

$$G_K = \sum_{i=0}^{K} P_{S_i}. \tag{12.3}$$

For test specimen sizes of practical interest it is reasonable to limit attention to a group of most serious flaws which are a small fraction of the total number. This means that, for S values of primary interest, $\frac{V}{\omega} P_S$ is small compared with unity. The difference expression, (12.2), may then be replaced by

$$\frac{V}{\omega} P_{S_K} (1 - G_{K-1})^{\left(\frac{V}{\omega} - 1\right)}. \tag{12.4}$$

For mathematical simplicity it will also be assumed G_K can be approximated as a continuous function of S, namely $G(S)$, and that

$$\frac{V}{\omega} P_{S_K} (1 - G_{K-1})^{\left(\frac{V}{\omega} - 1\right)} \approx -\frac{\partial}{\partial G}\left[\exp\left(-\frac{V}{\omega} G\right)\right] dG \tag{12.5}$$

where the right side is evaluated at S_K.

It is clear that $G(S)$ must be a positive increasing function of S which is zero when S is zero. Consequently, S, considered as a function of G, is a monotonically increasing function of that argument. From these comments, one can see that the assumptions made above, general as they are, have certain necessary implications regarding the effect of the size of the volume subjected to fracture testing. For example, the expected average failure stress S_A from a large number of trials is

$$S_A = \int_{G=0}^{\infty} S\, e^{-\frac{V}{\omega} G} \frac{V}{\omega}\, dG \tag{12.6}$$

and

$$\frac{\partial S_A}{\partial V} = \frac{1}{V} \int_0^{\infty} S\left(1 - \frac{V}{\omega} G\right) e^{-\frac{V}{\omega} G} \frac{V}{\omega}\, dG. \tag{12.7}$$

When plotted as a function of G, the expression

$$\left(1 - \frac{V}{\omega} G\right) e^{-\frac{V}{\omega} G} \tag{12.8}$$

has the same area under the positive portion, G less than ω/V, as is enclosed by the negative portion, G greater than ω/V. Since S always increases with G, the result of the integration by which $\partial S_A/\partial V$ is calculated must be different from zero and negative under all permissible choices of V, ω, and $G(S)$.

For illustration, consider the choice of $G(S)$ made by WEIBULL [4].

$$\left.\begin{array}{ll} G = 0 & \text{for } S < \sigma_0, \\ G = \left(\dfrac{S - \sigma_0}{\sigma}\right)^n & \text{for } S > \sigma_0. \end{array}\right\} \tag{12.9}$$

This corresponds to having σ_0 as a lower threshold of strength and to a flaw probability P_S of

$$P_S = \left(\frac{S - \sigma_0}{\sigma}\right)^{n-1} \frac{dS}{\sigma}. \tag{12.10}$$

With n greater than unity, this implies an ever greater probability as the supposed flaw becomes less serious in its weakening effect, an assumption which is plausible in its gross aspects. A frequency plot of test results falling within small, equal intervals of S is expected to resemble the function

$$f(S) = n\left(\frac{S-\sigma_0}{\sigma}\right)^{n-1} e^{-\frac{V}{\omega}\left(\frac{S-\sigma_0}{\sigma}\right)^n}. \tag{12.11}$$

If n is large, this function has negligible magnitude, except near the value of S for which it is maximum.

For the assumed case of uniform tension, we obtain for the average strength,

$$S_A = \sigma_0 + \sigma\left(\frac{\omega}{V}\right)^{1/n} \Gamma\left(1 + \frac{1}{n}\right) \tag{12.12}$$

and for the mean square relative deviation (relative variance)

$$\eta^2 = \left(\frac{S-S_A}{S_A}\right)^2_{\text{average}} = \left(1 - \frac{\sigma_0}{S_A}\right)^2 \frac{1.5}{n^2}. \tag{12.13}$$

If σ_0/S_A is negligible, the value of η is about $1.2/n$ and can be computed readily for comparison with observed scatter of experimental results when only n is known.

To attribute to a flaw population the lowering of fracture stress from theoretical estimates by a factor of 100 to 1000 has been criticized by some as inferring less uniformity in the results of fracture stress measurements than is actually found. However, measurements of the effect of size upon fracture stress provide a direct indication of the magnitude of expected scatter due to the flaws. Relation (12.13) gives the predicted theoretical scatter when P_S is represented by relation (12.10). The scatter actually found is invariably greater rather than less than the magnitude predicted by Eq. (12.13).

This point was discussed from a different aspect by FISHER and HOLLOMON[1]. Assuming a distribution of micro-cracks which is equivalent to a more realistic P_S than that of Eq. (12.10), FISHER and HOLLOMON concluded that the degree of uniformity of fracture stress experimentally found corresponded in theoretical terms to having a realistically large number of flaws in the test specimen sizes under consideration. The increase of η with decrease in number of flaws predicted by their analysis may help to explain the large variation of results which has been found in measurements of the strength of single crystal filaments or "whiskers" of metal[2].

In addition it was shown that this model gave essentially correct predictions of the influences upon fracture stress of flaw orientation during deformation of metals.

To assume σ_0/S_A negligible permits considerable simplification of mathematical analysis. It may be noted that one cost of this simplification is the fact that the corresponding value of η is insensitive to size of test volume whereas a review of strength measurement data shows a definite trend toward decrease of η with increase in size of the specimen tested as predicted by the Fisher-Hollomon statistical model and by Eq. (12.13).

13. Influences of growth of crack-life flaws. When a statistical analysis of fracture strength is employed with a view toward obtaining general trends of fundamental significance, it is helpful to keep in mind certain limitations of this

---

[1] J. C. FISHER and V. H. HOLLOMON: Metals Techn. **14**, No. 5 (1947).
[2] S. S. BRENNER: J. Appl. Phys. **27**, 1484 (1956).

approach. For example, to obtain determinations of a parameter such as σ_0 requires determination of S_A for scaled sets of strength measurements extending over an extremely large size range. Although in principal it is possible to do this without alteration of the material flaws, in practice it is difficult to avoid introduction of new classes of more serious flaws or increased seriousness of existing flaws in the largest sizes tested. Intuitively, one would expect some lower threshold to the weakening influence of the various possible material defects. However, if one studies graphs of log S_A as a function of log V from past experimental work, no convincing evidence for a σ_0 term of significant size is found. The previous comment may assist in reconciling this fact with one's intuitive idea that a limit of weakening influence nevertheless exists. In each set of experimental results on a particular material, a limited range and variety of largest flaws dominate the results. For consistency with general trends, it is necessary to choose for the flaw probability function a form which permits small flaws to be more numerous than large ones to a degree controllable by adjustment of the parameters of the function. The Weibull flaw probability function satisfies this requirement and permits convenient simplifications in the consideration of flaw growth and of non-uniform stress fields. For reasons apparent from the above comment, it is believed mathematical convenience also justifies dropping the σ_0 term of the Weibull formula in the analyses which follow.

Consider next the influence of flaw growth upon the parameter n. For illustration, it will be assumed that the controlling S-flaw of each volume element ω can be thought of as an embedded penny-shaped crack of radius r and that the growth rate of r under tensile stress S follows the trends shown in Fig. 4. The resistance to fracture extension at onset of unstable fast fracturing is assumed to be a constant. Since dG/dA is proportional to the product $S^2 r$, one has

$$S^2 r = S_0^2 r_0 \tag{13.1}$$

where S_0 and r_0 are any selected reference pair of values of stress and critical crack size. The S value of a volume element containing a crack of radius r then becomes

$$S = S_0 \left(\frac{r_0}{r}\right)^{\frac{1}{2}}. \tag{13.2}$$

Thus

$$\frac{dS}{S\,dt} = -\frac{1}{2}\left(\frac{dr}{r\,dt}\right). \tag{13.3}$$

The number of volume elements ω in V containing S-flaws in the range S to $(S+dS)$ is, with σ_0 omitted from equation (12.10)

$$N_s = \frac{V}{\omega}\left(\frac{S}{\sigma}\right)^{n-1}\frac{dS}{\sigma}. \tag{13.4}$$

This relation is represented by curve A of Fig. 8. If the flaw growth effect corresponds to a constant fractional change in diameter per unit time of each penny-shaped crack, equation (13.3) shows that each point of curve A may be thought of as moving to the left with constant velocity, so that, after some period of time under tensile loading, the frequency graph of S-flaws would be represented by a parallel curve such as curve C. However, this situation is most unlikely. Since our interest is primarily in representing the behavior of the largest flaws, it is expected that the fractional time rate of extension will be an increasing function of their length or a decreasing function of their S-value.

Thus the influence of flaw growth is expected to be represented by a curve of smaller slope, such as curve B of Fig. 8. There is no reason to draw curve B other than as a straight line since no more is attempted than an approximate representation of the largest flaws as allowed by a Weibull type probability function. A decrease of the parameter n with time under load is, therefore, to be expected. Results of size effect tests with varied loading time have consistently shown this trend. From the above considerations, one anticipates that damage to the strength properties of a material will cause an increase in the scatter of test data, a lowering of the average measured strength, and an increased sensitivity of the strength to size. These trends are in general agreement with engineering experience.

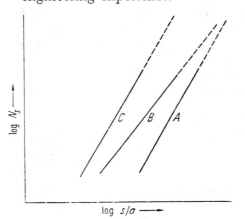

Fig. 8. Schematic graphs of the relation between $\log N_s$ and $\log(S/\sigma)$ from Eq. (13.4) showing influence of two hypothetical trends of flaw growth.

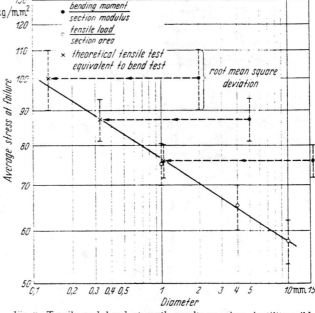

Fig. 9. Tensile and bend strength results on a low ductility mild steel at liquid air temperature. Replotted from data of DAVIDENKOV, SHEVANDIN, and WITTMAN.

14. Measurements of effects of size upon fracturing. When the statistical approach is applied to a non-uniformly stressed solid, for example, a bar subjected to bending moments, some modifications of the above analysis are required. It is, at the outset, necessary to consider whether the influence of stress gradient over regions comparable to the instability size of a crack may be neglected. With this condition satisfied, one may make a computation of the expected average value, S'_A, for a set of trials in which the strength value of each trial is given as the computed largest tensile stress, S_m, in the test piece corresponding to the loads just before onset of sudden fracture. The result is

$$S'_A = \left(\frac{1}{R}\right)^{1/n} S_A \tag{14.1}$$

where S_A is as given by Eq. (12.12) and

$$R = \frac{1}{V} \iiint \left(\frac{S(x, y, z)}{S_m}\right)^n dx\, dy\, dz. \tag{14.2}$$

Here the integral is taken to extend throughout that portion of the test piece subjected to tension and $S(x, y, z)$ is the maximum tensile stress at the point (x, y, z).

In the experiment to be discussed next, test conditions were selected for which very little flaw growth during time under load would be expected. The experiment provides the best known comparison of results of statistical fracture strength analysis for bend tests with those for tensile tests. Fig. 9 shows results replotted from DAVIDENKOV, SHEVANDIN, and WITTMAN[1]. It was the purpose of this work to investigate fracture size effect of a commercial metal under conditions of extreme brittleness. Thus the steel chosen had a high phosphorous content and the tests were done at liquid air temperature, the area reduction at this temperature being insignificant. The room temperature ultimate tensile strength was 62 kg/mm^2. Small specimens were machined from the ends of larger specimens after they had been broken. The tests were made in tension in one set of tests and in bending in a second set of tests. For each specimen size, sixteen to thirty results were obtained.

The results show areas of agreement with the statistical analyses discussed above. When the logarithm of the failure stress is plotted against the logarithm of specimen diameter as in Fig. 9, the slopes of the curves through the data are about the same for the bend tests as for the tension tests and suggest an average value of 25 for the constant n. Estimates of η were made from the published data and were found to vary in the range of 1.1 to 1.9 times the theoretical value. Each group of bend tests provides an average failure stress which is theoretically equal to the failure stress for a group of tensile bars which are of smaller size by an amount which can be calculated by computing R for the bend tests from Eq. (14.2). The close agreement with the tensile test results of the shifted bend test points as shown in Fig. 9 is somewhat surprising in view of uncertainties of calculation and measurement. The computation of R for the bend tests assumed the stress to be proportional to distance from a neutral plane through the axis of the cylindrical specimen and to fall off linearly with distance along the rod from the central point of load application. Due to the large value of n, the results are determined primarily by stresses in a small region at the position of greatest tension. Since the stress drop with distance along the bar from the midpoint is undoubtedly not linear in the region of the greatest stress, the computation of R using that stress gradient may be questioned. Possibly changes in the magnitude of the greatest tensile stress from that predicted by the simple beam formula compensate in some way so that a correct computation of R would not produce greatly different results.

Notched bar size effect studies in tension using a good quality structural low carbon steel were made by BROWN, LUBAHN, and EBERT[2]. The test conditions permitted considerable amounts of plastic deformation at the root of the notch. The bars broke suddenly, essentially at the maximum recorded load. The trend with size of the measured nominal stresses corresponded to $n = 50$. Accurately scaled bend tests of notched bars using tough heat-treated Ni—Cr alloy steels were reported by SHEARIN, RUARK, and TRIMBLE [6]. The values of n, again based upon maximum load, showed little consistency from one plate to another but were all in the range 60 to 100.

From experiments by ANDEREGG[3] on small diameter glass rods in tension, a value of n of about 8 may be deduced as characterizing the effect of changes in specimen length.

McVICKER and IRWIN[4] studied the effect of specimen length upon strength of glass fibers on a basis similar to that of ANDEREGG. A comparison was made of

---

[1] N. DAVIDENKOV, E. SHEVANDIN and F. WITTMAN: Trans. Amer. Soc. Mech. Engrs. **69**, A 63 (1947).
[2] W. F. BROWN, J. P. LUBAHN and L. J. EBERT: Welding J. Res. **26**, S. 554 (1947).
[3] F. O. ANDEREGG: Ind. and Eng. Chem. **31**, 290 (1939).
[4] NRL Laboratory Report in preparation.

fibers damaged by handling with others carefully protected from surface abrasion. The decrease of average strength for the damaged fibers was much greater for long than for short specimens suggesting the damage did not greatly alter the number of relatively small flaws. The corresponding n values were 3 for the damaged fibers and 12 for the undamaged fibers. Despite use of careful testing methods the measured η values exceeded the value $1.2/n$ by somewhat more than a factor of three. Use of a flaw distribution function more like that of the Fisher-Hollomon model discussed in Sect. 12 would have improved agreement of theoretical and measured η values.

In the report of their experiments by SHEARIN, RUARK, and TRIMBLE [6], attention was primarily upon total work done by the loading forces in causing fracture rather than on the value of the maximum load. Fig. 10 is typical of tests of this nature on steels which are tough in their resistance to crack initiation. Unstable fast fractures occurred in some, though not in all, of the largest size bars. The load-strain curve for the largest size bar of Fig. 10 was selected to show an example of a test which ended with a sudden fracture. Three prominent size effects were noted: as the size of the test piece was increased, the total work per unit volume to complete fracture became less, the initial crack at the root of the notch tended to become well developed at a smaller total strain, and the chance of sudden fracturing was increased. With respect to the first two effects, one may observe that the notch indentation is, in each specimen, the largest flaw and that for equal nominal stress the force tendency acting to extend the notch is proportional to the specimen size. For a more detailed view, one may observe that, during early stages of yielding, flaws in similar positions relative to the notch would have the same stress environment in the different test sizes. Individually, similar flaws in similar positions but in different size bars would develop into extending cracks at the same rate. However, the region subjected to large stresses increases with the specimen size. Thus, because of flaw probability considerations, a notch develops into a natural crack the more readily the larger the bar size. Even the largest bar size of Fig. 10 is too small for a pronounced influence of $\mathscr{G}_c$ upon the nominal or average stress. Thus for these tests the nominal stress at which the crack develops is controlled by plastic yielding and is reduced in the larger bars only by the relatively small amounts typical of size effects associated with plastic deformation.

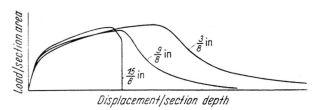

Fig. 10. Typical load-deformation curves for three sizes of notched bar bend tests of a steel. From SHEARIN, RUARK, and TRIMBLE.

15. Factors which control degree of brittleness. The zone of plastic strains in the region near the advancing edge of a running crack is influenced by the fact that propagation velocities for plastic strains are normally much less than the propagation velocities of elastic strains. For example, it was pointed out by VON KARMAN[1] and by TAYLOR[2] that the velocity, $V(\varepsilon)$, of a plastic strain, ε, under conditions of rapidly applied uniaxial tension or compression might be estimated from

$$V(\varepsilon) = \sqrt{S'/\varrho}$$

---

[1] T. v. KARMAN and P. DUVEZ: J. Appl. Phys. **10**, 987 (1950).
[2] G. I. TAYLOR: J. Inst. Civ. Engrs. **26**, 486 (1956).

where S' is the slope, $dS/d\varepsilon$, of a curve relating the force per unit original area, S, sometimes called "engineering stress", to the average sectional strain, ε; and where ϱ is the density of the material.

At the strain for which S' is zero, the plastic reduction of area of a tensile bar becomes locally concentrated because strengthening due to work hardening becomes at this point less than weakening due to reduction of section area. In consistency with this interpretation $V(\varepsilon)$ is zero when S' is zero.

In the tearing of a ductile metal foil the separations which compose the advance of the crack form in a zone of large plastic strains. A considerable fraction of the total deformation work is expended in this region which does not expand in proportion to crack length because the characteristic propagation speed for large plastic strains is less than even a slow time rate of motion of the crack. Thus the zinc foil fracture development model discussed in Sect. 7 corresponds roughly to the conditions of the modified Griffith theory even though lengthening of the zinc sheets by creep was observed along with the crack extension.

Consider next a cleavage crack rapidly traversing a large plate of structural grade low carbon steel, several centimeters thick. Crack velocities ranging from 0.9 to 1.5 km/sec have been observed. For thickness reduction by plastic strain to occur in advance of such cracks the slope, S', would need to be in the range of 3 to 9% or more of Young's modulus. One would estimate from inspection of static stress-strain curves that plastic strains greater than 2% would not occur, a result which agrees with what has been commonly observed. However, a dynamic rather than a static stress-strain relation should be used for this estimate. Reference [15] gives a brief summary of dynamic measurements for two steels.

In the first suggestions of a relation between plastic instability and brittleness[1] local adiabatic temperature rise was of equal importance to rate of strain hardening. This consideration as well as influence of strain rate upon plastic flow stress should be taken into account. A simple association of strain hardening with brittleness under various conditions of crack extension cannot therefore be expected. However, the general features of all these considerations are at least qualitatively known.

When a crack is subject only to two-dimensional considerations as for the severing of a thin ductile foil, the fact that plane strain conditions are necessary at the advancing edge is not relevant. However, in the case of cracks traversing thick sections the regions of the crack well removed from free surfaces must conform to conditions of plane strain. Calculations using Eqs. (5.1) through (5.4) show that, for fixed r and K, the ratio of greatest shearing stress for plane stress to greatest shearing stress for plane strain has values of approximately 1, 1.4, and 3 at the three ϑ values of $\pi/2$, $\pi/3$, and zero. Thus the zone of plastic strains near the advancing edge of a crack is substantially altered and restricted when the situation changes from one of plane stress to one of plane strain. The magnitude of this influence upon values of $\mathscr{G}_c$ can be estimated by comparing the plane strain $\mathscr{G}_c$ value shown on Fig. 5, 1.8×10^7 erg/cm, with the measurement for the same aluminum alloy under conditions of plane stress shown in Table 2, 5.2×10^7 erg/cm. The yield and fracture strengths of high strength aluminum alloys are only slightly influenced by temperature and strain rate for temperatures within 100° C of room temperature.

Observations on several steels indicate a condition which is predominately one of plane strain will prevail when

$$3 E \mathscr{G}_c \leq \sigma_Y^2 B \tag{15.1}$$

---

[1] See articles by ZENER and by READ in Ref. [6].

where σ_y is yield stress, B is the span of the crack between free surfaces, and E is Young's modulus. When this condition is met, a predominately cleavage crack can be made to progress slowly through a section of low carbon steel. Otherwise rapid motion of the crack is necessary and plane stress conditions accompanied by large plastic strains occur if the motion drops below a critical speed.

From the preceding discussion it is clear that tendencies toward embrittlement from lowering of temperature and from increase of strain rate should be abrupt in those situations where plastic flow stress changes abruptly with temperature and strain rate. In addition a pronounced decrease in the plastic strains near a crack results from the addition of tensile stress parallel to the leading edge.

These factors are of a kind which might be influenced beneficially by planned variations in composition and microstructure of the materials. They are superimposed upon the cracks, porosity, and unwanted constituents which normally occur due to manufacturing conditions in most commercial materials.

Other causes for embrittlement exist frequently in forms which are also influenced by temperature, strain rate, and state of stress. For example it has been known for many years that crack extension is assisted in the case of brass by slight amounts of mercury and in the case of glass by water[1]. KIES[2] has measured values of $\mathscr{G}_c$ for commercial lantern slide glass as influenced by water vapor. He found little strengthening of the glass with reduction of humidity except at a relative humidity of 2% or less. The results, which are included in Table 2 (end of Sect. 6), include a substantial segment of slow crack extension as a portion of the crack length used in the $\mathscr{G}_c$ computation. The observations by KIES are difficult to explain unless one assumes that $\mathscr{G}_c$, regarded as a function of time rate of crack extension, can have a maximum in the region of relatively low velocity. A relationship of this kind has in fact been found by BENBOW and ROESLER[3] for polymethylmethacrylate. Assuming such a maximum exists, it then appears the influence of moisture in the case of glass can be described as essentially a lowering of the $\mathscr{G}_c$ barrier to rapid crack extension. The strong affinity of water molecules for a clean surface of glass is well known. Presumably the increase of crack extension velocity at a given stress and crack length may be ascribed to a lowering of the surface tension of glass. This interpretation would be consistent with the fact that under rapid load application the strength of glass is not improved by removal of moisture.

An ambiguity of interpretation occurs in discussions of the embrittling influence of hydrogen in steel and other metals. It is known that molecular hydrogen films form on clean metal surfaces. PETCH and STABLES[4] suggested that the embrittling influence of hydrogen should be interpreted as lowering the effective surface tension of the metal. However, the older internal pressure theory appears to have at least equal justification. A large number of papers on hydrogen embrittlement have appeared[5]. As BALDWIN has suggested the major features can be simply understood in terms of internal pressure even though an additional surface tension type of influence is simultaneously present.

It is well known that very large pressures develop from diffusion of atomic hydrogen through steel and the entrapment of molecular hydrogen in internal cavities. Eq. (5.11) gives the crack-extension force for a penny-shaped internal

---

[1] See, for example, T. C. BAKER and F. W. PRESTON: J. Appl. Phys. **17**, 179 (1946).
[2] Unpublished laboratory notes.
[3] J. J. BENBOW and F. C. ROESLER: Proc. Phys. Soc. B **70**, 201 (1957).
[4] N. J. PETCH and P. STABLES: Nature, Lond. **169**, 842 (1952).
[5] e.g. see W. M. BALDWIN and J. T. BROWN: J. of Metals, Trans. Sect. **6**, 298 (1954).

crack subjected to remotely applied tension σ. For internal pressure p, the expression for $\mathscr{G}$ has the same form with σ replaced by p. Remembering that when stress systems are combined, the K values are additive, the crack-extension force for combined tension and internal pressure becomes

$$\mathscr{G} = \frac{2(1-\nu)}{\pi \mu}(\sigma + p)^2 a. \qquad (15.2)$$

Thus if p is an appreciable fraction of σ, say one-half, the presence of the internal pressure may increase the value of $\mathscr{G}$ by more than a factor of two.

Whether $\mathscr{G}$ increases or decreases when the crack extends quickly depends upon the ratio of p to σ and upon the initial non-elastic volume of the crack. A self-arresting behavior with crack extension is necessary in order to explain appearance features of the fracture and the influence of time upon stress rupture life. One finds that minimum internal pressures, p, ranging from $\sigma/4$ (for negligible non-recoverable crack volume) to $5\sigma/4$ (for relatively large non-recoverable crack volume) are required for a self-arresting behavior. These values p are theoretically possible at room temperature and at hydrogen contents commonly observed under conditions known to produce embrittlement. For rapidly applied loads no influence of hydrogen is expected because there is no time for restoration of pressure during the large cavity expansion which occurs as the starting crack develops to critical size. Influence of hydrogen pressure upon the deepening of a crack open to a low pressure region is not excluded because the new fracture elements, whose growth and joining compose the crack extension process, are internal. The uncertain effect of regions of large plastic strain upon collection of hydrogen in a cavity as well as the surface energy effect proposed by Petch and Stables provide ample substance for additional study.

16. Closing comments. The plan of this article has been to present the principal concepts needed for explanation of experimental observations. The experiments discussed were limited to those necessary for purposes of illustration. Among significant experiments which were not discussed are those using essentially single crystal whiskers[1]. The condition of this area of work currently lies intermediate between that of tensile strength of pure liquids and fracture strength of polycrystalline and otherwise flawed solids.

With regard to solids of this latter kind published experimental observations of fracturing and fracture strength are numerous, as might be expected for a topic of considerable practical interest. Among such papers, those which describe the processes associated with fracture and those which report measurements of fracture size effects are of most value. Noteworthy investigations of these features have been made by Bridgman [8], Tipper [9], de Leiris [10], Kies, Sullivan and Irwin [11], Stanton and Batson [12], and Docherty [13]. The fracture mechanics concepts now at hand were stimulated by the experimental fact that fracture size effects of large magnitude were found in spite of painstaking elimination of superficial causes for differences in results, such as specimen shape, surface finish, selection of material, and the testing equipment.

Bibliography references.

[1] Timoshenko, S. P.: History of the Strength of Materials, p. 51. New York: McGraw-Hill 1953.
[2] Nadai, A.: Theory of Flow and Fracture of Solids, pp. 207–228. New York: McGraw-Hill 1950.
[3] Griffith, A. A.: The Phenomenon of Rupture and Flow in Solids. Phil. Trans. Roy. Soc. Lond., Ser. A 221, 163–198 (1920).

---
[1] e.g. see S. S. Brenner: J. Appl. Phys. 27, 1484 (1956).

[4] WEIBULL, W.: A Statistical Theory of the Strength of Metals. Proc. Roy. Swed. Inst. Eng. Res. **1939**, No. 151.
[5] WESTERGAARD, H. M.: Bearing Pressure and Cracks. J. Appl. Mech. **6** (2) (1939).
[6] Fracturing of Metals (A.S.M.S., Oct. 1947), A.S.M., Cleveland, 1948. Contains, among other papers: C. ZENER: The Micro-Mechanism of Fracture; J. E. DORN: The Effect of Stress State on the Fracture Strength of Metals; G. R. IRWIN: Fracture Dynamics; P. E. SHEARIN, A. E. RUARK, R. M. TRIMBLE: Size Effects in Steels and other Metals from Slow Bend Tests; T. A. READ: Plastic Flow and Rupture of Steel at High Hardness Levels; J. H. HOLLOMON: Fracture and the Structure of Metals.
[7] Fatigue and Fracture of Metals (M.I.T. Symposium, June 1950), p. 139. New York: Wiley 1950. Contains, among other papers: E. OROWAN: Fundamentals of Brittle Behavior in Metals; W. WEIBULL: The Statistical Aspects of Fatigue Failures and its Consequences; P. L. TEED: The Influence of Metallographic Structure on Fatigue.
[8] BRIDGMAN, P. W.: Studies in Large Plastic Flow and Fracture. New York: McGraw-Hill 1952. Chapter 12 on volume changes in simple compression is of special interest in reference to the nature of fracture origins.
[9] TIPPER, C. F.: Dimensions in Testing (Conference on Brittle Fracture in Steel). J. West Scotland Iron a. Steel Inst. **60** (1953).
[10] LEIRIS, H. DE: L'Analyse Morphologique des Cassures. Métaux, Corrosion, Industries, No. 316, Dec. 1951.
[11] KIES, J. A., A. M. SULLIVAN and G. R. IRWIN: Interpretation of Fracture Markings. J. Appl. Phys. **21**, 716 (1950).
[12] STANTON, T. E., and R. G. C. BATSON: On the Characteristics of Notched-Bar Impact Tests. Proc. Inst. Civ. Eng. **211**, 67 (1920/21). The published discussions of this paper are also of interest.
[13] DOCHERTY, J. G.: Slow Bending Tests on Large Notched Bars, Engineering, Vol. 139, p. 211. 1935.
[14] WELLS, A. A., and D. POST: The Dynamic Stress Distribution Surrounding a Running Crack. Proc. Soc. for Exper. Stress Analysis **15**, No. 2 (1958).
[15] KRAFFT, J. M., A. M. SULLIVAN and G. R. IRWIN: J. Appl. Phys. **28**, 379 (1957).

A CONTINUUM-MECHANICS VIEW OF CRACK PROPAGATION

By G. R. IRWIN* and Professor A. A. WELLS†

I.—INTRODUCTION

WHEN a structural fracture failure occurs in normal service, the failure mechanism takes the form of progressive crack extension. Thus, the subject of this review has always possessed a direct practical appeal. Nevertheless, before the publication of the work by Griffith in 1920[1] and Weibull in 1939[2] appropriate analytical viewpoints were lacking and subsequent developments based on these ideas have been slow relative to rates of progress in other fields. In 1892 Love wrote in his book, "A Treatise on the Mathematical Theory of Elasticity",[3] that "the conditions of rupture are but vaguely understood". Love's comment was still valid when the fourth edition of his book was published in 1926. In terms of behaviour on an atomic scale, it is still valid today. However, present knowledge does permit a fairly comprehensive understanding of macroscopic fracture behaviour, which is of substantial practical value.

In support of Love's comment, a review of historical facts shows that knowledge of fracture behaviour has always embraced certain experimental observations that were basically inconsistent with the contemporary analytical treatments regarded as most acceptable. The nature of these observations suggested enhanced danger of brittle-fracture failure with increase in structural dimensions. Conventional design practice tended to ignore these inconsistencies, a fault which remains largely uncorrected at the present time.

For example, consider the measurements on the strength of iron wire reported by Leonardo da Vinci (1564–1642). A drawing from da Vinci's sketch book (Fig. 1) shows his apparatus. Fine sand fell into the basket from the hopper until the wire broke. Motion of the basket actuated a spring which closed the outlet opening of the hopper. Da Vinci's comments, which are in the form of testing instructions, were written alongside the sketch and are generally regarded as containing an implication of actual test results. From an authoritative translation[4]

---

* Superintendent, Mechanics Division, U.S. Naval Research Laboratory, Washington, D.C., U.S.A.
† Professor of Structural Science, Queen's University of Belfast; formerly at the British Welding Research Association, Abington.

the significant passage is: "Observe what the weight was that broke the wire, and in what part the wire broke ... Then shorten this wire, at first by half, and see how much more weight it supports; and then make it one quarter of its original length, and so on, making various lengths and noting the weight that breaks each one and the place in which it breaks." Evidently da Vinci's experiments indicated that short lengths of wire were, on the average, stronger than long lengths of the same wire. Indeed, considering the quality of wire available at that time and the flaw-probability aspects inherent in these tests, the average one-quarter-length results must have been of the order of 15–25% higher than the full-length results, an easily recognizable difference.

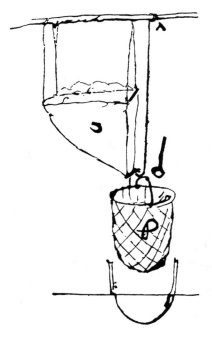

[*Courtesy Reynal and Co.*

FIG. 1.—The da Vinci apparatus for testing various lengths of iron wire. Redrawn from ref. (4).

On the other hand, Galileo* (1564–1642), who was primarily concerned with the mathematical aspects of mechanics, stated that the strength of a member in tension should depend only upon the area and not upon the length. Galileo's analytical approach dominated in the development of engineering design criteria. Mariotte (1620–1684) proposed that fracture occurs when the fractional elongation exceeds a certain limit, and

* For accounts of work by Leonardo da Vinci, Galileo, Mariotte, Coulomb, and Mohr, see Timoshenko.[5]

a criterion of the maximum-strain or -stress type for estimating fracture strength in tension has been conventional practice since that time. Coulomb (1736–1806) considered a shear-stress rule preferable for estimating limiting strength and added the thought that, in compression, the material would fail in shear with a strength enhancement due to sliding friction. Mohr (1860) provided a convenient diagrammatic formulation of what is sometimes referred to as the Coulomb–Mohr theory.

The summaries of work on strength of materials provided by Todhunter and Pearson[6] refer to two experimental findings of the da Vinci type. One reference is to measurements by Lloyd (about 1830), who determined that the average strength of short iron bars exceeded that of long iron bars. The second is a mémoire by Le Blanc (1839) who found long wires to be weaker than short iron wires of the same diameter. These results were properly interpreted as attributable to the increase in probability of a serious inhomogeneity with increase in dimensions.

After 1900, notched-bar impact testing became of increasing interest to metallurgists and engineers, primarily because of the sensitivity of such tests to the ductile–brittle transition. Ludwik (1909)[7] proposed an explanation of the transition behaviour. He regarded the plastic-flow stress and the fracture stress as separately determined. Thus, when the flow stress was raised above the fracture stress, brittle fracture without plastic flow was predicted. Subsequently, the Ludwik hypothesis lost favour because some plastic strain was observed to accompany initiation and propagation of a crack in structural metals, regardless of the degree of brittleness. However, if the unnecessary emphasis upon zero plastic strain is removed and the notch is a natural crack border, a close relationship will be seen between Ludwik's suggestion and the plane-strain crack-toughness concept discussed later.

An important programme of notched-bar impact tests was carried out at the National Physical Laboratory, Teddington, after the First World War. Stanton and Batson (1921)[8] reporting this work, emphasized the fact that, with increase in specimen dimensions, a substantial decrease in the fracture work per unit volume was observed. The serious implications of this finding as regards the general applicability of notched-bar testing, and upon structural-strength estimates based upon the usual scaling laws as well, were not overlooked in the published discussions of the Stanton–Batson paper. The static notched-bar bend tests conducted by Docherty[9] gave a similar indication that the increase, with test-piece dimensions, of the plastic work which precedes and accompanies fracture is smaller than would be predicted from geometrical similarity of the strain patterns. The situation was concisely summarized by Sachs

(1941)[10] in a review article which stated that plastic strain (without fracture) occurs with dimensional similitude, but that fracture tests sometimes show an increase of brittleness with specimen dimensions which, from similitude considerations, would not be anticipated.

In 1913 Inglis[11] had provided a theoretical analysis of the stresses around a two-dimensional elliptical opening of arbitrary eccentricity. Griffith formulated several proposals based upon the Inglis mathematical work. The idea proposed by Griffith[1] which remained of permanent interest was that the fracture strength of a brittle material (glass) would be limited by the largest of a distribution of tiny cracks always present in the material at the time of testing. Griffith suggested that the strength could be calculated from solid-state surface energy and crack size by a critical instability relationship. Instability was assumed to occur when the strain-energy release rate with crack extension exceeded the rate of increase in surface energy.

Weibull's statistical theory of strength[2] might be regarded as based upon a Griffith-type fracture theory. He considered that a test specimen or a structural component would behave as if composed of many equal units of volume, each possessing an individual intrinsic strength. The unit-volume strengths were assumed to act independently in limiting the strength of the total test volume. By assuming a specific frequency distribution of intrinsic strengths, Weibull predicted variations in strength with test volume and specimen shape which correspond at least qualitatively with observed fracture behaviour. Specific illustration of a Weibull–Griffith viewpoint was provided by Fisher and Hollomon,[12] who discussed a Weibull-type size distribution of Griffith cracks in relation to the fracture strength of metals.

A wide variety of fracture problems arising during World War II stimulated re-examination of these ideas, as well as the introduction of new ones. From the viewpoint of metal structures, progress in the understanding of fracture up to the end of 1947 is summarized in the proceedings of two conferences: the 1945 British Admiralty Conference on "Brittle Fracture of Mild-Steel Plate" at Cambridge[13] and the American Society for Metals Symposium on "Fracturing of Metals" at Chicago in 1947.[14] Also noteworthy are two 1947 review papers by Gensamer et al.,[15,16] citing more than 250 references. In the A.S.M. Symposium, Irwin[17] proposed that an understanding of the onset of crack propagation should be sought in terms of the development of unbalance between the Griffith-theory strain-energy release rate and the plastic-strain work rate required for crack extension. Zener's comments on micro-mechanisms provided the prologue for later work on the relation of fracture to dislocations. A large proportion of the material presented in the two conferences is still of interest and value.

Orowan (1949/1952)[18,19] suggested the application of the Griffith theory in fracture analysis along lines similar to those proposed by Irwin. The Griffith–Orowan theory was, however, less general and it was inferred that the analysis could be used only for relatively brittle metals. Subsequently, differences of opinion developed regarding the applicability of a modified Griffith-type theory. From Orowan's analysis, this conception was of value primarily for research guidance and applicable in practice only to brittle materials.[20] In the opinion of Irwin and Kies,[21] a modified Griffith theory could be employed widely in fracture-strength analysis in the presence of substantial amounts of plastic strain, so long as fracture occurred in advance of general yielding.

The lively interest in fracture, coupled with the lack of a dominating accepted viewpoint, resulted in a substantial confusion which is reflected in the 1948–1959 literature. For example, early in this period it was not clear what relationship, if any, existed between the Neuber notch-stress theory for a sharp notch with plastic strain, and the analysis of fracture in terms of energy rates. Metallurgists regarded the ductile–brittle transition behaviour as predictable in terms of a "transition temperature", but no general agreement existed on how to define and use this property in practical engineering applications. In the light of the well-established effect of dimensions, some regarded any "transition-temperature" concept as a dangerous oversimplification.

A forceful summary of serious fracture failures by Shank (1954),[22] coupled with the prospects for control of such failures, provided no assurance that disastrous fracture failures would not continue as in the past. Indeed, the de Havilland "Comet" fractures of 1953–1954, heavy-section fractures of large steam-turbine components (1954–1956), gas-transmission-pipe fractures of more than a mile in length, and additional welded-ship fractures, all occurred subsequent to Shank's paper.

We are too close to present times to know whether a gradient exists in the occurrence rate of such fractures. However, improved control methods applicable to the design and inspection of structures are available. These are closely related to recent clarifications of our knowledge of crack propagation. A review of the development of the clarifying ideas shows that the process consisted primarily in appropriate synthesis of viewpoints expressed in the period between 1900 and 1940. However, the desired clarifications were scarcely obvious. Comparatively recent research findings and careful fracture examinations have been exceedingly helpful.

II.—Development Aspects of Fracture Mechanics

Current fracture-mechanics concepts of macroscopic crack propagation are modifications of the ideas of Irwin,[17] Orowan,[18,19] Irwin and

Kies,[21] and Wells,[23] based on the Griffith theory. When using these it could be assumed that crack extension required a fixed rate of dissipation of strain energy, which would be constant at a given temperature for a given material. When the release of strain energy with crack extension exceeded this requirement, onset of rapid fracture followed. An excess release of strain energy (beyond the requirement for instability) developed with further crack extension, and—presumably—was used in particle accelerations incidental to the moving strain pattern of the rapid fracture.

Experimental facts demanded modifications of this concept. Wells[23] had reported that temperature measurements adjacent to the path of a running crack indicated that the local rate of dissipation of strain energy into heat was roughly in proportion to crack length. Irwin, Kies, and Smith[24] observed slow stable crack extension, under rising-load conditions, in high-strength aluminium alloy sheets before the onset of rapid fracture. Obviously, an equality between the strain-energy release rate and the fracture work rate must hold during any period of slow, stable crack extension.[25] From Wells's observations, this equality might be nearly maintained during rapid crack propagation.

These experimental observations[26-29] indicated that crack extension could be regarded as a rate-controlled process driven by a force. Irwin assumed that the appropriate crack-extension force was the Griffith-theory strain-energy release rate, which he designated by the symbol $\mathfrak{G}$. The critical value of this force for onset of rapid fracture, $\mathfrak{G}_c$, was expected to vary with elastic constraint (section thickness), temperature, and strain rate (for strain-rate-sensitive materials).

Schardin et al. in 1940[30] had found the limiting speed of crack extension in glass to be about equal to half the elastic shear-wave velocity, c_2. A summary of limiting crack-speed measurements prepared by Smith[29] indicated that this was a general characteristic of fast brittle fractures in a wide variety of solid materials. On the other hand, a mathematical analysis of a moving crack provided by Yoffe,[31] despite some lack of realism due to the fixed crack length, seemed adequate for the purpose of indicating that the limitation of crack speed from inertia alone would be the Rayleigh wave velocity, $\sim 0.9\, c_2$.

Saibel[32] pointed out that the velocity of the elastic stress wave signal through the disordered material in the fracture-process zone would scarcely exceed the speed of elastic waves pertaining to the liquid phase of the material, and concluded that a limiting speed of $\sim 0.5\, c_2$ would be in rough agreement with this consideration. Recently Barber[33] has suggested a viewpoint which might be regarded as equivalent to Saibel's analysis in other terms. The speed of a high-frequency elastic wave in a solid is reduced substantially by damping and dispersion as the wavelength becomes comparable to microstructural inhomogeneities. From considerations of this general nature, Barber concludes that the limitation imposed by inertia on crack speed, $0.9\,c_2$, should be multiplied by a factor of $2/\pi$. From this, a limiting velocity of $0.57\,c_2$ is predicted.

After attainment of the maximum crack speed, additional driving force causes division (branching) of a crack traversing a plate, or the roughening, termed "hackle", for a crack penetrating a thick section, and the average speed of the fracture is not further increased. Dynamic photoelastic studies by Wells and Post[34] and crack-speed studies by Kerkhof,[35] using ultrasonic irradiation, show that the acceleration in crack speed decreases steadily after a short initial period of rapid increase. The last factor-of-two increase in $\mathfrak{G}$, before the onset of crack division or hackle, increases the crack speed by no more than $\sim 10\%$.

The findings noted above support the concept of an equilibrium balance between crack-extension force and crack speed in the high-velocity range and are consistent with Wells's observations of heat flux from a running crack. For a crack extending in a slow stable manner, say, with the aid of stress-corrosion, it is clear that a rate-theory viewpoint would be appropriate. Other considerations are important to an understanding of the abrupt onset of rapid crack propagation and these are discussed later.

In a strict sense, fracture mechanics consists in determinations of the force $\mathfrak{G}$ from crack-stress-field analysis, and in observations of the influence of $\mathfrak{G}$ upon crack-extension behaviour. This includes, in particular, measurements of crack toughness in terms of a critical value either of $\mathfrak{G}$ or of a closely related crack-stress-field parameter indicative of the tensile stress on the fracture process zone.

Neuber[36] has suggested that plastic strain at a sharp notch would induce a behaviour equivalent to that resulting from having a larger notch-root radius, and that the maximum tensile stress at the notch would thus be reduced below the linear elastic value for the original root radius. From this analysis, the effective radius of curvature for a notch of extreme sharpness (or for a natural crack "border", see Fig. 2) is proportional in size to the local zone of plastic strain. Kuhn[37] discussed a procedure for representing crack-propagation conditions in high-strength sheet materials based upon Neuber's theoretical conception. In this procedure it is assumed that the results of tests with a sheet containing known crack lengths can be represented by taking the maximum stress at the crack border to be the ultimate tensile strength of the material. The effective root radius is then adjusted to correspond to the test results. Measurement, in this way, of an effective root radius as a toughness characteristic of the material is essentially equivalent to measurement of crack toughness in terms of $\mathfrak{G}_c$ except for matters of choice as regards stress-analysis methods. As the plastic zone adjacent to the crack border increases in size relative to crack length and net section, both procedures require a correction for plasticity. This correction is in the nature of a rough approximation, since a detailed plasticity analysis would render either procedure too complex for practical use.

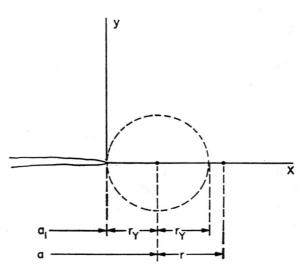

FIG. 2.—Schematic representation of crack-border plastic zone, showing the crack-size correction factor r_Y.

Wells[38] and Irwin[39,40] suggested that the correction for plastic strains adjacent to the crack could be assumed to be equivalent to adding to the crack length an increment of crack size proportional to $E\mathfrak{G}/(\sigma_Y)^2$, where E is Young's modulus and σ_Y is the tensile yield strength. A larger value of σ_Y was assumed for plane-strain than for plane-stress conditions at the crack border. With the aid of a suitable plasticity correction, the fracture-mechanics analysis has a useful degree of accuracy for any crack large enough to reduce the strength of a component section to less than the yield strength.

A solution for the plastic zone near a crack was obtained by McClintock and Hult,[41] for the simplified case in which all displacements are assumed parallel to the crack border (parallel shear) and the material behaves like a perfectly plastic solid (zero work-hardening). In this solution, when the plastic zone is small compared to the crack size and net section, the elastic stress field beyond the plastic zone is the same as that which would be present in the linear elastic solution, with the crack border advanced by half the diameter of the McClintock–Hult circular zone of plastic strain, as shown in Fig. 2. In the limiting case of small relative plastic zone size, the advancement of the crack border, r_Y, suggested by Irwin, corresponds exactly to the correction that would be anticipated from the mathematical solution of the parallel-shear-crack problem. Specifically, the value of r_Y thus suggested is

$$r_Y = \frac{1}{2\pi} \frac{E\mathcal{G}}{(\sigma_Y)} \qquad \ldots \ldots \quad (1)$$

The magnitude of the size of the plastic zone, as indicated by $2r_Y$, became of special importance with respect to the ductile–brittle transition in fracture behaviour.

After 1940 the terms "transition temperature" and "brittle temperature" began to appear as a matter of course in technical papers discussing results of notched-bar impact tests. The results of Stanton and Batson,[8] Docherty,[9] Shearin et al.,[42] and warning remarks in a well-known text-book by Salmon,[43] indicated that the conditions for ductile–brittle transition would depend upon specimen size and notch geometry, as well as upon testing speed. Nevertheless, a concept of the ductile–brittle transition temperature as a property of the material, determined by composition and microstructure, came into wide use. At the Conference on "Atomic Mechanisms of Fracture" held at Swampscott in 1959,[44] polycrystalline metals capable of slip and cleavage were regarded as having a "ductile–cleavage" transition temperature.

Irwin[45] pointed out that a well-defined ductile–brittle transition occurs in notched tensile plate tests of high-strength aluminium alloys (which have no "transition temperature"), in the thickness range where the plastic-zone size factor $2r_Y$ is comparable to the plate thickness. In this range the elastic constraint increases rapidly with decrease in the ratio of $2r_Y$ to plate thickness. The ductile–brittle transition in high-strength steels ($\sigma_{YS} \gtrsim 190$ kp/in^2) was similarly explained, whether it was produced by change in test temperature or by change in plate thickness. Thus it was learned (not for the first time) that separation, with relatively small plastic strain, normal to the applied tension, occurs in preference to oblique shear whenever the plastic zone is under sufficient elastic constraint for the stress conditions near the crack border to correspond to plane strain. Approximate estimates, as well as the reasonableness of the basic idea, suggested that this view of the ductile–brittle transition applied equally well to the strain-rate-sensitive steels of lower yield strength.

Because they antedate most of the confusion regarding the ductile–brittle transition, several comments from former years are of interest.

In 1862, Kirkaldy[6] stated that the appearance of a fractured iron bar could be changed from "wholly fibrous to wholly crystalline (aside from reduction of testing temperature)—in three different ways: (1) by altering the shape of the specimen so as to render it more liable to snap; (2) by treatment making it harder; (3) by applying the strain so suddenly as to render it liable to snap from having less time to stretch."

After directing attention to the arbitrary nature of a notched-bar impact test and to the lack of similitude in scaled experiments, Salmon,[43] in his text-book published in 1931, concluded: "In practice, owing to improper design or faulty manufacture, local concentrations of high stress may be set up. Hair-line cracks in forgings, flaws, inclusions, and other defects may exist. Such imperfections may and do imply the formation and spread of cracks during service. Certain materials will resist this tendency; others fail, usually with little or no deformation. The notched-bar test distinguishes between the two kinds of material." At a later point, Salmon's text-book summarized, in part, the view of "German metallurgists" on the effect of notches, in the following words: "In a notched-bar test a small volume of highly stressed material near the notch has to absorb the energy. This volume is surrounded by less highly stressed material which prevents plastic flow, i.e. it prevents the lateral distortion which accompanies the longitudinal strain, and, in consequence, a state of three-dimensional stress is set up—this will raise the resistance to shear, enhancing the tendency to cohesive failure. Apparently, brittle fractures in ductile materials are thus explained."

After a visit to Greenwich in 1935 to observe Docherty's slow-bend tests of notched bars of various sizes, an editor of *Engineering*[46] wrote: "The influence of size on cracking is of peculiar theoretical importance because it illustrates what now appears to be a fundamental restriction in the application of the Principles of Mechanical Similitude enunciated over 50 years ago by Professor James Thomson of Glasgow, and long used as the scientific basis of model testing. The application of these principles, only lightly shaken by Stanton and Batson, appears to be now temporarily overthrown—at least where cracking is in question; .. similar scale-effects have been established also in fatigue testing on notched bars... One of the conclusions.. is that the different steels might well be arranged in sequence of notch brittleness as judged.. by the size of test-piece at which the transition occurs from the more ductile to the more brittle type of failure.. but it must be borne in mind that the limiting size for any given quality of steel would depend on temperature and on the shape of the test-piece specified, as well as on many other factors."

Present-day concepts and procedures based on fracture mechanics consist largely in a return to the views quoted above, with more comprehensive understanding, and more advanced techniques for crack-toughness evaluation and methods for practical application. This represents substantial progress but why did it require twenty-five years? The following reasons for the slow progress made appear relevant:

(1) The convenience of testing notched bars of fixed size as a function of temperature tended to dim recognition of the influences of specimen shape and dimensions.

(2) When a mild-steel plate fractures from a sharp notch, the "thumbnail" of slow fracture adjacent to the notch usually differs markedly in appearance from the subsequent rapid portion of the fracture. It was noted that the difference in appearance was caused by a tendency towards cleavage in the rapid portion. A similar change in appearance occurred across the ductile–brittle transition range in notched-bar impact tests. The fact that cleavage in strain-rate-sensitive steels may be the result rather than the cause of rapid crack propagation could be argued from results of de Leiris on specimens of different lengths.[47] This question was discussed by Irwin[17] and by Felbeck and Orowan.[48] Nevertheless brittle-fracture research programmes strongly emphasized studies of cleavage. This emphasis distracted attention from proper understanding and control of low-stress fracture failures in actual structures.

(3) The tendency for materials experts to ascribe failures by fracture to material quality, without reference to crack size, stress level, and structural dimensions, interfered with the establishment of a sound linkage between structural mechanics and mechanical metallurgy.

(4) Examination and analysis of failure by fracture received insufficient attention. The fact that crack propagation from a starting crack of macroscopic size was responsible for most service fractures, was not generally recognized.

The importance of the fourth factor listed above can scarcely be overestimated. The present degree of acceptance of the fracture-mechanics viewpoint may be largely due to the intensive study devoted to three major fracture problems during the past ten years. In the first of these, failure of the de Havilland "Comet" commercial jet aircraft, the official report[49] was disappointing, in the sense that reference was made only to fatigue and inadequate window reinforcement as the factors responsible. However, the importance of crack propagation in the "Comet" failures was clearly acknowledged by the use of materials of greater crack-toughness and by the introduction of crack-arrest design features in subsequent commercial jet planes. The fracture safety of any aircraft structure rests primarily on the ability of components to resist propagation of such cracks as may, unavoidably, develop during service before replacement. The average stress level in the pressure hull of the "Comet" was about one-third of the yield strength of the material. Because of inadequate window reinforcement, the effective crack size determining the driving force for crack propagation was approximately the size of the window where fatigue cracking occurred.

In the period 1954–1956 a series of incidents occurred involving large steam turbines and generators in which heavy ($\sim 1{,}000{,}000$ lb) steel rotating components fractured unexpectedly.[50–52] The serious dangers, in terms both of damage and of financial loss, stimulated a very intensive study of the problem. The laboratory tests most closely simulating the service failures were those in which rotational speed was employed to fracture large notched discs.[53,54] A satisfactory correlation of these tests with other fracture tests, applying the concepts of mechanics, was provided by Wundt and his associates.[53,55] These same concepts assisted in arriving at an understanding of the service failures. In agreement with the laboratory fracture tests, examination of the service fractures suggested that the starting cracks had grown to a size of the order of several inches before the development of rapid crack propagation. Local concentrations of impurities were usually associated with these cracks and the stable growth possibilities were enhanced by the presence of hydrogen. The average stress on the region containing the primary origin was $\sim 30\%$ of the yield strength.

This fracture problem stimulated interest in ultrasonic inspection and a demand for steel components of greater toughness and cleanness. The desired degree of improvement of the steel was achieved in part by modifications in composition, but mainly through the enhanced soundness and cleanness of the forgings made possible by better steel-melting practice and by the installation of vacuum-pouring equipment during 1957–1959.[56]

The third major fracture problem was concerned with steel solid-propellant rocket chambers. This resulted in extensive investigations of crack propagation in very high-strength metals, beginning in 1958. Careful examinations of fracture failure coupled with realistic methods of laboratory crack-toughness evaluation, again proved most useful.[57] The rocket chambers which experienced serious cracking troubles were fabricated by welding and heat-treated, after welding, to a yield strength in the range 190,000–215,000 lb/in². Earlier cracks were always involved in these failures. With rare exceptions these occurred in welds or along weld borders. Large improvements in the accuracy of inspection, assisted by smooth finishing of the welds and the elimination of moisture-induced stress-corrosion, were necessary. When the crack-toughness of the material was maintained at a sufficiently high level for a small crack half through the cylindrical wall section to be stable with the normal tension equal to the yield stress, it was possible to effect the welding and inspection with enough care for reliable production chambers to be obtained.[58,59]

In 1959 the American Society for Testing Materials established a special committee for fracture testing of high-strength metals. The First Report of this committee[60] stated that "the validity of the analytical methods of fracture mechanics is sufficiently well established" to permit their use in determining whether a fracture test "is measuring the significant quantities governing performance" and the degree in which a fracture-test result "may be generalized to the more complex structure existing in service". The A.S.T.M. committee provided tentative recommendations on crack-toughness measurement procedures which have been extensively applied.

III.—Analytical Aspects of Crack-Extension Problems

For an understanding of fracture in terms of the mechanics of crack extension, the force acting to extend the crack is of major interest. The driving force for a crack is more sophisticated than a simple push or pull. However, the same basis for a force concept is present as exists for a crystalline dislocation line, and leads to definition of the crack-extension force as the irreversible strain-energy loss per unit area of crack extension. A generalized force can be calculated only in terms of a specific mathematical model. For the crack-extension force, as well as for the Peierls–Nabarro dislocation force, the model is one that assumes linear elastic behaviour outside a line-disturbance zone which is regarded as negligibly small. In both cases the force concept is valid only as an average along a disturbance-line length which is substantially greater than the size of the non-linear zone normal to the line.

Irwin[29] described the possible crack-border stress fields in terms of three crack-surface-displacement modes. The corresponding crack-extension forces were represented by $\mathfrak{G}_I$ (opening or tensile mode), $\mathfrak{G}_{II}$ (forward shear mode), and $\mathfrak{G}_{III}$ (parallel shear mode). Where the real crack situation differed substantially from any of these modes, the convention was to drop the fracture-mode subscript. Thus, the crack-extension force for a tensile crack penetrating a thick section, or for a circumferentially notched round tensile bar, is designated $\mathfrak{G}_I$ because the displacements near the crack border lie in planes normal to the crack surface, as in the plane-strain situation for the opening mode of crack-surface displacements. However, for a thorough-crack in a plate, such that the plastic zone at the crack border is at least comparable in magnitude to the plate thickness, the displacements near the crack border have components parallel to the crack surfaces. Although the stress field, from a generalized plane-stress viewpoint, does not differ from the corresponding two-dimensional stress field in plane strain, the crack-opening displacements are increased as a result of relaxation of the thickness-direction stress, and amounts of oblique shear fracture occur depending upon the size of the plastic zone relative to the plate thickness. The term used for the crack-extension force, in the case of a tensile crack which does not correspond to plain strain, is $\mathfrak{G}$. Except for the context, Irwin's practice in this regard could be ambiguous, since $\mathfrak{G}$ is also used[61] for the total energy release rate ($\mathfrak{G}_I + \mathfrak{G}_{II}$) in the case of an adhesive joint separating under combined tension and shear.

A general procedure for the determination of $\mathfrak{G}$ by experimental analysis has been discussed by Irwin and Kies.[21] Since this method is

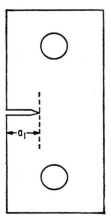

[Courtesy "Materials Research and Standards".

FIG. 3.—Single-edge-notch tensile specimen employed by Krafft and Sullivan. (Sullivan.[63])

quite simple and provides a clear picture of physical aspects, a brief outline is provided. Fig. 3 shows a single-notch, pin-loading type of crack-toughness test specimen introduced recently by Krafft[62] and Sullivan.[63] Suppose one regards this specimen simply as a general representation of a component containing a crack of size a. Fig. 4 shows schematically the load/elongation lines for two nearly equal crack lengths a and $(a + da)$. If the crack extension da occurred with no movement of the loading forces (l fixed), then the loss of strain energy is the triangular area of the figure with diagonal shading. If the same extension occurred with the load constant, this loss would be offset by a gain in strain energy represented by the rectangular area with horizontal shading. Neglecting a small area which can be regarded as an infinitesimal of higher order, the rectangular area is just twice as large as the triangular area with diagonal shading. Changes in strain energy introduced by movement of the loads, e.g., the rectangular area of the figure, play no part in calculation of the strain-energy loss attributable to the crack-extension process, since strain energy added in this way remains in the specimen and represents a reversible energy-change increment.

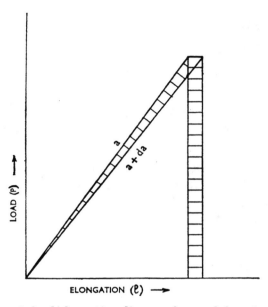

FIG. 4.—Schematic load/elongation diagram for crack lengths a and $a + da$.

On a unit-thickness basis, the area of the triangle with slant shading is $\mathfrak{G}da$. Let P represent the load per unit thickness, and define the compliance, C, by the equation:

$$l = CP \quad . \quad . \quad . \quad . \quad . \quad . \quad (2)$$

where l is the extension between the loading points. A simple calculation from Fig. 4 and equation (2) then shows that:

$$\mathfrak{G} = \frac{1}{2} P^2 \frac{dC}{da} \quad . \quad . \quad . \quad . \quad . \quad (3)$$

To find the value of $\mathfrak{G}$ for the specimen shown in Fig. 3 one needs measurements of C from load/elongation lines for a series of crack depths. Either a graphical or a curve-fitting procedure may then be used to obtain dC/da as a function of a (or C). After calibration in this way, equation (3) permits calculation of $\mathfrak{G}$ from given values of P and a (or P and C). This procedure is quite general in application. It is not limited to the tensile mode or to the particular specimen shape shown in Fig. 3.

Mathematical determinations of $\mathfrak{G}$ from crack-stress-field analyses were greatly simplified in the period 1955–1957 by papers which showed the close relationship of 𝔊 to the crack-border stress field.[26-28] Using x, y, z and r, θ co-ordinates, as shown in Fig. 5, the extensional stress, σ_y, near the border of a tensile crack and on the x, z plane is:

$$\sigma_y = \frac{K}{\sqrt{2\pi r}} \qquad \ldots \ldots \quad (4)$$

It was shown that:

$$E\,\mathfrak{G} = K^2 \qquad \text{(for plane stress)} \ldots \quad (5)$$

and

$$E\,\mathfrak{G} = K^2(1 - \nu^2) \qquad \text{(for plane strain)} \ldots \quad (6)$$

where E is Young's modulus and ν is Poisson's ratio. With this information one need only determine the limit of $\sigma_y\sqrt{2\pi r}$, with r approaching zero, to find K and thus 𝔊. Similar procedures using τ_{xy} and τ_{yz}, respectively, in place of σ_y, for Mode-II and Mode-III crack stress fields, result in values of K_{II} and K_{III}.

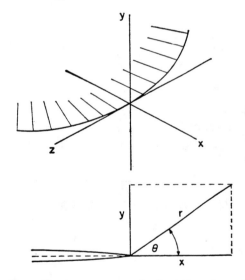

Fig. 5.—Co-ordinates x, y, z, and r, θ at the edge of a crack.

Application of this procedure with stress functions of types suggested by Westergaard[64] provided a large number of solutions for two-dimensional crack stress fields, including, of course, the value of 𝔊 derived by Griffith for normal tension on a central two-dimensional crack in an infinite plate.

A wide variation in symbols for the concepts represented by 𝔊 and K appears in past literature. For example, in discussing the Mode-I and -II crack-border stress fields, Williams[65] used stress-intensity parameters smaller by a factor of $\sqrt{2\pi}$ than K. Irwin, Paris,[66] and Erdogan[67] frequently use $\mathcal{K} = K/\sqrt{\pi}$. This review adheres to the K and $\mathfrak{G}$ nomenclature, which is standard practice in the reports of the A.S.T.M. fracture-testing committee.

Exact solutions in closed form are available for only a relatively small number of three-dimensional crack problems. Papers by Sneddon[68] and Sack[69] gave the stress distribution near a flat circular (or penny-shaped) crack in an infinite solid subjected to tension normal to the crack. Sneddon provided not only the value of $\mathfrak{G}$, as a function of crack size and applied stress, but also the limiting form of the stress equations close to the crack border. The inverse of this problem can be obtained from the sharp notch limit of Neuber's solution for a deep three-dimensional hyperbolic notch. Green and Sneddon solved the problem of a flat crack having an elliptical boundary.[70] The crack-border stress equations and values of $\mathfrak{G}$ around the elliptical crack border were not discussed.

Special practical and theoretical interest appertains to the problem of the elliptical crack. The practical interest arises from the fact that the starting cracks from fracture examinations so often take the form of surface cracks with a shape which can be approximately represented as half an ellipse. The theoretical interest concerns a critical aspect in the definition of a crack-extension force. The stress intensity K varies around the crack and one can ask whether the definition of $\mathfrak{G}$ is consistent with its representation as a point function of position along the crack border. These aspects of the elliptical-crack problem were discussed in a recent paper.[71] The analysis indicated that, for a three-dimensional crack problem, the crack-extension force $\mathfrak{G}$ has significance only as an average for a crack-border length substantially larger than r_Y. This question does not arise for two-dimensional crack problems, e.g., a crack extending through a plate analysed in terms of generalized plane stress.

To some extent, two-dimensional infinite-plate solutions for co-linear repeated cracks can be adapted to give approximate values of $\mathfrak{G}$ and K for plates of finite width. However, for the most part, unless the crack can be regarded as in an infinite solid or plate, compact analytical solutions of the crack stress fields are not available, and numerical methods are required to determine the relationship between crack size, component dimensions, and the applied stress. For example, calculations by Bowie[72] for a pair of cracks of equal length from a circular hole, were helpful in the study of the problem of fracture in a large steam-turbine rotor mentioned earlier. Several fractures started near the central bore-hole of the rotor. Wigglesworth's calculations of stresses near surface notches[73] permitted the determination of $\mathfrak{G}$ and K for a two-dimensional crack at a free surface boundary. Lachenbruch[74] has calculated K values for the two-dimensional surface crack with a superimposed temperature gradient normal to the free surface. Bueckner[75] provided numerical studies of several specific crack problems, as well as a careful mathematical check of the procedures suggested in refs. (21) and (28).

A series of reports, written by Sneddon and his co-workers, and distributed by the North Carolina State College, discuss advanced methods for linear elastic analysis of crack problems in isotropic bodies.[76-78] Investigations of crack problems for a homogeneous material with non-isotropic elastic properties have been conducted by Irwin[79] and Aug and Williams.[80] A paper by Williams[81] discusses a two-dimensional crack on the boundary between two materials of different elastic properties.

A recent paper by Barenblatt[82] reviews mathematical aspects of crack stress-field problems. The reasons for accepting the Rayleigh wave velocity as the maximum speed of a crack (purely from considerations of inertia) are given in detail. Barenblatt[83,84] prefers to remove the stress singularity at the crack border by the introduction of cohesive forces in a narrow strip adjacent to the border. For polycrystalline metals, removal of the stress singularity by introduction of a plastic zone, as discussed below, is more useful and appropriate, even though a strain singularity at the crack tip is still present. Barenblatt noticed that the equation for K values given by Irwin[29] for an infinite set of co-linear, two-dimensional cracks of arbitrary length is inaccurate. A short discussion of the degree of this inaccuracy is given in a University of Illinois report of 1962.[79]

As can be seen, linear elastic-fracture mechanics, although forming a portion of the larger domain of linear elasticity, constitutes essentially a complete analytical discipline in itself. The practical purpose is to establish an orderly descriptive procedure for representation of crack-extension behaviour in terms applicable both to laboratory tests and to service components.

It has been noted previously that an approximate correction for the local yielding near the crack border was desirable to ensure accuracy of this description in a close approach to conditions for general yielding. A detailed understanding of the shape and strain distribution of the local plastic zone is not necessary for the elementary tasks of fracture mechanics. However, available evidence suggests that, although a complete determination of crack-toughness from plastic-flow properties is unlikely, a close relationship to these properties exists nevertheless. To study such relationships it is necessary to associate plastic strain in a tensile or compression specimen with plastic strain in the fracture-process zone. More specific knowledge of the plastic-zone strain pattern would assist this task.

The parallel-shear mode of crack-surface displacement, which is rarely, if ever, observed in real fracturing, is by far the simplest of the three modes for mathematical treatment. McClintock[41] and his co-workers have discussed the plastic strains near a crack for the parallel-shear mode, assuming a perfectly plastic material. Neuber[36] provides an analysis of non-linear strains near a sharp notch or crack for the same type of deformation. The primary interest in these analyses stems from the fact that corresponding solutions for the opening mode (and forward-shear mode) are very complex. Thus, for current tasks which require a specific, albeit arbitrary, representation of the strains in the plastic zone, the parallel-shear analyses become of interest as convenient model-type representations. The McClintock model of the plastic zone is simpler than that of Neuber and has provided satisfactory estimates when used in a descriptive way with various sets of experimental data.

In the simple case where the net section and crack length are large compared to the plastic zone, the Hult–McClintock analysis predicts that the elastic/plastic boundary is a circle, as shown in Fig. 2. The elastic stresses outside this zone may be described in terms of linear elastic analysis, assuming the end of the crack to be located at the centre of the circle. The plastic strains, γ, are given by the equation

$$\gamma = \gamma_Y \frac{2r_Y \cos \varphi}{d} \quad \ldots \quad (7)$$

where d and φ are polar co-ordinates from the actual crack border and the angle is measured from the xz plane forward from the crack. The strain γ_Y is the elastic-limit shear strain at the boundary where $d = 2r_Y \cos \varphi$. Extensions of this type of solution to torsion of a grooved round bar of finite diameter were made by Walsh and MacKenzie[85] and of a grooved plate of finite thickness by Koskinen.[86] Experimental observations of torque and angular deflection by Walsh and MacKenzie showed reasonable agreement with the theory; the test bars were of 7075-T6 aluminium alloy. Koskinen's studies were in appropriate form for estimating the plastic-zone correction factor for notch tensile-test plates.

From the published discussion of Koskinen's results, these calculations indicate negligible inaccuracy of equation (1) until the net section stress exceeds $0.8 \sigma_Y$.[87] Koskinen's results have also been interpreted as indicating that the apparent K value for crack propagation will be low by ~ 10–15% when the net section stress equals σ_Y. Both these indications are in agreement with observations.[88] It should be noted, however, that Irwin's interpretation, though plausible, does not bear a simple relationship to plastic strains near the crack border.

The progressive extension of a crack during a many-cycle (low-stress-level) fatigue test can be regarded as a succession of low-cycle fatigue fractures of the fracture-process zone. A connection between many-cycle and low-cycle fatigue of the type thus suggested seems to exist.[89] For example, the relation for low-cycle fatigue life suggested by Coffin[90] estimates the life as being inversely proportional to the square of the plastic strain. If we use Coffin's estimate inside the small plastic zone of a crack extending slowly as a result of tension/zero/tension fatigue cycling, equation (7) suggests that segments of arbitrary fixed size (δ), adjacent to the crack border, would break in a number of cycles proportional to $(\delta/2r_Y)^2$. Thus, the rate of crack growth would be proportional to $(r_Y)^2$ or to $(K)^4$, where K is computed for the upper level of applied tension. A relationship between rate of crack growth and a K value of this kind has been reported by Paris and Erdogan.[66] A more general argument supporting the proportionality of crack-growth rate to $(K)^4$ appears in the doctoral thesis of Paris.[91]

McClintock[92] extended his studies of the parallel-shear plastic zone to construct a specific model for the development of unstable crack extension. A critical plastic-strain-fracture criterion and a structural size-length factor were involved, both suggestive of later developments. However, these are relatively new fields of investigation. More clarification is needed from experimental observations to appraise specific mathematical representations pertaining to the onset of fast fracture instability or to the growth rate of fatigue cracks.

IV.—INITIATION OF CRACK PROPAGATION

The unstable behaviour observed in the development of crack propagation can be regarded as closely related to the instabilities found in tensile testing of unnotched bars, as indicated by Tipper.[93,94] In a tensile test of a moderately tough metal bar, if the compliance is very large, both maximum load and fracture occur close together, when the conditions for a local necking type of plastic instability are satisfied. If the same bar is broken under low-compliance (stiff-machine) conditions, local necking occurs gradually. It is then possible to observe that the fracture instability starts from small cracks which form in central portions of the reduced section. The spreading and joining of these cracks tend to be self-catalysing to such a degree that the final fracture instability is quite sudden.

From studies of markings produced by rapid fractures in a wide variety of plastics and polycrystalline metals, Kies, Sullivan, and

Irwin[95] concluded that joining of locally discontinuous separations is characteristic of crack extension in tough plastics and in the polycrystalline metals. Initiation, spreading, and joining of advance separations appeared to form the basic mechanism of macroscopic crack extension, regardless of crack speed. Clearly, the advance-origin mechanism must be very rapid to be consistent with fast crack propagation. Correspondingly, when the advance separations near a stationary crack border start to join up, the inherent quickness of this type of tensile instability tends to produce an abrupt initial acceleration of the crack speed. Thus, for plane-strain tensile conditions at the crack border, normal behaviour at the onset of crack propagation consists in a sudden acceleration effectively equivalent to a jump in the crack velocity. One would expect to predict this behaviour in terms of the conditions necessary for the joining instability of advance origins. Gradual initial acceleration of crack speed, where observed, can be understood in terms of the stabilizing influences that block development of the rapid initial acceleration which would otherwise occur.

In the case of stable crack extension by fatigue, the maximum force, $\mathfrak{G}$, is not large enough to drive the crack through undamaged material. Thus, the incremental advances of the crack are limited to a small fraction of the zone of plastic-cycling damage adjacent to the crack border. Stable tensile crack extension occurs also, with the assistance of corrosion. As in the case of fatigue, instability is chiefly suppressed by the restricted size of the damaged zone at the crack border.

Gradual initial acceleration of the crack is readily observed in fracture tests on notched sheets of high-strength aluminium alloys, particularly when the size of the plastic zone is comparable to the plate thickness. For the usual edge-notched or centre-notched test specimens, assuming that the initial speed of separation between grips is maintained constant, a crack speed equal to the specimen width divided by the loading time is sufficient to cause a maximum load point.[88] This is a static-stress-pattern estimate based upon the influence of crack extension upon specimen compliance. For a 3-in.-wide plate loaded to fracture in 30 sec, this estimate corresponds to a speed of $\sim 0{\cdot}1$ in./sec. In initial work on crack-toughness testing at the U.S. Naval Research Laboratory, cinematograph film was used to determine the crack length at onset of fast fracture. These records showed that a small crack speed of the approximate size indicated above in fact developed at the maximum load and before the rapid final separation. The tendency towards gradual acceleration is quite clear visually, when one watches a notched wide-plate test of 2024-T3 aluminium. These observations, however, relate to plane-stress conditions near the crack border.

Despite the potential rapidity of the tensile-fracture instability under plane-strain conditions, slow or interrupted plane-strain tensile fracturing often occurs in tests on sharply notched plates, forming a crescent or "thumb-nail" of stable separation across the notch. Here, too, thickness-direction stress relaxation plays an important role. The region in which stress elevation is limited by plane-stress plastic flow tends to be quite large. The fracture in central regions is stabilized in addition by rapid load transfer to the unbroken sections adjacent to the free surfaces, where shear lips develop and may dominate in the rapid stage of the fracture. In tests on strain-rate-sensitive mild-steel plates, the dynamic elevation of the yield stress after the abrupt start of crack propagation restricts the size of the plastic zone, greatly reduces the size of the shear lips, and causes a distinct difference in appearance between the stable and fast portions of the plane-strain tensile-fracture surface.

The stabilizing influence of shear lips, discussed above, is essentially equivalent to conducting the fracture process of interest under very stiff loading conditions, and equivalent stabilizing tendencies can be obtained in comparatively brittle materials by specimen design. Examples of the latter type are the splitting tests used by Orowan for mica[96] and by Gilman[97] for single crystals, as well as various peel tests employed in the evaluation of adhesives.

Consider next the influence of notch sharpness in tensile-strength tests of notched specimens. This topic is discussed in the Third Report of the A.S.T.M. committee on fracture testing,[98] and it is concluded that the 0·001-in. notch-root radius suggested in the Committee's First Report[60] is too large to simulate a sharp natural crack (e.g., a fatigue crack) in certain low-toughness, high-strength alloy steels. As a result of this finding, current practice in crack-toughness tests on high-strength metals in the United States is tending towards the use of fatigue-crack extensions at the notch roots as standard practice in test-specimen preparation. In the tests discussed in the Third Report of the A.S.T.M. committee some slow crack extension was observed. However, this was primarily due to stress-corrosion associated with ink staining. The material, a hot-die steel with a yield strength of 235,000 lb/in^2, was quite brittle and unstable rapid crack extension occurred under nearly plane-strain conditions. From a graphical representation of the apparent crack toughness, K_c, as a function of $\sqrt{R}$ (where R is the notch root radius), the test results indicated that a root radius of ($r_Y/10$) would have been of adequate sharpness. However, preparation of a notch with the $\frac{1}{4}$-mil radius thus indicated would be impractical. Beyond this notch-radius size the apparent K_c value increased approximately in proportion to $\sqrt{R}$.

A similar result was found by Mulherin, Armiento, and Marcus[99] in notched-sheet tensile tests of 7075-76 aluminium performed under conditions that corresponded more nearly to those of plane stress. In these experiments the unstable region was on a larger scale, and the levelling off of the K_c value with reduction of R below a critical size was directly observable. The critical R value appeared to vary between $(r_Y/20)$ and $(r_Y/10)$. In the domain where the net section stress was $< 0\cdot 8\ \sigma_{YS}$ and R exceeded the critical value, the critical K value for onset of crack propagation, K_R, could be approximately represented by the expression

$$K_R = \tfrac{1}{2}\sigma_F\sqrt{\pi q R} \quad \ldots \quad (8)$$

where σ_F and q are adjustable constants. Fig. 6 contains an example of data from these experiments. As noted in ref. (35), the stress intensity K can be obtained from a Neuber-type analysis of notch stresses, assuming a small flank angle, with the expression,

$$K = \text{limit}\ (\tfrac{1}{2}\sigma_M\sqrt{\pi R}) \quad \ldots \quad (9)$$

where σ_M is the maximum stress at the notch root.

If we regard σ_F both as a fracture stress and as a maximum notch stress controlled by the effective root radius qR, then the reasonable assumption that qR is an order of magnitude greater than R brings the value of σ_F close to the ultimate tensile strength of the material.

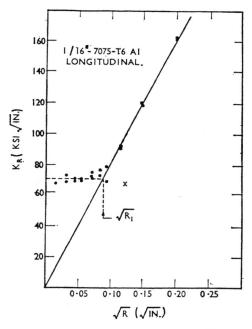

[*Courtesy Amer. Soc. Mech. Eng.*

Fig. 6.—Influence of notch sharpness upon the apparent critical values of K, here termed K_R. (*Mulherin* et al.[99])

In the above tests the amount of slow initial crack extension decreased with increase in notch-root radius and could be regarded as negligible compared to the size of the plastic zone when R was 20% above the critical value R_1 shown in Fig. 6. Thus, the fracture behaviour at larger values of R corresponded effectively to tensile instability, at a fixed stress level, of a plastic zone attached to the notch and proportional in size to the notch-root radius. The plane-strain data given in the Third Report of the A.S.T.M. committee are subject to similar interpretations.

The complexity of the effects of strain rate upon the initiation of crack propagation arises from the fact that the effective notch bluntness, as well as the elastic constraint, depends upon strain rate. A useful simplification is obtained if the investigation of strain-rate-sensitive materials is restricted to plane-strain conditions of fracture. Krafft and Sullivan[100] and Krafft[62] have reported rising-load/crack-toughness values, K_{Ic}, for various strain-rate-sensitive steels over a wide range of loading speeds. Because of the small size and the development of general yielding with increase in temperature, only results at low test temperatures were suitable for quantitative analysis. Measurements of plastic-flow properties were also made in uniaxial compression on the same materials and possible relationships between the fracture and flow properties are discussed. This work was facilitated by a novel, and relatively simple, testing machine developed by Krafft, which permits the maintenance of a wide range of nearly constant head speeds up to maximum loads of the order of 30,000 lb. The results are of exceptional interest.

For discussion of these results it is necessary to associate loading speed in the compression tests with stress or strain rate near the crack border of the fracture specimen, in an appropriate and systematic way. Observations of some interest result simply from the analysis of this work.[101]

Consider, first, that Fig. 2 represents the crack-border region of a specimen during a test in which the applied tensile load is proportional to time. The stress rate, $\dot{\sigma}_y = d\sigma_y/dt$, at a fixed point x, as shown in Fig. 2, can be found by differentiation of the stress equation

$$\sigma_y = K/\sqrt{2\pi r} \quad \cdots \cdots \quad (10)$$

The result is

$$\dot{\sigma}_y = \sigma_y\left(\frac{\dot{K}}{K} - \frac{\dot{r}}{2r}\right) \quad \cdots \cdots \quad (11)$$

Since

$$r = x - a_1 - r_Y \quad \cdots \cdots \quad (12)$$

it is found that

$$\dot{\sigma}_y = \sigma_y\left(\frac{\dot{K}}{K} + \frac{\dot{a}_1}{2r} + \frac{\dot{r}_Y}{2r}\right) \quad \ldots \quad (13)$$

To arrive at a measure of the stress rate governing the development of the plastic zone, it is appropriate to evaluate equation (13) at a point which is always in the same relative position close to the plastic zone. For simplicity, the governing stress rate, S, is defined as the value of $\dot{\sigma}_y$ at the point where $r = r_Y$.

Only a negligible error is made by inserting into the above equation

$$\dot{r}_Y = r_Y \frac{2\dot{K}}{K} \quad \ldots \quad (14)$$

neglecting, in this factor, the influence of time rate upon the yield stress σ_Y. The value of S is then found to be

$$S = \sigma_Y\left(\frac{2\dot{K}}{K} + \frac{\dot{a}_1}{2r_Y}\right) \quad \ldots \quad (15)$$

It is instructive to examine the influence of the crack speed upon the value of the stress rate at the crack border. For this purpose it is first assumed that the onset of crack propagation is so abrupt that, in determining the conditions leading to instability, the initial crack border may be considered as stationary and the term proportional to $\dot{a}_1$ disregarded. At the same time the hypothesized loading condition, $K = \dot{K}t$, can be inserted, thus obtaining

$$S = 2\sigma_Y/t \quad \ldots \quad (16)$$

For comparison, it is next assumed that movement of the crack border occurs and contributes to the conditions for onset of fast crack propagation. As a conservative and tentative value of $\dot{a}_1$, the W/t estimate (where $W = $ specimen width) corresponding to the small crack speed necessary to produce a maximum load point, is introduced. If this is done, the term of equation (15) proportional to $\dot{K}$ is negligible and the value for S becomes

$$S = \frac{2\sigma_Y}{t}\left(\frac{W}{4r_Y}\right) \quad \ldots \quad (17)$$

Estimates of the term in parentheses above range from 50 to 1000 in typical K_{Ic} tests. Thus, a very small crack speed causes an increase of 2–3 orders of magnitude in the stress rate.

From equation (16), the stress rate decreases with time during application of load. In a strain-rate-sensitive material, the magnitude of this stress-rate trend is enhanced by the corresponding decrease of σ_Y with decrease in stress rate. As a result, any tendency for the initial crack border to remain stationary is strengthened by plastic "blunting" of the crack and the plastic zone grows to a size much larger than the estimate of $2r_Y$ for the same driving force at stress rates corresponding to a moving crack. The abrupt initial acceleration normally expected in a plane-strain, tensile-fracture instability is, therefore, unusually pronounced in a strain-rate-sensitive material.

Krafft and Sullivan[100] used a stress-rate estimate essentially equivalent to equation (16) in establishing an empirical correlation between upper yield stress and K_{Ic} for three mild steels. The relationship found was of the form

$$K_{Ic} = M(\sigma_{UY})^{-3/2} \quad \ldots \ldots \quad (18)$$

where σ_{UY} is the upper yield stress. The proportionality constant, M, differed for each steel and was thought to represent crack-toughness factors in the materials that were not closely related to yield stress.

In later work, embracing a wider range of steels, Krafft[62] demonstrated that a direct proportionality between K_{Ic} and the strain-hardening exponent, n, gave a better correlation for all the data. Correlations with yield stress and plastic-flow stress, he explained, were to be expected in limited sets of data, because of the tendency of strain-hardening to decrease in a systematic way with increase of S and σ_Y.

Establishment of a relationship between K_{Ic} and strain-hardening introduces a new element. It is no longer sufficient to represent the time-rate influence by linear elastic analysis of the stress rate at the nominal border of the plastic zone. The plastic-flow tests at a given temperature provide a set of n values, each corresponding to a specific plastic strain rate. Within the plastic zone of the crack one has a wide range of strain rates. The task of correlation and its interpretations require representation of this range of strain rates in equation form as a function of distance from the crack border and of loading time.

For example, the McClintock–Hult model of the plastic zone might be used. In a manner similar to the derivation of equation (16), the plastic strain rate, $\dot{\varepsilon}$, at a distance d from the crack border is then found to be

$$\dot{\varepsilon} = \frac{2\varepsilon}{t} \quad \ldots \ldots \quad (19)$$

where

$$\varepsilon = \varepsilon_Y \frac{2r_Y}{d} \quad \ldots \ldots \quad (20)$$

and ε_Y is the strain at the elastic/plastic boundary. Actually, Krafft assumed that, because of elastic constraint and strain-hardening, the strain would be more nearly represented by

$$\varepsilon = \varepsilon_Y \left(\frac{r_Y}{d}\right)^{1/2} \quad \ldots \ldots \quad (21)$$

From this relationship

$$\dot{\varepsilon} = \varepsilon/t \quad \ldots \ldots \quad (22)$$

For the strain rate in flow-property tests an equation identical to equation (22) applies. Using test results on a low-strength mild steel, a sample of T-1 steel, and an 18% nickel maraging steel of very high strength, Krafft showed that the K_{Ic} values for each material varied with temperature and strain rate almost in direct proportionality to the strain-hardening exponent.

The success of Krafft's correlation is not affected by the introduction of a small numerical factor into the proportionality of $\dot{\varepsilon}$ to ε/t. Thus equations (19) and (20) might be used in place of equations (21) and (22). However, if this is done, it should be borne in mind that equation (19) must somewhat overestimate $\dot{\varepsilon}$ because no account has been taken of strain-hardening and plastic blunting of the crack. Thus Krafft's judgement that equation (22) provides a "best fit" to the data might reasonably be interpreted as indicating the approximate magnitude of this overestimate. Choices of this nature affect only interpretations and have no bearing on Krafft's observation that (K_{Ic}/n) tends to have a fixed value for each material.

Krafft assumed that his correlation meant that a plastic instability strain, ε, equal to the strain-hardening exponent, n, is dominant. From equation (21) he found the distance, d_T, from the crack border where $\varepsilon = n$ would occur. Krafft's d_T length factors are fixed values for each material independent of temperature and strain rate. The range of d_T is from $\sim 10^{-5}$ in. for the low-strength steels to 3×10^{-3} in. for the very high-strength maraging steel. As an alternative to Krafft's interpretation, equation (20) can be used in place of equation (21) to compute the values of d, where $\varepsilon = n$. These length factors d_n, lie in the same range as d_T but vary in proportion to $(2r_Y)^{1/2}$ with test temperature and loading speed.

The results of Krafft and his associates provide a greatly improved perspective on the factors that control rising-load crack-toughness in strain-rate-sensitive metals. With regard to the specific mechanism of the stationary crack-border instability, the results and their interpretations are suggestive rather than conclusive.

V.—General Behaviour Patterns

Fractures of glass and of glassy high-polymer plastics can be closely observed in various ways. Most fracture studies of these materials have been primarily concerned with rapid crack extension. If a glass rod is notched with a file and broken with a quick bending motion, a fracture surface is obtained with the typical appearance shown in Fig. 7

PLATE XIX

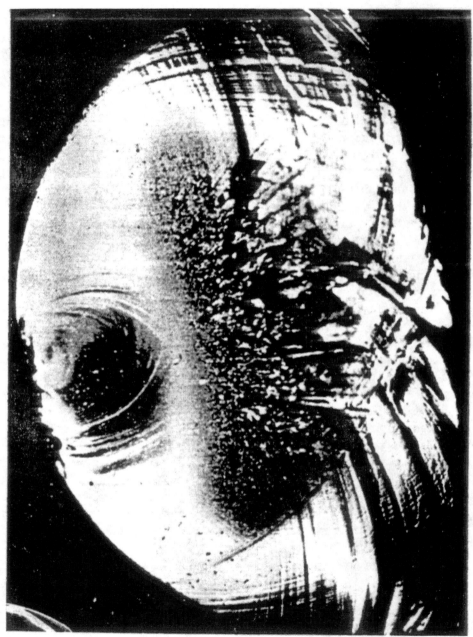

[*Courtesy U.S. Naval Research Lab.*

Fig. 7.—Fracture surface of a 0·3-in.-dia. glass rod, file-notched and broken in bending. Strong Wallner lines formed at irregularities produced by the notching and later by the hackle roughening. Surface was lightly silvered before photographing. (*Sullivan.*)

(Plate XIX). In the mirror region adjacent to the file-notch, with the aid of specular reflection lighting, a system of intersecting curves can be seen. These lines were first described by Wallner [102] and have been explained by Smekal.[103] Where the nearly circular leading border of the moving crack intersects the free surface of the rod, irregularities in the separation process generate elastic vibrations with shear component normal to the crack plane. The responsive tilt of the fracture surface at the interactions of the crack border with the largest of these elastic impulses traces out the Wallner lines. From measurement of the angle of intersection of two Wallner lines and a knowledge of the elastic-wave velocity, the crack speed can be computed.

Extensive crack-velocity studies of this general kind have been made by Kerkhof [104-106] and his associates. Kerkhof prefers to introduce the ripple marking by ultrasonic irradiation, a technique which adds substantially to the accuracy and completeness of the observations. From these investigations it is known that the crack speed increases steadily, though with decreasing acceleration, across the mirror region.

Ripple markings on a central segment of a fracture through an Araldite epoxy plate are shown in Fig. 8 (Plate XX) (Clark[107]). A crack was tapped through the plate using successive application of spreading-force impulses by means of a knife blade. Shear vibrations at 80,000 c/s were travelling nearly normal to the fracture segment shown. The ripples are of low amplitude and can be seen only with specular lighting. Owing to the curvature of the crack surface it is difficult to show as much in a photograph as can be observed visually. Nevertheless, one can see the tendency of the ripple spacings to increase with distance from the starting position and to assume an indistinguishably small spacing as the crack approaches the arrest point. This plate was carefully prepared in such a way as to be free from residual stresses and the ripple markings indicate negligible lag of the crack at the free surfaces of the plate. The maximum crack speed was ~ 400 ft/sec.

Fig. 9 (Plate XXI) from Kerkhof [105] shows the clarity with which the progress of the crack can be outlined by ripple markings. The general view (inset) shows the fracture surface of a glass tube broken by application of lateral pressure. The ends of the glass tube protruded from the pressure vessel and a crystal transmitted ultrasonic shear vibrations into the glass at one of the free ends of the tube. As can be seen, the cracks spreading around the central hole were so nearly coplanar that the difference in level at joining soon disappeared. A fracture of this kind is driven by the distributed spreading force pressure as the hydraulic fluid penetrates towards the leading border of the crack. Typically, the driving force remains below that necessary to produce hackle roughening.

PLATE XX

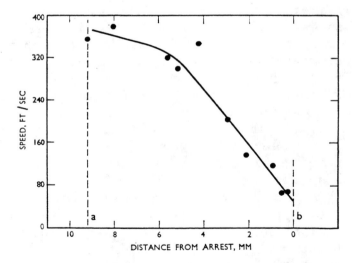

[*Courtesy U.S. Naval Research Lab.*

Fig. 8.—Ripple markings from 80,000 c/s vibrations for a start/arrest segment of fracture in a 3·5-mm-thick plate of Araldite. The graph shows estimates of crack speed during the approach towards the position of crack arrest. (*Clark.*[107])

PLATE XXI

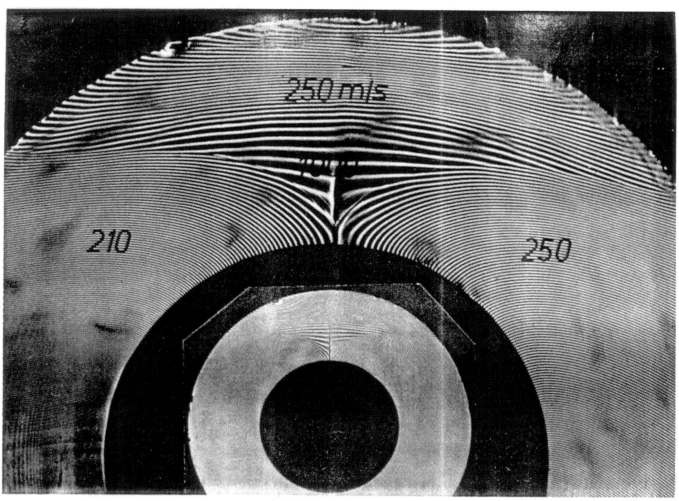

[*Courtesy Union Scientifique Continentale du Verre.*

FIG. 9.—Kerkhof's ripple markings on a glass-tube fracture. Larger photograph shows detail of inset.

Surrounding the inset portion of Fig. 9 is a photograph of the same fracture as taken at a higher magnification and with somewhat improved lighting. Kerkhof indicated the average speed of the crack in m/sec through selected regions by numbers on the photograph. It is clear that where local regions of elevated stress occurred, the corresponding increase of crack speed is similarly localized. Study of the available evidence from observations of types of ripple marking shows that the speed of a single running crack increases and decreases in phase with the tensile driving force.

Although gradual acceleration and deceleration of a running crack represents normal behaviour in response to gradual increase and decrease of the tensile force $\mathfrak{G}$, it will be recognized that circumstances favouring departures from normal behaviour often take place in the case of a long crack traversing a plate. For example, in a plate of glass there is frequently a state of residual compression in the surface layers. The crack border leads in the centre and lags at the free surfaces. Intermittent readjustments of the crack-border contour tend to change the net section (and stress level) controlling the crack speed adjacent to the free surfaces. Fluctuations in velocity occur in this way, particularly along the free-surface regions of the fracture. In fact, crack-speed measurements based upon micro-flash side views of fractures in glass plate have been reported as indicating that the crack speed always changes in a discontinuous manner,[108] whereas ripple markings (in central regions of a single running crack traversing a glass plate) indicate only gradual changes in the crack speed.

In the case of a long running crack in a steel plate, more pronounced free-surface effects occur. Rapid plane-stress yielding takes place in the surface layers. Fracture markings indicate that the crack is spreading outwards continuously from central regions towards the free surfaces, where final breakthrough normally takes place with a visible amount of oblique shear referred to by de Leiris as the "shear lip". Here the modulations in crack speed take the form of a tendency for the shear-lip fractures to occur in segments, thus producing an appearance of intermittent crack extension in microflash side views, as noted by van Elst, Korbee, and Verbraak[109] and others. Since crack extension is always locally discontinuous in the common structural metals and plastics, a certain amount of averaging across irregularities is implicit in the crack-speed concept itself. In this sense, apparent deviations of the crack speed from normal behaviour due to enhanced plastic yielding in the surface layers do not imply absence of an "in-phase" responsiveness of the fracturing process to the tensile force.

Although direct simultaneous observation of the force $\mathfrak{G}$ and the crack-speed response is the basic task of experimental fracture mechanics, the majority of such observations are concerned with the onset of rapid fracture rather than with changes in velocity of an already propagating crack. Fig. 10 (Plate XXII), taken from work by Wells and Post[34] shows four micro-flash views of the isochromatic fringes for the same crack traversing a plate of Columbia resin (CR-39). Approximately, the size of an isochromatic loop of fixed order number at the crack border is proportional to the crack-extension force $\mathfrak{G}$. Other similar photographic records were obtained and the average crack speed was measured across segments of the plate width.

The average values of tensile force in terms of K^2 and of crack speed are shown in Fig. 11. Although the force increased in proportion to the crack length throughout the region observed, the major part of the increase in velocity occurred before the crack was half-way across the plate. Beyond this half-way point the velocity was levelling off in its approach to the limiting crack speed for the material and, in a few cases, crack division was observed. In agreement with other work[110] the limiting speed approached was about half the elastic-shear-wave velocity ($\sim 0.5\ c_2$), a value well below the Rayleigh wave-velocity limit expected from considerations purely of inertia ($\sim 0.9\ c_2$).

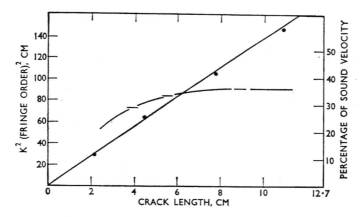

Fig. 11.—K^2 and crack speed as a function of crack length for cracks traversing CR-39 plates, from data of ref. (33).

Consider next the development of crack division and hackle. Yoffe concluded, from analytical studies of a running crack of constant length, that crack division was due to distortion of the crack-border stress field as a result of the inertia effects of the material. In reaching this conclusion, Yoffe neglected two factors. For the glassy and polycrystalline solids in which crack division is commonly observed, the formation and joining up of advance separations is the principal mechanism of progressive crack extension. Thus, it is desirable to explain crack division

PLATE XXII

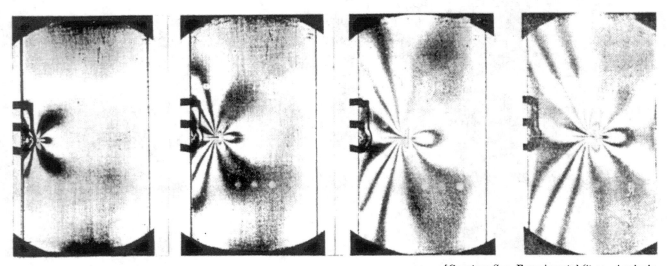

[*Courtesy Soc. Experimental Stress Analysis.*

FIG. 10.—The size of the crack-border stress field is indicated by the isochromatic loops adjacent to the leading edge of the crack, in four micro-flash views of the same crack traversing a plate of CR-39. (*Wells and Post*.[34])

in these terms. Secondly, both for the static[111] and the running crack, the largest tensile stress at a fixed distance from the crack border is nearly perpendicular to the crack plane and occurs at an angle of 60–70° from the plane of the crack.[29] The characteristic roughness of a plane-strain tensile fracture in materials of moderate toughness is due in part to the tendency for non-coplanar advance separations to develop in such regions. The ratio of the σ_y stress at the 60° angle to the σ_y stress at the 0° angle, both computed for the same value of r, increases from 1·3 to 1·45 as the crack speed increases from zero to $0·5\ c_2$.[61] This moderate alteration in stress distribution is too small to constitute the primary or controlling factor in crack division.

It has been noted previously that a limiting crack speed of $\sim 0·5\ c_2$ is anticipated from theoretical considerations, independent of the onset of crack division or hackle. Thus, as the crack speed approaches its limiting value with the driving force $\mathfrak{G}$ still increasing, both the distance from the crack border at which an advance separation can be activated by the local stress, and the time for spreading of such separations before arrival of the main crack, are continually increasing. The development of two advance separations on opposite sides of the crack plane, large enough for simultaneous propagation by a divided stress field, would necessarily result in crack division. This explanation of crack division is consistent with known facts bearing on fracture behaviour. In fact, fracture examinations show that crack division is preceded by numerous efforts at division which are unsuccessful because one of the two necessary new cracks is arrested. Evidently, instead of using onset of crack division as the reason for the limiting crack speed, one should regard the approach of the crack speed to a limiting value as the primary factor leading to the development of crack division.

When the tensile driving force on a deeply embedded crack exceeds the magnitude necessary for the limiting crack speed, an instability towards crack division develops. However, in typical examples, a local attempt at crack division can dominate only along one small segment of the crack border. Efforts at crack division, which would result in branching of the crack in a plate fracture, compete with other attempts at crack division which continually develop along various segments of the long crack border. The spreading and joining of sizable advance cracks, thus produced, cause some undercutting. Fragments termed "shards" are formed in this way and a pronounced roughening, termed "hackle", occurs. Hackle, for a crack spreading into a thick section of brittle material, is the counterpart of crack division or branching for a crack traversing a plate. The high-speed phenomena of crack division, hackle, and limiting velocity correspond in a consistent way to the view that a running crack is in stable balance with the driving force. A relationship of the low-speed crack-extension behaviour, including crack arrest, to the stable-behaviour viewpoint will be noted in subsequent discussion.

At the University of Illinois, studies by Hall and his co-workers of crack propagation in large mild-steel plates have resulted in a great amount and variety of data.[112,113] In one group of tests, welded slots were used to place the side quarters of a 5-ft-wide plate in residual tension.[114] Thus, when a small general tension was applied and a crack was started with wedge impact in a side notch, the crack ran rapidly at first, then slowed down as it entered the central regions of the plate, originally in a stress state of residual compression. Velocities observed ranged from ~5000 to ~300 ft/sec. Recordings were made of the signals from strain-gauges bonded to the plate at various distances from the crack. A recent re-examination of this work by Eftis and Krafft[115] has provided the data shown on the extreme right-hand side of Fig. 12.

The correlation between K value and crack speed, as illustrated in Fig. 13, exhibits a large scatter, particularly in the low-velocity region.

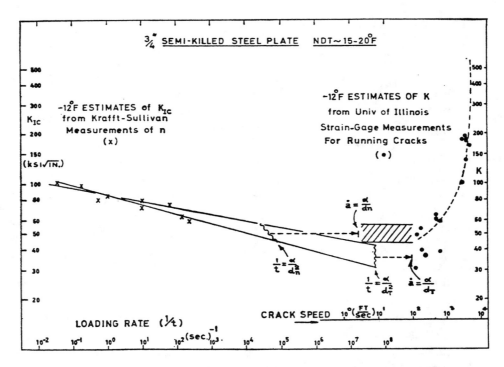

Fig. 12.—K as a function of reciprocal of loading time for rising-load K_{Ic} tests and as a function of crack speed for running cracks in a mild-steel plate. Data from refs. (112), (114), and (115).

Nevertheless, the data clearly show a general trend for the average crack speed to decrease with decrease in the crack-border tensile stress as measured by the K value. The decrease of K with decrease in average crack speed terminates when K becomes small enough for crack arrest.

In all the University of Illinois wide-plate tests, there was a general trend for the fracture-surface roughness, the value of K (or $\mathfrak{G}$) at the crack border, and the crack speed to increase or decrease together. On a size scale suggested by the chevron appearance, discontinuous velocity changes may well have occurred. However, the general trend seems to be acceptable as characteristic of normal behaviour.

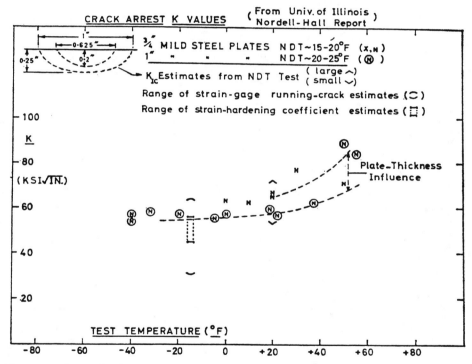

Fig. 13.—Critical values of K for the arrest of cracks in ¾- and 1-in. mild-steel plates from data in ref. (112). Estimated ranges of K_{Ic} are shown based on NDT measurements, on correlation with strain-hardening experiments (ref. 114), and on low-crack-speed estimates of K (ref. 114).

Observations of crack arrest in the University of Illinois tests were made, using tensile tests of 24-in.-wide mild-steel plates with a central longitudinal butt weld.[115] The tests were of the same type as the Wells (British Welding Research Association) B.W.R.A. wide-plate tests discussed later, in which crack arrest at small tensile loads has often been noted. Fig. 13 contains calculations by Nordell and Hall[113] of the K value* at crack arrest for $\frac{3}{8}$- and 1-in.-thick plates. The calculations were based upon the crack size at arrest, residual-stress measurements in the plate, and knowledge of the tensile load. The upward trend of the crack-arrest K values above 0°F is in part due to decrease of plane-strain constraint near the leading region of the crack. Rough estimates of the magnitude of the influence of plate thickness, shown in Fig. 13, are large enough to make it difficult to judge the magnitude of this upward trend in terms of actual K_I values for crack arrest from the present data.

The materials concerned in Fig. 13 were similar in regard to transition temperature to the plate that furnished the data in Fig. 12. One crack arrest was observed in a test which employed the Fig. 12 plate material. This point is indicated by a cross in Fig. 13 and agrees well with two crack-arrest calculations for one of the Fig. 13 steels. After allowances had been made for underestimates of the effective crack length at arrest and for the influence of plate thickness, the crack-arrest K_I value for the Fig. 12 steel at −15°F was reduced to fall within the range 44–55 lb/in². This region is shown with diagonal shading in Fig. 12. The unstable, low-velocity crack-extension behaviour can be represented in a nominal manner by a zero-slope extension of the trend line (K_1 vs. $\dot{a}$) through the crack-arrest zone.

Consider next the estimates of K_{Ic} from the Krafft–Sullivan tests shown in the left-hand portion of Fig. 12. Using specimens cut from the test plates which provided the K_I vs. $\dot{a}$ data, crack-toughness measurements under increasing load were conducted for a wide range of loading times at a series of low temperatures.[115] Measurements of flow properties were recorded over the same range of loading times and over an extended temperature range. Owing to the smallness of the test specimens, direct measurements of K_{Ic} at −15°F could not be made. However, a proportionality factor, relating K_{Ic} to the strain-hardening exponent, was established and this permitted each strain-hardening measurement to be given an equivalent K_{Ic} interpretation. The K_{Ic} values thus estimated for −15°F are shown plotted as a function of the reciprocal of the load-increase time, t.

For purposes of illustration assume that the actual values of K_{Ic} as a function of $1/t$ would lie in the zone defined by the trend lines through the K_{Ic} estimates on the left-hand side of Fig. 12. The correlation with strain-hardening and previous discussion suggest that crack-extension instability is governed by conditions closely related to the strain, $\varepsilon = n$, at a distance d_n or d_T from the crack border, depending upon whether equation (20) or equation (21) is used for computation of d. Among the

---

* The stress-intensity values given in ref. 112 are in terms of $\mathscr{K} = K/\sqrt{\pi}$. These are converted to values of K in Fig. 13.

factors which might stop the decrease of K_{Ic} with $1/t$ and introduce a rising trend corresponding to the stable portion of the K_I vs. $\dot{a}$ relationship, the most prominent is the development of an adiabatic thermal condition in the fracture-process zone. The locus for a change from isothermal conditions in a region of size d can be estimated from the equation

$$\frac{1}{t} = \frac{\alpha}{d^2} \qquad \ldots \ldots \quad (23)$$

where t is the load-increase time and α is the heat-diffusion coefficient (ratio of thermal conductivity to the product of density $\times$ specific heat).

In Fig. 12 the wavy line at $K_{Ic} = 50$ lb/in^2/in. and $1/t = 0.45 \times 10^4$/sec was computed from equation (23) assuming $d = d_n$. The vertical locus at $1/t = 5 \times 10^7$/sec was computed assuming $d = d_T$. As discussed earlier, a notch root radius of $0.05\ (2r_Y)$, in certain high-strength materials, results in the same crack-toughness as when the notch is fatigue-crack sharpened. From this one might plausibly use a small fraction of $(2r_Y)$ as a third estimate of the size of the process zone. For the data of Fig. 12, the estimate $d = 0.05\ (2r_Y)$ results in a locus for change in thermal conditions which crosses the zone of K_{Ic} values at about the same place as the locus computed for $d = d_n$ and with a smaller slope. Within the confines of present uncertainties, it is not possible to make a definite choice between the alternative assumptions for the fracture-process-zone size factor. However, the positions of the isothermal/adiabatic change loci relative to the crack-arrest K values suggest that a connection between K_{Ic} at increasing loads, measurements of K_I for crack arrest, and the subsequent stable high-velocity region, might be established on the basis of thermal conditions in an appropriately defined fracture-process zone at the crack border.

When $1/t$ has the value corresponding to equation (23), the strain rate at the border of the process zone corresponds to a crack speed in the range 2–50 ft/sec, depending upon the d value assumed. An attempt at crack extension which is unable to accelerate beyond this speed range would fail, since this is equivalent to a reduction in loading rate and thus to a higher crack-toughness.

From earlier discussion it has emerged that the stability of a high-velocity running crack is closely related to limitations on the speed of propagation of stress waves and plastic strains in the plastic zone leading the crack. The velocity of a uniaxial tensile plastic strain travelling along a rod decreases with increase in the plastic strain and is zero for a plastic strain large enough for instability. The conditions controlling propagation speed are, of course, much more complex at the border of a moving crack. However, it does not seem unreasonable to assume that, as the average crack speed in steel increases beyond some value of the order of 50 ft/sec, further increases in speed cannot develop without increase in the stress parameter, K_I. As the crack speed drops towards this value, intermittent arrest and an increasing degree of instability towards arrest are to be expected. The values of K_I for slow-moving cracks or crack arrest should be nearly the same as the minimum K_{Ic} approached by increasing the loading rate in a rising-load test. Estimates of K_{Ic} for Pellini–Puzak-type drop-weight tests support this conclusion.

The object of a small-flaw, drop-weight test is to determine a nil-ductility temperature termed the NDT. The inserted flaw consists of a notched segment of brittle weld bead on the tension surface of the test-plate. Particularly in tests above the NDT, indications of crack arrest appear on the specimens beneath the brittle weld flaw, apparently outlining the borders of the region embrittled by the weldment. If rapid crack extension occurs from this region through the test-plate, with yielding limited to the plane-strain crack-border plastic zone, the result is a flat break and the temperature is regarded as being below the NDT. However, if the stress at the tensile surface has to rise above the yield stress before crack propagation takes place, then a small bending yield of the plate is observed and the temperature is regarded as being above the NDT. Thus, at the NDT, the stress near the flaw must be nearly equal to the yield stress of the material corresponding to a loading time of the order of 10^{-4} sec, and the test is equivalent to a fast, rising-load K_{Ic} test for a specimen with a crack part way through it. Calculating the depth of this crack as being in the range 0·2–0·25 in. and the dynamic yield stress as being in the range 80,000–90,000 lb/in² led to estimates of K_{Ic} which bracket the Nordell–Hall crack-arrest values at the NDT temperature of the material. The estimates are shown in Fig. 13.

From these crack-arrest data, it can be deduced that the minimum K_{Ic} of the materials represented is relatively insensitive to testing temperature in the range from −40°F to +30°F (−40°C to 0°C). This is not inconsistent with the behaviour of the strain-hardening coefficient in the same temperature range. Further reductions of temperature would be expected to cause substantial decreases in the minimum crack-toughness. Judging from K_{Ic} tests of mild steels for 2×10^{-3} sec loading time, the decrease in crack-toughness with decrease in temperature can be expected to level off below −250°F (−155°C) in the region where twinning is the dominant mechanism of plastic strain.

An approximate proportionality of K_{Ic} to absolute temperature over a wide temperature range has been noted for several high-strength steels ($\sigma_{YS} \simeq 220$ lb/in²).[116] These K_{Ic} tests were made using a loading time of the order of 1 min. Since the strain-rate-sensitivity of very high-strength steels of this type is small, the minimum crack-toughness values might show a similar variation with temperature.

VI.—Crack Arrest in Mild Steels

The historical view of brittle-fracture studies interpreted in Sections I and II has indicated the extent to which certain ideas have been abandoned, only to be reclaimed at intervals of many years through the intervention of new evidence. One of the features in this interplay has undoubtedly been the conflicting requirements of the metallurgist and the engineer in the production of viable structures. The metallurgist has to deal with many variables in the development of tough materials and requires economical small-scale tests. Complete removal of uncertainties as regards specimen size and shape is an advantage here, for useful progress can then be made under strictly comparable conditions. The engineer, on the other hand, uses relatively few materials and is reasonably satisfied that they possess uniform properties in terms of his requirements. He is concerned with a wide variety of sizes and shapes which he expects to construe in terms of continuum mechanics. His natural preference is for tests on elements or components of structure reproduced at the full scale where necessary. It has been recognized by Boyd, in unrecorded discussions, that the object of much recent investigation has been to build a bridge between those two firm abutments. In the construction of this metaphorical bridge, attempts have been made to found a central pier, in terms of test procedures with specimens that are small enough to be duplicated in metallurgical research, yet large enough to illustrate full-scale behaviour. Examples of such are the U.S. Naval Research Laboratory drop-weight test and the various crack-arrest tests devised from the original of T. S. Robertson.

The drop-weight procedures have already been discussed in Section V. In now examining the developed range of crack-arrest tests it is appropriate to consider in some detail the circumstances that gave rise to them.[118] Naval vessels are subject to explosive attack in time of war and the resulting damage has to be minimized. Much experience has shown that the elimination of brittle cracking served well to localize damage under these conditions, where its origin could not, for obvious reasons, be controlled by the designer. Robertson's contribution was, therefore, to devise a test in which a brittle crack, started by wedging action in a full-thickness plate, could be arrested by running it into regions of increasing temperature. After much experiment he designed a specimen with an integral arch over the starting notch which could be struck at the crown in the direction of crack travel, to give a wedging force limited in magnitude by the yielding load in the arch member.

The space under the arch was ideally situated to confine liquid nitrogen as a local coolant and embrittling agent. In the next step Robertson noted that the stress system from the blow was insufficient in itself to project a crack uniformly through the plate in the absence of a temperature gradient, and so a uniform tension, transverse to the crack path, was added to the system. The condition for steady and repeatable cracking in the presence of possible parasitic thermal- and residual-stress effects was enhanced by application of the additional stress system through yielding strips.

Under these developed conditions the crack could be made to propagate uniformly through a plate specimen without a temperature gradient (except for the local region at the crack starting notch) subject only to the value of the applied tensile stress at a given plate temperature. The momentum of the blow was correspondingly established as a secondary variable controlling the duration, rather than the magnitude, of the wedging force. The value of the temperature-gradient test then emerged in terms, first, of its capability to demonstrate low applied-stress fracture below a certain arrest temperature; secondly, in the precision with which the critical stress for propagation could be defined below this temperature; and, thirdly, in the sharpness with which the transition temperature could be defined in terms of a sudden rise in crack-arrest applied stress.

In the version of this test that has now been used for many years by the British Admiralty in the qualification of steels for hull service, two further features may be mentioned. First, possible bending effects arising from the crack-starting blow are counteracted by a substantial mass in contact with the opposite specimen face. Secondly, the agreed interpretation of crack-arrest temperature in gradient tests relates to the position on the plate surfaces where the shear lips cease to be severed. Under these conditions of use at plate thicknesses up to $\sim 1\frac{1}{2}$ in. the specimens have proved to be sufficiently small to permit the test to form a useful part of programmes of actual steel development. A particularly valuable feature of the test, exploited originally in Admiralty programmes and later by the United Kingdom Atomic Energy Authority[119] in the development of thick steel plates for nuclear-reactor pressure vessels, has been to sound warnings with regard to higher yield points and larger plate thicknesses, in terms of the crack-arresting properties likely to be obtained. It has been found that corresponding Charpy V notch impact energies at the crack-arrest temperature rise substantially with these two variables. Similar indications, it must also be admitted, can be derived from the slow notched tension tests of Baker and Tipper[120] as these are judged on the appearance of fracture. But the latter test, in isolation, leaves room for doubt, since the material in the path of the fracture is first taken to its ultimate strength and extension, and the demonstration of low-stress fractures at somewhat lower temperatures in the Robertson tests furnishes complementary reasons for caution.

The attractiveness to engineers of the crack-arrest test is evident from the variants subsequently developed, most of which have contributed perspective and permanence to the original concept. The variants include the Japanese Double-Tension[121,122] and the United States SOD[123] (Standard Oil Developments) or Esso[124] tests. It is convenient to discuss these under the following headings: (1) specimen length and plate width, (2) generation of wedging force, and (3) temperature control.

The Robertson specimen is short in the direction of applied tension and the associated loading-jack system stores comparatively little energy. The effects of the intervening yielding strips are somewhat difficult to determine. By comparison, long specimens have been used in the SOD and Double-Tension tests, but without significant change of result. If anything, the crack is somewhat harder to arrest in long specimens when it extends beyond the centre of width of the plate, and this has given rise to stated limitations on acceptable crack-arrest positions.

When dealing with thicker plates, British work has created a preference for specimens that are larger, in order to contain full cracks with curved fronts at arrest. Testing machines have been built to examine full plate widths as great as 8 ft, keeping the aspect ratios of the specimens low and approximately constant.[119,125] Such tests can be usefully performed only under the more stringent isothermal conditions. They have demonstrated increases in crack-arrest temperature for the same material of the order of 30 degC in a few cases, as compared with temperature-gradient tests on smaller standard specimens.

The relationship between the impact momentum and the wedging force in the original Robertson test is not capable of succinct evaluation, and three other methods of crack generation may be briefly described. A high-speed wedge is driven into the initial slit in the SOD test.[123] A shaped ear, connected by an isthmus to the main specimen, is statically loaded to fracture in tension by an ancillary device in the Double-Tension test,[124] and provides the possibility of accurate measurement of the wedging force. In N. C. R. E. wide-plate tests the Robertson knob is omitted, but the initial slit is supplemented by a transverse hole of relatively small diameter, into which a cone-ended nail can be explosively driven.[125] Although it does not permit measurement of the wedging force, this device is among the simplest of the crack-starting devices.

Temperature gradient in crack-arrest tests is an important feature, creating the need for interpretation, as indicated above. Two extremes of behaviour may be noted by way of illustration. Under isothermal conditions well below the crack-arrest temperature a crack stopping under the action of a declining driving-stress system is mainly smooth and relatively square-ended, without substantial shear lips. Such crack arrests are sometimes observed in casualties and in wide-plate tests. Conversely, with a temperature gradient and a rising driving-stress characteristic, e.g. in a long test plate, or in a testing machine with ample stored energy, the arrested crack is deeply curved, with substantial shear lips.[126] Under such conditions the assessment of a crack-arrest temperature is necessarily subject to some convention, such that the value of a gradient test depends upon its economy in the number of specimens required to achieve a definite result. The conventions employed by others than Robertson have ranged, in terms of the temperature adopted, from the position of the tip of the crack to the severed shear lips at the plate surface.

The problem of thermal stress in temperature-gradient specimens is not as severe as it might seem, although differentials as large as 100 degC may exist. For plane heat flow with no surface exchanges, it may be proved that singly connected bodies are free from thermal stress when they have heat sources and sinks at their outer boundaries that are allowed to reach equilibrium with one another. This ideal has been approached by Akita and Ikeda in long Esso specimens, by arranging for simultaneous local cooling with liquid nitrogen at the notched edge and local hot-water heating at the opposite edge.[124]

Before going on to discuss the mechanical analysis of these tests, it is appropriate to describe briefly the notched and welded wide-plate test insofar as it may be used to demonstrate crack-arresting properties. Unlike the above, with the exception of the Double-Tension test, this is a wholly static test, and is mainly performed on square test plates with approximately 36-in. sides. The test configuration remains essentially the same as that devised by Greene,[127] with a central longitudinal butt weld containing jeweller's sawcut notches in the centres of length of the edges prepared for welding. In the British Welding Research Association tension test a balanced double-V weld preparation is used, and the sawcuts are uniformly 0·2 in. deep over the full thickness. Other notch variants have been employed at the University of Illinois.[115] In the absence of the weld metal, such specimens would retain high brittle strengths to very low temperatures, as in the Tipper test, but the effect of the weld is to embrittle the crack-starting region by heat and thermal plastic strain, and crack propagation is assisted by the longitudinal tensile residual stresses of yield-point magnitude. The test is conducted isothermally, but at suitably low temperatures fracture initiation takes place at low applied stresses, and symmetrical crack arrests occur. Re-initiation of these arrested cracks requires much higher applied stresses, if the cracks are long enough to have passed out of the residual stress field. Although the applied stress at initiation is uncontrolled, except in terms of test temperature and the severity of the initial defect (in terms of "hot" cracking of the weld metal), such tests also show a limiting applied stress for crack arrest. However, there is a tendency for arrests to be grouped either relatively near the weld or, at marginally higher applied stress, near the outer plate edges. Notched and welded wide-plate tests have been performed on steels of yield strengths from 35,000 to 110,000 lb/in^2, of thicknesses ranging from $\frac{1}{4}$ to 3 in., and at temperatures down to $-190°$C.[128] The crack-arresting behaviour described above is apparently universal, as it is in the other crack-arresting tests.

Mechanics of Crack-Arrest Tests

The feature of most crack-arrest tests which permits analysis of the results in terms of the mechanics methods of Section III, is the relative absence of yielding from the imposed loading when the specimen is below the crack-arrest temperature. Indeed, a superficial calculation of the concentrated stresses near the crack border shows that even the yielding there is confined to ranges mainly less than plate thickness, so that crack-length corrections are superfluous in the larger specimens.

The relevant stress-intensity terms for the combination of applied stress σ and crack-wedging force are, respectively:[28]

$$K_\sigma = \sigma \sqrt{W \tan \frac{\pi a}{W}} \quad \quad \quad (24a)$$

$$K_r = 2\sqrt{\frac{a}{\pi}} \int_0^a \frac{\sigma_r dx}{\sqrt{a^2 - x^2}} \quad \quad \quad (24b)$$

where W is plate total width and a is the half-length of a central crack or depth of each of a pair of edge cracks. The second expression is for similar geometries, with σ_r as the local value of crack-surface pressure at a distance x from the longitudinal plane of symmetry within the crack length. These expressions may be added algebraically to determine the total K value. The next important step is to determine the drop in σ that occurs as the plate is progressively severed, and this depends upon the lengthwise span L at which the plate grips are fixed in space, the resulting compliance of the whole plate, and its change during cracking. By use of equation (24(a)) above, transposed in terms of $\mathfrak{G}$, and equation (3) of Section III, it may be shown that the stress-reduction factor equal to the cracked and uncracked compliance ratios, is simply,

$$1 + \frac{4W}{\pi L} \ln \sec \frac{\pi a}{W} \quad \quad \quad (25)$$

This term illustrates the important effect of changing the plate length in crack-arrest tests. At a ratio of length to width of $8/\pi$, the crack-border stress intensity arising from applied stress is close, at all except the greatest possible crack lengths, to the value for an infinite plate, as noted by Irwin in relation to the photoelastic observations of Wells and Post.[33] With shorter plates the applied stress falls off more quickly, and it would seem that the plate shape was well chosen experimentally by Robertson to permit constancy of stress intensity over a substantial middle zone in his short specimens. It will also be seen from equation (24(b)), when it is evaluated for a concentrated wedging force, how the latter complements the applied stress in maintaining a uniform level of crack-border stress intensity from the start of cracking.

Successful quantitative analyses of results from crack-arrest tests, mainly in terms of crack-extension force, have been made.[129,130,124] The authors of the last-named paper, in detecting consistent effects of wedge momentum on the deduced values of fracture-toughness at arrest, chose to correlate the latter with the peak velocity of the brittle crack. However, they also had to neglect the K contribution from the initial blow, which they maintained to be negligible at the position of crack arrest. The latter assumption might now appear to be doubtful, tending to obscure the essential simplicity of behaviour of the material at crack arrest by attributing to the test a memory of already completed events.

The variation in K values over the whole range of possible crack speeds has also been discussed in Section V, with reference to the University of Illinois data obtained from wide plates of mild steel. In these tests a great deal of information was made directly available by dynamic strain measurements. It is also of interest to examine British low-stress, crack-arrest data from welded and notched wide plates, derived from applied loads at fracture and distribution of residual stresses in the weld; these, however, had to be interpreted through the above calculations of attenuation of applied stress during fracture, since corresponding dynamic measurements were not available. Crack-arrest K values for six steels of differing thickness and yield point are summarized from these determinations in Table I. The conclusions are broadly the same as those reached in Section V. In particular, the crack-arrest K values correspond clearly with expectations of plane-strain fracture.

Moreover, recent static-loading tests on full-thickness wide plates of Steel W (Table I), with symmetrical 3-in.-deep edge notches, revealed low-stress fractures and closely comparable crack-initiation K values at similar temperatures.[127] The notches were square-ended and finished with a 0·006-in. saw blade. Curiously enough, such easy crack initiations had not previously been obtained from many examples of arrested brittle cracks of similar length, and this may have been due to the

TABLE I.—*Calculated Crack-Stress Intensity Values at Arrest in Welded and Notched Wide Plates*

Material	Test Temp., °C	Applied Stress at Fracture, lb/in² × 10³	Half Crack Length at Arrest, in.	Stress Intensity (K), lb/in²√in. × 10³		
				Residual Stress	Applied Stress	At Arrest
Steel F 1 in. thick Y.P. 36 lb/in² × 10³	−5	0	2·6	56	0	56
	−4	0·7	2·15	60	2	62
	−1	2·7	2·9	52	8	60
	6	2·9	1·95	61	7	68
	−8	6·5	2·85	53	19	72
	−5	6·5	3·05	50	20	70
	−4	11·4	{4	42	41	83
			8	18	57	75
Steel P 1 in. thick, Y.P. 36 lb/in² × 10³	−28	7·8	14	5	51	56
	−25	5·2	3·9	43	19	62
	−25	4·3	4·4	38	17	55
	−8	3·4	2·6	55	9	64
	−8	6·1	4·7	36	24	60
	−8	9·9	3·05	46	30	76
	−8	13·4	5·25	33	55	88
	−7	4·0	3·6	45	14	59
	0	9·4	3·3	48	31	79
	1	0	2·0	61	0	61
	4	9·6	3·65	45	33	78
	10	10·5	2·25	59	62	121
Steel Q 1 in. thick Y.P. 40 lb/in² × 10³	−27	5·1	12·6	8	31	39
	−25	8·5	17·0	1	76	77
	−22	1·8	4·0	47	6	53
	−15	7·0	2·9	59	21	80
Steel S 1 in. thick Y.P. 43 lb/in² × 10³	−56	5·6	4·0	50	20	70
	−40	7·6	8·4	20	38	58
	−40	8·7	4·0	50	31	81
	−30	2·5	2·9	62	7	69
	−24	6·3	5·6	36	27	63
Steel T, 1 in. thick Y.P. 49 lb/in² × 10³	−65	14·6	9·0	21	77	98
	−50	7·4	2·6	77	21	98
Steel W 3 in. thick Y.P. 38 lb/in² × 10³	−45	5·2	2·3	63	14	77
	−20	8·5	2·45	61	23	84
	−10	13·7	3·25	52	66	118
	−10	8·5	2·55	60	14	74
	−8	{0	1·5	66	0	66
		12·9	7·0	24	61	85

Notes: 1 K_r corrected for yield-point changes from one steel to another.
        2 K_r not corrected for yielding when external stress is applied.

gathered shear lips and comparatively rough textures of the latter cracks. It would now appear that the majority of running brittle cracks in mild steels are manifestations of plane-strain fracture, such that arrests are simple reversals on the time scale of possible plane-strain initiations. Nevertheless, the question of detailed equivalence is still uncertain. Running cracks in mild steels need not always exemplify plane-strain fractures, since Japanese investigators[131] have shown, by progressive thinning of plates subjected to crack-arrest tests, that fracture-toughness values at arrest at particular temperatures become subject to transitional increases at $\sim \frac{1}{2}$ in. thickness. The latter can readily be interpreted as progression towards plane-stress fracture conditions.

Confirmation of these conclusions could clarify the restricted thinking that differentiates at present between the treatment of mild steels and that of materials of higher strengths and lower ductilities, in terms of a distinction between crack initiation and propagation in the former, and plane-stress and plane-strain fracture in the latter. The occurrence of an intermediate temperature range in mild steel where elevated-toughness fracture initiations and arrests of the plane-stress type are combined with sustained propagation in the flat-tensile or plane-strain mode, is in part a reflection of enhanced strain-rate-sensitivity. The apparent complexity, particularly of initiation behaviour in this temperature range, is probably to be attributed to boundary effects in specimens of restricted dimensions. Where these boundary effects are reduced in specimens of large dimensions the behaviour becomes simplified, and approaches that of brittle high-strength materials.

REFERENCES

1. A. A. Griffith, *Phil. Trans. Roy. Soc.*, 1920, [A], **221**, 163.
2. W. Weibull, *Ing. Vetenskaps Akad. Handlingar*, 1939, **(151)**.
3. A. E. H. Love, "A Treatise on the Mathematical Theory of Elasticity". **1892**: Cambridge (University Press).
4. Article by Arturo Uccelli in "Leonardo da Vinci". **1956**: New York (Reynal and Co.).
5. S. P. Timoshenko, "History of Strength of Materials". **1953**: New York and London (McGraw-Hill).
6. I. Todhunter and K. Pearson, "History of the Theory of Elasticity and of the Strength of Materials", Sections 1503 and 936. **1886**: Cambridge (University Press).
7. P. Ludwik, "Elemente der technologischen Mechanik". **1909**: Berlin (Julius Springer).
8. T. E. Stanton and R. G. C. Batson, *Proc. Inst. Civil Eng.*, 1921, **211**, 67.
9. J. G. Docherty, *Engineering*, 1932, **133**, 645; 1935, **139**, 211.
10. G. Sachs, *Trans. Amer. Inst. Min. Met. Eng.*, 1941, **143**, 13.
11. C. E. Inglis, *Trans. Inst. Naval Architects*, 1913, **55**, 219.
12. J. C. Fisher and J. H. Hollomon, *Trans. Amer. Inst. Min. Met. Eng.*, 1947, **171**, 546.

13. "B.I.S.R.A. Conference on Brittle Fracture of Mild-Steel Plate" (Cambridge). **1945**: London (The Association).
14. "Fracturing of Metals" (A.S.M. Symposium, Chicago 1947). **1948**: Cleveland, O. (Amer. Soc. Metals).
15. M. Gensamer, E. Saibel, J. T. Ransom, and R. E. Lowrie, "The Fracture of Metals: A Report to the Bureau of Ships, U.S. Navy". **1947**: New York (Amer. Weld. Soc.).
16. M. Gensamer, E. Saibel, and J. T. Ransom, *Weld. J.*, 1947, **26**, 443s.
    M. Gensamer, E. Saibel, and R. E. Lowrie, *ibid.*, 1947, **26**, 472s.
17. G. R. Irwin, ref (14), p. 147.
18. E. Orowan, *Rep. Progress Physics*, 1949, **12**, 185.
19. E. Orowan, "Fatigue and Fracture of Metals" (edited by W. M. Murray), p. 139. **1952**: New York (John Wiley); London (Chapman and Hall).
20. E. Orowan, *Weld. J.*, 1955, **34**, (3), 157s.
21. G. R. Irwin and J. A. Kies, *ibid.*, 1952, **31**, 95s.
22. M. E. Shank, "A Critical Survey of Brittle Fracture in Carbon Plate Steel Structures Other than Ships." **1954**: New York (Welding Research Council of the Engineering Foundation); "Symposium on the Effect of Temperature on the Brittle Behaviour of Metals, with Particular Reference to Low Temperatures" (Special Tech. Publ. No. 158), p. 45. **1954**: Philadelphia, Pa. (Amer. Soc. Test. Mat.).
23. A. A. Wells, *Weld. Research*, 1953, **7**, 34r.
24. G. R. Irwin, J. A. Kies, and H. L. Smith, *Proc. Amer. Soc. Test. Mat.*, 1958, **58**, 640.
25. F. A. McClintock and G. R. Irwin, "Plasticity Aspects of Fracture Mechanics", (Special Tech. Publ. No. 381), p. 84. **1965**: Philadelphia, Pa. (Amer. Soc. Test. Mat.)
26. G. R. Irwin, *U.S. Naval Research Lab. Rep.* (4763), 1956; "Proceedings of the 1955 Sagamore Research Conference on Strength Limitations of Metals" Vol. II, p. 289. **1956**: Washington, D.C. (U.S. Office of Technical Services).
27. G. R. Irwin, *Proc. 9th Internat. Congr. on Applied Mechanics* (Brussels, 1956), Paper No. **101(II)**.
28. G. R. Irwin, *J. Appl. Mechanics*, 1957, **24**, 361.
29. G. R. Irwin, "Fracture" in "Encyclopædia of Physics", Vol. VI, p. 551. **1958**: Berlin (Springer-Verlag).
30. H. Schardin, D. Elle, and W. Struth, *Z. techn. Physik*, 1940, **21**, 393.
31. E. H. Yoffe, *Phil. Mag.*, 1951, **42**, 739.
32. E. Saibel, ref. (14), p. 275.
33. S. W. Barber, "Symposium sur la Resistance mécanique du Verre" (Florence, 1961) p. 847. **1962**: Brussels (Union Scientifique Continentale du Verre).
34. A. A. Wells and D. Post, *Proc. Soc. Exper. Stress Analysis*, 1958, **16**, 69.
35. F. Kerkhof, *Naturwiss.*, 1953, **40**, 478.
36. H. Neuber, "Kerbspannungslehre". **1937**: Berlin (Julius Springer).
37. P. Kuhn, "Colloquium on Fatigue; Stockholm, May 1955" (edited by W. Weibull and F. K. G. Odqvist), p. 131. **1956**: Berlin (Springer-Verlag).
38. A. A. Golestaneh, *Brit. Weld. Research Assoc. Rep.* (**C33/1/57**), 1957.
39. G. R. Irwin, "Structural Mechanics", p. 557. **1960**: London (Pergamon Press).
40. G. R. Irwin, "Proceedings of the 1960 Sagamore Research Conference on Ordnance Materials", p. IV.63. **1961**: Washington, D.C. (U.S. Office of Technical Services).
41. F. A. McClintock and J. A. H. Hult, *Proceedings of the 9th International Congress on Applied Mathematics*, (Brussels 1956), Vol. VIII, p. 51.
42. P. E. Shearin, A. E. Ruark, and R. M. Trimble, ref. (14), p. 167.
43. E. H. Salmon, "Materials and Structures", Vol. I: "The Elasticity and Strength of Materials". **1931**: New York and London (Longmans, Green).
44. "Fracture" (edited by B. L. Averbach *et al.*). **1959**: New York (John Wiley); London (Chapman and Hall).
45. G. R. Irwin, *Trans. Amer. Soc. Mech. Eng.*, 1960, [D], **82**, 417.

46. Editorial Note, *Engineering*, 1935 (March 15).
47. H. de Leiris, *Bull. Soç. Franç. Mécaniciens*, 1952, **2**, (6).
48. D. K. Felbeck and E. Orowan, *Weld. J.*, 1955, **34**, 571s.
49. *DeHavilland Official Rep. on "Comet" Failure*, **1955**.
50. C. Schabtach, E. L. Fogleman, A. W. Rankin, and D. H. Winne, *Trans. Amer. Soc. Mech. Eng.*, 1956, **78**, 1567.
51. D. H. Emmert, *ibid.*, 1956, **78**, 1547.
52. D. R. De Forrest, L. P. Grobel, C. Schabtach, and B. R. Seguin, "Investigation of the Generator Rotor Burst at the Pittsburgh Station of the Pacific Gas and Electric Co.", paper (57-PWR-12) submitted to A.S.M.E. Power Division Conference, Allentown, Pa., **1957**.
53. D. H. Winne and B. M. Wundt, *Trans. Amer. Soc. Mech. Eng.*, 1958, **80**, 1643.
54. G. O. Sankey, *Proc. Amer. Soc. Test. Mat.*, 1960, **60**, 721.
55. J. D. Lubahn and S. Yukawa, *ibid.*, 1958, **58**, 661.
56. C. J. Boyle, R. M. Curran, D. R. DeForrest, and D. L. Newhouse, *ibid.*, 1962, **62**, 1156.
57. G. R. Irwin and J. A. Kies, "Fracture Theory as Applied to High-Strength Steel Pressure Vessels", paper submitted to Amer. Soc. Metals Golden Gate Conference, **1960**.
58. G. R. Irwin and J. E. Srawley, *Materialprüfung*, 1962, **4**, (1), 1.
59. G. R. Irwin, "Structural Aspects of Brittle Fracture", *Appl. Materials Research*, 1964, **3**, (2), 65.
60. "Fracture Testing of High-Strength Sheet Materials" (1st Rep. of Special A.S.T.M. Cttee.), *Amer. Soc. Test. Mat. Bull.*, 1960, **(243)**, p. 29.
61. G. R. Irwin, "Relatively Unexplored Aspects of Fracture Mechanics", *Univ. Illinois Rep.* **(TAM 240)**, 1963.
62. J. M. Krafft, *Appl. Materials Research*, 1963, **3**, 88.
63. A. M. Sullivan, *Mat. Research Stand.*, 1964, **4**, (1), 20.
64. H. M. Westergaard, *Trans. Amer. Soc. Mech. Eng.*, 1939, [A], **61**, 49.
65. M. L. Williams, *ibid.*, 1957, **79**, 104.
66. P. Paris and F. Erdogan, *ibid.*, 1963, [D], **85**, 528.
67. F. Erdogan and G. C. Sih, *ibid.*, 1963, [D], **85**, 519.
68. I. N. Sneddon, *Proc. Phys. Soc.*, 1946, **187**, 229.
69. R. A. Sack, *Proc. Roy. Soc.*, 1946, [A], **58**, 729.
70. A. E. Green and I. N. Sneddon, *Proc. Cambridge Phil. Soc.*, 1950, **46**, 159.
71. G. R. Irwin, *Trans. Amer. Soc. Mech. Eng.*, 1962, [E], **84**, 651.
72. O. L. Bowie, *J. Math. Phys.*, 1956, **25**, (1), 60.
73. L. A. Wigglesworth, *Mathematics*, 1957, **4**, 76.
74. A. H. Lachenbruch, *J. Geophys. Research*, 1961, **66**, 4273.
75. H. F. Bueckner, *Trans. Amer. Soc. Mech. Eng.*, 1958, **80**, 1225.
76. I. N. Sneddon, "Dual Series Relations", lectures delivered to Applied Mathematics Research Group, N. Carolina State College, March **1963**.
77. M. Lowengrub, "A Two-Dimensional Crack Problem", memorandum submitted to Applied Mathematics Research Group, N. Carolina State College, June **1963**.
78. M. Lowengrub, "An External Crack Problem with Asymmetrical Loading", *ibid.*, Sept. **1963**.
79. G. R. Irwin, "Analytical Aspects of Crack Stress-Field Analysis", *Univ. Illinois Rep.*, **(TAM 213)**, 1962.
80. D. D. Aug and M. L. Williams, *Trans. Amer. Soc. Mech. Eng.*, 1961, [E], **83**, 372.
81. M. L. Williams, *Bull. Seismological Soc. America*, 1959, **49**, 199.
82. G. I. Barenblatt, "Advances in Applied Mechanics", Vol. VII, p. 55, **1962**.
83. G. I. Barenblatt, *Priklad. Matem. i. Mekh.*, 1959, **23**, (3/5), 1.
84. G. I. Barenblatt, R. L. Salganik, and G. P. Cherepanov, *ibid.*, 1962, **26**, (2), 328.
85. J. B. Walsh, and A. C. Mackenzie, *J. Math. Phys. Solids*, 1958–60, **7–8**, 247.
86. M. F. Koskinen, *Trans. Amer. Soc. Mech. Eng.*, 1963, [D], **85**, 585.
87. G. R. Irwin, *ibid.*, 1963, [D], **85**, 593, (discussion).

88. G. R. Irwin, *Weld. J.*, 1962, **41**, (11), 519s.
89. G. R. Irwin, *Metals Eng. Quart.*, 1963, **3**, (1), 24.
90. L. F. Coffin, *Trans. Amer. Soc. Mech. Eng.*, 1960, [D], **82**, 671.
91. P. C. Paris, Ph.D. Thesis, Lehigh Univ., **1962**.
92. F. A. McClintock, *J. Appl. Mechanics*, 1958, **25**, 582; 1959, **26**, 467.
93. C. F. Tipper, *Admiralty Ship-Welding Cttee Rep.*, No. **R3**. 1948: London (H.M. Stationery Office).
94. C. F. Tipper, "The Brittle-Fracture Story". **1962**: Cambridge (University Press).
95. J. A. Kies, A. M. Sullivan, and G. R. Irwin, *J. Appl. Physics*, 1950, **21**, 716.
96. E. Orowan, *Z. Physik*, 1933, **82**, 235.
97. J. J. Gilman, "Fracture", p. 193. **1959**: New York (John Wiley); London (Chapman and Hall).
98. "Fracture Testing of High-Strength Sheet Materials", (3rd Rep. of Special A.S.T.M. Cttee.), *Mat. Research Stand.*, 1961, **1**, 877.
99. J. H. Mulherin, D. F. Armiento, and H. Marcus, "Fracture Characteristics of High-Strength Aluminium Alloys Using Specimens with Variable Notch-Root Radii", paper presented to Amer. Soc. Mech. Eng. meeting, Philadelphia, 1963.
100. J. M. Krafft and A. M. Sullivan, *Trans. Amer. Soc. Metals*, 1963, **56**, 160.
101. G. R. Irwin, "Crack-Toughness Testing of Strain-Rate-Sensitive Materials", paper presented to Amer. Soc. Mech. Eng. meeting, Philadelphia, **1963**.
102. H. Wallner, *Z. Physik*, 1939, **114**, 368.
103. A. Smekal, *Glastechn. Ber.*, 1950, **23**, 57.
104. F. Kerkhof, *Naturwiss.*, 1953, **40**, 478.
105. F. Kerkhof, "Symposium on the Mechanical Resistance of Glass" (Florence 1961), p. 779. **1962**: Brussels (Union Scientifique Continentale du Verre).
106. F. Kerkhof, *Glastechn. Ber.*, 1960, **33**, 456.
107. A. B. J. Clark, [*U.S.*] *Naval Research Lab. Progress Rep.*, p. 19, **1964**.
108. H. Schardin, "Fracture", p. 297. **1959**: New York (John Wiley); London Chapman and Hall).
109. H. C. van Elst, W. L. Korbee, and C. A. Verbraak, *Trans. Met. Soc. A.I.M.E.*, 1962, **224**, 1298.
110. H. L. Smith and W. J. Ferguson, "Crack-Velocity Measurements in Cellulose Acetate", [*U.S.*] *Naval Research Lab. Progress Rep.*, **1950**.
111. D. Post, *Proc. Soc. Exper. Stress Analysis*, 1954, **12**, 99.
112. F. F. Videon, F. W. Barton, and W. J. Hall, "Brittle-Fracture Propagation Studies", SSC-148, U.S. Dept. Commerce OTS (PB 181535), **1963**.
113. W. J. Nordell and W. J. Hall, *Weld. J.*, to be published.
114. F. W. Barton and W. J. Hall, "Studies of Brittle-Fracture Propagation in Six-Foot-Wide Steel Plates with a Residual Strain Field", U.S. Navy Bureau of Ships, SSC-130, NAS-NRC, Washington, D.C., **1961**.
115. J. Eftis and J. M. Krafft, *Trans. Amer. Soc. Mech. Eng.*, 1965, [B], **187**, 257.
116. J. E. Srawley, T. C. Lupton, and W. S. Kenton, "Crack Toughness of Two High-Strength Sheet Steels", *U.S. Naval Research Lab. Rep.* (**5895**), 1963.
117. G. R. Irwin, "Fracture by Progressive Crack Extension", paper presented to Symposium on "Structural Dynamics under High-Impulse Loading" (September 1962); ASD, Wright-Patterson Air Force Base, **1963**.
118. T. S. Robertson, *Engineering*, 1951, **172**, (Oct. 5), p. 445; *J. Iron Steel Inst.*, 1953, **175**, 361.
119. A. Cowan and H. G. Vaughan, *Nuclear Eng.*, 1962, **7**, 57, 61.
120. J. F. Baker and C. F. Tipper, *Proc. Inst. Mech. Eng.*, 1956, **170**, 65.
121. M. Yoshiki and T. Kanawaza, *J. Soc. Naval Architects, Japan*, 1957, (**102**), 39.
122. M. Yoshiki and T. Kanawaza, *Univ. Tokyo Rep.* (**SR-6004**), 1960.
123. F. J. Feely, M. S. Northup, S. R. Kleppe, and M. Gensamer, *Weld. J.*, 1955, **34**, 596s.
124. Y. Akita and K. J. Ikeda, *Transportation Tech. Research Inst. Rep.* No. (**56**), 1963.

125. T. S. Robertson and P. R. Christopher, [U.S.] *Naval Construction Research Estab. Rep.* (**R345**), 1956.
126. G. M. Boyd, *Engineering*, 1953, **175**, 65, 100.
127. T. W. Greene, *Weld. Research*, 1949, **14**, (5), 193s.
128. A. A. Wells, Houdremont Lecture to the Internat. Inst. Welding, *Brit. Weld. J.*, 1965, **12, 2**.
129. M. Kanamori, T. Kanazawa, M. Nakajima, and H. Itagaki, *Univ. Tokyo Rep.* (**6103**), 1961.
130. A. A. Wells, *U.S. Naval Research Lab. Rep.* (**4705**), 1956.
131. Rep. of Research Cttee. for Iron and Steel, Japan Welding Engineering Soc., **1963**.

© THE INSTITUTE OF METALS 1965

PLASTIC ZONE NEAR A CRACK AND FRACTURE TOUGHNESS

by

G. R. Irwin
U. S. Naval Research Laboratory
Washington 25, D. C.

Introduction

In order to discuss fracture in terms of stresses we need a descriptive terminology related to the stresses near the leading edge of the crack. One might term the simplest method for doing this, a one-parameter method. A natural and appropriate characterization of the one-parameter type is provided by the crack-extension force $\mathcal{G}$ or equivalently by the stress intensity factors K and $\mathcal{K}$. These terms are related as follows (1):

$$K^2 = \pi \mathcal{K}^2 = \begin{cases} \mathcal{G} E & \text{(plane stress)} \\ \dfrac{\mathcal{G} E}{1-\nu^2} & \text{(for plane strain)} \end{cases} \quad \text{- - - (1)}$$

E is Young's modulus and ν is Poisson's ratio.

As an example of the descriptive use of these stress field parameters, one can apply tension to a test specimen containing a starting crack and observe the value of the stress parameter for onset of rapid crack extensions. The special values of the parameters obtained in this way are designated K_c, $\mathcal{K}_c$, and $\mathcal{G}_c$. When carefully done, this testing procedure results in a useful, reproducible and quantitative measurement of fracture toughness (2).

Normally in fracture tests of this type the stress field parameter value is not obtained from an experimental stress analysis. Instead the length of the slow crack is marked with a straining procedure and an appropriate theoretical analysis in terms of crack length $2a$, gross section stress σ, and specimen width W is used to calculate the stress field parameter. For example, in the case of the centrally-notched specimen, Figure 1, the recommended computation procedure is based upon the equations:

$$K^2 = \sigma^2 W \tan u \quad \text{- - - - - - - - - - - - - - - - - (2)}$$

$$\text{and} \quad u = \frac{\pi}{W}\left(a + \frac{K^2}{2\pi\sigma_{YS}^2}\right) \quad \text{- - - - - - - - - - - (3)}$$

where σ_{YS} is the uniaxial tensile yield stress.

The stress analysis basic to equations (2) and (3) is derived from a mathematical model which differs from the real specimen principally in just two ways. One difference is that the mathematical model corresponds to a specimen subject to relatively small normal stresses on the side boundaries. In the test situations of practical interest this difference from the real test specimen has a negligible influence on the value of K and is not of interest in this discussion. The second, and more significant difference stems from the fact that the mathematical model is based entirely on equations of linear elasticity. The allowance for zones of plastic strains at the crack ends, illustrated by the term added to a in equation (3), merely assumes the influence of stress relaxation in the plastic zones is equivalent to additional crack length in the simple way shown.

The existence of a difference between the real test situation and the idealized mathematical model employed for test analysis purposes is a universal characteristic of physical measurements. Sometimes the difference is recognized but ignored, in other instances an approximate correction procedure is employed. An example like the crack stress field problem in some respects can be given from another field if one allows going back some years in time, back to the period when the study of resonance frequencies of open and closed tubes was of considerable interest. It was noted the tubes resonated as if the effective length exceeded the actual length. Intuitively it seemed that a length equal to half the tube diameter added at each open end would be about the right correction. Use of an end effect correction of this amount was suggested by such authorities as Rayleigh and Lamb and served a useful purpose in this application for many years. I have been informed some theoretical calculations during the past 15 years have been made which show the proper end effect correction is very nearly but not exactly equal to half the tube diameter.

As was true of the acoustic tube end-effect so also in the case of the plastic zone correction there are simple considerations which suggest an approximate answer to the problem.

Notched Sheets in Tension

The elastic theory stress analysis for the end region of a long crack in an infinite plate predicts the tension normal to the crack, σ_y, acting across the line of expected crack extension is given by

$$\sigma_y = \frac{K}{\sqrt{2\pi r}} \quad \text{---------- (4)}$$

where r is the distance from the end of the crack. By solving for r with σ_y put equal to the uniaxial yield stress σ_{YS} one obtains

$$r_{YS} = \frac{1}{2\pi} \left(\frac{K}{\sigma_{YS}}\right)^2 \quad \text{---------- (5)}$$

Since stress relaxation in the plastic zone means more stress must be carried beyond this zone it is clear the actual zone of plastic strains must extend to a greater distance than r_{YS} from the end of the crack. On the other hand our problem does not actually require knowing the size and shape of the plastic zone or the stresses within this zone. What we need is an estimate of the additional crack length such that the elastic stress field for the lengthened crack would correspond closely to the real stress field in regions beyond the zone of plastic strains. This added crack length should be smaller than the actual plastic zone size and might, in fact, be nearly equal to r_{YS}. The situation is shown schematically in Figure 1.

Consider next the net-section stress expected from equations (2) and (3). One can derive from equation (2) and (3) the relation

$$\frac{1}{W}\left(\frac{K}{\sigma_{YS}}\right)^2 = \left(\frac{\sigma_N}{\sigma_{YS}}\right)^2 \left(1 - \frac{2a}{W}\right)^2 \tan\left[\frac{\pi a}{W} + \frac{1}{2W}\left(\frac{K}{\sigma_{YS}}\right)^2\right] \quad - (6)$$

Taking the square root of both sides of equation (6) one has

$$\left(\frac{K}{\sigma_{YS}}\right)\frac{1}{\sqrt{W}} = \left(\frac{\sigma_N}{\sigma_{YS}}\right)\left(1 - \frac{2a}{W}\right)\left\{\tan\left[\frac{\pi a}{W} + \frac{1}{2W}\left(\frac{K}{\sigma_{YS}}\right)^2\right]\right\}^{1/2} \quad - - - (7)$$

Using the term on the left of equation (7) as the abscissa variable and $\left(\frac{\sigma_N}{\sigma_{YS}}\right)$ as the ordinate, Figure 2 shows essentially the predicted variation of net-section stress with inverse square root of specimen width. Curves for two fixed values of a/W are shown. As the specimen width becomes smaller, conditions on the net section approach those of general yielding. Because of this one expects measured values of net-section stress would fall below proportionality to $1/\sqrt{W}$ and approach some value in the yield-tensile strength range as specimen width is decreased. If we use only the region of positive slope, equation (7) predicts a behavior of the type expected.

Now, if the plastic zone correction factor had been taken to be

$$r_p = \frac{p}{2\pi}\left(\frac{K}{\sigma_{YS}}\right)^2 \quad - - - - - - - - (8)$$

the effect of the proportionality factor p is to multiply the number scales of ordinate and abscissa by $1/\sqrt{p}$. By taking p to be unity the net-section stress predicted by equation (7) has a maximum value of σ_N nearly equal to 1.15 σ_{YS}, a plausible value in the yield-tensile strength range. This maximum occurs when the size of r_p is nearly half of the unbroken ligament of the specimen.

From equations (4) and (5) one can see that, as an estimate of the plastic zone correction factor, r_{YS} is unlikely to be wrong by as much as a factor of two. From Figure 2 and equation (7) it is clear the proportionality factor, p, must be selected to be in the range 0.75 to 1.3 in order to predict a leveling off of net-section stress in the range of 1.0 to 1.3 σ_{YS}.

The value of K is determined by stress level and by crack length. In the above discussion the plastic zone correction was discussed only in terms of an adjustment to the crack length. A more accurate though more complex correction procedure can be developed

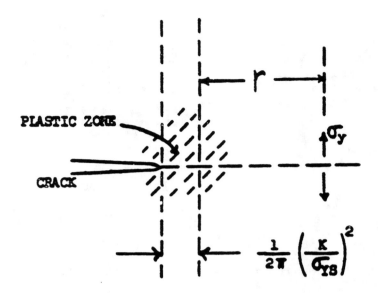

Fig. 1 - The distance r from the effective edge of the crack to the position of the normal tensile stress σ_y. The size of the plastic zone correction is shown schematically relative to the plastic zone size.

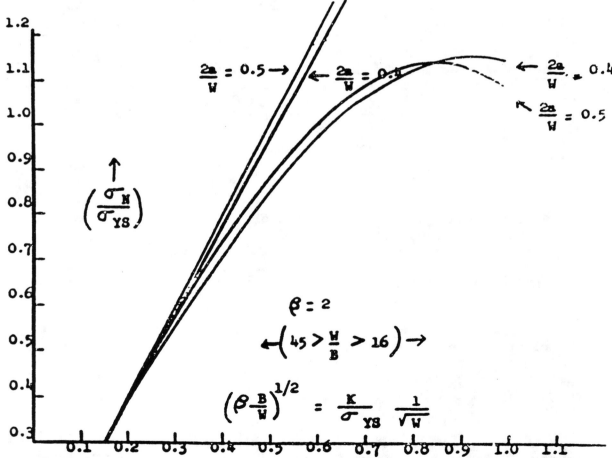

Fig. 2 - Influence of the added crack length r_{YS} upon the relation of net section stress to inverse square root of specimen width for geometrically similar centrally notched sheet tensile specimens as predicted by equation (7).

by use of a stress level correction term in addition to the crack length correction term. The latter is not, in this case, as large, and the relative magnitudes of the two correction terms can be adjusted to give an improved fit to the actual stress field in the specimen.

As an illustration consider again the centrally-notched sheet tensile specimen. The K value of equation (2) is derived from a stress analysis which predicts the stresses σ_y along the line of expected crack extension are given by

$$\sigma_y = \sigma \left\{ 1 - \left(\frac{\sin \frac{\pi x}{W}}{\sin u} \right)^{-2} \right\}^{-1/2} \quad \text{-------- (9)}$$

$$\text{with } u = \frac{\pi a}{W} + \frac{p_1}{2W} \left(\frac{K}{\sigma_{YS}} \right)^2 \quad \text{-------- (10)}$$

Assume that the crack length is half the specimen width, the value of β is 2π where

$$\beta = \frac{K^2}{B \sigma_{YS}^2}, \quad \text{----------- (11)}$$

and W/B is 16. In this event selection of a plastic strain correction equal to r_{YS} has the effect of increasing K by about 10 percent. A similar increase of K would occur by addition to σ of a stress correction term equal to 10 percent of σ. Alternatively one could add $r_{YS}/2$ to a and increase σ only 5 percent. All three corrective procedures would predict the same K value to within 1 percent. However, the agreement of equation (9) with an experimental stress analysis across the net section along the x-axis is sensitive to these alternatives.

A program of experimental stress measurements to explore this situation was initiated at the Naval Research Laboratory last year. The experiment employed a 12 inch wide sheet of 1/8-in thick 7075 T-6 Al alloy with a central slot and a photoelastic coating on each side. The central slot ended in 3/16-in diameter holes to deter crack formation. The experiment consisted in application of a series of increasing tensile loads and the determination, at each load, of the corrective factors to the applied cross section stress σ and to the crack length.

Replacing σ in equation (9) by $(\sigma + \sigma_1)$, for each of the tensile loads σ, it was possible to find a pair of values, σ_1 and p_1, such that σ_y from equation (9) (thus modified) fitted the measured values of σ_y along the x-axis throughout the region of elastic strains. Values of p were then found using the relation

$$\tan \left(\frac{\pi a}{W} + \frac{pK^2}{2W\sigma_{YS}^2} \right) = \left(1 + \frac{\sigma_1}{\sigma} \right)^2 \tan \left(\frac{\pi a}{W} + \frac{p_1 K^2}{2W\sigma_{YS}^2} \right) \text{--- (12)}$$

The experimental values of pr_{ys} are shown on figure 3 as a function of r_{ys}. At light loads the circular opening at each end of the slot had a strong influence on the results. However, the measured crack length correction appeared to approach r_{ys} as the load was increased and it was not possible within experimental uncertainties to conclude that a significant difference of p from unity was indicated by these measurements.

From the equation for K as influenced by σ, and p, r_{ys} the effect of these corrections decreases as specimen width increases. Thus a study of K as a function of specimen size should assist finding best values for the correction terms. Theoretically this procedure might seem the basic method for their determination. In practice there are many pitfalls. At specimen widths exceeding 50 times the sheet thickness (for steel) it is often necessary to use buckling restraints on the test sheet to insure a flat sheet test situation at onset of fast fracture. If an oblique shear mode of fracture occurs, too much restraint interferes with transverse displacements natural to the shear fracture and too little restraint allows an undesirable deviation from sheet flatness. For heat treated high strength sheet steels some residual stress may exist in the specimen and affect test results in a way dependent upon specimen size. The latter effect can be avoided by use of test specimens cut from a large sheet of 7075 T-6 Aluminum. Figure 4 shows results for such specimens. Restraints to prevent buckling were necessary for W/B greater than 80. In addition 24 frames per second movies instead of ink staining was the method used for estimating crack length at onset of rapid fracture in all but two of the results shown. The calculations assumed p equal to unity. From the appearance of the results a different choice of p would not have improved uniformity of K_c as a function of specimen width. In fact, as shown on Figure 4, the largest shift of a plotted point by removing the plastic strain correction is smaller than variation of K_c due to causes not related to plastic strain.

Plane Strain Stress Conditions

Measurements of crack toughness in terms of the K_c value designated K_{Ic} are conveniently made using a round tensile bar with a circumferential notch of sharp root radius (3). In several respects these tests are less complex than crack toughness measurements of sheet material. Because of notch sharpness there are no shear lips growing at the side boundaries of the initial crack to assist stable slow crack extension and the separation due to slow crack extension normally is so small it seems permissible to neglect this factor in calculating results. Thus ink staining of the slow crack is not required. In addition the plane strain stress conditions at the edge of the sharp notch result in an effective notch depth substantially less than r_{ys} because of the increased elastic constraint. Comparing plane stress and plane strain stress conditions from the crack edge elastic stress equations for equal K value, one finds the average of the square of the largest shearing stress is less for plane strain than for plane stress by a factor of 0.39. If

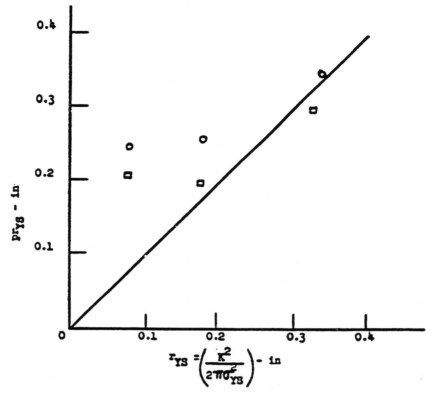

Fig. 3 - Experimental stress analysis measurements of pr_{YS} plotted against r_{YS} for a 1/8" thick centrally slotted sheet tensile specimen of 7075 T-6 aluminum.

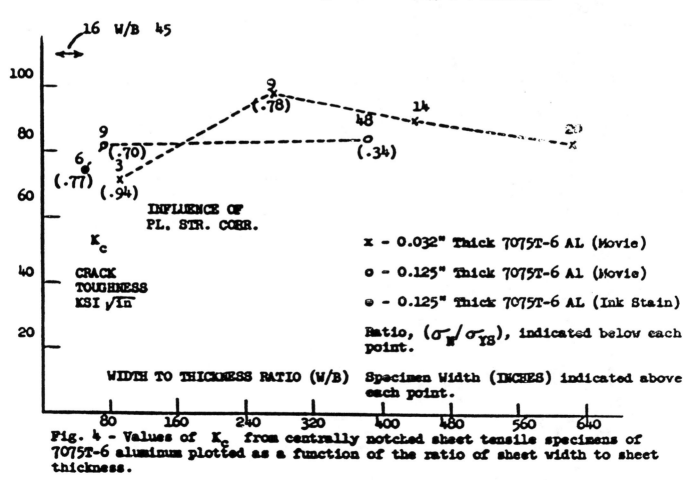

Fig. 4 - Values of K_c from centrally notched sheet tensile specimens of 7075T-6 aluminum plotted as a function of the ratio of sheet width to sheet thickness.

one chooses as the effective increase of notch depth the value $r_{YS}/2\sqrt{2}$ (= 0.35 r_{YS}), the use of this 10 percent smaller factor permits some simplification in the numbers appearing in the K value equation for a specimen notched so as the make the net section area half of the cross section area. The equation for K then becomes

$$K\left[1 - \frac{K^2}{2\pi D \, \sigma_{YS}^2}\right]^2 = 0.233 \sqrt{\pi D} \, \sigma_N \quad \text{------- (13)}$$

where D is the cross diameter of the specimen and σ_N is the net section stress at fracture.

Equation (13) is of the form

$$x\left[1 - (1/2) x^2\right] = 0.233 \frac{\sigma_N}{\sigma_{YS}} \quad \text{------ (14)}$$

where $x = \dfrac{K}{\sqrt{\pi D} \, \sigma_{YS}}$

Figure 5 shows the values of ratio σ_N/σ_{YS} as a function of x from equation (14). At values of x approaching 0.63 equation (14) predicts a pronounced departure of σ_N from proportionality to inverse square root of specimen size and a trend toward a maximum value of 1.74 σ_{YS}. Data points from tests of 7075 T-6 Aluminum (squares), and tests of a rotor steel (circles) are also shown.

The plastic zone correction factor of $r_{YS}/2\sqrt{2}$ is an intermediate choice relative to the data shown and is consistent with the plane stress correction factor from approximate analysis of the influence of elastic constraint. In view of the fact that plastic yielding near a crack is also influenced by strain hardening, stress gradient, and other significant details it seems pointless to attempt more than a rough estimate of the plastic strain corrections in terms of their proportionality to $(K/\sigma_{YS})^2$. From the considerations discussed above the plastic strain correction factor should be in the range $.8r_{YS}$ to $1.3r_{YS}$ for conditions of plane stress and should be in the range $0.3r_{YS}$ to $0.5r_{YS}$ for plane strain. Given any selections in these ranges, materials and test conditions could probably be found for which the selected corrections would seem correct.

Fracture Mode Transition and a Minimum Toughness Criterion

Plastic yielding near the leading edge of a crack was discussed above in terms of its influence upon the stress analysis applied to test specimens. For a crack traversing of a plate one can observe that a change from flat-tensile to oblique shear fracture can be introduced by reduction of plate thickness (4). This fracture mode transition correlates with the ratio of r_{YS} to plate thickness B in such a way that mid-range conditions of fracture transition occur when

$$2r_{YS} = B \quad \text{----------- (15)}$$

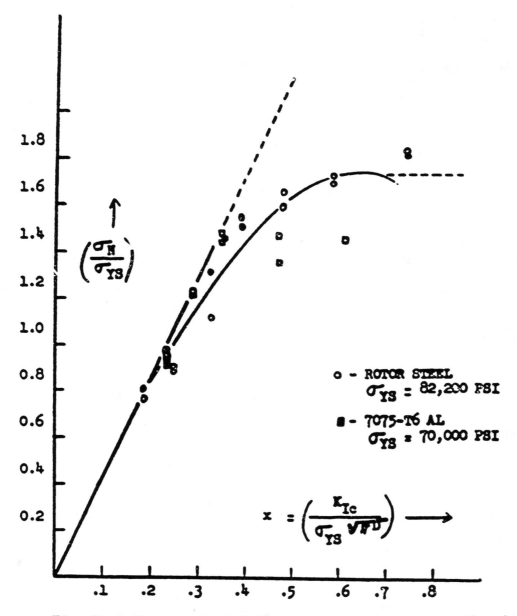

Fig. 5- Influence of the added effective notch depth ($x_{YS}/2\sqrt{2}$) upon the relation of net section stress to inverse square root of specimen diameter for circumferentially notched round tensile bars as predicted by equation (13). Shown on the figure are experimental points for a rotor steel and for 7075-T6 aluminum.

$$\beta_c = \frac{K_c^2}{B\,\sigma_{YS}^2} = \pi \quad \text{-------} \quad (16)$$

Figure 6 shows illustrative data for 7075 T-6 Al. The dashed line represents equation (16).

Intuitively one feels use of a sheet material tough enough so that full shear fracturing is observed should bring a substantial measure of protection from fracture failure due to small flaws. For high strength sheet steels the percent shear for a running crack is consistently above 50 percent for $\beta_c \geq 2\pi$. A β_c value of 2π has been suggested as a minimum toughness criterion for such structures (5). Figure (7) provides some support for this idea.

Commonly fracture failure of a high strength steel pressure vessel occurs by the spread through the wall thickness of a small crack and the subsequent lateral propagation of the crack after local severing through the wall has occurred. The computations shown on this figure are intended to represent the stress (in fraction of yield stress) required for crack propagation at various stages of this process. The movement of the leading edge of the crack assumed in the calculations is shown on the figure by dashed lines.

To the left the curves refer to extension of a partial crack from one surface through the plate. To the right the curves refer to extension of a through crack. Three levels of toughness are represented, as indicated by the β_c numbers on the curves to the right. The same three levels of toughness, but expressed in terms of β_{Ic} values, appear on the graph at the left. β_{Ic} is defined by the equation

$$\beta_{Ic} = \frac{K_{Ic}^2}{B\,\sigma_{YS}^2} \quad \text{-------------} \quad (17)$$

To obtain values of β_{Ic} corresponding to β_c use was made of the empirical relation

$$\beta_c = \beta_{Ic}(1 + 1.4\,\beta_{Ic}^2) \quad \text{--------} \quad (18)$$

In terms of percent shear the assumptions were intended to correspond approximately to 80, 50, and 5 percent shear in the appearance of the running crack. For the lower two toughness levels the stress required for propagation is generally less than 80 percent of the yield stress and decreases steadily with no indication of an arrest tendency as the crack completes through the thickness. For $\beta_c = 2\pi$ a stress equal to the yield stress is required both for the half depth crack and for the through crack of half length equal to the plate thickness. Hence there is a suggestion that fracture arrest may occur. With moderate additional toughness stresses greater than the yield stress would be required throughout the partial crack extension process.

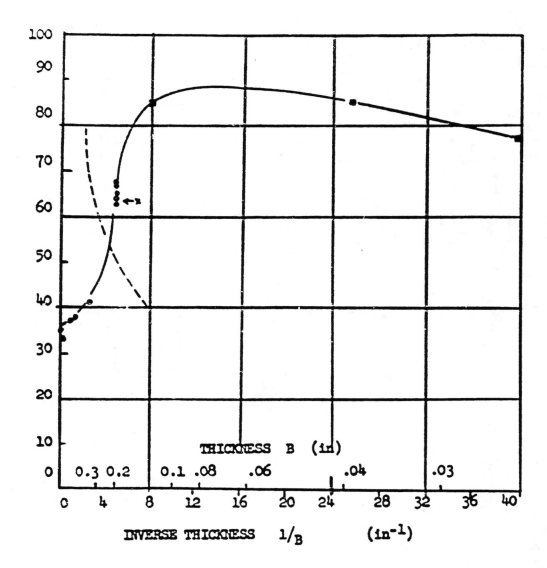

Fig. 6 - K_c as a function of reciprocal sheet thickness (1/B) for 7075-T6 aluminum alloy. Dashed line is the β_c = 2.4 locus

Fracture mode transition is a general thing occurring in low carbon steel and soft metals as well as in high strength metals. The meaning, however, of the fracture appearance is somewhat different because the running crack fracture appearance is determined by a relative plastic zone size based upon dynamic rather than the static yield properties. For high strength metals there is relatively little dynamic elevation of the yield stress. A mild steel structure would be expected to bear loads less than those required for general yielding with reference to the yield stress for slow loading, a stress level probably less than 40,000 psi. However, for rapid loading yielding occurs at a stress level higher by a factor of 2 to 3 depending upon loading speed. Thus in terms of percent of dynamic yield stress a stress level sufficient to arrest a crack in the first through-crack position shown in Figure 6 is, for low carbon structural steel, unlikely to exceed the lowest of the three curves and corresponds to less than 10 percent shear lip in the running crack fracture appearance. Assuming there is no dynamic overstressing and that engineers usually estimate design loads in a conservative manner for mild steel structures, a fracture toughness criterion of the nil-ductility type means, in most such structures, about the same degree of crack propagation safety as one has in a high strength rocket chamber for which the wall toughness corresponds to $\beta_c = 2\pi$ and the fracture appearance is nearly 100 percent shear.

Summary

The effect of plastic strains on the crack edge stress field, on fracture mode transition, and on the toughness necessary for arrest of a short through-crack can be estimated approximately in terms of a length r_{YS} given by

$$r_{YS} = \frac{1}{2\pi}\left(\frac{K}{\sigma_{YS}}\right)^2$$

In each case the estimate is only of the rough approximation type. However, it should be noted in support of these estimates that substantial improvements will require solution of complex multi-parameter problems. Within the framework of a simple one-parameter representation of the stress elevation near the leading edge of a crack, exact answers to these problems do not exist. In addition, each of the problems discussed is affected by other uncertainty factors of comparable magnitude which are not representable in terms of an elastic-plastic stress analysis problem.

References

(1) G. R. Irwin, J. A. Kies, H. L. Smith, "Fracture Strengths Relative to Onset and Arrest of Crack Propagation," Proc. ASTM, Vol. 58, p. 640, 1958

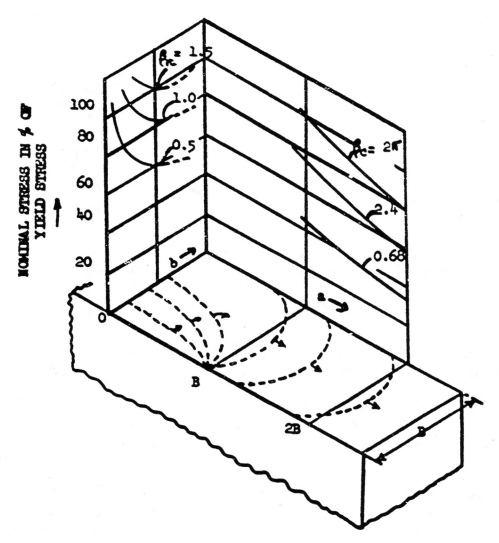

Fig. 7 - Right background graph shows relative stress for propagation of through-cracks. Three levels of toughness are represented corresponding to $\mathcal{G}_c$ values of 2π, 2.4, and 0.68. Left background graph shows relative stress for propagation of partial cracks. Computations were based upon plane strain $\mathcal{G}_{Ic}$ values estimated as equivalent to the through-crack $\mathcal{G}_c$ values.

(2) ASTM Committee Report, "Fracture Testing of High Strength Sheet Materials," ASTM Bulletin, p. 29-40, Jan. 1960

(3) G. R. Irwin, "Dimensional and Geometric Aspects of Fracture," ASM Conf. on Fracture of Eng. Mat'ls, Troy, N. Y., August 24-25, 1959

(4) G. R. Irwin, "Fracture Mode Transition for a Crack Traversing a Plate," Trans. ASME, Vol. 82, Series D, p. 417 (1960)

(5) G. R. Irwin, J. A. Kies, "Fracture Theory as Applied to High Strength Steels for Pressure Vessels," Golden Gate Metals Conf., San Francisco, Calif., Feb. 4-6, 1960

YIELDING OF STEEL SHEETS CONTAINING SLITS

By D. S. Dugdale

Engineering Department, University College of Swansea*

(*Received* 27th *November*, 1959)

Summary

Yielding at the end of a slit in a sheet is investigated, and a relation is obtained between extent of plastic yielding and external load applied. To verify this relation, panels containing internal and edge slits were loaded in tension and lengths of plastic zones were measured.

1. Introduction

Notches, such as holes and slits, generally reduce the strength of sheet material. If no inelastic deformation is permissible, stresses must nowhere exceed the initial yield stress, though this may be subject to a size effect. If large-scale yielding is to be avoided, the applied load divided by minimum area of section should not reach the plastic flow stress. It is the purpose of the present work to trace the spread of plasticity from a centre of stress concentration as loads are increased.

Relaxation methods can often be applied to a particular problem for determining the field of plastic strain produced by given loads (Allen and Southwell 1950). However, in the present work, a prototype problem is sought which will allow a direct calculation of extent of yielding as a function of external load. When a very thin sheet containing a straight cut is loaded in a direction perpendicular to the cut, it may be expected that yielding will be confined to a very narrow band lying along the line of the cut.

2. Analysis of the Problem

An ideal elastic–plastic material is considered which flows after yielding at a constant uniaxial tensile stress Y. Uniform tensile stress T is applied to the edges of an infinite sheet in a direction perpendicular to an internal cut of total length $2l$. Yielding must occur over some length s measured from the end of the cut, as shown in Fig. 1 (a). It is suggested that the sheet may be considered to deform elastically under the action of the external stress together with a tensile stress Y distributed over part of the surface of a hypothetical cut of length $2a$ as shown in Fig. 1 (b). Since internally applied forces are in static equilibrium, the distribution of stress in the sheet must be independent of elastic constants (Coker and Filon 1957, p. 518).

The problem of a straight cut loaded over part of its edge has been examined by Muskhelishvili (1953, p. 340). His stress functions were found to assume a

---

*Present address : Metallurgical Engineering Dept., Illinois Institute of Technology, Chicago 16, Illinois, U.S.A.

simple form when account was taken of the symmetry of the present configuration. It is convenient to introduce variables α and β defined by $x = a \cosh \alpha$, $l = a \cos \beta$ (see Fig. 1). The stress σ_y acting at points on $y = 0$ was determined in the form of a series in ascending powers of α having a leading term $\sigma_y = -2Y\beta/\pi\alpha$. The

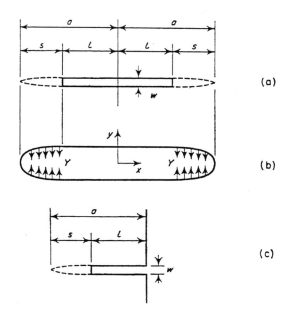

Fig. 1. Geometry of slit: (a) internal slit, (b) internal stresses acting on slit, (c) edge slit.

analogous expression for stress due to the external loading is $\sigma_y = T/\alpha$. When these stresses are superimposed, it is necessary that the stress at the point $\alpha = 0$ (i.e. $x = a$) should not be infinite, so the coefficient of $1/\alpha$ must vanish, viz. $T - 2Y\beta/\pi = 0$. This readily leads to the relation

$$\frac{s}{a} = 2 \sin^2 \left(\frac{\pi}{4} \frac{T}{Y} \right). \tag{1}$$

When T/Y approaches unity, this relation gives $s/a \simeq 1 - (\pi/2)(1 - T/Y)$. The last expression may be alternatively derived from stress functions given by WESTERGAARD (1939). By Saint-Venant's principle, stresses at $x = a$ should not depend on the exact distribution of stress over the segment $-l < x < l$ if l/a is small. Hence the partial loading required may be simulated by an inward tensile stress Y acting over the whole segment $-a < x < a$ together with central concentrated forces of magnitude $2lY$ acting outwards.

When T/Y is very small, equation (1) gives $s/l = 1.23 (T/Y)^2$. This relation gives the scale of plastic deformation if yielding actually occurs, but it need not imply that yielding must occur for any stress, however small.

3. TEST PANELS

The above analysis refers to an infinite sheet in a state of plane stress having a geometry defined by the length of slit. In practice, a sheet must have a definite thickness and a limited width. It was assumed that the sheet would be effectively of infinite size if the total length of slit never exceeded one fifth of the sheet width. This point has been discussed by FROST and DUGDALE (1958). If the sheet is to yield when the tensile stress at any point reaches the uniaxial yield stress, it must be possible for shearing to occur without constraint on planes at 45° to the plane of the sheet. With two exceptions indicated later, slits were made of width w equal to the sheet thickness (0·050 in.).

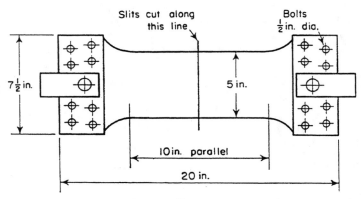

FIG. 2. Test panel.

Load was transmitted to the test panels through bolted joints as shown in Fig. 2. Measurements with extensometers on a panel without slits showed that stresses over the central section did not vary from the average value by more than 2 per cent. Steel produced by the Air Blown Bessemer process was selected, as this steel responds to strain-etching. The chemical analysis was: carbon 0·05 per cent, manganese 0·40 per cent, nitrogen 0·013 per cent. When cutting operations were completed, the panels were annealed at 900 °C, precautions being taken to avoid scaling.

TABLE 1. *Experimental values*

Internal slits			Edge slits		
Half-length of slit l (in.)	Applied stress T (tons/in²)	Plastic zone length s (in.)	Length of slit l (in.)	Applied stress T (tons/in²)	Plastic zone length s (in.)
0·50	3·3	0·042	0·50	3·7	0·057
0·40	5·0	0·088	0·50	4·6	0·093
0·25	6·2	0·087	0·40	5·7	0·116
0·25	6·9	0·122	0·40	7·0	0·197
0·25	8·0	0·185	0·25	8·3	0·204
0·25	10·0	0·470	0·25	9·5	0·350
0·10*	11·2	0·394	0·15*	10·8	0·448

*Width of slit $w = 0·10$ in.

4. EXPERIMENTAL RESULTS

Tensile load was applied to each panel for 2 min. The stress T was calculated from the total width of the test section (5 in.) taking no account of the presence of slits (FROST and DUGDALE 1958). After being loaded, panels were aged at 250 °C for a few hours and were dissected for examination. Portions including the slit and plastically deformed zone were polished on both sides and etched with Fry's reagent (JEVONS 1925). The plastic zone tapered to a fine point, and its length was often uncertain to the extent of 0.003 in. It was measured with a steel rule and magnifying glass. Average values of the four measurements for each panel are shown in Table 1.

A strip was cut from each panel for a tensile test. When results for all panels were averaged, lower yield stresses of 12.0, 12.5 and 13.0 tons/in² were obtained for strain rates of 10^{-6}, 10^{-5} and 10^{-4} per second. For plotting Fig. 3, a value of 12.5 tons/in² was selected. Strain hardening was found to begin at a strain of 1.5 per cent.

5. Discussion of Results

In Fig. 3 experimental values are shown in relation to the curve calculated from equation (1). It should be noted that an edge slit was considered to have a length equivalent to half the total length of an internal slit, as shown in Fig. 1. It is thought that the experimental points fall sufficiently near the calculated curve to

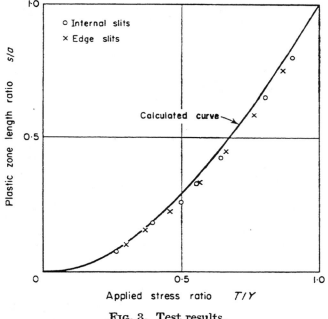

FIG. 3. Test results.

show that the analysis for internal slits is correct and that it applies also to edge slits. It can be seen that all experimental points fall slightly to the right of the curve. However, they may be brought into agreement with the curve by an adjustment of only about 3 per cent in the yield stress Y. As the yield stress corresponds to an arbitrarily chosen strain rate, it may be unwise to attach any significance to this discrepancy.

Acknowledgment

The writer is indebted to the Metallurgical Staff of Messrs. RICHARD THOMAS and BALDWINS, Ebbw Vale, for assistance in selecting steel of a suitable grade.

REFERENCES

ALLEN, D. N. DE G. and SOUTHWELL, R.	1950	*Phil. Trans.* A **242**, 379.
COKER, E. G. and FILON, L. N. G.	1957	*Photo-Elasticity* (Cambridge University Press)
FROST, N. E. and DUGDALE, D. S.	1958	*J. Mech. Phys. Solids* **6**, 92.
JEVONS, J. D.	1925	*J. Iron Steel Inst.* **111**, 191.
MUSKHELISHVILI, N. I.	1953	*Theory of Elasticity* (Noordhoff).
WESTERGAARD, H. M.	1939	*J. Appl. Mech.* **6**, A-49.

The Mathematical Theory of Equilibrium Cracks in Brittle Fracture

By G. I. BARENBLATT

Institute of Geology and Development of Combustible Minerals of the U.S.S.R. Academy of Sciences, Moscow, U.S.S.R.[*]

    I. Introduction
   II. The Development of the Equilibrium Crack Theory
  III. The Structure of the Edge of an Equilibrium Crack in a Brittle Body
      1. Stresses and Strains Near the Edge of an Arbitrary Surface of Discontinuity of Normal Displacement
      2. Stresses and Strains Near the Edge of an Equilibrium Crack
      3. Determination of the Boundaries of Equilibrium Cracks
  IV. Basic Hypotheses and General Statement of the Problem of Equilibrium Cracks
      1. Forces of Cohesion; Inner and Edge Regions; Basic Hypotheses
      2. Modulus of Cohesion
      3. The Boundary Condition at the Contour of an Equilibrium Crack
      4. Basic Problems in the Theory of Equilibrium Cracks
      5. Derivation of the Boundary Condition at the Contour of an Equilibrium Crack by Energy Considerations
      6. Experimental Confirmation of the Theory of Brittle Fracture; Quasi-Brittle Fracture
      7. Cracks in thin Plates
   V. Special Problems in the Theory of Equilibrium Cracks
      1. Isolated Straight Cracks
      2. Plane Axisymmetrical Cracks
      3. The Extension of Isolated Cracks Under Proportional Loading; Stability of Isolated Cracks
      4. Cracks Extending to the Surface of the Body
      5. Cracks Near Boundaries of a Body; Systems of Cracks
      6. Cracks in Rocks
  VI. Wedging; Dynamic Problems in the Theory of Cracks
      1. Wedging of an Infinite Body
      2. Wedging of a Strip
      3. Dynamic Problems in the Theory of Cracks

References

---

[*] Present address: Institute of Mechanics, Moscow State University, Moscow, USSR.

I. Introduction

In recent years the interest in the problem of brittle fracture and, in particular, in the theory of cracks has grown appreciably in connection with various technical applications. Numerous investigations have been carried out, enlarging in essential points the classical concepts of cracks and the methods of analysis. The qualitative features of the problems of cracks, associated with their peculiar non-linearity as revealed in these investigations, makes the theory of cracks stand out distinctly from the whole range of problems in the present theory of elasticity. The purpose of the present paper is to present a unified view of how the basic problems in the theory of equilibrium cracks are formulated, and to discuss the results obtained.

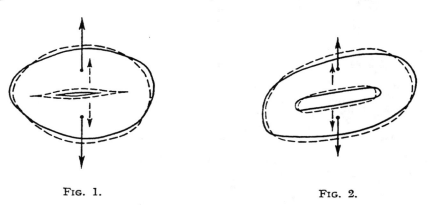

Fig. 1.   Fig. 2.

The object of the theory of equilibrium cracks is the study of the equilibrium of solids in the presence of cracks. Consider a solid having cracks (Fig. 1) which are in equilibrium under the action of a system of loads. The body, able to sustain any finite stresses, is assumed to be perfectly brittle, i.e. to retain the property of linear elasticity up to fracture. The possibility of applying the model of a perfectly brittle body to real materials will be discussed later.

The opening of a crack (the distance between the opposite faces) is always much smaller than its longitudinal dimensions; therefore cracks can be considered as surfaces of discontinuity of the material, i.e. of the displacement vector. Henceforth, unless the contrary is stated, plane cracks of *normal* discontinuity are considered, i.e. cracks are pieces of a plane bounded by closed curves (crack *contours*), at which only the normal component of the displacement vector has a discontinuity. The case when the tangential component of the displacement vector is discontinuous at the discontinuity surface can be treated in the same manner.

One might think that the investigation of the equilibrium of elastic bodies with cracks can be carried out by the usual methods of the theory of elasticity in the same way as it is done for bodies with cavities (Fig. 2).

However, there exists a fundamental distinction between these two problems. The form of a cavity undergoes only slight changes even under a considerable variation in the load acting upon the body, whereas cracks, whose surface also constitutes a part of the body boundary, can expand a good deal even with small increase of the load, to which the body is subjected. (In Figs. 1 and 2, dotted lines indicate additional loads and the corresponding positions of the body boundaries.)

Thus, one of the basic assumptions of the classical linear theory of elasticity is not satisfied in problems of the theory of cracks, namely the assumption about the *smallness* of changes in the boundaries of a body under loading, which permits one to satisfy the boundary conditions at the surface of the unstrained body. This fact makes the problem of the equilibrium of a body with cracks, unlike traditional problems of the theory of elasticity, essentially non-linear. In the theory of cracks one must determine from the condition of equilibrium not only the distribution of stresses and strains but also the boundary of the region, in which the solution of the equilibrium equations is constructed.

Non-linear problems of this type ("problems with unknown boundaries") have long been known in various fields of continuum physics. Suffice it to mention the theory of jets and the theory of finite-amplitude waves in hydrodynamics, the theory of flow past bodies in the presence of shock waves in gas dynamics, Stefan's problem of freezing in the theory of heat conduction, etc. The main difficulty in all these problems lies in the determination of the boundary of the region in which the solution is sought. Likewise, the basic problem in the theory of equilibrium cracks is the determination of the surfaces of cracks when a given load is applied.

The differential equations of equilibrium and the usual boundary conditions of the theory of elasticity cannot in principle give the solution of this problem without the introduction of some additional considerations. This may be seen from the fact that we can construct a formal solution of the equations satisfying the usual boundary conditions *no matter how we prescribe crack surfaces*. The analysis of these solutions shows that in general the tensile stress σ normal to the surface of a crack is infinite at the crack contour. More exactly, near an arbitrary point of the crack contour

$$(1.1) \qquad \sigma = \frac{N}{\sqrt{s}} + \quad \text{finite quantity,}$$

where s is the distance of a point of the body lying in the plane of a crack from the crack contour, N is the *stress intensity factor*, a quantity dependent on the applied loads, the form of the crack contour, and the coordinates of the point considered, but independent of s. The form of a normal section of the deformed crack surface near the contour appears in such cases unnaturally rounded (as in Fig. 3 or somewhat different; see details below).

Generally speaking, however, there exist such exceptional contours of cracks for which stresses at the edges are finite ($N = 0$) under a given load; at the same time the opposite faces of cracks close smoothly at the edges. The form of a section of the crack surface near the edge appears then as a cusp, cf. Fig. 4. It can be shown that for such contours, and only for them, the energy released by a small change in the contour of a crack is equal to zero. It follows that only such contours can bound equilibrium cracks.

FIG. 3.   FIG. 4.

Thus, when all loads acting upon a body are given, the problem of the theory of equilibrium cracks may be formulated as follows: for a given position of initial cracks and a given system of forces acting upon the body one requires the determination of the stresses, the strains, and the contours of cracks so as to satisfy the differential equations of equilibrium and the boundary conditions, and to insure finiteness of stresses (or, which is the same, a smooth closing of the opposing faces at the crack edges). If the position of the initial cracks is not given, then, since according to our model the body can sustain any finite stress, the solution of the problem formulated above is not unique. This is only natural because at one and the same load in one and the same body there need not be any cracks, or there may be one crack, or two, and so on.

In the general case of curved cracks, the shape is determined not only by the load existing at a given moment but also by the whole history of loading. If however, the symmetry of the body and the applied monotonically increasing loads assure the development of plane cracks, then the contours of cracks are determined by the current load alone. All the results at present available in the theory of cracks correspond to particular cases of this simplified formulation of the problem.

A given system of forces acting upon the body should in general include not only the loads applied to the body. The following example illustrates what is meant. Let us attempt to determine the contour of an equilibrium crack in the case of the loads represented in Fig. 1. If, in accordance with the usual approach in the theory of elasticity (as in the case of the cavity shown in Fig. 2), the surface of the crack is considered to be free of stresses, the result will be paradoxical: whatever contour of the crack we would take, the tensile stresses at its edge are always infinite. Consequently, there cannot exist an equilibrium crack; however small the force of extension may be, the body that has a crack breaks in two!

Such an obvious lack of agreement with reality can be easily explained: simply using the model of an elastic body, we have not taken into consideration all forces acting upon the body. It appears that — and this is also one of the main distinctions between the problems of the theory of cracks and the traditional problems of the theory of elasticity — for developing an adequate theory of cracks it is necessary to consider molecular forces of cohesion acting near the edge of a crack, where the distance between the opposite faces of the crack is small and the mutual attraction strong.

Although consideration of forces of cohesion settles the matter in principle, it complicates a great deal the analysis. The difficulty is that neither the distribution of forces of cohesion over the crack surface nor the dependence of the intensity of these forces on the distance between the opposite faces are known. Moreover, the distribution of forces of cohesion in general depends on the applied loads. However, if cracks are not too small, there is a way out of the difficulty: with increasing distance between the opposite faces the intensity of forces of cohesion reaches very quickly a large maximum, which approaches Young's modulus and then diminishes rapidly.

Therefore two *simplifying assumptions* can be made. The *first* is that the area of the part of the crack surface acted upon by the forces of cohesion can be considered as negligibly small compared to the entire area of the crack surface. According to the *second* assumption the form of the crack surface (and, consequently, the local distribution of forces of cohesion) near the adges, at which the forces of cohesion have the maximum intensity, does not depend on the applied load.*

The intensity of the forces of cohesion has the highest possible value for a given material under given conditions. This happens for instance at all edge points of a crack formed at the initial rupture of the material as the load increases. For most real materials cracks are *irreversible* under ordinary conditions. If an irreversible crack is produced by an artificial cut without subsequent expansion or is obtained from a crack that existed under a greater load by diminishing the load, then the intensity of forces of cohesion at the crack contour will be lower than the maximum possible one. The forces of cohesion that act at the surface of a crack compensate the applied extensional loads and secure finiteness of stresses and smooth closing of the crack faces. With an increase in extensional loads the forces of cohesion grow, thus adjusting themselves to the increasing tensile stresses, and the crack does not

---

* Sh. A. Sergaziev very neatly compared cracks for which these assumptions are satisfied with "zippers".

expand further until the highest possible intensity of forces of cohesion is reached. The crack starts expanding[+] only upon reaching the highest possible intensity of forces of cohesion at the edge.

Successive expansion of the crack edge under increasing extensional load is represented schematically in Fig. 5.

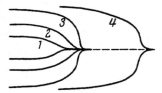

Fig. 5.

1,2. The intensity of forces of cohesion is less than the maximum.
3,4. The intensity of forces of cohesion is equal to the maximum.

If use is made of the first of the above assumptions, molecular forces of cohesion will enter in the conditions that determine the position of crack edges only in the form of the integral

$$(1.2) \qquad K = \int_0^d \frac{G(t)dt}{\sqrt{t}},$$

where $G(t)$ is the intensity of the forces of cohesion acting near the crack edge, t is the distance along the crack surface taken along the normal to the crack edge, and d is the width of the region subject to the forces of cohesion. For those contour points, at which the second assumption applies, this integral represents a constant of the given material under given conditions (temperature, composition, and pressure of the surrounding atmosphere, etc.), which determines its resistance to the formation of cracks. It can be shown that the quantity K is related to the surface tension of the material T_0, the modulus of elasticity E, and Poisson's ratio ν by means of the simple equation

$$(1.3) \qquad K^2 = \frac{\pi E T_0}{1 - \nu^2}.$$

Furthermore, for all points of the crack edge at which the intensity of forces of cohesion is a maximum, the stress-intensity factor N, entering in

---

[+] Quite a similar situation arises when a body moves over a rough horizontal surface under the action of a horizontal force. The motion of the body begins only after the force exceeds the highest possible value of the friction for the given body and the given surface.

(1.1) and calculated without taking into account the forces of cohesion should be equal to K/π. For all points of the edge at which the intensity of forces of cohesion is below the maximum, the stress intensity factor calculated without considering forces of cohesion is smaller than K/π.

The foregoing considerations elucidate sufficiently the nature of the forces of cohesion involved in this problem, and it is now possible to formulate the fundamental problem of the theory of equilibrium cracks.* When the symmetry of the body, of the initial cracks, and of the monotonously increasing forces insures development of a system of plane cracks, this problem can be stated as follows.

Let a system of contours of initial cracks be given in a body. It is required to find the stress and displacement fields corresponding to a given load as well as a system of contours of plane cracks surrounding the contours of the initial cracks (and perhaps coinciding with them partly).

Mathematically the problem consists in constructing such a system of contours that the factor of intensity of the tensile stress, calculated without taking into account the forces of cohesion, should be equal to K/π at all edge points, not lying on the contours of the initial cracks, and should not exceed K/π at all points of contours, lying on the contours of the initial cracks.

The foregoing formulation of the problem eliminates from our direct consideration the molecular forces of cohesion (they enter only through the constant K). Therefore stress and strain fields furnished by the solution of this problem will not be realistic in a small neighbourhood of the crack edges.

When cracks are reversible, or when the applied load is great enough to cause the contours of all the cracks to expand beyond the contours of the initial cracks, the form of the latter evidently is no longer of any importance.

The equilibrium state corresponding to the highest possible intensity of forces of cohesion at least at one point of the crack contour can be stable or unstable. Accordingly, further extension of the crack with increasing load proceeds in essentially different ways. In the case of stable equilibrium, a slow quasi-static transition of the crack from one equilibrium state to another takes place, when the load is increased gradually. If the equilibrium is unstable, the slightest excess over the equilibrium load is accompanied by a rapid crack extension that has a dynamic character. In some cases, when there exist no neighbouring stable states of equilibrium, this leads to the complete rupture of the body. The theory of cracks developed in such a way that problems of this latter type were mainly treated until recently.

---

* Such general formulations of problems seem advisable to us despite the fact that their general solution in effective form exceeds by far the possibilities of present mathematics. General statements of problems are a help in realizing the meaning of specific solutions and difficulties arising in developing the theory.

Sometimes the condition for the onset of crack extension is therefore identified with the condition for complete fracture of the body. It should be clearly understood, however, that this is only true in special cases, the practical significance of which must not be exaggerated.

Below, following a brief outline of the development of the mathematical theory of cracks, the fundamentals of the theory of equilibrium cracks are given as well as the results for the most typical special problems treated hitherto. At the end of this review dynamic problems in the theory of cracks are discussed briefly.

When writing this article the author endeavoured to avoid the repetition of available presentations of some aspects of brittle fracture. Thus the review deals with the theory of cracks proper, i.e. with the mathematical theory of brittle fracture. The numerous available experimental investigations are referred to only inasmuch as they are necessary for confirming the theory presented and establishing the limits of its applicability. Experimental investigations of brittle fracture, unlike the mathematical theory, were discussed more than once in reviews and monographs. At the same time, questions concerning exclusively mathematical techniques of solving the problems of elasticity theory are discussed only briefly, if at all. Also the question of the formation of the initial cracks will not be touched.

Trying to preserve a unified point of view in discussing certain results of other investigators, the author permitted himself sometimes a deviation from the original treatment.

II. The Development of the Equilibrium Crack Theory

Investigations in the field of the theory of cracks were started by C. E. Inglis [1] about fifty years ago. His paper presents the solution of a problem within the classical theory of elasticity concerning the equilibrium of an infinite body with an isolated elliptical cavity (in particular, with a straight-line cut) in a uniform stress field. N. I. Muskhelishvili [2] — also within the classical theory of elasticity — obtained in a simpler and more effective form the solution of a problem concerning the equilibrium of an infinite body having an elliptical cavity in an arbitrary stress field.

However, in spite of their outstanding significance for subsequent investigations, papers [1, 2] did not prepare the foundations for the theory of cracks proper. The fact is that the solutions obtained in these papers possess two properties which were difficult to explain. First, the length of a crack was found to be indefinite at a given load so that it was possible to construct a solution with an arbitrary value of this parameter. Everyday experience suggests nevertheless that the dimensions of cracks existing in a body should be connected somehow with the extensional loads applied to the body. As the load increases, cracks existing in the body do not expand at first when the load is small; upon reaching a certain load they begin to expand, the expansion depending on the manner in which the load is applied. In some cases cracks expand rapidly up to complete rupture of the body with the load maintained constant, in other cases they expand slowly, stopping as soon as the increase of the load is suspended. Since the opening of a crack is usually small compared to its longitudinal dimensions, it is natural to represent a crack as a cut; but then the tensile stresses at the crack edges in Inglis' problem are infinite, and *in general* the same thing happens in the problem treated by Muskhelishvili. Clearly solutions with infinite tensile stresses at the edges of a crack are unacceptable in a physically correct model of a brittle body. Thus, direct application of the classical scheme of the theory of elasticity to the problem of cracks leads to a problem which is incomplete and yields physically unacceptable solutions.

A. A. Griffith's papers [3, 4] are rightly considered fundamental for the theory of cracks of brittle fracture. The important idea, first advanced in these papers is that an adequate theory of cracks requires the improvement of the model accepted for a brittle body by the consideration of molecular forces of cohesion acting near the edge of a crack.

Griffith treated the following problem: An infinite brittle body stretched by a uniform stress p_0 at infinity has a straight crack of a certain size $2l$. It is required to determine the critical value of p_0 at which the crack begins to expand. The molecular forces of cohesion were considered as forces of surface tension being internal forces for the given body; their effect on the stress and strain field was neglected.

Under this condition the change ΔF of free energy ("total potential energy" in Griffith's terminology) of a brittle body with a crack, compared to the same body under the same loads but without a crack, is equal to the difference between the surface energy of the crack U and the decrease in strain energy of the body due to formation of the crack W. For the crack to expand, the change in free energy of the body must not grow with an increase in the size $2l$ of the crack. Thus, the parameters of the critical equilibrium state are obtained from the condition

(2.1) $$\frac{\partial (U - W)}{\partial l} = 0.$$

But the surface energy of the crack U is equal to the product of the surface area of the crack and the energy T_0 required to form the unit surface of the crack. Under certain sufficiently general assumptions, the quantity T_0, the surface tension, can be considered constant for a given material under given conditions. Therefore, according to Griffith, the determination of the critical load reduces to the determination of the quantity $\partial W/\partial l$, "the elastic energy release rate". Analysing the simplest case, Griffith calculated this quantity by using Inglis' results [1] and obtained relations determining the critical values of tensile stress in the forms

$$(2.2) \qquad p_0 = \sqrt{\frac{2ET_0}{\pi(1-\nu^2)l}}, \qquad p_0 = \sqrt{\frac{2ET_0}{\pi l}},$$

for plane strain and plain stress, respectively.

The theoretical part of Griffith's paper contains also the results of the investigation of the structure of a crack near its ends. This is carried out on the basis of the classical solution of elasticity theory, constructed without considering forces of cohesion, hence with infinite tensile stresses at the ends of the crack, if it has the shape of a cut. Griffith made an attempt to improve this description of a crack by considering it as an elliptical cavity with a finite radius of curvature ρ at the end (Fig. 3). However, according to his own estimate the magnitude of the radius of curvature at the end of the crack was of the order of the intermolecular distance, which clearly indicates the incorrectness of the approach: in any investigation based on the concept of a continuous medium distances of intermolecular order of magnitude cannot be considered as finite.

This part of Griffith's work is inadequate for the following reason. In determining the equilibrium size of a crack, the effect of molecular forces of cohesion on the stress and strain fields can be neglected, but this cannot be done in analysing the structure of a crack near its ends. The distance at which the effect of forces of cohesion is appreciable is comparable to the distance over which the form of a crack varies essentially. Therefore, to a considerable part, Griffith's analysis of the structure of crack edges cannot be accepted as correct, and in particular his conclusion concerning the rounded form of cracks near the ends is wrong, as will be shown in detail later. This aspect of the matter, obviously of prime importance, remained unclarified until recently and led in a number of cases to misinterpretations of Griffith's results [5].

In addition to the basic shortcoming pointed out here, there were some errors in calculations in the theoretical part of the paper [3]. Shortly after it had appeared, A. Smekal [6] published a detailed comment on it, containing also quite an interesting general discussion of the problem of brittle fracture and correcting the errors.

In a subsequent paper by K. Wolf [7] a more precise and simpler account of Griffith's results was given, and similar calculations were made for somewhat different (but also uniform) states of stress. In [7] the relation of Griffith's theory of fracture to previously proposed theories of strength was also discussed.

In connection with his experiments on the splitting of mica I. V. Obreimov investigated [8] the tearing-off of a thin shaving from a body by a splitting wedge that slides over its surface and has a single point of contact with the shaving. Using the approximate methods of thin-beam theory, Obreimov established the relation between the form parameters of a crack and the surface tension by means of an energy method similar to that used in Griffith's paper. The method of paper [8] was continued later by many investigators [9–12].

The determination of the elastic energy release rate $\partial W/\partial l$ for tensile stress fields more complex than a uniform one, as well as for other configurations of cracks encountered considerable mathematical difficulties. The investigations of H. M. Westergaard [13], I. N. Sneddon [14, 15], I. N. Sneddon and H. A. Elliot [16], M. L. Williams [17] clarified the distribution of stresses and strains near the discontinuity surfaces of the displacement. Together with the classical papers by Muskhelishvili [2, 18, 19] the investigations of Westergaard and Sneddon constitute the mathematical basis of subsequent works on the theory of cracks. However, the conditions of equilibrium for new particular cases and, still less, for a somewhat more general case of loading were not obtained in these papers.

In the papers by R. A. Sack [20], T. J. Willmore [21], and O. L. Bowie [22] the conditions of equilibrium were obtained for some new special cases of loading and position of cracks. The energy method was applied directly in these papers, and thus considerable difficulties in the calculations had to be overcome. In view of the fact that the equilibrium states in the problems treated in [20–22] are unstable and unique, the conditions of equilibrium are identical with those for complete fracture of the body.

The papers by G. R. Irwin [23] and E. O. Orowan [24], in which the concept of quasi-brittle fracture was developed, represent an important stage in the theory of cracks. Irwin and Orowan noticed that a number of materials, which behave as highly ductile in standard tensile tests, fracture by a *quasi-brittle* mechanism when cracks are forming. This means that the arising plastic deformations are concentrated in a very narrow layer near the surface of a crack. As was shown by Irwin and Orowan, it is possible in such cases to employ Griffith's theory of brittle fracture, introducing instead of surface tension the effective density of surface energy. This quantity, in addition to the specific work required to produce rupture of internal bonds (= surface tension), includes the specific work required to produce plastic deformations in the surface layer of a crack; it is sometimes several orders of magnitude larger than the surface tension.

The idea of quasi-brittle fracture extended considerably the range of applicability of the theory of brittle fracture and was undoubtedly one of the main reasons for reviving interest in this problem. Irwin, Orowan and other authors published a series of papers [23—32] devoted to the development of the generalized theory of brittle fracture, to the investigation of the limits of its applicability, and to the analysis of experimental data from the view point of this theory. Special notice deserves the paper by

H. F. Bueckner [33] in which a quite general energy analysis of brittle and quasi-brittle fracture was carried out on the basis of the Griffith-Irwin-Orowan scheme.

In all the foregoing papers the question of the structure of a crack near its edge remained without clarification. In a very interesting paper [34] devoted to the physico-chemical analysis of deformation processes, P. A. Rebinder first expressed the thought about the wedge-like form of a crack at its ends and about the necessity of a corresponding development of Griffith's theory. H. A. Elliot [35], N. F. Mott [36], and Ya. I. Frenkel [5], in analysing the form of a crack, proceeded from the idea of a crack of infinite length between two solid blocks of the material, which were at normal intermolecular distance from each other before formation of the crack.

In [35] the blocks were considered to be semiinfinite. Starting from the classical solution for a straight-line crack [1] and a disk-shaped crack [20] having a diameter $2c$ in a uniform tensile stress field p, the distributions of normal stresses σ_y and lateral displacements v were determined in [35] for points of the planes distant half the normal intermolecular distance from the crack plane. The function $\sigma_y(2v)$ containing p and c as parameters was identified with the relation between molecular forces of cohesion and the distance; by integrating this function, the surface tension was determined, which thus was found to be connected with p and c. The author identified this relation with the condition of fracture, which of course differed from Griffith's condition. The distribution of the lateral displacements so obtained was identified with the form of the crack.

Such an approach is inadequate for the following reasons. The formal application of the apparatus of classical elasticity for the determination of stresses and deformations near the edge of a crack is unjustifiable, since in applying this apparatus all distances (even those which are considered small) must be large compared to the intermolecular distance. Moreover, forces of cohesion act not only inside the body but also on a part of the crack surface. If this fact is taken into account, the edges of a crack have a pointed rather than a rounded shape, and there is no infinite stress concentration at the ends. This will be shown below in detail. Thus, stress and displacement distributions near the edge of the crack surface differ essentially from the corresponding distributions obtained according to the solutions of Inglis [1] and Sack [20], in which the surface of cracks was supposed to be free of stress. Note also that the decrease of $\sigma_y(2v)$ with increasing v is very slow in paper [35], much slower indeed than the natural velocity of diminution of the intensity of forces of cohesion.

Ya. I. Frenkel [5] treated the problem of a crack of infinite length cutting through a thin strip in longitudinal direction. The use of the approximate theory of thin beams, which is unsuitable for analysing the form of a crack near its ends, did not permit him to obtain an adequate result. Incidentally, the comments on Griffith's theory contained in this paper cannot be accepted as well justified either. Frenkel criticizes Griffith because of the instability of the equilibrium in the case of a straight crack in a uniform tensile stress field (considered by Griffith) and he ascribes this instability to Griffith's wrong idea about the form of the crack ends. This is not true. The conclusion about stability or instability of equilibrium of a crack does not depend on considerations concerning the structure of the crack ends. As will be shown later, instability of a crack in a uniform field occurs even when allowance is made for smooth closing of cracks at the ends; it is a part of the problem itself rather than a consequence of the peculiar crack shape assumed. Frenkel's conclusion concerning the existence of a stable state of equilibrium in addition to the unstable one is due to his incorrect replacement of the uniform state of stress by another one.*

In a paper by A. R. Rzhanitsyn [37] an attempt was made to solve the problem of a circular crack in a body subjected to a uniform tensile stress under consideration of the molecular forces of cohesion distributed over the crack surface and with smooth closing of the crack. Unfortunately the application of inadequate methods (averaging stresses and strains) did not allow the author to obtain the correct conditions of equilibrium.

An idea first suggested by S. A. Khristianovitch [38] was of great importance for the proper understanding of the structure of cracks near their ends. Khristianovitch considered, in connection with the theory of the so-called hydraulic fracture of an oil-bearing stratum, an isolated crack in an infinite body under a constant all-round compressive stress at infinity, maintained by a uniformly distributed pressure of a fluid contained inside the crack. The problem was treated in the quasi-static formulation. In solving it, Khristianovitch hit upon the indefiniteness of the crack length. He noticed, however, the following circumstance. Under the assumption that the fluid fills the crack completely, tensile stresses at the end of the crack are always infinite, whatever the size of the crack. But if the fluid fills the crack only partially, so that there is a free portion of the crack surface which is not wetted by the fluid, then at one exceptional value of the crack length tensile stresses at the ends of the crack are finite. It turned out that for this value of the crack length (and only for this one) the opposite faces of the crack close smoothly at its edges. Khristianovitch advanced a hypothesis of finiteness of stresses or, which is the same, of smooth closing of the opposite faces of a crack at its edges as a fundamental condition determining the size of a crack. The use of this hypothesis made it possible to solve a number of problems concerning formation and expansion of cracks in rocks [38—43].

---

* Besides these basic shortcomings there are some errors in calculations in [5] indicated in [37].

In all these papers, however, molecular forces of cohesion were not taken into account directly. Now in dealing with cracks in rock massifs it is quite permissible to neglect forces of cohesion. The estimates show that the effect of rock pressure is far greater here than the action of forces of molecular cohesion, particularly if the natural fissuring of rocks is taken into consideration. Under other conditions (in particular, in many cases when massifs are simulated in laboratories) forces of cohesion play an important part and their consideration is of great significance in analysing the conditions of equilibrium and expansion of cracks.

A very interesting early work by H. M. Westergaard [44] should be mentioned in connection with these investigations (see also [13]). On the basis of the analogy with the contact problem noted by the author, it is stated that there is no stress concentration at the end of a crack in such brittle materials as concrete. The same paper gives formulas which describe correctly stresses and strains near the ends of equilibrium cracks of brittle fracture in the absence of forces of cohesion. However, Westergaard did not connect the condition of finiteness of stress with the determination of the longitudinal dimension of a crack, which he assumed to be given.

In papers [45, 46] by G. R. Irwin (see also [47, 48, 49, 33]) an important formula was established that correlates the strain-energy release rate with the stress intensity factor near the ends of a crack in a problem of the classical theory of elasticity. On the basis of this formula the strain-energy release rate was determined, and the conditions of fracture were obtained for several new cases of loading and position of cracks [47, 50, 32, 51, 52].

Beginning with the work of Griffith, in most of the theoretical investigations problems of a similar type were treated: the equilibrium state, in which the intensity of forces of cohesion at the contour is a maximum, turns out to be unstable, and the condition for the onset of expansion of a crack coincides with the condition for the beginning of complete fracture of the body. Thus the condition for onset of the expansion is identified in some papers with the onset of rapid crack propagation and fracture for all cracks. In general, that is not true. Cracks actually may be stable so that the beginning of crack development is not necessarily connected with the fracture of a body; and one should not imagine that stable cracks are rare, that they are not encountered in practice and are difficult to produce experimentally. As the experimental investigations carried out by numerous authors beginning from I. V. Obreimov [8] show, the extension of cracks is stable in many cases throughout the greater part of the process of fracture. A. A. Wells [30] obtained stable cracks over a certain range of extensional forces in steel plates under combined external tensile stresses and internal stresses due to welded seams. F. C. Roesler [53] and J. J. Benbow [54] investigated stable conical cracks in glass and silica. The same authors [9] obtained stable cracks in wedging a strip of organic glass. Recently

J. P. Romualdi and P. H. Sanders [52] obtained stable cracks within certain limits of loads for a tensile plate stiffened by riveted ribs. References to other investigations in which stable cracks were obtained and analysed can be found in a monograph by B. A. Drozdovsky and Ya. B. Fridman [55]. All these papers confirm strongly the possibility of using the concept of brittle and quasi-brittle fracture for stable cracks.

Consideration of stable cracks greatly extends the problems that can be formulated in the theory of equilibrium cracks. Indeed, in the case of unstable cracks, only the determination of the load at which a crack begins to expand is of interest, since the process becomes dynamic upon reaching this load. In the case of stable cracks, however, one has to investigate the quasi-static expansion of cracks with change in loads.

In papers [56—61] the formulation of problems in the theory of equilibrium cracks of brittle fracture was improved and supplemented in accordance with the foregoing considerations. In these papers a new approach to problems of the theory of cracks was proposed, which is based on the general formulation of problems concerning elastic equilibrium of bodies in the presence of cracks, as it was given in [40]. The further discussions in this review are based on this approach, which is presented in the following chapter. A number of new problems of the theory of cracks were formulated and solved on that basis.

III. THE STRUCTURE OF THE EDGE OF AN EQUILIBRIUM CRACK IN A BRITTLE BODY

1. *Stresses and Strains Near the Edge of an Arbitrary Surface of Discontinuity of Normal Displacement*

As has already been pointed out, one can construct a formal solution of the differential equations of the theory of elasticity, which satisfies the boundary conditions corresponding to the applied load, if one prescribes arbitrarily a surface of discontinuity of the displacement. In the present section the behavior of the solutions of the equations of elasticity near the edge of a surface of discontinuity of displacement is investigated. For simplicity we shall restrict ourselves here to surfaces of discontinuity of normal displacement, appearing as plane faces bounded by closed curves (*contours*).

Near an arbitrary point O at the contour of such a surface, let us take a vicinity whose characteristic dimension is small compared to the radius of curvature of the contour at the point O. Deformation in this vicinity can be considered as plane and corresponding to a straight infinite cut in an infinite body subjected to a system of symmetrical loads (see Fig. 6; the plane of deformation is a plane normal to the contour of the discontinuity surface at the point O; the trace of the cut in the drawing is the intersection of that plane with the discontinuity surface). Loads can be applied at the surface of the cut and inside the body; the loads at the surface can be assumed to be normal without losing the generality of the further analysis. Consider now this configuration in more detail.

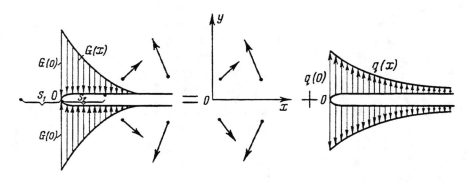

Fig. 6.

The stress and displacement fields can be presented as the sum of two fields (Fig. 6), the first of which corresponds to a continuous body under loads applied inside the body; the second belongs to a body with a cut, symmetrical loads being applied at the surface of the cut only. The shape of the deformed surface of the cut is determined by the second state of stress, since normal displacements at the place of the cut for the first state of stress are equal to zero by symmetry.*

The analysis of the first state of stress can be carried out by the usual methods of the theory of elasticity and is of no special interest; we shall consider this state of stress as given. Let us assume that the line of the cut corresponds to the positive semi-axis x; the normal stresses, $g(x)$, applied at the surface of the cut in the second state of stress, represent the difference between the stresses applied at the surface of the cut in the actual field, $G(x)$, and the stresses at the place of the cut, $p(x)$, corresponding to the first state of stress.

Applying Muskhelishvili's method [18] to the analysis of the second state of stress, we obtain the relations determining stresses and displacements

(3.1) $$\sigma_x^{(2)} + \sigma_y^{(2)} = 4 \operatorname{Re} \Phi(z),$$

(3.2) $$\sigma_y^{(2)} - i\sigma_{xy}^{(2)} = \Phi(z) + \Omega(\bar{z}) + (z - \bar{z})\overline{\Phi'(z)},$$

---

* This convenient method of reducing the load to a load distribution over the discontinuity surface was developed in the most general form by H. F. Bueckner [33].

(3.3) $$2\mu(u^{(2)} + iv^{(2)}) = \varkappa\varphi(z) - \omega(\bar{z}) - (z - \bar{z})\overline{\Phi(z)},$$

$$\varkappa = 3 - 4\nu,$$

where $z = x + iy$; $\sigma_x^{(2)}, \sigma_y^{(2)}, \sigma_{xy}^{(2)}$ are the components of the stress tensor of the second state of stress; $u^{(2)}, v^{(2)}$ are the displacement components along the x and y axes corresponding to the second state of stress; $\mu = E/2(1 + \nu)$ is the shear modulus, E is Young's modulus, and ν is Poisson's ratio. The analytical functions $\varphi, \omega, \Phi, \Omega$ are expressed by formulas

(3.4) $$\Phi(z) = \Omega(z) = \varphi'(z) = \omega'(z) = \frac{1}{2\pi i \sqrt{z}} \int_0^\infty \frac{\sqrt{t}\, g(t)\, dt}{t - z},$$

(3.5) $$\varphi(z) = \omega(z) = \frac{1}{2\pi i} \int_0^\infty g(t) \ln \frac{\sqrt{t} + \sqrt{z}}{\sqrt{t} - \sqrt{z}}\, dt.$$

At the cut ($x \geqslant 0$, $y = 0$) and its prolongation ($x \leqslant 0$, $y = 0$) the following relations hold:

(3.6) $$\sigma_x^{(2)} = \sigma_y^{(2)} = 2\,\text{Re}\,\Phi(z), \qquad \sigma_{xy}^{(2)} = 0, \qquad v^{(2)} = \frac{4(1-\nu^2)}{E}\,\text{Im}\,\varphi(z).$$

Using known formulas for limiting values of a Cauchy-type integral at the ends of the contour [19], we obtain an expression for the tensile stresses near the end of the cut along its prolongation,

(3.7) $$\sigma_y^{(2)} = -\frac{1}{\pi\sqrt{s_1}} \int_0^\infty \frac{g(t)\, dt}{\sqrt{t}} + g(0) + O(\sqrt{s_1}),$$

where s_1 is the small distance of the point considered from the end of the cut (Fig. 6). Similarly, we have for the distribution of normal displacements of points at the surface of the cut near its end

(3.8) $$v^{(2)} = \mp \frac{4(1-\nu^2)}{\pi E} \sqrt{s_2} \int_0^\infty \frac{g(t)\, dt}{\sqrt{t}} + O(s_2^{3/2}),$$

where s_2 is the distance of a surface point of the cut from its end, and negative and positive signs correspond to the upper and lower faces of the cut, respectively (Fig. 6).

This result also fully elucidates the distribution of normal tensile stresses and normal displacements near the contour of an arbitrary surface of normal discontinuity. Indeed, the following formulas are readily obtained from relations (3.7) and (3.8):

$$(3.9) \quad \sigma_y = \frac{N}{\sqrt{s_1}} + G(0) + O(\sqrt{s_1}), \quad v = \mp \frac{4(1-\nu^2)N\sqrt{s_2}}{E} + O(s_2^{3/2}),$$

where σ_y is the tensile stress at a point of the body a small distance s_1 away from the contour of the discontinuity surface, lying in the osculating plane to the contour of the discontinuity surface through the point O; N is *the*

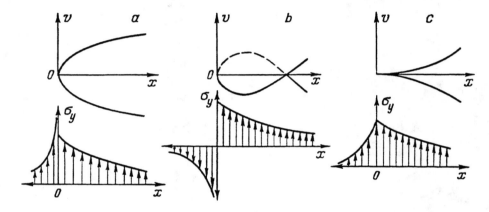

FIG. 7.

stress intensity factor, a quantity dependent on the acting loads, on the the configuration of the body and of the discontinuity surfaces in it, and on the coordinates of the point O considered; $G(0)$ is the magnitude of the normal stress applied to the discontinuity surface at O (Fig. 6); s_2 is the small distance of a point of the discontinuity surface from its contour. Depending on the sign of N, there are in general three possibilities.

If $N > 0$, an infinite tensile stress acts at the point O. The shape of the deformed discontinuity surface and the distribution of normal stresses σ_y near the point O are represented in Fig. 7a.

If $N < 0$, then an infinite compressive stress acts at the point O; the shape of the deformed discontinuity surface and the distribution of stresses near O are represented in Fig. 7b. The opposite faces of the crack overlap in this case, and it is quite evident that this case is physically unrealistic.

Finally, if $N = 0$, the stress acting near the contour is finite and tends to the normal stress applied at point O of the contour if O is approached. Thus the stress σ_y is continuous at the contour, and the opposite faces of the discontinuity surface close smoothly (Fig. 7c).

The investigation of the stress and strain distribution near the edge of the surface of normal discontinuity was begun by Westergaard [44, 13] and Sneddon [14, 15] and continued later by the author [40], by Williams [17], and by Irwin [45—47]. In view of the character of the stress states considered in [14, 15, 45—47] results were obtained only for the case $N > 0$.

2. Stresses and Strains Near the Edge of an Equilibrium Crack

The results obtained in the preceding section pertain to an arbitrary surface of discontinuity of normal displacement. We now show that, for an equilibrium crack, $N = 0$ at all points of its contour.

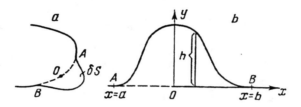

Fig. 8.

Consider a possible state of the elastic system, which differs from the actual state of equilibrium only by a slight variation in the form of the crack contour in a small vicinity of the arbitrary point O (Fig. 8). The new contour is a curve that encloses the point O lying in the plane of the crack. This curve is tangential to the former contour of the crack at points A and B close to O; everywhere else the contours of all the cracks remain unchanged. In view of the closeness of the points of tangency A and B to the point O, the initial contour of the crack at the portion AB can be considered as straight. The distribution of normal displacements of the points of the new crack surface and the distribution of tensile stresses at these points prior to the formation of the new crack surface are, according to the above, given, to within small quantities, by

$$(3.10) \qquad v = \mp \frac{4(1 - \nu^2) N}{E} \sqrt{h - y}, \qquad \sigma_y = \frac{N}{\sqrt{y}},$$

where N is the stress intensity factor at the point O.

The energy released in the formation of the new crack surface, which is equal to the work required to close this new surface, is given by

$$\delta A = \tfrac{1}{2} 2 \int\limits_{\delta S} \sigma_y |v| dS = \frac{4(1-v^2)N^2}{E} \int\limits_a^b dx \int\limits_0^h \sqrt{\frac{h-y}{y}} dy$$

(3.11)
$$= \frac{2(1-v^2)\pi N^2}{E} \int\limits_a^b h\,dx = \frac{2(1-v^2)\pi N^2 \delta S}{E},$$

where δS is the area of the projection of the new crack surface on its plane.

The condition of equilibrium of the crack requires that δA vanishes; this together with (3.11) implies that $N = 0$. Thus we arrive at a very important result characterizing the structure of cracks near their contours:

1. *The tensile stress at the contour of a crack is finite.*
2. *The opposite faces of a crack close smoothly at its contour.*

It appears, therefore, that contrary to Griffith's conception the form of a crack near its edge is as represented in Fig. 4. Since the only acting forces at the surface of a crack near its contour are forces of cohesion, it follows from (3.9) that the tensile stress at the crack contour is equal to the intensity of forces of cohesion at the contour. In particular, if there are no forces of cohesion, the tensile stress at the crack contour is equal to zero.

The condition of finiteness of stresses and smooth closing of the opposite faces at the edges of a crack was first suggested as a hypothesis by S. A. Khristianovitch [38], to serve as a basic condition that determines the position of the crack edge. The proof of this condition given above follows [60] mainly. Formula (3.11) for the case of plane stress was first proved by Irwin [45, 46] irrespective of finiteness of stresses and smoothness of closing (see also the review by Irwin [47] and the paper by Bueckner [33]). The early paper by Westergaard [44] contains a statement concerning the absence of stress concentration at the end of a crack in brittle materials like concrete, but the condition of finiteness of stress that appears in this work was not connected with the determination of the size of the crack.

We have confined ourselves here to the examination of cracks of normal discontinuity only for simplicity of treatment. Analogous reasoning, in particular the proof of finiteness of stress at the crack edge, can be extended without any substantial changes to cover the general case in which also the tangential displacement components have a discontinuity at the crack surface.

3. Determination of the Boundaries of Equilibrium Cracks

The conditions of finiteness of stresses and smooth closing of a crack at its contour permit us to formulate the problem of equilibrium cracks for a given system of loads acting upon the body: for a given position of initial cracks and a given system of forces acting upon the body, it is required to find stresses, deformations, and crack contours in the elastic body so as to satisfy the differential equations of equilibrium and the boundary conditions, and to insure finiteness of stresses and smooth closing of the opposite faces at the crack contours.

We shall illustrate the solution of this problem by an elementary example of an isolated straight crack in an infinite body under all-round compressive stress q at infinity and with concentrated forces P applied at opposite points of the crack surface (Fig. 9).

The solution of the equilibrium equations satisfying the boundary conditions can be obtained by Muskhelishvili's method [18] for an *arbitrary crack length $2l$*. Stresses and displacements are expressed by formulas (3.1)—(3.3) with

$$\Phi(z) = = \frac{2\zeta^2}{l(\zeta^2-1)}\left\{\frac{P}{\pi(\zeta^2+1)} - \frac{ql(\zeta^2+1)}{4\zeta^2}\right\},$$

(3.12)

$$z = \frac{l}{q}\left(\zeta + \frac{1}{\zeta}\right).$$

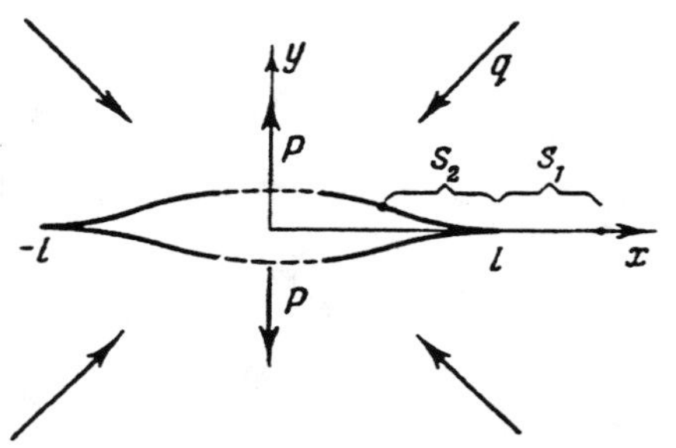

FIG. 9.

Evidently, equilibrium equations and boundary conditions do not determine the length of the crack. The distributions of stresses σ_y at the prolongation of the crack and normal displacement v of points of the crack surface near its edge are given by

$$\sigma_y = \left(\frac{P}{\pi l} - q\right)\sqrt{\frac{l}{8s_1}} + O(1), \qquad v = \mp \frac{(1-\nu^2)}{E}\left(\frac{P}{\pi l} - q\right)\sqrt{8s_2 l} + O(s_2^{3/2}).$$

(3.13)

Finiteness of stress and smooth closing of the crack at its ends are assured simultaneously by the condition

(3.14) $$l = \frac{P}{\pi q},$$

which determines the crack size under given loads P and q.

Let us now attempt to determine the size $2l$ of an isolated straight crack in an infinite body stretched by uniform stress p_0 at infinity in the direction perpendicular to the crack. If the crack surface is assumed to be free of stress, then one can easily show that the tensile stress at the prolongation of the crack near its edge depends on the distance s_1 as follows:

$$\sigma_y = \frac{p_0 \sqrt{l}}{\sqrt{2s_1}} ; \tag{3.15}$$

hence it appears that for no l the stress σ_y will be finite at the crack end and there does not exist an equilibrium crack! This paradoxical result is due to the fact that we did not take into account the molecular forces of cohesion acting near the crack edges and thus did not completely account for the loads acting upon the body. The consideration of these forces and the definitive formulation of problems in the theory of equilibrium cracks of brittle fracture are discussed in the following section.

IV. Basic Hypotheses and General Statement of the Problem of Equilibrium Cracks

1. *Forces of Cohesion; Inner and Edge Regions; Basic Hypotheses*

In order to construct an adequate theory of cracks of brittle fracture, it is necessary to supplement the model of a brittle body by considering the molecular forces of cohesion acting near the edge of a crack at its surface. It is known that the intensity of forces of cohesion depends strongly on the distance. Thus, for a perfect crystal the intensity f of forces of cohesion acting between two atomic planes at the distance y from each other is zero if y is equal to the normal intermolecular distance b. With y increasing up to about one and a half of b, the intensity f grows and reaches a very high maximum $f_m \sim \sqrt{ET_0/b} \sim E/10$; after that it diminishes rapidly with further increase of y (Fig. 10). Here E is Young's modulus, and T_0 is the surface tension related to $f(y)$ by the formula

$$2T_0 = \int_b^\infty f(y) dy. \tag{4.1}$$

The maximum intensity f_m defines the theoretical strength, i.e. the strength of a solid if it were a perfect crystal. The actual strength of solids is usually several orders of magnitude lower because of defects of crystal structure. For amorphous bodies the relation between the intensity of forces of cohesion and the distance has qualitatively the same character.

Data at present available, which confirm the above character of the relation between the intensity of forces of cohesion and the distance, lead to the following conclusion. It has long been known that the strength of thin fibers exceeds considerably that of large specimens of the same material [62, 63]. Experiments carried out recently with filamentary crystals of some metals revealed an exceptionally high strength approaching the theoretical value [63]. It is supposed that this phenomenon is due to the relatively small amount of structural defects in thin fibers and filamentary crystals. Furthermore, numerous direct measurements of the intensity of molecular forces of cohesion for glass and silica [64—66] were made recently. The

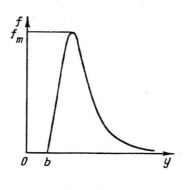
FIG. 10.

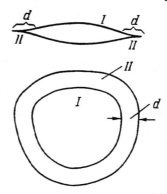
FIG. 11. I-inner region, II-edge region.

very elegant method used in this kind of measurements is based on the application of a regenerative microbalance and was suggested and employed by B. V. Deryagin and I. I. Abrikosova [64, 65]. However, these direct measurements deal with very great distances y compared to the normal intermolecular distance and thus determine only the end of the falling branch of the curve $f(y)$. A macroscopic theory for forces of cohesion at such distances was developed by E. M. Lifshitz [64] and was found in good agreement with the results of these aforementioned measurements. The relation $f(y)$, if y equals several normal intermolecular distances, seems to be beyond any strict quantitative theory and difficult for the direct experimental determination at present. A description of available attempts to estimate the relation $f(y)$ at such distances and, consequently, the theoretical strength can be found in [67, 63, 68].

The distance between the opposite faces of a crack varies from magnitudes of the order of the intermolecular distance near the crack edge to sometimes rather great magnitudes far from the edge. It is therefore convenient to divide the crack surface into two parts (Fig. 11). The opposite faces in the first part — *the inner region of the crack* — are a great distance apart, hence their interaction is vanishingly small, and the crack surface can be considered free of stresses caused by the interaction of the opposite faces. The opposite faces of a crack in the second part are adjacent to the crack contour — *the edge region of the crack* — and come close to each other so that the intensity of the molecular forces of cohesion acting on this part of the surface is great. Of course, the boundary between the edge and inner region of the crack surface is conventional to a certain extent. For very small cracks there may be no inner region of the crack at all.

Since the distribution of the forces of cohesion over the surface of the edge region is not known beforehand, a substantial part of the loads applied to the body is not known. It is thus impossible to handle the problem of cracks directly in the way it was stated in Chapter III. But the following method of solving problems of cracks is possible in principle: the distance between the opposite faces of a crack is found at each surface point as a function of the unknown distribution of forces of cohesion over the surface. Assuming the relation $f(y)$ between forces of cohesion and distance as given, a relationship can be obtained which determines the distribution of forces of cohesion over the crack surface.

Such an approach is not practicable. First, the relation $f(y)$ is not known to a sufficient extent for a single real material. Even if it were known, the problem would constitute a very complex non-linear integral equation, the effective solution of which presents great difficulties even in the simplest cases.*

Attempts were made to prescribe the distribution of forces of cohesion over the crack surface in a definite manner, but these attempts cannot be considered sufficiently well founded.

For sufficiently large cracks, consideration of which is of principal interest, the difficulty connected with our lack of knowledge of the distribution of forces of cohesion can be avoided without making any definite assumptions concerning this distribution. In this case the general properties of the relation between forces of cohesion and distance allow the formulation of two basic hypotheses which not only simplify essentially the further analysis, but permit the determination of contours of cracks, although the forces of cohesion are finally altogether excluded from consideration as loads acting upon the body.

First hypothesis: The width d of the edge region of a crack is small compared to the size of the whole crack.

This hypothesis is acceptable because of the rapid diminution of forces of cohesion with the increase in the distance between the opposite faces of

---

* In papers of M. Ya. Leonov and V. V. Panasyuk [69, 70] the relation $f(y)$ is approximated by a broken line, and on the basis of this approximation a linear integral equation for the normal displacements of the crack surface points is derived. It is solved approximately, the representation of the solution being not quite successfully selected so that the form of the crack at its end appears wedge-shaped with a finite edge angle. In fact, as was shown above, the edge angle must be zero. The shortcoming of these papers lies also in the application of the results obtained by the methods of mechanics of continua to cracks whose longitudinal dimensions are only of the order of several intermolecular distances.

a crack. Of course, there exist micro-cracks to which this hypothesis cannot be applied. However, as the width d of the edge region is quite small, the hypothesis is already valid for very small cracks and certainly for all macro-cracks. Nevertheless, the width d is considered to be sufficiently great compared to micro-dimensions (for instance, compared to the lattice constant in a crystalline body), so that it is permissible to employ the methods of continuum mechanics over distances of the order of d.

Second hypothesis: The form of the normal section* of the crack surface in the edge region (and consequently the local distribution of the forces of cohesion over the crack surface) does not depend on the acting loads and is always the same for a given material under given conditions (temperature, composition and pressure of the surrounding atmosphere and so on).

When the crack expands, the edge region near a given point, according to the second hypothesis, moves as if it had a motion of translation, and the form of its normal section remains unchanged. This hypothesis is applicable only to those points of the crack contour where the maximum possible intensity of forces of cohesion is reached; an expansion of the crack occurs then at this point with an arbitrarily small increase in the loads applied to the body.

Equilibrium cracks, on whose contour is at least one such point, will be called *mobile-equilibrium* cracks to distinguish them from *immobile-equilibrium* cracks which do not possess this property, i.e. do not expand with an infinitesimal increase in the load. Thus the second hypothesis and all conclusions based on it are applicable to reversible cracks as well as to irreversible equilibrium cracks, which formed at the initial rupture of a brittle body in the process of increasing the load. It is not applicable to cracks which result from equilibrium cracks existing at some greater load by diminishing that load; nor can it be applied, to artificial cuts made without subsequent expansion.

The second hypothesis is suggested by the fact that the maximum intensity of the forces of cohesion is so very great and exceeds by several orders of magnitude the stresses which would arise under the same loads in a continuous body without a crack. Therefore it is possible to ignore the change of stress in the edge region when loads vary and, consequently, the corresponding variation of the normal sections.

These two hypotheses reformulate the results of the qualitative analysis of the brittle-fracture phenomenon carried out by a number of investigators beginning with Griffith. They are the only assumptions concerning the forces of cohesion which underlie the theory presented below and appear in this explicit form in [56, 57].

---

* = intersection with a plane normal to the crack contour.

2. Modulus of Cohesion

The body considered is assumed to be linearly elastic up to fracture. The elastic field in the presence of cracks can then be represented as the sum of two fields: a field evaluated without taking into account forces of cohesion and a field corresponding to the action of forces of cohesion alone. Therefore the quantity N entering in formulas (3.15) and, as was proved, equal to zero can be written as $N = N_0 + N_m$, where the stress intensity factor N_0 corresponds to the loads acting upon the body and to the same configuration of cracks *without* considering forces of cohesion, and the stress intensity factor N_m corresponds to the same configuration of cracks and forces of cohesion only.

According to the first hypothesis the width d of the edge region acted upon by forces of cohesion is small compared to the crack dimensions on the whole and, in particular, to the radius of curvature of the crack contour at the point considered. In determining the value of N_m we may thus assume that the field belongs to the configuration discussed in Section III,1, i.e. to an infinite body with a semi-infinite cut, with symmetrical normal stresses being applied to the surface of the cut. Hence it follows from (3.7) that

$$(4.2) \qquad N_m = -\frac{1}{\pi}\int_0^\infty \frac{G(t)dt}{\sqrt{t}} = -\frac{1}{\pi}\int_0^d \frac{G(t)dt}{\sqrt{t}},$$

where $G(t)$ is the distribution of forces of cohesion different from zero only in the edge region $0 \leqslant t \leqslant d$.

According to the second hypothesis, the distribution of forces of cohesion and the width d of the edge region at those points of the contour, where the intensity of forces of cohesion is a maximum, do not depend on the applied load; the integral in (4.2) represents then a constant characterizing the given material under given conditions. This constant will be denoted by K:

$$(4.3) \qquad K = \int_0^d \frac{G(t)dt}{\sqrt{t}}.$$

It was termed *the modulus of cohesion* since this quantity characterizes the resistance of the material to an extension of its cracks, caused by the action of forces of cohesion. As will be shown below, the quantity K is the only characteristic of the forces of cohesion, that enters in the formulation of the problem of cracks.

The dimension of the modulus of cohesion is:

$$(4.4) \qquad [K] = [F][L]^{-3/2} = [M][L]^{-1/2}[T]^{-2},$$

where $[F]$, $[L]$, $[M]$, and $[T]$ denote the dimensions of force, length, mass, and time, respectively. Constants of a similar dimension are encountered in the contact problem of the theory of elasticity [71, 72, 73]. It is no coincidence, that there exists a profound connection between the contact problem and problems in the theory of cracks of brittle fracture; it seems that this was first pointed out in the papers of Westergaard [44, 13].

3. *The Boundary Condition at the Contour of an Equilibrium Crack*

For points of the contour of an equilibrium crack, at which the maximum intensity of cohesion is reached, the second hypothesis is applicable, and (4.2) may be written as

$$(4.5) \qquad N_m = -\frac{1}{\pi} K;$$

considering that $N = 0$, we obtain

$$(4.6) \qquad N_0 = \frac{1}{\pi} K.$$

The boundary condition at contour points of an equilibrium crack, at which the intensity of forces of cohesion is maximal, can also be formulated as follows: on approaching these points, the normal tensile stress σ_y at the points of the body lying in the crack plane, if calculated without taking into account forces of cohesion, tends to infinity according to the law

$$(4.7) \qquad \sigma_y = \frac{K}{\pi \sqrt{s}} + O(1),$$

where s is the (small) distance from the contour point considered. Satisfying (4.6) at least at one point of the contour is the condition that the crack is in the state of mobile equilibrium.

One should not connect, in general, the reaching of the state of mobile equilibrium by the crack with the onset of its unstable rapid growth and still less with complete fracture of the body. A mobile-equilibrium crack may be either stable or unstable. Only in case of instability is the condition for the onset of rapid crack propagation given by (4.6). However, not even in this case is complete fracture of the body unavoidable, since the transition from the unstable state of equilibrium to the other, stable one, is possible. Numerous examples illustrating various possibilities will be discussed in the following chapter.

If a crack is irreversible and there are points on its contour where the intensity of forces of cohesion is less than maximal,* then the second hypothesis is not applicable at such points. Since cohesive forces that act in the edge region of the crack surface are smaller near such points than those acting near points of the type considered above, it follows from (4.2) that $-N_m < K/\pi$; and since $N_0 = -N_m$, we have for these points

$$N_0 < \frac{K}{\pi}. \tag{4.6a}$$

As the load increases, forces of cohesion in the edge region grow; they compensate the increase in the load and insure finiteness of stress and smooth closing at the crack contour. However, the crack does not expand at a given contour point until the forces of cohesion become maximal. The second hypothesis now becomes applicable, and condition (4.6) is satisfied.

In determining the form of contours of equilibrium cracks, conditions (4.6) and (4.6a) permit us to exclude the forces of cohesion altogether from the consideration of the loads acting upon the body. Instead, we work with their overall integral characteristic, the modulus of cohesion. Special estimates show [57, 58] that the influence of molecular forces of cohesion on the stress and displacement field is essential only in the neighbourhood of the edge in a region of the order of magnitude d. Forces of cohesion thus determine the structure of cracks near their ends, and the forms of crack contours depend on them only through the integral characteristic K.

4. Basic Problems in the Theory of Equilibrium Cracks

The basic problem in the theory of equilibrium cracks can be stated in its most general form as follows. A certain system of initial cracks and a process of loading the body, i.e. a system of loads acting upon the body, dependent on one monotonously increasing parameter λ, are given. The value of λ for the initial state may be assumed as zero. It is required to determine the form of the crack surfaces and to find the distribution of stresses and strains in the body corresponding to any $\lambda > 0$. The process of varying the load is supposed to be sufficiently slow so that dynamic effects need not be considered.

When the symmetry of body, loads, and initial cracks insures the possibility of developing a system of plane cracks and the extensional loads grow monotonously with increasing λ, the configuration of cracks in the body is determined by the current load only and not by the whole history of the

---

* For instance, contour points of non-expanded cuts or of cracks formed from cracks which existed under a greater load when the load is diminished.

process of loading, as it is in the general case. In this case the problem is formulated as follows (it will be called problem A). In a body bounded by a surface Σ contours of an initial system of plane cracks Γ_0 are given (Fig. 12; the plane of the drawing is the plane of the cracks). It is required to find the elastic field and the contours of a system of plane cracks Γ enclosing the contours Γ_0 (and perhaps coinciding with them partially) corresponding to a given load, i.e. to a given value of λ.

This problem reduces mathematically to the following one. It is required to construct the solution of the differential equations of equilibrium of elasticity theory in the region bounded by plane cuts with contours Γ and by the body boundary Σ under boundary conditions corresponding to the given load. The contours Γ must be determined so that condition (4.6) is satisfied at points of these contours not lying on Γ_0, and condition (4.6a) at points of Γ lying on Γ_0.

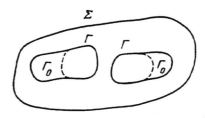

Fig. 12.

If the cracks are reversible or if the applied loads are sufficiently great so that the contours Γ do not coincide with Γ_0 at any point, then the form of the initial contours is of no importance. It is then possible, without prescribing the initial cracks, to formulate directly the problem of determining the contours Γ of equilibrium cracks of a given configuration so that condition (4.6) is satisfied at each point of Γ. Here we assume that the initial cracks are such that they are compatible with the realization of the given configuration of cracks when the load increases. This problem will be called problem B.

It may happen that a solution of either of the above stated problems does not exist. If this happens, it has quite a different significance for the problems A and B. If no solution of problem A exists this means that the applied load exceeds the breaking load, hence its application causes fracture of the body. The limiting value of the parameter λ up to which the solution of problem A exists, corresponds to the breaking load. The determination of the breaking load for a given configuration of the initial cracks and a given process of loading presents an important problem in the theory of cracks. Non-existence of the solution of problem B signifies that, whatever initial cracks may be within a given configuration, they will not expand under a given load, hence the applied load is too small. In such cases the conventional description of the state would be that mobile-equilibrium cracks do not form under the given load.

5. Derivation of the Boundary Condition at the Contour of an Equilibrium Crack by Energy Considerations

Molecular forces of cohesion so far have been considered as external forces applied to the surface of the body. This was necessary for analysing the structure of cracks near their ends.

If only boundary conditions are to be obtained, another approach can be employed which considers the forces of cohesion as internal forces of the system. On the basis of this approach, the idea of which goes back to Griffith [3, 4], a relation between the modulus of cohesion and other characteristics of the material will be obtained.

As before, let there be a certain configuration of equilibrium cracks in a brittle body and consider as in Section III,2 a possible state of the elastic system, which differs from the real one only by a variation in the crack contour near a certain point O (Fig. 8). However, unlike Section III,2, the characteristic size of the new area of the crack surface is assumed to be large compared to the dimension d of the edge region, though small compared to the size of the crack as a whole; according to the first hypothesis (Section IV,1) such an assumption is permissible. Under this assumption forces of cohesion can be considered merely as forces of surface tension, and a certain amount of work must be done to overcome these forces in increasing the crack surface. The influence of forces of cohesion on the stress and strain fields can be neglected since it is essential only in the neighbourhood of the crack edge, whose dimension is of the order of the width of the edge region.

The work δA required for the transition from the actual state to a virtual one is equal to the difference between the corresponding increment in surface energy δU and released elastic energy δW:

(4.8) $$\delta A = \delta U - \delta W.$$

For the actual state of an elastic system to be an equilibrium state, δA must vanish, hence

(4.9) $$\delta U = \delta W.$$

Quite similarly to Section III,2 an expression for δW is obtained:

(4.10) $$\delta W = \frac{2(1-\nu^2)\pi N_0^2 \delta S}{E},$$

where N_0 is the value of the stress intensity factor at the point O calculated without taking into consideration the forces of cohesion. Formula (4.10) in a somewhat different form was established by Irwin [45–47].

If the form of the edge region near a given point of the contour corresponds to the maximum intensity of forces of cohesion, then, according to the above, in forming a new crack surface the edge region is displaced without deformation; the work against the forces of cohesion per unit of newly formed surface is then constant and equal to the surface tension T_0. Therefore, $\delta U = 2T_0 \delta S$, because two surfaces form in rupture. Together with (4.9) and (4.10), we have

$$(4.11) \qquad N_0 = \sqrt{\frac{ET_0}{\pi(1-\nu^2)}}.$$

Comparing (4.11) and (4.6), we obtain a relationship correlating the modulus of cohesion K, defined independently by (4.3), with the surface tension T_0 and the elastic constants of the material E and ν:

$$(4.12) \qquad K^2 = \frac{\pi E T_0}{1-\nu^2}.$$

6. Experimental Confirmation of the Theory of Brittle Fracture; Quasi-Brittle Fracture

After Griffith's work [3, 4] many investigators attempted to carry out experimental verifications of the theory of brittle fracture. We cannot

Spherical bulbs				Cylindrical tubes			
$2l$ inches	D inches	p_0 psi	$p_0\sqrt{l}$	$2l$ inches	D inches	p_0 psi	$p_0\sqrt{l}$
0.15	1.49	864	237	0.25	0.59	678	240
0.27	1.53	623	228	0.32	0.71	590	232
0.54	1.60	482	251	0.38	0.74	526	229
0.89	2.00	366	244	0.28	0.61	655	245
				0.26	0.62	674	243
				0.30	0.61	616	238

analyse all this work here in detail and shall dwell only on several of the most characteristic papers, referring for details and discussion of other numerous investigations to the special publications [62, 55, 74-78].

Griffith's paper [3] gives descriptions and results of the following experiments. Cracks of various length $2l$ were placed on spherical glass bulbs and cylindrical tubes, whose diameter D was sufficiently great so that a special verification showed no influence of the diameter on the experimental results. After the tubes and bulbs had been annealed to relieve residual stresses produced by making the cracks, they were loaded from the inside by hydraulic pressure up to fracture. The breaking stress p_0 corresponding to each crack length $2l$ was measured.

According to the foregoing theory it appears that the breaking stress p_0 at which a given crack becomes unstable (onset of mobile equilibrium) can depend only on the crack length $2l$ and the modulus of cohesion K. From dimensional analysis [79] it follows that $p_0 = \alpha K/\sqrt{l}$, where α is a dimensionless constant. Consequently, $p_0\sqrt{l}$ must be constant for a given material (in full accord with (2.1)).

Griffith's experiments, which are tabulated here, confirm the constancy of this quantity and thus the foregoing theoretical scheme.

The remarkably elegant experiments of Roesler [53] and Benbow [54], in which stable conical cracks were produced, are of special interest for the confirmation of the theory of brittle fracture. The scheme of these experiments is presented in Fig. 13; the photograph of conical cracks in fused silica, borrowed

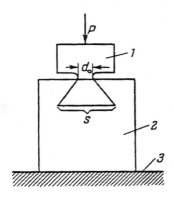

Fig. 13.
1. Steel indentor.
2. Specimen.
3. Steel support.

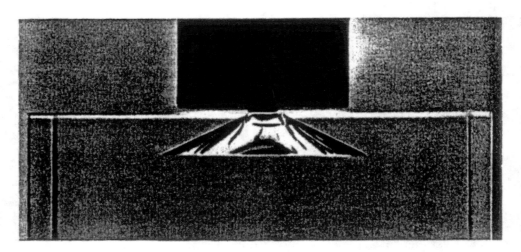

Fig. 14.

from Benbow's paper [54], is given in Fig. 14. The cracks were formed by the penetration into a specimen of glass [53] and fused silica [54] of a cylindrical steel indentor with a flat end. In accordance with the above, the diameter s of the base of a conical crack can depend only on the diameter d_0 of the indentor base, the force P pressing the indentor, the modulus of cohesion K, and Poisson's ratio ν. Since the correct formulation of the corresponding problem of elasticity theory does not include Young's modulus, it should not be included in the number of determining parameters of the crack problem. Dimensional analysis yields

$$(4.13) \qquad s = \left(\frac{P}{K}\right)^{2/3} \varphi\left[\frac{K^{2/3} d_0}{P^{2/3}}, \nu\right],$$

where φ is a dimensionless function of its arguments.

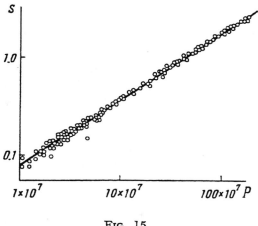

Fig. 15.

Experiments carried out with indentors of three diameters on eleven glass specimens [53] confirm well the existence of the universal relation (4.13). At large values of P, when the first argument of the function φ becomes vanishingly small, *self-similarity* takes place, and the following relationship holds:

$$(4.14) \qquad s = \left(\frac{P}{K}\right)^{2/3} \varphi_1(\nu).$$

Fig. 15 represents a graph, taken from Benbow's paper [54], of the $s(P)$ relation according to data from experiments with fused silica carried out under conditions corresponding to the self-similar regime. As can be observed, these experiments give a conclusive proof of the validity of relation (4.14) and confirm thereby the above scheme.

The experiments described were carried out with materials which can be considered as perfectly brittle. This refers especially to fused silica. Benbow [54] presents certain facts indicating that the mechanism of crack formation in fused silica is closer to being perfectly brittle than it is in glass: cracks in glass grow for a long time under constant load, whereas in fused silica their size is established quickly and then remains unchanged; after removal of the load, cracks in glass remain distinctly visible, but in silica they are imperceptible, etc. However, the significance of the theory of brittle fracture greatly exceeds what should be the limits of its applicability to those comparatively rare materials that are perfectly brittle. Experimental investigations show that when cracks appear some materials, which behave as highly plastic bodies in common tensile tests, fracture in such a way that plastic deformations, though present, are concentrated in a thin layer near the crack surface.

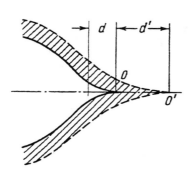

Fig. 16.

D. K. Felbeck and E. O. Orowan [28] carried out experiments on fracture of low-carbon steel plates with a saw-cut crack under conditions corresponding to Griffith's scheme of uniform extension. Experimental results are in good agreement with Griffith's formula, but the surface-energy density exceeds by about three orders of magnitude the surface tension of the material investigated. It was found in good agreement with the specific work of plastic deformations in the layer near the crack surface, which was determined by independent measurements.

On the basis of this and similar experimental results Irwin [23] and Orowan [24] advanced the concept of quasi-brittle fracture, which permitted an important extension of the limits of applicability of the theory of brittle fracture. Here the theory of brittle fracture covers the case when the plastic deformations are concentrated in a thin layer near the crack surface. The energy T required to form the unit surface of a crack is expressed as the sum of the specific work against the forces of molecular cohesion T_0 ($=$ surface tension) and the specific work of plastic deformation T_1:

$$(4.15) \qquad T = T_0 + T_1.$$

A formal extension to quasi-brittle fracture is made as follows (Fig. 16, the plastic deformation zone near the surface is shaded). Imagine the whole plastic region cut out and shift the crack end to the end of the plastic region. This can be done, if the forces exerted by the plastic zone upon the elastic zone are considered as external forces applied to the crack surface. After that the previous reasoning remains unchanged, if the plastic zone is assumed as thin and use is again made of the hypothesis concerning the invariability of the edge region (which here includes the boundary of the elastic and plastic zones). The modulus of cohesion is now expressed as

$$(4.16) \qquad K = \int_0^{a+a'} \frac{G_1(t)dt}{\sqrt{t}} = \frac{\pi E T}{1-\nu^2},$$

where $G_1(t)$ is the distribution of normal stresses acting on the boundary of the elastic and plastic zones.

When the contribution of molecular forces of cohesion to integral (4.16) can be ignored in comparison to the contribution of stresses that act in the region ahead of the actual crack end and have the order of magnitude of the yield point stress σ_0, we obtain an estimate for the modulus of cohesion:

$$(4.17) \qquad K = \frac{\pi E T_1}{1-\nu^2} \sim 2\sigma_0 \sqrt{d'}.$$

Note that the value of σ_0 at the yield point near the crack end may differ from that at the yield point obtained in tensile tests with large specimens.

The concept of quasi-brittle fracture is somewhat related to the concept of the "plastic particle" at the ends of notches with a zero radius of curvature, advanced in a classical monograph by H. Neuber [80].

In the following we shall speak of cracks of brittle fracture, bearing in mind the possibility of extending the results to the case of quasi-brittle fracture. Of course, in this latter case it is necessary to take into consideration the irreversibility of cracks of quasi-brittle fracture.

7. Cracks in Thin Plates

If the state of stress can be assumed to be plane, then all relations derived previously for the case of plane strain hold also for thin plates, if only E is replaced by $E(1-\nu^2)$ and the modulus of cohesion is assumed to have some other value K_1. Repeating the derivation of formula (4.12) for the plane stress state we obtain

$$(4.18) \qquad K_1^2 = \pi E T.$$

The experiments show that the surface energy density T in the case of quasi-brittle fracture increases with a reduction in the plate width [48], which is due to a broadened plastic-strain zone near the crack surface. An approximate theoretical analysis of this phenomenon was attempted by I. M. Frankland [81].

Bearing in mind the complete analogy of the analysis of plane stress and plane strain we shall in the following consider only plane strain.

V. Special Problems in the Theory of Equilibrium Cracks

This Chapter deals with solutions of special problems in the theory of cracks available at present. A few of the examples have illustrative character, but most problems presented are interesting in themselves.

1. Isolated Straight Cracks

In this and the following section isolated mobile-equilibrium cracks are examined, and all along the contour the maximum intensity of forces of cohesion is assumed to prevail. The problem reduces here to the determination of crack contours corresponding to a given load so that condition (4.6) is satisfied at these contours, and it represents a particular case of problem B formulated above. It is supposed that the initial cracks guarantee the possibility of producing such cracks; the necessary requirements for the initial cracks in the cases of reversible and irreversible cracks follow readily from the solutions obtained.

Let us consider an isolated straight mobile-equilibrium crack extending along the x-axis from $x = a$ to $x = b$ in an infinite body subject to plane strain. Let $p(x)$ be the distribution of normal stresses, which arise at the place of the crack in a continuous body under the same loads. This distribution is computed by the usual methods of elasticity, and we may consider it as given. It may be shown by using Muskhelishvili's solution [2, 18] that tensile stresses near the crack ends calculated without taking into account forces of cohesion become infinite according to the law $\sigma_y = N/\sqrt{s} + \ldots$, where

$$N_a = \frac{1}{\pi \sqrt{b-a}} \int_a^b p(x) \sqrt{\frac{b-x}{x-a}} dx, \qquad N_b = \frac{1}{\pi \sqrt{b-a}} \int_a^b p(x) \sqrt{\frac{x-a}{b-x}} dx$$

(5.1)

are the values of the stress intensity factors for points a and b, respectively. Satisfying condition (4.6) at these points, we obtain relations that determine the coordinates of the crack ends a and b:

$$(5.2) \quad \int_a^b p(x) \sqrt{\frac{b-x}{x-a}} dx = K\sqrt{b-a}, \qquad \int_a^b p(x) \sqrt{\frac{x-a}{b-x}} dx = K\sqrt{b-a}.$$

In particular, if the applied load is symmetrical with respect to the crack middle, where we place the origin of coordinates, then $-a = b = l$, and Eqs. (5.2) become one relation determining the half-length of the crack l:

(5.3)
$$\int_0^l \frac{p(x)dx}{\sqrt{l^2-x^2}} = \frac{K}{\sqrt{2l}}.$$

Note that (5.2) and (5.3) represent finite equations, since $p(x)$ is a given function. These equations determine the position of the ends of an isolated straight-line mobile-equilibrium crack under a given load, if this load guarantees that such a crack can exist.

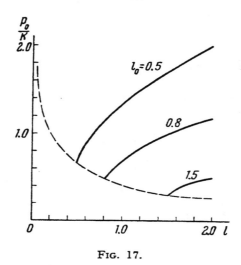

Fig. 17.

A method to calculate the strain-energy release rate $\partial W/\partial l$ for a symmetrical isolated crack was indicated by K. Masubuchi [82]. He proposed a trigonometrical representation of stresses $p(x)$ and displacements v of points of the crack surface,

(5.4) $\quad p(x) = \dfrac{E}{4l} \sum\limits_{n=1}^{\infty} nA_n \dfrac{\sin n\theta}{\sin \theta}, \quad v = \tfrac{1}{2} \sum\limits_{n=1}^{\infty} A_n \sin n\theta, \quad x = l \cos \theta.$

As was shown by Masubuchi,

(5.5)
$$\frac{\partial W}{\partial l} = \frac{\pi E}{8(1-\nu^2)l} \sum_{n=1}^{\infty} (nA_n)^2.$$

Equating this expression to $4T$, where T is the surface energy density, a relation between the applied stresses and the crack size can be obtained, though in a form far more complicated than (5.3).

Let us now look at a few examples. A crack may be kept open by a uniform tensile stress p_0 applied at infinity. As already pointed out, this problem was first treated by Griffith [3, 4]. In this case $p(x) \equiv p_0$ and equation (5.3) yields

$$(5.6) \qquad l = \frac{2K^2}{\pi^2 p_0^2}.$$

Relation (5.6) appears in Fig. 17 as the dotted line. One sees that the size of a mobile-equilibrium crack diminishes with increasing tensile stress, which is indicative of the instability of mobile equilibrium in this case. Despite this instability the size l defined by (5.6) has a physical meaning: If there is a crack of length $2l_0$ in a body, to which constant tensile stress p_0 is applied at infinity, then at $l_0 < l$ this crack does not expand (and closes if it is a reversible crack) while at $l_0 > l$ it grows indefinitely. Thus, the equilibrium size is in a certain sense critical (this will be discussed in more detail in Section V,3). It is obvious that instability of mobile equilibrium in this case fully corresponds to the substance of the matter and, contrary to the opinion expressed by Frenkel [5], is not connected with Griffith's incorrect ideas about the geometry of the crack ends.

If stresses vanish at infinity, and if a crack is maintained by a uniformly distributed pressure applied over a part of its surface ($0 \leqslant x \leqslant l_0$) while the remaining part of the crack surface ($l_0 \leqslant x \leqslant l$) is free of stress, then the half-length of the mobile-equilibrium crack l is given by the relation [58]

$$(5.7) \qquad \sqrt{\frac{l}{l_0}} \arcsin\left(\frac{l_0}{l}\right) = \frac{K}{p_0 \sqrt{2l_0}}.$$

This relation is shown in Fig. 17 by the solid lines which may be obtained from each other by a similarity transformation. It is evident that the opening of a crack, i.e. the appearance of a free segment, is possible provided l_0 is not less than the corresponding size of a mobile-equilibrium crack kept open by a uniform tensile stress at infinity, p_0, which is determined by (5.6). Therefore all the solid lines (Fig. 17) must start from the dotted line.

A limiting case of (5.7) is of interest. It occurs when p_0 tends to infinity and l_0 tends to zero, while $2p_0 l_0 \equiv \text{Const} = P$. This corresponds to a crack kept open by concentrated forces applied at opposite points of its surface. The half-length of the crack is then given by

$$(5.8) \qquad l = \frac{P^2}{2K^2}.$$

Note that (5.6) and (5.8) may be obtained, disregarding the value of the numerical factor, by dimensional analysis. For example, the size of a crack maintained by concentrated forces is determined only by the magnitude P of these forces and the overall characteristic of the forces of cohesion, K. It is obvious that the modulus of elasticity and Poisson's ratio do not enter in the number of determining parameters, since the corresponding problem of the theory of elasticity is naturally formulated only in terms of stresses. Considering the dimensions of P and K, we see that it is possible to set up only one combination with the dimension of length from these quantities, namely the ratio P^2/K^2, and no dimensionless combination exists. Thus the length of a mobile-equilibrium crack must be proportional to P^2/K^2, and the coefficient of proportionality a universal constant [cf. 79].

Let now a crack be maintained by two equal and opposite concentrated forces P, whose points of application are separated by L along the common line of action of the forces; the crack is supposed to be perpendicular to the line of action of the forces and located symmetrically [58].

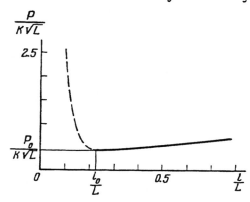

Fig. 18.

The distribution of tensile stresses at the place of the crack in a continuous body is in this case given by

$$(5.9) \qquad p(x) = \frac{PL}{2\pi(x^2 + L^2)} \left[1 - \nu + 2(1 + \nu) \frac{L^2}{x^2 + L^2} \right]$$

(the origin of coordinates is taken in the middle of the crack). Using (5.3), we obtain the relation determining the crack size in the form

$$(5.10) \qquad \frac{P}{K\sqrt{L}} = \left(1 + \frac{L^2}{l^2} \right)^{3/2} \frac{\sqrt{2}}{[2 + (3 + \nu)L^2/l^2]\sqrt{L/l}}.$$

A plot of $P/K\sqrt{L}$ versus the relative length of the crack l/L for $\nu = 0.25$ is shown in Fig. 18. As can be seen, at $P > P_0$ two lengths of a mobile-equilibrium crack correspond to each value of P, the smaller decreasing and the greater increasing with increasing P. States of mobile equilibrium corresponding to the smaller equilibrium length are unstable; the corresponding branch of the load-length diagram in Fig. 18 is shown by the dotted line. States corresponding to the greater length are stable (solid line in Fig. 18).

The smaller size l_1 is the critical size at a given load P; initial cracks present in the body and smaller than $2l_1$ do not expand under the action of applied loads of magnitude P (in case of reversible cracks they close), and those which are greater expand until the crack reaches the second (stable) equilibrium size.* At $P < P_0$ equation (5.10) has no solution. This means that, whatever length of the initial crack we take, it will not develop into a mobile-equilibrium crack at the given load. The size of a mobile-equilibrium crack l_0 different from zero corresponds to the critical value of forces P_0.

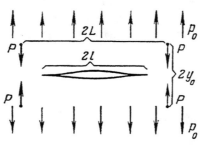

FIG. 19.

An interesting problem concerning the influence of riveted stiffeners on crack propagation was treated by J. P. Romualdi and P. H. Sanders [52]. This problem is schematized by the authors as follows (Fig. 19). An infinite plate is stretched by a uniform stress p_0 in the direction perpendicular to a crack. The action of the rivets and the stiffeners is represented by two symmetrically located pairs of opposite concentrated forces equal in magnitude to P; they are considered as given.

Substituting the corresponding stress distribution in (5.3) and working out† the elementary though somewhat cumbersome integrals, we obtain the relation between the applied load and the half-length of an equilibrium crack l:

$$\frac{p_0 \sqrt{L}}{K} = \frac{\sqrt{2}}{\pi} \frac{P}{K\sqrt{L}} \bar{y}_0 \left[\frac{1-\nu}{A\sqrt{A-B+2}} + \frac{12(1+\nu)\bar{y}_0^2}{A^2(A+B-2)\sqrt{A-B+2}} + \right.$$

(5.11) $$\left. \frac{2(1+\nu)(2B-A-4)}{A^2\sqrt{A-B+2}} + \bar{y}_0^2 \frac{(1+\nu)(B+A)(2B-A-4)}{A^3(A+B-2)\sqrt{A-B+2}} \right] + \frac{\sqrt{2}}{\pi\sqrt{\bar{l}}};$$

$$\bar{y}_0 = \frac{y}{L}, \quad \bar{l} = \frac{l}{L}, \quad B = \bar{y}_0^2 + \bar{l}^2 + 1, \quad A = \sqrt{B^2 - 4\bar{l}^2}.$$

---

* Note that, because of dynamic effects accompanying the expanding of the initial cut, the crack actually may "overshoot" the stable equilibrium state to some extent. This will be discussed later in more detail.

† Computation of the integrals and numerical calculations for the graphs in Fig. 20 were made by V. Z. Parton and E. A. Morozova.

The results of the calculations are plotted in Fig. 20 for $\nu = 0.25$, $P/K\sqrt{L} = 0.2$ and for several values of the parameter y_0/L. As is seen, mobile-equilibrium cracks are unstable in the absence of stiffeners. The influence of the stiffeners shows itself first of all in an increase of the size of a mobile-equilibrium crack at a given load and, as an especially important feature, in the appearance of stable states of mobile equilibrium at sufficiently small y_0/L, i.e. when rivets are spaced closely enough. The appearance of stable states of mobile equilibrium changes considerably the character of the crack expansion (see details below).

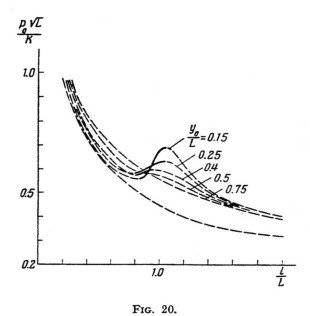

Fig. 20.

The authors observed experimentally the transition of cracks from unstable mobile-equilibrium states to stable ones; their experiments, carried out with aluminium alloy plates in the presence and absence of stiffeners, reveal a considerable increase in size of mobile-equilibrium cracks in the presence of stiffeners at the same value of p_0. In [52] the stress intensity factor at the crack ends was also determined experimentally for several stable and unstable mobile-equilibrium states. In the absence of stiffeners, measurements of the stress intensity factor were made by the direct method, i.e. by diminution of tensile stresses near the crack ends (at distances obviously large compared to the size of the crack-edge region). In the presence of stiffeners the stress intensity factors were measured indirectly. The values of these factors were found to coincide except in two cases when they were smaller by approximately 15 per cent. However, these two tests carried out with one and the same specimen, with a stable crack in one case and unstable crack in the other, gave values of the stress intensity factor close to each other. (A somewhat lower value of this factor at the end of the stable crack can be explained by the considerable dynamic effects which, according to the authors, occur in the transition from the unstable state to the stable one.) Thus it may be supposed that the deviation observed is due to some peculiarity of the specimen. Altogether, these experiments confirm directly the proposed general scheme.

This discussion can be readily extended to straight cracks in an anisotropic medium, placed in the planes of elastic symmetry of the material. The problem of a straight crack in an orthotropic infinite body subjected to a uniform stress field was treated by T. J. Willmore [21] and A. N. Stroh [83]. In [83], the results of [16] were also extended to cover the case of a straight crack in an anisotropic body under an arbitrary stress field, and the stress intensity factors at crack ends were found for this problem. Paper [84] brings the solution of the general problem concerning a straight mobile-equilibrium crack in an orthotropic body subjected to an arbitrary stress field symmetrical with respect to the line of the crack.

2. Plane Axisymmetrical Cracks

If a disk-shaped mobile-equilibrium crack of radius R is maintained in an infinite body by an axisymmetrical load, tensile stresses near the crack contour calculated without taking into account forces of cohesion tend to infinity according to the law

$$(5.12) \quad \sigma_y = \frac{N}{\sqrt{s}}, \quad N = \frac{1}{\pi\sqrt{R/2}} \int_0^R \frac{rp(r)dr}{\sqrt{R^2 - r^2}},$$

where $p(r)$ is the tensile-stress distribution at the place of the crack in a continuous body subjected to the same loads. According to the general condition (4.7), the equation determining the radius of a mobile-equilibrium crack R is

$$(5.13) \quad \int_0^R \frac{rp(r)dr}{\sqrt{R^2 - r^2}} = K\sqrt{\frac{R}{2}}.$$

This equation was established in [56, 57]. Its derivation is based on the application of the method of Fourier-Hankel transforms, developed by I. N. Sneddon [14, 15] for solving axisymmetrical problems of elasticity. In particular, if a mobile-equilibrium crack is kept open by a uniform tensile stress at infinity p_0, then $p(r) \equiv p_0$ and the radius of an equilibrium crack is given by

(5.14) $$R = \frac{K^2}{2p_0^2}.$$

This problem was first solved by R. A. Sack [20] by the energy method; his method is quite similar in principle to Griffith's [3, 4] treatment of the corresponding plane problem.

If there is no tensile load at infinity, and if the crack is kept open by a uniformly distributed pressure p_0 over a part of its surface ($0 \leqslant r \leqslant r_0$) while the remaining part of the crack surface ($r_0 \leqslant r \leqslant R$) is free, then the radius of the mobile-equilibrium crack is found from the relation

(5.15) $$\frac{p_0 \sqrt{r_0}}{K} = \frac{1}{\sqrt{2}} \left(\frac{r_0}{R}\right)^{-3/2} \left[1 + \sqrt{1 - \left(\frac{r_0}{R}\right)^2}\right].$$

Here, just as in the plane case, the radius of the loaded part of the crack surface r_0 must not be less than the critical radius for a given pressure p_0, which is defined by (5.14). In particular, if a disk-shaped crack is maintained open by equal and opposite concentrated forces P applied at its surface, then the radius of a mobile-equilibrium crack is determined by the formula

(5.16) $$R = \left(\frac{P}{\sqrt{2}\,\pi K}\right)^{2/3}.$$

Relations (5.14) and (5.16) can be obtained, except for the numerical factor, from dimensional analysis (cf. (5.6) and (5.8)).

If a disk-shaped crack is kept open by equal and opposite forces P whose points of application are $2L$ apart along the common line of action, then the radius of a mobile-equilibrium crack R is determined from the equation

(5.17) $$\frac{P}{KL^{3/2}} = \pi \sqrt{2} \left(\frac{L}{R}\right)^{-3/2} \left(1 + \frac{L^2}{R^2}\right)^2 \left(1 + \frac{2-\nu}{1-\nu} \frac{L^2}{R^2}\right)^{-1}$$

The above solutions were obtained in [56, 57]; the interpretation of the relations obtained is quite similar to the corresponding cases for a straight crack.

3. *The Extension of Isolated Cracks Under Proportional Loading; Stability of Isolated Cracks*

The problem of this section is a special case of problem A. A complete investigation is carried out for symmetrical loading and initial cracks, straight and disk-shaped cracks being considered simultaneously. An example of a problem concerning the growth of an unsymmetrical initial crack is given, which illustrates the general procedure of solving this problem. Under proportional loading the tensile stresses at the place of the crack, but in a *continuous* body subjected to the same load, are proportional to the loading parameter λ; hence $p(x) = \lambda f(x)$ and $p(r) = \lambda f(r)$ in the cases of straight and disk-shaped cracks, respectively. Introducing the dimensionless variable ξ equal to x/l and r/R in these cases, respectively, one obtains relations (5.3) and (5.12) in the form

(5.18) $$\frac{\sqrt{2}\,\lambda}{K} = \varphi(c),$$

where $\varphi(c)$ is defined respectively by

(5.19) $$\varphi(c) = \left[\sqrt{c} \int_0^1 \frac{f(c\xi)d\xi}{\sqrt{1-\xi^2}}\right]^{-1}, \quad \varphi(c) = \left[\sqrt{c} \int_0^1 \frac{f(c\xi)\xi d\xi}{\sqrt{1-\xi^2}}\right]^{-1}$$

and c denotes, respectively the half-length $l/2$ or the radius R.

Thus the relation of the crack length to the parameter λ of proportional loading is completely determined by the length of the initial crack and by the function $\varphi(c)$, corresponding to a given load distribution. Certain properties of the function $\varphi(c)$ can be obtained under the most general assumptions. Omitting the case of a crack maintained by concentrated forces applied at its surface, let us suppose that the crack is kept open by any loads, in particular, by concentrated loads applied inside the body and perhaps by distributed loads applied at the crack surface. In this case the functions $p(x)$, $p(r)$, and, consequently, $f(c\xi)$ are obviously bounded. For small c we obtain from (5.19), respectively:

(5.20) $$\varphi(c) = \frac{2}{\pi f(0)\sqrt{c}} + \ldots, \quad \varphi(c) = \frac{1}{f(0)\sqrt{c}} + \ldots.$$

Suppose that the tensile loads applied to the body on each side of the crack are bounded and, for definiteness, equal to λP. Then the following relations are valid:

(5.21)
$$\int_{-\infty}^{\infty} p(x)dx = \lambda P, \quad \int_0^{\infty} f(c\xi)d\xi = \frac{P}{2c},$$

$$\int_0^{\infty} p(r)r\,dr = \frac{\lambda P}{2\pi} \quad \int_0^{\infty} f(c\xi)\xi d\xi = \frac{P}{2\pi c^2}.$$

Eqs. (5.21) and (5.19) yield asymptotic representations for the functions $\varphi(c)$ when $c \to \infty$:

$$(5.22) \qquad \varphi(c) = \frac{2\sqrt{c}}{P} + \ldots, \qquad \varphi(c) = \frac{2\pi c^{3/2}}{P} + \ldots.$$

Thus, under the assumptions made, $\varphi(c)$ tends to infinity when $c \to 0$ and $c \to \infty$. Owing to the boundedness of $f(c\xi)$, the integrals in expressions (5.19) do not become infinite at any c, therefore $\varphi(c)$ vanishes nowhere and, consequently, has at least one positive minimum, one falling branch, and one rising branch. If the forces applied to the body on either side of the crack are not bounded, then the function $\varphi(c)$ may not have rising branches and, consequently, minima. This happens in particular in case of a uniform tensile stress field when $p = \lambda p_0$ and

$$(5.23) \qquad \varphi(c) = \frac{2}{\pi p_0 \sqrt{c}}, \qquad \varphi(c) = \frac{1}{p_0 \sqrt{c}}$$

for a straight and axisymmetrical crack, respectively.

By definition, an equilibrium crack is stable if no (sufficiently small) change in its contour produces forces which tend to move the crack further away from the disturbed state of equilibrium. It is evident that immobile-equilibrium cracks are always stable. For stability of a mobile-equilibrium crack it is necessary that its size should grow with an increase of the loading parameter λ. Suppose indeed that the corresponding size of a mobile-equilibrium crack c grows with increasing load. If the crack size is diminished without changing the load ($\lambda = $ const), the crack extension force will be greater than it was in equilibrium. Therefore the equilibrium is disturbed, and the crack tends to widen under the action of the excess force. Conversely, if the crack size is slightly increased compared with its equilibrium size, then the equilibrium is disturbed in the opposite direction, and the crack tends to close, if it is reversible.* If near a given equilibrium state the equilibrium crack size c diminishes with an increase of λ, then it is obvious that its small change under a constant load will produce forces favouring further departure from the equilibrium state. The corresponding equilibrium state will be unstable. Hence the equilibrium state of a crack is stable, if for given c and λ the following condition is satisfied:

$$(5.24) \qquad \frac{dc}{d\lambda} > 0.$$

---

* If the crack is irreversible, then with an increase in its size no reverse closing takes place, but no further expansion of the crack takes place either. Equilibrium is attained in this case because of diminution of forces of cohesion acting in the edge region of the crack.

Differentiating (5.18) with respect to λ, we find

(5.25) $$\frac{dc}{d\lambda} = \frac{\sqrt{2}}{K\varphi'(c)}.$$

Thus the condition for stability of the state of mobile equilibrium is

(5.26) $$\varphi'(c) > 0,$$

and only those states of mobile equilibrium are stable which correspond to rising portions of the curve $\varphi(c)$.

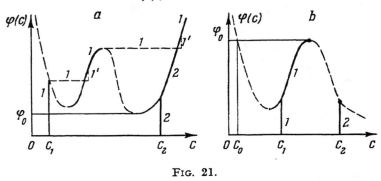

Fig. 21.

Now we have everything that is necessary for the complete investigation of the extension of an isolated symmetrical crack under proportional loading. Let a function $\varphi(c)$, such as shown in the graph of Fig. 21, correspond to a given system of loads applied to the body and consider first the case when $\varphi(c) \to \infty$ as $c \to \infty$ (Fig. 21a). Such a case occurs in particular when the loads applied on both sides of the crack are bounded. Let the dimension of the initial crack $2c_1$ correspond to an unstable brach of $\varphi(c)$. Then the crack length remains constant with increase of λ, until λ reaches the magnitude, for which the initial crack of size $2c_1$ becomes one of mobile equilibrium. Since the mobile equilibrium is unstable, the crack begins to expand under constant load, until it reaches the nearest stable mobile-equilibrium state. With further increase of λ the crack size grows continuously, until the load corresponding to a maximum of $\varphi(c)$ is reached, then changes again in a stepwise manner when the transition to another stable branch takes place, after which it grows continuously with increasing λ. The path of the point representing the change of the crack is indicated by the number 1 in Fig. 21a.* Let now the size of an initial crack $2c_2$ correspond to a stable

---

* Owing to dynamic effects that occur in this transition, the crack may overexpand a little beyond the size of the stable mobile-equilibrium crack corresponding to the given load (apparently that happened in the experiments described in paper [52]). In this case, a further increase in the load leaves the length unchanged up to reaching mobile equilibrium, after which the crack starts to lengthen further. Naturally, the purely static theory considered here cannot describe these dynamic effects; the corresponding parts of the graph in Fig. 21a are dotted and designated by the number 1′.

branch of $\varphi(c)$. The crack size now remains unchanged up to the load at which it reaches mobile equilibrium, after which it increases continuously. The path of the representative point is indicated by the number 2 in Fig. 21a. In the case considered, no fracture of the body occurs for any values of the parameter λ. If λ is less than its critical value (corresponding to the lowest of the minima of $\varphi(c)$), then great as the size of the initial crack may be, it does not expand under a given load. The size of the mobile-equilibrium crack corresponding to this critical value of λ is finite.

This means in particular: if a crack is kept open by forces applied inside the body and perhaps by loads distributed over the crack surface, and if the forces applied on each side of the crack are bounded, then there exists a critical value of the parameter λ; for all values of λ greater than the critical one there exists at least one stable and one unstable state of mobile equilibrium.

Let us now turn to the case when $\varphi(c) \to 0$ as $c \to \infty$ (Fig. 21b). If the size of an initial crack $2c_1$ corresponds to a stable branch of $\varphi(c)$, then the crack does not expand until a load is reached at which its state becomes a mobile equilibrium. After that, the crack grows continuously with increasing λ, until a value of λ is reached that corresponds to a maximum. If this λ-value is exceeded, the solution of the problem does not exist any longer, and fracture of the body occurs. The path of the representative point is indicated by the number 1 in Fig. 21b. If the size of an initial crack $2c_2$ corresponds to the right-hand unstable branch of $\varphi(c)$, then no expansion of the initial crack occurs with increasing λ, until a value of λ is reached for which the state of the initial crack becomes a mobile equilibrium. The slightest exceeding of this value of λ causes complete fracture of the body. If the size of an initial crack $2c_3$ corresponds to the left-hand unstable branch of the curve $\varphi(c)$, then for $c_3 < c_0$ the crack develops in the same manner as in case 2; for $c_3 > c_0$ the development of the crack is similar to case 1 in Fig. 21a before reaching a maximum, after which the body fractures.

The investigation of other forms of the curve $\varphi(c)$ can easily be carried out by combining the cases considered. We see that the knowledge of the function $\varphi(c)$ makes it possible to describe completely the behavior of a symmetrical isolated crack in an infinite body under proportional loading. In the case of reversible cracks, a change in the crack size can be traced by means of the graph of $\varphi(c)$ also for a non-monotonous variation in the load. It is of interest to note that in this case a decrease in the load produces a stepwise diminution of the crack size, but this happens, in general, when critical equilibrium states are passed that are different from those corresponding to an increase in the load.

Recently, L. M. Kachanov [84a] carried out an investigation generalizing the previous treatments so as to cover the case of the time-dependent modulus of cohesion. This investigation is of basic importance in connection with the problems of so-called "stress rupture".

The analysis carried out in the present section is based on [59].

Consider now the solution of a problem concerning the extension of an unsymmetrical initial crack in one simple case. Let a straight initial crack with the end coordinates $x = -a_0$ and $x = b_0$ be given in an infinite unloaded body (for definiteness assume $b_0 < a_0$) and let equal and opposite concentrated forces P be applied at opposite points of the crack surfaces, say, at $x = 0$. The magnitude of the force P plays the role of the loading parameter. According to (5.1), the values of the tensile-stress intensity factors N_0 at $x = -a$ and $x = b$ are, respectively,

$$(5.27) \qquad N_a = \frac{P}{\pi \sqrt{b+a}} \sqrt{\frac{b}{a}}, \qquad N_b = \frac{P}{\pi \sqrt{b+a}} \sqrt{\frac{a}{b}}.$$

When $P < P_1$, where

$$(5.28) \qquad \frac{P_1^2}{K^2} = \frac{(b_0 + a_0)b_0}{a_0},$$

both factors N_a and N_b are less than K/π so that the crack expands neither to the right nor to the left. At $P = P_1$ the factor N_b becomes equal to K/π,

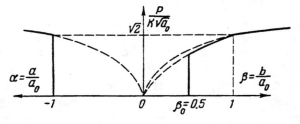

FIG. 22.

mobile equilibrium is reached and the end b begins to move to the right. The advance depends on the magnitude of the applied force according to the relation

$$(5.29) \qquad \frac{P^2}{K^2} = \frac{b(a_0 + b)}{a_0}.$$

As long as $P < P_2$, where

$$(5.30) \qquad \frac{P_2^2}{K^2} = 2a_0,$$

we have $N_a < K/\pi$, and the left end does not move. At $P = P_2$, we have $b = -a_0$, a symmetrical crack in mobile equilibrium, and at $P > P_2$ the development of the crack continues according to (5.8). The development of the initial crack with changing P is plotted in Fig. 22.

4. Cracks Extending to the Surface of the Body

If a crack extends to the surface of the body, it becomes difficult to obtain effective analytical solutions. Mapping of the corresponding region on a half-plane cannot be carried out by means of rational functions, and Muskhelishvili's method does not make it possible to obtain solution in finite form. Therefore it is necessary to resort to numerical methods in analysing such problems.

A number of numerical solutions have been derived up to now; the mobile-equilibrium states are unstable in all analysed cases.

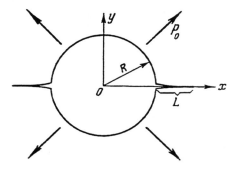

Fig. 23.

O. L. Bowie [22] treated the problem of a system of k symmetrically located cracks of equal length extending to the free surface of a circular cut in an infinite body (Fig. 23). The body is stretched at infinity by the all-round stress p_0. Bowie employed Muskhelishvili's method for calculating stresses and strains. To obtain the solution in effective form, the author used a polynomial approximation to the analytical function mapping the exterior of the circle with adjacent cuts on the exterior of the unit circle. For the determination of the dimensions of mobile-equilibrium cracks Bowie used directly Griffith's energy method and computed the strain-energy release rate. Numerical calculations were made for cases of one crack and two diametrically opposite cracks. To obtain sufficient accuracy of calculations it proved necessary to retain about thirty terms in the polynomial representation of the mapping function. The numerical results for the cases $k = 1$ and $k = 2$ obtained by Bowie are shown in Fig. 24. It follows from these computations that at $L/R > 1$ the tensile stress for two cracks with a circular cavity is very close to the tensile stress for one crack of length

$2(L + R)$, so that the influence of the cavity proper is almost unnoticeable. Furthermore, in the case of small crack lengths the conditions of mobile equilibrium are obviously determined by the tensile stresses directly at the

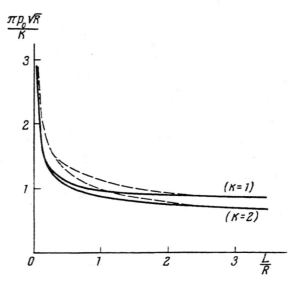

FIG. 24. ———uniaxial tension, - - - all-round tension.

surface of the circular cavity. As is known, in case of uniaxial extension the highest tensile stress at the boundary of the cavity is equal to $3p_0$ and in case of all-round extension $2p_0$. Thus the ratio of equilibrium loads in these cases should approach 2/3, and this is found in agreement with Bowie's calculations.

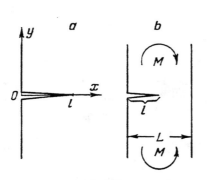

FIG. 25.

The problem of a straight crack ending on a straight free boundary of the half-space (Fig. 25) was treated independently by L. A. Wigglesworth [85] and G. R. Irwin [51] using different methods.

Wigglesworth [85] investigated the case of an arbitrary distribution of normal and shearing stresses over the faces of the crack. For a symmetrical distribution of stresses he reduced the problem to an integral equation for the complex displacement $w(x) = u(x) + iv(x)$ of points of the crack surface:

$$(5.31) \qquad \int_0^l L(x,t)w(t)dt = -\frac{4(1-v^2)}{E}\int_0^x p(x)dx.$$

Here $L(x,t)$ is a singular integral operator and $p(x) = \sigma(x) + i\tau(x)$; $\sigma(x)$ is the distribution of normal stresses; $\tau(x)$ is the distribution of shearing stresses. Equation (5.31) is solved in the paper by an integral-transform method. Detailed calculations are made for the case when the surfaces of the crack and boundary are free of stresses, the tensile stress p_0 being applied at infinity parallel to the boundary of the half-space.

For stresses near the crack end the author obtains in this special case the following relations:

(5.32)
$$\sigma_x + \sigma_y = 1.586 \sqrt{\frac{l}{s}} p_0 \sin\frac{\phi}{q},$$

$$\sigma_x - \sigma_y + 2i\sigma_{xy} = -0.793 \sqrt{\frac{l}{s}} p_0 \sin\phi \exp\left(\frac{3i\phi}{q}\right),$$

hence we find at the prolongation of the crack ($\Phi = \pi$)

(5.33)
$$\sigma_x = \sigma_y = 0.793 \, p_0 \sqrt{\frac{l}{s}}, \qquad \sigma_{xy} = 0,$$

which together with (4.6) gives the expression for the length of the mobile-equilibrium crack in the form

(5.34)
$$l = \frac{K^2}{\pi^2 (0.793)^2 p_0^2} \approx 1.61 \frac{K^2}{p_0^2}.$$

Irwin [51] investigated only the last special case. He represented the unknown solution as the sum of three fields. The first field corresponds to a crack ($-l \leqslant x \leqslant l$, $y = 0$) in an infinite body subjected to constant tensile stress p_0 at infinity, the second field corresponds to the same crack under normal stresses $Q(x)$ symmetrical with respect to the x and y axes and applied at the crack surface, the third field corresponds to a half-space $x \geqslant 0$ without crack, at the boundary of which ($x = 0$) the distribution of normal stresses $P(y)$, symmetrical with respect to the x axis, is given. Satisfying the boundary conditions at the free boundary and the crack surface, Irwin obtained for $P(y)$ and $Q(x)$ the system of integral equations

(5.35)
$$\int_0^l Q(x) \frac{2\sqrt{l^2 - x^2}\, y}{\pi(y^2 + x^2)\sqrt{y^2 + l^2}} \left[\frac{2y^2}{y^2 + x^2} + \frac{y^2}{y^2 + l^2} - 2\right] dx +$$

$$+ p_0 \left[\frac{2y}{\sqrt{y^2 + l^2}} - \frac{y^3}{(y^2 + l^2)^{3/2}} - 1\right] = P(y),$$

$$-4 \int_0^\infty P(y) \frac{xy^2}{\pi(x^2 + y^2)^2} dy = Q(x),$$

which he solved by the method of successive approximations. The first approximation yields a relation for the length of the mobile-equilibrium crack l:

$$(5.36) \qquad l = \frac{2K^2}{\pi^2 1.095^2 p_0^2} = 1.69 \frac{K^2}{p_0^2},$$

which differs, as is seen, insignificantly from the more exact relation (5.34).

H. F. Bueckner [50] treated a problem of one straight crack reaching the boundary of a circular cavity in an infinite body. No stress is applied at infinity and at the boundary of the cavity, the surface of the crack is free of shearing stresses, normal stresses are applied symmetrically and vary according to a given law — $p(x)$. Such a form of the problem arises in the analysis of rupture of rotating disks. Like Wigglesworth [85], Bueckner proceeds independently from a singular integral equation for the lateral displacements of points of the crack surface. He considers a one-parameter family of particular solutions of this equation, corresponding to certain special distributions $p_n(x)$. In the general case it is recommended to represent $p(x)$ as a linear combination of $p_n(x)$:

$$(5.37) \qquad p(x) = \sum_{n=0}^{n=m} \alpha_n p_n(x);$$

the coefficients α_n are determined by the least-square method or by collocation. The factor of stress intensity at the crack end N_0 is expressed in terms of the coefficients α_n.

If the length of the crack is far less than the radius of the circular cavity, then we have in the limit the previous particular case of a straight boundary. As it follows from Bueckner's calculations in this particular case when $p = p_0 = \text{Const}$, the expression for the length of a mobile-equilibrium crack is

$$(5.38) \qquad l = \frac{2K^2}{\pi^2 1.13^2 p_0^2} = 0.159 \frac{p_0^2}{K^2},$$

which is in good agreement with (5.34) and (5.36).

In [50] Bueckner also treated a problem of a crack reaching the surface of an infinitely long strip of finite width under an arbitrary load, symmetrical with respect to the line of the crack (Fig. 25b). He showed that it is possible to replace with a high degree of accuracy the integral equation occurring in this case by one with a degenerated kernel. The numerical solution obtained by Bueckner in the special case when the load is produced by couples M, applied on both sides of the crack at infinity, gives the relation between the length of a mobile-equilibrium crack and the load; it is represented by the curve in Fig. 26.

As has already been pointed out, in all cases discussed in the present section mobile-equilibrium cracks are unstable. Thus, when loads increase, extension of an initial crack does not take place until it reaches mobile equilibrium, after which the body fractures. In these problems the load at which an initial crack reaches mobile equilibrium coincides with the breaking load, which is in general not true.

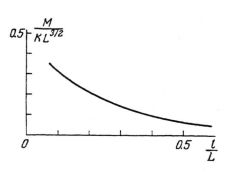

FIG. 26.

In the paper by D. H. Winne and B. M. Wundt [32] some of the solutions presented in this section were employed for the analysis of fracture of rotating notched disks, and of notched beams in bending. The experiments conducted by Winne and Wundt, analysed on the basis of these calculations, revealed close coincidence of the values of surface-energy density T (or, which amounts to the same, of the moduli of cohesion K) determined from the angular speed, at which fracture of rotating notched disks occurs, and from the loads at which fracture of notched beams in bending occurs. This confirms that the quantities T and K are characteristics of the material and do not depend on the nature of the state of stress.

5. Cracks near Boundaries of a Body; Systems of Cracks

Crack development in bounded bodies possesses some characteristic peculiarities. Difficulties of a mathematical character do not allow us to

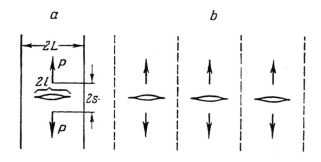

FIG. 27.

carry out here as complete an investigation as in the case of isolated cracks. However, the qualitative features and some of the quantitative characteristics of this phenomenon can readily be elucidated in connection with the simplest problems that yield to analytical solution. Let us examine first of all the problem of a straight crack in a strip of finite width (Fig. 27a). The crack is assumed to be symmetrical with respect to the middle line of the strip, and the direction of its propagation is normal to the free boundary. The load keeping the crack open is considered symmetrical with respect to the line of the crack and the middle line of the strip.

In solving the problem we use the method of successive approximations developed by D. I. Sherman [86] and S. G. Mikhlin [87]. As the first approximation we take the solution of a problem in the theory of elasticity for the exterior of a periodical system of cuts (Fig. 27b). Denoting again by $p(t)$ the distribution of tensile stresses, which would be at the place of the cracks in a continuous body under the same loads, we obtain the equation determining the half-length of a mobile-equilibrium crack l in the form

$$(5.39) \quad \int_{-m}^{m} p[t_0(t)] \sqrt{\frac{m+t}{m-t}} dt = K \sqrt{\frac{\pi m}{2L}},$$

where $t = \sin(\pi t_0/2L)$, $m = \sin(\pi l/2L)$. In the particular case represented in Fig. 27, when the crack is maintained by equal and opposite concentrated forces P with points of application $2s$ apart along their common line of action, (5.39) becomes

$$(5.40) \quad \frac{P}{K\sqrt{L}} = \frac{\sqrt{8(\alpha^2 + 1) \sin(\pi l/L)}}{\pi \cosh \sigma \left[1 - \nu + (1 + \nu) \dfrac{\sigma(2\alpha^2 + 1) \cosh \sigma}{\alpha(\alpha^2 + 1)m}\right]},$$

where $\alpha = \sinh \sigma/m$, $\sigma = \pi s/2L$. When $s = 0$ (concentrated forces applied at the crack surface), (5.40) reduces to

$$(5.41) \quad \frac{P}{K\sqrt{L}} = \sqrt{\frac{2}{\pi} \sin \frac{\pi l}{L}}.$$

Let us also quote the relation between the size of a mobile-equilibrium crack and the load for the case of a uniform tensile stress at infinity, $P/2L$,

$$(5.42) \quad \frac{P}{K\sqrt{L}} = \sqrt{\frac{2}{\pi} \cot \frac{\pi l}{2L}}.$$

Relation (5.40) for various σ is presented in Fig. 28. The solid and dotted lines denote, as usual, stable and unstable branches. As is seen, for $\sigma \geqslant \sigma_c \approx 0.5$ there are no stable branches, hence for distances between points of application of forces exceeding $2L/\pi \approx 0.64 L$ mobile-equilibrium cracks are always unstable. Quite similarly to the analysis in Section V.3 (extension of an isolated crack under proportional loading) the graph in Fig. 28 makes it possible to describe completely the extension of any symmetrical initial crack when the load increases.

The present analysis is based on papers [58, 88]. The solution of the corresponding problem in the theory of elasticity for the case $s = 0$ was obtained by Irwin [45]. The problem of a periodical system of cracks under uniform loading at infinity was solved by Westergaard [13] and independently by W. T. Koiter [89].

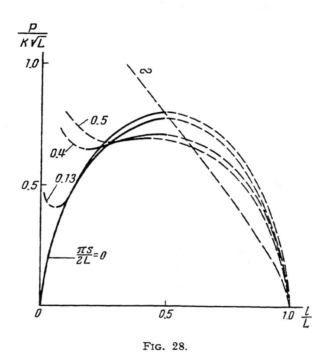

Fig. 28.

In the first approximation only the shearing stresses vanish at the lines of symmetry (shown by the dotted lines in Fig. 27b, which correspond to the boundaries of the strip); the normal stresses are different from zero. To obtain the second approximation, the first approximation is added to the solution for an uncracked strip, at the boundaries of which the normal stresses are given; their distribution is chosen in such a manner as to compensate the normal stresses at the boundary obtained in the first approximation. Now the boundary condition is no longer satisfied at the crack surface. To obtain the third approximation, the second approximation is added to the solution for the exterior of a periodical system of cuts, at the surface of which the distribution of normal stresses is equal to the difference between the given stresses and those obtained in the second approximation, and so on.

Special estimates obtained in [88] show that for stable mobile-equilibrium states the considerations of the second and subsequent approximations leads to corrections of the order of 2.5—3 per cent in the above relations. This permits us to confine ourselves to the first approximation.

In addition to these problems (the periodical system of cracks and the system of radial cracks ending in a circular cavity), several other problems of systems of cracks have been treated; they deal with straight cracks located along one straight line. Mathematical methods developed by Muskhelishvili [90, 18], D. I. Sherman [91], and Westergaard [13] permit

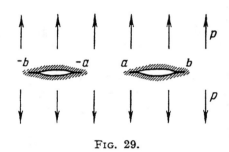

Fig. 29.

the reduction of any such problem to quadratures. Let us here consider the simplest example: it is the problem of the extension of two collinear straight cracks of the same length in an infinite body, stretched by a uniform stress p at infinity (Fig. 29). This problem was treated by Willmore [21]; it also occurs in a paper by Winne and Wundt [32] (the authors refer to a private communication by Irwin). According to the solution presented in [21], the sizes of the cracks remain unchanged at $p < p_1$, where

(5.43) $$p_1 = \sqrt{\frac{2}{b}\frac{K}{\pi}\left\{\frac{K'(\alpha)\sqrt{\alpha(1-\alpha^2)}}{E'(\alpha)-\alpha^2 K'(\alpha)}\right\}}, \qquad \alpha = \frac{a}{b} < 1.$$

Here K', E' are standard notations of elliptic integrals.

At $p = p_1$ the cracks attain an unstable state of mobile equilibrium, after which the inside edges of the cracks join and form a crack of length $2b$. The further extension of the crack depends on whether the bracketed expression in (5.43) is greater or less than unity. If it is less than unity, which happens for $\alpha < 0.027$, the size of the crack resulting from the joining of the inside edges is less than the size of the mobile-equilibrium crack corresponding to the load p_1. In this case the crack remains unchanged up to the load $p_2 = \sqrt{2}\,K/\pi\sqrt{b}$, after which the body fractures. If it is greater than unity, complete fracture of the body occurs immediately upon reaching the load p_1. Assuming $b - a = 2l$ and making $b \to \infty$ in (5.43), we obtain in the limit (5.6), as expected. The solution given in [32] leads to the same qualitative results. However, it cannot be accepted as correct because it is based on the erroneous expressions of the stress intensity factors given in [47].

The case of two identical cracks maintained open by concentrated forces applied at their surface was treated in [88]. A complete investigation of the general case of symmetrical loading for a system of two cracks can be carried out quite similarly, with expressions for the stress intensity factors at the crack ends $x = a$ and $x = b$

$$N_b = \frac{\sqrt{2}}{\pi \sqrt{b(b^2 - a^2)}} \left[\int_a^b p(t) t \sqrt{\frac{t^2 - a^2}{b^2 - t^2}} dt + C \right]$$

$$N_a = \frac{\sqrt{2}}{\pi \sqrt{a(b^2 - a^2)}} \left[\int_a^b p(t) t \sqrt{\frac{b^2 - t^2}{t^2 - a^2}} dt - C \right]$$

$$C = \frac{b}{K'\left(\frac{a}{b}\right)} \int_a^b \frac{dt}{\sqrt{(b^2 - t^2)(t^2 - a^2)}} \int_a^b \frac{p(t_0) \sqrt{(b^2 - t_0^2)(t_0^2 - a^2)} dt_0}{t_0 + t}$$

(5.44)

As is seen from these examples, collinear cracks "weaken" each other and reduce their stability. Ya. B. Zeldovitch noticed that in the case

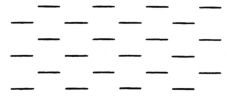

FIG. 30.

of a "chess-board" pattern of cracks (Fig. 30) the inverse phenomenon occurs. As the calculations show, even for uniform normal loads at the crack surfaces, mobile-equilibrium cracks may become stable for a certain mutual position.

We consider briefly the so-called "size effect" in the brittle fracture of bounded bodies. Take similarly shaped bodies, which differ only in the characteristic size d and in the characteristic scale of the applied extensional load S (it is supposed that macroscopic cracks present in the bodies are also geometrically similar). In brittle fracture, the value $S = S_0$ that corresponds to fracture depends only on the characteristic size of the body d and the modulus of cohesion K. There is only one way to form a characteristic having the dimension S from the quantities K and d, and it is impossible to make any dimensionless combinations. Therefore the following simple relations govern the magnitude of the breaking load:

$$(5.45) \qquad S_0 = \varepsilon_1 K d^{3/2}, \qquad S_0 = \varepsilon_2 K d^{1/2}, \qquad S_0 = \varepsilon_3 K d^{-1/2},$$

where S_0 has the dimension of a force, of a force distributed along a line (as, for instance, a concentrated force in plane strain), and of a stress, respectively. The quantities ε_i are constants for a given geometrical configuration of the body. About the fracture of geometrically similar bodies a great deal of experimental data is at present available, which permits clarification of the limits of applicability of the theory of brittle fracture. Detailed information on this topic can be found in a paper by B. M. Wundt [92], and some new results have been presented by S. Yusuff [93].

6. Cracks in Rocks

The investigation of crack extension in rock massifs is of great interest in theoretical geology. Cracks can form in rocks because of various causes of tectonic character, but also because of some artificial actions (mining excavations, hydraulic fracture of oil-bearing strata, etc.).

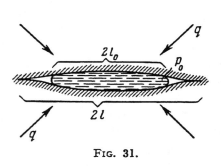

Fig. 31.

In connection with the theory of hydraulic fracture of an oil-bearing stratum a number of problems of the theory of cracks have been treated, among them *the problem of the vertical crack:* A crack in an infinite space subjected to all-round compressive pressure q at infinity is maintained open by a flowing viscous fluid injected into it (Fig. 31). The main peculiarity of the problem is that the fluid does not fill the crack completely: there is always a free part of the crack on both sides of the wetted area. The pressure p_0 in the flowing fluid troughout the wetted area of the crack can be considered constant in first approximation. Indeed, at the end of the wetted area an abrupt narrowing of the crack takes place, and almost all of the pressure drop will occur there. The problem is called so, because the actual fissure, idealised by this problem, is located in a vertical plane, and q represents the lateral pressure of the rocks. In comparison with the action of lateral rock and fluid pressures the action of the forces of cohesion may be neglected, as estimates show.* Condition (5.3) determining crack sizes becomes

(5.46) $$\int_0^l \frac{p(x)dx}{\sqrt{l^2-x^2}} = 0, \quad p(x) = \begin{cases} p_0 - q, & 0 \leqslant x \leqslant l_0; \\ -q, & l_0 < x \leqslant l; \end{cases}$$

hence

(5.47) $$l = l_0 \left[\sin \frac{\pi q}{2p_0}\right]^{-1}.$$

The expression for the maximum half-opening of the crack v_0 is

(5.48) $$v_0 = \frac{8(1-v^2)p_0 l_0}{\pi E} \ln \cot \frac{\pi q}{4p_0}.$$

As calculations show, for values l_0/l close to unity which are usually encountered in practice, the opening of the crack is almost constant all along the wetted area of the crack; the crack closes rapidly along the free part. — This problem of the vertical crack was first stated and solved in a paper by Zheltov and Khristianovitch [38].

The problem of the horizontal crack [40] is stated as follows. In a heavy half-space at a certain depth H a horizontal disk-shaped crack is formed by injecting viscous fluid as before; the surface of the crack is again divided into a wetted part ($0 \leqslant r \leqslant R_0$) and a free part ($R_0 < r \leqslant R$), and the fluid pressure p in the wetted part may again be considered as constant. Forces of cohesion, as in the preceding case, are neglected. Under the assumption that the depth of the crack position H is sufficiently great, the boundary condition at the boundary of the half-space need not be taken into account. The condition of finiteness of stresses at the crack contour yields in this case

(5.49) $$\frac{p-\gamma H}{p} = \sqrt{1 - \left(\frac{R_0}{R}\right)^2}.$$

where γ is the specific weight of the rock. For the volume of the injected fluid one obtains

(5.50) $$V = \frac{4(1-v^2)pR^3}{E} \varphi\left(\frac{R_0}{R}\right), \quad \varphi(z) = z^3\left[\frac{2}{3} - \frac{z}{3} - \frac{z}{3(1+\sqrt{1-z^2})}\right].$$

---

* The condition that forces of cohesion be negligibly small is $K/q\sqrt{l} \ll 1$. It is in general not satisfied in laboratory scaling.

In practice, $z = R_0/R$ is close to unity so that it is possible to use the asymptotic form of (5.50)

$$(5.51) \qquad V = \frac{4(1-\nu^2)pR^3}{3E} \sqrt{2(1-z)}\, [1 + \sqrt{2(1-z)} - 3(1-z)].$$

The maximum half-opening of the crack is determined by the formula

$$(5.52) \qquad v_0 = \frac{8(1-\nu^2)pR_0}{\pi E} \arccos\left(\frac{R_0}{R}\right).$$

Thus, if the depth of the crack position, the fluid pressure, and the specific weight of the rock are known, R_0/R can be found according to (5.49). Then the crack radius is obtained from (5.51) and a knowledge of the total volume of the injected fluid V, after which the determination of the remaining parameters does not encounter any difficulties.

In [40, 41] problems were also treated concerning horizontal cracks in a radially varying pressure field caused by the higher lying rocks. Under certain conditions a complete wetting of the crack surface (i.e. the absence of a free part) may in this case occur.

Yu. P. Zheltov [43] proposed an approximate method for solving the problem of the horizontal crack in a radially varying vertical pressure field. A comparison between the results obtained by this method and the exact solutions for certain cases showed quite satisfactory agreement.

By using the method of successive approximations Yu. A. Ustinov [94] estimated the influence of the free boundary in the problem of the horizontal crack. If the depth is larger than twice the crack radius, the influence of the free boundary is negligibly small.

The problem of a crack formed by driving a horizontal wedge of constant thickness into a heavy space was treated in [39].

The solution of the problem of the vertical crack was extended by Zheltov [42] to cover the case when the rock is permeable and the injected fluid flows through the rock.

VI. Wedging; Dynamic Problems in the Theory of Cracks

1. *Wedging of an Infinite Body*

Wedging is formation of a crack in a solid by driving a rigid wedge into it. The most characteristic property of the wedging of a brittle body is that the wedge surface never comes in complete contact with the body: there is always a free portion in the front part of the wedge; ahead of the wedge a free crack forms, which closes at some distance from the edge of the wedge (Fig. 32).

It appears that the problem of wedging of an infinite body by a fixed wedge [39, 58, 95] is the simplest to formulate among problems of this kind; it yields to an effective exact solution by the methods of elasticity theory and gives a qualitative idea of wedging under more complex conditions.

Let a uniform, isotropic brittle body be wedged by a thin, symmetrical, perfectly rigid semi-infinite wedge with thickness $2h$ at infinity (Fig. 32). In front of the wedge a free crack forms, which closes smoothly at a certain

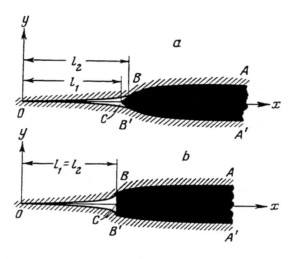

Fig. 32.

point O; the position of the point O with respect to the front point of the wedge C is not known beforehand and must be determined in the course of solving the problem. If the wedge has a rounded front part (Fig. 32a), the position of the points of departure of the crack surface from the wedge, B and B', is not prescribed and must also be determined in the course of solving the problem. If the wedge has a truncated front part (Fig. 32b) as e.g. in the case of a wedge of constant thickness, the position of the points of contact is quite definite; they coincide with the corners of the wedge front. It is evident that the stress at the points of departure is in this case infinite. We shall at first assume that there is no friction at the surface of contact between wedge and body.

The field of elastic stresses and strains satisfies the usual equations of static elasticity in the exterior of the crack. In view of the assumed slenderness of the wedge, the boundary conditions may be transferred from the crack surface proper to the x-axis. Without considering forces of cohesion, the boundary conditions are

(6.1)
$$\sigma_{xy} = 0, \quad \sigma_y = 0 \quad (0 \leqslant x < l_2, \ y = 0),$$
$$v = \pm f(x - l_1), \quad \sigma_{xy} = 0 \quad (l_2 \leqslant x < \infty, \ y = 0);$$

here σ_y, σ_{xy} are the stress-tensor components; l_1 and l_2 are the distances of the point O from the edge of the wedge and from the points of departure B, B'; $f(t)$ defines the wedge surface in a system of coordinates with origin at the front point of the wedge; the positive and negative signs correspond to the upper and lower faces of the cut, respectively.

As is seen, the problem of wedging is a peculiar combination of the contact problem in the theory of elasticity [18, 72, 73] and the problem of the theory of cracks.

The position of the points of departure of the crack surface from the wedge in the case of a wedge with rounded edge, and the position of the point of closing with respect to the edge are determined from the following conditions:

1. *Stresses at the points of departure must be finite.* For the contact problem a similar condition was first suggested as a hypothesis by Muskhelishvili [96, 18] and independently by A. V. Bitsadze [97]; it was proved in [61].

2. *Stresses at the crack edge are finite or, which is the same, the opposite faces of a crack close smoothly at its end.* Since the intensity of forces of cohesion at the crack edge is maximal, stresses near the crack edge calculated without taking into account forces of cohesion must tend to infinity according to (4.7).

The problem of wedging is a mixed problem of the theory of elasticity. For its solution it is convenient to consider the singular integral equation for the compressive force acting on the face of the wedge, $\sigma_y = -\phi(x)$. If $\phi(x)$ is known, the determination of the elastic field obviously reduces to the solution of the first boundary-value problem in the theory of elasticity for the exterior of a semi-infinite straight-line cut, which can be found by Muskhelishvili's method ([18], § 95). This solution gives an expression for the lateral displacements at points of contact between wedge and crack surface:

$$(6.2) \qquad v = \frac{4(1-\nu^2)}{\pi E} \int_{l_2}^{\infty} \phi(\sigma^2)\sigma \ln\left|\frac{\sigma+\zeta}{\sigma-\zeta}\right| d\sigma,$$

where $\zeta = \sqrt{x}$, and the root may assume positive and negative values for displacements of the upper and lower face. The second condition (6.1) yields the fundamental integral equation of the problem:

$$(6.3) \qquad \int_{l_2}^{\infty} \phi(\sigma^2)\sigma \ln\left|\frac{\sigma+\zeta}{\sigma-\zeta}\right| d\sigma = \pm \frac{\pi E}{4(1-\nu^2)} f(\zeta^2 - l_1),$$

which can be shown to be equivalent to the singular integral equation obtained from (6.3) by differentiation with respect to ζ:

$$(6.4) \qquad \int_{|\sigma|>l_{\mathbf{2}}} \frac{\phi(\sigma^2)\sigma d\sigma}{\sigma-\zeta} = \pm \frac{\pi E}{2(1-\nu^2)} \zeta f'(\zeta^2-l_1),$$

and the condition

$$(6.5) \qquad \phi(x) = \frac{Eh}{2\pi(1-\nu^2)x} + O\left(\frac{1}{x^2}\right) \quad (x \to \infty),$$

where $h = f(\infty)$. By using the methods for singular integral equations developed in the monograph by Muskhelishvili ([19], Chapter 5) the solution of equation (6.4) can be found in the form:

$$(6.6) \qquad \phi(x) = \frac{1}{\pi\sqrt{x(x-l_2)}} \left[A - \frac{E}{2(1-\nu^2)} \int_{l_\mathbf{2}}^{\infty} \frac{f'(t-l_1)\sqrt{t(t-l_2)}\,dt}{t-x}\right],$$

where A is an indefinite constant.

The integral in (6.6) does exist in view of the finiteness of $f(\infty) = h$, and it tends to zero as $x \to \infty$; this together with (6.5) determines the value of the constant A:

$$(6.7) \qquad A = \frac{Eh}{2(1-\nu^2)}.$$

For finiteness of stress at the points of departure $x = l_2$ in case of a wedge with rounded edge, it is necessary and sufficient that the bracketed expression in (6.6) vanishes at $x = l_2$. This gives one equation for the determination of l_1 and l_2:

$$(6.8) \qquad h = \int_{l_\mathbf{2}}^{\infty} f'(t-l_1) \sqrt{\frac{t}{t-l_2}}\, dt.$$

Now the following expression for the tensile stresses at the prolongation of the cut results from the solution:

$$(6.9) \qquad \sigma_y = \frac{E}{2\pi(1-\nu^2)\sqrt{(l_2-x)(-x)}} \left[h - \int_{l_\mathbf{2}}^{\infty} \frac{f'(t-l_1)\sqrt{t(t-l_2)}\,dt}{t-x}\right].$$

Together with (4.7) it leads to

$$\text{(6.10)} \qquad h - \int_{l_2}^{\infty} f'(t - l_1) \sqrt{\frac{t - l_2}{t}}\, dt = \frac{2K\sqrt{l_2}(1 - \nu^2)}{E}$$

Relations (6.8) and (6.10) are finite equations which determine the unknown constants l_1 and l_2.

In the particular case of constant wedge thickness $f(t) \equiv h$, condition (6.8), which is no longer valid, is replaced by the relation $l_1 = l_2$, and (6.10) gives the following expression for the length of a free crack in front of a "square" wedge:

$$\text{(6.11)} \qquad l_1 = l_2 = \frac{E^2 h^2}{4(1 - \nu^2)^2 K^2}.$$

In [95] other special forms of the wedge are also treated such as a wedge rounded-off with a small radius of curvature and a wedge rounded-off according to a power law. Investigation of the first example shows that

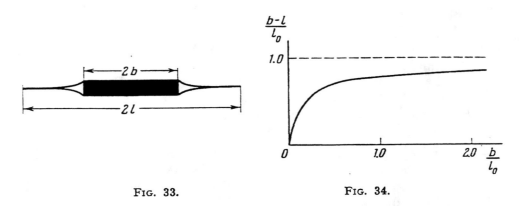

Fig. 33. Fig. 34.

roundness affects slightly the length of the free crack in front of the wedge. In [95] also a case when Coulomb friction acts on the faces of the wedge is treated.

In [84] wedging of an anisotropic body by a semi-infinite rigid wedge is studied.

I. A. Markuzon [98] treated a problem of wedging an infinite body by a wedge of finite length $2b$ (Fig. 33). In case of constant thickness of the wedge $2h$, the relation between crack length $2l$ and wedge length $2b$, other things being equal, is as represented in Fig. 34 (l_0 is the length of a free crack for an infinite wedge defined by (6.11)).

In [98] the influence of a uniform compressive or tensile stress at infinity on the length of a free crack, when the wedge is of finite length, was also investigated.

Relation (6.11) can be used for the experimental determination of the modulus of cohesion K. For that purpose a wedge is driven into a plate of the testing material, the wedge being substantially more rigid than the plate. The length L of the resulting free crack is measured. The modulus of cohesion can then be found according to the formula

$$(6.12) \qquad K = \frac{Eh}{2(1-\nu^2)\sqrt{L}}.$$

The wedge must be sufficiently long in order to eliminate the influence of the plate boundary, and it should be driven in, until the distance between the wedge end and the crack end stops varying with further displacement of the wedge. The plate must be wide and sufficiently thick so that the state of stress essentially corresponds to plane strain. To insure a straight-line form of the crack, it is necessary to compress the specimen in the direction of crack propagation. This is recommended by Benbow and Roesler [9]. (It can be shown that (6.11) and (6.12) remain unchanged in this case.)

2. Wedging of a Strip

In strict formulation, problems concerning the wedging of bounded bodies are very difficult to solve. Up to now there are but a few approximate solutions, based on the application of the approximations of simple beam theory.

The first of these solutions was obtained by I. V. Obreimov [8]; as a matter of fact, this work was the first investigation in which wedging was considered. In connection with his experiments on the splitting of mica, Obreimov examined the case when a strip being torn off has small thickness and only one-point contact with the wedging body (Fig. 35). In order to establish a relation between the surface tension of mica and the parameters of the crack shape, Obreimov applied to this problem the methods of strength of materials, considering a shaving as a thin beam. The theoretical part of the work of Obreimov is not free from shortcomings. Later, corrections were introduced into these calculations in the book by V. D. Kuznetsov [99] as well as by M. S. Metsik [10] and N. N. Davidenkov [12]. In addition Metsik improved the experimental procedure of [8]. Application of the approximations of thin-beam theory for the determination of the crack length is justifiable in some cases. However, these approximations cannot be applied to describe the form of the crack surface in the immediate vicinity of its edge, even if the distribution of forces of cohesion in the edge region is explicitly considered, as it was done by Ya. I. Frenkel [5]. The fact is that the longitudinal dimension of the edge region cannot be assumed to be large compared to the shaving thickness; hence a shaving cannot be considered as a thin beam in the region where the forces of cohesion are acting.

To illustrate the approximate approach based on the methods of simple beam theory, we discuss the paper by Benbow and Roesler [9] in more detail. Note that in this work possibilities and limits of applicability of the above approach are most clearly pointed out.

The following statement of the problem is considered (Fig. 36). A strip of finite width b is wedged symmetrically so that the crack passes along the

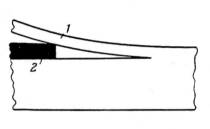
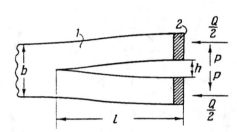

FIG. 35.
1. a Body being wedged.
1. a Wedge.

FIG. 36.
1. a Body being wedged.
2. Grips.

middle line of the strip. At the end of the strip, compressive forces $Q/2$ are applied to insure straight crack propagation; the wedging force P produces a crack length l and initial width h.

Having obtained an expression for the strain energy from dimensional considerations, the authors write the equilibrium condition for the crack in the form

$$\frac{T}{E} = \frac{h^2}{l} \phi\left(\frac{b}{l}\right), \tag{6.13}$$

so that for a given material the quantity h^2/l is uniquely determined by the quantity b/l. The experiments made with specimens of two different plastics [9] give a conclusive proof of the existence of such a one-one relation.

For small b/l, i.e. for long cracks, it is possible to obtain an asymptotic form of relation (6.13) by considering both halves of the strip as thin beams fixed at the section corresponding to the crack end. The expression for the strain energy of the strip is in this case

$$U = 3h^2 B/l^3, \tag{6.14}$$

where $B = EJ$ is the stiffness of the beam, $J = nb^3/96$, and n is the transverse thickness of the beam. The surface energy of the crack is, evidently, $2Tnl$.

In the mobile-equilibrium state the variation of surface energy corresponding to a small variation of the crack length δl is equal to the corresponding variation of strain energy of the strip. Hence it follows that

$$(6.15) \qquad -\frac{\partial U}{\partial l} = 2Tn \quad \text{or} \quad \frac{T}{E} = \frac{3h^2 b^3}{64 l^4}.$$

By comparing the second formula (6.15) with (6.13), an asymptotic expression for $\Phi(b/l)$ can be found as $b/l \to 0$:

$$(6.16) \qquad \phi = \frac{3}{64}\left(\frac{b}{l}\right)^3.$$

From (6.15) an expression for the length of an equilibrium crack is obtained:

$$(6.17) \qquad l = \sqrt[4]{\frac{3h^2 b^3 E}{64 T}} = \sqrt[4]{\frac{3h^2 b^3 E \pi}{64 K^2 (1-\nu^2)}}.$$

Thus in this case the length of the crack is proportional only to $\sqrt{h}$, whereas in the wedging of an infinite body by a semi-infinite wedge the length of the crack is proportional to h^2 (cf. (6.11)).

Relation (6.15) was used by Benbow and Roesler for the determination of the surface-energy density of the plastics investigated. The careful experimentation and the scrupulous evaluation of the sources of possible errors and of their magnitude are remarkable.

In the recent review of J. J. Gilman [11] a detailed summary and a bibliography of experimental investigations on wedging can be found.

3. Dynamic Problems in the Theory of Cracks

Considerable attention is nowadays given to questions of dynamics of cracks. A detailed consideration of these questions is beyond the scope of the present review; we shall confine ourselves here to a brief information about basic results achieved in theoretical investigations of dynamics of cracks.

In the paper by N. F. Mott [36] the crack-expansion process is treated in the case of an isolated straight crack in an infinite body subjected to a uniform field of tensile stresses p_0.* On the basis of dimensional analysis Mott obtained an expression for the kinetic energy of a body,

$$(6.18) \qquad \mathscr{E} = k\rho l^2 V^2 p_0^2 / E^2,$$

---

* Unlike [36], plane strain is here considered rather than the state of plane stress.

where ρ is the density of the body, l the half-length of the crack, V the rate of crack expansion, and k a dimensionless factor which Mott considered constant and left indefinite. Adding to the static-energy equation (2.1) the derivative with respect to l of the kinetic energy (6.18) and assuming the remaining terms in (2.1) to be the same as in Griffith's static problem, Mott found the rate of crack expansion

$$(6.19) \qquad V = \left[\frac{\pi(1-\nu^2)}{k}\right]^{1/2} \left(\frac{E}{\rho}\right)^{1/2} \left(1 - \frac{l_*}{l}\right),$$

where l_* is the half-length of the mobile-equilibrium crack defined by (5.4). Thus, as the crack propagates, its extension rate increases, approaching the limit

$$(6.20) \qquad V_0 = \left[\frac{\pi(1-\nu^2)}{k}\right]^{1/2} \left(\frac{E}{\rho}\right)^{1/2}.$$

The ultimate rate constitutes, according to Mott, a certain part of the longitudinal wave propagation velocity. In this reasoning the use of the static expression for the decrease of the strain energy W remains unfounded. Moreover, the quantity k in (6.18) and (6.19) need not be constant in general; it may depend on l/l^*, V/c_1 and other dimensionless combinations.

E. Yoffe [100], using the exact formulation of dynamic elasticity theory, investigated the problem of a straight crack of constant length, moving with constant velocity in an infinite body stretched by uniform stress at infinity. Notwithstanding the somewhat artificial character of the problem, an important result was obtained in this paper, which has quite a general meaning: If the crack propagation rate becomes greater than a certain critical rate, the direction of crack propagation is no longer the direction of maximum tensile stress, and the crack begins to curve. The magnitude of the critical rate V_1 is about $0.4\, c_1$, where c_1 is the longitudinal wave propagation velocity in the given material (the ratio V_1/c_1 depends slightly on Poisson's ratio ν of the material).

D. K. Roberts and A. A. Wells [101] made an attempt to evaluate the constant k, which remained indefinite in [36]. Using the value of k obtained, they found the ultimate crack expansion rate close to that found by Yoffe. However, their estimate, based as it is on the solution of a static problem, is too rough; and since the straight-line direction of crack propagation in [101] was assumed as certain, the close agreement between the critical rate found by Yoffe [100] and the ultimate rate obtained in [101] must be considered as incidental.

If the straight-line direction of crack propagation is somehow insured (for instance, by a large compression of the body in the direction of crack propagation or by the anisotropy of the material), then the maximum rate of crack propagation coincides with the velocity of propagation of Rayleigh surface waves in the given material, which is about $0.6\,c_1$.

The fact that the ultimate rate of crack propagation coincides with the Rayleigh velocity was first stated by A. N. Stroh [102]. The heuristic proof given in that paper amounts to the following. Stroh correctly notes that the ultimate rate of crack propagation does not depend on the surface energy of the body, and he assumes the surface energy to be zero. Proceeding from this, Stroh is led by energy considerations to the conclusion that the tensile stress near the crack end (on its prolongation) is equal to zero. Thus the crack may be thought of as a disturbance moving on a stress-free surface, which can propagate only with the Rayleigh velocity. In fact, from Stroh's reasoning it may only be concluded that the tensile stress at the very contour of the crack is equal to zero. From this fact, however, it does not follow that the rate of crack propagation is equal to the Rayleigh velocity, as can be seen from the following simple example. Take a body subjected to all-round compressive stress at infinity and wedged by a semi-infinite wedge as in Fig. 32, moving with infinitely small velocity. Forces of cohesion and, consequently, surface energy are assumed to be zero. In view of the infinitesimal velocity of the wedge, dynamic effects are insignificant, hence, according to Section III,2, the tensile stress at the crack end must vanish. At the same time, the rate of crack propagation is equal to the velocity of the wedge, i.e. it is also infinitesimal.

By arguments based on the analysis of exact solutions of the dynamic equations of elasticity, the conclusion that the ultimate rate of crack propagation is equal to Rayleigh velocity was drawn independently and simultaneously by several authors. I. W. Craggs [103] considered steady propagation of a semi-infinite straight crack with symmetrically distributed normal and shearing stresses applied on a part of the crack surface adjacent to the edge. In a paper by Dang Dinh An [104] a non-steady field of stresses and strains was investigated, acting in an infinite body with a semi-infinite crack, along the surface of which symmetrical concentrated forces normal to the crack surface begin to move suddenly away from the edge with constant velocity. Paper [95] examines the wedging of an infinite isotropic brittle body by a semi-infinite rigid wedge of arbitrary form, moving with constant velocity. In [84] a similar problem is treated for a case of an anisotropic body. B. R. Baker [105] considers a non-steady distribution of stresses and strains in a solid with a semi-infinite crack, at the surface of which constant normal stress is applied at the initial moment, after which the crack begins to expand with constant velocity.

From the various problems treated in these papers the following general result was obtained which led to our earlier conclusion: when the characteristic rate involved in the problem approaches the Rayleigh velocity, peculiar resonance phenomena arise. Note that the appearance of resonance when the Rayleigh velocity is approached is not specific for the problems of cracks: the investigation of the problem of a punch moving along the boundary of a half-space, carried out by L. A. Galin [72] and J. R. M. Radok [106], reveals [95] that the same resonance phenomena occur, when the velocity of the punch approaches the Rayleigh velocity. It appears that the limiting character of the Rayleigh velocity is most directly illustrated by a problem of wedging. Obviously the maximum possible rate of crack propagation can be reached in wedging a body by a moving wedge. The analysis of this problem shows [95] that with increasing velocity of the wedge the length of the free crack in front of the wedge decreases and tends to zero when the Rayleigh velocity is approached. For larger wedge velocity a free crack does not form in front of the wedge. Hence the maximum rate with which a crack can expand is equal to Rayleigh velocity.

K. B. Broberg [107, 108] treated the problem of a uniformly expanding crack of finite length in an infinite body subjected to a uniform tensile stress field. The solution obtained by Broberg is an asymptotic representation for great values of time of the solution of the problem treated by Mott [36] and Roberts and Wells [101]. However, unlike [101], Broberg's solution was obtained on the basis of the exact methods of the dynamic theory of elasticity. Independently of [102–104, 57, 95, 105] and in full accord with the results of these investigations, Broberg obtained that the rate of crack expansion in his problem, equal to the ultimate rate of crack expansion in the problem considered in [36, 101], coincides with the Rayleigh velocity.

Note the papers by B. A. Bilby and R. Bullough [109] and F. A. McClintock and S. P. Sukhatme [110] which treat uniformly moving cracks of finite and infinite length, respectively, at the surface of which symmetrical shearing stresses parallel to the crack edge were applied. Instead of plane strain we have in this problem what is often called *anti-plane strain*: one displacement component, parallel to the crack edge, is different from zero. The investigation of such cracks reduces to the solution of a single wave equation (reducing to Laplace's equation for equilibrium cracks). Cracks under anti-plane strain conditions are of considerable interest, being the simplest model for which an effective solution is possible for many problems, which are intractable for cracks under plane-strain conditions because of the great mathematical difficulties.

An analysis of the dynamics of crack propagation on the basis of the approximations of the simple beam theory was carried out by J. J. Gilman [11] and J. C. Suits [111].

ACKNOWLEDGEMENT

The author is very grateful to Prof. Ya. B. Zeldovitch and Prof. Yu. N. Rabotnov (USSR Academy of Sciences) and Dr. S. S. Grigorian for the invariable interest and attention given to his work on cracks and for a number of valuable advices. He recalls with appreciation the valuable discussions with Prof. S. A. Khristianovitch (USSR Academy of Sciences). The author considers it his pleasant duty to express his sincere thanks to Prof. G. Kuerti (USA) and Prof. G. G. Chernyi for the amiable assistance in writing this review. Credit is also given to I. A. Markuzon who assisted the author in compiling the bibliography.

References

1. INGLIS, C. E., Stresses in a plate due to the presence of cracks and sharp corners, *Trans. Inst. Nav. Arch.* 55, 219–230 (1913).
2. MUSKHELISHVILI, N. I., Sur l'intégration de l'équation biharmonique, *Izvestiya Ross. Akad. nauk* 13, 6 ser., 663–686 (1919).
3. GRIFFITH, A. A., The phenomenon of rupture and flow in solids, *Phil. Trans. Roy. Soc.* A 221, 163–198 (1920).
4. GRIFFITH, A. A., The theory of rupture, *Proc. 1st Intern. Congr. Appl. Mech.*, Delft, pp. 55–63 (1924).
5. FRENKEL, YA. I., The theory of reversible and irreversible cracks in solids, *Zhurn. tekhn. fiz.* 22, 1857–1866 (1952) (in Russian).
6. SMEKAL, A., Technische Festigkeit und molekulare Festigkeit, *Naturwiss.* 10, 799–804 (1922).
7. WOLF, K., Zur Bruchtheorie von A. Griffith, *Zeitschr. ang. Math. Mech.* 3, 107–112 (1923).
8. OBREIMOV, I. V., The splitting strength of mica, *Proc. Roy. Soc.* A 127, 290–297 (1930).
9. BENBOW, J. J., and ROESLER, F. C., Experiments on controlled fractures, *Proc. Phys. Soc.* B 70, 201–211 (1957).
10. METSIK, M. S., A photoelastic measurement of the splitting work of mica crystals, *Izvestiya VUZ, ser. fiz.* No. 2, 58–63 (1958) (in Russian).
11. GILMAN, J. J., Cleavage, ductility and tenacity in crystals, *in* "Fracture" (B. L. Averbach et al., eds.), pp. 193–221. Wiley, New York, 1959.
12. DAVIDENKOV, N. N., On the surface energy of mica, *Prikladna mekhanika* 6, No. 2, 138–142 (1960) (in Ukrainian).
13. WESTERGAARD, H. M., Bearing pressures and cracks, *J. Appl. Mech.* 6, No. 2, A–49–A–53 (1939).
14. SNEDDON, I. N., The distribution of stress in the neighborhood of a crack in an elastic solid, *Proc. Roy. Soc.* A 187, 229–260 (1946).
15. SNEDDON, I. N., "Fourier Transforms". McGraw Hill, New York, 1951.
16. SNEDDON, I. N., and ELLIOT, H. A., The opening of a Griffith crack under internal pressure, *Quart. Appl. Math.* 4, 262–267 (1946).
17. WILLIAMS, M. L., On the stress distribution at the base of a stationary crack, *J. Appl. Mech.* 24, 109–114 (1957).
18. MUSKHELISHVILI, N. I., "Some Basic Problems of the Mathematical Theory of Elasticity", 4th ed. Izd. AN SSSR, M.-L., 1954 (Transl. from 3rd ed., Nordhoof, Holland, 1953).
19. MUSKHELISHVILI, N. I., "Singular Integral Equations". GTTI, M.-L., 1946 (Transl. Nordhoof, Holland, 1953).

20. SACK, R. A., Extension of Griffith theory of rupture to three dimensions, *Proc. Phys. Soc.* 58, 729–736 (1946).
21. WILLMORE, T. J., The distribution of stress in the neighborhood of a crack, *Quart. Journ. Mech. Appl. Math.* 2, 53–64 (1949).
22. BOWIE, O. L., Analysis of an infinite plate containing radial cracks originating at the boundary of an internal circular hole, *J. Math. and Phys.* 25, 60–71 (1956).
23. IRWIN, G. R., Fracture dynamics, *in* "Fracturing of Metals", pp. 147–166. ASM, Cleveland, Ohio, 1948.
24. OROWAN, E. O., Fundamentals of brittle behavior of metals, *in* "Fatigue and Fracture of Metals" (W. M. Murray, ed.), pp. 139–167. Wiley, New York, 1950.
25. IRWIN, G. R., and KIES, J. A., Fracturing and fracture dynamics, *Weld. Journ. Res. Suppl.* 31, 95s–100s (1952).
26. IRWIN, G. R., and KIES, J. A., Critical energy rate analysis of fracture strength, *Weld. Journ. Res. Suppl.* 33, 193s–198s (1954).
27. OROWAN, E. O., Energy criteria of fracture, *Weld. Journ. Res. Suppl.* 34, 157s–160s (1955).
28. FELBECK, D. K., and OROWAN, E. O., Experiments on brittle fracture of steel plates, *Weld. Journ. Res. Suppl.* 34, 570s–575s (1955).
29. WELLS, A. A., The mechanics of notch brittle fracture, *Weld. Research* 7, 34r–56r (1953).
30. WELLS, A. A., The brittle fracture strength of welded steel plates, *Quart. Trans. Inst. Nav. Arch.* 98, 296–311 (1956).
31. WELLS, A. A., Strain energy release rates for fractures caused by wedge action, *NRL Rept.* No. 4705 (1956).
32. WINNE, D. H., and WUNDT, B. M., Application of the Griffith-Irwin theory of crack propagation to the bursting behavior of disks including analytical and experimental studies, *Trans. ASME* 80, 1643–1658 (1958).
33. BUECKNER, H. F., The propagation of cracks and the energy of elastic deformation, *Trans. ASME* 80, 1225–1230 (1958).
34. REBINDER, P. A., Physico-chemical study of deformation processes in solids, *in* "Yubileinyi Sbornik, Posviashchennyi XXX-letiu Velikoi Oktyabr'skoi Revolutsii", pp. 533–561. Izd. AN SSSR, M., 1947 (in Russian).
35. ELLIOT, H. A., An analysis of the conditions for rupture due to Griffith cracks, *Proc. Phys. Soc.* 59, 208–223 (1946).
36. MOTT, N. F., Fracture of metals; theoretical consideration, *Engineering* 165, 16–18 (1948).
37. RZHANITSYN, A. R., On the fracture process of the material in tension, *in* "Issledovaniya po Voprosam Stroitel'noi Mekhaniki i Teorii Plastichnosti" (A. R. Rzhanitsyn, ed.), pp. 66–83. Stroiizdat, M., 1956 (in Russian).
38. ZHELTOV, YU. P., and KHRISTIANOVITCH, S. A., On the mechanism of hydraulic fracture of an oil-bearing stratum, *Izvestiya AN SSSR, OTN*, No. 5, 3–41 (1955) (in Russian).
39. BARENBLATT, G. I., and KHRISTIANOVITCH, S. A., On roof stoping in mine excavations, *Izvestiya AN SSSR, OTN*, No. 11, 73–86 (1955) (in Russian).
40. BARENBLATT, G. I., On some problems of the theory of elasticity arising in investigating the mechanism of hydraulic fracture of an oil-bearing stratum, *Prikl. matem. i mekhan.* 20, 475–486 (1956) (in Russian).
41. BARENBLATT, G. I., On the formation of horizontal cracks in hydraulic fracture of an oil-bearing stratum, *Izvestiya AN SSSR, OTN*, No. 9, 101–105 (1956) (in Russian).
42. ZHELTOV, YU. P., An approximate evaluation of the size of a crack forming in hydraulic fracture of a stratum, *Izvestiya AN SSSR, OTN*, No. 3, 180–182 (1957) (in Russian).

43. ZHELTOV, YU. P., On the formation of vertical cracks in a stratum by means of a filtrating fluid, *Izvestiya AN SSSR, OTN*, No. 8, 56–62 (1957) (in Russian).
44. WESTERGAARD, H. M., Stresses at a crack, size of the crack and the bending of reinforced concrete, *J. Americ. Concrete Inst.* 5, No. 2, 93–102 (1933).
45. IRWIN, G. R., Analysis of stresses and strains near the end of a crack traversing a plate, *J. Appl. Mech.* 24, 361–364 (1957).
46. IRWIN, G. R., Relation of stresses near a crack to the crack extension force, *Proc. 9th Intern. Congr. Appl. Mech., Brussels* pp. 245–251 (1957).
47. IRWIN, G. R., Fracture, *in* "Handbuch der Physik", B. VI (S. Flügge, ed.), pp. 551–590. Springer, Berlin, 1958.
48. IRWIN, G. R., KIES, J. A., and SMITH, H. L., Fracture strength relative to onset and arrest of crack propagation, *Proc. Americ. Soc. Test. Mater.* 58, 640–657 (1958/1959).
49. KIES, J. A., SMITH, H. L., IRWIN, G. R., La mécanique des ruptures et son application aux travaux de l'ingénieur, *Mém. scient. rev. métallurgie* 57, No. 2, 101–117 (1960).
50. BUECKNER, H. F., Some stress singularities and their computation by means of integral equations, *in* "Boundary Problems in Differential Equations" (R. E. Langer, ed.), pp. 215–230. Univ. Wisconsin Press, 1960.
51. IRWIN, G. R., The crack-extension-force for a crack at a free surface boundary, *NRL Rept.* No. 5120 (1958).
52. ROMUALDI, J. P., and SANDERS, P. H., Fracture arrest by riveted stiffeners, *Proc. 4th Midwest. Conf. Solid Mech.*, Univ. Texas Press pp. 74–90 (1959/1960).
53. ROESLER, F. C., Brittle fracture near equilibrium, *Proc. Phys. Soc.* B 69, 981–992 (1956).
54. BENBOW, J. J., Cone cracks in fused silica, *Proc. Phys. Soc.* B 75, 697–699 (1960).
55. DROZDOVSKY, B. A., and FRIDMAN, YA. B., "The Influence of Cracks on the Mechanical Properties of Structural Steels". Metallurgizdat, M., 1960 (in Russian).
56. BARENBLATT, G. I., On the equilibrium cracks due to brittle fracture, *Doklady AN SSSR* 127, 47–50 (1959) (in Russian).
57. BARENBLATT, G. I., On the equilibrium cracks due to brittle fracture. Fundamentals and hypotheses. Axisymmetrical cracks, *Prikl. matem. i mekhan.* 23, 434–444 (1959) (in Russian).
58. BARENBLATT, G. I., On the equilibrium cracks due to brittle fracture. Straight-line cracks in flat plates, *Prikl. matem. i mekhan.* 23, 706–721 (1959) (in Russian).
59. BARENBLATT, G. I., On the equilibrium cracks due to brittle fracture. Stability of isolated cracks. Relation to energy theories, *Prikl. matem. i mekhan.* 23, 893–900 (1959) (in Russian).
60. BARENBLATT, G. I., On conditions of finiteness in mechanics of continua. Static problems in the theory of elasticity, *Prikl. matem. i mekhan.* 24, 316–322 (1960) (in Russian).
61. BARENBLATT, G. I., On some basic ideas of the theory of equilibrium cracks forming during brittle fracture, *in* "Problems of Continuum Mechanics" (M. A. Lavrent'ev, ed.), pp. 21–38. SIAM, Philadelphia, 1961.
62. ALEXANDROV, A. P., ZHURKOV, S. P., "The Phenomenon of Brittle Fracture". GTTI, M., 1933 (in Russian).
63. GARBER, R. I., and GUINDIN, I. A., Strength physics of crystalline bodies, *Uspekhi fizicheskikh nauk* 70, 57–110 (1960) (in Russian).
64. DERYAGIN, B. V., ABRIKOSOVA, I. I., and LIFSHITZ, E. M., Molecular attraction of condensed bodies, *Uspekhi fizicheskikh nauk* 64, 493–528 (1958) (in Russian).
65. DERJAGUIN, B. V., and ABRIKOSSOVA, I. I., Direct measurements of molecular attraction of solids, *J. Phys. Chem. Solids* 5, No. 1/2, 1–10 (1958).

66. JONGH, J. G. V., DE, "Measurements of Retarded van der Waals-Forces". Thesis, Abels, Utrecht, 1958.
67. BORN, M. and KUN HUANG, "Dynamical Theory of Crystal Lattices". Clarendon Press, Oxford, 1954.
68. LEIBFRIED, G., Gittertheorie der mechanischen und thermischen Eigenschaften der Krystalle, in "Handbuch der Physik", B. VII, I (S. Flügge, ed.), p. 104–324. Springer, Berlin, 1955.
69. LEONOV, M. YA., and PANASYUK, V. V., Development of the finest cracks in a solid, *Prikladna mekhanika* 5, 391–401 (1959) (in Ukrainian).
70. PANASYUK, V. V., Determination of stresses and strains near the finest crack, *Nauchn. zapiski In-ta mashinovedenia i avtomatiki AN USSR* 7, 114–127 (1960) (in Russian).
71. LANDAU, L. D., and LIFSHITZ, E. M., "Theory of Elasticity". Pergamon Press, London, 1960.
72. GALIN, L. A., "Contact Problems of the Theory of Elasticity". GITTL, M., 1953 (in Russian).
73. SHTAERMAN, I. YA., "Contact Problem of the Theory of Elasticity". GITTL, M., 1949 (in Russian).
74. PARKER, E. R., "Brittle Behavior of Engineering Structures". Wiley, New York, 1957.
75. DAVIDENKOV, N. N., "Impact Problem in Metallography". Izd. AN SSSR, M.-L., 1938 (in Russian).
76. PASHKOV, P. O., "Fracture of Metals". Sudpromgiz, L., 1960 (in Russian).
77. POTAK, YA. M., "Brittle Fracture of Steel and Steel Parts". Oborongiz, M., 1955 (in Russian).
78. AVERBACH, B. L., and oth. (eds.), "Fracture". Wiley, New York, 1959.
79. SEDOV, L. I., "Similarity and Dimensional Methods in Mechanics", 4th ed. GITTL, M., 1957 (Transl. from 4th ed., Acad. Press, New York, 1959).
80. NEUBER, H., "Kerbspannungslehre. Grundlagen für genaue Spannungsrechnung". Springer, Berlin, 1937.
81. FRANKLAND, I. M., Triaxial tension at the head of a rapidly running crack in a plate, *Paper ASME*, No. APM-11 (1959).
82. MASUBUCHI, K., Dislocation and strain energy release during crack propagation in residual stress field, *Proc. 8th Japan Nath. Congr. Appl. Mech.* pp. 147–150 (1958/1959).
83. STROH, A. N., Dislocations and cracks in anisotropic elasticity, *Phil. Mag.*, VIII ser. 3, 625–646 (1958).
84. BARENBLATT, G. I., and CHEREPANOV, G. P., On the equilibrium and propagation of cracks in an anisotropic medium, *Prikl. matem. i mekhan.* 25, 46–55 (1961) (in Russian).
84a. KACHANOV, L. M., On kinetics of crack growth, *Prikl. matem. i mekhan.* 25, 498–502 (1961) (in Russian).
85. WIGGLESWORTH, L. A., Stress distribution in a notched plate, *Mathematika* 4, 76–96 (1957).
86. SHERMAN, D. I., A method of solving the static plane problem of the theory of elasticity for multiconnected regions, *Trudy Seismologicheskogo Instituta Akademii Nauk SSSR*, No. 54 (1935) (in Russian).
87. MIKHLIN, S. G., The plane problem of the theory of elasticity, *Trudy Seismologicheskogo Instituta Akademii Nauk SSSR*, No. 65 (1935) (in Russian).
88. BARENBLATT, G. I., and CHEREPANOV, G. P., On the influence of body boundaries on the development of cracks of brittle fracture, *Izvestiya AN SSSR, OTN, ser. mekh. i mash.*, No. 3, 79–88 (1960) (in Russian). Correction. The same journal, No. 1 (1962).

89. KOITER, W. T., An infinite row of collinear cracks in an infinite elastic sheet, *Ingenieur-Archiv* 28, 168–172 (1959).
90. MUSKHELISHVILI, N. I., Basic boundary-value problems in the theory of elasticity for a plane with straight-line cuts, *Soobschenia AN Gruz. SSR* 3, 103–110 (1942) (in Russian).
91. SHERMAN, D. I., A mixed problem in the theory of potential and in the theory of elasticity for a plane with a finite number of straight-line cuts, *Doklady AN SSSR* 27, 330–334 (1940) (in Russian).
92. WUNDT, B. M., A unified interpretation of room-temperature strength of notched specimens as influenced by their size, *Paper ASME*, No. 59-MET-9 (1959).
93. YUSUFF, S., Fracture phenomena in metal plates, *Paper presented at the Xth Intern. Congr. Appl. Mech., Stresa* (1960).
94. USTINOV, YU. A., On the influence of the free boundary of a half-space on the crack propagation, *Izvestia AN SSSR, OTN, ser. mekh. i mash.*, No. 4, 181–183 (1959) (in Russian).
95. BARENBLATT, G. I., and CHEREPANOV, G. P., On the wedging of brittle bodies, *Prikl. matem. i mekhan.* 24, 667–682 (1960) (in Russian).
96. MUSKHELISHVILI, N. I., "Some Basic Problems of the Mathematical Theory of Elasticity", 2nd ed. Izd. AN SSSR, M.-L., 1935 (in Russian).
97. BITSADZE, A. V., On local deformations of elastic bodies in compression, *Soobschenia AN Gruz. SSR* 3, 419–424 (1942) (in Russian).
98. MARKUZON, I. A., On the wedging of a brittle body by a wedge of finite length, *Prikl. matem. i mekhan.* 25, 356–361 (1961) (in Russian).
99. KUZNETSOV, V. D., "Surface Energy of Solids". GITTL, M., 1954 (Transl., H. M. Stat. Office, London, 1957).
100. YOFFE, E., The moving Griffith crack, *Phil. Mag.*, VII ser. 42, 739–750 (1951).
101. ROBERTS, D. K., and WELLS, A. A., The velocity of brittle fractures, *Engineering* 178, 820–821 (1954).
102. STROH, A. N., A theory of the fracture of metals, *Advances in Physics* 6, 418–465 (1957).
103. CRAGGS, I. W., On the propagation of a crack in an elastic-brittle material, *J. Mech. Phys. Solids* 8, 66–75 (1960).
104. DANG DINH AN, Elastic waves by a force moving along a crack, *J. Math. and Phys.* 38, 246–256 (1960).
105. BAKER, B. R., Dynamic stresses created by a moving crack, *Paper presented at the Xth Intern. Congr. Appl. Mech., Stresa* (1960).
106. RADOK, J. R. M., On the solutions of problems of dynamic plane elasticity, *Quart. Appl. Math.* 14, 289–298 (1956).
107. BROBERG, K. B., The propagation of a brittle crack, *Paper presented at the Xth Intern. Congr. Appl. Mech., Stresa* (1960).
108. BROBERG, K. B., The propagation of a brittle crack, *Arkiv för Fysik* 18, 159–129 (1960).
109. BILBY, B. A., and BULLOUGH, R., The formation of twins by a moving crack, *Phil. Mag.*, VII ser. 45, 631–646 (1954).
110. MCCLINTOCK, F. A., and SUKHATME, S. P., Travelling cracks in elastic materials under longitudinal shear, *J. Mech. Phys. Solids* 8, 187–193 (1960).
111. SUITS, J. C., Cleavage, ductility and tenacity in crystals. Discussion, *in* "Fracture" (B. L. Anderson, and oth., eds.), pp. 223–224. Wiley, New York, 1959.

ON THE WESTERGAARD METHOD OF CRACK ANALYSIS*

G. C. Sih**

ABSTRACT

The Westergaard method of crack analysis, published almost thirty years ago, is shown to be invalid for a class of crack problems dealing with the infinite medium with cracks under applied loads at infinity. The necessary modifications of the Westergaard method are derived from the complex potential formulation of Muskhelishvili. The examples of a single line crack in an infinite plate owing to biaxial tension and pure shear are discussed.

INTRODUCTION

A survey of the literature on the analysis of crack problems shows that the Westergaard method[1] has been most frequently quoted and used by the practitioners in fracture mechanics for nearly thirty years. Surprisingly enough it has yet to be pointed out that the method in ref. 1 suffers severe restrictions for a class of problems dealing with the infinite medium with a crack (or cracks) subjected to external loads at infinity. These restrictions will be derived in the work to follow from the more general consideration of complex potentials originated by Muskhelishvili.[2]

In the theory of two-dimensional isotropic elasticity, the stresses and displacements may be expressed in terms of two complex functions $\phi(z)$ and $\psi(z)$ of the variable $z = x + iy$. They are

$$\sigma_x + \sigma_y = 4 \, \text{Re} \, [\phi'(z)]$$
$$\sigma_y - \sigma_x + 2i\tau_{xy} = 2[\bar{z}\phi''(z) + \psi'(z)] \qquad (1)$$

and

$$2\mu(u+iv) = \kappa\phi(z) - z\overline{\phi'(z)} - \overline{\psi(z)} \qquad (2)$$

where κ takes the value $3-4\nu$ for plane strain and $(3-\nu)/(1+\nu)$ for generalized plane stress and ν is the Poisson's ratio. The shear modulus is denoted by μ. Eqs. (1) and (2) may be simplified by introducing symmetry conditions as follows:

SYMMETRIC PROBLEMS

If the external loads are placed symmetrically with respect to the x-axis along which the cracks are situated, then the shearing stress τ_{xy} must vanish at $y=0$, i.e.,

$$\text{Im} \, [\bar{z}\phi''(z) + \psi'(z)] = 0, \qquad \text{for } y = 0 \qquad (3)$$

Eq. (3) can be satisfied by taking

$$\psi'(z) + z\phi''(z) + A = 0 \qquad (4)$$

where A is a real constant depending upon the applied load. Making use of eq. (4) and letting $\kappa = 3 - 4\nu$ for plane strain, the stresses take the form

---

* The results presented in this paper were obtained in the course of an investigation carried out under Contract Nonr-610(06) with the Office of Naval Research in Washington, D.C.
** Professor of Mechanics, Lehigh University, Bethlehem, Pa.

$$\sigma_x = 2 \text{ Re}\left[\phi'(z)\right] - 2y \text{ Im}\left[\phi''(z)\right] + A$$
$$\sigma_y = 2 \text{ Re}\left[\phi'(z)\right] + 2y \text{ Im}\left[\phi''(z)\right] - A$$
$$\tau_{xy} = -2y \text{ Re}\left[\phi''(z)\right] \tag{5}$$

and the displacements are

$$2\mu u = 2(1-2\nu) \text{ Re}\left[\phi(z)\right] - 2y \text{ Im}\left[\phi'(z)\right] + Ax$$
$$2\mu v = 4(1-\nu) \text{ Im}\left[\phi(z)\right] - 2y \text{ Re}\left[\phi'(z)\right] - Ay \tag{6}$$

Hence, the problem is reduced to the determination of a single complex function $\phi(z)$ satisfying the necessary boundary conditions. Eqs. (5) and (6) agree with eqs. (4-6) and (9-10) in ref. 1, respectively, only if

$$2 \phi'(z) = Z, \quad A = 0$$

In general, the constant A cannot be neglected arbitrarily. To illustrate this point, consider the problem of an infinite medium with a central crack of length 2a along the x-axis. The boundary conditions are

$$\sigma_y = \tau_{xy} = 0, \quad y = 0, \quad -a < x < a$$
$$\sigma_x = \epsilon\sigma^\infty, \quad \sigma_y = \sigma^\infty, \quad \tau_{xy} = 0, \text{ as } (x^2+y^2)^{1/2} \to \infty \tag{7}$$

The solution to this problem is given by (2)

$$\phi'(z) = (\sigma^\infty/2)\left[z/(z^2-a^2)^{1/2}\right] - (1-\epsilon)(\sigma^\infty/4)$$
$$\psi'(z) = (a^2\sigma^\infty/2)\left[z/(z^2-a^2)^{3/2}\right] + (1-\epsilon)(\sigma^\infty/2) \tag{8}$$

Inserting eq. (8) into (4), A is found to be

$$A = -(1-\epsilon)(\sigma^\infty/2)$$

Note that A vanishes only in the special case of $\epsilon=1$ corresponding to the case of uniform tension at infinity. The same applies to the problem of an infinite row of collinear cracks spaced periodically in an infinite medium[3].

SKEW-SYMMETRIC PROBLEMS

For loads applied skew-symmetrically with respect to the crack line, say along the x-axis, the normal stress σ_y is required to vanish at $y=0$, or

$$\text{Re}\left[2\phi'(z) + \bar{z}\phi''(z) + \psi'(z)\right] = 0, \text{ at } y = 0 \tag{9}$$

It follows that

$$\psi'(z) + 2\phi'(z) + z\phi''(z) + iB = 0 \tag{10}$$

where B is a real constant. Substituting eq. (10) into eqs. (1) and (2) and separating the real and imaginary parts give

$$\sigma_x = 4 \text{ Re}\left[\phi'(z)\right] - 2y \text{ Im}\left[\phi''(z)\right]$$
$$\sigma_y = 2y \text{ Im}\left[\phi''(z)\right] \tag{11}$$

$$\tau_{xy} = -2 \text{ Im}[\phi'(z)] - 2y \text{ Re}[\phi''(z)] - B$$

and

$$2\mu u = 4(1-\nu) \text{ Re}[\phi(z)] - 2y \text{ Im}[\phi'(z)] - By$$
$$2\mu v = 2(1-2\nu) \text{ Im}[\phi(z)] - 2y \text{ Re}[\phi'(z)] - Bx$$
(12)

The Westergaard version of eqs. (11) and (12) may be obtained by selecting an Airy stress function of the form $y \text{ Im } Z$. The results are the same as those given above if

$$2 \phi'(z) = Z, \quad B = 0$$

The restriction of B=0 leads to a trivial solution for the problem of uniform in-plane shear applied to an infinite medium containing a crack. For this problem, the conditions are

$$\sigma_y = \tau_{xy} = 0, \quad y = 0, \quad -a < x < a$$
$$\sigma_x = \sigma_y = 0, \quad \tau_{xy} = \tau^\infty, \quad \text{as } (x^2 + y^2)^{1/2} \to \infty$$
(13)

From ref. 2, the complex functions are

$$\phi'(z) = -(i\tau^\infty/2)[z/(z^2-a^2)^{1/2}] + i\tau^\infty/2$$
$$\psi'(z) = i\tau^\infty[z/(z^2-a^2)^{1/2}] - i(a^2\tau^\infty/2)[z/(z^2-a^2)^{3/2}]$$
(14)

The constant B may thus be found from eqs. (10) as

$$B = -\tau^\infty$$

Hence, B cannot vanish for a non-trivial solution.

It should be mentioned that the Westergaard method is valid for loads applied to the crack surfaces since in such cases the constants A and B have no contribution.

Received May 9, 1966

REFERENCES

1. H. M. Westergaard — J. Appl. Mech., 6, A49-53 (1937).
2. N. I. Muskhelishvili — Some Basic Problems of Mathematical Theory of Elasticity, P. Noordhoff Ltd., Groningen, Holland (1953).
3. I. N. Sneddon — Private communication.

RÉSUMÉ - La méthode de Westergaard d'analyse de la fissuration, qui fut publiée voici quelque trente années, n'est pas applicable à une classe de problèmes de fissuration où des charges à l'infini agissent sur un milieu infini comportant des fissures.

Les modifications qu'il convient d'apporter à la méthode de Westergaard sont déduites de la formule des potentiels complexes due à Muskhelishvili.

A titre d'exemples, on discute les cas d'une fissure simple, ou d'une linge de fissures simples, dans une plaque infinie, soumise à tension biaxiale d'une part, et à cisaillement pur d'autre part.

ZUSAMMENFASSUNG - Die Westergaard Methode der Rissanalyse, welche vor beinahe 30 Jahren veröffentlicht wurde, wird ungültig bewiesen für eine Klasse der Rissprobleme, die sich um das unendliche Mittel mit Anrissen unter unendlich entfernten Belastungen behandeln. Die nötigen Modifizierungen der Westergaard Methode sind von der komplex-potentiellen Formulation des Muskhelishvili abgeleitet. Die Beispiele eines Anrisses einer einzelnen Linie in einer unendlichen Platte infolge von zweiachsiger Spannung und reiner Scherbelastung werden besprochen.

On the Modified Westergaard Equations for Certain Plane Crack Problems

J. EFTIS AND H. LIEBOWITZ

School of Engineering and Applied Science, The George Washington University, Washington, D.C. (U.S.A.)

(Received March 31, 1972)

ABSTRACT

An error in Westergaard's equation for a certain class of plane crack problems, originally pointed out by Sih, is briefly discussed anew. The source of the difficulty is traced to an oversight in an earlier work by MacGregor, upon whose work Westergaard based his equations. Several examples of interest illustrating the consequences of the necessary correction to these equations are given.

1. Introduction

The Westergaard equations, which apply for a certain class of plane problems in linear elasticity, were shown to be generally incorrect by Sih in 1966, [1]. Specifically, by use of the well known Goursat–Kolosov complex representation of the plane problem, it was shown that the stress and displacement field equations appropriate to the restricted class of problems alluded to above include a real constant term which is lacking in the Westergaard equations.

In this paper the constant term which, according to Sih's analysis, should be appended to Westergaard's equations, is shown to be the result of an oversight in a lesser known work of MacGregor [2], upon whose work Westergaard based his formulations [3]. The consequences of the corrected equations are then demonstrated for several familiar plane crack problems, and for the approximate plane crack-tip stress and displacement field equations.

The problem of the centrally cracked strip of finite width loaded unaxially in uniform tension is also discussed. A Westergaard type stress function is introduced which provides an approximate closed form solution. This approximate solution has the merits of yielding the Fedderson secant formula for the crack-tip stress intensity factor, and for providing an analytical expression for the crack opening displacement which closely matches experimental data and which is a considerable improvement over the calculation first introduced by Irwin [4].

2. Modified Westergaard Equations

In MacGregor's complex characterization of the plane problem (omitting body force) the holomorphic functions

$$J(z) = \theta(x, y) + i\Omega(x, y)$$
$$H(z) = \theta_0(x, y) + i\Omega_0(x, y) \tag{1}$$

are introduced together with their derivatives

$$iJ'(z) \equiv iW(z) = \Phi(x, y) + i\Psi(x, y)$$
$$H'(z) \equiv -K(z) = -\Gamma(x, y) - i\Pi(x, y) . \tag{2}$$

The bi-harmonic Airy stress function $U(x, y)$ is represented as a linear combination of the single-valued harmonic functions θ and θ_0 by

$$U(x, y) = y\theta + \theta_0 = U(z, \bar{z}) = \frac{i(\bar{z}-z)}{2} \text{Im}\,[iJ(z)] + \text{Re}\,[H(z)] \tag{3}$$

The complex representation of the plane stress field is then readily shown to be

$$\sigma_{xx} = 2\Phi + y\frac{\partial \Phi}{\partial y} + \frac{\partial \Gamma}{\partial x} = 2\,\text{Re}\,[iW(z)] - y\,\text{Im}\,[iW'(z)] + \text{Re}\,[K'(z)]$$

$$\sigma_{yy} = -y\frac{\partial \Phi}{\partial y} - \frac{\partial \Gamma}{\partial x} = +y\,\text{Im}\,[iW'(z)] - \text{Re}\,[K'(z)]$$

$$\sigma_{xy} = -\Psi - y\frac{\partial \Phi}{\partial x} + \frac{\partial \Gamma}{\partial y} = -\text{Im}\,[iW(z)] - y\,\text{Re}\,[iW'(z)] - \text{Im}\,[K'(z)]. \tag{4}$$

For the restricted class of plane problems for which $\sigma_{xy}=0$ at all points along the line $y=0$, which includes plane crack problems in which the internal crack (or cracks) is situated along the x-axis and where the applied loads are symmetrically located with respect to the crack plane, it follows from (4) that

$$\Psi - \frac{\partial \Gamma}{\partial y} = \text{Im}\,[iW(z) + K'(z)] = 0. \tag{5}$$

Consequently

$$\Psi = \frac{\partial \theta}{\partial x} = \frac{\partial \Gamma}{\partial y} = -\frac{\partial \Pi}{\partial x}$$

from which it necessarily follows that

$$\frac{\partial \Phi}{\partial x} = -\frac{\partial}{\partial x}\left(\frac{\partial \Gamma}{\partial x}\right)$$

which is satisfied in the most general sense if one chooses

$$\Phi + A \equiv -\frac{\partial \Gamma}{\partial x} \tag{6}$$

or

$$\text{Re}\,[iW(z) + K'(z)] \equiv -A \tag{7}$$

everywhere. Here A is a real constant. The oversight in MacGregor's work rests in the fact that A was omitted or, put another way, was necessarily presumed to be zero. Substituting equations (5) through (7) into (4) and introducing

$$Z(z) \equiv iW(z) \tag{8}$$

one obtains

$$\sigma_{xx} = \text{Re}\,[Z(z)] - y\,\text{Im}\,[Z'(z)] - A$$

$$\sigma_{yy} = \text{Re}\,[Z(z)] + y\,\text{Im}\,[Z'(z)] + A$$

$$\sigma_{xy} = -y\,\text{Re}\,[Z'(z)] \tag{9}$$

which are the equations obtained by Sih when $Z(z) \equiv 2\phi'(z)$. Because the stress components are required to satisfy given boundary conditions the constant A will in general depend on the manner of the applied loading and will vanish only for rather special loading conditions.

The displacement field equations must likewise be corrected. In the Goursat–Kolosov representation the displacement field is specified by the well known form [5]

$$2\mu(u+iv) = \kappa\phi(z) - z\,\overline{\phi'(z)} - \overline{\psi(z)} \tag{10}$$

where $u(x,y)$ and $v(x,y)$ are respectively the x- and y-components of the displacement vector, $\mu = E/2(1+v)$ is the shear modulus, E and v are Young's Modulus and Poisson's Ratio respectively, and $\kappa = [3-v/1+v]$ for plane stress and $\kappa = [3-4v]$ for plane strain. The holomorphic functions $\phi(z)$ and $\psi(z)$ can be shown to be related to those introduced in equations (1), (2) and (3) by the relations

$$iW(z) = Z(z) = 2\phi'(z)$$
$$H(z) = X(z) + z\phi(z)$$
$$H'(z) = -K(z) = \psi(z) + \phi(z) + z\phi'(z)$$
$$X'(z) = \psi(z).\tag{11}$$

Adding to (10) its complex conjugate in the case of plane stress, one obtains

$$Eu = (3-v)\operatorname{Re}[\phi(z)] - (1+v)\{x\operatorname{Re}[\phi'(z)] + y\operatorname{Im}[\phi'(z)] + \operatorname{Re}[\psi(z)]\}.$$
$$Ev = (3-v)\operatorname{Im}[\phi(z)] + (1+v)\{x\operatorname{Im}[\phi'(z)] - y\operatorname{Re}[\phi'(z)] + \operatorname{Im}[\psi(z)]\}.\tag{12}$$

The Goursat–Kolosov equivalent of equations (5) and (7), with the help of (11), read

$$\operatorname{Im}[z\phi''(z) + \psi'(z)] = 0$$
$$\operatorname{Re}[z\phi''(z) + \psi'(z)] \equiv A$$

or

$$z\phi''(z) + \psi'(z) = A\tag{13}$$

everywhere. Integrating

$$Z\phi'(z) - \phi(z) + \psi(z) = Az + B$$

which is equivalent to the pair of equations

$$x\operatorname{Re}[\phi'(z)] - y\operatorname{Im}[\phi'(z)] - \operatorname{Re}[\phi(z)] + \operatorname{Re}[\psi(z)] = Ax$$
$$y\operatorname{Re}[\phi'(z)] + x\operatorname{Im}[\phi'(z)] - \operatorname{Im}[\phi(z)] + \operatorname{Im}[\psi(z)] = Ay.\tag{14}$$

The constant B, which must be real, can be omitted because its retention merely serves to add to the displacement field a term which represents a rigid body displacement. Upon combining (12) with (14)

$$Eu = 2(1-v)\operatorname{Re}[\phi(z)] - (1+v)2y\operatorname{Im}[\phi'(z)] - (1+v)Ax$$
$$Ev = 4\operatorname{Im}[\phi(z)] - (1+v)2y\operatorname{Re}[\phi'(z)] + (1+v)Ay.\tag{15}$$

To avoid confusion with the bar symbol used to denote complex conjugation let

$$2\phi(z) = \int Z(z)\,\mathrm{d}z \equiv \tilde{Z}(z)\tag{16}$$

where upon

$$Eu = (1-v)\operatorname{Re}[\tilde{Z}(z)] - (1+v)y\operatorname{Im}[Z(z)] - (1+v)Ax$$
$$Ev = 2\operatorname{Im}[\tilde{Z}(z)] - (1+v)y\operatorname{Re}[Z(z)] + (1+v)Ay\tag{17}$$

emerge as the modified Westergaard field equations for plane stress.

3. Applications

To illustrate use of the modified Westergaard equations it is worth while to treat anew the familiar problem of the infinite plate with colinear periodic cracks as shown in Fig. 1. The factor k is any real number.

Using the Kolosov equations [5]

$$\sigma_{xx} + \sigma_{yy} = 2\{\phi'(z) + \overline{\phi'(z)}\} = 4\operatorname{Re}[\phi'(z)]$$
$$\sigma_{yy} - \sigma_{xx} + 2\mathrm{i}\sigma_{xy} = 2\{\bar{z}\phi''(z) + \psi'(z)\}.\tag{18}$$

The boundary conditions can be expressed as follows: For all points situated on any crack border

$$\sigma_{yy} + \mathrm{i}\sigma_{xy} = 2\operatorname{Re}[\phi'(z)] + \{\bar{z}\phi''(z) + \psi'(z)\} = 0\tag{19}$$

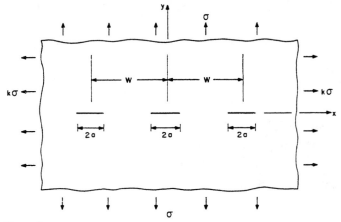

Figure 1.

At $|z| = \infty$

$$\sigma_{yy}(\infty) = \sigma, \quad \sigma_{xx}(\infty) = k\sigma, \quad \sigma_{xy}(\infty) = 0. \tag{20}$$

Due to the symmetry of the loading relative to the x-axis (13) must be satisfied. With $\bar{z}=z$ at $y=0$, (13) reduces (19) to

$$2\,\text{Re}[\phi'(z)] = -A \tag{21}$$

for all points on any crack border. In semi-inverse fashion, owing to the periodic and symmetric nature of the crack spacing, the function $2\phi'(z)$ can be chosen to have the form

$$2\phi'(z) \equiv \frac{g(z)}{\left\{\sin^2\left(\frac{\pi z}{W}\right) - \sin^2\left(\frac{\pi a}{W}\right)\right\}^{\frac{1}{2}}} - A \tag{22}$$

where the denominator of the first term has no real part along the crack borders. The function $g(z)$ is presumed to be holomorphic in the region of definition, except possibly at the point $z=\infty$, and must be such that $\text{Im}[g(x)]=0$ along the crack borders. The function $2\phi'(z)$ so defined satisfies boundary condition (21).

From boundary condition (20)

$$\sigma_{yy}(\infty) - \sigma_{xx}(\infty) + 2i\sigma_{xy}(\infty) = (1-k)\sigma = 4y\,\text{Im}[\phi''(z)] + 2A - 4iy\,\text{Re}[\phi''(z)]$$

from which

$$(1-k)\sigma = (\bar{z}-z)2\phi'(z) + 2A, \quad |z| \to \infty. \tag{23}$$

Inasmuch as $2\phi'(z)$ must be holomorphic throughout, including the point at infinity, it will therefore be continuous at and in the neighborhood of this point, and for $|z|$ arbitrarily large

$$2\phi'(z) \to \frac{g(z)}{\sin\left(\frac{\pi z}{W}\right)} - A$$

where upon

$$(\bar{z}-z)\left\{\frac{g'(z)}{\sin\left(\frac{\pi z}{W}\right)} - \frac{g(z)}{\sin^2\left(\frac{\pi z}{W}\right)} \cdot \frac{\pi}{W}\cos\frac{\pi z}{W}\right\} + 2A = (1-k)\sigma \tag{24}$$

which can be identically satisfied by choosing

$$g(z) = \sigma\sin\left(\frac{\pi z}{W}\right); \quad A = \tfrac{1}{2}(1-k)\sigma. \tag{25}$$

The condition that Im $[g(x)] = 0$ along the crack borders is also seen to be satisfied.

The stress function which solves this problem is thus

$$Z(z) = \frac{\sigma \sin\left(\frac{\pi z}{W}\right)}{\left\{\sin^2\left(\frac{\pi z}{W}\right) - \sin^2\left(\frac{\pi a}{W}\right)\right\}^{\frac{1}{2}}} - \tfrac{1}{2}(1-k)\sigma. \tag{26}$$

For uniaxial uniform tension applied in the y-direction, $k=0$ and $A = \sigma/2$. The stress function (26) then assumes a form equivalent to that given by Sanders [6]. When $k=1$, $A=0$, which corresponds to loading by equal uniform biaxial tension. The stress function introduced by Westergaard for this problem in reference [3] is therefore a solution only for this special loading condition.

As another illustration of consequence concerning this particular class of plane crack problems, consider the so-called crack-tip stress and displacement field equations. These can be obtained for opening mode crack surface displacements (mode I) by consideration of the problem of Fig. 1, modified to a single centrally located crack of length $2a$. A stress function which will satisfy the boundary conditions along such a cut has the form

$$2\phi''(z) = Z(z) = \frac{g(z)}{\{z^2 - a^2\}^{\frac{1}{2}}} - A.$$

Proceeding as in the previous example, it will turn out that $g(z) = \sigma z$ and $A = (1-k)\sigma/2$ so that

$$Z(z) = \frac{\sigma z}{\{z^2 - a^2\}^{\frac{1}{2}}} - (1-k)\frac{\sigma}{2}. \tag{27}$$

Introducing crack-tip polar coordinates (r, θ) through the coordinate transformation $\zeta = (z - a) = r e^{i\theta}$

$$Z(\zeta) = \frac{\sigma(\zeta + a)}{\{(\zeta + a)^2 - a^2\}^{\frac{1}{2}}} - (1-k)\frac{\sigma}{2}.$$

For $|\zeta|$ very small, i.e., $|\zeta| \ll a$

$$Z(\zeta) \cong \frac{K_I}{\{2\pi\zeta\}^{\frac{1}{2}}} - (1-k)\frac{\sigma}{2} \tag{28}$$

where

$$K_I \equiv \sigma\{\pi a\}^{\frac{1}{2}} \tag{29}$$

is the crack-tip stress intensity factor. Substituting (28) into (9), (16) and (17) one obtains for the plane stress crack-tip stress and displacement fields the approximations

$$\sigma_{xx} \cong \frac{K_I}{\{2\pi r\}^{\frac{1}{2}}} \cos\left(\frac{\theta}{2}\right)\left[1 - \sin\left(\frac{\theta}{2}\right)\sin\left(\frac{3\theta}{2}\right)\right] - (1-k)\sigma$$

$$\sigma_{yy} \cong \frac{K_I}{\{2\pi r\}^{\frac{1}{2}}} \cos\left(\frac{\theta}{2}\right)\left[1 + \sin\left(\frac{\theta}{2}\right)\sin\left(\frac{3\theta}{2}\right)\right]$$

$$\sigma_{xy} \cong \frac{K_I}{\{2\pi r\}^{\frac{1}{2}}} \sin\left(\frac{\theta}{2}\right)\cos\left(\frac{\theta}{2}\right)\cos\left(\frac{3\theta}{2}\right)$$

$$u \cong \frac{K_I}{\mu}\left\{\frac{r}{2\pi}\right\}^{\frac{1}{2}} \cos\left(\frac{\theta}{2}\right)\left[\left(\frac{1-v}{1+v}\right) + \sin^2\left(\frac{\theta}{2}\right)\right] - \frac{\sigma}{E}(1-k)\, r \cos\theta$$

$$v \cong \frac{K_I}{\mu}\left\{\frac{r}{2\pi}\right\}^{\frac{1}{2}} \sin\left(\frac{\theta}{2}\right)\left[\left(\frac{2}{1+v}\right) - \cos^2\left(\frac{\theta}{2}\right)\right] + \frac{v\sigma}{E}(1-k)r \sin\theta. \tag{30}$$

Again only when $k = 1$, i.e., equal uniform biaxial tensile loading, do these equations reduce to the form currently found in the literature [7].

To further illustrate use of the modified Westergaard equations consider the centrally cracked strip (plate) of finite width loaded uniaxially in uniform tension, Fig. 2, of great interest in fracture toughness testing, and which has not been given an exact closed form solution.

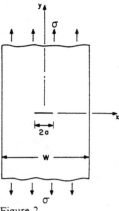
Figure 2.

A widely used approximate solution to this problem was first introduced by Irwin [4], by means of the stress function (26) with $k = 1$, which, as has been shown, is the exact solution to the periodic colinear crack problem in an infinite sheet loaded in uniform biaxial tension. To the stress field associated with this stress function Irwin adds a uniform horizontal compressive stress of magnitude σ along the vertical edges of the strip which, interestingly, has the effect of compensating for the missing A term. This combination partially satisfies boundary conditions along the vertical edges, leaving a horizontal stress of varying magnitude which depends on the relative crack size. The crack tip stress intensity factor emanating from this stress function is the so-called tangent formula

$$K_1 = \sigma \left\{ W \tan \left(\frac{\pi a}{W} \right) \right\}^{\frac{1}{2}} \tag{31}$$

Subsequently more accurate truncated series (polynomial) representations for K_1 have obtained by Isida [8] and Scrawley et al. [9], which show the tangent formula to be in varying degree of small error, depending on the crack size. Recently a secant formula has been proposed by Fedderson [9]

$$K_1 = \sigma \left\{ \pi a \sec \left(\frac{\pi a}{W} \right) \right\}^{\frac{1}{2}} \tag{32}$$

which matches almost identically Isida's K_1 values, deemed to be the most accurate. Having the added virtue of being concise and therefore relatively simple to use, Fedderson's secant formula has now in some quarters replaced the tangent formula in fracture toughness testing.

There will be some practical interest then in obtaining the corresponding stress function, that is, one which comes acceptably close to solving the problem of Fig. 2 and which yields the secant formula for K_1.

It is convenient to let

$$Z(z) \equiv Z^*(z) - A . \tag{33}$$

Then

$$\sigma_{xx} = \text{Re}[Z^*(z)] - y \, \text{Im}[Z^{*\prime}(z)] - 2A$$
$$\sigma_{yy} = \text{Re}[Z^*(z)] + y \, \text{Im}[Z^{*\prime}(z)] \tag{34}$$
$$\sigma_{xy} = -y \, \text{Re}[Z^{*\prime}(z)]$$

$$\tilde{Z}(z) = \int (Z^*(z) - A)\, dz = \tilde{Z}^*(z) - Az \tag{35}$$

and

$$Eu = (1-v)\,\text{Re}\,[\tilde{Z}^*(z)] - (1+v)\,y\,\text{Im}\,[Z^*(z)] - 2Ax$$
$$Ev = 2\,\text{Im}\,[\tilde{Z}^*(z)] - (1+v)\,y\,\text{Re}\,[Z^*(z)] + 2vAy. \tag{36}$$

A stress function which satisfies the crack border boundary condition, partially satisfies the vertical edge boundary condition and yields the secant formula for K_I has the form

$$Z(z) = Z^*(z) - A = \frac{\sigma\left\{\frac{\pi a}{W}\csc\left(\frac{\pi a}{W}\right)\right\}^{\frac{1}{2}}\sin\left(\frac{\pi z}{W}\right)}{\left\{\sin^2\left(\frac{\pi z}{W}\right) - \sin^2\left(\frac{\pi a}{W}\right)\right\}^{\frac{1}{2}}} - \tfrac{1}{2}\sigma\left\{\frac{\pi a}{W}\csc\left(\frac{\pi a}{W}\right)\right\}^{\frac{1}{2}} \tag{37}$$

For $|\zeta| \ll a$, where $\zeta = z - a = r\,e^{i\theta}$

$$Z^*(\zeta) \cong \frac{\sigma\left\{\frac{\pi a}{W}\csc\left(\frac{\pi a}{W}\right)\right\}^{\frac{1}{2}}}{\left\{\dfrac{\frac{2\pi\zeta}{W}\sin\left(\frac{\pi a}{W}\right)\cos\left(\frac{\pi a}{W}\right)}{\sin^2\left(\frac{\pi a}{W}\right) + \frac{2\pi\zeta}{W}\sin\left(\frac{\pi a}{W}\right)\cos\left(\frac{\pi a}{W}\right)}\right\}^{\frac{1}{2}}}$$

from which

$$Z^*(\zeta) \cong \frac{\sigma\left\{\pi a \sec\left(\frac{\pi a}{W}\right)\right\}^{\frac{1}{2}}}{\{2\pi\zeta\}^{\frac{1}{2}}} = \frac{K_I}{\{2\pi\zeta\}^{\frac{1}{2}}}. \tag{38}$$

Using (38) and

$$2A = \sigma\left\{\frac{\pi a}{W}\csc\left(\frac{\pi a}{W}\right)\right\}^{\frac{1}{2}} \tag{39}$$

in (34) through (36) will give the crack tip stresses and displacements as in equations (30), except that $(1-k)\sigma$ is replaced by $2A$ as given by (39).

The crack border condition

$$\sigma_{yy} = \sigma_{xy} = 0, \quad y = 0, \quad |x| < a$$

is seen to be satisfied by inspection. At $z = \tfrac{1}{2}W + iy$

$$Z^{*\prime}\left(\frac{W}{2} + iy\right) = i2A\frac{\pi}{W}\tanh\left(\frac{\pi y}{W}\right)\left[\frac{\sin\left(\frac{\pi a}{W}\right)}{\cosh\left(\frac{\pi y}{W}\right)}\right]^2\left\{1 - \left[\frac{\sin\left(\frac{\pi a}{W}\right)}{\cosh\left(\frac{\pi y}{W}\right)}\right]^2\right\}^{-\frac{1}{2}}$$

which has no real part. Thus $\sigma_{xy}(\tfrac{1}{2}W, y) = 0$ for all y. On the other hand

$$\sigma_{xx}\left(\frac{W}{2}, y\right) = 2A\left\{\frac{1 - \left[\dfrac{\sin\left(\frac{\pi a}{W}\right)}{\cosh\left(\frac{\pi y}{W}\right)}\right]^2\left[1 + \frac{\pi y}{W}\tanh\left(\frac{\pi y}{W}\right)\right]}{\left\{1 - \left[\dfrac{\sin\left(\frac{\pi a}{W}\right)}{\cosh\left(\frac{\pi y}{W}\right)}\right]^2\right\}^{\frac{1}{2}}} - 1\right\}. \tag{40}$$

(37) would be an exact solution if the right side of (40) were to vanish for all values of y, all other boundary conditions having been satisfied. Results of calculation of (40) are shown in Fig. 3. For small crack sizes, $(\pi a/W) \leq 0.3$, the right side of (40) gives values very close to zero along the entire vertical edge, having a maximum value of about four percent of the applied load at the crack plane when $(\pi a/W) = 0.3$. For $(\pi a/W) > 0.5$ the resulting horizontal boundary stress exceeds fifteen percent of the applied load at the crack plane. The pattern of this boundary stress distribution is interesting in that through Poisson's Ratio effects it tends to suppress vertical displacement of points situated just above and below the crack plane.

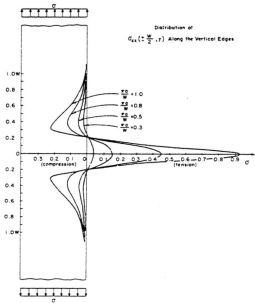

Figure 3.

Owing to the greater relative accuracy of the secant formula for K_I, one might expect that for small to moderate crack sizes, e.g., $(\pi a/W) < 0.5$, the stress function (37) will yield good estimates for other centrally located quantities such as the crack opening displacement, of interest in elastic compliance calibrations. For displacement gage points located along the plate center line, it follows from (36), after some calculation, that

$$\frac{E}{\sigma W} v(o, y) = \left\{ \frac{\pi a}{W} \csc\left(\frac{\pi a}{W}\right) \right\}^{\frac{1}{2}} \left[\frac{2}{\pi} \cosh^{-1} \left[\frac{\cosh\left(\frac{\pi y}{W}\right)}{\cos\left(\frac{\pi a}{W}\right)} \right] \right.$$

$$\left. - \frac{(1+v)}{W} y \left\{ 1 + \left[\frac{\sin\left(\frac{\pi a}{W}\right)}{\sin\left(\frac{\pi y}{W}\right)}\right]^2 \right\}^{-\frac{1}{2}} + \frac{v}{W} y \right]. \quad (41)$$

Calculation of (41) is compared with experimental data obtained from Alum. 7075–T6 center cracked sheets, reported in reference [10], and shown in Fig. 4. The data points defining the experimental curve were obtained in the low load or elastic range. The predicted crack opening displacement, eq. (41), is a considerable improvement over Irwin's calculation, and is surprisingly close to the experimental curve in the large crack size range where the vertical edge boundary condition is poorly approximated. The fact that the predicted compliance curve lies

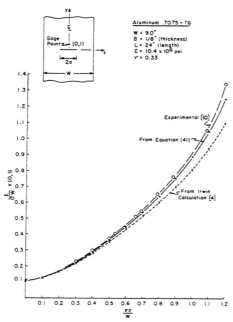

Figure 4.

entirely below the experimental curve appears to be explainable by the particular nature of the distribution of the excess of vertical edge boundary stress shown in Fig. 3. Imposition of an identical distribution along the vertical edges, but reversed in sense, (leaving those edges free of traction as they should be) would tend to increase somewhat the vertical displacement from that given by (41) for all points a little above and below the crack plane and would thereby elevate the curve of (41).

Acknowledgment

The authors wish to acknowledge support for this work extended by the NASA-Langley Research Center through NASA Grant NGR 09–010–053.

REFERENCES

[1] G. C. Sih, On The Westergaard Method of Crack Analysis, *Journ. Fract. Mech.*, 2 (1966).
[2] C. W. MacGregor, The Potential Function Method for The Solution of Two-Dimensional Stress Problems, *Trans. Am. Math. Soc.*, 38 (1935).
[3] H. M. Westergaard, Bearing Pressures and Cracks, *Journ. Appl. Mech.*, 6 (1939).
[4] G. R. Irwin, *Fracture Testing of High-Strength Sheet Materials Under Conditions Appropriate for Strength Analysis*, NRL Report 5486, 1960.
[5] I. S. Sokolnikoff, *Mathematical Theory of Elasticity*, McGraw-Hill, New York (1956).
[6] J. L. Sanders, Jr., On The Griffith-Irwin Fracture Theory, *Journ. Appl. Mech.*, (27) 1960.
[7] P. C. Paris and G. C. Sih, *Stress Analysis of Cracks*, STP 381, ASTM (1965).
[8] M. Isida, The Effect of Longitudinal Stiffness in a Cracked Plate Under Tension, *Proc. 4th U.S. Cong. Appl. Mech.*, (1962).
[9] W. F. Brown, Jr., and J. E. Srawley, *Plane Stress Crack Toughness Testing of High Strength Metallic Materials*, STP 410, ASTM, 1969.
[10] H. Liebowitz and J. Eftis, On Nonlinear Effects in Fracture Mechanics, *Journ. Engr. Fract. Mech.*, 3 (1971).

RÉSUMÉ

On discute brièvement une erreur, qui a été signalée à l'origine par Sih, dans l'équation de Westergaard relative à une certaine catégorie de problèmes de fissures planes.

L'origine de la difficulté réside dans une surestimation faite par MacGregor dans un travail antérieur, sur lequel Westergaard a basé ses équations.

On donne plusieurs exemples intéressants qui illustrent les conséquences des corrections qu'il est nécessaire d'apporter à ces équations.

ZUSAMMENFASSUNG

Ein Fehler in Westergaard's Formel für eine gewisse Klasse von Flächenrißproblemen, auf den schon Sih hingewiesen hatte, wird besprochen.

Der Ursprung dieser Schwierigkeit liegt in einem Versehen das McGregor in einer früheren Arbeit unterlaufen ist, auf welcher Westergaard's Formeln aufgebaut sind.

Die Konsequenzen die sich aus den für diese Formeln erforderlichen Korrekturen ergeben, werden an Hand verschiedener interessanter Beispiele nachgewiesen.

A CRITICAL RE-EXAMINATION OF THE WESTERGAARD METHOD FOR SOLVING OPENING-MODE CRACK PROBLEMS

R. J. Sanford
Ocean Technology Division
Naval Research Laboratory
Washington, D.C., 20375

(Received and accepted for print 17 July 1979)

Introduction

The Westergaard [1] complex stress function technique for solving opening mode crack problems has played an important role in the development of linear elastic fracture mechanics. Irwin [2] used Westergaard's solution for an internal crack subjected to uniform biaxial stress at infinity to derive approximate equations for the stresses in the neighborhood of a crack tip in which the "stress intensity factor" concept was introduced. Following this development, there was a period of activity in which the Westergaard method was used extensively to derive stress intensity factors for various geometries (see, for example, the compilation by Paris and Sih [3] and references therein). Early on, Irwin [4] modified the Westergaard equations to include an arbitrary uniform stress in the direction parallel to the crack. An analytical review of this modification was later provided by Sih [5] and Eftis and Liebowitz [6] by relaxing a boundary condition imposed by Westergaard. In the sequel it will be shown that these "modified Westergaard equations" are still too restrictive to solve a broad class of opening mode crack problems of technical interest; however, there is a further relaxation of the boundary conditions which does permit solution of these problems.

History of the Method

In his original paper, Westergaard demonstrated that "in a restricted but important group of cases the normal stresses and the shearing stress in the directions x and y can be stated in the form" [1]:

$$\sigma_x = \text{Re} Z(z) - y \text{Im} Z'(z) \tag{1}$$

$$\sigma_y = \operatorname{Re} Z(z) + y \operatorname{Im} Z'(z) \qquad (2)$$

$$\tau_{xy} = -y \operatorname{Re} Z'(z) \qquad (3)$$

where

$$Z = Z(z) = Z(x+iy) = \operatorname{Re} Z + i \operatorname{Im} Z \qquad (4)$$

and

$$Z'(z) = \frac{dZ}{dz} = \operatorname{Re} Z' + i \operatorname{Im} Z' \qquad (5)$$

For traction-free cracks located along the x-axis and subjected to symmetric loads, the stress function, Z, is chosen such that;

$$\operatorname{Re} Z(z) = 0 \qquad (6)$$

over the domain of the crack (to ensure that $\sigma_y = 0$). Note that the other boundary condition, $\tau_{xy} = 0$, along the entire x-axis is automatically satisfied from the form of eq. (3).

In 1958, Irwin in a discussion [4] of the experimental work of Wells and Post [7] noted that the equations of Westergaard could be modified to include an arbitrary uniform stress in the x-direction, σ_{ox}; thus, the modified Westergaard equations became

$$\sigma_x = \operatorname{Re} Z - y \operatorname{Im} Z' - \sigma_{ox} \qquad (7)$$

$$\sigma_y = \operatorname{Re} Z + y \operatorname{Im} Z' \qquad (8)$$

$$\tau_{xy} = -y \operatorname{Re} Z' \qquad (9)$$

The inclusion of the non-singular stress, σ_{ox}, was necessary to explain the tilt of the isochromatic fringe loops away from the normal in the work of Wells and Post [7]. The Irwin modification, in conjunction with the near field stress equations, has formed the basis for the analysis of photoelastic fringe patterns in the neighborhood of a crack tip from which the stress intensity factor has been obtained for a variety of geometries. A review of some of these methods can be found in references [8, 9, 10]. Sih [5] starting with the Goursat-Kolosov complex representation of the plane problem [11], i.e.:

$$\sigma_x + \sigma_y = 4 \operatorname{Re}[\phi'(z)] \qquad (10)$$

$$\sigma_y - \sigma_x + 2i\tau_{xy} = 2[\bar{z}\phi''(z) + \psi'] \qquad (11)$$

showed that the symmetry condition $\tau_{xy} = 0$ on $y = 0$ could be satisfied by a less restrictive assumption than employed by Westergaard and obtained the following expressions for the stresses:

$$\sigma_x = 2\operatorname{Re}[\phi'(z)] - 2y \operatorname{Im}[\phi''(z)] + A \qquad (12)$$

$$\sigma_y = 2\text{Re}[\phi'(z)] + 2y\text{Im}[\phi''(z)] - A \qquad (13)$$

$$\tau_{xy} = -2y\text{Re}[\phi''(z)] \qquad (14)$$

Where, according to Sih's analysis, A is a real constant depending on the applied load. Here, as in the modified Westergaard equations, the problem is reduced to finding a single stress function which satisfies the boundary conditions. Eftis and Liebowitz [6] showed that Sih's equations (12-14) are equivalent to the modified Westergaard equations (7-9) if

$$2\phi'(z) = Z(z) + A \qquad (15)$$

where

$$2A = -\sigma_{ox} \qquad (16)$$

Recently Tada, Paris and Irwin [12] suggested that a mode I stress function of the form:

$$Z(z) = \frac{K}{\sqrt{2\pi z}} + \sum_{n=1}^{N} A_n z^{n-1/2} \qquad (17)$$

be used in conjunction with the boundary collocation method to solve opening mode crack problems with finite boundaries. At this point it is instructive to derive the equation for the maximum shear stress from the modified Westergaard equations, eq. (7-9). Recall that

$$(\tau_m)^2 = \left(\frac{\sigma_x - \sigma_y}{2}\right)^2 + (\tau_{xy})^2 \qquad (18)$$

thus

$$(\tau_m)^2 = y^2 Z' \cdot \overline{Z}' + y\sigma_{ox}\text{Im}Z' + \left(\frac{\sigma_{ox}}{2}\right)^2 \qquad (19)$$

where $\overline{Z}$ = complex conjugate of Z. In particular, along the axis of symmetry, $y=0$, eq. (19) reduces to:

$$2\tau_m = |\sigma_{ox}| \qquad (20)$$

Thus, if the modified Westergaard eqs. are valid, the photoelastic fringe order (proportional to $2\tau_m$) ahead of the crack is constant FOR ANY STRESS FUNCTION Z. Clearly, this is not always the case. For example, for cracks approaching a boundary as in Figure 1 of a compact tension specimen isochromatic fringe loops typically form ahead of the crack. This counter-example raises questions about the validity of the modified Westergaard eqs. in general, and on the use of the series stress function (eq. 17) in particular, for solving problems in which the boundary or stress gradient ahead of the crack can be expected to play a significant role.

Re-examination of the Problem

With these historical results and observations as background it is instructive to re-examine the derivation of the original Westergaard equations and subsequent modified Westergaard equations. For this purpose the Goursat-Kolosov formulations used by Sih will be used (eq. 10-11). Separating eq. (11) into real and imaginary parts yields:

$$\tau_{xy} = x\text{Im}\phi'' - y\text{Re}\phi'' + \text{Im}\psi' \qquad (21)$$

Imposing the symmetry condition on $y=0$ results in the following relation which must be satisfied:

$$x\text{Im}\phi'' + \text{Im}\psi' = 0 \quad \text{on} \quad y = 0 \qquad (22)$$

Let:

$$\eta(z) = z\phi''(z) + \psi'(z) \qquad (23)$$

then, the symmetry condition of eq. (22) is equivalent to;

$$\text{Im}\eta(z) = 0 \quad \text{on} \quad y = 0 \qquad (24)$$

Substituting eq. (23) into eq. (11) yields:

$$\sigma_y - \sigma_x + 2i\tau_{xy} = 2[(\bar{z} - z)\phi''(z) + \eta(z)] \qquad (25)$$

and

$$\sigma_x = 2\text{Re}\phi' - 2y\,\text{Im}\phi'' - \text{Re}\eta \qquad (26)$$
$$\sigma_y = 2\text{Re}\phi' + 2y\,\text{Im}\phi'' + \text{Re}\eta \qquad (27)$$
$$\tau_{xy} = -2y\,\text{Re}\phi'' + \text{Im}\eta \qquad (28)$$

The symmetry condition of eq. (24) can be satisfied by any one of three conditions dependent upon the degree of constraint to be placed on $\eta(z)$. Each case will be examined individually in order of increasing generality.

Case 1:
$$\eta(z) = 0 \quad \text{for all } z \qquad (29)$$

Setting $2\phi' = Z(z)$ yields:

$$\sigma_x = \text{Re}Z - y\text{Im}Z'$$
$$\sigma_y = \text{Re}Z + y\text{Im}Z'$$
$$\tau_{xy} = -y\,\text{Re}\,Z'$$

These are the Westergaard equations.

Case 2: $\quad\quad \eta(z) = A$, a real constant for all z $\hfill (30)$

Setting $2\phi' = Z(z) - A$, yields

$$\begin{aligned}\sigma_x &= \text{Re } Z - y\text{Im} Z' - 2A \\ \sigma_y &= \text{Re} Z + y\text{Im} Z' \\ \tau_{xy} &= -y \text{ Re} Z'\end{aligned} \quad\quad (31)$$

which can be recognized as Irwins's modified Westergaard equations with $\sigma_{ox} = 2A$. Note that this case is equivalent to the condition: $\text{Im}\eta(z) = 0$ for all z.

Case 3: $\quad\quad \text{Im}\eta(z) = 0 \quad \text{on } y = 0 \hfill (32)$

Setting $2\phi' = Z(z) - \eta(z)$

yields
$$\sigma_x = \text{Re} Z - y\text{Im} Z' + y\text{Im}\eta' - 2\text{Re}\eta \quad\quad (33)$$
$$\sigma_y = \text{Re} Z + y\text{Im} Z' - y\text{Im}\eta' \quad\quad (34)$$
$$\tau_{xy} = -y\text{Re} Z' + y \text{ Re}\eta' + \text{Im}\eta \quad\quad (35)$$

For each of the above cases, the stress function $Z(z)$ is the familiar Westergaard stress function for the geometry under consideration and must satisfy the condition of eq. (6). Similarly, $\eta(z)$ must satisfy eq. (24). As a complement to the series stress function, eq. (17), a suitable function for $\eta(z)$ is:

$$\eta(z) = \sum_{m=0}^{M} \alpha_m z^m \quad\quad (36)$$

With this choice of $\eta(z)$, the photoelastic fringe order ahead of the crack is of the form:

$$2\tau_m = 2\left|\sum_{m=0}^{M} \alpha_m x^m\right| \quad\quad (37)$$

Clearly, of the three possible constraint conditions, only the "generalized" Westergaard Eq. (33-35) can provide the flexibility to solve plane crack problems such as that shown in Fig. 1 in which the photoelastic fringe order varies ahead of the crack. Moreover, the order of the polynomial, M, necessary to adequately describe the stress state can be estimated from a comparison of a plot of the fringe order ahead of the crack with eq. (37).

Figure 1 — Isochromatic pattern for a crack approaching a boundary in a compact tension specimen ($a/W = 0.75$).

Acknowledgments

The research reported herein was supported by the Office of Naval Research, Mechanics subelement, project RR023-03-45. During the period in which this research was performed the author was a Visiting Research Associate in the Mechanical Engineering Department, University of Maryland, under the terms of the Advanced Graduate Training Program administered by the Naval Research Laboratory. The author wishes to express his thanks to Dr. G. R. Irwin for stimulating the author's interest in this problem.

References

1. H.M. Westergaard, J. Appl. Mech., **6**, A-49 (1939).
2. G.R. Irwin, Handbuch der Physik, Vol. VI, p. 558, Springer, Berlin (1958).
3. P.C. Paris and G.C. Sih, Fracture Toughness Testing and its Applications, STP 381, p. 30. ASTM, Phil. PA (1964).
4. G.R. Irwin, discussion of ref. 7, Proc. SESA, **16**, 93 (1958).
5. G.C. Sih, Int. J. Fract. Mech., **2**, 628 (1966).
6. J. Eftis and H. Liebowitz, Int. J. Fract. Mech., **8**, 383 (1972).
7. A.A. Wells and D. Post. Proc. SESA, **16**, 69 (1958).
8. A.S. Kobayashi, Experimental Techniques in Fracture Mechanics, SESA Monograph #1, Westport, Conn. (1973).
9. A.S. Redner, Fracture Mechanics and Technology, p. 607, Noordhoff Int. (1978).
10. J.M. Etheridge and J.W. Dally, Exp. Mech., **17**, 248 (1977).
11. N.I. Muskhelishvili, Some Basic Problems of the Mathematical Theory of Elasticity, p. 112, Noordhoff Ltd., Netherlands (1953).
12. H. Tada, P. Paris and G. Irwin, The Stress Analysis of Cracks Handbook, p. 1.27, Del Research Corp., Hellertown, PA (1973).

INFLUENCE OF NON-SINGULAR STRESS TERMS AND SPECIMEN GEOMETRY ON SMALL-SCALE YIELDING AT CRACK TIPS IN ELASTIC–PLASTIC MATERIALS

By S. G. Larsson[†] and A. J. Carlsson

Department of Strength of Materials and Solid Mechanics,
The Royal Institute of Technology, Stockholm, Sweden.

(*Received 23rd October* 1972)

Summary

The plane strain elastic–plastic state at a crack tip is determined for compact tension, bend, double edge-cracked and centre-cracked specimens using a finite element method with triangular constant-strain elements. The solutions are found to differ by 10 to 30 per cent at the ASTM-limit as regards fracture surface displacement, normal stress and plastic zone size. In order to bring the boundary layer solution for the crack problem into agreement with the solution for a specific specimen one has to modify this solution. The modification consists of an addition to the boundary tractions for the boundary layer problem of tractions corresponding to the non-singular, constant second term in a series expansion of the normal stress parallel to the crack plane.

1. Introduction

One of the basic assumptions behind the application of linear elastic fracture mechanics to elastic–plastic materials is that plastic deformation at the crack tip is governed by the intensity of the elastic stress singularity, that is, the stress intensity factor K_I. This is considered to be true if the plastic zone size is small compared to other geometric dimensions of the problem, such as, for example, crack length a. The extent of the plastic zone, $2r_p$, is approximately $2r_p = (K_I/Y)^2/3\pi$ for plane strain, where Y is the tensile yield strength. A requirement of a maximum allowable K_I, for example, $K_I < Y\sqrt{a}/\sqrt{2.5}$ as in ASTM (1970), thus guarantees that the plastic zone size is much smaller than the crack length.

If it is true that the state at the crack tip in an elastic–plastic material at low load levels is determined by the stress intensity factor, then the crack problem can be solved by using a boundary layer approach; that is, assuming that the boundary value stresses of the elastic–plastic crack problem are given by the extension of the validity of the singular term in the elastic stress solution,

$$\sigma_{ij} = \frac{K_I}{(2\pi r)^{1/2}} f_{ij}(\Theta), \qquad (1)$$

to large values of r and small-scale yielding (Rice, 1968). In (1), r and Θ are polar co-ordinates referred to the crack tip, and the functions f_{ij} are given by the elastic solution.

---

[*] Now at Saab-Scania AB, Trollhättan, Sweden.

In the present investigation the elastic–plastic problem has been solved for cracks in different types of specimens using the actual boundary conditions. It is then found that these solutions for the state at the crack tip cannot be related to the boundary layer solution via the stress intensity factor K_I alone even at load levels significantly below the ASTM-limit. Already, initiation of yielding (that is, in this case plastic yielding of the first element) occurs at different K_I-values for different specimens. This shows that in a series expansion of plastic zone size R in terms of the ratio of stress intensity factor to yield strength,

$$R = \alpha_1(K_I/Y)^2 + \ldots,$$

terms other than the leading one become important already when R is of the order of magnitude of the element dimension.† In order to get good agreement between the boundary layer solution and the actual solutions one has to modify the boundary layer solution. The modification consists of an addition to the boundary tractions of the boundary layer problem of tractions corresponding to the non-singular term of the x-direction stress for the actual geometry. The magnitude of this stress is found from the solution obtained by the elastic finite element method for that geometry. Throughout a large region behind and in front of the crack tip, this stress is independent of the x-coordinate. In this way, using a two-parameter description of the problem, the solutions for the different specimens can be brought into exact agreement up to load levels slightly above the ASTM-limit.

The solutions considered here are restricted to a linearly elastic, ideally-plastic material obeying the von Mises yield condition and flow rule.

2. Finite Element Method and Element Model

A finite element method program developed by Härkegård and Larsson (1973) at the Royal Institute of Technology in Stockholm for the analysis of elastic–plastic structures loaded in plane strain has been used in the present investigation. The program is based on an elastic–plastic constitutive matrix obtained through inversion of the Prandtl–Reuss equations for a material obeying the von Mises yield condition and flow rule (Yamada, Yoshimura and Sakurai, 1968). This makes an incremental treatment of elastic–plastic problems possible. Triangular constant-strain elements are used in the computation.

The computational procedure is as follows. First, the elastic solution which corresponds to incipient plastic yielding of the most highly-stressed element is determined. Then, the load is changed incrementally in such a way that at most one element becomes plastic at each load increment. A second restriction on the load increments is that they must not be larger than that corresponding to $\Delta K_{I\max} = 0\cdot 01\ Y\sqrt{a}$. The procedure is the same for the computations with actual geometries and with the modified boundary layer problems. As shown in Fig. 1, good correspondence is obtained between values of effective stress and effective plastic strain determined in this way (open circles) and the stress–strain curve of the material (drawn curve). The values given by the half-filled circles were obtained by choosing the load increments in such a way that one element became plastic at each load increment. For the

† That is, about one-half per cent of crack length a.

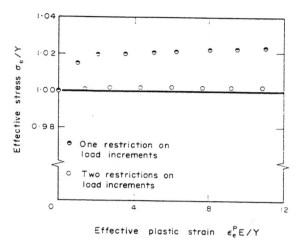

Fig. 1. Correlation between stress–strain relation as assumed and as determined from computed effective stress and strain.

individual stress and strain values the insertion of a maximum allowable load increment meant a change of at most 3 per cent.

The geometries studied are shown in Fig. 2; due to symmetry only the shaded parts need to be considered. The specimens are characterized by the dimensionless parameters $W/a = 2.0$, $v = 0.3$, $E/Y = 400.0$, where W is a relevant width dimension according to Fig. 2, a is crack length or half crack-length, v is Poisson's ratio, and E is Young's modulus. In all cases the finite element method representation shown in Fig. 3 was used for the region close to the crack tip. The region inside $r_0 = 0.024\,a$ is made up of 48 elements of approximately equal size. For larger radii the elements are determined by 15 concentric circles and 13 radial rays at equal angular intervals $\alpha = 15°$ so that the elements are almost equilateral. The radii of the first 14 circles are $r_i = (1+\alpha)^i r_0$ and the radius of the last is $r = 0.8\,a$. The number of elements in the region common to the different cases is 423 and the number of degrees of freedom is 476. In addition, up to 140 elements are used to form the remaining parts of the plates. The complete element configuration for the CT-specimen is shown in Fig. 4.

The computations were made on an IBM 360/75 and each load increment required an average computing time of 6.3 s. The number of load increments was about 200 which corresponds to a total computing time of 21 min.

3. Elastic Solution

3.1 Actual specimen geometries

For all specimens and for the boundary layer case yielding begins in the element situated just above the crack tip, element b in Fig. 3. This occurs at the values of the stress intensity factor given in Table 1, where for the different specimens K_I is determined according to Isida (1971) and Bowie (1964). Area-average stresses obtained from the analytical elastic solution of (1) by integration over an area corresponding to the elements in Fig. 3 show that the above-mentioned element should yield first, although for a somewhat higher value of the stress intensity factor, namely,

$$K_I = 0.174\, Y\sqrt{a}.$$

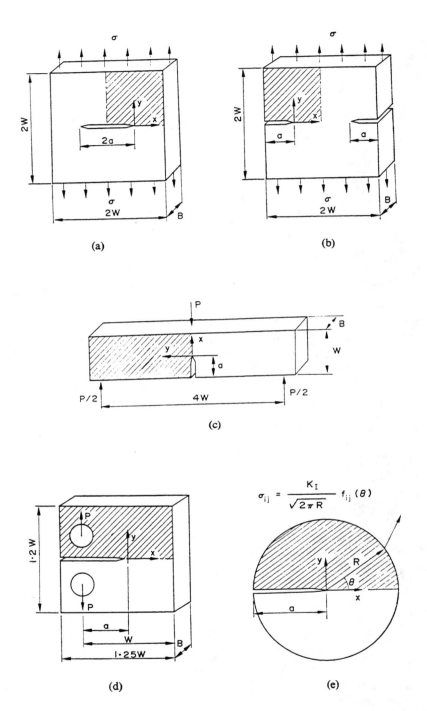

FIG. 2. Geometries studied (for symmetry reasons only shaded parts are considered): (a) Centre-cracked specimen; (b) double edge-cracked specimen; (c) bend specimen; (d) compact tension specimen; and (e) boundary layer approach.

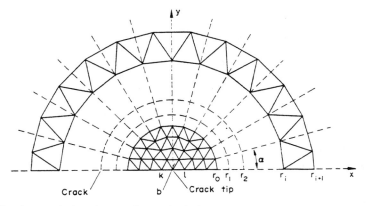

Fig. 3. Finite element design close to the crack tip for all geometries. $r_0 = 0.024\,a$; 423 elements inside $r = 0.8\,a$.

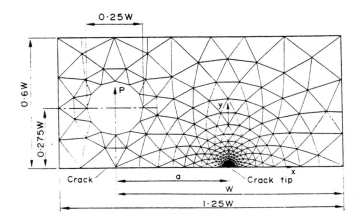

Fig. 4. Element configuration for CT-specimen.

TABLE 1. *Values of stress intensity factor for first plastic yielding for different cases*

Cases	$\dfrac{T_z}{(K_I/\sqrt{a})}$	Notation	$\dfrac{K_I}{Y\sqrt{a}}$	Reference
Centre-cracked specimen		────────	0·151	ISIDA (1971)
Double edge-cracked specimen		─────	0·153	BOWIE (1964)
Bend specimen		─·─·─	0·166	ASTM (1970)
Compact tension specimen		─×─×─	0·170	ASTM (1970)
Boundary layer approach		─────	0·161	
Modified boundary layer approach (CC)	−0·589	○	0·151	
Modified boundary layer approach (DEC)	−0·144	◐	0·158	
Modified boundary layer approach (Bend)	0·033	□	0·162	
Modified boundary layer approach (CT)	0·291	◩	0·167	

3.2 Modified boundary layer problem

Analytically, the state of stress at the crack tip in the elastic case is given by the singular stress components of (1) and by non-singular stress terms, and one may write

$$\sigma_{ij} = \frac{K_I}{(2\pi r)^{1/2}} f_{ij}(\Theta) + T_{ij0} + T_{ij\infty}(r). \qquad (2)$$

By separating the non-singular stress components into two parts it is possible to choose these so that T_{ij0} is constant over a large distance in front of and behind the crack tip whereas $T_{ij\infty}(r)$ is non-zero only close to the outer boundaries of the specimen and $\lim_{r \to 0} T_{ij\infty}(r) = 0$. Of the T_{ij0}-components only the normal stress parallel to the crack plane, the x-direction stress, is different from zero, and this component is denoted by T_x.

An analytical eigen-value expansion leading to a representation of the stress components equivalent to (2) was given by WILLIAMS (1957) for external cracks. In such an expansion the term independent of r would correspond to T_{ij0} in (2) while the terms of power larger than or equal to one-half in r would correspond to $T_{ij\infty}(r)$.

In the present work the total x-direction stress σ_x is determined from the solutions obtained by the elastic finite element method for the different specimens and for the boundary layer problem for elements with one side on the crack surface. The results are shown in Fig. 5 for a load level just below yielding: $K_I = 0.15\ Y\sqrt{a}$ and for a distance 0.02–$0.24\ a$ behind the crack tip. From this plot the non-singular stress for each specimen is obtained as the average difference between $\sigma_x(r, \Theta)$ for the corresponding specimen and boundary layer solutions:

$$T_x = \sigma_x(r, \pi)\bigg|_{\text{SPEC.}} - \sigma_x(r, \pi)\bigg|_{\text{B.L.}}$$

Since T_x varies linearly with applied load it is proportional to $K_I/\sqrt{a}$. In this way, the values of T_x given in Table 2 are obtained for the different specimens.

The stress T_x as determined for a certain specimen geometry is added to the singular stresses of (1) which specifies the boundary values of the boundary layer problem. In this way, a modified boundary layer solution is obtained which is expected to agree with the solution for the actual geometry. So modified, the boundary layer solutions give a first plastically-yielding element at the K_I-levels shown in Table 1. The agreement with first plastic-yield K_I-values for the actual geometries is very good, and it would have been exact had the average T_x been determined from element stresses over a more narrow range close to the crack tip.

TABLE 2. *Values of* T_x *for different cases*

Case	$T_x/(K_I/\sqrt{a})$
Centre-cracked specimen	-0.589
Double edge-cracked specimen	-0.144
Bend specimen	0.033
Compact tension specimen	0.291

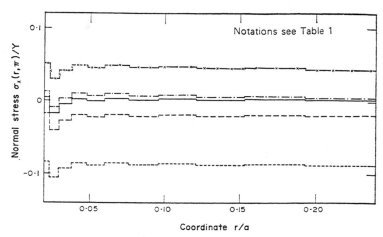

Fig. 5. The x-direction stress in the crack plane for elements behind the crack tip at $K_I = 0.15\ Y\sqrt{a}$.

4. Elastic–Plastic Solution

Once yielding has started in one element the plastic zones grow by at most one element per load increment. The elastic–plastic boundary assumes, due to the element configuration, a very irregular form. Figure 6 shows smoothed estimates of the boundaries for the different specimens at a load level just below the highest allowable one according to ASTM for small-scale yielding, that is, $K_I = 0.6\ Y\sqrt{a}$. A closer study of the growth of the plastic zones at increasing load shows that in a certain load range they grow linearly with K_I^2. For the compact tension, bend and double edge-cracked specimens this interval is $0.4 < K_I/Y\sqrt{a} < 1.0$, whereas for the centre-cracked specimen it is $0.4 < K_I/Y\sqrt{a} < 0.55$. In Fig. 7 this fact has been used to

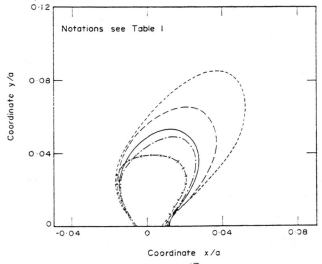

Fig. 6. Plastic zones at a load level $K_i = 0.6\ Y\sqrt{a}$, that is, just below the ASTM-limit.

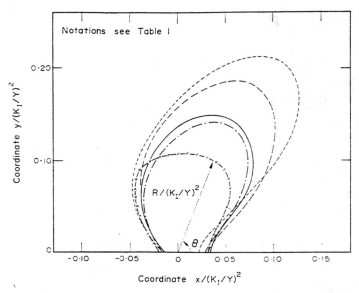

Fig. 7. Plastic zones in a coordinate system, non-dimensionalized with respect to the characteristic length parameter $(K_I/Y)^2$.

show the elastic–plastic boundaries in a coordinate system in which x and y are non-dimensionalized with respect to the characteristic length parameter $(K_I/Y)^2$. These curves are obtained by determining for each case in the above load interval the average extent $R(\Theta)/(K_I/Y)^2$ of the plastic zone. The maximum deviation of $R/(K_I/Y)^2$ was 6 per cent, except for $\Theta = 0$ when it was somewhat larger. Figures 6 and 7 also show the plastic zone for the pure boundary layer problem. With the scaling of coordinates chosen in Fig. 7 the plastic zones for the different cases would concide if the plastic state were determined by K_I alone. The plastic zones of the modified boundary layer solutions are shown in Fig. 8 together with the zones for the actual geometries. The difference in size is at most three elements. The plot is for a load level $K_I = 0.6\ Y\sqrt{a}$.

The displacement of the crack surfaces at $K_I = 0.6\ Y\sqrt{a}$ is shown in Fig. 9. It is of special interest to consider the displacement very near the crack tip. Figure 10 shows the variation of displacement with load in the first nodal point behind the crack tip, that is, point k in Fig. 3. The coordinates of point k are $x = -0.006\ a$, $y = 0$. In the Figure, the corresponding displacement values for the modified boundary-value problems are also given. As is the case with the plastic zones the plastic part of the fracture surface displacement is expected to grow linearly with K_I^2, when it is measured at a point $r/(K_I/Y)^2 = $ const., that is, at a point moving with increasing load in the physical coordinate system. On the other hand, the elastic part of the fracture surface displacement increases linearly with $K_I\sqrt{r}$. For points close to the crack tip this part can be neglected except for small values of K_I. In the present work it is found that the plastic part of the displacement at a fixed point on the fracture surface varies linearly with K_I^2/EY. In Fig. 11 the total displacement δ at the point

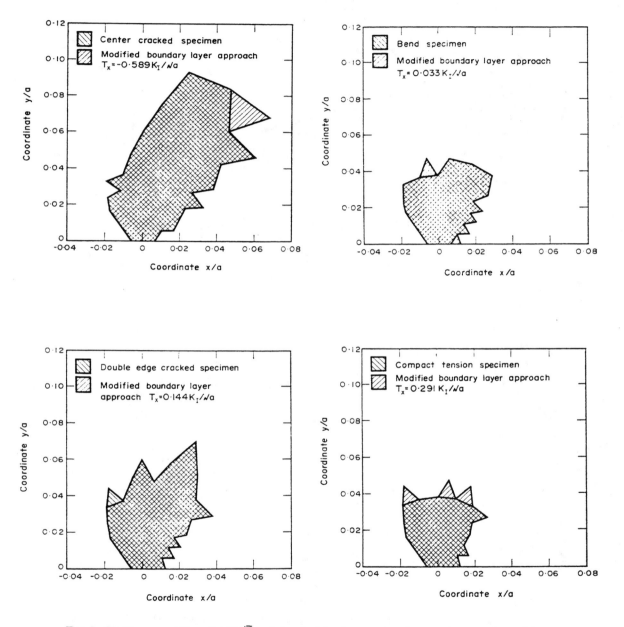

FIG. 8. Plastic zones at $K_I = 0.6\, Y\sqrt{a}$ as determined by the computer for actual geometries and for corrected boundary-value solution.

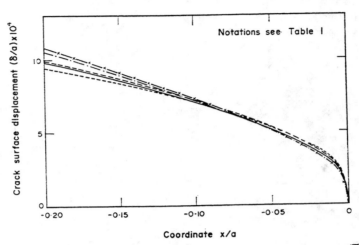

Fig. 9. Displacement of the crack surfaces at a load level $K_I = 0.6\, Y\sqrt{a}$.

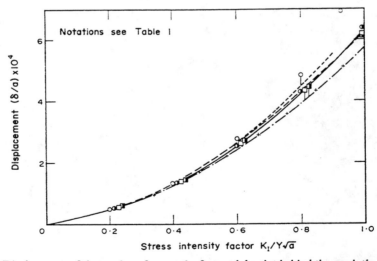

Fig. 10. Displacement of the crack surfaces at the first nodal point behind the crack tip as a function of load.

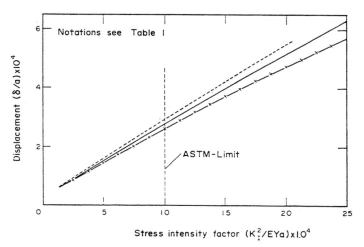

FIG. 11. Displacements in Fig. 10 as a function of K_I^2/EY.

$x = -0{\cdot}006\,a$ is shown as a function of K_I^2/EY; up to the ASTM-limit the relation is of the form

$$\delta = c \cdot \frac{K_I^2}{EY} + d \cdot \frac{K_I Y \sqrt{r}}{EY},$$

where c and d are constants. The constant c varies between 0·24 (CT-specimen) and 0·27 (CC-specimen). At the ASTM-limit there is a knee in the curves. For larger values of K_I^2/EY the CC-curve is linear with slope 0·26 whereas the CT-curve has a continuously decreasing slope. The corresponding values for the slope found by RICE and TRACEY (1973) in the boundary layer solution for the displacement measured at

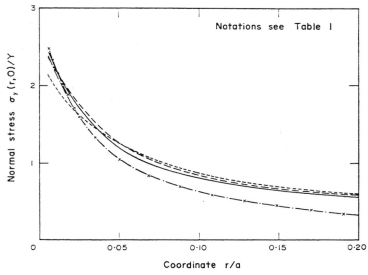

FIG. 12. Stress normal to the crack plane as a function of distance r from the crack tip for $K_I = 0{\cdot}6\,Y\sqrt{a}$.

the crack tip is $c = 0.25$. Their finite element design allows for a discontinuous displacement at the crack tip which the element configuration of the present work does not.

The stress distribution σ_y in front of the crack is shown in Fig. 12 for $K_I = 0.6\, Y\sqrt{a}$.

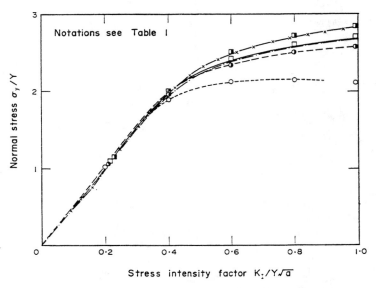

Fig. 13. Stress normal to the crack plane in the first nodal point in front of the crack as a function of load.

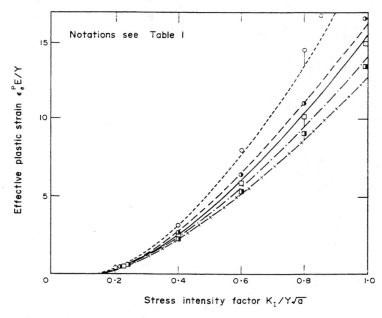

Fig. 14. Effective plastic strain in the first plastically-yielding element as a function of load.

The distribution was obtained by fitting a curve to average nodal stresses of nodal points situated on the x-axis. The stress at the crack tip is excluded for obvious reasons. Figure 13 shows the stress σ_y at the first nodal point in front of the crack ($x = 0.006\,a$, $y = 0$ or point l in Fig. 3) as a function of K_I. It seems to approach asymptotically the value $\sigma_y = 2.97\,Y$ as predicted by Prandtl's slip-line field solution (see ASTM (1970)), except in the centre-cracked specimen where it is considerably lower. Finally, Fig. 14 shows the effective plastic strain in the first plastically-yielding element as a function of K_I. In Figs. 13 and 14 values are also given for the corrected boundary layer problems.

5. Discussion

The results obtained show that a boundary layer solution of the elastic–plastic crack problem is not a meaningful representation of a real crack problem, at least not for materials with low strain-hardening. However, the boundary layer solution can be modified or corrected to give a true picture of the state at the tip of a crack in any specimen or structure. The modification amounts to an addition to the boundary values of the boundary layer problem of the tractions corresponding to the non-singular x-direction stress. This stress is constant along the crack plane over a large distance in front of and behind the crack tip. It can be determined from the elastic solution of the corresponding crack problem.

For the four different specimens studied there is a large difference in the stress, strain and crack surface displacement parameters describing the state at the crack tip already at load levels far below the ASTM-limit. The crack surface displacement seems to be the parameter which correlates best to the stress intensity factor K_I (see Figs. 9, 10 and 11).

In linear elastic fracture mechanics evaluation of fracture toughness is often based on displacement criteria, the secant slope evaluation method. According to the results given in Fig. 10 for the displacement as a function of the stress intensity factor this will lead to very good correlation between fracture toughness values determined with bend and compact tension specimens. Also, values determined with centre-cracked and double edge-cracked specimens would differ very little up to stress levels equal to the ASTM-limit. However, the difference between the bend and CT values on the one hand and the CC and DEC values on the other is 7 per cent at $K_I = 0.6\,Y\sqrt{a}$. Most valid K_{Ic}-values are determined by using CT-specimens. For comparison sometimes bend specimens have been used. It would be more decisive to use a centre-cracked or edge-cracked specimen for comparison.

The corresponding difference between the specimens at the ASTM-limit is 18 per cent for effective plastic strain in the most highly strained element and 26 per cent for normal stress σ_y at the first nodal point in front of the crack. Thus, if the real fracture criterion is one of critical stress or strain, then one would expect very large geometry effects in addition to those included in the stress intensity factor. The picture may, however, be different from that found in the present work if one takes the stable growth of the crack into account. Some preliminary results indicate that strain-hardening of the material does not change the main features.

Acknowledgement

The writers wish to acknowledge that the idea of relating the differences found in small-scale yielding behaviour between different specimens to the non-singular stress terms arose from discussions between them and Professor J. R. Rice. The work has been financially supported by the Swedish Board for Technical Development.

REFERENCES

ASTM	1970	Annual Book of ASTM Standards, Pt. 31, p. 911. American Society for Testing Materials, Philadelphia.
BOWIE, O. L.	1964	*J. appl. Mech.* **31**, 208.
HÄRKEGÅRD, G. and LARSSON, S. G.	1973	*Computers and Structures*, Vol. 3. In press.
ISIDA, M.	1971	*Int. J. Fracture Mech.* **7**, 301.
RICE, J. R.	1968	*Fracture* (edited by LIEBOWITZ, H.), Vol. 2, *Mathematical Fundamentals*, p. 191. Academic Press, New York.
RICE, J. R. and TRACEY, D. M.	1973	*Numerical and Computer Methods in Structural Mechanics* (edited by FENVES, S. J., PERRONE, N., ROBINSON, A. R. and SCHNOBRICH, W. C.). In press. Academic Press, New York.
WILLIAMS, M. L.	1957	*J. appl. Mech.* **24**, 109.
YAMADA, Y., YOSHIMURA, N. and SAKURAI, T.	1968	*Int. J. mech. Sci.* **10**, 343.

LIMITATIONS TO THE SMALL SCALE YIELDING APPROXIMATION FOR CRACK TIP PLASTICITY

By J. R. Rice

Division of Engineering, Brown University, Providence, R.I., U.S.A.

(*Received 6th March* 1973)

Summary

Recent finite-element results by S. G. Larsson and A. J. Carlsson suggest a limited range of validity to the 'small scale yielding approximation', whereby small crack tip plastic zones are correlated in terms of the elastic stress intensity factor. It is shown with the help of a model for plane strain yielding that their results may be explained by considering the non-singular stress, acting parallel to the crack at its tip, which accompanies the inverse square-root elastic singularity. Further implications of the non-singular stress term for crack tip deformations and fracturing are examined. It is suggested that its effect on crack tip parameters, such as the opening displacement and *J*-integral, is less pronounced than its effect on the yield zone size.

1. Introduction

RECENTLY, LARSSON and CARLSSON (1973) have shown that the range of validity of the 'small scale yielding' approximation for crack tip plastic zones is substantially more limited than previous analyses had suggested (see, for example, RICE [1967a, 1968a]). To describe the approximation, let r, θ be polar coordinates centred at the tip of a crack in a body under plane strain deformations. The small-displacement-gradient linear elastic solution results in stresses of the form

$$\sigma_{ij} = Kr^{-1/2}f_{ij}(\theta) + \text{non-singular terms} \qquad (1.1)$$

near the crack tip, where K is the stress intensity factor and where the set of universal functions f_{ij} is normalized so that the singular part of the stress acting ahead of the tip, normal to the plane of the crack, is $K(2\pi r)^{-1/2}$. The small scale yielding approximation then incorporates the notion that, even though (1.1) is inaccurate within and near a small crack tip yield zone, its dominant singular term should in some sense still govern the deformation state within that zone. Hence, the actual elastic–plastic problem is replaced by a problem formulated in boundary layer style, whereby a semi-infinite crack in an infinite body is considered and the actual conditions of boundary loading are replaced by the asymptotic boundary conditions that

$$\sigma_{ij} \to Kr^{-1/2}f_{ij}(\theta) \qquad \text{as } r \to \infty. \qquad (1.2)$$

Hence, as is often said, the small yield zone is 'surrounded' by the dominant elastic singularity, and the applied loadings and geometric shape of the body influence conditions within the plastic region only insofar as they enter the formula for K, as computed elastically.

A consequence of this formulation is that the plastic zone dimension r_p and the crack tip opening displacement δ_t, when definable, are given by formulae of the type

$$r_p = \alpha K^2/\sigma_0^2, \qquad \delta_t = \beta K^2/E\sigma_0, \tag{1.3}$$

where E is elastic tensile modulus, σ_0 is yield strength, and α and β are dimensionless factors which may, for example, depend on Poisson's ratio, strain-hardening exponent, etc., but are independent of the applied load and specimen geometry. Now, by comparing equations such as (1.3), generated by the boundary layer formulation, to available complete elastic–plastic solutions, RICE (1967a, 1968a) found that the approximation was valid up to substantial fractions of the loads corresponding to general yielding. Of course, in the limit of very small load levels, the solutions coincide exactly. It turns out to be important that such complete solutions were, however, available only for the anti-plane strain case and for the Barrenblatt–Dugdale–BCS (Bilby–Cottrell–Swinden) yield model.

By contrast, LARSSON and CARLSSON (1973) performed plane strain elastic–plastic calculations, by the finite element method, for a variety of specimen geometries, and found significant discrepancies with the boundary layer formulation, even within the rather small range of yield zone sizes allowed by the ASTM limits for fracture test correlation in terms of K-values. For example, by fitting their numerical results to (1.1) they found that at loads corresponding to the ASTM limit, α would have to differ by a factor of two between the compact tension and center cracked specimens.

Larsson and Carlsson were able to explain their results in terms of a suggestion by the present writer that differences from specimen to specimen in the 'non-singular terms' of (1.1) could be responsible for the discrepancies. Indeed, from the analyses of WILLIAMS (1957) and IRWIN (1960), a more detailed form than (1.1) for the in-plane stress components is

$$\begin{bmatrix} \sigma_{xx} & \sigma_{xy} \\ \sigma_{yx} & \sigma_{yy} \end{bmatrix} = \frac{K}{\sqrt{r}} \begin{bmatrix} f_{xx}(\theta) & f_{xy}(\theta) \\ f_{yx}(\theta) & f_{yy}(\theta) \end{bmatrix} + \begin{bmatrix} T & 0 \\ 0 & 0 \end{bmatrix} + \text{terms which vanish at crack tip.} \tag{1.4}$$

Here, (x, y) is the plane of straining and the crack coincides with the x-axis, so it is seen that the portion of the non-singular stress field which does not vanish at the tip amounts to a uniform stress $\sigma_{xx} = T$ acting parallel to the crack plane. Thus, by first determining T in terms of the applied load for each of their specimens, LARSSON and CARLSSON (1973) were able to verify that a two-parameter boundary layer formulation, in which (1.2) is replaced by the requirement of an asymptotic approach to the field given by the two leading terms of (1.4), could closely match their results for the different specimens.

The aim in the present paper is to study further this T-effect, and to clarify the manner in which it results in deviations from (1.3) at such substantially lower levels of applied load than had been expected from earlier studies. Much of this discussion is given in terms of a simple model for plane strain yielding, consisting of two slip bands emanating symmetrically from the crack tip. It is also shown that there is no similarly strong T-effect on formulae for the value of the J-Integral. This and related implications of the T-effect for fracture are discussed.

2. A Model for Plane Strain Yielding

The model for plane strain yielding is illustrated in Fig. 1. Plastic relaxation occurs by sliding on two bands at angles $\pm\phi$ with the crack plane. These bands sustain a

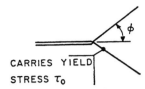

Fig. 1. Crack tip yield model.

yield stress τ_0 in shear, and their length r_p is determined by the following approximate argument (RICE, 1967a).† Consider first a mode II shear crack under stress intensity factor $K^{(s)}$, so that the elastic field analogous to (1.1) results in

$$\sigma_{yx}^{(s)} = \frac{K}{(2\pi r)^{1/2}} + \cdots \qquad (2.1)$$

for the shear stress exerted directly ahead of the crack, in its own plane. If this is relaxed through sliding in the crack plane under a yield stress τ_0, the small scale yielding estimates of the extent of the plastic zone and the crack tip sliding displacement are

$$r_p^{(s)} = (\pi/8)[K^{(s)}/\tau_0]^2, \qquad \delta_t^{(s)} = (1-\nu^2)[K^{(s)}]^2/E\tau_0. \qquad (2.2)$$

Now, for the mode I tensile case, the elastic field (1.1) results in a shear stress

$$\sigma_{\phi r} = \frac{\sin\phi \cos(\tfrac{1}{2}\phi) K}{2(2\pi r)^{1/2}} + \cdots \qquad (2.3)$$

along the planes at angles $\pm\phi$ where sliding is presumed to take place. By comparing this to (2.1), we can identify $K^{(s)}$ as

$$K^{(s)} = \tfrac{1}{2}\sin\phi \cos(\tfrac{1}{2}\phi) K \qquad (2.4)$$

and, as an approximation, estimate the extent of the plastically relaxed zones and crack tip sliding displacement in each from (2.2). Hence, for small scale yielding,

$$\left.\begin{array}{l} r_p \approx r_p^{(s)} = (\pi/64)\sin^2\phi\,(1+\cos\phi) K^2/\tau_0^2, \\ \delta_t \approx 2\delta_t^{(s)}\sin\phi = \tfrac{1}{4}(1-\nu^2)\sin^3\phi\,(1+\cos\phi)\,K^2/E\tau_0 \end{array}\right\} \qquad (2.5)$$

where a trigometric identity is used and where δ_t is the total opening at the tip between upper and lower crack surfaces. In fact, this expression for δ_t differs from that given originally (RICE, 1967a) in that the $\sin\phi$ multiplying $\delta^{(s)}$, and giving its projection onto the y-direction, had been omitted.

If we choose the value of ϕ as that which maximizes the extent of the yielded zone, then $\cos\phi = \tfrac{1}{3}$, so that $\phi = 70\cdot 6°$, and (2.5) become

$$\left.\begin{array}{l} r_p = \dfrac{\pi}{18}\left(\dfrac{K}{\sqrt{3}\tau_0}\right)^2 \approx 0\cdot 17\left(\dfrac{K}{\sqrt{3}\tau_0}\right)^2, \\[2mm] \delta_t = \dfrac{16}{27}\sqrt{\dfrac{2}{3}}\,\dfrac{(1-\nu^2)K^2}{E(\sqrt{3}\tau_0)} \approx 0\cdot 44\,\dfrac{K^2}{E(\sqrt{3}\tau_0)} \quad \text{(for } \nu = 0.3\text{)}, \end{array}\right\} \qquad (2.6)$$

† A numerical solution of this model for $\phi = 45°$ has been reported by BILBY and SWINDEN (1965), who modelled the crack and yield bands by a finite set of discrete dislocations, having fixed positions but variable Burgers vectors.

where the results are given in terms of the equivalent tensile strength $\sigma_0 = \sqrt{3}\,\tau_0$ for purposes of comparison with (1.3) and with numerical finite-element solutions to the full elastic–plastic equations for a non-hardening von Mises material. The most accurate of such solutions for the small scale yielding formulation is probably that of RICE and TRACEY (1973), employing singular elements. They reported a maximum plastic zone extent at 71° with numerical coefficients of 0·152 for r_p and 0·493 for δ_t. Similar results were obtained by LARSSON and CARLSSON (1973) and also by LEVY, MARCAL, OSTERGREN and RICE (1971), in an earlier implementation of singular elements, except that the latter obtained a numerical coefficient about 14 per cent lower for δ_t. Thus, the simple model seems to be in fair agreement with more accurate solutions. Indeed, if we thought of the yield zone as not being confined to discrete bands, but rather as a diffuse zone, and used (2.5_1) to predict the distance to the elastic–plastic boundary, then a yield zone shape in good agreement with that of the numerical solutions results over the 'centred fan' range of ϕ from 45° to 135°.

Now let us consider the effect of the T-term on this model. Evidently, a uniform stress field $\sigma_{xx} = T$ creates a uniform shear stress

$$\sigma_{\phi r} = -T \sin \phi \cos \phi \qquad (2.7)$$

along a plane at angle ϕ with the crack plane. Since it is uniform, a solution to the yield model for the case of $T = 0$ also provides the solution when $T \neq 0$ if we make the replacement

$$\tau_0 \to \tau_0 + T \sin \phi \cos \phi. \qquad (2.8)$$

Hence, the solution for the modified boundary layer formulation, in which T is accounted for as discussed earlier, is given directly from (2.5) as

$$\left.\begin{array}{l} r_p = (\pi/64) \sin^2 \phi\, (1+\cos \phi)\, K^2/(\tau_0 + T \sin \phi \cos \phi)^2, \\ \delta_t = \tfrac{1}{4}(1-\nu^2) \sin^3 \phi\, (1+\cos \phi)\, K^2/[E(\tau_0 + T \sin \phi \cos \phi)]. \end{array}\right\} \qquad (2.9)$$

To see the real significance of these results, let us keep in mind that K and T are directly proportional to the applied loadings. For example,

$$K = \sigma_{yy}^\infty (\pi a)^{1/2}, \qquad T = \sigma_{xx}^\infty - \sigma_{yy}^\infty \qquad (2.10)$$

for the Inglis–Kolosov configuration of a crack of length $2a$ under remotely uniform biaxial stressing (Fig. 2). Now, if (2.9) is expanded in a series about $T = 0$, using the value of $\phi = 70.6°$ which maximizes r_p in that case, then

$$\left.\begin{array}{l} r_p = \dfrac{\pi}{18}\left(\dfrac{K}{\sqrt{3}\tau_0}\right)^2 \left[1 - \tfrac{4}{3}\sqrt{\dfrac{2}{3}}\left(\dfrac{T}{\sqrt{3}\tau_0}\right) + \cdots\right], \\ \delta_t = \dfrac{16}{27}\sqrt{\dfrac{2}{3}}\dfrac{(1-\nu^2)K^2}{E(\sqrt{3}\tau_0)}\left[1 - \tfrac{2}{3}\sqrt{\dfrac{2}{3}}\left(\dfrac{T}{\sqrt{3}\tau_0}\right) + \cdots\right], \end{array}\right\} \qquad (2.11)$$

where by comparison with (2.6), the bracketed terms represent the deviation from the small scale yielding approximation. Thus,

$$r_p = \dfrac{\pi^2}{18} a \left(\dfrac{\sigma_{yy}^\infty}{\sqrt{3}\tau_0}\right)^2 \left[1 + \tfrac{4}{3}\sqrt{\dfrac{2}{3}}\left(\dfrac{\sigma_{yy}^\infty - \sigma_{xx}^\infty}{\sqrt{3}\tau_0}\right) + \cdots\right] \qquad (2.12)$$

for the configuration of Fig. 2.

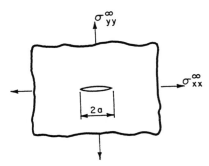

FIG. 2. Crack under biaxial tension.

There is a fundamental difference between these expansions and similar expansions as carried out from the available complete elastic–plastic solutions for anti-plane strain or for the Barrenblatt–Dugdale–BCS model. Namely, for every such solution the corresponding series for r_p and δ_t are of the form

$$r_p = \alpha(K/\sigma_0)^2[1 + \lambda(\sigma_{\mathrm{appl}}/\sigma_0)^2 + \ldots], \qquad (2.13)$$

where α and λ are constants and where σ_{appl} is some nominal applied stress. A typical example for the latter type of model, wherein yield is supposed to be confined to plastic zones sustaining the tensile yield strength σ_0 and lying in the plane of the crack, is

$$r_p = a\{[\cos(\pi\sigma_{yy}^\infty/2\sigma_0)]^{-1} - 1\} = (\pi^2/8)\,a(\sigma_{yy}^\infty/\sigma_0)^2[1 + (5\pi^2/48)(\sigma_{yy}^\infty/\sigma_0)^2 + \ldots] \qquad (2.14)$$

for the configuration of Fig. 2. The feature of interest for all such solutions is that the deviation from the small scale yielding approximation is *quadratic* in the applied load, whereas for the present inclined shear band model (and, by implication, for the exact elastic–plastic plane strain solution) the deviation is *linear* in the applied load. Indeed, this difference would seem to be at the root of the Larsson–Carlsson observation of a substantially more limited range of validity to the small scale yielding approximation than had been evident from the earlier solutions.

In retrospect, it is easy to see how this distinction comes about: the non-vanishing but non-singular T-terms are the source of the linear deviation in (2.11) and (2.12). This term is completely without effect on the Barrenblatt–Dugdale–BCS model. For example, changing T by changing σ_{xx}^∞ in Fig. 2, or by alterations of boundary conditions in other cases which would induce a uniform σ_{xx} if the response were elastic, has no effect on the solutions for this model. Of course, the same is not true for the plane strain model of Fig. 1, as (2.7) to (2.9) show. In anti-plane strain, there is a similar possibility of a non-vanishing but non-singular term amounting to a uniform shear stress σ_{xz}, and this would presumably result also in a linear deviation. But this term exists only when loadings are unsymmetrical relative to the crack line, and its effect has been undetected simply because solutions have been done only for symmetrical loadings.

It is interesting to examine further the predictions of the simple model as given by (2.9). The *apparent* α-value, in the notation of (1.3), will be called R_0 and is given by

$$R_0 = r_p/(K/\sqrt{3}\tau_0)^2 = (3\pi/64)\sin^2\phi\,(1+\cos\phi)/[1+(T/\tau_0)\sin\phi\cos\phi]^2. \qquad (2.15)$$

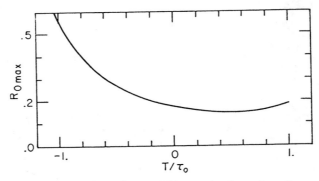

Fig. 3. Effect of T on yield zone size for a given K.

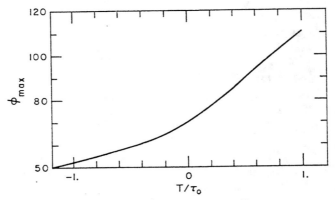

Fig. 4. Angle ϕ at which yield zone size is maximum.

Figures 3 and 4 show the variation, with T/τ_0, of the value $R_{0,\max}$ resulting when this is at a maximum, and the corresponding values, $\phi_{\max}$, of ϕ. Now consider the results summarized in Table 1 for four specimens analyzed by LARSSON and CARLSSON (1973). The first column shows T in ratio to K for each specimen, with a being crack depth. The data are taken from solutions for loadings at the ASTM limit, $K = 0.6\,\sigma_0\sqrt{a}$, and corresponding ratios of T/τ_0 are listed. Next is shown $R_{0,\max}$ from the numerical solution and from (2.15). Finally $\phi_{\max}$ is shown from the same sources. It is given as a

TABLE 1. *T-effect on yield zones for fracture specimens.*

Specimen	$T\sqrt{a}/K$	T/τ_0	$R_{0,\max}$ LARSSON and CARLSSON (1973)	Equations (2.9) and (2.15)	$\phi_{\max}$ (deg.) LARSSON and CARLSSON (1973)	Equations (2.9) and (2.15)
Center cracked	−0.59	−0.57	0.23	0.29	58–63–69	58
Doubled edge cracked	−0.14	−0.13	0.20	0.19	64–67–73	67
Bend	0.03	0.03	0.15	0.17	63–71–78	71
Compact tension	0.29	0.28	0.11	0.15	68–83–105	81

range for the numerical solution because of the typically broad and not precisely defined maximum for R_0, and the angles such as 58°–63°–69° for the center cracked specimen correspond to those for which the yield zone extent is 0·98 $R_{0,\,\text{max}}$–$R_{0,\,\text{max}}$–0·98 $R_{0,\,\text{max}}$, respectively. The agreement is remarkably good. Also, as remarked earlier with $T = 0$, a reasonable indication of the shape of the plastic zone is given by regarding (2.9) as being the distance to the elastic–plastic boundary at any angle ϕ. The writer is pleased to acknowledge the assistance of Professor A. J. Carlsson in providing the numerical evaluation of the formula (2.9) and the comparison with his finite-element data as discussed here.

It is also worthy of note that the variation of crack tip opening displacement between specimens at the same K is considerably less than that in plastic zone size. For example, if the values of T/τ_0 and ϕ_{max} at the ASTM limits are taken from the Table, then (2.9) with $v = 0.3$ gives *apparent* β-values, in the notation of (1.3), which range from 0·49 for the center cracked to 0·41 for the compact tension specimen. By comparison, 0·44 is the actual β-value, from (2.6) for $T = 0$. The numerical solutions of LARSSON and CARLSSON (1973) showed a similarly small variation in apparent β-values.

3. ABSENCE OF T-EFFECT ON THE J-INTEGRAL

Recently BROBERG (1971) and BEGLEY and LANDES (1972) have suggested that failure criteria could be based on the J-Integral, in the sense of its use by RICE (1968a, b) as a measure of the intensity of crack tip deformations in non-linear materials. By its compliance interpretation, it is known that

$$J = (1-v^2)\frac{K^2}{E} \tag{3.1}$$

for small scale yielding. It is of interest to know if there is a linear deviation of J from this formula, due to a T-effect, as is the case with plastic zone size (equations (2.11) and (2.12)). In fact, it is shown here that there is no T-effect on J, so that

$$J = (1-v^2)\frac{K^2}{E}\left[1+\mu\left(\frac{\sigma_{\text{appl}}}{\sigma_0}\right)^2+\ldots\right], \tag{3.2}$$

i.e. the deviation is *quadratic* in applied load.

To see this, consider the modified boundary layer formulation for a semi-infinite crack in an infinite body, with boundary conditions of asymptotic approach to the first two terms of (1.4). There will be a yield zone at the crack tip, but we focus on the nature of the solution in elastically deformed material outside of a circle of radius greater than the greatest extent of the non-linear zone. By doing a 'WILLIAMS (1957) expansion' of the stress field for this *outer* region we arrive at the representation

$$\sigma_{ij} = T\delta_{ix}\delta_{jx}+Kr^{-1/2}f_{ij}(\theta)+r^{-1}A_{ij}(\theta)+r^{-3/2}B_{ij}(\theta)+r^{-2}C_{ij}(\theta)+\ldots. \tag{3.3}$$

Of course, the terms of exponent more negative than $-\tfrac{1}{2}$ are customarily deleted because if this were written for the *inner* region, they would result in unbounded energy.

By seeking the most general r^{-1} stress field corresponding to symmetrical loadings about the x-axis, one easily shows from the equilibrium equations and compatibility equation $\nabla^2(\sigma_{rr}+\sigma_{\theta\theta}) = 0$, for an isotropic material, that

$$A_{rr} = \xi \cos \theta, \qquad A_{\theta\theta} = \eta \cos \theta, \qquad A_{r\theta} = \eta \sin \theta, \qquad (3.4)$$

where ξ and η are constants and where $\theta = \pm\pi$ are the crack surfaces. In fact, these expressions when multiplied by r^{-1} correspond to the stress field due to an edge dislocation with Burgers vector in the y-direction plus a concentrated line force pointing in the x-direction. If the A_{ij}-terms are to satisfy traction-free boundary conditions on the crack surfaces, $\eta = 0$, and if there is to be no net force transmitted across a circuit surrounding the crack tip, $\xi = 0$. Hence, there is no r^{-1} term in (3.3):

$$A_{ij} = 0. \qquad (3.5)$$

Choosing the path Γ in the definition of J to be a large circle of radius r surrounding the tip,

$$\begin{aligned} J &= \int_\Gamma [W\, dy - T_i(\partial u_i/\partial x)\, ds] \\ &= r \int_{-\pi}^{+\pi} [\tfrac{1}{2}\sigma_{ij}\varepsilon_{ij} \cos \theta - \sigma_{ix}(\partial u_i/\partial x) \cos \theta - \sigma_{iy}(\partial u_i/\partial x) \sin \theta]\, d\theta. \end{aligned} \qquad (3.6)$$

The displacement derivatives and strains will have expansions in powers of r identical to those for the stresses. Further, since J is path-independent, r may be chosen as large as we wish. By considering the different powers of r which remain in (3.3) when the r^{-1} term is deleted, one sees that in the limit $r \to \infty$, only the first two terms of (3.3) will contribute to J. But this means that J takes on the same value which it would have if there were no plastic zone and the material responded elastically everywhere, and this value is well-known to be that given by (3.1) independently of T. Hence there is no T-effect on the J-integral.

The significance of this is made evident by the work of CHEREPANOV (1967), HUTCHINSON (1968), and RICE and ROSENGREN (1968) on crack tip singularities in 'power-law' strain hardening materials. The strength of their leading singular term, which dominates the deformation field near the crack tip, well within the plastic zone, is expressed solely in terms of J. From this we conclude that there is no T-effect on the dominant singularity, although there will of course be an effect on the overall shape of the plastic zone. But this needs two qualifications. First, as remarked by RICE and ROSENGREN (1968) and MCCLINTOCK (1971), and as is also evident from an earlier anti-plane strain analysis by RICE (1967b), the question as to whether the 'dominant' singularity really governs over physically significant size scales for fracturing depends on how strongly the material strain-hardens. Indeed, with the non-hardening idealization there is no such one-parameter characterization and different specimens may have different near tip fields, at the same J-value, when load magnitudes are beyond the range of validity of the unmodified boundary layer formulation (1.2). For example, the T-effect leads to slight differences in a crack tip parameter such as the opening displacement in non-hardening calculations, as remarked earlier. The second point is that the dominant singularity, when present, is parameterized in terms of the value, J_{tip}, of J on a contour of vanishing radius about the tip. This will equal the value (3.1) as computed on contours in the elastic region only to the extent that a 'total strain' formulation of plasticity if appropriate. It is likely that the development of pointed vertices on small-offset yield surfaces makes this a good approximation to actual behavior for cases without substantially non-radial loading. However, some approximation is involved and J_{tip} may therefore differ from (3.1) by an amount which depends on conditions in plastically deformed regions away from the tip, where there *is* a T-effect. For example, the incremental, small scale yielding, non-hardening solutions by LEVY *et al.* (1971) and RICE and TRACEY (1973), which take no account of vertex formation, result in a J_{tip} value about 20 per cent less than that of (3.1). This percentage reduction could be affected linearly by T, although the net effect on J_{tip} would seem small.

4. DISCUSSION AND SUMMARY

The model discussed here shows deviations from the small scale yielding solution at relatively low levels of applied load, in agreement with the results of LARSSON and CARLSSON (1973). Their cause is evidently due to the non-singular stress term T acting parallel to the crack plane. The effect on the plastic zone size is quite pronounced at load levels corresponding to the ASTM limit, although there seems to be less effect on near-tip parameters such as the crack tip opening displacement and J-integral. As regards the near-tip stress state, recall that the Prandtl field, which is thought to give the stress state as $r \to 0$ for a non-hardening material (RICE, 1967a, 1968a, b), involves a positive σ_{xx} both ahead and behind the tip. This suggests that specimens with a negative value of T would tend to show a more rapid fall-off, with increasing r, from the hydrostatically elevated stresses of the Prandtl field than would be the case for those with positive T. Indeed, this agrees with the numerical results of LARSSON and CARLSSON (1973) who find, for example, that the center-cracked specimen exhibits a considerably more rapid stress fall-off than do the other specimens listed in Table 1.

This latter kind of T-effect is likely to be important for stress induced fracture mechanisms, such as cleavage micro-cracking, whereas the T-effect on crack tip deformation parameters would seem more relevant to cases of ductile void-growth. Effects of both kinds could be involved when high stress levels are important to void nucleation by the cracking or de-cohesion of second-phase particles (RICE and JOHNSON, 1970). As for experimental studies which might reveal a T-effect on critical K-values for fracture, HALL (1971) has compared four crack test specimen designs for 2219-T87 aluminum and 5Al–2·5Sn ELI titanium alloys. Two of his specimens, namely the bend and compact tension, coincide with those of Table 1. With aluminum, Hall finds a critical K-value for the compact tension specimen which is typically about 25 per cent lower than that for the bend specimen, when comparison is made at the maximum load allowed in the ASTM procedures. This, incidentally, corresponds to a plastic zone extent which is only 6 to 7 per cent of the crack depth a. On the other hand, with titanium Hall finds a less definitive effect, and the bend specimen results instead in the lower critical K-value, by about 10 per cent.

The inclusion of T as a second crack tip parameter was shown to characterize suitably small *plane strain* yield zones, when K alone becomes inadequate. More generally, for actual three-dimensional tensile mode crack tip stress states, it would seem necessary to supplement K with two parameters, say S and T. Here, T is the non-singular σ_{xx} introduced earlier whereas S represents a similar non-singular σ_{zz}, acting perpendicular to the principal plane of deformation. For plane strain, $S = vT$, but this will not be so in general. Just as T influences yielding in the plane, S would seem to influence the transition to a non-plane-strain yielding mode involving through-thickness deformation as observed, for example, in thin notched sheets with 'plane stress' yielding.

Acknowledgment

This study was supported by the United States National Aeronautics and Space Administration under grant NGL–40–002–080. I am grateful to Professor A. J. Carlsson and to Mr. G. Harkegaard of the Royal Institute of Technology, Stockholm for bringing the problem to my attention and for several helpful discussions.

References

BEGLEY, J. A. and LANDES, J. D.	1972	*Fracture Toughness*, Part II, ASTM–STP–514, p. 1. American Society for Testing and Materials, Philadelphia.
BILBY, B. A. and SWINDEN, K. H.	1965	*Proc. Roy. Soc. Lond.* **A285**, 22.
BROBERG, B.	1971	*J. Mech. Phys. Solids* **19**, 407.
CHEREPANOV, G. P.	1967	*Prik. Mat. Mekh.* **31**, 476.
HALL, L. R.	1971	*Fracture Toughness Testing at Cryogenic Temperatures*, ASTM–STP–496, p. 40. American Society for Testing and Materials, Philadelphia.
HUTCHINSON, J. W.	1968	*J. Mech. Phys. Solids* **16**, 13, 337.
IRWIN, G. R.	1960	*Structural Mechanics* (edited by GOODIER, J. N. and HOFF, N. J.), p. 557. Pergamon, New York.
LARSSON, S. G. and CARLSSON, A. J.	1973	*J. Mech. Phys. Solids* **21**, 263.
LEVY, N., MARCAL, P. V., OSTERGREN, W. J. and RICE, J. R.	1971	*Int. J. Fracture Mech.* **7**, 143.
MCCLINTOCK, F. A.	1971	*Fracture: An Advanced Treatise* (edited by LIEBOWITZ, H.), Vol. III, *Engineering Fundamentals and Environmental Effects*, p. 47. Academic Press, New York.
RICE, J. R.	1967a	*Fatigue Crack Propagation*, ASTM–STP–415, p. 247. American Society for Testing and Materials, Philadelphia.
	1967b	*J. Appl. Mech.* **34**, 287.
	1968a	*Fracture: An Advanced Treatise* (edited by LIEBOWITZ, H.), Vol. II, *Mathematical Fundamentals*, p. 191. Academic Press, New York.
	1968b	*J. Appl. Mech.* **35**, 379.
RICE, J. R. and JOHNSON, M. A.	1970	*Inelastic Behavior of Solids* (edited by KANNINEN, M. F., ADLER, W., ROSENFIELD, A. and JAFFE, R.), p. 641. McGraw-Hill, New York.
RICE, J. R. and ROSENGREN, G. F.	1968	*J. Mech. Phys. Solids* **16**, 1.
RICE, J. R. and TRACEY, D. M.	1973	*Numerical and Computational Methods in Structural Mechanics* (edited by FENVES, S. J., PERRONE, N., ROBINSON, A. R. and SCHONOBRICH, W. C.). Academic Press, New York. In press.
WILLIAMS, M. L.	1957	*J. Appl. Mech.* **24**, 109.

BIAXIAL LOAD EFFECTS IN FRACTURE MECHANICS[†]

H. LIEBOWITZ, J. D. LEE and J. EFTIS

School of Engineering and Applied Science, The George Washington University, Washington, DC 20052, U.S.A.

Abstract—Our investigation into the effects of load biaxiality thus far, has produced several findings which, in our opinion, are deemed to be important.

(a) The standard expressions for elastic stress and displacement in the crack-tip region, i.e. the so-called "singular-solution', cannot be considered to be approximations that are acceptable in a completely general sense.

(b) This conclusion is best illustrated in the instance of a biaxially loaded infinite sheet with a flat (horizontal) central crack, wherein the effect of the load applied parallel to the plane of the crack appears entirely in the second terms of the series representations for local stress and displacement. Omission of these contributions, which is the usual practice, is tantamount therefore to denial of the physical presence of the horizontal load. Thus, in calculations of stress, displacement and related quantities of interest in the crack border region by means of the standard expressions, no biaxial load effects will appear, leading thereby to the erroneous impression that load applied parallel to the plane of the crack can have no influence with regard to the fracture problem.

(c) For the infinite sheet problem with a horizontal central crack, our analytical analysis shows significant biaxial load effect on crack border region and crack edge displacement, on local maximum shear stress, on the pattern of maximum shear isostats, on the angle of initial crack extension, and on local elastic strain energy density and strain energy rate. On the other hand, both the elastic stress intensity factor (as to be expected) and the J-integral show no sensitivity whatsoever to the presence of the horizontal load.

(d) The analytical results referred to above for the infinite sheet are also seen in the results obtained for a finite sheet using finite element numerical analysis.

(e) A nonlinear finite element analysis of the same biaxially loaded finite specimen geometry, designed to simulate elastic–plastic material behavior under conditions of no unloading, shows that the global energy rate, the J-integral, the plastic stress and strain intensity factors (in the sense of Hilton and Hutchinson), and the size of the crack border region plastic yield, all have pronounced biaxial load dependency.

1. INTRODUCTION

FOR A plane cracked-body of arbitrary size and shape, with arbitrarily applied loads, the notion that the state of plane elastic stress and displacement in the neighborhood of crack tip can be adequately and generally specified by the so-called "singular solution", has achieved the status of a truism in contemporary fracture mechanics. Eftis *et al.*[1, 2] have shown that this proposition which appears to be reasonable on face value, quantitatively speaking, is nevertheless unacceptable due to the quite arbitrary practice of omitting the second term of the series representation for the stresses, a contribution which is independent of distance from the crack tip. Such an omission can lead to errors of both quantitative and qualitative nature in the prediction of stress and displacement related quantities of interest.

The simplest way to demonstrate the general inadequacy of the one-term representation of the crack tip stress and displacement field is through analysis of a plate, finite or infinite in size, with a centered flat crack subjected to symmetric biaxial loading applied to its outer boundaries (see Fig. 1). For this problem the effects of the horizontal load (which is parallel to the plane of the crack) shows up solely and entirely in the second term of the series representations. Thus, failure to include the second term, which is the general practice, in effect denies the physical presence of the load applied parallel to the crack, and misleads one into thinking that load biaxiality has no significant bearing on fracture problem whatsoever.

For an infinite center-cracked specimen subjected to symmetric and uniform biaxial loading, one is able to obtain an exact solution in a relative simple form, on the basis of which it is possible to obtain expressions for the crack tip stress and displacement field in terms of series representations (see Section 2). The importance of the second term in the series representation and the biaxial effects on isostats of maximum shear, angle of initial crack extension, and local

---

[†]A shorter version of this paper has been presented at the International Conference On Fracture Mechanics and Technology, held in Hong Kong from 21–25 March 1977.

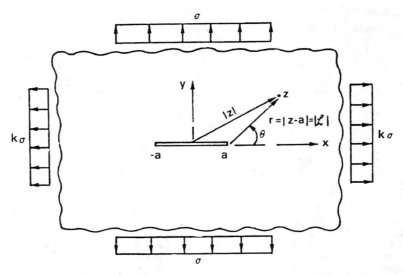

Fig. 1. Plane biaxially loaded center-cracked geometry.

energy rate are discussed in Sections 3–5. The stress intensity factor and the J-integral are shown to be independent of loading applied parallel to the crack.

For a finite center-cracked specimen, Lee and Liebowitz[3], combining finite-element analysis with series-type analytic solution around the crack tip, are able to prove both numerically and analytically that the stress intensity factor remains independent of the biaxial load factor k. In Section 6 we generalize the procedure and indicate the importance of the second term in this modified finite-element analysis.

Although there are significant biaxial effects on maximum shear isostats, angle of initial crack extension, local energy rate, etc., in the linear analysis our calculations show that J-integral is independent of biaxial load factor k. It is of interest to note that Kibler and Roberts[4] have observed experimentally an increase in the apparent fracture toughness with increasing biaxial load.

In the area of nonlinear finite-element analysis of the crack problem, we mention the works by Hutchinson[5, 6], Rice and Rosengren[7], Goldman and Hutchinson[8], Shih[9, 10], Hilton and Hutchinson[11], and especially that of Hilton[12] who analyzed an infinite center-cracked specimen by dividing the specimen into three parts, an elastic region, in which an analytic solution can be represented in the form of a Laurent series expansion, outside of a circular arc centered at origin with sufficiently large radius, a small region around crack tip, in which Hutchinson's solutions are valid, and a region in between analyzed by finite-element method. Hilton[12] reported the biaxial effect on strain intensity factor. Lee and Liebowitz[13] performed a nonlinear finite-element analysis on finite center-cracked specimen subjected to biaxial loading and found that there are significant biaxial effects on energy rate (global), J-integral, stress and strain intensity factor. In Sections 7 and 8, we briefly describe the procedures and show some of the numerical results. We also include our recent calculations showing the biaxial effect on the size of the plastic zone.

2. WILLIAMS EIGENFUNCTION EXPANSION[14]

The plane problem of elasto-statics in the absence of body force for the homogeneous isotropic solid reduces to the specification of Airy's stress function $U(r, \theta)$ which determines the stress components according to the relations

$$t_{rr} = r^{-1}U_{,r} + r^{-2}U_{,\theta\theta}, \tag{2.1}$$

$$t_{\theta\theta} = U_{,rr}, \tag{2.2}$$

$$t_{r\theta} = r^{-2}U_{,\theta} - r^{-1}U_{,r\theta}. \tag{2.3}$$

The condition for stress compatibility requires U to satisfy the biharmonic equation

$$\nabla^4 U = 0. \tag{2.4}$$

Let $U(r, \theta)$ have the form

$$U(r, \theta) = r^{\lambda+1} F(\theta) \qquad (2.5)$$

where λ is an unspecified parameter greater than or equal to 0.5 in order to fulfil the requirement that the strain energy in a finite region around crack tip should be finite. Substituting (2.5) into eqn (2.4), one obtains

$$F(\theta) = \bar{A} \cos(\lambda+1)\theta + \bar{B} \sin(\lambda+1)\theta + \bar{C} \cos(\lambda-1)\theta + \bar{D} \sin(\lambda-1)\theta. \qquad (2.6)$$

For a center-cracked specimen subjected to symmetric biaxial loading, it is required that

$$t_{\theta\theta}(r, \pm\pi) = t_{r\theta}(r, \pm\pi) = 0, \qquad (2.7)$$

$$t_{rr}(r, \theta) = t_{rr}(r, -\theta), \qquad (2.8)$$

$$t_{\theta\theta}(r, \theta) = t_{\theta\theta}(r, -\theta), \qquad (2.9)$$

$$t_{r\theta}(r, \theta) = -t_{r\theta}(r, -\theta), \qquad (2.10)$$

which in turn requires that

$$F(\pi) = F'(\pi) = 0 \qquad (2.11)$$

and

$$F(\theta) = F(-\theta). \qquad (2.12)$$

Equations (2.11) and (2.12) lead to an infinite set of eigenvalues

$$\lambda_n = n/2, \qquad n = 1, 2, 3, \ldots \qquad (2.13)$$

and corresponding eigenfunctions

$$F_n(\theta) = \begin{cases} \cos(\lambda_n+1)\theta - \cos(\lambda_n-1)\theta, & \text{if } n \text{ is even} \\ \cos(\lambda_n+1)\theta - (\lambda_n+1)\cos(\lambda_n-1)\theta/(\lambda_n-1), & \text{if } n \text{ is odd.} \end{cases} \qquad (2.14)$$

It is straight-forward to show that the series representation of the stress and displacement fields, in the case of generalized plane stress, are

$$t_{xx} = \sum_{n=1}^{\infty} C_n r^{n/2-1} \{-(1+\beta_n)\cos(n-2)\theta/2 + \alpha_n \cos(n-6)\theta/2\} \qquad (2.15)$$

$$t_{yy} = \sum_{n=1}^{\infty} C_n r^{n/2-1} \{(1-\beta_n)\cos(n-2)\theta/2 - \alpha_n \cos(n-6)\theta/2\} \qquad (2.16)$$

$$t_{xy} = \sum_{n=1}^{\infty} C_n r^{n/2-1} \{\sin(n-2)\theta/2 - \alpha_n \sin(n-6)\theta/2\} \qquad (2.17)$$

$$E u_x = \sum_{n=1}^{\infty} C_n r^{n/2} \{(\gamma_n - \delta_n)\cos n\theta/2 + \eta_n \cos(n-4)\theta/2\} + B \qquad (2.18)$$

$$E u_y = \sum_{n=1}^{\infty} C_n r^{n/2} \{(\gamma_n + \delta_n)\sin n\theta/2 - \eta_n \sin(n-4)\theta/2\} \qquad (2.19)$$

where E is the Young's modulus, ν is the Poisson's ratio and

$$\alpha_n \equiv \begin{cases} 1, & \text{if } n \text{ is odd,} \\ (n-2)/(n+2), & \text{if } n \text{ is even,} \end{cases}$$

$$\beta_n \equiv 4\alpha_n/(n-2),$$

$$\gamma_n \equiv 4(-3+\nu)\alpha_n/[n(n-2)],$$

$$\delta_n \equiv 2(1+\nu)/n,$$

$$\eta_n \equiv 2(1+\nu)\alpha_n/(n-2),$$

and B, C_n are unknown constants to be determined by boundary conditions, other than those specified along the line of crack. For the stresses, if the terms of order $r^{1/2}$ and above are omitted, and correspondingly the terms of order $r^{3/2}$ and above for the displacements are omitted, then in the neighborhood of crack tip we have[1]

$$t_{xx} \cong \frac{K}{\sqrt{(2\pi r)}} \cos\frac{\theta}{2}\left[1 - \sin\frac{\theta}{2}\sin\frac{3\theta}{2}\right] + A, \tag{2.20}$$

$$t_{yy} \cong \frac{K}{\sqrt{(2\pi r)}} \cos\frac{\theta}{2}\left[1 + \sin\frac{\theta}{2}\sin\frac{3\theta}{2}\right], \tag{2.21}$$

$$t_{xy} \cong \frac{K}{\sqrt{(2\pi r)}} \sin\frac{\theta}{2}\cos\frac{\theta}{2}\cos\frac{3\theta}{2}, \tag{2.22}$$

$$u_x \cong \frac{K}{\mu}\sqrt{\left(\frac{r}{2\pi}\right)} \cos\frac{\theta}{2}\left[(\kappa-1)/2 + \sin^2\frac{\theta}{2}\right] + A\frac{\kappa+1}{8\mu}r\cos\theta + \frac{B}{E}, \tag{2.23}$$

$$u_y \cong \frac{K}{\mu}\sqrt{\left(\frac{r}{2\pi}\right)} \sin\frac{\theta}{2}\left[(\kappa+1)/2 - \cos^2\frac{\theta}{2}\right] - A\frac{3-\kappa}{8\mu}r\sin\theta, \tag{2.24}$$

where

$$K = 4\sqrt{(2\pi)}C_1,$$

$$A = -2C_2,$$

$$\mu = \frac{E}{2(1+\nu)}, \tag{2.25}$$

$$\kappa = \begin{cases} (3-\nu)/(1+\nu), & \text{plane stress.} \\ 3-4\nu, & \text{plane strain.} \end{cases}$$

Eftis *et al.*[1] further obtained the holomorphic function $2\phi'(z)$, specifying the solution for the center-cracked specimen of infinite size subjected to uniform and biaxial loading (see Fig. 1), as

$$2\phi'(z) = \sigma\{z/(z^2-a^2)^{1/2} - (1-k)/2\}, \tag{2.26}$$

where $z \equiv x + iy$, from which it can be shown that

$$K = \sigma(\pi a)^{1/2}, \tag{2.27}$$

$$A = -(1-k)\sigma, \tag{2.28}$$

$$B = -\frac{(1-k)(\kappa+1)}{8\mu}\sigma a. \tag{2.29}$$

One can readily see that the stress intensity factor K is independent of the biaxial load factor k, and that A, B are increasing linearly with respect to k. Also, one is able to prove by several different means that all coefficients in the series representations (2.15)–(2.19) except C_2 (being equivalent to A) and B are independent of biaxial factor k. In other words, the uniform biaxial load applied parallel to the crack does not affect the stress intensity factor; it enters into the picture only through the coefficients A and B exclusively. Therefore, to emphasize once again, omission of the second term in the series representation is not just a matter of accuracy, on the contrary, it is equivalent to denying biaxial effects due to the physical presence of the load applied parallel to the crack. The qualitative and quantative difference between a one-term representation and a two-term representation will be described in Sections 3–5.

3. ISOCHROMATIC FRINGE PATTERNS

It is generally agreed among those engaged in photoelastic analysis of crack border stress patterns that a two parameter characterization of the crack tip region stress components is necessary to reproduce experimental results. This idea was first introduced by Irwin[15], and has been used by other investigators[16, 17]. Cotterell[18] called attention to the influence that the coefficient of the second term of the Williams eigenfunction expansion has on the shape and orientation of the isostatic loops. The procedure adopted by the photoelastic analysts is in the right direction, so to speak, however, the procedure itself is empirical and fails to make apparent, in any general sense, the fact that the boundary loading has a pronounced influence on the isochromatic patterns close to the crack tip.

In the crack tip region, $0 < r/a \ll 1$, the maximum shear stress τ_m, according to the expression

$$\tau_m^2 = (t_{yy} - t_{xx})^2/4 + t_{xy}^2, \tag{3.1}$$

is obtained, by using the two-term representations (2.20)–(2.22), to be [1]

$$\tau_m = \sigma\left\{\frac{a \sin^2 \theta}{8r} + \frac{(1-k)^2}{4} + (1-k) \sin \theta \sin \frac{3\theta}{2} \sqrt{\left(\frac{a}{8r}\right)}\right\}^{1/2} \tag{3.2}$$

The isostats, contours of constant shear stress, in the vicinity of crack tip are shown in Figs. 2–8, for different biaxial load factor k. The pattern of change from biaxial tension–tension to biaxial tension–compression loading is clearly illustrated. The symmetric loop occurs only for the equal biaxial tension–tension loading which is the only case for which one-term approximation gives correct results. In case of uniaxial loading, $k = 0$, the forward tilt of the loops shown in Fig. 5 compares well with experimental results.

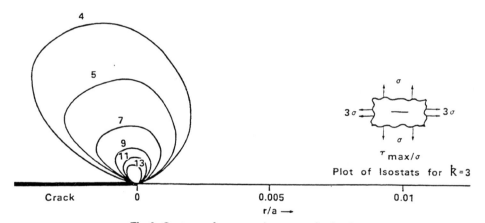

Fig. 2. Contours of constant shear stress for $k = 3$.

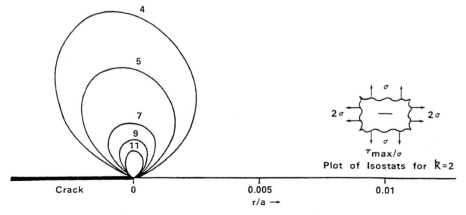

Fig. 3. Contours of constant shear stress for $k = 2$.

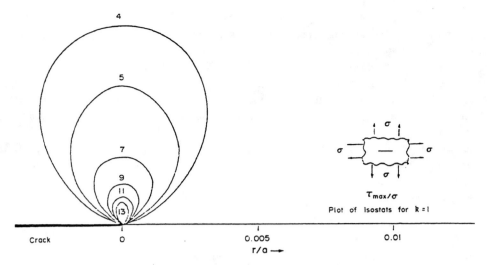

Fig. 4. Contours of constant shear stress for $k = 1$.

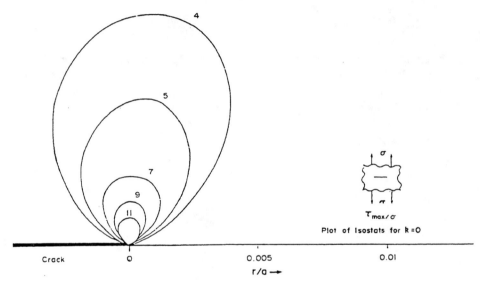

Fig. 5. Contours of constant shear stress for $k = 0$.

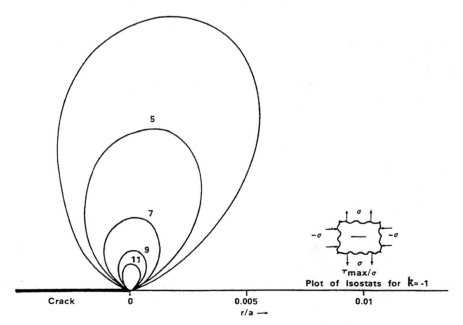

Fig. 6. Contours of constant shear stress for $k = -1$.

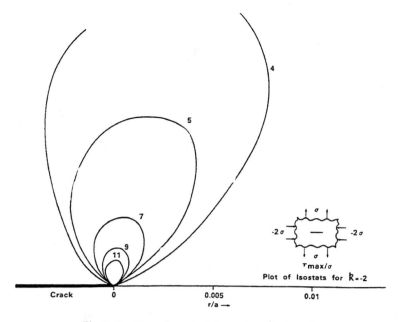

Fig. 7. Contours of constant shear stress for $k = -2$.

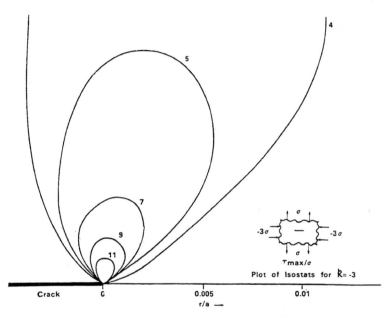

Fig. 8. Contours of constant shear stress for $k = -3$.

When the $A = -(1-k)\sigma$ term is disregarded, instead of eqn (3.2), one obtains

$$\tau_m^* = \sigma \sin \theta \sqrt{\left(\frac{a}{8r}\right)}. \tag{3.3}$$

A measure of the error for the maximum shear stress associated with the use of (3.3), as against eqn (3.2), can be obtained from the ratio τ_m/τ_m^* which is shown in Figs. 9–11 for illustrative purposes.

Since plastic deformation is connected with maximum shear, the above results imply that the extent of plastic yield is significantly affected by the nature of the boundary loading. From the isochromatic patterns shown in Figs. 4 and 6, one expects a larger yield region for tension–compression loading compared to tension–tension loading. Indeed, this prediction is verified by the finite-element solutions in the nonlinear analysis which will be discussed in later sections.

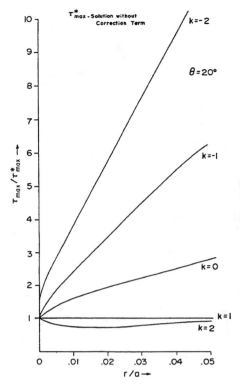

Fig. 9. Error in the conventional shear stress calculation near the crack tip at $\theta = 20°$.

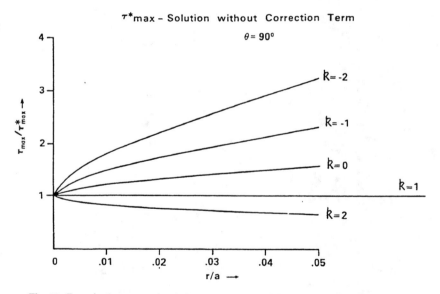

Fig. 10. Error in the conventional shear stress calculation near the crack tip at $\theta = 90°$.

4. ANGLE OF INITIAL CRACK EXTENSION

Another aspect of the crack problem for which a one-term characterization of the crack border region stress appears to be insufficient, reveals itself in the prediction of the angle of initial crack extension for the biaxially loaded specimen with a centered flat crack. The maximum normal stress criterion maintains that a crack will begin to extend radially along the plane on which the stress normal to this plane attains a positive (tensile) maximum value [19–21]. Sih's hypothesis for the prediction of angle of initial crack extension maintains that crack growth will initiate along the radial direction with orientation θ_0 relative to the x-axis along which the local elastic strain energy density attains a minimum [22]. In other words, for some very small fixed, but unspecified, radial distance r_0 from the crack tip, $0 < (r_0/a) \ll 1$, the

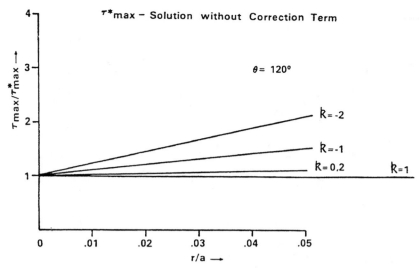

Fig. 11. Error in the conventional shear stress calculation near the crack tip at $\theta = 120°$.

maximum normal stress criterion requires that

$$[t_{\theta\theta}]_{\theta_0} > 0, \qquad \left[\frac{\partial t_{\theta\theta}}{\partial \theta}\right]_{\theta_0} = 0, \qquad \left[\frac{\partial^2 t_{\theta\theta}}{\partial \theta^2}\right]_{\theta_0} < 0, \tag{4.1}$$

while Sih's hypothesis requires that

$$\left[\frac{\partial \phi}{\partial \theta}\right]_{\theta_0} = 0, \qquad \left[\frac{\partial^2 \phi}{\partial \theta^2}\right]_{\theta_0} > 0. \tag{4.2}$$

We do not necessarily agree or disagree with the general adequacy of these criteria; we employ them nevertheless because they afford a basis for comparison, illustrating the point that we wish to make. At $r = r_0$, the two-term representation for stress gives[1, 2],

$$[t_{\theta\theta}]\theta_0 = \sigma\left[\frac{8}{3}\alpha_0 \cos^3 \frac{\theta_0}{2} - (1-k)\sin^2 \theta_0\right], \tag{4.3}$$

$$\left[\frac{\partial t_{\theta\theta}}{\partial \theta}\right]_{\theta_0} = -2\sigma\left[\alpha_0 \cos \frac{\theta_0}{2} + (1-k)\cos \theta_0\right]\sin \theta_0, \tag{4.4}$$

$$\left[\frac{\partial^2 t_{\theta\theta}}{\partial \theta^2}\right]_{\theta_0} = \sigma\left\{\alpha_0\left[\sin \theta_0 \sin \frac{\theta_0}{2} - 2\cos \theta_0 \cos \frac{\theta_0}{2}\right] - 2(1-k)\cos 2\theta_0\right\}, \tag{4.5}$$

$$[\phi]_{\theta_0} = \beta_0\left\{-2\sin^4 \frac{\theta_0}{2} + (3-\kappa)\sin^2 \frac{\theta_0}{2} + (\kappa-1) - (1-k)\sqrt{\left(\frac{2r_0}{a}\right)}\cos \frac{\theta_0}{2}\left[(\kappa-1) + 8\sin^4 \frac{\theta_0}{2}\right.\right.$$
$$\left.\left. - 6\sin^2 \frac{\theta_0}{2}\right]\right\} + (1-k)^2(\kappa+1)\sigma^2/8\mu, \tag{4.6}$$

$$\left[\frac{\partial \phi}{\partial \theta}\right]_{\theta_0} = \beta_0\left\{\sin \theta_0\left[-2\sin^2 \frac{\theta_0}{2} + (3-\kappa)/2\right] - (1-k)\sqrt{\left(\frac{2r_0}{a}\right)}\sin \frac{\theta_0}{2}\right.$$
$$\left. \times \left[-(\kappa+11)/2 + 5\sin^2 \theta_0 + 5\sin^2 \frac{\theta_0}{2}\right]\right\}, \tag{4.7}$$

$$\left[\frac{\partial^2 \phi}{\partial \theta^2}\right]_{\theta_0} = \beta_0\left\{\left(\frac{3-\kappa}{2} - 2\sin^2 \frac{\theta_0}{2}\right)\cos \theta_0 - \sin^2 \theta_0 - (1-k)\sqrt{\left(\frac{2r_0}{a}\right)}\left[-(\kappa+11)\cos \frac{\theta_0}{2}\right.\right.$$
$$\left.\left. + 15 \sin \theta_0 \sin \frac{\theta_0}{2} + 10\sin^2 \theta_0 \cos \frac{\theta_0}{2} + 20 \sin 2\theta_0 \sin \frac{\theta_0}{2}\right]\bigg/4\right\}, \tag{4.8}$$

where

$$\alpha_0 \equiv \frac{3}{8}\sqrt{\left(\frac{a}{2r_0}\right)},$$

$$\beta_0 \equiv \sigma^2 a/8\mu r_0.$$

If the second term is not included in the series representation, one can readily see that, at $\theta_0 = 0$,

$$t_{\theta\theta} = \sigma \sqrt{\left(\frac{a}{2r_0}\right)} > 0,$$

$$\frac{\partial t_{\theta\theta}}{\partial \theta} = 0,$$

$$\frac{\partial^2 t_{\theta\theta}}{\partial \theta^2} = -2\sigma\alpha_0 < 0,$$

$$\frac{\partial \phi}{\partial \theta} = 0,$$

$$\frac{\partial^2 \phi}{\partial \theta^2} = \beta_0(3-\kappa)/2 > 0,$$

and this implies the crack will *always* extend along x-axis regardless of the applied load condition according to either the maximum normal stress criterion or the minimum strain energy density criterion. We note, on the contrary that the biaxial load tests of Kibler and Roberts[4] show that the angle of initial crack extension is well above the value of zero when the biaxial load factor $k \geq 1.3$, and increases as k increases. However, if we do have the second term included, according to eqn (4.5), when

$$k - 1 > \alpha_0, \tag{4.9}$$

$[\partial^2 t_{\theta\theta}/\partial \theta^2]_{\theta_0} = 0$ is no longer negative, and hence $\theta_0 = 0$ is no longer a solution to the problem of initial crack extension. Also one can show that, if inequality (4.9) is satisfied[3],

$$\theta_0 = 2\cos^{-1}\{[\alpha_0 + \sqrt{(\alpha_0^2 + 8(k-1)^2)}]/4(k-1)\} \tag{4.10}$$

satisfies all three requirements in (4.1). For example, at $\alpha_0 = 1$ which gives the acceptable value $r_0/a = 0.070$, the variation of $t_{\theta\theta}$ with θ plotted for different values of k is shown in Fig. 12[1]. The positive maximum of $t_{\theta\theta}$ can be seen to occur for nonzero values of θ_0, which increases as k increases, starting in the vicinity of $k = 2$. These curves, although only in qualitative agreement with Kibler and Roberts experimental results, nevertheless demonstrate once again

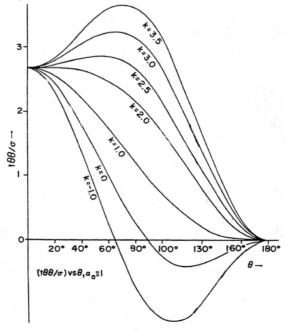

Fig. 12. Variation of normalized tensile stress with orientation and biaxial load factor.

the importance of second term in the expression (2.20). Similarly, according to eqn (4.8), when

$$k - 1 > \alpha_0 \frac{16(3-\kappa)}{3(\kappa+11)}, \tag{4.11}$$

$[\partial^2 \phi/\partial \theta^2]_{\theta_0} = 0$ is no longer positive and hence $\theta_0 = 0$ is no longer a solution to the problem of initial crack extension. The experimental data, reported in Ref. [4, 23] for plexi-glass (PMMA, $\nu \simeq 0.4$), showed that the angle of initial crack extension has already begun to turn from the initial crack plane for $k = 1.3$. This is qualitatively in agreement with what is obtained from inequality (4.11), assuming $r_0 = 0.07a$.

5. LOCAL ENERGY RATE AND J-INTEGRAL

The elastic strain energy per unit thickness over a circular region centered at the crack tip with radius r_0, $0 < r_0/a \ll 1$, can be obtained as [2],

$$\begin{aligned}\Phi &= \int_0^{r_0} \int_{-\pi}^{\pi} \phi(r,\theta) r\, dr\, d\theta \\ &= \frac{\sigma^2 \pi r_0^2}{16\mu} \left\{ (2\kappa - 1)a/r_0 - \frac{32(1-k)(5\kappa - 7)}{15\pi} \sqrt{\left(\frac{a}{2r_0}\right)} + (1-k)^2(\kappa+1) \right\}. \end{aligned} \tag{5.1}$$

The local energy rate, defined as the change of local strain energy with crack size, is obtained as

$$\frac{\partial \Phi}{\partial a} = \frac{\sigma^2 \pi r_0}{16\mu} \left\{ (2\kappa - 1) - \frac{16(1-k)(5\kappa - 7)}{15\pi} \sqrt{\left(\frac{r_0}{2a}\right)} \right\}. \tag{5.2}$$

The first term on the right side of eqn (5.2) is the contribution to the energy rate based on use of only the singular parts of the expressions for local crack tip stresses and displacements. The presence of the second term in the series representations for stress and displacement naturally give rise to the second contribution to the right side of eqn (5.2), which expresses the influence of the biaxial loading on the *local* energy rate. The extent of this influence is shown in Fig. 13, for r_0/a values ranging from 0.01 to 0.07 and for Poisson's ratio $\nu = 0.3$, in the case of plane stress.

At this point, one may ask whether the value of *J*-integral will be affected by using two-term representation for stress and displacement field? Eftis *et al.*[2] examined the *J*-integral along a circular contour centered at the crack tip with radius r by direct calculation involving not only

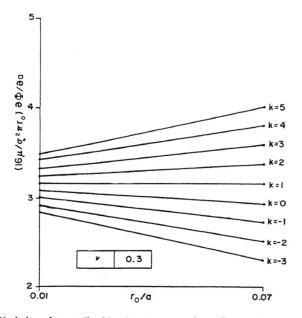

Fig. 13. Variation of normalized local energy rate with radius and biaxial load factor.

the K-term but also the A-term for stresses and displacements. The result obtained was

$$J = \frac{\kappa+1}{8\mu}K^2$$
$$= \frac{\pi a \sigma^2(\kappa+1)}{8\mu}, \tag{5.3}$$

where although calculated with two-term representations, J is independent of k, which indicates the j-integral is insensitive to the presence of the load $k\sigma$ parallel to the crack.

6. FINITE CENTER-CRACKED SPECIMEN

The plane stress problem of a finite rectangular plate, length = $2L$ and width = $2W$, with a centered crack, crack size = $2a$, subjected to uniform biaxial loading, $t_{yy} = \sigma$ along $y = \pm L$ and $t_{xx} = k\sigma$ along $x = \pm W$, has been solved by finite-element analysis combined with a series-type analytical solution around the crack tip[3].

The procedure will be briefly described as follows. Referring to Fig. 14, in region R_1 series-type analytical solutions for stresses (N terms) and displacements ($N+1$ terms) can be written as:

$$t_{ij} = \sum_{n=1}^{N} C_n r^{n/2-1} f_{ij(n)}(\theta), \tag{6.1}$$

$$u_i = \sum_{n=1}^{N} C_n r^{n/2} g_{i(n)}(\theta) + B\delta_{i1}, \tag{6.2}$$

where f_{ij} and g_i are as in eqns (2.15)–(2.19), and where, at this stage, C_n, $n = 1, 2, \ldots N$, and B are the unspecified coefficients to be determined. Let region R_2 be divided into triangular finite elements. If there are M nodal points along semi-circle S_1 and $2M > N+1$, then, at each of the M nodal points, the x and y components of nodal force can be written as

$$F_x = \sum_{n=1}^{N} C_n F_{x(n)},$$
$$F_y = \sum_{n=1}^{N} C_n F_{y(n)}, \tag{6.3}$$

where $F_{x(n)}$ and $F_{y(n)}$ are obtainable directly from the stress distribution specified by $f_{ij(n)}(\theta)$. Since so far the analysis is linear and principle of superposition applies, one is able to obtain the displacements at each nodal point as:

$$u_x = \hat{u}_x + \sum_{n=1}^{N} C_n u_{x(n)},$$
$$u_y = \hat{u}_y + \sum_{n=1}^{N} C_n u_{y(n)}, \tag{6.4}$$

where $\hat{u}_x$ and $\hat{u}_y$ are the contribution to the displacement field due to the applied stresses and $u_{x(n)}$, $u_{y(n)}$ are the contribution to the displacement field due to the nth term of the stresses along semi-circle S_1. At each of the M nodal points along S_1, the displacements calculated from eqns (6.4) should match those calculated from expressions (6.2) with r being specified as r_0. In other words, we have $2M$ equations

$$\hat{u}_x + \sum_{n=1}^{N} C_n u_{x(n)} = \sum_{n=1}^{N} C_n r_0^{n/2} g_{1(n)} + B,$$
$$\hat{u}_y + \sum_{n=1}^{N} C_n u_{y(n)} = \sum_{n=1}^{N} C_n r_0^{n/2} g_{2(n)}, \tag{6.5}$$

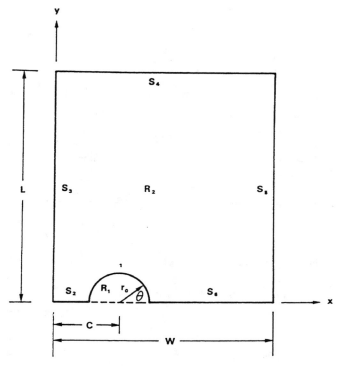

Fig. 14. First quadrant of center-cracked specimen.

and $N+1$ unknowns, C_n, $n = 1, 2, \ldots N$, and B, which will be determined by method of least squares.

Following this procedure, Lee and Liebowitz[3] have proved *analytically* and *numerically* that

$$K = \sigma(\pi a)^{1/2} h_1\left(\frac{L}{a}, \frac{W}{a}\right), \tag{6.6}$$

$$A = -\left[h_2\left(\frac{L}{a}, \frac{W}{a}\right) - k\right]\sigma, \tag{6.7}$$

$$B = -\left[h_3\left(\frac{L}{a}, \frac{W}{a}\right) - k\right]\frac{(\kappa + 1)E}{8\mu}\sigma a, \tag{6.8}$$

and moreover, that all the other coefficients C_n, $n \geq 3$, are independent of the biaxial load factor k. In eqns (6.6)–(6.8), h_1, h_2, and h_3 are correction factors depending on the dimensionless geometric parameters L/a and W/a. The similarity between eqns (2.27)–(2.29) and (6.6)–(6.8) are remarkable. All the results mentioned in Sections 2–5 can be reproduced for the finite rectangular plate and the difference, if there is any, is only quantitative in nature. However, we take this opportunity to make the following point. The stress intensity factor K being independent of biaxial load factor k, means that the biaxial effect reveals itself *only* through the coefficients A and B. This is proved both analytically and numerically, provided the A and B-terms are included in the analysis. On the other hand, if the A-term is not included as part of the analytic solution for region R_1, then the apparent values of stress intensity factor, K^*, will vary with the biaxial load factor k. If the B-term is further excluded from the analysis, the outcome will be even worse. The apparent value K^* normalized by the correct value K is plotted against k in Fig. 15, which once again indicates the importance of the A-term in another aspect.

7. NONLINEAR ANALYSIS

Lee and Liebowitz[13] performed a nonlinear finite-element analysis for a finite center-cracked specimen, of which the material property is characterized by a Ramberg–Osgood type stress–strain relation, subjected to biaxial loading. It was found that there are significant biaxial effects on energy rate (global), J-integral, and stress (strain) intensity factor.

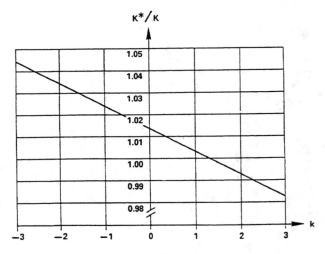

Fig. 15. The apparent stress intensity factor as function of biaxial load factor $c = 0.5$, $l = 2.5$.

Here in this section we briefly describe the procedures used in the nonlinear analysis, which is detailed in Ref. [13]. The Ramberg–Osgood type stress-strain relation can be written as:

$$E\epsilon_{ij} = (1+\nu)s_{ij} + \frac{1-2\nu}{3}t_{kk}\delta_{ij} + \frac{3}{2}\alpha\sigma_e^{n-1}s_{ij}, \qquad (7.1)$$

where effective stress σ_e and stress deviator s_{ij} are defined as:

$$\sigma_e^2 \equiv \frac{3}{2}s_{ij}s_{ij}, \qquad (7.2)$$

$$s_{ij} \equiv t_{ij} - \frac{1}{3}t_{kk}\delta_{ij}. \qquad (7.3)$$

In the case of generalized plane stress, eqn (7.1) reduces to the following matrix form:

$$\begin{vmatrix} \epsilon_x \\ \epsilon_y \\ \gamma_{xy} \end{vmatrix} = \frac{1}{E} \begin{vmatrix} 1+g & -\nu-0.5g & 0 \\ -\nu-0.5g & 1+g & 0 \\ 0 & 0 & 2(1+\nu)+3g \end{vmatrix} \begin{vmatrix} t_{xx} \\ t_{yy} \\ t_{xy} \end{vmatrix}, \qquad (7.4)$$

where

$$g \equiv \alpha\sigma_e^{n-1}, \qquad (7.5)$$

$$\sigma_e^2 = t_{xx}^2 + t_{yy}^2 - t_{xx}t_{yy} + 3t_{xy}^2. \qquad (7.6)$$

One sees that the matrix linking the stresses and strains in eqn (7.4) depends on the effective stress σ_e, as does the stiffness matrix of each triangular finite element. Therefore it is necessary to employ an iteration process to solve this nonlinear problem. The iteration process adopted can be described as follows. For each element and for a specifically given applied stress σ and biaxial load factor k, a trial value g^* is assigned and, after the governing matrix equation is solved, one may obtain the calculated value g^{**}, and this process will be continued until, for each element, the percentage difference between g^* and g^{**} is below some allowable value of error. After the iteration process is completed, it is straightforward to calculate the strain energy density ϕ_I and the complementary energy density ψ_I of the Ith element as follows:

$$\phi_I = \frac{1}{E}\left\{\frac{1+\nu}{3}\sigma_e^2 + \frac{1-2\nu}{6}t_{kk}^2 + \frac{\alpha n}{n+1}\sigma_e^{n+1}\right\}_I, \qquad (7.7)$$

$$\psi_I = \frac{1}{E}\left\{\frac{1+\nu}{3}\sigma_e^2 + \frac{1-2\nu}{6}t_{kk}^2 + \frac{\alpha}{n+1}\sigma_e^{n+1}\right\}_L. \qquad (7.8)$$

Then the global strain energy and complementary energy, per unit thickness, are

$$\Phi = \sum A_I \phi_I, \tag{7.9}$$

$$\Psi = \sum A_I \psi_I, \tag{7.10}$$

where A_I is the area of the Ith element, and the summation is performed over all the elements. Since one can easily prove that

$$\Psi = \oint t_{ij} n_j u_i \, ds - \Phi, \tag{7.11}$$

where the first term of the right hand side of (7.11) is nothing but the line integral of the inner product of stress vector and displacement vector over the entire exterior boundary, the nonlinear energy rate **G** in fixed load and fixed grip situations can be obtained as:

$$\mathbf{G} = \frac{\partial \Psi}{\partial (2a)}\bigg|_{\text{fixed load}} = -\frac{\partial \Phi}{\partial (2a)}\bigg|_{\text{fixed grip}}. \tag{7.12}$$

It is also straightforward to calculate the J-integral

$$J = \oint_\Gamma (\phi \, dy - t_{ij} n_j u_{i,x} \, ds) \tag{7.13}$$

along any closed contour Γ surrounding the crack tip.

As far as the crack tip solution is concerned, we outline some of the Hutchinson analytical solutions in plane stress as follows[5]:

$$t_{ij} = K_\sigma r^{-1/(n+1)} \mathbf{t}_{ij}(\theta), \tag{7.14}$$

$$\epsilon_{ij} = K_\epsilon r^{-n/(n+1)} \boldsymbol{\epsilon}_{ij}(\theta), \tag{7.15}$$

where K_σ and K_ϵ are stress intensity factor and strain intensity factor respectively, $\mathbf{t}_{ij}$ and $\boldsymbol{\epsilon}_{ij}$ are dimensionless functions of θ being detailed in [5, 6], and K_σ and K_ϵ are related as

$$K_\epsilon = \frac{\alpha}{E} K_\sigma^n. \tag{7.16}$$

Moreover, Hutchinson, utilizing the path independency of the J-integral, is able to establish the following J–K_σ relation

$$\frac{\alpha}{E} K_\sigma^{n+1} c_n = J, \tag{7.17}$$

where c_n depends on material hardening coefficient n. In case of plane stress, c_n takes the typical values of 3.86, 3.41, 3.03, 2.87 for $n = 3, 5, 9, 13$ respectively. Equation (7.17) will be used to calculate the stress intensity factor K_σ from the value of J-integral and eqn (7.16) will be used to calculate strain intensity factor K_ϵ.

Yield stress σ_Y, according to usual engineering definition[24], is obtained as

$$\sigma_Y = (0.002 E/\alpha)^{1/n}, \tag{7.18}$$

which means in a simple tension test, when the stress reaches σ_Y, the corresponding strain reaches $\sigma_Y/E + 0.002$. In other words, material loaded to the defined yield stress will be found, upon release of the load, to have suffered a permanent deformation of 0.2%. Now, in the fracture problem, if one considers that a certain part of the specimen has yielded when the effective stress σ_e of that part of the specimen is larger than or equal to the yield stress, then one may find the plastic zone size A_p for a given biaxial loading.

8. NUMERICAL RESULTS

For convenience sake, we define the following dimensionless quantities:

$$\begin{aligned} c &\equiv a/W, \\ l &\equiv L/W, \\ \bar{\alpha} &\equiv \alpha \sigma_Y^{n-1}, \\ \bar{\sigma} &\equiv \sigma/\sigma_Y, \\ \hat{G} &\equiv GE/W\sigma_Y^2, \\ \hat{J} &\equiv JE/W\sigma_Y^2, \\ \hat{K}_\sigma &\equiv K_\sigma/\sigma_Y W^{1/(n+1)}, \\ \eta &\equiv A_p/4LW. \end{aligned} \qquad (8.1)$$

For illustrative purpose, in this work, we fixed

$$\bar{\alpha} = 0.02, \quad n = 13, \quad \nu = 0.33, \quad l = 2.5. \qquad (8.2)$$

In Fig. 16, the dimensionless energy rate $\hat{G}$ is plotted as a function of the dimensionless applied stress $\bar{\sigma}$ for two cases, one of which is uniaxial, i.e. $k = 0$, and the other tension–compression, with $k = -3$, at $c = 0.5$. It is noticed that the biaxial effect on energy rate is enormous and that it is increasing as applied stress increases. In Fig. 17, nonlinear energy rate G, for different values of k, normalized by the linear and uniaxial energy rate $\bar{G}$, is plotted as function of $\bar{\sigma}$. It is noticed that, first, as $\bar{\sigma} \to 0$, $G/\bar{G}$ for different value of k approaches zero, which means not only the nonlinear effect but also the biaxial effect on energy rate disappears; second, for applied stress σ less than $0.3\sigma_Y$, G decreases as biaxial load factor k increases, which means that the

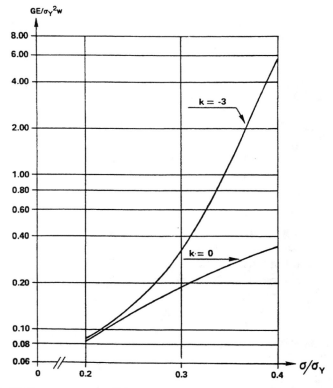

Fig. 16. Energy rate vs applied stress: $\nu = 0.33$, $\bar{\alpha} = 0.02$, $n = 13$, $l = 2.5$, $c = 0.5$.

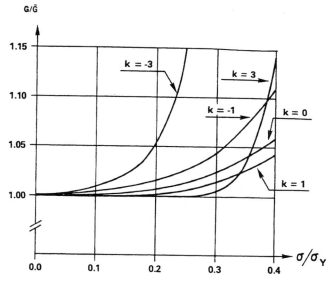

Fig. 17. Biaxial and nonlinear effects on energy rate: $\nu = 0.33$, $\bar{a} = 0.02$, $n = 13$, $l = 13$, $c = 0.5$.

tensile (compressive) stress applied parallel to the crack tends to strengthen (weaken) the specimen. However, at higher value of $\bar{\sigma}$, $G(k = 3)$ becomes even larger than $G(k = -1)$. We believe that $G(\bar{\sigma} = 0.4, k = 3) > G(\bar{\sigma} = 0.4, k = -1)$ is due to the nonlinearity caused mainly by the large stress $k\sigma$ ($= 1.2\sigma_Y$) applied parallel to the crack. The values of the J-integral as functions of the biaxial load factor k, for different values of $\bar{\sigma}$, are plotted in Figs. 18 and 19, from which it may be seen that the general characteristics and the numerical values of the J-integral are similar to those of the energy rate. However, we do detect that the numerical difference between G and J increases as applied stress increases, especially at larger k values (positive or negative). From Fig. 20, it could be seen that G is about 7.8% lower than J when $k = -3$, and it

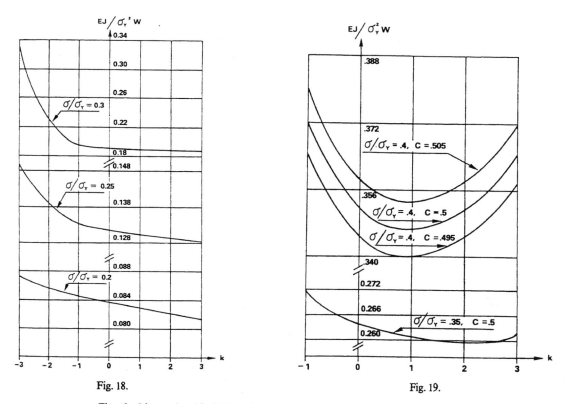

Fig. 18. J-integral vs biaxial load factor: $\nu = 0.33$, $\bar{a} = 0.02$, $n = 13$, $l = 2.5$, $c = 0.5$.

Fig. 19. J-integral vs biaxial load factor: $\nu = 0.33$, $\bar{a} = 0.02$, $n = 13$, $l = 2.5$.

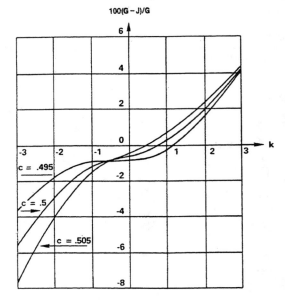

Fig. 20. Percentage difference between energy rate and J-integral: $\nu = 0.33$, $\bar{\alpha} = 0.02$, $n = 13$, $l = 2.5$, $\sigma/\sigma_Y = 0.4$.

is about 4.4% higher than J when $k = 3$ at $c = 0.505$, $\bar{\sigma} = 0.4$. The biaxial effect on stress intensity factor is shown in Fig. 21, from which one can see once again the qualitative difference between linear and nonlinear analyses. However, the biaxial effect on K_σ is less than that on the J-integral and the energy release rate because, according to eqn (7.17), we have the following relation:

$$\frac{K_\sigma(k)}{K_\sigma(k=0)} = \left[\frac{J(k)}{J(k=0)}\right]^{1/(n+1)}. \tag{8.3}$$

The biaxial effect on the strain intensity factor K_ϵ is shown in Fig. 22. Because the dimension of linear stress intensity factor is different from that of the nonlinear stress intensity factor, one cannot compare these two quantities directly. Therefore we make use of the small scale yield stress intensity factor (dimensionless)[5]

$$\bar{K}_{ssy} = [(\sigma/\sigma_Y)^2 c\pi/\bar{\alpha} c_n]^{1/(n+1)} \tag{8.4}$$

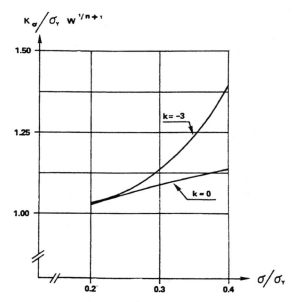

Fig. 21. Stress intensity factor vs applied stress: $\nu = 0.33$, $\bar{\alpha} = 0.02$, $n = 13$, $l = 2.5$, $c = 0.5$.

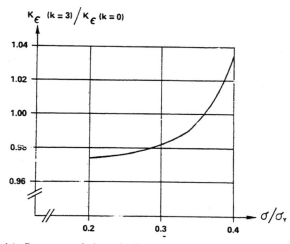

Fig. 22. Biaxial effects on strain intensity factor: $\nu = 0.33$, $\bar{a} = 0.02$, $n = 13$, $l = 2.5$, $c = 0.5$.

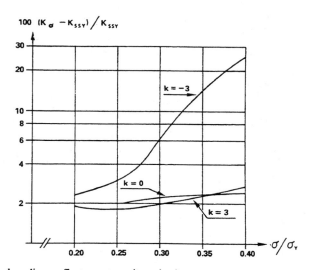

Fig. 23. Biaxial and nonlinear effects on stress intensity factor: $\nu = 0.33$, $\bar{a} = 0.02$, $n = 13$, $l = 2.5$, $c = 0.5$.

and plot the percentage difference between K_σ and K_{ssy} against σ/σ_Y for $k = -3, 0, 3$ in Fig. 23. It was also found that even when $\sigma/\sigma_Y \to 0$, there is still a 2% difference between K_σ and K_{ssy}. This difference is due to the fact that the linear value of the J-integral for an infinite center-cracked specimen has been used for the right hand side of eqn (7.17). In Fig. 24, the normalized size of the plastic zone, in which the effect stress σ_e is larger than or equal to the yield stress σ_Y, is plotted as function of biaxial load factor k for different values of σ/σ_Y. One notices the rapid rise of plastic zone size as the absolute value of k increases.

9. CONCLUSION

In linear elastic fracture mechanics, the importance of the second term in the series representation for crack tip stresses has been demonstrated on the basis of an exact analytical solution for an infinite center-cracked specimen subjected to symmetric and uniform biaxial loading. It has been proved that the stress intensity factor K is independent of biaxial load factor k and only the second term in the series representation of stresses accounts for the uniformly applied stress $k\sigma$ parallel to the crack. Therefore the arbitrary practice of omitting the second term is in effect denying the biaxial effects whatsoever. On the other hand, with the second term being included, one is able to show that:

(1) The orientation of isostatic loop in the first quadrant of the specimen is tilting clockwise as biaxial load factor k is decreasing (see Figs. 2–8).

(2) There are significant biaxial effects on the maximum shear stress near the crack tip (see Figs. 9–11).

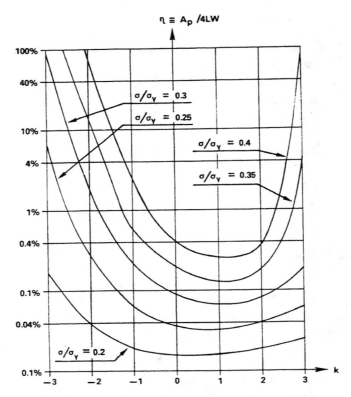

Fig. 24. Biaxial effect on plastic zone size: $\nu = 0.33$, $\bar{a} = 0.02$, $n = 13$, $l = 2.5$, $c = 0.5$.

(3) When the biaxial load factor k becomes larger than a certain value, $\theta_0 = 0$ is no longer the angle of initial crack extension, to be specific, according to maximum normal stress criterion,

$$\theta_0 = \begin{cases} 0, & k - 1 \leq \alpha_0 \equiv \frac{3}{8} \sqrt{\left(\frac{a}{2r_0}\right)} \\ 2\cos^{-1}\{[\alpha_0 + \sqrt{(\alpha_0^2) + 8(k-1)^2}]/4(k-1)\}, & k - 1 > \alpha_0 \end{cases} \quad (9.1)$$

and, according to minimum strain energy density criterion, there is no (relative) minimum strain energy density between zero and 90 degrees if $k - 1 > 16(3 - \kappa)\alpha_0/3(\kappa + 11)$.

(4) The *local* strain energy rate increases linearly with biaxial load factor.

(5) The value of J-integral is independent of biaxial load factor.

For a finite specimen, utilizing series-type analytical solution in the crack tip region and finite-element analysis, numerical results indicate that the stress intensity factor is independent of biaxial load factor. However, if the second term is not included in the analysis, one will mistakenly obtain apprarent values of stress intensity factor K^*, which decreases as k increases.

In the nonlinear analysis, a generalized Ramberg–Osgood stress–strain relation is adopted to characterize the material nonlinearity. The finite center-cracked specimen is analyzed by finite-element method. Numerical results indicate that:

(1) When the applied stress σ, perpendicular to the crack, is small, relatively speaking, the *global* energy rate **G** and J-integral increase rapidly as biaxial load factor k decreases. However, when σ becomes larger, larger tensile stress $k\sigma$ is in correspondence with larger **G** and **J** (see Figs. 16–19).

(2) The numerical difference between energy rate and J-integral, **G**–**J**, increases (decreases) as biaxial load factor k increases (decreases). The absolute value of the difference increases as applied stress σ increases.

(3) The relations between J-integral, stress intensity factor, and strain intensity factor may be expressed as

$$J = \frac{\alpha}{E} K_\sigma^{n+1} c_n = K_\sigma K_\epsilon c_n. \quad (9.2)$$

Therefore, qualitatively speaking, the biaxial effects on stress (strain) intensity factor is similar to that on J-integral.

(4) The plastic zone size, as expected, increases as applied stress σ increases. Moreover, for a fixed σ, plastic zone size increases very rapidly as the absolute value of biaxial load factor increases (see Fig. 24).

Acknowledgement—The authors wish to acknowledge financial support for this work from the Office of Naval Research (N00014-75-C-0946), NASA–Langley Research Center (NGL 09-010-053), and Air Force Office of Scientific Research (AFOSR-76-3099).

REFERENCES

[1] J. Eftis, N. Subramonian and H. Liebowitz, *Engng Fracture Mech.* **8**(4) (1976).
[2] J. Eftis, N. Subramonian and H. Liebowitz, *Engng Fracture Mech.* **9**(4), 753 (1977).
[3] J. D. Lee and H. Liebowitz, *Technical Report*, School of Engineering and Applied Science, The George Washington University, submitted to ONR, 1976.
[4] J. J. Kibler and R. Roberts, *J. Engng Indust.* **92**, 727 (1970).
[5] J. W. Hutchinson, *J. Mech. Phys. Solids* **16**, 13 (1968).
[6] J. W. Hutchinson, *J. Mech. Phys. Solids* **16**, 337 (1968).
[7] J. R. Rice and G. F. Rosengren, *J. Mech. Phys. Solids* **16**, 1 (1968).
[8] N. L. Goldman and J. W. Hutchinson, *Int. J. Solids Struct.* **11**, 575 (1975).
[9] C. F. Shih, Harvard University *Report DEAP S*-10, 1974.
[10] C. F. Shih, *Fracture Analysis, ASTM STP* **560**, 187 (1974).
[11] P. D. Hilton and J. W. Hutchinson, *Engng Fracture Mech.* **3**, 435 (1971).
[12] P. D. Hilton, *Int. J. Fracture* **9**, 149 (1973).
[13] J. D. Lee and H. Liebowitz, *Technical Report*, School of Engineering and Applied Science, The George Washington University, submitted to ONR, 1977.
[14] M. L. Williams, *J. Appl. Mech.* **24**, 109 (1957).
[15] G. R. Irwin. *Proc. Soc. Expl. Stress Analysis*, Vol. 16, p. 69, 1958.
[16] D. G. Smith and C. W. Smith, *Engng Fracture Mech.* **4**, 357 (1972).
[17] P. S. Theocares and E. Gdoutos, *Engng Fracture Mech.* **7**, 331 (1975).
[18] B. Cotterell, *J. Fracture Mech.* **2**, 526 (1966).
[19] F. Erdogen and G. C. Sih, *J. Bas. Engng* **85**, 1 (1963).
[20] V. V. Panasyuk, L. T. Berezhnitskiy and S. Ye. Kovehik, *Prikladnaga Mekhanika* **1**, 1 (1965).
[21] J. G. Williams and P. D. Ewing, *J. Fracture Mech.* **8**, 441 (1972).
[22] G. C. Sih, *Mechanics of Fracture* (Ed. G. C. Sih), Vol. 1. Noordhoff, Leyden (1973).
[23] P. S. Leevers, J. C. Radon and L. E. Culver. *Polymer* **17**, 627 (1976).
[24] E. E. Sechler, *Elasticity in Engineering*. Wiley, New York (1952).

(*Received* 30 *May* 1977)

A Novel Principle for the Computation of Stress Intensity Factors

By H. F. Bueckner*)

Ein ebener Verzerrungszustand in einem elastischen Bereich mit Kerbe oder Riß wird betrachtet. Der Bereich steht unter der Einwirkung von Randkräften; es wird gezeigt, daß der Faktor K der Spannungsintensität an der Wurzel einer Kerbe als Mittelwert der Randkräfte dargestellt werden kann und daß die hierbei auftretenden Gewichtsfunktionen von den Randverschiebungen zweier spezieller Spannungsfelder abgeleitet werden können, deren jedes durch eine „fundamentale Singularität" an der Wurzel der Kerbe und durch die Abwesenheit von äußeren eingeprägten Kräften gekennzeichnet werden kann.

A state of plane strain in a notched or cracked elastic domain under the action of boundary tractions is considered. It is shown that the stress intensity factor K at a root of a notch can be represented in the form of a weighted average of the tractions, and that the weight functions involved can be derived from the boundary displacements of two special stress fields, each of which is characterized by a "fundamental singularity" at the root and by the absence of externally impressed forces.

В упругой области с надрезом или трещиной рассматривается состояние плоской деформации. На кромки области действуют силы. Показано, что в впадине надреза фактор К, определяющий интенсивность напряжения, может быть выражен средним значением сил на кромках. Встречающиеся при этом весовые функции определяются с помощью смещений границы двух специальных полей напряжения, из которых каждое может быть охарактеризовано через „Фундаментальную особенность" в впадине надреза и через отсутсвие действуюйих внешних сил.

1. Introduction

We shall be concerned with stress concentrations and in particular with the intensity factor K which governs the distribution of stresses near the root of a crack or a notch and which is of significance in the mechanics of brittle fracture. We confine our considerations to states of plane strain within the realm of linear elasticity. Reference will be made to the complex plane of $z = x + iy$ as well as to holomorphic functions of z. The following examples are to illustrate important cases of stress concentrations. In Fig. 1a—c a slab of infinite

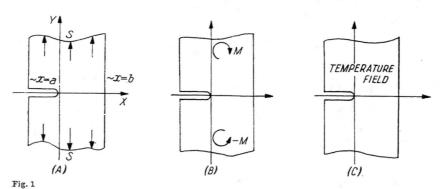

Fig. 1

extension represents the elastic specimen. It is bounded by the lines $x = a$, $x = b$, where $a < 0$, $b > 0$. The slab is notched along the interval $\langle a, 0 \rangle$ of the x-axis. In the case Fig. 1a the slab is under uniform tension at infinity, while Fig. 1b shows the slab exposed to two moments M, $-M$ at $y = \infty$, $y = -\infty$ respectively. In the case of Fig. 1c a temperature field creates the stresses; in order to fix the ideas here, let us assume that the slab is at temperature $T = 0$ for all times $t < 0$, and that thereafter the temperature of the side $x = a$ is forced to be $T = 1$, while the side $x = b$ is kept at $T = 0$. It is fair to assume that the crack will not disturb the flow of heat in the slab, so that $T = T(x, t)$ may be written. In all three cases a stress concentration at the root of the notch ($z = 0$) appears. The following asymptotic relations hold as we approach the root along the x-axis:

$$\lim x^{1/2} |\sigma_y| = \frac{K}{\sqrt{2}}, \qquad x \to +0 \tag{1.1}$$

$$\lim |x|^{-1/2} |v| = \frac{2\sqrt{2}(1-\nu^2)}{E} K, \qquad x \to -0. \tag{1.2}$$

---

*) Engineering Mathematician, General Electric Co., Schenectady, N. Y. Adjunct Professor of Mechanics, Rensselaer Polytechnic Institute, Troy, N. Y. USA.

Here E ist YOUNG's modulus, ν POISSON's parameter, v is the displacement in y-direction along one side of the notch and K is a certain constant, also known as the stress intensity factor, [1]—[6]. K depends on the shape of the specimen and on the impressed load (or temperature field) of the slab. In the case of Fig. 1c K is even time-dependent. Since information about K has a bearing on materials testing and engineering structural applications, the literature of the computation of K shows an abundance of efforts to obtain K one way or the other. Unfortunately the simplicity of the laws (1.1), (1.2) is not reflected by what it takes to compute K reliably and accurately. It is therefore desirable to develop methods of computation, which combine simplicity with scientific acceptability.[*] This is the basic concern of this paper.

Certain simplifications were introduced by the author some years ago [4], [5]. It is useful to mention them again at this place.

1. It is generally much simpler to compute the stress field in the unnotched specimen. In the case of Fig. 1 the stresses in the slab without a notch cannot depend on y. Moreover the shearing stress τ_{xy} must vanish everywhere. From the conditions of equilibrium at $x = a$, $x = b$ it follows that $\sigma_x \equiv 0$. This leaves us with σ_y, and we have

$$\sigma_y = S \text{ (tension at infinity)} \qquad \text{for Fig. 1a} \tag{1.3}$$

$$\sigma_y = 12\, M\, d^{-3}\,(x - x_0), \quad 2x_0 = a + b, \quad d = b - a \qquad \text{for Fig. 1b} \tag{1.4}$$

$$\sigma_y = -2\,G\,\alpha\,\frac{1+\nu}{1-\nu}\left[T(x,t) + \frac{x-b}{d}\right] \qquad \text{for Fig. 1c} \tag{1.5}$$

Here G is the shear modulus and α the thermal expansion coefficient. The field in the unnotched specimen shows no stress concentration at $z = 0$. Let us subtract this field from the one in the notched specimen. The difference field can be interpreted as the response to normal tractions along the notch and to nothing else. Thus the conditions

$$\sigma_y = -p(x), \qquad \tau_{xy} = 0 \qquad \text{along the notch}, \tag{1.6}$$

no tractions along $x = a$, $x = b$ and vanishing stresses at infinity determine the difference field. The function $-p(x)$ is identical with the right hand side of (1.3), (1.4), (1.5) as the case may be. The new field yields the K we want.

2. In view of (1.2) ist suffices to establish a relation between $p(x)$ and the normal displacement $v(x)$ along one side of the notch. In many cases it is easier to describe the relation by assuming $v(x)$ and deriving $p(x)$ from it. For the notched slab this leads to

$$p(x) = \frac{d}{dx}\int_a^0 L(x,t)\,v(t)\,dt \tag{1.7}$$

with $L(x,t)$ representing some singular linear integral operator. The integral must be taken as a CAUCHY principal value. Some details about $L(x,t)$ are in [4]. If $p(x)$ is given, equation (1.7) can be solved for $v(x)$, which in turn, with the aid of (1.2), permits to find K.

The steps so described have the advantage that data along the notch only have to be dealt with. Nevertheless we have to compute $v(x)$ as often as we prescribe $p(x)$. The question arises whether or not another reduction of computational efforts is still possible. After all we do not need all of $v(x)$ but the quantity K, which is nothing more but a functional on some space of functions $v(x)$. Reinterpreting the functional with the aid of the associated functions $p(x)$, we should expect a relation of the form

$$K = \frac{\sqrt{2}}{\pi}\int_a^0 p(x)\,m(x)\,dx \tag{1.8}$$

where $m(x)$ is some weight function which does not depend on $p(x)$. The value $m(t)$ has a simple interpretation. It represents the value of $\pi K/\sqrt{2}$ for that field, which is generated by two concentrated unit forces at $x = t$, the forces acting on opposite faces of the notch and in the direction of the inner normal of the elastic domain. But this interpretation is unattractive, because it makes the function $m(x)$ an abstract of an infinity of fields. It will be shown in this paper, that there is a second interpretation of the function $m(x)$ as normal displacement of one special field. If $m(x)$ can be computed for this field, then (1.8) points the way for an advantageous method to find K. Our paper will pursue this very aspect.

2. The first fundamental problem

Consider some domain A in the z-plane, e. g. the interior of a JORDAN curve B as shown in Fig. 2a. In the absence of body forces, any state of plane strain in A can be expressed by the formulas of MUSKHELISHVILI [7] as follows:

$$2\mu w = \varkappa\,\varphi(z) - z\cdot\overline{\varphi'(z)} - \overline{\psi(z)}; \qquad w = u + iv \tag{2.1}$$

$$\sigma_x = \operatorname{Re}\left[\varphi'(z) + \overline{\varphi'(z)} - z\cdot\overline{\varphi''(z)} - \overline{\psi'(z)}\right] \tag{2.2}$$

$$\sigma_y = \operatorname{Re}\left[\varphi'(z) + \overline{\varphi'(z)} + z\cdot\overline{\varphi''(z)} + \overline{\psi'(z)}\right] \tag{2.3}$$

$$\tau_{xy} = \operatorname{Im}\left[\bar{z}\,\varphi''(z) + \psi'(z)\right]. \tag{2.4}$$

Here

$$\mu = E/2(1+\nu), \qquad \varkappa = 3 - 4\nu; \tag{2.5}$$

u, v are the displacements in x-direction and y-direction respectively. The functions $\varphi''(z), \psi'(z)$ are holomorphic in A. If A is bounded and simply connected, then φ, ψ are also holomorphic in A. If φ, ψ yield a stress field $\sigma_x, \sigma_y, \tau_{xy}$ then all other pairs $\varphi^* = \varphi + i\alpha z + \beta$, $\psi^* = \psi + \gamma$ with arbitrary complex constants β, γ and

---

[*] FICHERA [9] presents such a method in order to deal with the stress concentration at a concave corner.

an arbitrary real constant α yield the same stress field. The constants α, $\varkappa \beta - \gamma$ represent the family of rigid body motions. It should be emphasized, that φ, ψ depend on the choice of the coordinate system x, y. If we change from z to

$$\zeta = \varepsilon (z - a), \qquad |\varepsilon| = 1 \tag{2.6}$$

then the pair of functions

$$\varphi_1(\zeta) = \varepsilon\, \varphi(z), \qquad \psi_1(\zeta) = \bar{\varepsilon}\, [\psi(z) + \bar{a}\, \varphi'(z)] \tag{2.7}$$

yields the same field of stresses and displacements. We observe that

$$\bar{\zeta}\, \varphi_1'(\zeta) + \psi_1(\zeta) = \bar{\varepsilon}\, [\bar{z}\, \varphi'(z) + \psi(z)]. \tag{2.8}$$

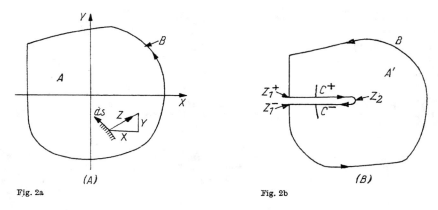

Fig. 2a Fig. 2b

If ds is an oriented arc-element in A, then the components X in x-direction, Y in y-direction of the traction which attacks the elastic material to the left of ds are given by the formula

$$Z\, ds = (X + i\, Y)\, ds = - i\, dP \qquad \text{(Fig. 2a)} \tag{2.9}$$

where

$$P = P(z) = \varphi(z) + z\, \overline{\varphi'(z)} + \overline{\psi(z)}. \tag{2.9'}$$

The function $P(z)$ is not analytic in general. Note that (2.1) can also be written in the form

$$2\, \mu\, w = (\varkappa + 1)\, \varphi(z) - P(z). \tag{2.1'}$$

The strain energy density W per unit length can be expressed in the form

$$\begin{aligned} 2\, E\, W &= (1 - \nu^2)\, (\sigma_x^2 + \sigma_y^2) + 2\, (1 + \nu)\, (\tau_{xy}^2 - \sigma_x\, \sigma_y) \\ &= 8\, (1 + \nu)\, (1 - 2\, \nu)\, (\operatorname{Re} \varphi')^2 + 2\, (1 + \nu)\, |\bar{z}\, \varphi''(z) + \psi'(z)|^2. \end{aligned} \tag{2.10}$$

Of practical importance are those states of plane strain for which the integral

$$H = \int_A W\, dx\, dy \tag{2.11}$$

exists. This means that finite energy (per unit length) is stored in A. In most engineering applications only such states are encountered. Yet this investigation will lead to special fields of infinite energy in A, even if A is of bounded area. For this reason the distinction $H < \infty$ and $H = \infty$ has to be kept in mind.

Throughout this paper we have to refer to the first fundamental problem of elasticity. The problem requires to determine fields of displacements and stresses in a domain A whose boundary is exposed to a self-equilibrated system of tractions. For domains and boundary tractions of a very general nature K. O. FRIEDRICHS [12] and later G. FICHERA [10], [11] have shown the unique existence of a stress field of finite energy which responds to the tractions. More precisely the stresses approach limit values in some sense as the boundary is approached from within A, and the limit values are compatible with the prescribed tractions. Naturally the field of displacements is not unique, but its ambiguity is within the group of rigid body motions. The results in [12] cover in particular states of plane strain in domains with a notch or crack. While the questions of existence and uniqueness are answered, some details remain for further investigation, e. g. the way the stresses approach limit values as one moves to the boundary. In this context the work by Russian authors has to be mentioned. Let the domain A be simply connected and bounded by a smooth JORDAN curve B (Fig. 2a) whose curvature exists as a HÖLDER-continuous*) function of arclength s. Let $f(s)$ be a complex-valued function, continuously defined on B and such that the derivative $df/ds = \dot{f}$ is H-continuous on B. The quantity $Z = - i\, \dot{f}$ can be interpreted as traction on B. The continuity of $f(s)$ implies that the tractions distributed

---

*) A function $h(s)$, defined in $\langle a, b \rangle$ is called HÖLDER-continuous in that interval, if $|h(s) - h(t)| \leq h_0\, |s - t|^\lambda$, where $h_0 \geq 0$ and $\lambda > 0$ are certain constants. More generally a function $g(z)$, defined on some set $\mathfrak{M}$ of the z-plane is called HÖLDER-continuous on $\mathfrak{M}$, if $|g(z) - g(z')| \leq g_0\, |z - z'|^\lambda$ with certain constants $g_0 \geq 0$ and $\lambda > 0$. Unless we say otherwise, we assume $0 < \lambda \leq 1$. We shall use the abbreviation H-continuous for HÖLDER-continuous.

over B have a vanishing resultant. In order to obtain a vanishing moment of them, it is necessary and sufficient to impose the condition

$$\text{Re} \int_B \bar{z}\, df = 0. \tag{2.12}$$

Under these assumptions MUSKHELISHVILI [7] shows the existence of two functions φ, ψ, each holomorphic in A, such that the associated $P(z)$ can be continuously extended to the closure $\bar{A}$ of A and such that P coincides with f on B. This basic result can be extended to domains which are bounded by a finite number of JORDAN curves with H-continuous curvature; here reference is made to SHERMAN, whose work has found a sufficiently detailed description in [7]. Another extension is due to MAGNARADZE [13], [14], [15]. He admits boundaries B with a countable number of corners, but not with cusps. In view of these results and also with regard to the phenomena of stress concentration at the root of a crack or a notch, we shall introduce some general concepts and statements, which will serve as a basis for the investigations in this paper.

We begin with geometric considerations. Take an oriented JORDAN curve B and one of the two domains of the z-plane, separated by B, e. g. the domain of its interior. Let B be so oriented that the domain (A) is to its left. From some point z_1 of B (see Fig. 2b) we draw a JORDAN arc into A. Let z_2 be its other endpoint. Orienting this arc (C) such that it goes from z_1 to z_2 we denote it by C^+; C^- will be the same arc but with opposite orientation. The points $z \in C$, $z \neq z_2$ will be counted twice, and we set $z = z^+$, $z = z^-$ in order to indicate that the point belongs to C^+ or to C^- respectively. We consider now the closed and oriented curve G, which follows the route from z_1^- along B to z_1^+, from z_1^+ to z_2 along C^+ and finally from z_2 along C^- to z_1^-. G is the boundary of a new domain A', which consists of all those points of A which are not on C. G leaves A' to the left. The construction of G can be generalized by cutting the domain A along a finite number of mutually nonintersecting JORDAN arcs each of which goes from some point of B into A. The outcome is a new domain A', consisting of all points of A not on a cut, and an oriented closed curve G which appears as boundary of A'. G has doublepoints along the cuts. From now on, such a closed curve G, in one orientation or the other, will be called a finite contour. We also admit C^+, C^- as a contour if z_1 is counted once. A finite contour has at least one point z_0 which is not a doublepoint. Consider the map $z' = 1/(z - z_0)$. It transforms G into some infinite curve G', which carries an orientation due to that of G and to the map. We consider $z' = \infty$ as point of G', and we count it once only. We denote G' as infinite contour; it appears as boundary of a certain domain.

An oriented JORDAN arc will be called regular, if it possesses a curvature which is an H-continuous function of arclength s. If a finite contour can be decomposed into a finite number of regular arcs, each with the orientation induced by the contour (G), then we call G a finite regular contour. The image of G under the map $z' = 1/(z - z_0)$ will be referred to as an infinite regular contour.

In the sequel the word contour will always mean a finite or an infinite regular contour. Any contour can be parametrized with respect to the arclength s in the form $z = z(s)$, where s varies in some closed interval $-a \leq s \leq a$ ($a = \infty$ is admitted) and where $z(a) = z(-a)$; also $z(a)$ is counted once only. If $z(t) = z(t')$, we count $z(t)$, $z(t')$ as different if only $0 < |t - t'| < 2a$. The orientation shall be that of increasing s.

Let a domain A in the z-plane be bounded by a finite number of contours $G_1, G_2, \ldots, G_n$, each so oriented that A is to its left. Any two of them shall have no point $z \neq \infty$ in common. Along G_k the derivative dz/ds cannot have more than a finite number of points of discontinuity, the corners of A. We measure the angle at a corner as a material one, indicating the angular area of A at the corner; thus the angle is 2π at $z = 0$ in Fig. 1. We shall call A a regular domain, if the angle at any corner $z \neq \infty$ is greater than zero. The reader checks easily that Fig. 1 and Fig. 3 show regular domains.

We turn to the first fundamental problem for a regular domain. In this case the tractions on the various contours G_k are prescribed, subject to the condition that they form a self-equilibrated system with respect to the boundary of A as a whole. It has been shown [7] that the problem can be reduced to one, for which the tractions on each G_k have neither resultant force nor moment. Consequently it suffices to prescribe $P(z)$ on the boundary of A such that $P(z)$ is continuous on each G_k and that $\text{Re} \int_{G_k} \bar{z}\, dP = 0$ for $k = 1, 2, \ldots, n$.

In this context we shall impose some conditions on the behavior of $P(z)$ on the contours. Reference will be made to certain open arcs on a contour G, where open arc means a JORDAN arc without its endpoints. A function $f(z)$, defined on G, will be called a load function on G, if it meets these conditions:

(1) f is continuous on G; an at most finite number of exceptional points on G exists (including the corners and, if G is infinite, the point $z = \infty$) such that $\dot{f} = df/ds$ is H-continuous on any open arc not containing an exceptional point; moreover, if G is infinite, $\dot{f}$ for sufficiently large $|s|$ shall satisfy an H-condition of the form

$$\left| s^2 \dot{f} \Big|_{s_1}^{s_2} \right| \leq f_0 \cdot \left| \frac{1}{s_1} - \frac{1}{s_2} \right|^\lambda \tag{*}$$

with certain constants $f_0 \geq 0$, $\lambda > 0$; furthermore $s^2 \dot{f} \to 0$ as $|s| \to \infty$.

(2) $\quad \text{Re} \int_G \bar{z}\, df = 0.$

If G is infinite, then condition (1) implies that $|\dot{f}| \leq f_0 |s|^{-2-\lambda}$ for large $|s|$, $\bar{z} \dot{f}$ is L-integrable on G, and the moment condition (2) makes sense. We introduce two more definitions:

Definition 1: Let A be a regular domain. A state of plane strain in A, described with the aid of two functions φ, ψ will be named **regular**, if
  (1) φ, ψ are holomorphic in A and φ, P are continuous in the closure $\bar{A}$ of A;
  (2) $P(z)$ is a load function on any of the contours which bound A;
  (3) φ', ψ are continuous in $\bar{A}$ with the exception of at most a finite number of points $z' \neq \infty$ on the boundary, near which $\varphi'(z) = O(|z - z'|^{-1/2})$;
  (4) $\varphi'(z) = O(|z|^{-1-p})$ with some constant $p > 0$, as $z \to \infty$;
  (5) the state has finite energy in A.

It should be emphasized that distinct points of the boundary are also considered as different in A. The relation $f(z) = O(g(z))$ means that f/g stays bounded as $z \to z_0$, where z_0 is a point in whose vicinity f, g are compared with one another; the relation $f(z) = o(g(z))$, which will be found elsewhere in this paper, means $f/g \to 0$ as $z \to z_0$.

Definition 2: A regular domain A will be called **regular elastic**, if (rigid body motions disregarded) a regular state φ, ψ responds uniquely to any set of load functions $f(z)$ on the various contours, which form the boundary of A, in the sense that $\dot{P} = \dot{f}$ on the boundary.

If $z = \infty$ is on the boundary of A, we can modify φ and ψ by additive constants, such that $P(\infty) = 0$ for a regular state. This does not change the stress field. For this reason we can and we shall assume that $f(\infty) = 0$ for the load function.

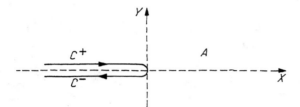

Fig. 3

Let us now describe at least one example of a regular elastic domain. We take the z-plane with a cut along the negative real axis (Fig. 3). Its boundary consists of an infinite notch or crack with the faces C^+, C^-. It is so oriented that A is to its left. The most general load function takes the form

$$f(x) = \pm Q(x) + R(x) \qquad \text{on } C^+, C^- \text{ respectively.} \tag{2.13}$$

Q, R must be continuous in $-\infty < x \leq 0$; in addition $Q(0) = 0$, $Q \to 0$, $R \to 0$ as $x \to -\infty$. With the exception of a finite number of points $x_n < x_{n-1} < \cdots < x_1 < 0$ the derivatives $Q'(x), R'(x)$ exist everywhere, and they are H-continuous on any interval which does not contain an abscissa x_k. The functions show the behavior

$$\begin{aligned} Q'(x) &= O(|x|^{-2-\lambda}), & R'(x) &= O(|x|^{-2-\lambda}) \\ Q(x) &= O(|x|^{-1-\lambda}), & R(x) &= O(|x|^{-1-\lambda}) \end{aligned} \qquad \text{as } x \to -\infty; \quad \lambda > 0. \tag{2.14}$$

The functions $Q_*(x) = Q(1/x)$, $R_*(x) = R(1/x)$ have H-continuous derivatives in $1/x_n < x < 0$, and we may set

$$Q_*(0) = Q'_*(0) = R_*(0) = R'_*(0) = 0 \tag{2.15}$$

in accord with that continuity. The relations (2.14) are equivalent with

$$\begin{aligned} Q'_*(x) &= O(|x|^\lambda), & R'_*(x) &= O(|x|^\lambda) \\ Q_*(x) &= O(|x|^{1+\lambda}), & R_*(x) &= O(|x|^{1+\lambda}) \end{aligned} \qquad \text{as } x \to 0. \tag{2.14*}$$

The moment condition (2) above takes the special form

$$\mathrm{Re} \int_{-\infty}^{0} x\, Q'(x)\, dx = -\mathrm{Re} \int_{-\infty}^{0} Q(x)\, dx = 0.$$

We shall not have to invoke it. — In what follows $\sqrt{z}$ represents the main branch of the square root in A, i.e. $\sqrt{z}$ is positive on the positive real axis. All integrals below go from $-\infty$ to zero, unless other bounds are indicated. For $z \in A$ the following functions $q, r, \varrho, \varphi, \psi$ are well-defined and holomorphic in A:

$$q(z) = \frac{1}{2\pi i} \int \frac{Q(t)\, dt}{t - z}, \qquad r(z) = \frac{-\sqrt{z}}{2\pi} \int \frac{R(t)\,|t|^{-1/2}\, dt}{t - z}, \tag{2.16}$$

$$\varphi(z) = q(z) + r(z); \qquad \varrho(z) = -\overline{q(\bar{z})} + \overline{r(\bar{z})}; \qquad \psi(z) = \varrho(z) - z\,\varphi'(z). \tag{2.17}$$

We have

$$P(z) = \varphi(z) + \overline{\varrho(z)} + (z - \bar{z})\,\overline{\varphi'(z)}. \tag{2.18}$$

The following formulas are consequences of (2.16):

$$\sqrt{z}\, r(z) = S + \frac{1}{2\pi} \int \frac{R(t)\,|t|^{1/2}\, dt}{t - z} \qquad \text{with } S = \frac{1}{2\pi} \int R(t)\,|t|^{-1/2}\, dt \tag{2.19}$$

$$q'(z) = \frac{1}{2\pi i} \int \frac{Q(t)\, dt}{(t - z)^2} = \frac{1}{2\pi i} \int \frac{Q'(t)\, dt}{t - z}; \qquad \sqrt{z} \cdot r'(z) = \frac{1}{2\pi} \int \frac{R'(t)\,|t|^{1/2}\, dt}{t - z} \tag{2.20}$$

$$z^2\, q'(z) = \frac{1}{2\pi i} \int \frac{Q_*(\tau)\, d\tau}{\tau - \xi}, \qquad z \cdot \sqrt{z} \cdot r'(z) = \frac{-1}{2\pi} \int \frac{R_*(\tau)\,|\tau|^{1/2}\, d\tau}{\tau - \xi} \qquad \text{where } \xi = 1/z. \tag{2.21}$$

We must now consider the behavior of the functions in (2.16)–(2.21) as $z \to x^+$, $z \to x^-$ from within A, where x^+, x^- are points of C^+, C^- respectively. In particular we must investigate the approaches $z \to 0$ and $z \to \infty$ ($\xi \to 0$). In this context well-known theorems about fundamental properties of Cauchy integrals will have to be applied. For details the reader is referred to the literature and with emphasis to Muskhelishvili's book "Singular integral equations" [8], where chapter 4 offers all necessary information. The numerators of the integrands in (2.16) are H-continuous on any closed interval of the negative t-axis. Therefore $q(z)$, $r(z)$ can be continuously extended to the closure $\overline{A}$ of A, the points $z = 0$, $z = \infty$ possibly excluded. The boundary values of q, r on C^+, C^- are given by the well-known Plemelj-formulas. These formulas yield

$$q(x^+) - q(x^-) = Q(x); \qquad r(x^+) - r(x^-) = R(x) \tag{2.22}$$

for opposite points x^+, x^- on the crack. — The derivatives $q'(z)$, $r'(z)$ can be continuously extended to $\overline{A}$, the points $z = x_k$, $z = 0$, $z = \infty$ possibly excluded. The existence of this extension is due to the circumstance, that the functions $Q'(t)$, $|t|^{1/2} R'(t)$ in (2.20) are H-continuous on any finite interval (a, b) of the negative t-axis, which does not contain a point x_k. Turning to $q'(z)$, we derive from (2.20)

$$2\pi i q'(z) = \int_{-\infty}^{x_1} \frac{Q'(t)\, dt}{t - z} + \int_{x_1}^{0} \frac{Q'(t) - Q'(0)}{t - z}\, dt + Q'(0) \log \frac{z}{z - x_1}. \tag{2.23}$$

The first integral on the right represents a function of z which is holomorphic in $|z| < x_1$, while the second integral is a continuous function of z in the intersection of $\overline{A}$ with the disk $|z| < x_1$. The latter is related to the H-continuity of the numerator in the second integral and to the vanishing of that numerator at $t = 0$. From (2.23) it follows that

$$q'(z) = L \log z + q_1(z), \qquad L = Q'(0)/2\pi i \tag{2.24}$$

where $q_1(z)$ is continuous in $\overline{A}$ for $|z| < |x_1|$. The function $\log z$ represents the main branch of the log-function in A, i. e. it takes real values on the positive real axis. The reasoning leading to (2.24) is easily extended. In particular we find the function $\sqrt{z}\, r'(z)$ to be continuous in $\overline{A}$ for $|z| < |x_1|$; moreover

$$\lim_{z \to 0} \sqrt{z}\, r'(z) = T = -\frac{1}{2\pi} \int R'(t) \cdot |t|^{-1/2}\, dt. \tag{2.25}$$

A more refined result can be obtained. We write

$$r'(z) - \frac{T}{\sqrt{z}} = -\frac{\sqrt{z}}{2\pi} \int \frac{R'(t)\, |t|^{-1/2}\, dt}{t - z}; \tag{2.26}$$

the H-continuity of $R'(t)$ in $(x_1, 0)$ implies (see [8], p. 74)

$$-\frac{1}{\pi} \int \frac{R'(t)\, |t|^{-1/2}\, dt}{t - z} = \frac{R'(0)}{\sqrt{z}} + O(|z|^{p'-1/2}), \qquad p' > 0, \tag{2.26'}$$

so that

$$r'(z) - \frac{T}{\sqrt{z}} = \frac{1}{2} R'(0) + O(|z|^{p'}). \tag{2.26''}$$

The behavior of q', r' near a point $x_k \in C^+$ or $x_k \in C^-$ can be found in analogy to (2.23), (2.24). Leaving details to the reader we simply state here that

$$q'(z) = O(\log d), \qquad r'(z) = O(\log d), \qquad d^{-1} = |z - x_k|; \qquad \text{as } z \to x_k. \tag{2.27}$$

From (2.24), (2.26) the continuity of q, r at $z = 0$ follows. As for $z \to \infty$, we use (2.21). The functions Q_*, R_* vanish at $x = 0$ and are H-continuous in $(1/x_n, 0)$. This implies that $z^2 q'(z)$, $z \sqrt{z}\, r'(z)$ are continuous at $z = \infty$ on $\overline{A}$. What we have ascertained so far, permits us to state already, that φ, ϱ, P are continuous in $\overline{A}$, and that furthermore φ', ψ satisfy conditions (3), (4) of definition 1. One may set $p = 1/2$ as regards condition (4). The function P takes the form

$$P(x) = \varphi(x) + \overline{\varrho(x)} = q(x) - q(\overline{x}) + r(x) + \overline{r(x)} \quad \text{on } B, \tag{2.28}$$

where x, $\overline{x}$ represent opposite points on the crack. A comparison with (2.13), (2.22) yields $P(x) = f(x)$ on B.

It remains to show that the state φ, ψ has finite energy and that it responds uniquely to (2.13). The expression (2.10) for EW contains the terms $\mathrm{Re}\, \varphi'$ and $\overline{z}\, \varphi'' + \psi' = \varrho' - \varphi' + (\overline{z} - z)\, \varphi''$. We must show that these functions are square integrable over A. But our results on q', r' imply already that φ', ϱ' are square integrable, and it remains to show the same for the function $(z - \overline{z})\, \varphi''$. It is possible to find a common upper bound M to the functions $|Q'(t)|$, $|Q'_*(\tau)|$, $|R'(t)|$, $|R'_*(\tau)|$. Equations (2.20) on $q'(z)$ yield, after differentiation,

$$(z - \overline{z})\, q''(z) = \frac{1}{2\pi i} \int \frac{Q'(t)\, (z - \overline{z})\, dt}{(t - z)^2} \tag{2.29}$$

and

$$|(z - \overline{z})\, q''(z)| < \frac{M}{\pi} \int_{-\infty}^{\infty} \frac{|y|\, dt}{(x - t)^2 + y^2} = M. \tag{2.30}$$

In the same vein we find with the aid of (2.20), (2.21)

$$\left| (z - \overline{z})\, \frac{d}{dz} \left(\sqrt{z}\, r'(z) \right) \right| < M, \tag{2.31}$$

$$\left| (z - \overline{z})\, \frac{d}{dz} \left(z^2\, q'(z) \right) \right| = \left| (\xi - \overline{\xi})\, \frac{d}{d\xi} \left(z^2\, q'(z) \right) \right| < M, \tag{2.32}$$

$$\left| (z - \overline{z})\, \frac{d}{dz} \left(z \sqrt{z}\, r'(z) \right) \right| < M. \tag{2.33}$$

Inequalities (2.30)–(2.33) imply

$$|(z - \overline{z})\, \varphi''(z)| \begin{cases} \leq c_1 |z|^{-1/2} & \text{for } 0 < |z| \leq c_3 \\ \leq c_2 |z|^{-3/2} & \text{for } |z| > c_3 \end{cases} \tag{2.34}$$

with suitable constants c_1, c_2, c_3. From (2.34) it follows that $(z - \bar{z}) \cdot \varphi''(z)$ is square integrable over A. At this juncture we can state that φ, ψ is regular.

Finally we deal with the question of uniqueness. While this can be decided on the basis of energy considerations, we can also settle the question by using the properties of continuity, as they are associated with a regular state. Let us assume that we have a regular state in A, say φ_1, ψ_1, for which $P_1(z)$ vanishes on the boundary. We write $\varrho_1(z) = \psi_1(z) + z\,\varphi_1(z)$. $P_1 = 0$ on B leads to

$$\varphi_1(z) + \overline{\varrho_1(z)} = 0, \qquad z \in C^+, z \in C^-. \tag{2.35}$$

Let us introduce the function

$$\varrho_2(z) = \overline{\varrho_1(\bar{z})}. \tag{2.36}$$

It is holomorphic in A and continuous in $\overline{A}$. Condition (2.35) can be rewritten as

$$\varphi_1(x) + \varrho_2(\bar{x}) = 0 \quad \text{on } C^+, C^-. \tag{2.37}$$

Here $x, \bar{x}$ denote opposite points on the boundary. From (2.37) it follows that $F(z) = \varphi_1(z) - \varrho_2(z)$ takes the same value at $x, \bar{x}$. Since $F(z)$ is continuous in $\overline{A}$, $F(z)$ is necessarily an entire function. The continuity of F at $z = \infty$ implies boundedness of F and, by LIOUVILLE's theorem, $F(z) \equiv \text{const}$. Without loss of generality we may assume $\varphi_1(0) = 0$; from (2.37) it follows that $\varrho_2(0) = 0$. Therefore $F(z) \equiv 0$, hence $\varphi_1 \equiv \varrho_2$. Condition (2.37) leads now to

$$\varphi_1(x) + \psi_1(\bar{x}) = 0. \tag{2.38}$$

The function $\varphi_2(z) = \sqrt{z}\,\varphi_1(z)$ is holomorphic in A, continuous in $\overline{A}$ for $z \neq \infty$, and it takes, by virtue of (2.38) the same value on opposite points $x, \bar{x}$ of B. For this reason $\varphi_2(z)$ is an entire function. But $\varphi_2(0) = 0$, therefore $\varphi_1^2(z) = \varphi_2(z)/z$ is also an entire function. Its continuity in $\overline{A}$, leads, (as before with $F(z)$), to the result $\varphi_1(z) \equiv 0$. Altogether we have found $\varphi_1(z) \equiv 0$, $\varrho_2(z) \equiv 0$, $\varrho_1(z) \equiv 0$; this establishes the uniqueness of the state φ, ψ by (2.17).

3. Normal and fundamental singularities of fields in a regular elastic domain

We turn to a description of phenomena of stress concentration near the root of a crack. We shall assume that the root is at $z = 0$ and that the crack leaves the origin in the direction of the negative real axis. Fig. 1a shows a typical configuration. As in the case of the domain of Fig. 3, it is useful to express the field not with the aid of φ, ψ but rather with φ and ϱ, where $\varrho(z) = \psi(z) + z\varphi'(z)$. Consequently the formulas (2.1)–(2.4) take the form

$$2\mu w = \varkappa \varphi(z) - \overline{\varrho(z)} + (\bar{z} - z)\,\overline{\varphi'(z)}, \tag{3.1}$$

$$\sigma_x = \text{Re}\,[\varphi'(z) + 2\,\overline{\varphi'(z)} - \overline{\varrho'(z)} + (\bar{z} - z)\,\overline{\varphi''(z)}], \tag{3.2}$$

$$\sigma_y = \text{Re}\,[\varphi'(z) + \overline{\varrho'(z)} - (\bar{z} - z)\,\overline{\varphi''(z)}], \tag{3.3}$$

$$\tau_{xy} = -\,\text{Im}\,[\varphi'(z) - \varrho'(z) + (z - \bar{z})\,\varphi''(z)]. \tag{3.4}$$

Before going into general considerations we return to the example of Fig. 3. Let us introduce the following functions

$$\varphi_0(z) = 2T\sqrt{z}, \qquad \varrho_0(z) = 2\overline{T}\sqrt{z}, \tag{3.5}$$

$$\varphi_1(z) = Lz\log z, \qquad \varrho_1(z) = -\overline{L}z\log z. \tag{3.6}$$

From (2.24), (2.26) it follows that

$$\varphi = \varphi_0 + \varphi_1 + \varphi_2; \qquad \varrho = \varrho_0 + \varrho_1 + \varrho_2 \tag{3.6}$$

where φ_2', ϱ_2' are continuous in $\overline{A}$ for $|z| < x_1$. Writing $g(t) = \varphi_2'(t^+) - \varphi_2'(t^-)$ for $x_1 < t \leq 0$, we derive from CAUCHY's integral formula for φ_2' and the subdomain $|z| < c = \left|\frac{1}{2}x_1\right|$ of $\overline{A}$ the representation

$$\varphi_2'(z) = \frac{1}{\pi i}\int_{-c}^{0}\frac{g(t)\,dt}{t - z} + h(z), \tag{3.7}$$

where $h(z)$ is holomorphic in $|z| < c$. We may apply the technique of (2.29), (2.30) to (3.7). Result:

$$y\,\varphi_2''(z) = O(1) \qquad \text{as } z \to 0. \tag{3.8}$$

The same holds verbally for $y\,\varrho_2''(z)$. — Formulas (3.6), (3.8) provide the basis for asymptotic representation of stresses and displacements near $z = 0$. It is seen that φ_2, ϱ_2 contribute to the order $O(1)$ to the stresses; they contribute to the order $O(z)$ to the quantity $w(z) - w(0)$, where $w(z)$ denotes the complex displacement. Neglecting φ_2, ϱ_2 we obtain expressions for stresses and displacements which characterize the behavior of the field quantities near $z = 0$. The expressions contain two complex coefficients T, L. If $T = 0, L \neq 0$ the stresses are unbounded like $\log|z|$. If $T \neq 0$ the functions φ_1, ϱ_1 can be neglected versus φ_0, ϱ_0. In this case one finds asymptotic laws of the form

$$w(z) - w(0) \cong r^{1/2}\,F_1(\vartheta); \qquad \sigma_{ik} \cong r^{-1/2}\,F_{ik}^1(\vartheta) \qquad \text{where} \quad z = r\,e^{i\vartheta},$$

$$\sigma_{11} = \sigma_x, \qquad \sigma_{12} = \tau_{xy}, \qquad \sigma_{22} = \sigma_y. \tag{3.9}$$

The functions F_1, F_{ik} refer to the angle ϑ only; they depend on T as a parameter. Since the functions are extensively described in the vast literature of fracture mechanics, we need not reproduce them at this place. Among the angles ϑ the two cases $\vartheta = \pm\pi$, $\vartheta = 0$ are distinguished by the geometry of the crack. The angles $\vartheta = \pm \pi$ pertain to the crack itself (or rather to the tangent of it at $z = 0$), while $\vartheta = 0$ represents the straight line extension of the crack into the elastic material. Associated with these angles are two complex-valued quantities of particular interest. One of them is the complex crack-opening, defined by

$$\Delta w = -i [w(z^+) - w(z^-)], \tag{3.10}$$

where z^+, z^- represent opposite points on C^+, C^- respectively. The other one is the traction

$$Z = \tau_{xy} + i \sigma_y \qquad \text{at } y = 0, \quad x > 0, \tag{3.11}$$

which attacks the lower half-plane. The formulas above yield, with $K = 2\sqrt{2}\, T$,

$$\lim_{x \to +0} x^{1/2}(-iZ) = \frac{1}{2}\sqrt{2}\, K, \tag{3.12}$$

$$\lim_{x \to -0} (-x)^{1/2} \frac{1}{2} \Delta w = \frac{2\sqrt{2}\,(1-\nu^2)}{E} K. \tag{3.12'}$$

As one can see the formulas (3.12), (3.12') generalize (1.1), (1.2) towards a situation in which the stress intensity factor K must be allowed to take a complex value. From now on we admit a complex K, and we broaden our objective accordingly.

It is not by accident that the coefficients of $\sqrt{z}$ in the formulas (3.5) for φ_0 and ϱ_0 take conjugate complex values $2T$ and $2\overline{T}$. This is due to the two circumstances that the crack leaves $z = 0$ in the direction of the negative x-axis and that $P(z)$ must be a load function on the crack. Were $\varrho_0(z) = 2T^* \sqrt{z}$ and $T^* \neq \overline{T}$, then $P(z)$ could not have an H-continuous derivative on C^+, C^- at $z = 0$. The reader is invited to check this in more detail on his own. — We possess in (2.25) a formula for T and thus for K. It is now the time to interpret it. By (2.9), (2.18) we have

$$Z = -i(Q' + R') \quad \text{on} \quad C^+, \qquad Z = i(-Q' + R') \quad \text{on} \quad C^-, \tag{3.13}$$

(2.25) and (3.13) yield

$$K = 2\sqrt{2}\, T = -\frac{\sqrt{2}}{\pi} \int_{-\infty}^{0} R'(t)\, |t|^{-1/2}\, dt = \frac{\sqrt{2}}{2\pi} \int_B \frac{Z\, ds}{\sqrt{z}} \tag{3.14}$$

where the integration in the last term follows the oriented crack. In words: A complex weight, namely $(\sqrt{2}\,\pi \sqrt{z})^{-1}$, is attached to the complex traction, and K is determined as the boundary integral over the weighted traction. This is but one way of interpretation of (3.14). It is more illuminating to imagine that the weight is the complex displacement of some other field, whereupon K becomes a "mixed energy". We shall see in the sequel, that the second point of view shows the path to the computation of K in more general cases. Before we leave the example of Fig. 3 let us take a brief look at the special case $Q = 0$, $\operatorname{Im} R = 0$. Here the notch is loaded by normal tractions, which act in opposite directions at opposite points. Set $p(t) = -R'(t)$ and $m(t) = |t|^{-1/2}$; it turns out that (3.14) takes the special form (1.8).

Consider now the general case of a regular state φ, ϱ in the domain A. By definition $\varphi'(z) = O(|z|^{-1/2})$ near $z = 0$. This implies the continuity of $y\,\varphi'(z)$ in $\overline{A}$ for some neighborhood $r < \varepsilon$ of the origin. Since the same holds for P, φ, we find, that the function $\varrho(z)$ must be continuous in $\overline{A}$ for $r < \varepsilon$.

One can always write

$$\varphi'(z) = \varphi_0'(z) + g(z)\, z^\tau; \qquad \varrho'(z) = \varrho_0'(z) + h(z)\, z^\tau \tag{3.15}$$

for some neighborhood $r < \varepsilon$ of $z = 0$ and $z \in A$. Here φ_0, ϱ_0 refer to (3.5) and some value of T; τ is some real constant. For certain regular states it will be possible to choose T and $\tau > -1/2$ such that $g(z)$, $h(z)$ are continuous in $\overline{A}$ for $r < \varepsilon$ and some $\varepsilon > 0$. The example of Fig. 3 is one of them. If a choice of T and $\tau > -1/2$ is possible, then T is unique. We introduce

Definition 3: A regular state in A for which T und $\tau > -1/2$ can be so chosen that $g(z)$, $h(z)$ are continuous in $\overline{A}$ for $r < \varepsilon$ and some $\varepsilon > 0$ will be called a regular state with a **normal singularity** at $z = 0$. We denote T as the coefficient and $K = 2\sqrt{2}\, T$ as the intensity of the singularity.

Assuming a normal singularity, we may apply CAUCHY's integral formula to $g(z)$, and

$$g(z) = \frac{1}{2\pi i} \int_{B'} \frac{g(t)\, dt}{t - z}, \qquad z \in A, \quad |z| < \eta \tag{3.16}$$

where B' is the boundary of the intersection of A with some disk $r < \eta < \varepsilon$; B' is to be so oriented that it leaves the intersection to the left. From (3.16) one can derive, in analogy to the step from (3.7) to (3.8), that

$$y\, g'(z) = O(1) \qquad \text{as } z \to 0. \tag{3.17}$$

If $T \neq 0$, then (3.15), (3.17) show that the asymptotic behavior of stresses and displacements near $z = 0$ is that of the state φ_0, ϱ_0. It is useful to note that (3.10), (3.11), (3.12), (3.12') retain their meaning and hold verbally for a regular state with a normal singularity of intensity K at $z = 0$.

One would conjecture that any regular state in A must have a normal singularity at $z = 0$. While we shall not pursue this idea in all generality, we shall prove a little more under supplementary conditions on crack form and the load function on the crack. — A JORDAN arc will be called analytic, if it can be embedded in some domain D, such that Im $\omega(z) = 0$ along the arc for some function $\omega(z)$ which is holomorphic and schlicht in D. Any analytic arc is regular. The function $\omega(z)$ maps D 1:1 onto a domain Δ of the complex ω-plane. The image of the arc is some interval $\langle a, b \rangle$ on the real ω-axis. Let $z = p(\omega)$ be the inverse map; $p(\omega)$ is holomorphic and schlicht in Δ. If $z = 0$ is one of the endpoints of the analytic arc, and if the arc leaves that point in the direction of the negative x-axis, we may assume without loss of generality that $\omega(0) = 0$, $\omega'(0) = 1$; this implies $p(0) = 0$, $p'(0) = 1$; we may also assume $a < 0$ and $b = 0$. The analytic arc is the image of $\langle a, 0 \rangle$ under the map $p(\omega)$.

Theorem 1: *Let the crack be analytic in some neighborhood $r = |z| < \varepsilon$. Any regular state φ, ϱ in A, for which the analytic part of the crack is load-free, has a normal singularity at $z = 0$. Moreover φ, ϱ admit expansions of the form*

$$\varphi(z) = \sum_{k=1}^{\infty} a_k z^{k/2}, \qquad \varrho(z) = \sum_{k=1}^{\infty} b_k z^{k/2} \qquad \text{with } b_1 = \bar{a}_1 \tag{3.18}$$

for some neighborhood $r < \varepsilon'$.

Proof: Referring to the functions $\omega(z)$, $p(\omega)$ above, we may assume that $p(\omega)$ is holomorphic and schlicht in some domain $\Delta = \{\omega : |\omega| < 2\eta\}$ where $\eta > 0$ is suitably chosen; we may also set $p(0) = 0$, $p'(0) = 1$, $a = -\eta$, $b = 0$; the image of $\langle -\eta, 0 \rangle$ under $p(\omega)$ is an analytic part of the crack. We choose η so small that the image D of Δ under $p(\omega)$ belongs to $\overline{A}$ and that $\varphi'(z)$ is continuous in $D \cap \overline{A}$ for $z \neq 0$. We determine $\varepsilon' > 0$ such that $|\omega| < \eta$ if only $|z| < \varepsilon'$. Associated with $p(\omega)$ is the function $q(\omega) = \overline{p(\bar{\omega})}$. It is also holomorphic and schlicht in Δ; moreover $q(0) = 0$, $q'(0) = 1$. Therefore $q(\omega) - p(\omega) = O(\omega^2)$ as $\omega \to 0$. The function $q(\omega)$ has the feature that $q(\omega(z)) = \overline{p(\omega(z))} = \bar{z}$ on the analytic part of the crack, which corresponds to $\langle -\eta, 0 \rangle$.

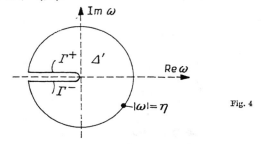

Fig. 4

Let Δ' be the domain of the points $0 < |\omega| < \eta$, the points of the negative real ω-axis excluded. It is a disk with a cut, as shown in Fig. 4. We denote upper and lower side of the cut by Γ^+, Γ^- respectively. We introduce the following functions: $\varphi_1(\omega) = \varphi(p(\omega))$, $\varrho_1(\omega) = \varrho(p(\omega))$ and

$$\varphi_2(\omega) = \varrho_1(\omega) + (q(\omega) - p(\omega))\frac{\varphi_1'(\omega)}{p'(\omega)}. \tag{3.19}$$

These functions (not to be confused with what appears in (3.6)) are holomorphic in Δ'. They are also continuous in the closure of Δ'. This is obvious for φ_1 and ϱ_1. As regards $\varphi_2(\omega)$ we have $\varphi'(z) = \varphi_1'(\omega)/p'(\omega) = O(|z|^{-1/2}) = O(|\omega|^{-1/2})$ as $z \to 0$, $\omega \to 0$. But the factor $q(\omega) - p(\omega)$ makes $\varphi'(z)$ continuous at $\omega = 0$, and φ_2 is continuous in the closure of Δ' as asserted. The function $p(\omega)$ maps Γ^+ into C^+ and Γ^- into C^-. We have

$$\varphi_2(\omega) = \varrho(z) + (\bar{z} - z)\varphi'(z); \qquad z \text{ on the crack}, -\eta \leq \omega < 0, \tag{3.20}$$

due to the characteristic feature of the function $q(\omega)$. The condition of an unloaded crack can therefore be expressed in the form

$$\varphi_1(\omega) + \overline{\varphi_2(\omega)} = \text{const.} \qquad \text{on } \Gamma^+, \Gamma^-. \tag{3.21}$$

The implications of (3.21) are partially those of (2.35). In particular the function $\varphi_1(\omega) - \overline{\varphi_2(\bar{\omega})}$ is holomorphic in the disk $|\omega| < \eta$. Feeding this result back into (3.21) one can conclude that

$$\sqrt{\omega}\,\varphi_1(\omega) = \sqrt{\omega}\,K(\omega) + L(\omega); \qquad \sqrt{\omega}\,\varphi_2(\omega) = -\sqrt{\omega}\,K^*(\omega) + L^*(\omega) \tag{3.22}$$

$$K^*(\omega) = \overline{K(\bar{\omega})} + \text{const.}, \qquad L^*(\omega) = \overline{L(\bar{\omega})};$$

$\sqrt{\omega}$ is meant to represent the main branch of the square-root, while K, L are certain holomorphic functions of ω in $|\omega| < \eta$. The continuity of φ_1, φ_2 implies $L(0) = 0$. Therefore

$$\sqrt{\omega}\,\varrho_1(\omega) = \sqrt{\omega}\,K_1(\omega) + L_1(\omega), \qquad L_1(0) = 0, \tag{3.23}$$

where K_1, L_1 are holomorphic in $|\omega| < \eta$. As we return to the functions $\varphi(z)$, $\varrho(z)$, we observe that the expansions (3.22), (3.23) yield immediately the expansions (3.18) of theorem 1. This in turn implies that the state φ, ϱ has a normal singularity at $z = 0$. The assertion $b_1 = \bar{a}_1$ comes from the argument about $T^* = \overline{T}$ above. Theorem 1 is thus proved.

If the weight function in (3.14) is to be interpreted as complex displacement of some other state, then that state cannot be regular. We shall therefore consider states in A with a "stronger" than normal singularity. To this end we introduce functions of the form

$$\varphi_s(z) = \frac{a}{\sqrt{z}}q_{11}(z) + q_{12}(z); \qquad a \neq 0, \quad q_{11}(0) = 1,$$

$$\varrho_s(z) = \frac{b}{\sqrt{z}}q_{21}(z) + q_{22}(z); \qquad q_{21}(0) = 0 \quad \text{or} = 1, \tag{3.24}$$

where the $q_{ik}(z)$ are holomorphic in some neigborhood of $z = 0$. Special cases of such functions are

$$\varphi_s(z) = \frac{a}{\sqrt{z}} + a' \sqrt{z}; \qquad \varrho_s(z) = \frac{b}{\sqrt{z}} + b' \sqrt{z}; \qquad a \neq 0 \,. \tag{3.25}$$

To these functions we assign $\psi_s(z) = \varrho_s(z) - z\, \varphi_s'(z)$. The functions φ_s, ψ_s are essential to

Definition 4: A state φ, ψ in A has a **fundamental singularity** at $z = 0$ (a root of a crack) if
(a) φ, ψ represent a regular state in almost all subdomains A_n of A, defined by $|z| > 1/n$, $n = 1, 2, \ldots$.
(b) $\dot{P} = 0$ on B for $z \neq 0$;
(c) $\varphi = \varphi_s + \varphi_r, \psi = \psi_s + \psi_r$ for some region $A - A_n$ where φ_s, ψ_s are of type (3.24) while φ_r, ψ_r describe a regular state.

We remark that the domains A_n are well-defined and regular for sufficiently large n. An example of a state with a fundamental singularity is easily found. Take the domain of Fig. 3 and the functions (3.25) with $a' = b' = 0$, $b = \bar{a}$. For simplicity we shall say that the state φ, ψ of definition 4 or the state φ, ϱ (with $\varrho = \psi + z\, \varphi'$) is fundamental. For fixed A the linear combination $\varphi = \alpha_1 \varphi_1 + \alpha_2 \varphi_2$, $\psi = \alpha_1 \psi_1 + \alpha_2 \psi_2$ of fundamental states φ_1, ψ_1 and φ_2, ψ_2 with real combination coefficients α_1, α_2 is again fundamental. It will be shown in section 4, that the complex displacements of certain fundamental states furnish the weight function for the computation of K of a state with a normal singularity. We proceed now in order to establish features of uniqueness and existence of fundamental states.

Let the coefficient a in (3.24) be prescribed. Writing $P = P_s + P_r$, where P_s, P_r refer to the states φ_s, ψ_s and φ_r, ψ_r respectively, we observe that $\dot{P}_s = - \dot{P}_r$ on B for $z \neq 0$. Since P_r must exhibit the properties of regularity of a load function on B, the same must hold for P_s. In particular P_s must go to a limit as $z \to 0$ along the crack. In order to check on this point, it will suffice to consider the functions in (3.25). For these

$$P_s = \frac{a}{\sqrt{z}} + \frac{\overline{b}}{\sqrt{z}} + a' \sqrt{z} + \overline{b'\sqrt{z}} + \frac{1}{2}(z - \bar{z})\overline{\left[\frac{-a}{z\sqrt{z}} + a' \sqrt{z}/z\right]} =$$
$$= \frac{e^{-i\vartheta/2}}{\sqrt{r}} \left\{ a + \overline{b}\, e^{i\vartheta} - \frac{1}{2} \bar{a}\, e^{i\vartheta}(e^{2i\vartheta} - 1) + a' r\, e^{i\vartheta} + \overline{b'} r\, e^{-i\vartheta} + \frac{1}{2} \bar{a}' r\, (e^{2i\vartheta} - 1) \right\}. \tag{3.26}$$

Now $y = O(x^2) = O(z^2)$ as $z \to 0$ along the crack. Furthermore $\vartheta \to \pm \pi$ as $z \to 0$ along C^+, C^- respectively. In order to make P_s going to a limit it is necessary that $b = \bar{a}$ in (3.25) and $b = \bar{a}$, $q_{21}(0) = 1$ in (3.24). Let two states with a fundamental singularity and the same coefficient a exist. The difference of these two states is necessarily of the form

$$\varphi = \sqrt{z}\, p_{11}(z) + p_{12}(z) + \varphi_r; \qquad \varrho = \sqrt{z}\, p_{21}(z) + p_{22}(z) + \varrho_r \tag{3.27}$$

where the $p_{ik}(z)$ are holomorphic in some neigborhood $r < \varepsilon$. This state has finite energy. It satisfies the regularity conditions of a regular state in A and on B, and $\dot{P} = 0$ along B. The state therefore is regular; its unique existence implies that its stress field vanishes everywhere. Consequently the coefficient a in (3.24) determines the fundamental state uniquely, if rigid body motions are disregarded. It is now evident that the fundamental states form a real linear space of dimension ≤ 2.

It will be useful to study (3.26) in more detail for special cases and with $b = \bar{a}$ from here on. Since C^+, C^- represent the same regular arc in different orientations for some neigborhood of $z = 0$, they must satisfy the differential equations

$$\dot{z} = - e^{-i\omega}, \qquad \dot{\omega} = \gamma \tag{3.28}$$

where $\gamma = \gamma(s)$ is the curvature as a function of arclength, $- \omega$ the tangent angle, and where (as before) the dot denotes differentiation with respect to arclength. We may assume $z = \omega = 0$ for $s = 0$. By definition of a regular arc the function $\gamma(s)$ is H-continuous in some interval, say $0 \leq s \leq h$, i.e.

$$|\gamma(s) - \gamma(t)| \leq \Gamma |s - t|^\lambda; \qquad \Gamma \geq 0, \quad 0 < \lambda \leq 1 \quad \text{for } 0 \leq s, t \leq h \tag{3.29}$$

with certain constants Γ, λ. This implies

$$\gamma(s) = \gamma_0 + O(s^\lambda); \qquad \gamma_0 = \gamma(0) \quad \text{as } s \to 0 \tag{3.30}$$
$$\omega(s) = \gamma_0 s + O(s^{1+\lambda}) \quad \text{as } s \to 0 \,. \tag{3.30'}$$

From (3.28) it follows

$$\dot{x} = -\cos \omega, \qquad \dot{y} = \sin \omega, \qquad \ddot{x} = \gamma \sin \omega, \qquad \ddot{y} = \gamma \cos \omega \,. \tag{3.31}$$

(3.30), (3.30') and (3.31) yield

$$x = -s + \gamma_0 s^3/6 + O(s^{3+\lambda}); \qquad y = \gamma_0 s^2/2 + O(s^{2+\lambda}) \,. \tag{3.32}$$

Turning now to the polar coordinates r, ϑ, we set $\theta = \vartheta - \pi$ on C^+, $\theta = \vartheta + \pi$ on C^- and $\theta = 0$ at $s = 0$. Disregarding the different orientations of these arcs we assign the same s to opposite points on the crack. From (3.32) we derive

$$r = s - \gamma_0 s^3/24 + O(s^{3+\lambda}), \qquad \theta = -\gamma_0 s/2 + O(s^{1+\lambda}) \,. \tag{3.33}$$

From (3.28)
$$\dot{r} = \cos(\omega + \theta), \qquad r\dot{\theta} = -\sin(\omega + \theta),$$
$$\ddot{r} = -(\dot{\omega} + \dot{\theta})\sin(\omega + \theta); \qquad r\ddot{\theta} = -(\dot{\omega} + 2\dot{\theta})\cos(\omega + \theta). \tag{3.34}$$

Altogether (3.30), (3.30'), (3.33), (3.34) lead to
$$\dot{\theta} = -\gamma_0/2 + O(s^\lambda), \qquad r\ddot{\theta} = O(s^\lambda), \qquad \ddot{\theta} = O(s^{\lambda-1}), \qquad \ddot{r} = O(s). \tag{3.35}$$

To these preparations we add the remark that any function $G(s)$, defined in $\langle 0, h\rangle$ and differentiable in $(0, h\rangle$ is H-continuous on $(0, h\rangle$ if $|\dot{G}(s)| \leq c s^\alpha$ where $c \geq 0$ and $\alpha > -1$ represent some constants. Indeed for $t < s$:
$$|G(s) - G(t)| \leq c \int_t^s x^\alpha\, dx = \frac{c}{\alpha + 1}(s^{\alpha+1} - t^{\alpha+1}). \tag{3.36}$$

Since $s^{1+\alpha}$ is H-continuous, so is $G(s)$.

Example: $G(s) = \dot{\theta}(s)$.

We shall now establish conditions under which the real linear space of the fundamental states has dimension 2. Necessary and sufficient to achieve dimension 2 is the existence of fundamental states for $a = 1$ and $a = i$. We begin with a discussion of P_s for the functions (3.25) with these coefficients: $b = \bar{a}$, $a' = 0$ and $b' = \gamma_0 \operatorname{Im} a$. This leads to

$$P_s = \frac{1}{2\sqrt{r}} e^{-i\vartheta/2}(1 + e^{i\vartheta})^2(2 - e^{i\vartheta}) = \frac{2\,e^{i\vartheta/2}}{\sqrt{r}}(2 - e^{i\vartheta})\cos^2\vartheta/2 \qquad \text{if } a = 1, \tag{3.37}$$

$$P_s = \frac{e^{-i\vartheta/2}}{2\sqrt{r}}\{2i(1 + e^{i\vartheta}) + i e^{i\vartheta}(e^{2i\vartheta} - 1) + 2\gamma_0 r\}$$
$$= \frac{e^{-i\vartheta/2}}{2\sqrt{r}}\{i(1 + e^{i\vartheta})^2(e^{i\vartheta} - 2) + 4i(1 + e^{i\vartheta}) + 2\gamma_0 r\}$$
$$= \frac{e^{-i\vartheta/2}}{\sqrt{r}}\left\{2i(e^{2i\vartheta} - 2e^{i\vartheta} + 2)\cos^2\frac{1}{2}\vartheta - \frac{2y}{r} + \gamma_0 r\right\} \qquad \text{if } a = i. \tag{3.38}$$

In either case P_s is twice continuously differentiable in the half-open interval $(0, h\rangle$. The important terms in both cases are

$$F_1(s) = \frac{\cos^2\dfrac{1}{2}\vartheta}{\sqrt{r}} = \frac{\sin^2\dfrac{1}{2}\theta}{\sqrt{r}}, \tag{3.39}$$

$$F_2(s) = \frac{\gamma_0 r - 2y/r}{\sqrt{r}} = \frac{\gamma_0 r + 2\sin\theta}{\sqrt{r}}. \tag{3.40}$$

By (3.33) we have $F_1(s) = O(s^{3/2})$, $F_2(s) = O(s^{(1/2)+\lambda})$ as $s \to 0$. This makes P_s continuous in the closed interval $\langle 0, h\rangle$ if we define $P_s = 0$ at $s = 0$. We turn to $\dot{P}$; $\dot{\theta}$ being H-continuous on $\langle 0, h\rangle$, it will suffice to consider the function F_1 in the case $a = 1$ and F_1, F_2 in the case $a = i$. We differentiate these functions twice with respect to s and obtain, with the aid of (3.33), (3.34), (3.35), the following order-relations:

$$\ddot{F}_1 = O(s^{-1/2}); \qquad \ddot{F}_2 = O(s^{\lambda-3/2}). \tag{3.41}$$

By the remark on the function $G(s)$ above, the function $\dot{F}_1$ can be considered as H-continuous in $\langle 0, h\rangle$. The behavior of $\dot{F}_2$ depends on λ. Obviously $\dot{F}_2$ is H-continuous in $\langle 0, h\rangle$ if $\lambda > 1/2$. The H-continuity of $\dot{F}_1$ implies the H-continuity of $\dot{P}_s$ in the case $a = 1$; the H-continuity of $\dot{F}_1$, $\dot{F}_2$ implies the H-continuity of $\dot{P}_s$ for the case $a = i$. Assuming $\lambda > 1/2$ in (3.29) from here on, we may sum up the preceding results in this statement: The function P_s satisfies the regularity conditions of a load function on C^+, C^- in some neigborhood $r < \varepsilon$ of $z = 0$.

We are now going to construct states with a fundamental singularity at $z = 0$. To this end we pick some point z_0 at arclength s_0 on the crack (Fig. 5), where $0 < s_0 < h$, so that the preceding considerations are valid for the crack between z_0 and $z = 0$. By C_0^+, C_0^- we shall denote the parts of C^+, C^- on that particular section of the crack, the points $z = 0$, $z = z_0$ to be included. We introduce

$$H(z) = 2\sqrt{z(1 - z/z_0)}\,(1 - z/z_0)^3(1 + cz) - p(z); \qquad c = 7/2 z_0, \tag{3.42}$$

where $p(z)$ represents a polynomial. The function $H(z)$ is holomorphic outside of C_0^+, C_0^-; $p(z)$ can be and is so chosen that $H(z)$ is holomorphic at infinity with $H(\infty) = 0$. We observe that $H'(z) = O(z^{-2})$, $zH'(z) + H(z) = O(z^{-2})$ as $z \to \infty$. Near $z = 0$ we may write

$$H(z) = 2\sqrt{z}\, q(z) - p(z), \tag{3.43}$$

where $q(z)$ is holomorphic in some neigborhood of $z = 0$. Moreover $q(0) = 1$, $q'(0) = 0$. The latter comes from the choice of the coefficient c in (3.42). We set up:

$$\varphi_s(z) = a\, H'(z)\,, \qquad \varrho_s(z) = \bar{a}\, H'(z) + \frac{b'}{3}[H(z) + z\, H'(z)]\,; \qquad b' = \gamma_0 \operatorname{Im} a\,. \tag{3.44}$$

The functions φ_s, ϱ_s are of the type (3.24). They are holomorphic in A and even beyond A, since $H(z)$ is holomorphic outside the cut C_0^+, C_0^-. Let us consider them in the domains A_n (definition 4) for sufficiently large n. If n is large enough the circle $\mathfrak{L}$ of radius $1/n$ around the origin (Fig. 5) will intersect with the boundary B of A at precisely two points, namely a point ζ^+ on C_0^+ and its opposite ζ^- on C_0^-. This makes the exterior of $\mathfrak{L}$ in A a well-defined domain A_n. If A is bounded by the contours $G_1, G_2, \ldots, G_n$ ($n = 2$ in Fig. 5) and if G_1 contains C_0^+, C_0^-, then A_n is bounded by $G_2, \ldots, G_n$ and a certain contour G_1', the latter consisting of $\mathfrak{L}$ and that part of G_1 which runs outside of $\mathfrak{L}$. The functions $\varphi_s, \varrho_s, \varphi_s', \psi_s, \varphi_s'', \varrho_s', P_s(z)$ are continuous in the closure of A_n. The behavior of $H(z)$ at infinity enforces that the state φ_s, ϱ_s has finite energy in A_n even if A were of unbounded area. P_s satisfies the regularity conditions of a load function on the boundary of A_n. Notice here that the factor $(1 - z/z_0)^3$ in (3.42) makes $\dot{P}_s$ H-continuous on any open arc of G_1 which contains z_0. We assert that φ_s, ϱ_s is a regular state in A_n; this statement needs verification in only one respect: We must show that P_s satisfies the moment condition on the contours $G_1', G_2, \ldots, G_n$. Let $\mathfrak{C}$ be an oriented rectifiable curve outside of some

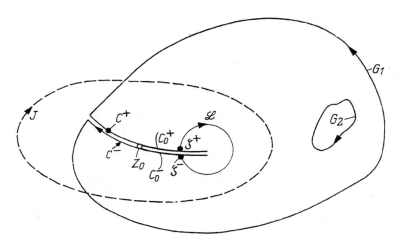

Fig. 5

neigborhood of $z = 0$. We write

$$M(\mathfrak{C}) = \int_{\mathfrak{C}} \bar{z}\, dP_s(z) \tag{3.45}$$

and consider various cases of $\mathfrak{C}$. Let $\mathfrak{C}$ represent some JORDAN curve J (Fig. 5) which has C_0^+, C_0^- either in its exterior or in its interior. In either case CAUCHY's integral theorem applies to the functions φ_s, ϱ_s in the form

$$\int_J \varphi_s\, dz = 0\,, \qquad \int_J \varrho_s\, dz = 0\,, \tag{3.46}$$

since φ_s, ϱ_s are either holomorphic inside J or outside J and since $\varphi_s = O(z^{-2})$, $\varrho_s = O(z^{-2})$ as $z \to \infty$. With the aid of (3.46) the quantity $M(J)$ can be given a simple form. We write

$$M(J) = -\int_J P_s\, d\bar{z} = -\int_J (\varphi_s + \bar{\varrho}_s + (z - \bar{z})\, \overline{\varphi}_s')\, d\bar{z} = -\int_J (\varphi_s + z\, \overline{\varphi}_s')\, d\bar{z} =$$
$$= -\int_J \varphi_s\, d\bar{z} - \int_J z\, d\overline{\varphi}_s = \int_J (\overline{\varphi}_s\, dz - \varphi_s\, d\bar{z})\,. \tag{3.47}$$

It is obvious that $\operatorname{Re} M(J) = 0$, i.e. P_s satisfies the moment condition on J. This basic result is easily generalized. Both (3.46) and the steps in (3.47) stay valid, if J is replaced by a finite contour in $\overline{A}_n$. Due to the behavior of φ_s, ϱ_s at $z = \infty$ we may even replace J by an infinite contour in $\overline{A}_n$, if such a contour exists. Consequently we can state, that P_s satisfies the moment condition on $G_1', G_2, \ldots, G_n$; φ_r, ϱ_s then define a regular state in A_n.

From (3.43) it follows that

$$\varphi_s(z) = \frac{a}{\sqrt{z}} + \varphi^*(z)\,; \qquad \varrho_s(z) = \frac{\bar{a}}{\sqrt{z}} + b'\sqrt{z} + \varrho^*(z)\,, \tag{3.48}$$

with

$$\varphi^*(z) = z^{3/2} Q_1(z) + R_1(z)\,; \qquad \varrho^*(z) = z^{3/2} Q_2(z) + R_2(z)\,. \tag{3.48'}$$

Here R_1, R_2 are polynomials while Q_1, Q_2 are holomorphic in some neighborhood of $z = 0$. The function $P^*(z)$, associated with φ^*, ϱ^* is continuous in $\overline{A}$; it is especially continuous on C_0^+, C_0^-, and it admits an H-continuous derivative with respect to arclength on these arcs. This can be easily verified by using the tools that led us to the result that P_s in (3.37), (3.38) satisfies the regularity conditions of a load function on C_0^+, C_0^-. We leave details to the reader. The formulas (3.48), (3.48'), the behaviour of $P^*(z)$ and our result about (3.37), (3.38) establish this: The function P_s, associated with (3.44), satisfies the regularity conditions of a load function on C_0^+, C_0^-. But then P_s satisfies the regularity conditions of a load function on all contours $G_1, G_2, \ldots, G_n$. Equations (3.48), (3.48') imply

$$M(\mathfrak{L}) = O(n^{-1/2}) \quad \text{as} \quad n \to \infty . \tag{3.49}$$

Combining (3.49) with $\operatorname{Re} M(G_1') = 0$ we obtain the result that $\operatorname{Re} M(G_1) = 0$. In words: The function P_s by (3.44) satisfies the moment condition on G_1. Since it satisfies the moment condition on the other contours G_k, we may now conclude our considerations of (3.44) with

Lemma 1: *P_s, as associated with φ_s, ϱ_s in (3.44), represents a load function on each of the contours which bound A.*

Since A is assumed to be a regular elastic domain, there must exist a regular state φ_r, ϱ_r in A, such that $\dot{P}_r = -\dot{P}_s$ on B, where P_s is the quantity of lemma 1. But then the state $\varphi = \varphi_s + \varphi_r$, $\varrho = \varrho_s + \varrho_r$ is fundamental in accordance with definition 4. The fundamental singularity has the coefficient a. Therefore

Theorem 2: *For any $a \neq 0$ in (3.24) and for $\lambda > 1/2$ in (3.29) there exists a fundamental state in A. Rigid body motions disregarded, the state is unique.*

We add

Theorem 3: *If the crack is analytic near $z = 0$, then for $|z| < \varepsilon'$ and some $\varepsilon' > 0$ the functions φ_s, ϱ_s, of a fundamental state in A admit expansions*

$$\varphi_s = \sum_{k=-1}^{\infty} a_k z^{k/2} , \qquad \varrho_s = \sum_{k=-1}^{\infty} b_k z^{k/2} \qquad \text{with} \quad b_{-1} = \bar{a}_{-1} .$$

Theorem 3 can be proved in analogy to the proof of theorem 1. We do not go into more detail.

4. Weight Functions

We assume two states φ_1, ϱ_1 and φ_2, ϱ_2 in a regular elastic domain A. All quantities associated with them will be distinguished by subscripts 1,2 respectively. State φ_1, ϱ_1 shall be regular with a normal singularity of intensity $K = 2\sqrt{2}\,T$ at $z = 0$. At the same point state φ_2, ϱ_2 shall have a fundamental singularity with coefficient a. Take now an oriented, piecewise smooth curve $\mathfrak{C}$ in A, going from some point ζ' to another point ζ''. Along $\mathfrak{C}$ the two states exhibit tractions which attack the material to the left. We wish to consider the work which the tractions of state φ_k, ϱ_k accomplish through the displacements of state φ_m, ϱ_m. This work is

$$W_{km} = \int_{\mathfrak{C}} (X_k u_m + Y_k v_m)\, ds = \operatorname{Re} \int_{\mathfrak{C}} (X + iY)_k (u - iv)_m\, ds =$$
$$= -\operatorname{Re} i \int_{\mathfrak{C}} \bar{w}_m\, dP_k = \operatorname{Im} \int_{\mathfrak{C}} \bar{w}_m\, dP_k ; \qquad k, m = 1, 2 . \tag{4.1}$$

We are particularly interested in

$$W^*(\mathfrak{C}) = W_{12} - W_{21} = \operatorname{Im} \int_{\mathfrak{C}} (\bar{w}_2\, dP_1 - \bar{w}_1\, dP_2) . \tag{4.2}$$

Formulas (2.1'), (4.2) yield

$$2\mu\, W^*(\mathfrak{C}) = \operatorname{Im} \int_{\mathfrak{C}} (\bar{P}_1\, dP_2 - \bar{P}_2\, dP_1) + (\varkappa + 1) \operatorname{Im} \int_{\mathfrak{C}} (\bar{\varphi}_2\, dP_1 - \bar{\varphi}_1\, dP_2) . \tag{4.3}$$

Integration by parts leads to

$$\operatorname{Im} \int_{\mathfrak{C}} (\bar{P}_1\, dP_2 - \bar{P}_2\, dP_1) = \operatorname{Im} \bar{P}_1 P_2 \Big|_{\zeta'}^{\zeta''} - \operatorname{Im} \int_{\mathfrak{C}} (\bar{P}_2\, dP_1 + P_2\, d\bar{P}_1) = \operatorname{Im} \bar{P}_1 P_2 \Big|_{\zeta'}^{\zeta''} , \tag{4.4}$$

$$\operatorname{Im} \int_{\mathfrak{C}} (\bar{\varphi}_2\, dP_1 - \bar{\varphi}_1\, dP_2) = \operatorname{Im} (\bar{\varphi}_2 P_1 - \bar{\varphi}_1 P_2) \Big|_{\zeta'}^{\zeta''} + \operatorname{Im} \int_{\mathfrak{C}} (P_2\, d\bar{\varphi}_1 - P_1\, d\bar{\varphi}_2) . \tag{4.5}$$

Furthermore

$$\int_{\mathfrak{C}} (P_2\, d\bar{\varphi}_1 - P_1\, d\bar{\varphi}_2) = \int_{\mathfrak{C}} (\varphi_2\, d\bar{\varphi}_1 - \varphi_1\, d\bar{\varphi}_2) + \int_{\mathfrak{C}} (z - \bar{z}) \overline{(\varphi_2'\, d\varphi_1 - \varphi_1'\, d\varphi_2)} - \int_{\mathfrak{C}} \overline{g(z)}\, d\bar{z} , \tag{4.6}$$

where $g(z) = \varrho_1 \varphi_2' - \varrho_2 \varphi_1'$. The second integral in (4.6) has a vanishing integrand; the first one, in analogy to (4.4), can be given the form

$$\operatorname{Im} \int_{\mathfrak{C}} (\varphi_2\, d\bar{\varphi}_1 - \varphi_1\, d\bar{\varphi}_2) = -\operatorname{Im} \varphi_1 \bar{\varphi}_2 \Big|_{\zeta'}^{\zeta''} . \tag{4.7}$$

Putting all the pieces together, we obtain

$$2\mu\, W^*(\mathfrak{C}) = (\varkappa + 1) \operatorname{Im} \int_{\mathfrak{C}} g(z)\, dz + \operatorname{Im} R \Big|_{\zeta'}^{\zeta''} \tag{4.8}$$

with

$$R = \bar{P}_1 P_2 + (\varkappa + 1) [\bar{\varphi}_2 (P_1 - \varphi_1) - \bar{\varphi}_1 P_2] .$$

We turn to special cases of $\mathfrak{C}$. If $\mathfrak{C}$ is closed then R does not contribute and

$$2\mu W^*(\mathfrak{C}) = \operatorname{Im}(\varkappa + 1) \int_\mathfrak{C} g(z)\, dz; \qquad \mathfrak{C} \text{ is closed}. \tag{4.9}$$

The integral over $g(z)$ in (4.9) does not necessarily vanish, since A need not be simply connected. However if $\mathfrak{C}$ is a Jordan curve whose interior belongs to A, then by Cauchy's integral theorem $W^*(\mathfrak{C}) = 0$. This is a special case of Betti's theorem, according to which the work done by the tractions of state (1) through the displacements of state (2) equals the work which the tractions of state (2) accomplish through the displacements of state (1). Let $G_1^*, G_2^*, \ldots, G_m^*$ be a set of finite contours inside A, bounding a subdomain A' of A. Let the contours be so oriented that A' is to their left. Since Cauchy's integral theorem applies to the integral $g(z)\, dz$ over the boundary of A', we may write

$$\sum_{k=1}^m W^*(G_k^*) = 0. \tag{4.10}$$

Our next aim is to extend (4.10) by replacing the G_k^* by essential parts of the boundary of A. We observe that $W^*(\mathfrak{C}')$ is well-defined, if $\mathfrak{C}'$ is a Jordan arc on B which does not contain the point $z = 0$. Formula (4.8) stays valid if $\mathfrak{C}'$ does not contain a point of discontinuity of the functions $\varrho_1, \varphi_1', \varrho_2, \varphi_2'$. Of such points there is at most a finite number. Let $z' \neq 0$ be one of them. We have $\varphi_k'(z) = O(|z - z'|^{-1/2})$ as $z \to z'$. Since φ_k, P_k are continuous in $\overline{A}$ at z', it follows that $\varrho_k^*(z) = \varrho_k(z) + (\bar{z} - \bar{z}')\, \varphi_k'(z)$ is also continuous at z'; but

$$g(z) = \varrho_1 \varphi_2' - \varrho_2 \varphi_1' = \varrho_1^* \varphi_2' - \varrho_2^* \varphi_1' = O(|z - z'|^{-1/2}) \qquad \text{as } z \to z'. \tag{4.11}$$

For this reason the integral in (4.8) makes sense for any Jordan arc $\mathfrak{C}'$ on B, if only $z = 0$ is not on the arc. Furthermore the steps leading to (4.8) retain their validity for such an arc, and (4.8) holds for $W^*(\mathfrak{C}')$. Next we consider the case $\mathfrak{C} = \mathfrak{L}$, where $\mathfrak{L}$ is the circle from ζ^+ to ζ^- in Fig. 5. That circle has radius $1/n$. We choose n so large that the points $0 < |z| < 2/n$ are in $\overline{A}$ and that $\varrho_1, \varphi_1', \varrho_2, \varphi_2'$ are continuous at those points. The quantity $W^*(\mathfrak{L})$ is well-defined by (4.2) as either a proper or an improper Riemann-integral; (4.8) stays valid for $\mathfrak{C} = \mathfrak{L}$. Finally we turn again to the subdomain A_n of A, as referred to in definition 4 and Fig. 5. Its boundary B_n consists of the contours $G_1', G_2, \ldots, G_n$, each so oriented, that A_n is to the left. We assert the validity of (4.10) for $m = n$ and $G_1^* = G_1', G_2^* = G_2, \ldots, G_n^* = G_n$. Indeed W^* is well-defined for any of the new contours, and (4.8) applies in any of these cases. Cauchy's integral theorem holds for $g(z)$ and B_n. The possibility $z = \infty$ on B_n is of no consequence, since $g(z) = O(|z|^{-1-p})$ with some constant $p > 0$, as $z \to \infty$. We prefer to write (4.10) for the boundary B_n of A_n in the form

$$W^*(\mathfrak{L}) = -W^*(B') \tag{4.12}$$

where B' is the complement of $\mathfrak{L}$ on B_n; B' can also be interpreted as that part of B which is outside of $|z| = 1/n$. Since $P_2 = 0$ on B', we have

$$-W^*(B') = -\operatorname{Im} \int_{B'} \bar{w}_2\, dP_1. \tag{4.13}$$

Along $\mathfrak{L}$ the behavior of the two states is of the asymptotic character for small $|z|$, as we have described it above. In particular both states give rise to

$$z\, g(z) = -a\,\overline{T} - \bar{a}\, T + O(|z|^q) \qquad \text{as} \quad |z| = 1/n \to 0; \tag{4.14}$$

here $q > 0$ is some constant; it has been assumed that $\varrho_1(0) = 0$. The latter is no essential restriction of generality, since the right hand side of (4.8) does not change if ϱ_1 is altered by an additive constant. From (4.14) it follows that

$$\lim_{n \to \infty} \int_\mathfrak{L} g(z)\, dz = 2\pi i\, (a\,\overline{T} + \bar{a}\, T). \tag{4.15}$$

We assert

$$R \Big|_{\zeta^-}^{\zeta^+} \to 0 \qquad \text{as} \quad n \to \infty. \tag{4.16}$$

This follows from

$$\overline{\varphi_2(z)}\, (P_1(z) - \varphi_1(z)) \Big|_{\zeta^-}^{\zeta^+} = \bar{\varphi}_2 \left(\bar{\varrho}_1 + (z - \bar{z})\,\bar{\varphi}_1'\right) \Big|_{\zeta^-}^{\zeta^+} = \left(2\, \bar{a}\, \overline{T} + o(1)\right) \Big|_{\zeta^-}^{\zeta^+} = o(1) \quad \text{as } n \to \infty, \tag{4.16'}$$

as well as from the continuity of P_1, P_2, φ_1 on the crack. We may now combine (4.15), (4.16) into

$$\lim_{n \to \infty} 2\mu W^*(\mathfrak{L}) = 2\pi\, (\varkappa + 1)\, (a\,\overline{T} + \bar{a}\, T). \tag{4.17}$$

Since $w_2 = O(|z|^{-1/2})$ as $z \to 0$, the integral (4.13) goes to a limit as $n \to \infty$. Putting the preceding results together we obtain, as $n \to \infty$, a limit relation from (4.12). It can be given the form

$$\operatorname{Re}(\bar{a}\, T) = -\frac{\mu}{2\pi\,(\varkappa + 1)} \int_B (X_1 u_2 + Y_1 v_2)\, ds. \tag{4.18}$$

Let u, v represent the displacements for the fundamental state with coefficient $a = 1$, and let u^*, v^* be those of the fundamental state with coefficient $a = i$. Setting $U = u + i u^*$, $V = v + i v^*$, we may combine formulas (4.18) for the two cases into

$$T = - \frac{\mu}{2 \pi (\varkappa + 1)} \int_B (X_1 U + Y_1 V) \, ds \, . \tag{4.19}$$

Changing from T to K and expressing $\varkappa, \mu$ in terms of E, ν we rewrite (4.19) as

$$K = - \frac{\sqrt{2} \, E}{8 \pi (1 - \nu^2)} \int_B (X_1 U + Y_1 V) \, ds \, . \tag{4.20}$$

This is the formula for K which we announced in the introduction.

If A is symmetric with respect to the x-axis (Fig. 6) then the crack forms an interval $(-l, 0)$ on that axis. The point $-l$ may be either a single point or a doublepoint. If the crack only is loaded, and if the tractions are normal, i.e. $\sigma_y = -p(t)$, $\tau_{xy} = 0$ on upper and lower side of the crack, then

$$\varphi(z) = \overline{\varphi(\bar z)}; \qquad \varrho(z) = \overline{\varrho(\bar z)} \, . \tag{4.21}$$

The intensity K and the coefficient T are real, and it suffices to identify U, V with the displacements u, v of the fundamental state with coefficient $a = 1$. Formula (4.20) can be rewritten in the form (1.8), i.e.

$$K = \frac{\sqrt{2}}{\pi} \int_{-l}^{0} p(t) \, m(t) \, dt \, ; \tag{1.8}$$

$m(t)$ is defined by

$$m(t) = - \frac{2 \mu}{\varkappa + 1} \, v(t) \, , \tag{4.22}$$

where $v(t)$ denotes the normal displacement of the fundamental state along the upper side of the crack. From $a = 1$ and (4.22) it follows that

$$m(t) \, (-t)^{1/2} \to 1 \qquad \text{as } t \to -0 \, . \tag{4.23}$$

Theorem 4: *The intensity K of a state φ_1, ϱ_1 with a normal singularity at $z = 0$ can be represented in the form (4.20); the functions U, V are the combinations $U = u + i u^*$, $V = v + i v^*$, where $w = u + i v$, $w^* = u^* + i v^*$ are the complex displacements along the boundary of the fundamental states with coefficients $a = 1$, $a = i$ respectively. In the case of a symmetric domain A, a crack $(-l, 0)$ on the x-axis and normal tractions $\pm p(t)$ on C^+, C^- respectively, the intensity K is real and given by (1.8) where $m(t)$ is the displacement on C^+ and normal to it for a certain fundamental state with a real coefficient a; $m(t)$ is normalized by the condition (4.23).*

Let us describe some examples. We reconsider the domain A of Fig. 3. The fundamental states were already mentioned. They are $\varrho_2 = \varphi_2 = 1/\sqrt{z}$ for $a = 1$ and $-\varrho_2 = \varphi_2 = i/\sqrt{z}$ for $a = i$. Hence

$$U = - \frac{\varkappa + 1}{2 \mu \sqrt{z}}, \qquad V = i \, U \, . \tag{4.24}$$

(4.19) and (4.24) lead to the formula (3.14).

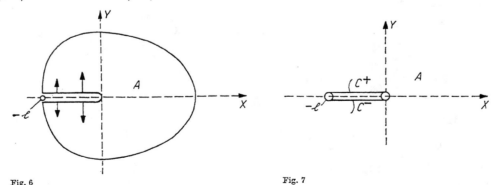

Fig. 6　　　　　　　　　　　　　　Fig. 7

Next we deal with the socalled GRIFFITH-crack. Here A is the z-plane with a cut along the interval $(-l, 0)$ along the x-axis, as shown in Fig. 7. The fundamental states of this case are obtained from those of Fig. 3, by replacing $1/\sqrt{z}$ by the function

$$F(z) = \left(\frac{l + z}{l \, z} \right)^{1/2} . \tag{4.25}$$

Therefore

$$K = \frac{\sqrt{2}}{2 \pi} \int_B F(z) \, Z_1 \, ds \, . \tag{4.26}$$

In the particular case $Z_1 = \pm i\, p(t)$, $p(t)$ = real-valued, on C^+, C^- respectively, formula (1.8) with $m(t) = |F(t)|$ applies; explicitly

$$K = \frac{\sqrt{2}}{\pi} \int_{-l}^{0} \left(\frac{l+t}{-lt}\right)^{1/2} p(t)\, dt. \tag{4.27}$$

The same formula was given by BARENBLATT [3], who derived the value of K from a direct representation of that stress-field which responds to the actual tractions on the crack.

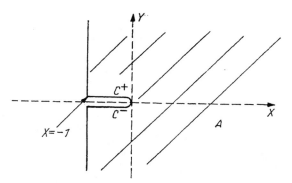

Fig. 8

Our last example is the case of the edge-crack in the half-plane $x > -1$ (see Fig. 8). The crack occupies the interval $(-1, 0)$ of the x-axis. We confine our efforts to the case of symmetry. As shown in [4], the state with a normal singularity at $z = 0$ can be represented by means of the formulas

$$\varphi_1(z) = Q(z) - Q(-2-z) - 2(1+z) Q'(-2-z),$$
$$\varrho_1(z) = \varphi_1(z) + 4(1+z)^2 Q''(-2-z); \tag{4.28}$$

$$Q(z) = \int_{-1}^{0} \frac{q(t)\, dt}{t-z}, \qquad q(t) = \frac{2\mu}{\varkappa(x+1)\pi} v_1(t), \tag{4.29}$$

where $v_1(t)$ stands for the normal displacement along the upper side C^+ of the crack. Along the crack $P_1(z) = \varrho_1(z) + \overline{\varphi_1(z)}$, and in detail.

$$P_1(x) = 2\,\text{Re}\, Q(x) - 2 Q(-2-x) - 4(1+x) Q'(-2-x) + 4(1+x)^2 Q''(-2-x). \tag{4.30}$$

We have $p(x) = -dP_1(x)/dx$. Differentiating (4.30) and expressing Q with the aid of (4.29) we obtain

$$p(x) = \frac{d}{dx} \int_{-1}^{0} L(x,t)\, q(t)\, dt \quad \text{with} \quad L(x,t) = \frac{-2}{t-x} + \frac{2}{t+x+2} + \frac{4(1+x)}{(t+x+2)^2} - \frac{8(1+x)^2}{(t+x+2)^3}. \tag{4.31}$$

The integral in (4.31) must be interpreted as CAUCHY principal value. Given $p(x)$ relation (4.31) represents an integral equation for the unknown displacement $v_1(x)$. The integral equation was first derived by WIGGLESWORTH [16]; somewhat later and independently of WIGGLESWORTH, the author derived the same integral equation as the limit case of an integral equation of a rotor with a radial crack. The integral equation was used [4], [5] in order to determine $v_1(x)$ numerically for the case of $p(x) \equiv 1$. The following approximation was found:

$$v_1(x) = \frac{\pi(\varkappa+1)}{2\mu} q(x); \qquad q(x) = q_0(x) \sqrt{-x},$$
$$8 q_0(x) \approx 1.87 - 0.12(1+x) + 0.58(1+x)^2 - 0.30(1+x)^3. \tag{4.32}$$

From (1.2) and (4.32) the value $K = 1.13$ was derived. It is remarkable that the function $q(x)$ in (4.32) contains already enough information for the construction of $m(x)$. Let us introduce

$$\varphi_2 = -\varphi_1 + (1+z)\varphi_1'; \qquad \varrho_2 = -\varrho_1 + (1+z)\varrho_1'. \tag{4.33}$$

These functions yield

$$P_2 = -P_1 + (1+x)\frac{\partial P_1}{\partial x} + y\frac{\partial P_1}{\partial y}; \qquad w_2 = -w_1 + (1+x)\frac{\partial w_1}{\partial x} + y\frac{\partial w_1}{\partial y}. \tag{4.34}$$

Now if $P_1 = cx + d$ with certain constants c, d along the crack, we have $p(x) \equiv -c$ and $P_2 \equiv c - d$ on C^+, C^-. We have $P_2 = -P_1 = \text{const.}$ along $x = -1$. Altogether: $P_2 = 0$ on the boundary, and we may now state that φ_2, ϱ_2 describes a fundamental state with a real coefficient a. The weight $m(t)$ can be derived from v_1 with the aid of the second relation (4.34). Result:

$$m(t) = \frac{1}{q_0(0)\sqrt{|t|}} \{(1-t) q_0(t) + 2t(1+t) q_0'(t)\}; \qquad q_0(t) \text{ defined by (4.32)}. \tag{4.35}$$

There is another way of finding $m(t)$. The integral equation (4.31), although derived under the assumptions of a regular state applies to fundamental states as well, and one would expect

$$\frac{d}{dx} \int_{-1}^{0} L(x,t)\, m(t)\, dt = 0. \tag{4.36}$$

The weight $m(t)$ can be determined as that solution of (4.36) which also satisfies the normalizing condition (4.23).

We conclude this paper with some general remarks about $m(t)$. By (4.23) this function is positive for small $|t|$; is it positive throughout $(-l, 0)$? The answer is affirmative for the examples above. It is desirable to deal with the question in all generality. Linked to this problem and even more important is this question: If $m_1(t)$, $m_2(t)$ pertain to domains A_1, A_2 with the same crack and if $A_1 \subset A_2$, would then $m_1(t) \geqq m_2(t)$? (Fig. 9).

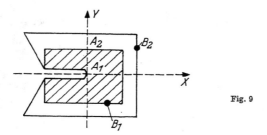

Fig. 9

A positive answer to this question would permit to find upper and lower bounds for $m(t)$. The practical importance of such bounds needs hardly be underlined. We shall now give a partial answer to the second question. The crack being analytic, the expansions in theorem 3 become applicable, and we have in particular the representation

$$\varphi_k = \frac{1}{\sqrt{z}} F_k(z) + G_k(z) + H_k; \qquad \varrho_k = \frac{1}{\sqrt{z}} F_k(z) - G_k(z); \qquad k = 1, 2, \tag{4.37}$$

for any fundamental state φ_k, ϱ_k in A_k with a real coefficient; the functions F_k, G_k are holomorphic in some neighborhood of $z = 0$; their expansions in powers of z have real coefficients; H_1, H_2 are real constants. Let $a = 1$; then $F_k(0) = 1$. Set $F_k'(0) = -c_k$. The normal displacement $v_k(t)$ on C^+ of the state φ_k, ϱ_k is given by

$$v_k(t) = -\frac{\varkappa + 1}{2\mu \sqrt{|t|}} F_k(t). \tag{4.38}$$

The associated function $m_k(t)$ is

$$m_k(t) = \frac{F_k(t)}{\sqrt{|t|}} = \frac{1}{\sqrt{|t|}} + c_k \sqrt{|t|} + \cdots. \tag{4.39}$$

Lemma 2: $c_1 > c_2$.

This means that $m_1(t) > m_2(t)$ for small $|t|$. The proof is easy. We form the state $\varphi = \varphi_2 - \varphi_1$, $\varrho = \varrho_2 - \varrho_1$ and observe that it is regular in A_1 with a normal singularity of coefficient $T = 1/2\,(c_1 - c_2)$. By (4.19) we have

$$\frac{\pi(\varkappa + 1)}{\mu}(c_1 - c_2) = -\int_{B_1} [(X_2 - X_1) u_1 + (Y_2 - Y_1) v_1]\, ds, \tag{4.40}$$

where $X_1 + i Y_1$, $X_2 + i Y_2$ are the complex tractions of states φ_1, ϱ_1 and φ_2, ϱ_2 respectively, the tractions attacking the material in A_1 along B_1. We have of course $X_1 = Y_1 = 0$ along B_1. The integral in (4.40) can be split:

$$\frac{\pi(\varkappa + 1)}{\pi}(c_1 - c_2) = \int_{B_1} [(X_2 - X_1)(u_2 - u_1) + (Y_2 - Y_1)(v_2 - v_1)]\, ds +$$
$$+ \int_{B_1} [(X_1 - X_2) u_2 + (Y_1 - Y_2) v_2]\, ds;$$

here the first integral on the right is twice the energy of φ, ϱ in A_1, while the second integral can be interpreted as twice the energy of φ_2, ϱ_2 in $A_2 - A_1$. Both energies are positive so that $c_1 > c_2$ as asserted.

References

[1] G. R. Irwin, J. A. Lies and H. L. Smith, Fracture strength relative to onset and arrest of crack propagation, Proceedings of American Society for Testing Materials, Vol. 58, p. 640–657 (1958/59).
[2] A. A. Griffith, The phenomenon of rupture and flow in solids, Philosophical Transactions of the Royal Society of London, Series A, Vol. 221, p. 163–198 (1920).
[3] G. I. Barenblatt, Mathematical theory of equilibrium cracks, Advances in Applied Mathematics, Academic Press, New York and London, Vol. 7, pp. 55–129.
[4] H. F. Bueckner, Some stress singularities and their computation by means of integral equations, Boundary Problems in Differential Equations, The University of Wisconsin Press, Madison 1960, p. 215–230.
[5] H. F. Bueckner and I. Giaever, The stress concentration of a notched rotor subjected to centrifugal forces, ZAMM 46, p. 265–273 (1966).
[6] H. F. Bueckner, The propagation of cracks and the energy of elastic deformation, Transactions of the ASME, p. 1225 to 1230 (1958).
[7] N. I. Muskhelishvili, Some basic problems of the mathematical theory of elasticity, Groningen, Holland 1953, P. Noordhoff, Ltd.
[8] N. I. Muskhelishvili, Singular integral equations, Groningen, Holland 1953, P. Noordhoff Ltd.
[9] G. Fichera, Sulla torsione elastica dei prismi cavi, INAC 389.
[10] G. Fichera, Sull' esistenza e sul calcolo delle soluzioni dei problemi al contorno, relativi all' equilibrio di un corpo elastico, INAC 248.

11 G. Fichera, Sui problemi analitici dell'elasticità piana, INAC 286. "INAC" refers to Pubblicazioni dell'Istituto per le Applicazioni dell Calcolo, Consiglio Nazionale delle Ricerche, Italy.
12 K. O. Friedrichs, On the boundary-value problems of the theory of elasticity and Korn's inequality, Ann. of Math. 48, p. 441–471 (1947).
13 L. G. Magnaradze, Les problèmes fondamentaux de la théorie de l'élasticité à deux dimensions pour les contours a points anguleux, Comptes Rendus (Doklady) de l'Académie des Sciences de l'URSS, Vol. **XVI**, No. 8, p. 149–153 (1937).
14 L. G. Magnaradze, Solution of the fundamental problems of plane theory of elasticity in the case of contours with corners, Comptes Rendus (Doklady) de l'Académie des Sciences de l'URSS, Vol. **XIX**, No. 9, p. 673–676 (1938).
15 L. G. Magnaradze, The fundamental problems of the theory of plane elasticitie of contours with corners (in Russian), Travaux de l'Institut Mathématique de Tbilissi, Vol. **IV**, p. 43–76 (1938).
16 L. A. Wigglesworth, Stress distribution in a notched plate, Mathematica 4, p. 76–96 (1957).

Manuskripteingang: 20. 3. 1969

Anschrift: Dr. Hans F. Bueckner, Schenectady, N.Y., 1184 Bellemead Court, Zipcode 12309

SOME REMARKS ON ELASTIC CRACK-TIP STRESS FIELDS

James R. Rice

Division of Engineering, Brown University, Providence, Rhode Island

Abstract—It is shown that if the displacement field and stress intensity factor are known as functions of crack length for any symmetrical load system acting on a linear elastic body in plane strain, then the stress intensity factor for any other symmetrical load system whatsoever on the same body may be directly determined. The result is closely related to Bueckner's [1] "weight function", through which the stress intensity factor is expressed as a sum of work-like products between applied forces and values of the weight function at their points of application. An example of the method is given wherein the solution for a crack in a remotely uniform stress field is used to generate the expression for the stress intensity factor due to an arbitrary traction distribution on the faces of a crack. A corresponding theory is developed in an appendix for three-dimensional crack problems, although this appears to be directly useful chiefly for problems in which there is axial symmetry.

INTRODUCTION

Consider a two-dimensional linear elastic body containing a straight crack under conditions of plane strain or of generalized plane stress. Both the body and all applied load systems to be considered are assumed symmetrical about the crack line so that only the tensile opening mode of crack tip deformation may result. Two distinct load systems are shown in Fig. 1 and denoted by Q_1 and Q_2. The displacement field and stress intensity factor are assumed known as a function of crack length l for one of the load systems, say Q_1.

The principal result of this study is in showing that this information is sufficient to determine the stress intensity factor for the other load system Q_2. Of course, the 1 and 2 systems may represent any arbitrarily chosen load systems and thus it is being shown that if a solution for the displacement field and stress intensity factor is known for any particular load system, then this information is sufficient to determine the stress intensity factor for any other load system whatsoever.

Bueckner [1] has presented a similar result, basing his argument on analytic function representations of elastic fields for isotropic materials. Here we see that this dependence between solutions for different load systems arises as a consequence of what is known on the relation between stress intensity factors and strain energy variations [2, 3] and of the properties of point functions. To develop the argument, consider the following preliminary remarks:

(a) Q_1 and Q_2 are considered as "generalized forces" in the sense that the stress vector **t** on the boundary Γ and body force **f** within region A resulting from, say, load system 1 are expressible in the form

$$\mathbf{t} = Q_1 \mathbf{t}^{(1)} \quad \text{on } \Gamma \quad \text{and} \quad \mathbf{f} = Q_1 \mathbf{f}^{(1)} \quad \text{in } A, \tag{1}$$

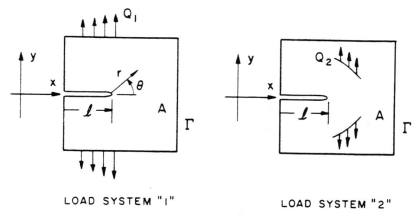

Fig. 1.

with similar expressions for load system 2, where $\mathbf{t}^{(1)}$, $\mathbf{t}^{(2)}$ and $\mathbf{f}^{(1)}$, $\mathbf{f}^{(2)}$ are functions of *position only*.

(b) If $\mathbf{u}$ is any displacement field in the body, then with it we may associate "generalized displacements" q_1 and q_2 by

$$q_i = \int_\Gamma \mathbf{t}^{(i)} \cdot \mathbf{u} \, d\Gamma + \int_A \mathbf{f}^{(i)} \cdot \mathbf{u} \, dA. \tag{2}$$

Thus if a variation $\delta \mathbf{u}$ is given to the displacement field, $Q_1 \delta q_1$ will be the work (per unit thickness) of load system 1 and $Q_2 \delta q_2$ of 2. We shall write $Q_1 \mathbf{u}^{(1)}$ for the elastic displacement field induced by load system 1 and $Q_2 \mathbf{u}^{(2)}$ for that by 2. Hence if both load systems are *simultaneously* applied to the body, then by superposition

$$\mathbf{u} = Q_1 \mathbf{u}^{(1)} + Q_2 \mathbf{u}^{(2)} \quad \text{and} \quad q_i = C_{ij} Q_j \tag{3}$$

(summing on repeated indices) where the compliances are

$$C_{ij} = \int_\Gamma \mathbf{t}^{(i)} \cdot \mathbf{u}^{(j)} \, d\Gamma + \int_A \mathbf{f}^{(i)} \cdot \mathbf{u}^{(j)} \, dA. \tag{4}$$

Here $C_{ij} = C_{ij}(l)$ because $\mathbf{u}^{(i)} = \mathbf{u}^{(i)}(x, y, l)$.

(c) The stress intensity factor K is defined by

$$K = \lim_{r \to 0} (2\pi r)^{\frac{1}{2}} \sigma_{yy}, \tag{5}$$

where σ_{yy} is the y directed tensile stress acting at distance r along the line directly ahead of the crack tip. We denote $Q_1 K^{(1)}$ as the factor induced by load system 1 and $Q_2 K^{(2)}$ as that by 2, where $K^{(i)} = K^{(i)}(l)$.

(d) It is known that if W is the elastic strain energy (per unit thickness) of a loaded, cracked body, then

$$(\partial W / \partial l)_{\text{fixed displ.}} = -K^2/H, \tag{6}$$

where "fixed displacements" means that the derivative is taken under conditions for which loaded portions of the body are constrained against working displacements. H is an appropriate elastic modulus: for an isotropic material it is $E/1-\nu^2$ for plane strain and E for generalized plane stress; for anisotropic materials the modulus may be chosen from the work of Sih et al. [4]. Thus when both load systems are simultaneously applied to the same body, we may think of representing W as a function of q_1, q_2 and l, and write

$$\partial W(q_1, q_2, l)/\partial l = -K^2/H, \quad \text{where } K = Q_1 K^{(1)}(l) + Q_2 K^{(2)}(l). \tag{7}$$

Of course, $\partial W(q_1, q_2, l)/\partial q_i = Q_i$.

These last remarks enable us to write, in the case of simultaneous action of both load systems, the perfect Pfaffian differential form

$$Q_1 \delta q_1 + Q_2 \delta q_2 - (K^2/H)\delta l = \delta W, \tag{8}$$

and this may be transformed to

$$q_1 \delta Q_1 + q_2 \delta Q_2 + (K^2/H)\delta l = \delta(Q_1 q_1 + Q_2 q_2 - W). \tag{9}$$

Clearly, the left side of this equation is a perfect differential and this has very important consequences: for if we consider q_i and K as functions of Q_1, Q_2 and l, as in equations (3), (7), then

$$\partial q_i/\partial l \equiv (dC_{ij}/dl)Q_j = \partial(K^2/H)/\partial Q_i \equiv 2K^{(i)}K^{(j)}Q_j/H. \tag{10}$$

Since this holds for all values of Q_1 and Q_2, we have

$$dC_{ij}(l)/dl = 2K^{(i)}(l)K^{(j)}(l)/H, \tag{11}$$

and this is seen to be a generalization of Irwin's [2] relation between compliance variations with crack length and stress intensity factors. A similar result has been derived in a special case by Rice and Levy (equations (9)–(14) of Ref. [5]); their application involved deriving cross terms analogous to C_{21}, given $K^{(1)}$ and $K^{(2)}$.

Here our viewpoint is different: it is assumed that we know the intensity factor $K^{(1)}$ and displacement field $\mathbf{u}^{(1)}$ associated with load system 1. This means that we also know C_{21}, as is clear from equation (4). Hence, we find that equation (11), written for $i = 2$ and $j = 1$, enables us to solve for $K^{(2)}$ solely from a knowledge of the solution for load 1:

$$K^{(2)}(l) = \frac{H}{2K^{(1)}(l)} \frac{dC_{21}(l)}{dl} \equiv \frac{H}{2K^{(1)}(l)} \left\{ \int_\Gamma \mathbf{t}^{(2)} \cdot \frac{\partial \mathbf{u}^{(1)}}{\partial l} d\Gamma + \int_A \mathbf{f}^{(2)} \cdot \frac{\partial \mathbf{u}^{(1)}}{\partial l} dA \right\}. \tag{12}$$

[Here we pause to recall that $\mathbf{u}^{(1)}$ will be non-unique to within rigid-body displacements and since these may be chosen arbitrarily for each crack length, $\partial \mathbf{u}^{(1)}/\partial l$ must be considered similarly non-unique. This has no effect on equation (12) since load set 2 is self-equilibrating and hence does no work on a rigid motion of the body.]

THE WEIGHT FUNCTION

Now, it is obvious that the stress intensity factor for load system 2 can in no way depend on the *particular* load system represented by 1. Hence the function

$$\mathbf{h}^{(1)} = \frac{H}{2K^{(1)}(l)} \frac{\partial \mathbf{u}^{(1)}(x, y, l)}{\partial l} \tag{13}$$

which, following Bueckner [1], we refer to as the *weight function*, must be essentially independent of the nature of load system 1.

To study its uniqueness consider another load system denoted by 3. This must give the same result for $K^{(2)}$ when substituted for 1 in equation (12) and thus

$$\int_\Gamma \mathbf{t}^{(2)} \cdot [\mathbf{h}^{(1)} - \mathbf{h}^{(3)}] \, d\Gamma + \int_A \mathbf{f}^{(2)} \cdot [\mathbf{h}^{(1)} - \mathbf{h}^{(3)}] \, dA = 0 \tag{14}$$

for all symmetrical self-equilibrating load systems 2. Since both 1 and 3 correspond to deformation fields that are symmetrical about the crack line we must have

$$h_x^{(i)}(x, -y) = h_x^{(i)}(x, y) + \omega^{(i)} y; \qquad h_y^{(i)}(x, -y) = -h_y^{(i)}(x, y) + \alpha^{(i)} + \omega^{(i)} x \tag{15}$$

where $\alpha^{(i)}$ and $\omega^{(i)}$ depend at most on crack length and reflect the rigid motion indeterminacy of $\mathbf{u}^{(i)}$. We now require that equation (14) hold for: (a) a unit point force in the y direction at an arbitrary point (x, y), with a symmetrical equilibrating force at $(x, -y)$ and (b) a unit point force in the x direction at an arbitrary point (x, y), a unit point force in the $-x$ direction at (x', y') and symmetrical forces at $(x, -y)$ and $(x', -y')$. This is readily shown to imply that

$$h_x^{(1)} - h_x^{(3)} = \lambda - \Omega y; \qquad h_y^{(1)} - h_y^{(3)} = \mu + \Omega x, \tag{16}$$

where λ, μ and Ω are constants. Thus $\mathbf{h}^{(1)} = \mathbf{h}^{(3)}$ to within rigid-body motions which are, in any event, arbitrary and inconsequential.

We therefore conclude that for any symmetrical load system leading to stress intensity factor K and displacement field $\mathbf{u}$, the function

$$\mathbf{h} = \mathbf{h}(x, y, l) = \frac{H}{2K} \frac{\partial \mathbf{u}}{\partial l} \tag{17}$$

(the derivative being taken at fixed values of the applied loads) is a universal function for a cracked body of any given geometry and composition, regardless of the detailed way in which the body is loaded. Once $\mathbf{h}$ is determined from the solution for any particular load system, the stress intensity factor induced by any other symmetrical load system $\mathbf{t}$ and $\mathbf{f}$ is, from equation (12),

$$K = \int_\Gamma \mathbf{t} \cdot \mathbf{h} \, d\Gamma + \int_A \mathbf{f} \cdot \mathbf{h} \, dA. \tag{18}$$

It should also be noted that once K is known we may go back to equation (17) and, by integrating $\partial \mathbf{u}/\partial l$ with respect to l, construct the entire displacement field provided it is known for one value of l (say, $l = 0$). Hence any one elastic crack solution is seen to contain a remarkable store of information. This information is most succinctly given through the weight function itself and it is of interest that the weight function may be determined directly in view of the following properties noted by Bueckner [1] and summarized briefly here:

Note that $\mathbf{h}$ satisfies the same differential equations as the displacement $\mathbf{u}$ and that when $\mathbf{h}$ is viewed as a displacement, the stresses which it produces require no body forces in A or boundary surface forces on Γ for their equilibration (this results, of course, since $\mathbf{h} \propto \partial \mathbf{u}/\partial l$). Ordinarily, these conditions would be interpreted as requiring that $\mathbf{h}$ be a rigid motion, since a state of zero stress satisfies the null loading conditions. However, the elastic uniqueness theorem applies to uniqueness on the class of crack tip displacement fields carrying bounded total energy and $\mathbf{h}$ is not of this class. The strongly singular part of $\mathbf{h}$ can be determined by recalling that the displacement fields to all (bounded energy) solutions for cracks take the form

$$\mathbf{u} = H^{-1}Kr^{\frac{1}{2}}\mathbf{g}(\theta) + \ldots \qquad (19)$$

where the dots stand for terms resulting in non-singular stresses, where $r\theta$ are polar coordinates centered at the crack tip and where $\mathbf{g}$ is a universal function of θ (and also of ratios of elastic moduli) [2–4].

Since $\partial r/\partial l = -\cos\theta$ and $\partial\theta/\partial l = \sin\theta/r$, where $\theta = 0$ along the line ahead of the crack, we have

$$\partial \mathbf{u}/\partial l = H^{-1}Kr^{-\frac{1}{2}}[(\sin\theta)(d\mathbf{g}/d\theta) - (\cos\theta)(\mathbf{g}/2)] + \ldots \qquad (20)$$

where now the dots represent all terms which are bounded at the tip and which, if taken individually, would correspond to bounded total energy. Hence we see from equation (17) that

$$\mathbf{h} = r^{-\frac{1}{2}}[2(\sin\theta)(d\mathbf{g}/d\theta) - (\cos\theta)\mathbf{g}]/4 + \mathbf{h}^* \qquad (21)$$

where $\mathbf{h}^*$ is a displacement field of the usual bounded energy class for an elastic crack problem.

From this point of view it is easy to see that $\mathbf{h}$ is a universal function for a given geometry and composition: its strongly singular ($r^{-\frac{1}{2}}$) part is universal and $\mathbf{h}^*$ is chosen so that it, together with the prescribed $r^{-\frac{1}{2}}$ term, results in zero surface tractions on Γ and zero body forces in A. Since the $r^{-\frac{1}{2}}$ term creates no tractions on the crack surfaces, the problem of determining $\mathbf{h}^*$ falls into the class of bounded energy problems for which there is uniqueness. Clearly, the result for $\mathbf{h}$ bears no relation to any particular load system to which the body may be subjected.

An example of the method follows. Also, a three-dimensional theory for a weight function is developed in the Appendix, although this theory appears at present to be less directly useful in the determination of stress intensity factors.

AN EXAMPLE

To illustrate the procedure let us follow Bueckner [1] in considering the Inglis problem of a crack of length l in an infinite body subject to a remotely uniform tensile stress σ. We take the origin of the xy coordinate system at one tip of the crack, the other being at $(l, 0)$, so that the y displacements along the upper and lower crack surfaces ($y = \pm 0$) and the stress intensity factor are

$$u_y = \pm 2\sigma x^{\frac{1}{2}}(l-x)^{\frac{1}{2}}/H, \qquad K = \sigma(\pi l/2)^{\frac{1}{2}}, \qquad (22)$$

where $H = E/(1-\nu^2)$ for plane strain. Thus

$$\left. \begin{array}{l} \partial u_y/\partial l = \pm \sigma x^{\frac{1}{2}}(l-x)^{-\frac{1}{2}}/H, \\ \\ \text{and} \\ \\ h_y(x, \pm 0, l) = (H/2K)(\partial u_y/\partial l) = \pm(2\pi l)^{-\frac{1}{2}}x^{\frac{1}{2}}(l-x)^{-\frac{1}{2}}, \end{array} \right\} \qquad (23)$$

from equation (17). We may therefore employ equation (18) to write the expression for K at the $(l, 0)$ crack tip due to a traction distribution $t_y = \pm p(x)$ along the surfaces of the crack and the result is

$$K = \left(\frac{2}{\pi l}\right)^{\frac{1}{2}} \int_0^l p(x) \left(\frac{x}{l-x}\right)^{\frac{1}{2}} dx. \tag{24}$$

This checks with known results (e.g. [3]) derived independently through Muskhelishvili's analytic function methods.

All problems of symmetrical loading may be reduced, by superposition of a solution for a crack-free body under the same loads, to a similar problem of prescribed normal tractions along the crack. Hence it would seem advisable that, to the extent possible, displacements of the crack surfaces [or better the weight function $h_y(x, \pm 0, l)$] as well as stress intensity factors be reported when crack problems are solved. Bueckner [1], for example, shows how his previous solution for the edge-cracked half-plane in tension may be employed to obtain the weight function for that case.

It is also worthy of note that the weight function at points remote from the crack tip could be determined with great accuracy by finite difference or finite element methods applied to the determination of $\mathbf{h}^*$ in equation (21). The inaccuracy of such methods near the tip would then be irrelevant. This procedure seems, in fact, to be closely related to that proposed by Barone and Robinson [6] for numerical determination of coefficients in eigenfunction expansions of stress fields about corner singularities.

Acknowledgement—This study was supported by the National Aeronautics and Space Administration under Grant NGL-40-002-080 to Brown University.

REFERENCES

[1] H. F. BUECKNER, A novel principle for the computation of stress intensity factors. *Z. angew. Math. Mech.* **50**, 529–546 (1970).
[2] G. R. IRWIN, Fracture Mechanics, in *Structural Mechanics*, edited by J. N. GOODIER and N. J. HOFF, pp. 557–591. Pergamon Press (1960).
[3] J. R. RICE, Mathematical Analysis in the Mechanics of Fracture, in *Fracture*, edited by H. LIEBOWITZ, Vol. II, pp. 191–311. Academic Press (1968).
[4] G. C. SIH, P. C. PARIS and G. R. IRWIN, On cracks in rectilinearly anisotropic bodies. *Int. J. Fracture Mech.* **1**, 189–203 (1965).
[5] J. R. RICE and N. LEVY, The part-through surface crack in an elastic plate. *J. Appl. Mech.* in press.
[6] M. R. BARONE and A. R. ROBINSON, Approximate Determination of Stresses near Notches and Corners in Elastic Media by an Integral Equation Method, University of Illinois Civil Engineering Studies, Structural Research Series No. 374 (1971).

APPENDIX

Three-dimensional bodies

An analogous theory may be developed for three-dimensional crack problems: let V be the volume and S the bounding surface of a body containing a planar crack, with both the body and all load systems under consideration being symmetrical about the plane of the crack. The contour lying along the tip of the crack is denoted by L; this is assumed smooth. Generalized forces Q_1 and Q_2 are defined so that, for example,

$$\mathbf{t} = Q_1 \mathbf{t}^{(1)} \quad \text{on } S \quad \text{and} \quad \mathbf{f} = Q_1 \mathbf{f}^{(1)} \quad \text{in } V \tag{A1}$$

for load system 1. Associated generalized displacements q_1 and q_2 may be defined through replacing Γ by S and A by V in equation (2). Further, equations (3) and (4) may be written with these same replacements when both load systems act simultaneously.

We shall wish to characterize energy changes when the crack surface is advanced by an infinitesimal amount δl, where δl is a smooth function of position along L marking the advance of the crack in a direction locally normal to L. The notation $\delta_l(\ldots)$ will denote the first order variation in the quantity $(\ldots)$, viewed as a function of crack position and some other variables, when only the crack position is varied. Thus we write in analogy to equation (6)

$$(\delta_l W)_{\text{fixed displ.}} = -\int_L [(K^2/H)\delta l]\, dL, \tag{A2}$$

where W is the total strain energy of the body and H has its plane strain value. For simultaneous action of both load systems,

$$K = Q_1 K^{(1)} + Q_2 K^{(2)} \tag{A3}$$

where $K^{(1)}$ and $K^{(2)}$ are functions of position along L the first assumed known and the latter to be found.

Since $\partial W/\partial q_i = Q_i$, a general variation in the strain energy may be written as

$$Q_1 \delta q_1 + Q_2 \delta q_2 - \int_L [(K^2/H)\delta l]\, dL = \delta W, \tag{A4}$$

and if we rearrange this in analogy to equation (9), viewing q_i and K as being dependent on Q_i and the crack position, then we may write in analogy to equation (10) that

$$\delta_l q_i \equiv (\delta_l C_{ij}) Q_j = \int_L \left[\frac{\partial}{\partial Q_i}\left(\frac{K^2}{H}\right) \delta l \right] dL \equiv \int_L \frac{2}{H}[K^{(i)} K^{(j)} Q_j \delta l]\, dL. \tag{A5}$$

Thus the three-dimensional version of the relationship between compliance variations and stress intensity factors is

$$\delta_l C_{ij} = \int_L \frac{2}{H}[K^{(i)} K^{(j)} \delta l]\, dL. \tag{A6}$$

When $i = 2$ and $j = 1$ this becomes, in analogy to equation (12),

$$\int_L \frac{2}{H}[K^{(1)} K^{(2)} \delta l]\, dL = \int_S \mathbf{t}^{(2)} \cdot \delta_l \mathbf{u}^{(1)}\, dS + \int_V \mathbf{f}^{(2)} \cdot \delta_l \mathbf{u}^{(1)}\, dV \tag{A7}$$

and this allows a (rather complete) knowledge of the solution for load system 1 to serve as a basis for determining the stress intensity factor for load system 2.

The difficulty is, of course, that three-dimensional solutions for any load system 1 will not be known with such complete generality that the first order variation $\delta_l \mathbf{u}^{(1)}$ can be determined for completely arbitrary variations δl along L. If the result were known, say as an equation of the form

$$\delta_l \mathbf{u}^{(1)}(P) = \int_L [U^{(1)}(P, P') \delta l(P')]\, dL(P'), \tag{A8}$$

where P denotes a general point of the body and P' a point along the crack tip, then we could solve for $K^{(2)}$ at any point P' along L as

$$K^{(2)}(P') = \frac{H(P')}{2K^{(1)}(P')} \left\{ \int_S \mathbf{t}^{(2)}(P) \cdot U^{(1)}(P, P') \, dS(P) + \int_V \mathbf{f}^{(2)}(P) \cdot U^{(1)}(P, P') \, dV(P) \right\} \quad (A9)$$

which is the general three-dimensional version of equation (12).

In similar fashion, a three dimensional weight function may be defined as

$$\mathbf{h}(P, P') = H(P') U^{(1)}(P, P') / [2K^{(1)}(P')], \quad (A10)$$

and this is a unique (to within rigid motions) function of P and P' for a given crack geometry in a body of given overall geometry and composition, being completely independent of the way in which the body is loaded. If it is determined from the solution for any particular load system, then the solution for K at P' induced by any other load system $\mathbf{t}$ and $\mathbf{f}$ may be obtained from

$$K(P') = \int_S \mathbf{t}(P) \cdot \mathbf{h}(P, P') \, dS(P) + \int_V \mathbf{f}(P) \cdot \mathbf{h}(P, P') \, dV(P). \quad (A11)$$

Of course, there will exist cases for which knowledge of an integrated average of the intensity factor, as in equation (A7), is sufficient and this presents less stringent requirements as to the generality in which $\delta_l \mathbf{u}^{(1)}$ must be known. An example is the case of a penny-shaped crack in an axially symmetric body. Then knowledge of the solution, as a function of crack radius, for any one axially symmetric load system would allow in an obvious way for the determination of the intensity factor for any other axially symmetric load system.

(*Received* 18 *June* 1971; *revised* 9 *November* 1971)

Абстрвкт—Оказывается, что если известны поле перемещений и фактор интенсивности напряжений, в качестве функций длины щели, для любой симметрической системы нагрузки, действующей на линейное упругое тело в плоском дефррмационном состоянии, тогда можно непосредственно определить фактор интенсивности напряжений для другой какой либо симметрической системы нагрузки, действующей на тоже самое тело. Решение тесно связанр с "функцией веса" Бюкнера /1/, вследствие которой представляется фактор интенсивности напряжений в виде суммы похожих работе пройзведений из приложенных усилий и значений функции веса в точках их приложения. Дается пример этого метода, в котором используется решения для щели под влиянием отдаленного однородного поля напряжений, для обобщения выражений фактора интенсивности напряжений, вследствие произвольного распределения тяговых усилий на поверхностях щели. В приложении, определяется соответсвующая теория для трехмерных задач щели, хотя это оказывается непосредственно полезным, главным образом, для задач с осевой симметрией.

Section Three
Subcritical Crack Growth Analysis

RECENT OBSERVATIONS ON FATIGUE FAILURE IN METALS

By W. A. Wood[1]

Synopsis

The mechanism of fatigue failure is viewed in the light of recent observations obtained by a modified taper-sectioning technique which reveals slipband contours at optical magnifications of 20,000 to 30,000 times. The observations show how the fatigue slipbands produce in the metal surface sharp notches, peaks, and intermediate contours and cause related forms of surface disintegration. Typical effects are illustrated by examples from copper and brass subjected to alternating torsion.

Early work on fatigue, summarized in the standard works by Gough (1)[2] and Moore (2), showed that the essential requirement for fatigue failure is a stress cycle that continually imposes a small plastic strain. Later work, notably by Gough and his co-workers (3), then showed that, of the disturbances this strain may cause, the ones responsible for failure are the slip movements; these movements in cubic metals concentrate in bands, and particularly intense bands, after enough cycles, turn into cracks. Thus fatigue raises two problems: (a) the special way in which metal structures may respond to a plastic strain that is small in magnitude and continually reversed in direction, (b) the way in which the resulting slipbands may turn into fine cracks.

These problems are often complicated in practice by concurrent structural changes not essential to the fatigue process. For example, nonsymmetrical stress-cycles superimpose the effects of a unidirectional deformation. Again, impurities and alloying elements by diffusing to the deformation zones may alter the local structure. In cold-worked metals recovery effects can modify local internal stresses. Accordingly in discussion of essential causes it is helpful first to confine attention to "pure fatigue"—for instance, to fatigue as it might occur under symmetrically alternating cycles in polycrystalline metal that is normally ductile, reasonably pure, and initially annealed. It is then possible as a result of recent work to suggest interpretations of both the special deformation and the starting of cracks.

The Deformation Process

The conditions for pure fatigue that the plastic strain should be small and alternately reversed have the following implications:

The *smallness* means that each application of strain need excite only "fine" slip, slip movements limited to a few

---

[1] Baillieu Laboratory, University of Melbourne, Australia.
[2] The boldface numbers in parentheses refer to the list of references appended to this paper, see p. 119.

lattice spacings at one time. This distinguishes fatigue from large-amplitude or static deformation, which evokes "coarse" slip, spontaneous avalanches of slip movements through hundreds of spacings. The fatigue strains can therefore be accommodated by movement of existing or easily-created dislocations, whereas the larger strains require wholesale multiplication of dislocations from (16) or similar sources. This is in accord with observation. Fatigue slipbands are at first faint and then grow slowly in strength. Static slipbands appear suddenly, each virtually at its final strength.

The *reversal* means that the small plastic strain can proceed by fine slip always. There is not the need that there is in static straining to carry on deformation by multiplication of dislocations in order to supplement dislocations which may become held-up at obstacles. For dislocations held-up in one half-cycle may still contribute to the reverse half-cycle by just turning back.

These distinctions may be illustrated by a comparison between the effects of a small plastic strain applied (*a*) in alternate directions, and (*b*) in unidirectional steps, the strain being small enough at each application to need only a few fine slip movements. Observations show that whereas the alternate applications produce the effects of fine slip, the unidirectional applications have all the effects of coarse slip, just as if they had been applied as a single large strain. They give rise to coarse slipbands and to the heavy disorientation and fragmentation shown always in X-ray diffraction patterns of statically deformed metal but not in metal subjected to pure fatigue. It appears therefore that the unidirectional applications, though individually needing only fine slip movements, still trigger avalanches of coarse slip. This they can do only if numbers of them are held-up at obstacles, and then at intervals released in bursts. Accordingly on this direct evidence unidirectional deformation is essentially a "stop-go" process; dislocations pile up at one moment and discharge the next. The distinguishing feature of pure fatigue is that comparable piling-up and outbursts are normally avoided.

These implications have the following practical consequences, by which also they may be tested:

1. Since continued reversals of a small plastic strain need not greatly increase the dislocation density, they need not cause much strain-hardening. This is readily confirmed (4–7). Strain-hardening is significantly additive only with reversals of large strains.

2. Because of this inefficient strain-hardening, continued reversal of the small strain can impose unusually high totals of plastic deformation, higher than the totals obtainable in large-amplitude or static straining. This also is readily shown, totals of several thousand per cent being obtainable before fracture (4–7). Thus if N cycles each with plastic component e are applied to a metal, it appears that the total plastic strain, which is proportional to Ne and which ought to determine the "cumulative damage," is disproportionately large when e is small.

3. From this it follows that continued reversal can give rise to an unusually high number of slip movements. Though some of these may merely retrace their paths, many must escape to the surface of the metal, where they are observed to build up the surface disturbances of continually increasing strength. Therefore it follows that these disturbances may become unusually strong.

Accordingly it is possible to see why pure fatigue deformation should be of a special kind; evidence that this is so, because of accompanying fine slip, has been summarized in a previous discussion (4). It is also possible to see that the unusually strong surface disturbances due to the fine slip are themselves capable of causing cracks, since because of the to-and-fro movement they can build up across a slipband irregular contours that may include sharp notches. This possibility has been the subject of subsequent work which is discussed next.

Observation of Slipband Disturbances

Interesting observations on the disturbances in fatigue slipbands have been made in recent years by Thompson (8), Craig (9), Forsyth (10), and Smith (11), using standard metallographic methods. However, though these methods are adequate for resolving the surface disturbances laterally, as the observations have shown, they do not so easily show the disturbances in depth. This information, which would show more directly how a disturbance might initiate a crack, is only obtainable by observing the disturbances in section. The following discussion is based on observations of this kind, obtained by a convenient technique.

The general procedure has been to prepare finely-ground and electropolished specimens with initially smooth surfaces, then to protect slipbands formed in fatigue tests by an electrodeposited coating, and finally to section the specimens at particularly small angles to the surface. A surface disturbance lying in a plane normal to the intersection of the surface and sectioning plane is then magnified in the section by the cosecant α, where α is the small-angle taper. With α about 2 or 3 deg, preliminary magnifications of 20 or 30 times are obtainable. The sections, after metallographic polishing, are then examined by optical microscope at 500 to 1000 magnifications. In this way the disturbances can be viewed in section at total effective magnifications up to 30,000 times.

It was also found that a simple way of making the small-angle sections was to use a cylindrical test specimen and merely grind on it a narrow longitudinal flat (12). A radial disturbance at the surface is then magnified in section by the cosecant $\alpha = R/x$, where R is the specimen radius and x the half-width of the flat, which at any point can be found accurately during the examination by microscope. The observations used below to illustrate the main findings were made in this way.

These refer to annealed OFHC copper and 70-30 brass subjected to alternating torsion in a Chevenard-type machine, all specimens having a test length of 1.75 in. and diameter of 0.2 in., one degree thus giving surface shear strain of 10^{-3}. Resulting slipbands were protected by silver plating. Sections were polished mechanically and finally etched in 10 per cent ammonium persulfate. This etchant leaves the silver untouched, so that the microstructure of the specimens and their contours show clearly against a white background. Magnifications of contours are given with the photographs in the form $t \times m$, where t is the preliminary taper magnification and m the subsequent optical magnification.

Etching also brings up in the polished section lines of deformation where heavy slip has been taking place, a feature established by Kemsley (13) in ordinary sections of copper fatigue specimens. The feature is of special interest here because it allows the internal slipbands to be related to the disturbances which they produce at the surface. The bands themselves, however, show significant variations with conditions of testing and it will be convenient to summarize these first, before describing the contour changes.

(a) Only limited lengths of a slipband

may respond to etching. This implies that the intensity of deformation along the band is correspondingly limited. Figures 1 and 2 illustrate extreme cases. Figure 1 taken from brass after 5×10^5 cycles at ± 5 deg, approximately half the expected life, shows bands which are long and fairly uniform in strength. Figure 2 from brass after 7×10^5 cycles at ± 5 deg shows typically limited bands. This photograph also serves to show a common occurrence in grains capable of cross-slip; the active slip concentrates and then take quite irregular paths, which may or may not follow grain boundaries. Figure 3 from brass after 10^4 cycles at the relatively large amplitude of ± 15 deg, which caused failure in another part of the specimen, illustrates both the absence of the etched-up slipbands and the appearance of the irregularly branching crack characteristic of the large amplitude.

This observation is of interest in confirming by an independent observation the point made earlier that excessive

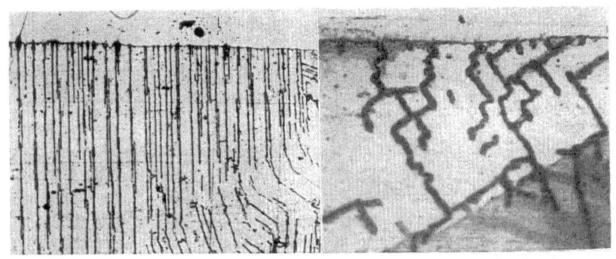

FIG. 1.—Long Etched-up Slipbands ($\times 200$) (Section parallel to surface).

FIG. 2.—Limited Etched-up Slipbands—Crazy Slip ($\times 1000$).

on a short length of one slip system and then jumps for a short distance to the intersecting system, so creating a curious pattern of "crazy" slip.

(b) The limited slipbands on the average become shorter as the amplitude of the fatigue cycle is larger; at sufficiently large amplitudes, therefore, no bands at all etch up. This means that the mechanism of cracking itself must change as the amplitude is larger, for cracks at large amplitudes cannot form by the opening-up of concentrated slipbands when such bands are not there. This was confirmed. It was found that cracks at large amplitudes begin at isolated points slipband deformation and slipband cracking is a feature only of cycles at small amplitudes. It is also of interest because the transition from small-amplitude to large-amplitude cracking appears to coincide with the knee of the S-N curve, suggesting that the curve owes its characteristic shape to the superposition of the two distinct mechanisms of deformation and cracking.

(c) The etched-up bands at small amplitudes in general can be readily observed at about one tenth the expected life of a specimen. At later stages they are detected without any difficulty. This will be evident from various photo-

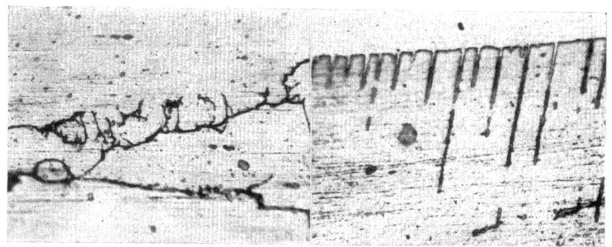

Fig. 3.—Irregular Cracking When Etched-up Slipbands Absent at Large Strain Amplitudes (× 1000).

Fig. 4.—Notch Formations (20,000).

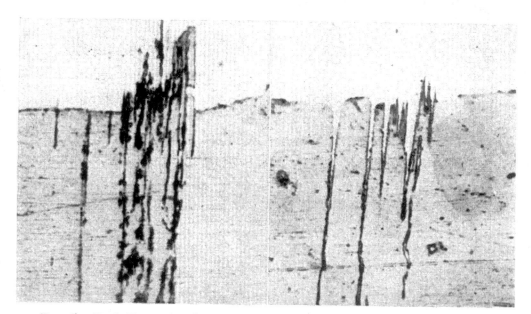

Fig. 5.—Peak Formation (12,500).

Fig. 6.—Combined Notches and Peaks (20,000).

graphs illustrating the slipband contours, which are described next.

Types of Slipband Disturbance

The higher magnifications showed that the etched-up slipbands give rise to characteristic disturbances wherever they meet the surface. The main forms which these disturbances take are illustrated below by photographs from the brass tested at ±5 deg and from the copper tested at ±2.5 deg, each amplitude giving an expected life of 10^6 cycles.

Notches and Peaks.—One form is an opening up of the head of a slipband into a notch; typical examples appear in Fig. 4, taken from copper after 10^5 cycles. A second form is the converse one of a band building up a peak; this is shown in Fig. 5, from brass after 5×10^5 cycles. A form combining both notches and peaks in an intermediate jagged contour is shown by Fig. 6, from copper after 5×10^5 cycles. Quite early in the life of a specimen a notch may penetrate deeply into its slipband, so turning it into a long narrow fissure. This form is illustrated by Fig. 7, from

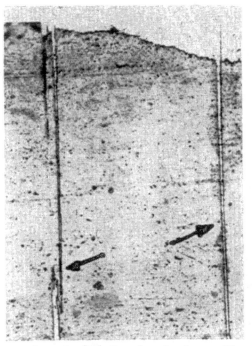

Fig. 7.—Notch Developed into Fissure (20,000).

copper after 5×10^4 cycles, only one twentieth of the expected life.

Surface Disintegration.—The disturbances may take special forms that accentuate a surface disintegration. One example is the peak formation itself which, as shown by Fig. 5, disintegrates when it protrudes too far. A second example occurs in grains capable of cross-slip; then fissures on intersecting may loosen a block of metal in the surface, an illustration being given in Fig. 8 from copper after 10^5 cycles. A third occurs when slipbands and the resulting fissures lie almost parallel to the surface. Then fine strips of metal, being in consequence loosely attached to the surface, are easily torn away; this is shown in Fig. 9, from brass after 10^5 cycles.

Block-Movements.—Fissures along the slipbands divide a grain into blocks between which the coherence is reduced. Relative movement of blocks as a whole is therefore to be expected. Typical relative movements, giving the surface

Fig. 8.—Intersecting Fissures Loosening Surface Block (20,000).

a serrated contour, are shown in Fig. 10, taken from brass after 5×10^5 cycles. In grains permitting cross-slip, the movements may detach a block from the surface, as illustrated by Fig. 11 from the same specimen. Accordingly the block-movement, though a secondary effect of the fissures, virtually constitutes a further mechanism of local failure.

Thus the observations suggest that the slip movements in a band may create a surface disturbance strong enough to start a crack. The observed disturbances on the average, however, are small, being estimated from the magnifications to be about 10^{-4} to 10^{-5} cm. The question next arises whether they cause

Fig. 9.—Fissures Nearly Parallel to Surface, with Local Disintegration (20,000).

Fig. 10.—Block Movements (20,000).

Fig. 11.—Cross-Slip Detaching Surface Block (Crack Forming) (12,500).

macroscopic cracking by propagating individually or by increasing in number.

Crack Formation

Observations on the specimens used in the work described here suggest that failure occurs when disturbances in the form of fissures become numerous enough to weaken the metal. The process appears to occur in the following stages.

The first is appearance of the slipband with a notch at its head, as already illustrated. At this stage the notch in effect is a small crack, but the slipband is merely a zone of heavy deformation. This follows from experiments in which a specimen was heated after the fatigue test and before the sectioning. The etched section then shows many notches with no slipbands attached, a condition not met in comparable specimens before heating. The deformation responsible for some of the slipbands can therefore be dispersed by heating. Since other evi-

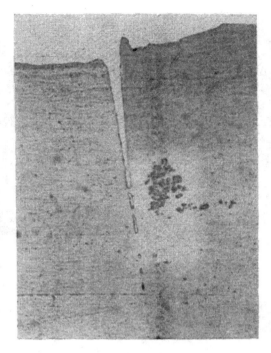

FIG. 12.—Fissure Sintering into Cavities (25,000).

in the heated specimens did retain their slipbands; but these now etched up as dotted lines, many of which were resolvable into links of cavities. Examples are shown by Figs. 12 and 13, from a copper specimen subjected to 10^5 cycles and then heated *in vacuo* for 1 hr at 800 C. This was taken to indicate that some of the slipbands had already opened up into fissures, which during heating had sintered at intervals to form cavities. This penetration of the notch into a slipband to form persisting fissures evidently institutes the second stage.

The next is an increase in number of fissures with continued cycling, rather than propagation of individual ones. That fissures may have difficulty in propagating is to be expected from the earlier observation that etched-up slipbands are usually limited in length, for

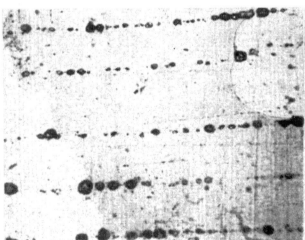

FIG. 13.—Balling-up of Etched-up Slipbands After Heating ($\times$ 1000) (Section parallel to surface).

FIG. 14.—Fissure Turning Through Angle in Cross-Slip System (25,000).

dence has shown that the fatigue slipbands are not zones of heavy strain-hardening (4) it would appear that at this early stage the bands which can be dispersed are zones of concentrated point-defects.

At the same time it was found in these heating experiments that other notches

fissures would penetrate readily into the lengths of heavy deformation but less readily beyond these limits. This difficulty is especially evident in grains exhibiting cross-slip. Then the fissures prefer to turn sharply from one slip system to the other rather than penetrate new ground in one direction. An example of sharply turning fissures is shown in Fig. 14, from brass after 10^5 cycles.

The final stage is then a macroscopic crack that originates at one fissure and traverses the weakened metal by jumping from one fissure to another. A typical example is to be seen in Fig. 15, from

directional deformation. Failure results from development of high local stresses.

At small amplitudes, in contrast, strain-hardening is of little importance. The main factor is the excessive plastic strain that can be imposed in consequence. This appears to produce strong concentrations of point defects in the

FIG. 15.—Crack Jumping from Fissure to Fissure (15,000).

brass after 10^6 cycles. In this stage general deterioration of the metal structure is accentuated. Thus block movements become enhanced, as also shown by Fig. 15; and the initial cracks may leave crystallographic directions for more irregular paths, as shown by Fig. 16 from the same specimen.

Conclusions

This and work previously discussed (4) show that the mechanism of failure under cyclic stress depends first on amplitude. At large amplitudes the early theories of Gough (3) and Orowan (14) might hold. Strain-hardening, and the corresponding internal stresses, increase significantly at each reversal; and after relatively few cycles they reach values comparable with those producing fracture in uni-

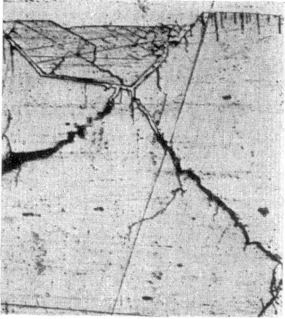

FIG. 16.—Crack Starting at Fissures, Then Developing Irregularly (10,000).

slipbands which respond to etching and abnormal disturbances where the bands meet the metal surface. It is unlikely that the point defects are responsible directly for the fissures forming in the slipbands, for, as emphasized by Mott (15), fatigue still occurs at low temperatures where diffusion of point defects is retarded. But it is reasonable to suppose that they weaken a band sufficiently to allow a fissure to form easily by extension of the notch produced at its head.

The observations accord with the working hypothesis previously put forward (4) that failure at low amplitudes begins as a simple geometrical consequence of the fine slip movements in a fatigue band, which can build up notches or peaks or intermediate contours simply according to their distribution across the band. It was desirable to show, however, that the disturbances thus predicted do in fact occur and that they are big enough to be significant. This the present observations seem to do.

References

(1) H. J. Gough, "Fatigue of Metals," D. Van Nostrand & Co., New York, N. Y. (1924).

(2) H. F. Moore and J. B. Kommers, "Fatigue of Metals," McGraw-Hill Book Co., Inc., New York, N. Y. (1927).

(3) H. J. Gough, "Crystalline Structure in Relation to Failure of Metals—Especially by Fatigue," *Proceedings*, Am. Soc. Testing Mats., p. 3. (Edgar Marburg Lecture 1933.)

(4) W. A. Wood, "Mechanism of Fatigue," Conference on Fatigue in Aircraft Structures, Academic Press Inc., New York, N. Y., p. 1 (1956).

(5) W. A. Wood and R. L. Segall, "Annealed Metals Under Alternating Plastic Strain," *Proceedings*, Royal Soc. London, Series A, Vol. 242, p. 180 (1957).

(6) W. A. Wood and R. L. Segall, "Softening of Cold-Worked Metal Under Alternating Strain," *Journal*, Inst. Metals (in press).

(7) W. A. Wood, "Cracking of α-Brass Under Alternating Strain," *Journal*, Inst. Metals (in press).

(8) N. Thompson, "Experiments Relating to the Origin of Cracks," Conference on Fatigue in Aircraft Structures, Academic Press Inc., New York, N. Y., p. 43 (1956).

(9) W. J. Craig, "Electron Microscope Study of the Development of Fatigue Failures," *Proceedings*, Am. Soc. Testing Mats., Vol. 52, p. 877 (1952).

(10) P. J. E. Forsyth, "Slipband Damage and Extrusion," *Proceedings*, Royal Soc. London, Series A, Vol. 242, p. 198 (1957).

(11) G. C. Smith, "The Initial Fatigue Crack," *Proceedings*, Royal Soc. London, Series A, Vol. 242, p. 189 (1957).

(12) W. A. Wood, "Formation of Fatigue Cracks," *Philosophical Magazine* (in press).

(13) D. S. Kemsley, "Interior Deformation Markings in Copper Fatigue Specimens," *Nature*, Vol. 178, p. 653 (1956).

(14) E. Orowan, "Theory of Fatigue," *Proceedings*, Royal Soc. London, Series A, Vol. 171, p. 79 (1939).

(15) N. F. Mott, "Work-Hardening and Initiation and Spread of Fatigue Cracks," *Proceedings*, Royal Soc. London, Series A, Vol. 242, p. 145 (1957).

(16) E. Orowan, "Dislocations and Mechanical Properties," Dislocations in Metals, Am. Inst. Mining and Metallurgical Engrs. (1954).

DISCUSSION

Mr. John J. Gilman.[1]—The very excellent photographs of this paper were quite impressive to me. I think they confirm a point of view that I have had and perhaps others here have had. This is that extrusions and intrusions as such do not seem to play a primary role in fatigue failure. In the author's figures we see large extrusions and adjacent intrusions, but the failures seem to have occurred between these points. I would like to draw attention to a more general bit of evidence that seems to bear on this point.

Coffin[2] has extensively investigated fatigue by means of constant strain amplitude tests. He plots the logarithm of the plastic strain amplitude *versus* the logarithm of the numbers of cycles to failure and finds that the data fall on a straight line. The point that is important to this discussion is that the plastic strain at failure in a simple tension test (one quarter cycle of loading) falls on the same curve as the multi-cycle data. Thus, failure during monotonic loading seems to be closely related to failure during cyclic loading. Since, in monotonic loading one has no opportunity to develop extrusions or intrusions but simply "staircases" the surface of the material, it may be concluded that intrusions are not a direct cause of fatigue failure. Of course, intrusions are a manifestation of the damage that occurs inside the material and for that reason they have an important secondary role.

Chairman T. J. Dolan.[3]—This observation probably relates primarily to conditions for which large cyclic plastic strains occur that lead to a limited life of, say, something less than 100,000 cycles. I think the author's paper deals primarily with the lower stress levels near the fatigue limit where the inelastic strains which occur are on a microscopic level (rather than observable as large scale plastic deformation). There seems to be a departure from this simple mathematical relationship for conditions involving failure in large numbers (millions) of cycles.

Mr. Gilman.—Mr. Coffin's results seem to prove the point for large cyclic strains. The author's results seem to prove the point for small cyclic strains, so it seems to be generally true that the microscopic shape changes are not of primary importance.

Mr. W. A. Backofen.[4]—My comment has to do with this question of a geometrical effect in the production of a fatigue crack. This is something the author has written about for a number of years and it seems rather important. There have been some experiments made in our laboratory recently that bear on this. Briefly, they consisted of bending

---

[1] Metallurgist, General Electric Co., Schenectady, N. Y.

[2] J. F. Tavernelli and L. F. Coffin, Jr., "A Compilation and Interpretation of Cyclic-Strain Fatique Test," *Transaction*, Am. Soc. Metals (1959).

[3] Head, Theoretical and Applied Mechanics Department, University of Illinois, Urbana, Ill.

[4] Associate Professor of Metallurgy, Massachusetts Institute of Technology, Cambridge, Mass.

experiments on large single crystals of copper in which the orientations were so determined that in certain crystals the active slip direction poked through the surface of observation, and when it operated it would give a step on the surface; there were other crystals which were similarly oriented in the axial sense but different in a rotational sense so that the slip direction happened to lie in the surface of observation.

Now when one compared identically oriented crystals except for this rotational difference, one could so pick his samples that the stresses on the active systems were identical. However, in those crystals with slip direction poking through the surface one did find cracking in due time, whereas in those crystals with slip direction lying in the surface, even though this was the primary system, the cracking could be effectively retarded, and in many instances, just about eliminated. Then one would force less highly stressed and less favorably disposed systems to come into play and become the cracks in time. Therefore, the matter of notches and crevices forming across the surface as a result of slip movement certainly seems to be important in crack production.

Mr. W. A. Wood (*author's closure*).—I agree with Mr. Gilman that extrusions and intrusions alone need not cause failure. But it should be added that these peaks and notches are surface expressions of abnormal disturbances which form internally along the slip planes and which certainly are concerned with failure. Mr. Gilman's further point regarding the significant work by Coffin is of special interest to me because of similar work in this laboratory. This work suggests, however, that the linear relation which he mentions ceases to hold at very small amplitudes of plastic strain. That is in accord with the comment made by Chairman Dolan on the same point. The experiment described by Mr. Backofen is also of great interest. I think this too shows that surface disturbances made by slip movements cannot be disregarded when they are taken in conjunction with the internal distortions created along the associated slip planes.

A Two Stage Process of Fatigue Crack Growth

- by -

P. J. E. Forsyth
Royal Aircraft Establishment

SUMMARY

A study of the initiation and growth of fatigue cracks has indicated that several complex processes are involved. The observations suggest why the present mathematical theories based on simple models are likely to be only of very limited application to the practical state of affairs.

The extension of a fatigue crack can occur in several ways. Initially it may grow by the process of its formation, that is, by an 'unslipping' or reverse glide mechanism forming a surface crevice which deepens with time. This unslipping mechanism has been designated Stage I fatigue growth. The circumstances favourable to its active continuance will be outlined in the paper. It is characterised by crystallographic fracture facets changing angle with orientation at grain boundaries. Because it is a shear stress dependent process these facets lie on or near the planes of maximum shear stress. In the general case, Stage I fatigue is superseded by a second stage. This second stage cracking occurs in a plane perpendicular to the maximum tensile stress. In most light alloys the fracture shows characteristic striations which are of diagnostic value. Because it can be demonstrated that these striations delineate successive positions of the crack tip, and their spacing represents the crack growth for each cycle of stress, they are additionally useful in determining crack growth history in this stage. By various observational techniques it has been possible to analyse the nature of these striations, and present a tentative explanation of their formation. Their appearance, and possibly the detail of the mechanism whereby they are formed, differs for various materials, and with both environment and frequency of cyclic stress. In the technologically important materials such as the high strength aluminium alloys the extension of the crack tip in one cycle of stress involves both a brittle and plastic growth period. The conditions favouring each of these components are considered.

Introduction

The prime evidence of fatigue failure is a crack, and the fatigue endurance is commonly divided into (a) the crack initiation period and (b) the crack growth period. Evidence for the early formation of fatigue cracks continues to accumulate and therefore this arbitrary division is being forced back further and further in time towards the origin of the process. If we accept that the common mode of crack formation is by the deepening of a slip band groove by an atomic process of dislocation movement, then the continued use of this division presents certain difficulties, and leads to the absurdity of trying to specify a minimum crack size in a process starting on an atomic scale. The test applied in these circumstances is to try to prise open the questionable slip band with a tensile stress, and if it gapes open then the band must have contained a fatigue crack.

This paper presents evidence for what is considered a rational division which has some physical meaning based on the mode of crack growth. It will be shown that there are basically two modes of crack growth, and that the change from one to the other is a natural division in behaviour. As we are not concerned in this context with safe fatigue conditions, although it is known that non-propagating cracks sometimes exist in this state, we can label the two behaviours stage I and stage II crack growth. Precracking damage will be discussed only in context with the process of growth of an existing crack.

Fatigue cracks, as a general rule, originate in slip bands, although both grain boundary and sub grain boundary initiation are not uncommon. These exceptions to the rule are fairly well understood and are explicable in terms of known grain boundary behaviour and environmental effects. Their importance is not denied, but in order to make a general classification of behaviour they will be discussed separately as subsidiary effects.

Stage I fatigue crack growth

The criterion for slip band cracking is the range of resolved shear stress on the slip plane. Thus cracks will form on those planes most closely aligned with the maximum shear stress directions in the component or fatigue specimen. For various reasons the surface grains behave in a soft manner, and in most cases the stress is a maximum at the surface. Therefore it is not surprising that the surface grains deform more than those in the interior of the metal. Furthermore, fatigue cracks originate at the free surface of these grains. The surface is not only a favoured place for dislocation movement, but also for the irreversible action of forming slip band grooves. The most important difference between slip bands formed by steady stress and those formed by cyclic stress is that the cyclic stress produces a slip band groove. The groove deepens with continued cyclic stress to form a crevice or intrusion. There is no evidence that the mechanism of crack extension changes while it follows the active slip plane. This mechanism of crack growth can persist for an appreciable proportion of the endurance, but whether this proportion is large or small it will be designated as stage I fatigue crack growth because it differs in an essential manner from later behaviour.

The crack can be studied in this early stage by means of various metallographic techniques such as taper sectioning, replication methods, or by studying the fracture surface itself. The intrusion or crevice forming process is often accompanied by extrusion of thin metallic slivers from the slip bands, sometimes in a closely associated manner as though some conjoint action was involved. There are many clear indications that the intrusion or crevice is formed by reverse slip, i.e. slip occurring in a preferred direction in neighbouring packets of slip planes. The behaviour envisaged is similar to that suggested by Shanley in his so-called 'unbonding' process of crack formation and demonstrated by various other workers (1) (2) (3). We know by direct observation that extrusion occurs in the direction of slip, and it has been assumed that the crevice or crack also grows in the slip direction. Direct evidence has now been obtained from the study of the crack surface which substantiates this point. Fig. 1 shows an extrusion in an aluminium-zinc alloy where the surface markings indicate the direction of slip and extrusion, and Fig. 2 shows features on the surface of an aluminium-zinc-magnesium alloy in the first stage region. These features which seem to be the traces of jogged slip again indicate the direction of slip, and consequently the direction of crack growth. The shape of the crack front in this stage of growth is difficult to ascertain, but the evidence from a transparent non-metallic material such as silverchloride shows that it can be sharply serrated, pushing forward at many places in a spear-head attack. Fig. 3 illustrates in a diagrammatic form, this type of cracking. Although extrusions rarely exceed 20 microns in length in the direction of growth, the crevices may extend crystallographically across many grains with tilt of the fracture path as the crack crosses a boundary from one grain to its neighbour. This produces a facetted texture, the general fracture plane being one of maximum shear. These facets may be only slightly disoriented in aluminium alloys which is probably related to the fact that fatigue slip can occur in these materials not only on the (111) but also on (110) and (100) planes (4). With this selection of planes the final crack path may be relatively planar.

If the two dislocation processes of extrusion and intrusion are similar, but working in opposite directions, it may be questioned why the former process always quickly ceases, whereas the latter may continue for a relatively long time. We know that in aluminium-4% copper alloy the extrusion process starts suddenly, and stops after only a few cycles, in most cases less than 100. This effect has been observed directly, and there are certain features in the structure of the extrusion which seem to confirm this. It seems that while the crystal surface is essentially plain, extrusion and intrusion have similar opportunities for operation, but the formation of an extrusion effectively notches the material and 'unlinks' the extruded region from the stress. The tip of the crevice becomes a very active dislocation source, and these sources feeding the extrusion process cease to operate. It is interesting to note that in the case of torsion, filamentary extrusions can occur many hundreds of microns in length. In torsion the shear can be transmitted during both halves of the stress cycle by frictional forces developed between the crack faces, there being practically no tension component to part them. Similarly a directly stressed specimen, if given a compressive mean stress will produce more marked extrusion than one loaded with zero mean stress. The newly formed crevices now grow by the operation of dislocations at the tip. This is probably enhanced by the cessation of operation of the other sources in the vicinity. As there is a valid reason for the cessation of extrusion, and an enhancement of the intrusion process, there seems to be no reason why the latter should not continue indefinitely. It will be shown that it eventually ceases in favour of a second stage growth process, usually when it meets a slip obstacle such as a grain boundary. In single crystals or in cold worked pure metals whose preferred orientation makes the grain boundaries less formidable dislocation barriers, this Stage I process may continue to final fracture. These crack surfaces are often remarkably free from fretting which suggests that the crack walls gape and the stress concentration operates quite freely in compression as well as in tension. This may be why it is the resolved shear stress range that is the important criterion in this stage.

The effectiveness of reverse slip processes in extending the crevice will depend on the operation of some mechanism whereby reversal of dislocation movement is blocked. This might happen, as has been suggested by several people; by surface contamination by gas atoms, the formation of oxide, and subsequent locking of dislocation sources. As this surface locking process would be expected to work equally over the whole crack tip surface, it would seem more likely that it is achieved by the interaction of dislocations themselves.

FIG. 1. EXTRUSION FROM SLIP BAND IN AN ALUMINIUM - 10% ZINC ALLOY.
(x 2,500) [Arrow A indicates markings on extrusion surface revealing slip direction]

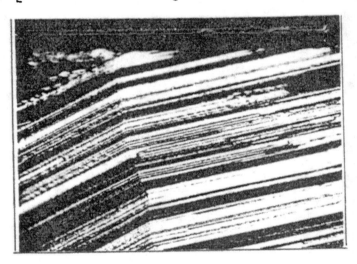

FIG. 2. TORSIONAL FRACTURE SURFACE OF AN ALUMINIUM 7.5% ZINC 2.5% MAGNESIUM ALLOY. (x 240)

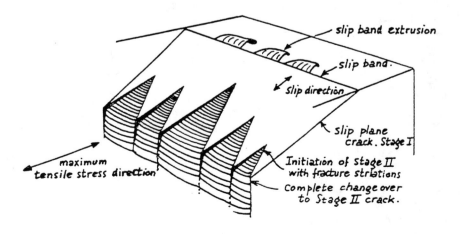

FIG. 3. SCHEMATIC DIAGRAM SHOWING SLIP PLANE CRACK, STAGE I, THE ONSET OF STAGE II CRACKING, AND THE NEW FRACTURE PLANE

A mechanism similar to that suggested by Cottrell & Hull (5) for the extrusion effect, seems to operate at the crack tip of pure aluminium growing by the Stage I process. This is illustrated in Figs. 4 and 5. There is no doubt that the crack tip is an active dislocation source, and it can be demonstrated that slip operates in two sets of planes originating from the crack tip. One set operates under the tension half cycle. The other set operates under the compressed half cycle i.e. when the shear directions are reversed.

Conditions favouring stage I crack growth

Certain conditions of stressing and environment favour this form of crack growth sometimes to the complete exclusion of the 2nd stage. Low stresses resulting in slow growth favour stage I behaviour. Mean tension stress will encourage the changeover to stage II which suggests that there is a minimum tensile stress necessary for stage II growth. Corrosion-fatigue conditions at stresses below the air fatigue limit may cause almost complete failure by slip plane cracking as shown in Fig. 6. This suggests that the effect of corrosion is to aid the unbonding process probably by film removal, or by the dissolution of crack tip material containing dislocation obstacles such as locked groups of dislocations. In the aluminium alloys considered in this context, the active slip may be accompanied by local chemical changes such as precipitation destruction, and solute disordering. The slip band crack usually changes from the slip plane to stage II growth i.e. in a plane perpendicular to the maximum tensile stress, when it reaches a grain boundary. The slip band cracks grow essentially on the single slip plane or planes containing a common direction, the slip direction. Occasionally it deviates in an alternating manner between two favourable oriented planes as shown in Fig. 7. Although a second plane is shown to be operative in the dislocation blocking process, it would seem where easy glide can occur the cracks grow a fair distance along a plane of one particular index. If the material and the conditions of stressing favour duplex slip then the stage I process gives way to the stage II process. If the material itself particularly favours duplex slip deformation e.g. an overaged aluminium alloy containing microscopically visible precipitate, then stage I may never develop in detectable form even under corrosion fatigue conditions. Stage II cracking will occur if the stress conditions favour duplex slip. These conditions are periodic high stress cycles, notched conditions, obstacles to easy glide blocking the crack tip, and the growth of the crack into the depths of the specimen where conditions of constraint exist, and the ratio of shear stress to tensile stress is low. The changeover is not necessarily sudden, the slip band cracking sometimes decreasing in a gradual manner. Fig. 8 shows the changeover (arrow A). It can be seen that although a weakened path has been produced by the slip process the crack still deviates to the normal plane.

Stage II fatigue crack growth

Whereas stage I growth is governed by the local shear stresses and the freedom of the dislocations, the criterion for growth in the second stage is the value of the maximum principal tensile stress operating in the component or specimen in the region of the crack tip. Fig. 8 showed that the change over could occur even when an apparently easy path for slip plane cracking existed. This apparent anomaly may be because of the changing value of β the shear stress/tensile stress ratio with depth in the specimen. This indicates that the peak tensile stress developed across the crack tip causes deviation of the crack from its slip path. If this is so it suggests that stage II cracking probably contains an element of cleavage. It will be shown later that certain observations have been made which confirm this view. The stage II mode of crack growth is characterised by microscopic features, the fracture striations, which may become visible as soon as the crack starts to deviate from the crystallographic path as

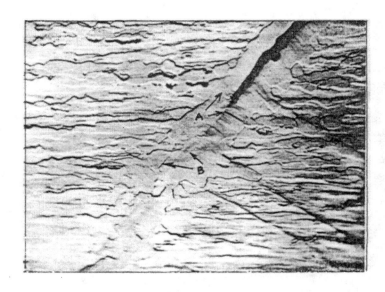

FIG. 4. A FATIGUE CRACK SUBSEQUENTLY SECTIONED, POLISHED AND ETCHED. The material was cold rolled pure aluminium. A zone of subgrain growth surrounds the crack and extends ahead of the crack tip. (x 1500) [Slip bands indicated by arrows A and B were produced by a subsequent stress cycle.]

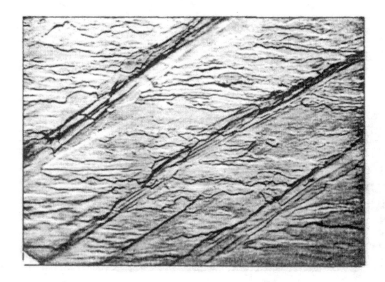

FIG. 5. THE SAME SPECIMEN AS ILLUSTRATED IN FIG. 4. Heavy localized slip has been produced in recrystallized zones by subsequent tensile deformation. (x 1500).

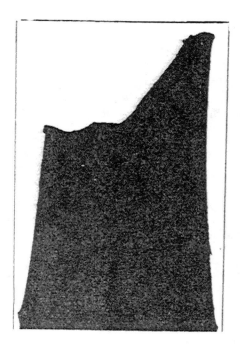

FIG. 6. SHADOWGRAPH OF FRACTURED ROLLS ROYCE CORROSION FATIGUE SPECIMEN. (x 12)

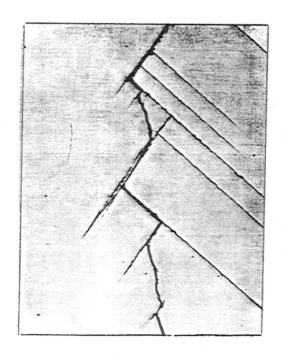

FIG. 7. FATIGUE CRACKS IN AN ALUMINIUM 7.5% ZINC 2.5% MAGNESIUM ALLOY. (x 750)

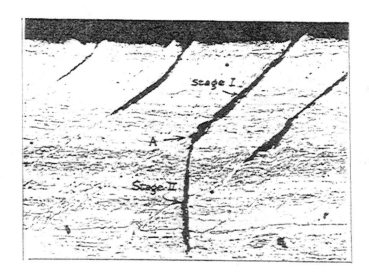

FIG. 8. SPECIMEN AS FOR FIG. 4, SHOWING CHANGEOVER OF MODE OF CRACKING, ARROW A. (x 200).

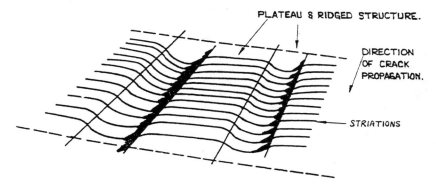

FIG. 9. A SCHEMATIC ILLUSTRATION OF STAGE II FRACTURE SURFACE.

shown diagramatically in Fig. 3. These features have proved to be of considerable diagnostic use in investigating service failures, and have been described elsewhere (6). Their appearance varies in detail from one material to another and these variations can be related in a qualitative way to the plastic behaviour of the material. Because the crack in that stage is governed by the direction of the maximum tensile stress, and there is an element of cleavage, this part of the fatigue fracture will be fairly smooth compared with the final tensile part of the fracture. This smoothness and reflectivity is the result of the crack growing across many small plateaux, and although the levels of the plateaux may differ they are substantially co-planar. Because their levels differ, a considerable area of the fracture may consist of fairly steep cliff edges joining up the plateaux. The nature of the surface is illustrated in a schematic form in Fig. 9.

The striations are ridges in the surface of the fracture, and it has been proved that each striation results from one stress cycle and the spacing of the striations represents the local crack growth during a stress cycle. Crack fronts do not extend at a constant rate over their whole length, local differences in rate are caused by straightening and local advancement. The evidence for this 'one for one' relationship between striations and stress cycles, is best seen in programme loaded specimens where the discrete cycles of the programme can often be indentified directly from the fracture surface, as shown in Figure 10. Because the striations are ridges on the surface, and delineate the successive positions of the crack tip, it has been assumed that they are the result of local plastic deformation at the crack tip with a resultant periodic deviation of the crack path. This being so it is not surprising that they are most clearly revealed in ductile materials. Again, because their formation depends on local plastic deformation the crack tip stress will govern the amount of deformation. It has been mentioned that there is a component of cleavage present in the striation, and the proportion of cleavage to ductile growth will affect the appearance of the striation as illustrated in Fig. 11. A high tensile mean stress superimposed on the peak positive half of the alternating stress will encourage the cleavage component, whereas a large alternating stress range will favour the plastic deformation component. The two types of behaviour are shown in Figs. 12 and 13.

There is probably no basic difference in behaviour between stage I cracking and the ductile component of the stage II striations. The important difference is the presence of the cleavage component which alters and governs the crack path. Not only does stress level control the amount of deformation at the crack tip, but it also controls the distance the cleavage crack grows each cycle. However, these two components can be made to operate independently by changing stressing and environmental conditions. Thus a direct linear relationship between spacing and stress has been obtained based on statistical counts taken on appropriate positions on fracture surfaces produced at different nominal stress levels (7). Local variations in spacing will occur depending on local stress concentrations resulting from the irregular crack front, thus the local stress concentration could be deduced from the striation spacing in the region. At low stresses nearer the crack origin or the change from stage I growth, the striations may not be resolved, either because of their close spacing, or because of the virtual absence of deformation at the crack root, and therefore the absence of a fracture surface ridge. During this stage of growth the fracture surface may appear featureless except for river patterns characteristic of cleavage. It therefore seems that at these early stages the fracture has all the characteristics of cleavage which can again be identified in the cleavage component observed as a separate part of the striations formed in the later stages of crack growth. The general conclusion from this must be that slip is virtually absent at the stage that cleavage sets in, dying out as the crack deepens along the slip plane. The cleavage component then predominates and causes growth for a period with the later appearance of a ductile component which continues to increase to final fracture. These changes in crack behaviour could be related entirely to changes in

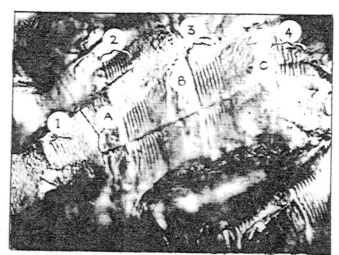

FIG. 10. FATIGUE FRACTURE SURFACE OF PROGRAMME LOADED SPECIMEN. (x 2,000) 1, 2, 3 and 4 show the repeated groups of 7 cycles of a particular load level. A, B and C are areas of low load level growth, where the fracture mode has changed to Stage 1 with a tilt of the fracture surface.

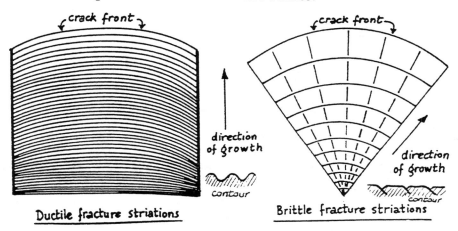

FIG. 11. A SCHEMATIC DIAGRAM ILLUSTRATING THE DIFFERENCES BETWEEN DUCTILE AND BRITTLE STRIATIONS.

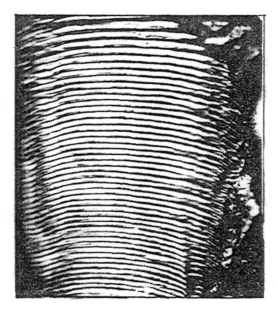

FIG. 12. DUCTILE STRIATIONS. (x 1500)

FIG. 13. BRITTLE CLEAVAGE STRIATIONS. A indicates crack front, B, river markings. (x 1500)

the shear stress to tensile stress ratio β. The tensile stress will increase to the fracture stress but the effective shear stress value will be $q_{max} - q_{yield}$. The crack starts from the surface where the grains will have a lower than average shear strength then grows into the constrained region of the bulk of the specimen where brittle cleavage conditions are favoured. The crack then grows to a stage where the stress has risen to a value at which deformation again occurs effectively, plastic flow occurs at the crack tip, and the conditions of constraint continue to decrease. In materials where brittle boundary fracture is common this cleavage period is replaced by a brittle boundary fracture period with stage II striations recurring at a later stage. The brittle boundary facets themselves may contain faintly marked striations indicating that a slight amount of plastic deformation is occurring at the crack tip even during this stage. This is discussed in the paper.

Frequency effects in stage II cracking

Reverse plane bend fatigue tests on an aluminium-7.5% zinc-2.5% magnesium alloy have shown that if the test frequency is changed from 1800 cycles per minute to 20 cycles per minute, the striation spacing increases by about a factor of 10. This change in spacing represents an increase in crack tip speed of the same order. A comparison of the two sets of striations showed that the low frequency ones contained more cleavage component including the characteristic river patterns which were absent on the high frequency striations. These features are all illustrated in Fig. 14 which shows a specimen that had been alternately fatigued at the two frequencies. It is well known that there is no such marked frequency effect on endurance, a decrease of life of 30% being reported by Schijve (8) for a change in frequency of about the same ratio. However, the fractographic effects were observed at an advanced crack stage, and are likely to be most active under conditions of triaxial stress where cleavage would be encouraged. Thus the frequency effect might only operate over a limited part of the life history of a specimen, and may have a minimal effect in sheet materials. The importance of this observation on the effect of frequency seems to lie in the fact that a few low frequency cycles at an advanced crack stage may be particularly damaging because of the initiation of the cleavage type striations.

Corrosion fatigue

The importance of corrosion fatigue lies in the fact that failures occur, under the conjoint action of corrosion and fatigue, at stresses very much lower than the air fatigue limit. A comparison of fracture behaviour of air and corrosion fatigue specimens showed that a continuity of behaviour existed which was related to the stress level of the test. For equivalent stresses, the air and the corrosion fatigue fractures showed a similar mode of fracture although the microscopic features were somewhat different. At very low corrosion fatigue stresses stage I cracking predominated, and as the stress was raised stage II cracking set in earlier. Thus a much greater proportion of stage I cracking can be obtained under corrosion-fatigue because it allows cracking to occur at very low stresses which are unfavourable for the stage II process.

At the higher stresses where the air and the corrosion fatigue S/N curves approach one another the fracture modes are similar but the striations formed in stage II are noticeably different. The striations produced under air are much more regular than those produced under corrosion the corrosion fatigue striations are characteristically brittle in nature with many river markings as previously described. They are also remarkably similar to the low frequency fracture striations. Thus it would seem that the action of the corrodant and the slow application of stress similarly encourage cleavage fracture in these aluminium alloys. These cleavage facets have been identified as (100) planes both by back reflection Xray and etch pit techniques. Fig. 15 shows local

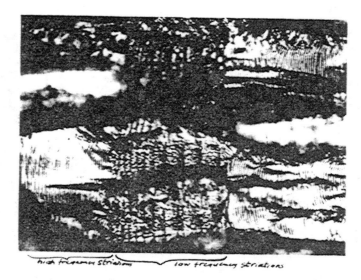

FIG. 14. ALUMINIUM 7.5% ZINC 2.5% MAGNESIUM ALLOY FATIGUED AT 1800 CYCLES PER MINUTE AND 20 CYCLES PER MINUTE. (x 600)

FIG. 15. LOCAL CHANGES IN MODE OF CRACKING UNDER CORROSION FATIGUE. (x 600)

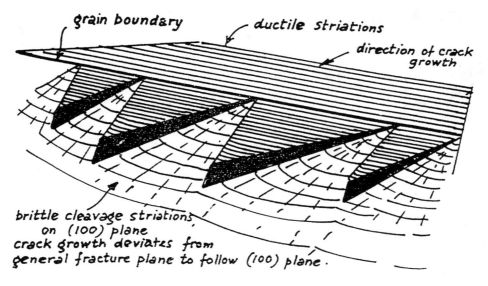

FIG. 16. A SCHEMATIC ILLUSTRATION OF FIG. 15.

changes occurring on the fracture surface as the crack deviates from the normal fracture plane to the cleavage plane. This is also illustrated schematically in Fig. 16. Fig. 17 shows the cleavage striations in more detail.

Fatigue crack growth in sheet materials

The growth of cracks in sheet materials is a subject of particular interest to the designers of aircraft structures. The conditions of working are those of relatively high mean stress with a superimposed small cyclic stress. The test pieces used to simulate working conditions are usually centrally slotted panels with the loads applied directly across the slot. Under the conditions of loading commonly used the crack starts very early in the fatigue life, most of which is taken up with its growth. The loads used may be applied at a constant level or applied as a repeated programme of varying loads. The combined effect of a notch and a high mean stress usually encourages the immediate onset of stage II growth with no indication of initial slip plane growth, thus the fracture striations may often be traced almost to the origin of the fracture. Most of the materials of interest to the aircraft designer are clad with pure aluminium or an aluminium-zinc alloy and the striations are particularly well defined in this layer. It has been possible to count the striations and construct a crack growth curve which agrees well with the curve constructed from surface observations. Apart from the local jumps in crack growth resulting from local stress concentrations or irregularities in the crack front, larger scale jumps are sometimes observed, particularly in the very strong aluminium-alloy to specification DTD687. This intermittent crack growth is the result of alternating tensile fracture jumps and periods of more gradual stage II fatigue growth. These separate periods of growth are revealed on the fracture as differences of texture as shown in Fig. 18. It can be seen that as the crack grows the front bulges forward advancing more rapidly in the central zone of the sheet. The bright textured areas are fatigue growth, and the dull areas are tensile fracture. The tensile crack growth has a tendency to tunnel down the centre of the sheet whereas the fatigue crack growth favours the surface layers. This is illustrated schematically in Fig. 19. For a considerable period the fatigue crack continues to run in a plane at right angles to the plane of the sheet, as the crack tunnels, the front becoming an acute arrow head form, a double $45°$ fracture surface appears which eventually changes over to single $45°$ fracture. Stage II striations can be detected on the early part of the $45°$ face, which is in agreement with surface observations of the crack growth rate. The crack may still grow in a non catastrophic manner, the fracture showing no clear change in appearance on attaining the critical crack length. Less ductile materials may grow fatigue cracks in the normal mode, and fail catastrophically with no change of fracture angle, except perhaps in the surface layers of the sheet. In the materials with less ductile behaviour where there is a particular tendency for the intermittent behaviour with periodic changes from fatigue to tensile fracture, advanced damage in the form of cracked particles is often detected in the boundary zones where the fatigue crack is becoming unstable and changing to the tensile mode. This precrack damage is naturally enough associated with the central zone of the material where the tensile crack is tunnelling. However, the tensile crack growth once started does not necessarily extend catastrophically, apparently re-establishing a stable crack front shape from which only the fatigue stress can cause further growth. The explanation for this may be that as the tensile crack tunnels forward it is retarded by the plastic deformation it has to achieve in the surface layers, the more acutely shaped it becomes the greater the length of crack front involved in this plastically deforming layer. Thus an energy balance is achieved.

Precracking damage

Various forms of metallurgical change have been observed to occur, usually

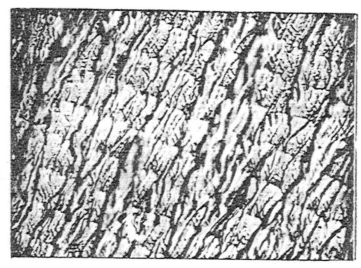

FIG. 17. BRITTLE CLEAVAGE STRIATIONS ON A CORROSION FATIGUE FRACTURE. (x 2,000)

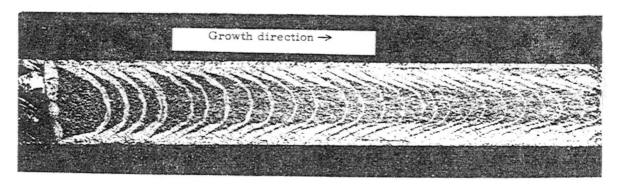

FIG. 18. FRACTURE GROWTH PATTERN ON DTD.687 SHEET SPECIMEN. (x 7)

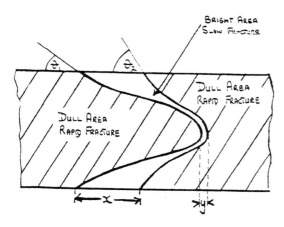

FIG. 19. SCHEMATIC ILLUSTRATION OF THE FEATURES OF THE GROWTH PATTERN SHOWN IN FIG. 18. x represents fatigue growth at surface and y represents fatigue growth in centre of sheet. The changes in crack front angle are represented by θ_1 and θ_2.

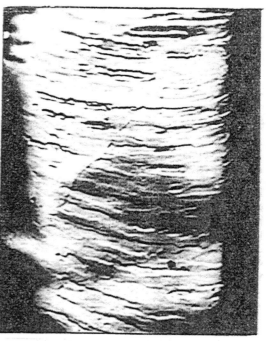

FIG. 20. THE REPOLISHED AND ETCHED FRACTURE SURFACE OF A PURE ALUMINIUM FATIGUE SPECIMEN STAGE II CRACKING. The substructure reveals the traces of the striations. (x 1500)

in the vicinity of the fatigue striations or particularly active slip planes. The more plastic deformation induced by the fatigue stress the more widespread the precracking damage is likely to be. Experiments on the effect of removing surface layers on residual fatigue strength, confirmed by direct metallographic observations, show that most of the damage, even in homogeneously stressed specimens, occurs in the surface layers of uncracked specimens. When slip plane cracks form then the damage may continue to develop in a narrow zone ahead of the crack tip. This zone may sometimes grow deeper than its associated slip plane crack, the crack deviating from its damage plane when influenced by changing stress conditions. No widespread zone of damage has ever been observed in stage II cracking which is in any way comparable with that found in stage I. However, stage II produces damage not ahead of the crack, but at the tip of the crack associated with the plastic deformation, and this damage does not appear to differ in any way from that which can be observed at the point of fracture of a tensile specimen in the same material. These observations are reconcilable with the fact that in stage I which may be a relatively long period, dislocation movement is more intense and continuously acting on a few small slip zones. In the stage II growth process the crack is continually growing into unaffected material, and each increment of growth is a small tensile test on practically unaffected material. Figs. 20 and 21 show stage II damage revealed in two different materials with different techniques. Fig. 20 illustrates sub-structure revealed on the fracture surface of a pure aluminium fatigue specimen by a very light electropolish followed by etching. This layer was only several microns deep and microsections prepared to show the material ahead of the crack tip revealed no advanced damage. The fact that the sub-structure delineates the stage II striations confirms that the two features are closely associated with one another. Similarly Fig. 21 shows the localised nature of the deformation in an aluminium alloy. The slip deformation has been revealed by a subsequent ageing treatment which causes slip band precipitation. It can be seen that the only appreciable slip is that associated with the striations themselves.

In most complex commercial alloys such as those used for aircraft skinning, the larger intermetallic particles may fracture in the vicinity of the crack tip. This form of damage is associated with high mean stresses, and large amounts of crack tip deformation. The stress raisers so formed may initiate fatigue cracks in advance of the main crack, but there is as yet no evidence that this behaviour occurs at a greater depth than a few tens of microns ahead of the main crack tip. Fig. 22 shows evidence for advance cracks in DTD683 alloy. As final fracture is approached the advanced damage becomes more extensive and the process of void formation sets in. Fig. 23 illustrates an aluminium-4% copper alloy tested in the solution treated condition, the observed features being near the point of final fracture. Arrow A indicates deep voids formed from cracked particles. The stage II crack growth striations show geometric growth which would appear to be related to the brittle fracture component.

Non-propagating cracks

It is of some interest to see if these observations on two stage crack propagation assist in any way to explain the phenomenon of the non-propagating crack. This form of crack occurring as it does in practice from the root of a notch, would be expected to start almost immediately as a stage II crack. An inspection of the various micrographs presented in various papers by Frost, (9) (10) suggests that this is so; it also reveals that these cracks usually stop on turning into the 45° mode i.e. they change to growth in the stage I mode, and then effectively become non-propagating. Orowan (11) has pointed out the importance of the straightness of the crack along its tip as regards its effectiveness as a stress raiser, and the fractographic evidence shows that the growth mode may change from place to place along the crack front. The behaviour of

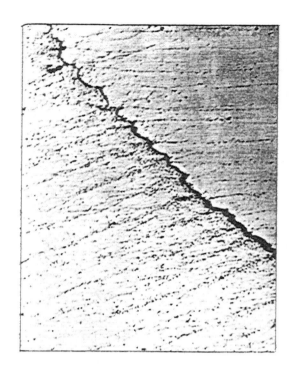

FIG. 21. A TAPER SECTION THROUGH A STAGE II FATIGUE CRACK IN AN ALUMINIUM-ZINC-MAGNESIUM ALLOY. The slip markings have been revealed by subsequent ageing after fatigue. (x 200)

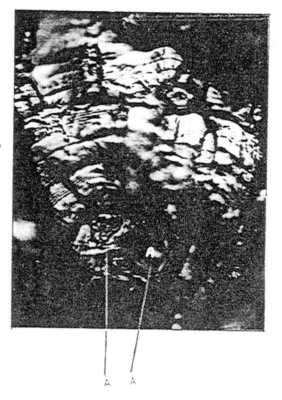

FIG. 22. A FATIGUE FRACTURE SURFACE IN DTD.683 ALLOY SHOWING PITS (A) WHERE PARTICLES HAVE CRACKED AND GROWN INTO FATIGUE CRACKS AHEAD OF THE MAIN CRACK. (Programme loaded specimen) (x 2000)

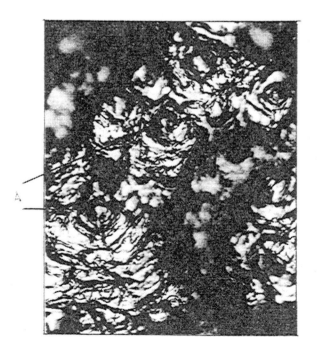

FIG. 23. FINAL FRACTURE ZONE OF A FATIGUED ALUMINIUM 4% COPPER ALLOY SHOWING PITS (A) AND GROWTH STRIATIONS. (x 1000)

the crack is therefore governed by the state of the majority of the crack front.*
If it can be established that a crack becomes non-propagating when most of the
length of its front has turned into the stage I mode, then we have a useful indication of the local conditions existing in the material. It is very probable
that service fractures continuously change their fracture mode with the various
stress levels encountered, because it has often been noticed that the fracture
surface of programmed specimens changes its angle with stress level taking up
a shear type path at the lowest stress levels of the programme. In both the
programme and the service case the recurrence of sufficiently high stresses will
restart the crack growing in the stage II mode.

It has been stated elsewhere that the changeover from stage I to stage II
occurs when the shear stress/tensile stress ratio reduces sufficiently to discourage the effective dislocation movement required for slip plane cracking and at the
same time encouraging cleavage conditions. For this reason a crack growing in
an un-notched specimen having once started to grow in the stage II mode is unlikely to revert to stage I again, although as the section is reduced a rapid shear
type growth may occur. We are faced with not only variations in the ratio, but
in the value of the shear stress, and it seems that the mode of growth the crack
will follow depends on both factors. When the crack, growing initially from a
notch, reaches a certain depth, the tensile component which had encouraged it to
start as a stage II crack is now reduced and this form of growth cannot be maintained. However, because conditions of constraint are such that easy slip plane
growth is impossible the crack becomes virtually non-propagating. It has been
suggested that these cracks do in fact grow very slowly although for all practical
purposes they have stopped. This is very likely to be true because they do seem
to turn over to the shear mode where very small dislocation movements could
maintain growth, although this is less likely with a strain ageing material. Again,
any atmospheric corrosion would encourage the slip plane growth as a corrosion-fatigue crack.

Fatigue fracture through grain boundary paths

In certain circumstances grain boundaries may behave in a particularly weak
manner, and cracks may originate in them. This form of boundary cracking has
been studied in an aluminium-zinc-magnesium ternary alloy heat treated to a
particularly susceptible grain boundary condition. Even at room temperature
boundary sliding could be induced in this alloy which developed into complete boundary fracture with time under load. Similarly fatigue cracks could, if the stress
was high enough, start in the grain boundaries, and extend to almost complete intercrystalline fracture. Stage I fracture has been defined as that resulting from
the movement of dislocations, commonly along a slip plane. It seems very likely
that a stage I type of cracking can occur at an obliquely inclined grain boundary
by reverse boundary sliding. This idea is substantiated by the fact that grain
boundary extrusion has been observed in an aluminium-4% copper fatigued at an
elevated temperature. Conditions in these materials favour stage II cracking because
local high tensile stress concentrations are likely to result from boundary sliding.

---

*Microsections suggest that most cracks occasionally travel for short distances in
the stage I mode, and it might be argued that such occasions existed in Frost's
specimens several times before they eventually become non-propagating. However, it must be remem   that a microsection only reveals two dimensions,
and we are only observing the nature of the crack tip at one particular position.

Therefore it is unlikely that stage I cracking will persist long before stage II takes over. It has already been stated that striations have been observed on boundary facets, the boundary fracture now taking the place of the cleavage component. A certain amount of plastic deformation is still occurring at the root of the growing crack, the amount depending on the local ductility i.e. on the state of the grain boundary resulting from ageing. Under these conditions it would seem that the rate of crack growth is particularly dependent on the metallurgical state of the grain boundary rather than that of the bulk material. There should be interesting differences in behaviour between a material cracking by cleavage, and another by an intercrystalline path, when fatigued in various active environments. The two types of behaviour could probably be achieved in the same material by varying the heat treatment.

Discussion

The relative importance of the two fatigue stages depends on the geometry of the specimen or component, the conditions of stressing, and the environment. The stage I process is important at low stresses with aggravating conditions such as corrosion, because it prepares the material for stage II cracking producing the right sort of stress concentration to the required depth in the material.

In steels that show air fatigue limits, it is reported that slip band cracks develop at stresses in the safe range, but they do not grow. Thus the fatigue limit appears to be that stress below which fatigue cracks do not propagate. However, in the presence of a corrodant these fatigue cracks will grow at extremely low stresses, and the material will show no true fatigue limit. This suggests that stage I cracking can be maintained in the presence of a corrodant, whereas in air dislocation movement ceases, and the cracks become non-propagating. This fact is substantiated by the observation that in some aluminium alloys a corrodant encourages the advancement of stage I cracks, at low stresses to the complete exclusion of the stage II growth. Stage II growth contains a component of plastic deformation, and therefore conditions which affect stage I might produce a similar if limited effect in the second stage. The limitation will be that the plastically deformed zone is formed in one cycle so that any damaging process must, to be effective, act very rapidly, because after the increment of cracking that element is no longer concerned with the growth process. However, the brittle cleavage component is affected in a marked manner by the environment. Observations on an aluminium-zinc-magnesium alloy fatigued in 3% NaCl solution, showed that the cleavage component predominated and dictated the path of fracture, along the (100) planes, with occasional excursions back to the normal fracture plane. Fatigue at low frequencies produced a similar effect, the fracture path often followed (100) planes along which the growth for each cycle of stress increased considerably. The association of cleavage with high tensile stresses and with triaxial stress conditions is not surprising. However, the fact that for similar stressing conditions it occurs only in the presence of a corrodant, or in air with slowly applied stresses, suggests the action of some time dependent phenomenon that encourages cleavage. There is no indication from electron micrographs of any electro-chemical removal of material from the crack tip, and it seems probable that cleavage is being encouraged by the formation of hydrogen ions, and the diffusion of these or others already existing in the material, to the crack tip. There is a strong analogy here with a hydrogen charged steel which behaved in the same way when fatigued in air at a frequency of 1800 cycles per minute. (12)

It is of interest to apply the ideas presented in this paper to the important practical effect of occasional overloads. It is generally agreed that occasional high positive overloads are beneficial to the fatigue life of a cracked component because they induce residual compressive stresses around the crack tip. Occasional high negative loads are fortunately not particularly detrimental because the crack tip is a less effective stress raiser in compression. It would appear that any residual stress would affect the cleavage component of crack growth, and a high residual compressive stress might inhibit it altogether. If this happened growth would continue by the stage I process. Eventually the residual stress field would be relaxed and stage II growth would continue again. In this respect we cannot wholly ignore the Bauschinger effect which may well encourage stage I growth. Stage I growth which involves the movement of dislocations will always tend to relax residual stresses of either sign, but because the induced stresses are usually beneficial, the stage I process is acting in a detrimental manner in preparing the crack tip for continuance of stage II growth.

Conclusions

A study of various aluminium alloy fracture surfaces shows that a general two stage classification can be made.

Stage I cracking is primarily the result of dislocation movement and fracture occurs along the slip plane.

Stage II cracking contains a component of cleavage.

Stage I cracking will proceed under conditions favouring single slip, whereas Stage II cracking occurs under conditions where slip becomes more complex and appreciable triaxial stresses exist.

The component of cleavage in stage II cracking varies in importance. The basic increment of fracture, the striation, will show a greater or less degree of brittle behaviour depending on the mechanical properties of the material, the environment, and the stressing conditions.

Cleavage is encouraged by a corrodant or by slow application of stress. It has also been reported that it is encouraged in a mild steel by electrolytically charging with hydrogen before fatiguing.

The phenomenon of non propagating cracks can be explained with reference to the two stages of crack growth. The cracks show features that suggest that they start almost immediately as stage II cracks, then grow to a length at which the crack tip is subject to too small a tensile stress to continue to grow in this mode. The cracking mode may change to stage I, but the rate of growth will be very small because of the constrained conditions at the crack tip.

Acknowledgements

The author gratefully acknowledges useful discussions with Mr. D. A. Ryder and Mr. C. A. Stubbington who have also worked on this investigation.

Published by permission of H. M. Stationery Office - Crown Copyright Reserved.

References

1. Shanley, R. F. — Proc. Inter. Union Theoretical and Applied Mech. Colloquim on Fatigue, Stockholm (1955) Springer Berlin, 251 (1956)

2. Forsyth, P. J. E. — Fatigue in Aircraft Structures, Columbia University Conference - 1956 - Academic Press Inc. N.Y.

3. Wood, W. A. — ibid

4. Broom, T, and Whittaker, V. N. — Nature, 177, 486-487 March, 1956.

5. Cottrell, A. H. and Hull, D. — Proc. Roy. Soc. (London) A, 242, (1957).

6. Forsyth, P. J. E., and Ryder, D. A. — "Metallurgia" Vol. 63, No. 377, p. 117, March, 1961.

7. Forsyth, P. J. E. and Ryder, D. A. — "Aircraft Engineering" April 1960.

8. Schijve, J. — 2nd Congress of the International Council of Aeronautical Sciences, September 1960 (Zurich).

9. Frost, N. E. — Proc, Inst. Mech. Eng. London, Vol. 173, No. 35.

10. Frost, N. E. — J. Mech. Eng. Science, June, 1960.

11. Orowan, E. — Fatigue in Aircraft Structures, Columbia University Conference 1956 (Discussion on C. E. Phillips paper, p. 121). Academic Press, Inc. N.Y.

12. Ryder, D. A. — Private Communication.

Mechanisms and Theories of Fatigue

CAMPBELL LAIRD

Department of Metallurgy and Materials Science
University of Pennsylvania
Philadelphia, Pennsylvania

Abstract

This paper presents an overview of the phenomena and mechanisms of fatigue. The cyclic deformation that occurs from the beginning of fatigue life is described, and the similarities and differences of this type of deformation with respect to monotonic deformation are briefly indicated. Emphasis is placed on pure metals, but more complicated materials are also treated. The effects of the magnitude of the strain amplitude are explored, and the regimes of high- and low-strain fatigue are defined. The evolution of fatigue damage from the dislocation structures formed by cyclic deformation is described, and theories of crack initiation are classified and discussed critically for both high- and low-strain fatigue. The cracks so formed initially propagate in the mode known as Stage I; but later, as the stress intensity increases with increase of crack length, the Stage II mechanism becomes dominant. The fractographic phenomena and mechanisms of fatigue-crack propagation associated with the different modes are briefly described. No attempt is made to describe the many theories of crack propagation now extant, but they are shown to be classifiable into two types: (a) those associated with plastic rupture at the crack tip and with attempts to quantify its relation to stress amplitude, and (b) those based on "exhaustion of ductility" at the crack tip followed by intermittent and incremental fracture. These types of mechanisms (and therefore the associated theories) are shown to apply to a wide range of materials for different regimes of stress intensity.

Introduction

The three opening papers of this seminar[1-3] have emphasized the serious problems and therefore the economic penalties associated with fatigue failures in energy and transportation systems, but fatigue is well-known to reach into many other areas of present-day technology.

Although the fatigue-monitoring techniques described by Buck and Alers[4] are clearly important in limiting the catastrophic effects of fatigue damage, it has long been recognized that a proper understanding of fatigue processes should provide an efficient route both to improving the design of components for resisting fatigue and also to better application of methods for monitoring fatigue damage. The fact that about half the papers of the present seminar are devoted to the fundamental aspects of fatigue reflects this recognition.

The purpose of the present paper is to provide an overview of the mechanisms of fatigue, which can be regarded as encompassing (*a*) the cyclic deformation of the material when subjected to fluctuating stresses, (*b*) the initiation of cracks in the material conditioned by the cyclic deformation, and (*c*) the subsequent propagation of the fatal cracks. These mechanisms are affected by microstructural and external variables in complex fashion. Since many of these variables are explored in the papers that follow, it seemed most appropriate here to deal with mechanisms in the simplest systems and to speculate briefly about the effect of the more complicated microstructural variables upon those mechanisms. In the final section of this paper, an attempt is made to show how a fundamental understanding of mechanisms can be used to understand the fatigue properties of nonferrous materials and notched members.

1. Mechanisms of Cyclic Deformation

(a) Low-Strain Fatigue

Let us consider an annealed copper single crystal that is subjected to a cyclic plastic strain of the order of 0.005. Because the crystal is very soft and oriented for single slip, the stresses necessary to enforce the strain are low, and this ensures that the plastic strain will be carried almost exclusively by the primary dislocations. As the crystal is cycled, it hardens; typical hysteresis loops associated with this hardening are shown in Fig. 1. Although the hardening per cycle is relatively small in the first cycles of life, as compared to monotonic hardening for equivalent cumulative plastic strains, it is nevertheless termed "rapid hardening" because the amount of hardening per cycle is initially large but declines with cycling. Eventually, the hardening rate falls to zero, and the crystal is then considered to be "in saturation." For example, the saturated flow stress for the crystal shown in Fig. 1 is approximately 28 MPa.

The details of this hardening have been very thoroughly studied by both mechanical tests and transmission electron microscopy,[5-20] and they have repeatedly been reviewed.[21-25] No attempt will be made here to recover this ground but simply to state the mechanistic position reflected in recent work,[16-20] on which there is a large, but not complete, measure of agreement. The salient facts about the hardening are as follows.

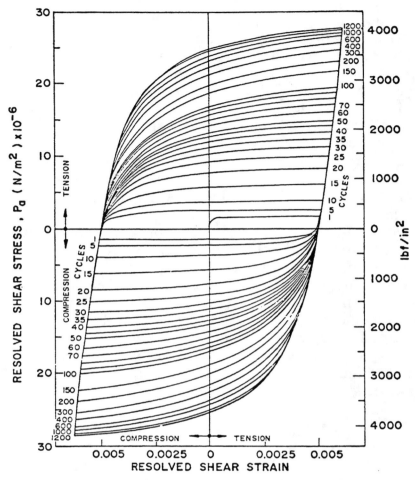

Fig. 1. The cyclic hardening of an annealed copper single crystal oriented for single slip and cycled through a plastic shear-strain amplitude of ±0.005. *(Courtesy of Finney, Ref 135)*

One can describe the rapid hardening and saturation stages by plotting the peak stress per cycle against the number of cycles, or accumulated strain summed without respect to sign, to generate a "cyclic-hardening" curve, of which typical examples are shown in Fig. 2. It will be noted that the saturated flow stresses, for three of the strains shown, are independent of strain. A plot of saturation stress versus the applied plastic-strain amplitude generates the "cyclic stress-strain curve," which for copper single crystals oriented for single slip appears as shown in Fig. 3. Thus there is a range of strains (region B in Fig. 3) corresponding to the strain-independent saturation stress shown by the cyclic-hardening curves of Fig. 2. This strain-independent region is known as the "plateau," and cyclic deformation can thus be classified into three regions: A, below the plateau (plastic-strain amplitude $\gamma_{pl} \leq 6 \times 10^{-5}$), B ($6 \times 10^{-5} \leq \gamma_{pl} \leq 7.5 \times 10^{-3}$), and C ($\gamma_{pl} > 7.5 \times 10^{-3}$); γ_{pl} is the plastic, resolved shear-strain amplitude. Regions A and B correspond to low-strain

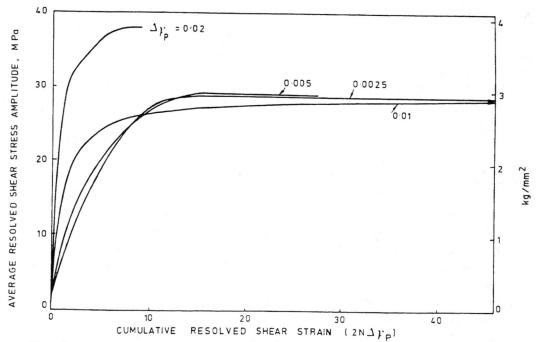

Fig. 2. Cyclic-hardening curves for copper single crystals cycled at the indicated plastic resolved shear-strain ranges. *(Courtesy of Finney, Ref 135)*

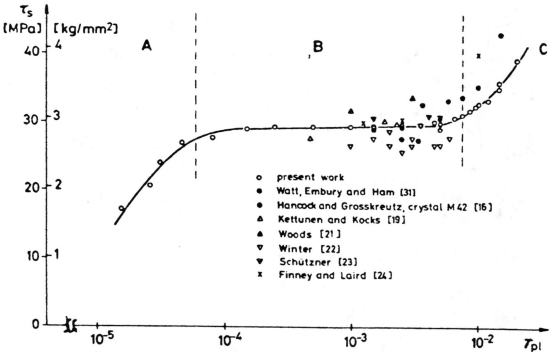

Fig. 3. Cyclic stress-strain curve for copper single crystals: resolved shear stress versus plastic resolved shear-stress amplitude. *(Courtesy of Mughrabi, Ref 19)*

fatigue where fatigue lives are generally in excess of 10^5 cycles to failure, and region C corresponds to high-strain fatigue.* The bulk of recent studies has been concerned with low-strain cyclic deformation, and the description that follows is derived largely from Ref 18 and 20.

The TEM observations of Hancock and Grosskreutz[12] and of Basinski et al.[13] make it clear that initial hardening occurs by the mutual trapping of edge dislocations. Typical dislocation structures associated with relatively few cycles are shown in Fig. 4 and 5. The trapped dislocations gather in bundles; between the bundles are relatively clear spaces in which screw dislocations will find it easy to glide.[18] They then drag out edge dislocations at the periphery of the bundles. The constant bombardment of the bundles by the screw dislocations as they shuttle to and fro in response to the applied stress has the effect of hammering the bundles into quite accurate edge orientation. It is clear, also, from TEM contrast information, that the signs of the dislocations are accurately balanced; the bundles thus consist of dense loop patches. With accumulation of strain in region A, or in region B prior to the onset of saturation, the loop patches gradually grow until they occupy a volume fraction of the order of 50%. For strains in the A region, a balance between the edge loop patches and intervening matrix channels plied by screw dislocations is eventually achieved. The deformation is thus quite homogeneous; the slip lines observed on free surfaces normal to the primary Burgers vector are very fine; and the slip is largely reversible in a mechanical sense. Under these conditions, the cyclic plastic strain is not damaging, and the life of the specimen could effectively be infinite.

Two parameters that have been widely recognized as of central significance for understanding the dislocation behavior of fatigued metals are the friction stress and the back stress.[26,27] Whereas the friction stress is independent of straining direction, the back stress changes sign during each half-cycle. It always reaches its maximum value at maximum applied strain, thus acting to lower the yield stress in the reversed direction. As it is of elastic nature, it soon decreases on straining in the direction favored by it, and then reverses so as again to oppose the imposed strains. Kuhlmann-Wilsdorf and Laird[20] have recently obtained measurements of the friction stress and back stress by analyzing cyclic hysteresis loops through a very simple scheme derived from the remote literature. This they used on the assertion that the hysteresis

---

*The distinction between high- and low-strain fatigue is conventionally different from that described here, however, and is as follows. If the plastic strain is greater than the elastic part of the total strain, the fatigue is considered "high strain." If the elastic strain is greater than the plastic strain, the fatigue is considered low-strain fatigue. The fatigue life associated with that total strain in which the magnitudes of the elastic and plastic strain are equal is termed the "transition life." Such a definition brooks no argument. However, when mechanisms of fatigue are also taken into account, this conventional definition does not describe the fatigue of single crystals very well, because of their low stresses and elastic strains.

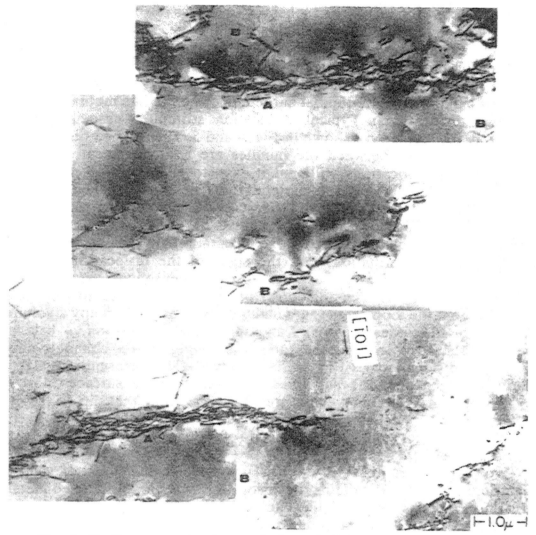

Fig. 4. Earliest stage in the evolution of dislocation structures in low-strain fatigue, showing edge dipoles assembled into loose groups. Copper single crystal cycled 9 reversals of ±0.0075 strain amplitude at 300 K. Foil oriented parallel to the primary slip plane; primary Burgers vector is [101]. *(Courtesy of Hancock and Grosskreutz, Ref 12)*

loops may be considered "reversible," meaning that the dislocation behavior is substantially unchanged from forward to reverse half-cycle, except for the sign change.

The well-known method is illustrated in Fig. 6.[20] At the start of the plastic deformation in the cycle (at τ_s), the back stress, τ_B, which was generated in the preceding half-cycle, acts in the same direction as the (now reversed) stress. Therefore, in the presence of a friction stress, τ_F, acting on the dislocations, Kuhlmann-Wilsdorf and Laird write[20] $\tau_s = \tau_F - \tau_B$. At the end of the forward cycle, when the

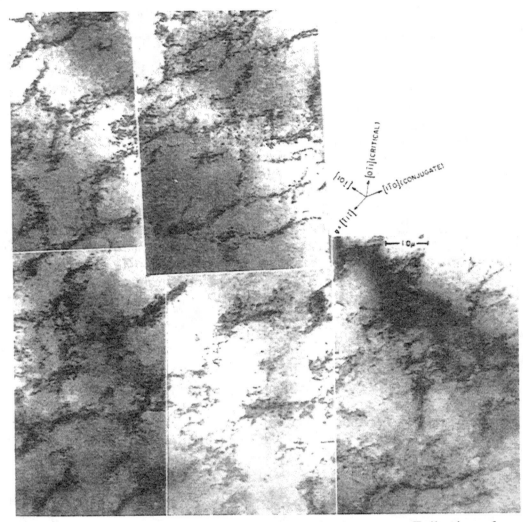

Fig. 5. Dislocation structure after 100 cycles in copper. Foil oriented parallel to the primary slip plane; however, dislocations with primary Burgers vector are not in contrast except near the lower-left corner of the micrograph. The strain here is a little high to produce a "typical" vein structure. *(Courtesy of Hancock and Grosskreutz, Ref 12)*

back stress has again reached its maximum value—but now opposed to the deformation—the applied stress is the sum of the friction stress and back stress: $\tau_E = \tau_F + \tau_B$. From these two equations, it follows:

$$\tau_F = (\tau_E + \tau_s)/2, \quad \tau_B = (\tau_E - \tau_s)/2$$

The friction stress and the back stress measured from the hysteresis loops of three different investigators, obtained for copper, are shown in Fig. 7 as a function of cycles. This result applies to plastic-strain amplitudes typical of the plateau. It will be noted that the friction stress and the back stress rise in parallel. From this and related analysis,

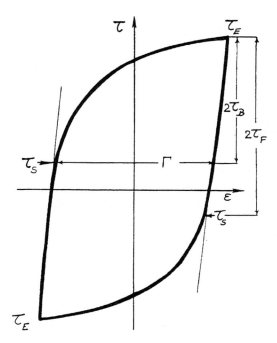

Fig. 6. The connection between the yield stress (τ_s) and maximum cyclic stress (τ_E), and the frictional stress (τ_F) and back stress (τ_B) derivable therefrom. *(Courtesy of Kuhlmann-Wilsdorf and Laird, Ref 20)*

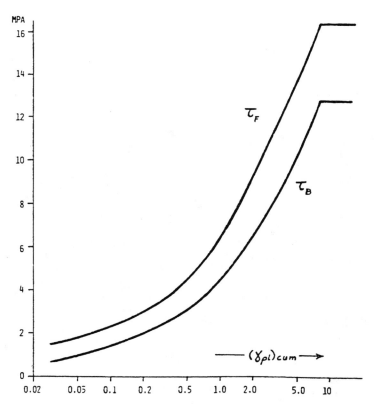

Fig. 7. The values of the friction stress, τ_F, and back stress, τ_B, derived in accordance with Fig. 6 for copper; $\gamma_{pl} = \pm 0.003$. *(Courtesy of Kuhlmann-Wilsdorf and Laird, Ref 20)*

Kuhlmann-Wilsdorf and Laird separate the friction stress into two parts, one equal in magnitude to the back stress (and presumed to have the same physical cause), the other, smaller, part with a different dependence on the number of cycles and saturating earlier than the former.[20] The smaller part, the "true" friction stress, is considered to be due to jog dragging by the screw dislocations shuttling in the channels. Dispersed point defects may also make a contribution—but a very small one, which saturates only after several cycles. It is further suggested that the channel width between the loop patches is self-adjusting so as always to let only one jog reside on a typical screw-dislocation segment shuttling to and fro in the channel. From this, one can predict that the channel width decreases with cycles in the same way that the jog-dragging part of the friction stress increases. Unfortunately, there are insufficient TEM micrographs of loop patches observed early enough in life to test this prediction. Typical views of "saturated" loop patches and concomitant channels are shown in Fig. 8.

Incidentally, Kuhlmann-Wilsdorf and Laird are able to deduce the rate of point-defect generation from this jog-dragging model[20] and find it in accord with other, indirect measures of fatigue-enhanced point-defect concentrations.

Further theoretical interpretation of the friction and back stresses depends on the behavior of the dislocations in the loop patches—whether they fully participate in the cyclic deformation via loop-flipping (the dipoles flip-flop from one 45° equilibrium position to the other in the manner shown schematically in Fig. 9 (after Feltner[10]) or whether the

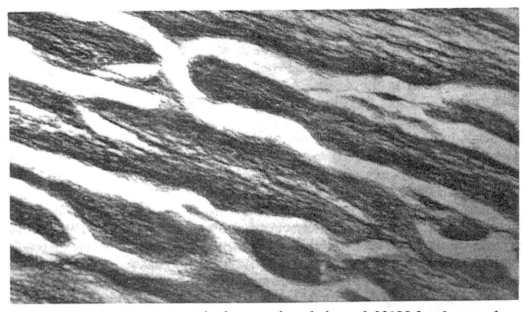

Fig. 8. Veins in a copper single crystal cycled at ±0.00125 for thousands of cycles. Approximately the same foil conditions as shown in Fig. 4. *(Courtesy of Finney, Ref 135)*

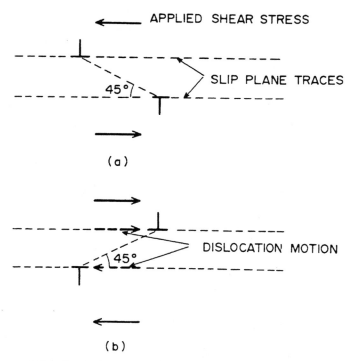

Fig. 9. Schematic representation of a dislocation dipole flipping from one 45° equilibrium position in tension (a) to compression (b).

loop patches are essentially rigid, acting as a nondeformable second phase. Kuhlmann-Wilsdorf and Laird[28] conclude in favor of complete participation of the loop patches in the cycling. This requires that the loop density be large enough to accommodate the plastic-strain amplitude and that the applied stress be large enough, in relation to the average loop size, to flip the loops. Quantitative considerations suggest that both conditions are fulfilled, provided that the aggregated dipolar loops behave like a Taylor dislocation lattice.[28] If so, the dislocation density in the loop patches must be in the $10^{11}/cm^2$ range. In this connection, it is most encouraging to report the recent measurement by Rapps et al.[29] These workers employed a description of the diffraction of fast electrons in crystals containing a high density of lattice defects, worked out for the case of narrow dislocation dipoles. The results were applied to the diffraction patterns observed on a cyclically deformed copper crystal containing loop patches, and the analysis yielded a dislocation density of $3.4 \times 10^{11}/cm^2$.

Now, the deformation of the loop patches requires a stress inversely proportional to the loop spacing. Since the loop patches are presumed to be stronger than the channels,[28] their deformation would lag slightly behind the deformation in the channels (which is being accomplished by the screw dislocations). This sets up a back stress in proportion

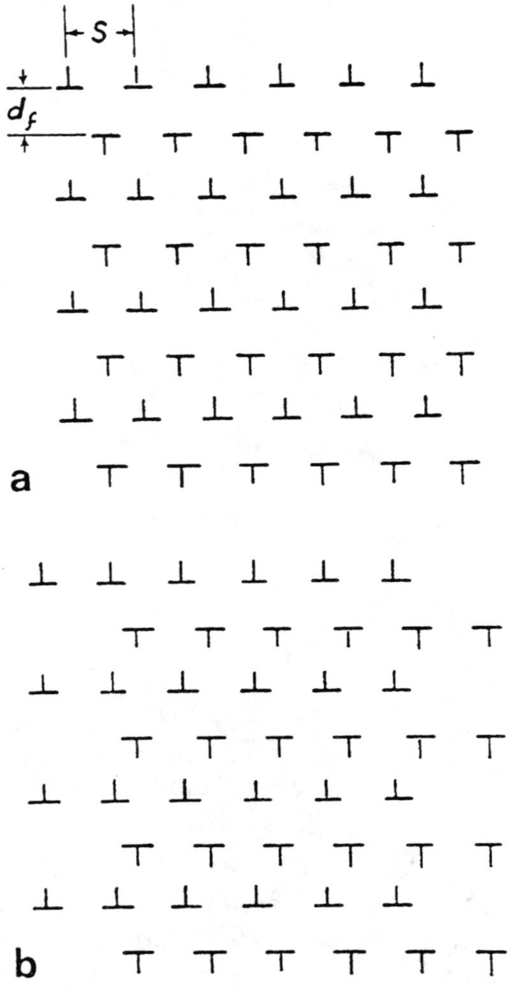

Fig. 10. The Taylor dislocation lattice corresponding to an idealized description of the dislocation structure in the loop patches: (a) in the symmetric configuration and (b) after shearing has produced polarization. The strength of the polarization, and hence the back stress produced by it, rises monotonically with strain.

to the dipole spacing, d_f, shown in Fig. 10. As the applied stress acts on the channel dislocations and the loop patches early in the half-cycle, the back stress will rise rapidly and loops will be forced to flip. After the requisite number of loops has flipped, the back stress rises more slowly, due to the buildup of a polarization stress. The action of polarization is illustrated schematically in Fig. 10. An interaction is now set up between the dislocations laid down by the channel screw dislocations and the edge-glide dislocations at the loop-patch surfaces. Locally, about these dislocations, the loops are forced to reorient in response to the dislocation stress fields. This reorientation is suggested to be different from that in the remainder of the loop patches,[28] because they screen the stresses of these dislocations. To pull the edge dislocations out of the energy wells thus generated for them requires a stress equal in magnitude to the back stress. Namely, removal of the edge dislocations restores the polarization stress, which their presence eliminated. It is pointed out that moving the dislocations through the polarized edges of the loop patches is the same motion as the relative shifting of the sublattices in the Taylor lattice, and it requires an equivalent stress, which can be computed.[28] As the motions involved are nonconservative, the corresponding stress is frictional. This establishes the magnitude of τ_F and τ_B, as well as the essential equality of the dominant part of τ_F with τ_B.[28] Many further details of cyclic deformation can be deduced from this model.[28]

The foregoing description applies either to the cyclic-deformation mechanism that occurs during rapid hardening or to the saturation condition associated with applied strains in the A region. If the applied strain is in the plateau region, such dislocation mechanisms terminate as saturation is approached, and the general fine slip gives way to the localized slip that produces the "persistent slip band." This phenomenon has major consequences for fatigue and thus deserves the emphasis of a separate section, which follows.

(b) Persistent Slip Bands

Persistent slip bands have long been recognized.[21] During early studies on the fatigue of polycrystalline metals, it was found that certain grains with the "softest" orientation for plastic deformation formed intense bands, which "persisted" when a thin layer of the surface was removed by electropolishing. This observation was the origin of the name, and for brevity, persistent slip bands will hereafter be called PSB's. Recent monocrystalline studies have gone a long way toward providing an understanding of their behavior.[16-19] For example, Finney and Laird[17] prepared monocrystalline fatigue specimens oriented in such a fashion that the primary slip lines were normal to the applied stress (tension-compression) and the primary Burgers vectors were oriented at as large an angle to the surface as possible consistent with the orientation requirement of the highest stress on the active slip plane. Therefore, when a specimen cycled at a strain in the plateau region entered saturation and the slip was localized in the PSB's, Finney and Laird were able to measure the amount of the plastic strain in the bands by interferometric techniques and compare the amount of plastic strain to the over-all applied strain. By interrupting their tests, repolishing the surfaces of the specimens, and then subjecting them to single excursions of tensile or compressive plastic strain, these workers were also able to show the deformation behavior at any point of the life. The optical appearance of a typical PSB is shown in Fig. 11, along with an interferogram that demonstrates the ease of measuring the slip offsets. The typical behavior of a PSB when subjected to several strain cycles is shown in Fig. 12. From results such as these, Finney and Laird concluded the following:[17]*

1 During saturation in the plateau region, essentially all the applied strain is carried by the PSB's. This means that a slight cyclic softening is associated

---

*Many of these conclusions are consistent with those of Winter,[16] who used a different technique. Mughrabi has been especially productive in defining the plateau.[19]

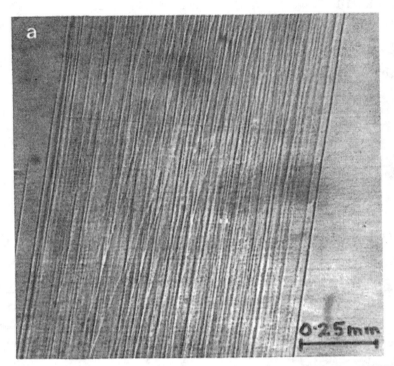

Fig. 11. (a) Optical micrograph of a discrete PSB during saturation. Specimen tested at $\gamma_{pl} = \pm0.0025$, repolished, and given a quarter cycle of plastic strain. (b) Monochromatic-light interferogram of coarse slip bands during saturation. Specimen tested for 30,000 cycles at ±0.00125, repolished, and then given a quarter-cycle plastic-strain increment. *(Courtesy of Finney, Ref 135)*

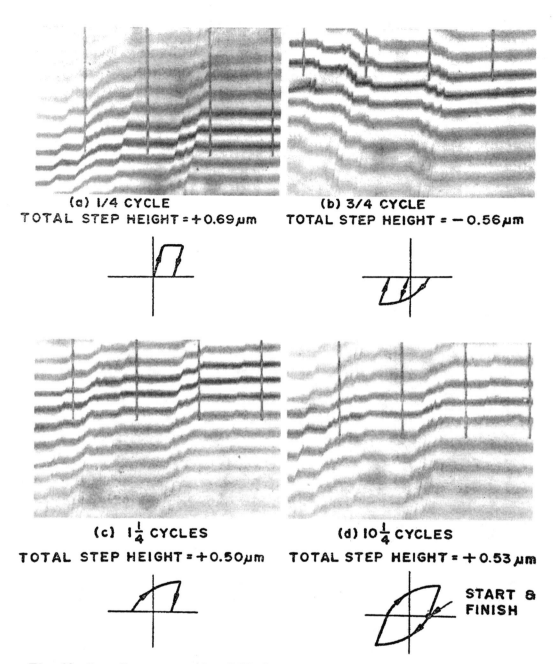

Fig. 12. Interferograms of a PSB in a copper single crystal cycled at a plastic-strain amplitude of 0.0025 for 2750 cycles, repolished, and subjected to the strain increments indicated. *(Courtesy of Finney and Laird, Ref 17)*

with the formation of the PSB's. As they form, they deprive the remainder of the gage section, containing loop patches and open channels, of the strain it previously carried. Such softening has frequently been observed as a yielding effect, a typical example of which is shown in Fig. 13.

2 In the plateau region, the PSB's deform under a stress that is independent of strain. In order to accommodate larger applied strains, the volume fraction

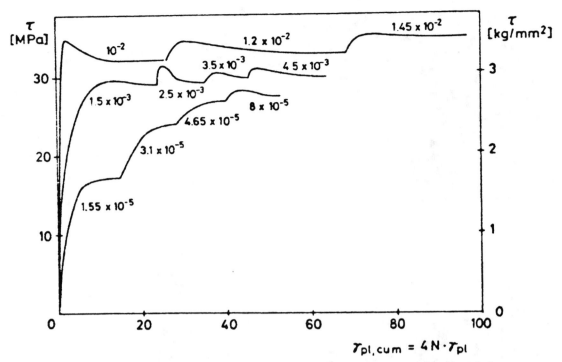

Fig. 13. Cyclic-hardening curves of three copper monocrystals, which were deformed at successively increasing values of γ_{pl}, denoted by numbers adjacent to the curves. Note the softening associated with the stresses where PSB's are known to form. *(Courtesy of Mughrabi, Ref 19)*

of the PSB's is forced to increase until, at the large-strain end of the plateau, essentially the whole of the gage section is a single PSB. Thus, the strain amplitude at the low end of the B region, 6×10^{-5}, is that required to form a single narrow PSB, whereas that at the high end represents the *average* strain of the PSB's.

3. The strain carried by groups of PSB's is macroscopically reversible, in that the local strain at the tensile maximum is identical to that at the compressive maximum. Also, the local strain within a band appears to increase linearly with the over-all strain during a typical excursion. Only at the finest distribution of slip within a PSB is the deformation not strictly reversible. This can be seen in Fig. 12, where a slight roughening of the surface is associated with the action of several cycles, although the slip offsets (representative of the local strain) continue to be mechanically reversible.

4. The PSB's are observed to penetrate right through a monocrystalline sample, and thus they act like a soft sandwich-filling within a harder matrix. It is not clear exactly at what point they pass right through the sample. It is probable that they nucleate near the surface of the specimen and spread during subsequent cycles. Certainly they have completed their penetration within a few hundred cycles from the beginning of saturation.

5. The dislocation structure of the PSB is significantly different from that of the matrix with its loop patches. It essentially consists of a cleared channel interrupted by rather thin—but dense—hedges of dislocations uniformly separated by approximately 1.5 μm (at least at room temperature).

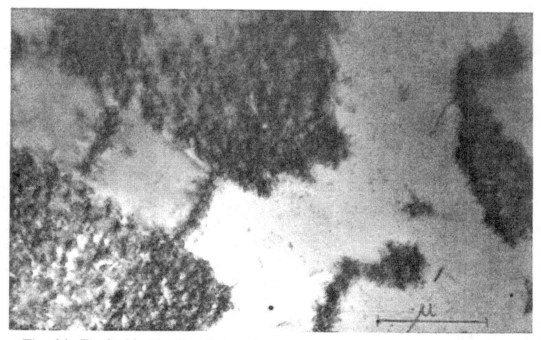

Fig. 14. Typical ladderlike dislocation structure of a PSB in a copper single crystal, viewed along the [1$\bar{2}$1] direction. The "rungs" of the ladder are the hedges seen in elevation. *(Courtesy of Finney, Ref 135)*

The typical appearance of a PSB, observed in elevation by TEM, is shown in Fig. 14. A schematic three-dimensional perspective of the PSB structure is shown in Fig. 15, and the dependence of the over-all structure on strain amplitude is also shown.

This concentration of strain in the PSB's is critical for the production of fatigue damage, and it is therefore important to understand their

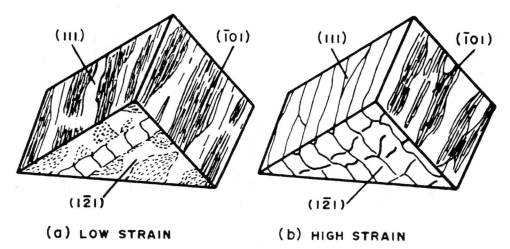

Fig. 15. Three-dimensional configuration of dislocations in copper single crystals strain-cycled to saturation. *(Courtesy of Finney and Laird, Ref 17)*

behavior. Unfortunately, we still are not sure how they form. For a while,[18] it seemed possible that they were formed from the loop patches by the agency of the free-channel screw dislocations bursting through the loop patches and creating an avalanche of mobile dislocations that cleared neighboring loop patches from their path. The PSB was conceived as originating out of the debris of this event.[18] Under some fatigue conditions, it may well occur in this fashion. For example, Neumann has shown that periodic strain bursts occur in a single crystal if the applied stress is gradually increased during cycling.[30] A typical result of his is shown in Fig. 16, and it may have the following interpretation. During the periods of quiescence, the dislocation structure will be of the loop-patch variety, but as the stress is increased, the particular array of dipole sizes becomes unstable, collapses and is remade of appropriate structure to carry the increased stress. Apparently at the plateau stress, the PSB structure is better capable of carrying large amounts of strain. Such a collapse of the loop patches from "outside" may well occur in situations where the stress is varying.

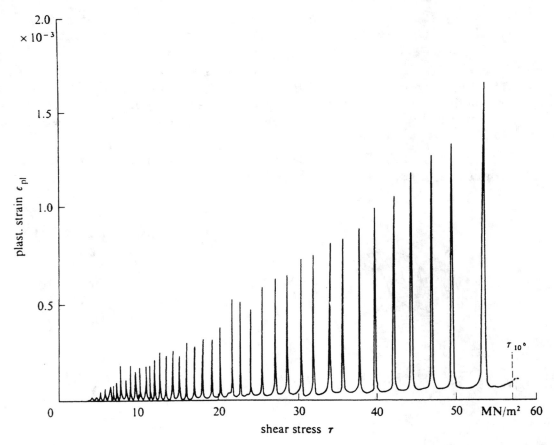

Fig. 16. Strain bursts in an aluminum single crystal, which occur during the gradual increase of the stress amplitude from zero. *(Courtesy of Neumann, Ref 30)*

In constant-strain conditions, however, the situation appears different, because Winter has taken a transmission electron micrograph that shows the apparent nucleation of a PSB from *inside* a loop patch. The nucleation appears to occur by rearrangement, and clearing, of the loops inside the patch, the exterior shell of the patch thus constituting two hedges of the newly forming PSB. Unfortunately, the structure he has observed may in fact be a PSB sectioned at a glancing angle and thus not representative of a true nucleation event. Thus the question of the formation mechanism of PSB's remains open.

However, our understanding of the deformation mechanism, once the PSB is formed, appears to be much more advanced.[18] The structure of the "hedges" in the PSB is considered to be dipolar walls consisting, typically, of an inner and an outer pair of dipolar tilt walls, having a density of 3×10^{10} cm^{-2} (see Ref 18). Since Antonopoulos and Winter have measured a dislocation density in the hedges an order of magnitude greater,[32] the dipolar walls are mixed with debris, consisting primarily of loops and dipoles with either the primary Burgers vector or a Burgers vector derived from it.[32]

Once the hedges have been established, the high strain of the PSB's is carried by glide dislocations moving in the clear channels between them in the form of mobile screw segments, laying down dipolar edge dislocations on them or taking dipolar edge dislocations up from the dipolar walls. Since the tilt-wall arrangement is even-numbered, one half of the edge dislocations is being taken up when the other half is deposited, in effect leaving the total dislocation content unchanged and simply transferring the edge dislocations from wall to opposite wall across the channels.[18]

Figure 17 illustrates the dislocation mechanism by which the transfer of edge dislocations from one side of the channels to the other is most likely to take place.[18] The plane of the drawing is meant to be the active slip plane, with the primary Burgers vector normal to the hedges; the stress is assumed to be applied so as to move the material above the drawing to the right of that below it. In that case, the downward-point-

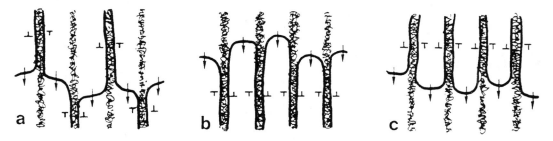

Fig. 17. Possible coordinated motion of screw dislocations along the channels of a PSB so as to maintain balanced dipolar walls in the hedges. *(Courtesy of Kuhlmann-Wilsdorf and Laird, Ref 18)*

ing arrows, indicating left-handed screw dislocations, also indicate their direction of motion if the usual symbols for positive ($\perp$) and negative ($\top$) edge dislocations signify that the extra half-plane is above the plane of the drawing (positive) and below the plane of the drawing (negative). Note the close coordination of the screw dislocations across the hedges. This occurs because a smaller shear stress is required to remove an edge dislocation from a tilt wall whose nearest neighbor is missing than if it is removed from a perfect wall. Thus Kuhlmann-Wilsdorf and Laird concluded that similarly signed screw dislocations on staggered atomistic planes, corresponding to the average distance of the dislocations in the tilt-wall layers, are likely to sweep through any one channel in loosely coordinated groups,, and that similarly there is a loose coordination among the screw dislocations in neighboring channels.[18] An idealized schematic representation of the dislocation arrangement and dipolar walls within a PSB, showing interchannel coordination, is shown in Fig. 18 (via an elevation of the PSB). The dislocations sketched may be regarded as those composing the innermost dipolar tilt-wall arrangement with entrapped debris, onto which one or more similar dipolar walls could be layered.[18] Many further details of PSB deformation have been explored.[18]

At an applied strain of $\gamma_{pl} \sim 7.5 \times 10^{-3}$, as noted above, the whole gage section of the specimen becomes a giant PSB. If the strain is greater than this, significant amounts of slip begin to occur on systems other than the primary, and the hedge structure of the PSB is broken down into a more complex cell structure, which is the object of study in high-strain fatigue.

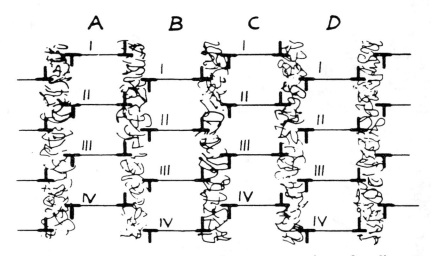

Fig. 18. Idealized schematic representation of a dislocation arrangement and dipolar walls within a PSB, viewed in elevation. For clarity, the wall width has been exaggerated. *(Courtesy of Kuhlmann-Wilsdorf and Laird, Ref 18)*

(c) High-Strain Fatigue

The over-all phenomenology of high-strain cyclic deformation is similar to that at low strain: in a pure material, rapid hardening first occurs and saturation subsequently sets in. At low strain, the kind of hardening is widely viewed as broadly similar to that which occurs in Stage I monotonic deformation of a single crystal.[24,33-36] The type of hardening that occurs at high strain, involving cells, is associated with Stage II monotonic deformation of a single crystal, and the onset of saturation is akin to Stage III. Many detailed similarities between the behaviors of cyclic and monotonic deformation have been pointed out.[34]

Monocrystalline studies of high-strain fatigue are rare because the experimental difficulties are severe. Accordingly, most of the work has been done on polycrystalline material and has emphasized the effects of different microstructural and external variables. For example, it has been shown that the saturation condition of wavy-slip materials is independent of the history of the material (grain size, subgrain structure, cold work).[33] This result can be illustrated by means of stress-strain curves. As shown in Fig. 19, the cyclic stress-strain curve of pure iron is considerably different from the monotonic stress-strain curves of the iron in either of two virgin states, annealed and cold worked, and it is independent of those states.[33,37]* However, in solid-solution alloys where the slip mode is planar, the saturation condition does depend

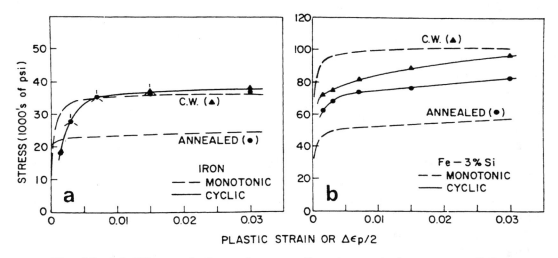

Fig. 19. (a) History-independent cyclic stress-strain curve of iron compared to the monotonic curves of the uncycled material; (b) history-dependent response of Fe-3% Si alloy compared to the monotonic response of the uncycled material. *(Courtesy of Feltner and Laird, Ref 37)*

*The cyclic deformation of monocrystalline bcc material is complex and most interesting[25,39-44] but for lack of space cannot be treated here. The reader is specially directed to the work of Mughrabi and his coworkers.[42-44]

strongly on the history of the material. This result is also shown in Fig. 19, where it will be noted that the cyclic stress-strain curves of Fe–3% Si alloy differ appreciably, depending on whether or not the alloy had been initially cold worked.[33,37] It will be noted from Fig. 19 that the flow stress of the initially cold worked material can be greater than that of the material after cycling. This is the result of cyclic softening, where an initially densely dislocated structure can be rearranged by the cycling into a form in which both the density and long-range stresses of the dislocations are much reduced. More details about cyclic softening can be found in Ref 24 and 35. Even in wavy-slip materials, if the prestrain is sufficiently severe (typically, above 50%), the rate of cyclic softening can be sufficiently small that the specimen fractures by fatigue before saturation is reached.[38] Under these conditions, history independence of the cyclic stress-strain curve no longer obtains.[38]

For those histories in which the cyclic stress-strain curve is independent of history, the dislocation cells that form on cycling are also independent of history.[34] When the prestrain is very severe, the dislocation cells subsequently produced are more misoriented, more angular and smaller, for a given strain amplitude, than those obtained with an initially annealed material (see Fig. 20 for typical cells produced in annealed and cold worked material by cycling). The mechanism of the cyclic deformation associated with such cell structure is as follows. The cell walls appear to accommodate the applied strain by providing a source of dislocation links.[34] These links bow out and shuttle across the cells to be captured, and partly annihilated, in the opposing walls. A mechanism similar in detail to that in PSB's is not envisaged here, because the cells, at high strain, are usually equiaxed, and thus the long channels of the PSB are not available for shuttling screw dislocations. However, the ways in which dislocations interact with the cell walls may well be similar to those that occur in PSB hedges. When the strain is reversed, the links bow in reverse fashion. Thus the flow stress can be explained by that necessary to force the links from the walls, and such a calculation yields reasonable agreement with experiment.[24,34] The saturation state is brought about by a balance between dislocation bowing and dislocation annihilation in the walls. This model was criticized by workers using single crystals at a time when our understanding of monocrystalline behavior was much less complete. Nowadays, the model seems more valid.

(d) Cyclic Deformation in Polycrystals and Complex Alloys

It is interesting to speculate on the extent to which the fundamental mechanisms described above apply to more complex materials. The first important point to note is that the cyclic-deformation mechanisms explored in recent work[18,20] apply not only to fcc materials, but, under appropriate conditions, to metals of other crystal structures as well—for

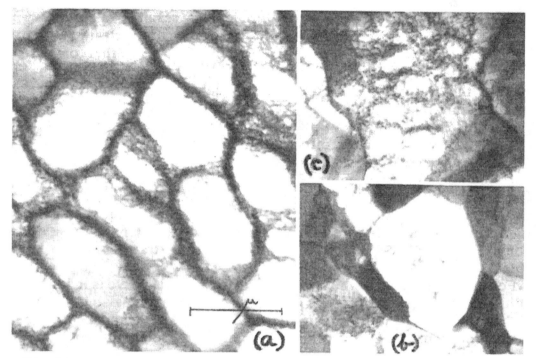

Fig. 20. (a) Typical uniform cells in annealed polycrystalline copper cycled at a plastic-strain amplitude of 0.0033 into saturation. (b) Dislocation structures in a specimen of copper initially cold swaged to a true strain of 2.10 and then cycled at a plastic-strain amplitude of 0.0031. Note the considerable variations in the structures, unlike the uniformity of those in initially annealed copper, because the prestrain is severe. (c) Typical dislocation structure associated with the prestrain of 2.10 before cycling. *(Courtesy of Laird et al., Ref 38)*

example, hcp[35] and bcc.[44] Moreover, the mechanisms advanced to explain cyclic deformation and the quantities that have been derived from them[18,20] depend on such parameters as the shear modulus, Poisson's ratio, and magnitude of the Burgers vector. Except for the shear modulus, most of these parameters do not vary much among different metals having a similar crystal structure. Consequently, it is no surprise to find, as recently shown by Mughrabi et al.,[44] that the cyclic stress-strain curves for monocrystalline copper and nickel effectively overlie one another when the flow stress is normalized against the shear modulus (see Fig. 21). Minor differences among metals can, however, be expected, insofar as stacking-fault energy influences the cyclic-deformation processes (see, for example, the behavior of silver[44]). Much more work is necessary to elucidate these differences.

The question of the relation between monocrystalline- and polycrystalline-fatigue results has been a recurring one. Kettunen showed an interesting correlation between the Wohler curves (plots of applied stress versus life) of polycrystals and single crystals.[45] Now Bhat and Laird

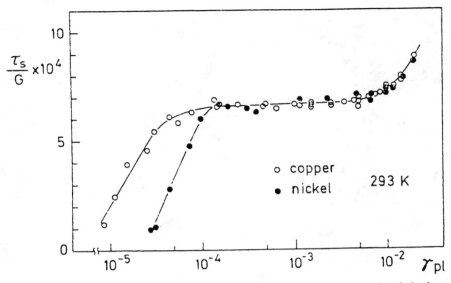

Fig. 21. Cyclic stress-strain curves for copper and nickel single crystals in which the stress has been normalized against the shear modulus. The logarithmic scale on the abscissa has been chosen for convenient presentation only. *(Courtesy of Mughrabi et al., Ref 44)*

have demonstrated a similar connection between the behaviors in cyclic deformation of the two kinds of materials.[46] The comparison is shown in Fig. 22. The polycrystal measurements of three sets of authors are plotted in a single curve;[33,47-49] the results of the different authors overlap and agree well in the regions of overlap. The single-crystal measurements of Mughrabi[19] and Finney and Laird[17] are also shown in Fig. 22. For ease of comparison with the monocrystal behavior, the polycrystal data are expressed in shear stress and shear strain using the Taylor factor appropriate for fcc crystals.[46] As is evident from Fig. 22, the cyclic stress-strain curve of the polycrystals definitely shows a plateau, although no measurement exists at low enough strain amplitudes to determine the lower end of the plateau. Nevertheless, the Coffin-Manson plots obtained by Lukáš and Klesnil,[48] which extend to long life, and the fatigue limit considerations of Laird[50] suggest that the lower end of the plateau should not differ much between monocrystals and polycrystals.[46]

It is noteworthy that the high-strain end of the plateau in polycrystals begins at a strain amplitude roughly an order of magnitude smaller than in single crystals.[46] The factors that control the plateau region in single crystals can be identified with those that control the first stage of monotonic deformation. It is reasonable to expect that in polycrystals, Stage I deformation is much more restricted than it is in single crystals.[46] Indeed, with respect to the role of alloying, the same conclusion should apply. Since Taylor's factor successfully brings the results of monocrys-

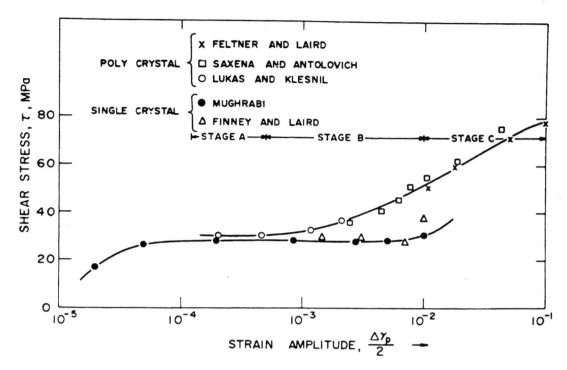

Fig. 22. Cyclic stress-strain curves for single-crystal and polycrystalline copper. Both curves show plateau behavior at low strain amplitudes. *(Courtesy of Bhat and Laird, Ref 46)*

talline and polycrystalline tests into coincidence at the plateau region, there are two interesting implications: (*a*) the deformation by PSB's is "homogeneous" among the grains and permeates the cross-section of the specimen, and (*b*) the high-strain end of the plateau coincides with the transition from high-strain fatigue (where the slip is general) to low-strain fatigue (where the slip is localized). To judge from the life measurements of Lukáš and Klesnil,[48] this appears to be the case.

Many of the fundamental mechanisms described above can be seen to apply to more complex materials if appropriate allowance is made for their special conditions. Indeed, for substitutional alloys, many of the details apply exactly.[15,35,46] Even in very complex alloys, broad similarities exist. Consider two examples: Al–4% Cu alloy aged to contain θ'' precipitates that are small, densely distributed and easy to cut by dislocations;[51] and Waspaloy containing equiaxed γ' precipitates 8 nm in diameter.[52] Both of these alloys show cyclic hardening followed by softening (see Fig. 23), but Stoltz and Pineau point out interesting differences between them.[52] In the Waspaloy, the number of cycles to maximum hardening [indicated by arrows in Fig. 23(a)] is about 25 and is constant regardless of strain amplitude. In the Al-Cu alloy, on the other hand, the number of cycles to maximum hardening increases with decrease in strain [Fig. 23(b)] and is found to occur at a critical *cumulative* strain, regardless of strain amplitude.[52] Thus, in Al-Cu-θ'',

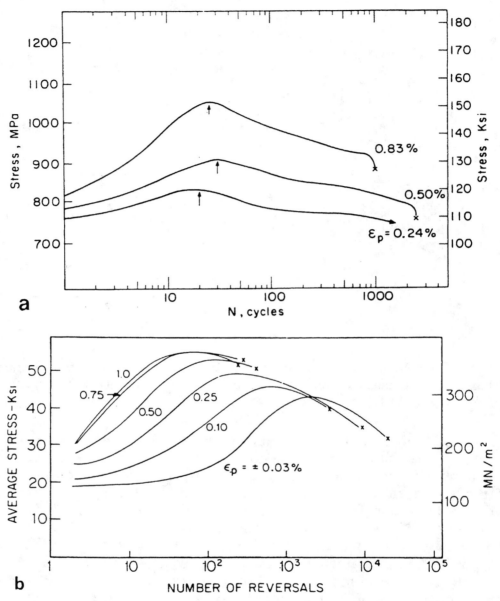

Fig. 23. (a) Cyclic response curves for Waspaloy with 8 nm γ' precipitates indicating hardening followed by softening. The maximum stress occurs at $N \approx 25$ regardless of plastic strain. *(Courtesy of Stoltz and Pineau, Ref 52)* (b) Cyclic response curves for Al–4% Cu alloy aged at 160 °C for 5 hr to produce a microstructure containing the metastable θ'' precipitate. The crosses at the ends of the curves denote fracture. *(Courtesy of Calabrese and Laird, Ref 51)*

the behavior has a similarity to a pure-metal single crystal, in that hardening first occurs by dislocation accumulation throughout the material, and softening occurs when the slip is subsequently localized, yielding PSB's. Since repeated shearing of the θ'' particles causes softening by a mechanism of particle disordering, at least in part,[51] the onset of cyclic softening is suggested to occur at a constant cumulative plastic strain.[52]

In Waspaloy, which is hardened much more relative to Al-Cu-θ″, all the plastic strain is localized in slip bands from the start of cycling. If the kinetics of the hardening and softening processes that govern the cyclic deformation are the same in each slip band regardless of plastic strain, then in order to accommodate the higher strains, Stoltz and Pineau argue, the number of slip bands must increase linearly with strain.[52] By sectioning cycled specimens and measuring the inter-slip-band spacing, they proved this to be the case.[52] Although there is an analogy here with the behavior of PSB's in the plateau region of the copper cyclic stress-strain curve, reference to Fig. 23(a) shows that the peak stress increases with plastic strain. This behavior is caused by the facts that Waspaloy is polycrystalline, has a planar slip mode, and was cycled at quite high strains. True plateau behavior may well be found at much lower strains, insofar as a material subject to cyclic softening will ever show a cycle-independent stress for a given applied strain.*

2. Mechanisms of Crack Nucleation

(a) Low-Strain Fatigue

In low-strain fatigue, the slip lines produced by the first few thousand cycles are very fine, consistent with general slip during rapid hardening and the associated vein production. As cycling continues, the PSB's form and intensify if the stress is above the plateau stress of the cyclic stress-strain curve (and therefore is above the fatigue limit[50]). Cracks ultimately form in the PSB's and, with continued cycling, multiply and link up to form the fatal crack. By testing single crystals, Gough showed that crack nucleation depends on the reversed shear-stress *range* acting on the active slip system, and not upon the normal stress component even when the values of these stresses were roughly equal.[55] The nucleation process has thus been viewed as a "pure" slip process.[56]

From the end of World War II to the beginning of the 1960's, long-life crack nucleation was intensively studied and has been repeatedly reviewed.[21,56-58] Many detailed and apparently conflicting observations have been reported,[56] and there is no intention of repeating these reviews here. Associated with these results, many mechanisms of crack nucleation, involving either single-dislocation reactions repeated every cycle

---

*There are other differences between the behaviors of pure metals and alloys. For example, the orientation of the crystal within a PSB in a pure metal is the same as that in the surrounding matrix containing veins. However, in a precipitation-strengthened material,[51,53] the volume of the PSB is misoriented with respect to the matrix by a few degrees. An explanation of this behavior has been advanced by Asaro[54] on a mechanical basis.

or group-dislocation processes occurring cyclically or occasionally, have been suggested.[21,56,57] There was a major debate about the role of cross-slip in crack nucleation. Nine, on the basis of single-crystal torsion tests,[59] demonstrated convincingly that repeated cross-slip is not basic to the formation of PSB's (and thus cracks) but rather seemed to inhibit the process. Other workers,[60] however, claimed that crack-nucleation phenomena were stimulated when the resolved shear stress on the cross-slip plane was increased, but Nine has argued that this need not be contradictory to his position.[59] It is interesting that Laird and Duquette,[56] reviewing this large body of work, were unable to conclude definitely in favor of any particular mechanism of crack nucleation. They tentatively suggested that a "Wood-type" mechanism, operating in a systematic fashion due to repeated deformation irreversibilities in a PSB, was the most likely choice for most materials. In this mechanism, alternating stressing results in different amounts of net slip on different planes and causes a general roughening of the surface.[61] Wood suggested that the valleys so formed act as stress raisers and promote further slip. This mechanism was criticized by Kennedy,[57] since it is unclear why, under alternating stress, the slip should continue to deepen the valleys, when the stress-concentration effect acts similarly on slip in either direction. However, May subsequently developed a purely statistical formalism[62,63] showing that with continued cycling, progressively deeper valleys would result from random slip.

The difficulties and confusion alluded to above resulted from the fact that even as recently as 1972,[56] our understanding of the mechanism of cyclic deformation in PSB's was very imperfect. Now, however, the situation is greatly improved;[17-20] the PSB behavior described in section 1(b) can be used as the basis of a crack-nucleation mechanism. Note that the development of the random slip within a PSB, shown in Fig. 12, is a demonstration of a Wood mechanism, operating locally. Moreover, the model described above provides a natural explanation[18] for the phenomenon of extrusions and intrusions, which leads to crack nucleation. This phenomenon was first recognized by Forsyth[64,65] in Al-4% Cu alloy, which, on fatiguing, showed PSB's and thin ribbons of metal subsequently extruded from them. A schematic of a slip-band extrusion and an optical micrograph of actual extrusions are given in Fig. 24. Typically, the extrusions were 10 μm high and about 100 mm thick; although it was initially believed that extrusions were confined to age-hardened materials, they were subsequently found to occur in almost every material examined, including AgBr,[66] and considerable morphological variation has been observed. (See Ref 56 for the many different circumstances under which extrusions can be found.) A model of extrusion formation that considers our improved understanding of PSB behavior has been offered,[18] as follows.

Figure 18 represents a cut normal through a PSB and parallel to the slip direction, and it shows schematically the idealized arrangement of

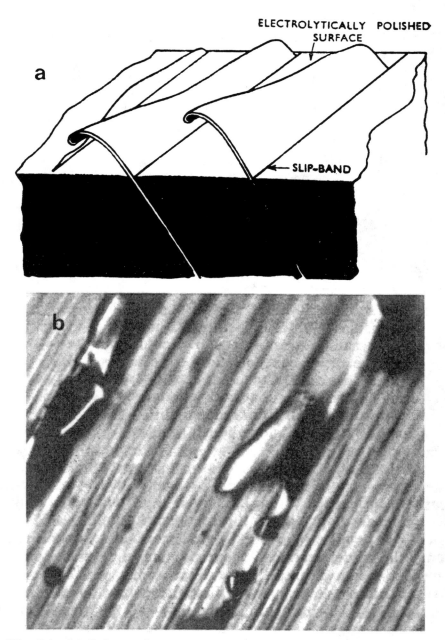

Fig. 24. (a) Schematic representation of slip-band extrusions. (b) Slip-band extrusion in age-hardened Al-Cu alloy fatigued at low strain. *(Courtesy of Forsyth, Ref 136)*

PSB walls and active slip-plane segments. As discussed in Ref 18, the cyclic strain will be accommodated by the interlinking screw-dislocation segments, all of the same sign within local areas. When cyclic strain is applied, waves of screw dislocations will sweep through the PSB somewhat as indicated in Fig. 25, in which the plane of the drawing is parallel to the slip plane.[18] The screw dislocations will eventually reach free surfaces in monocrystals and will be annihilated. In polycrys-

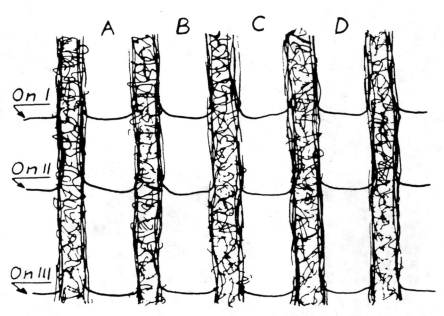

Fig. 25. Schematic view of advancing screw dislocations in the channels of a PSB projected into the primary glide plane. The displacement of the dislocations is greatly exaggerated in comparison with the width of the channels. The dislocations glide on planes separated according to Fig. 18, to which the numbering refers. *(Courtesy of Kuhlmann-Wilsdorf and Laird, Ref 18)*

tals, screw dislocations will reach grain boundaries and will be trapped. In single crystals they must be re-formed on the return cycle by mechanisms still unknown in detail. In polycrystals, the same screw dislocations, held up at the boundary, can be forced to retrace their paths, roughly, and to return to the opposite end of the PSB channel. Intrusions and extrusions would result if, for a packet of screw-dislocation segments, the return occurs on the next layer up or down.[18] For example, in channel B in Fig. 18 and 25, screw segments might travel along levels I, II and III in one half-cycle and return on levels II, III and IV on the return half-cycle, or this same kind of alternating slip might take place in a group of adjoining channels such as A, B and C but not D, and so forth.[18]

Kuhlmann-Wilsdorf and Laird have shown the deformation that would result from repetition of such displacements after several half-cycles (see Fig. 26). This figure traces the deformation of a surface initially planar, and normal to the primary Burgers vector at the beginning of the sequence, assuming the above path discrimination in tension and compression. At an external surface, the deformation would be seen as an extrusion if the initial plane had been bounded by the fatigued material on its right, or as an intrusion if it had been bounded at its left.[18] It is interesting to note that this mechanism is broadly similar to a path-discriminatory, dislocation-gating mechanism originally offered

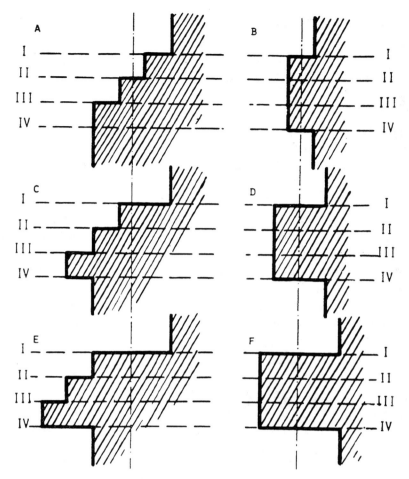

Fig. 26. Plastic deformation of a plane surface normal to the primary Burgers vector resulting from path discrimination in tension and compression. *(Courtesy of Kuhlmann-Wilsdorf and Laird, Ref 18)*

by Kennedy[57] long before the details of PSB behavior were established.

The dipolar walls within the PSB and directly adjacent to the surface will be forced to conform to the morphology of the extrusion. Since the PSB hedges contain many sessile dislocations among the debris, they can be expected not to be very mobile;[18] thus, reactive stresses due to distorting dipolar walls will limit the development of extrusions and intrusions to rather shallow profiles.[18] This explains why the "saturation" height of extrusions is of the order of microns. Presumably, the stress concentration of an intrusion gradually overcomes the resistance of the PSB hedges, and cracks ultimately form.

Once PSB behavior is understood in detail in other classes of material, particularly precipitation-strengthened alloys, detailed crack-nucleation models can be developed. In the meantime, it is wise to reiterate the warning of Laird and Duquette,[56] who, after reviewing the available evidence, concluded that fatigue-crack nucleation, in either the presence or the absence of aggressive environments, does not occur by the same mechanisms for every material.

(b) High-Strain Fatigue

At sufficiently high cyclic strains, dislocation-cell structures are general throughout a specimen, and the localized deformation associated with PSB's no longer occurs. Instead, the deformation is broadly homogeneous, and under such circumstances, cracks rarely occur transgranularly. Instead, notches develop at grain boundaries, and cracks form in them,[67-71] even in the absence of any inherent grain-boundary weakness. Laird and Duquette,[56] reviewing high-strain crack nucleation, understood the mechanism to be purely geometrical, but they deemphasized the crystallographic aspects of the slip involved and concluded that cracks formed as a result of a puckering effect produced by the varying, localized plastic deformation. For certain polymeric materials, direct evidence of such surface puckering had been obtained.[56] However, since then, a single effort at exploring the mechanistic details of crack nucleation in a pure metal has been completed by Kim and Laird,[72-75] and this has considerably improved our understanding. The details follow.

The technique newly applied to studying high-strain crack nucleation was interferometry,[74] used in conjunction with attempts to characterize the boundaries in which the cracks form.[75] Figure 27 shows the typical development of a grain-boundary crack at high strain (plastic strain = $\pm 0.57\%$) in pure copper. In Fig. 28, two-beam-interference fringes formed from white light show the gradual development of a notch at a grain boundary. With reversed cycling, at the beginning of life, a small step is formed in tension, but at the level of resolution obtainable with two-beam interferometry, it is cancelled by the compression stroke. With continued cycling, the amplitude of the boundary step slowly increases and resists complete cancellation in compression. Eventually a step 1.5 μm high, having a very sharp root radius, develops, and the crack grows along the grain boundary into the material from this step.[74] Detailed analysis of the sensitive boundaries and their adjacent grains established the requirements for nucleation by this mechanism,[75] namely:

1. The boundary has to be a high-angle boundary.
2. The slip on the most active slip system in either one or both of the adjoining grains should be directed at the intersection of the boundary with the specimen surface, in the manner shown schematically in Fig. 29; that is, the slip should operate over a long distance. Thus, a step can form most effectively at the grain boundary 1/2, because the dominant slip in grain 2 is directed at this boundary. However, no significant surface step can occur at boundary 2/3, because the slip requirements are not met.
3. The trace of the boundary in the free surface should be at an angle in the range 30° to 90° with respect to the stress axis.

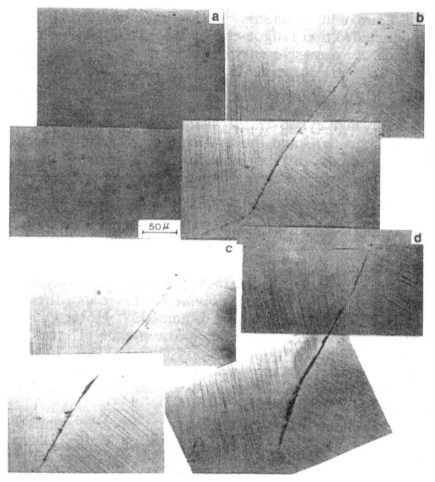

Fig. 27. Development of a crack nucleus in a polycrystalline specimen of copper cycled at a plastic strain of ±0.0057. The micrographs were taken at the tension strain limit **(a)** after 21 cycles (21 T); **(b)** 52 T; **(c)** 81 T; **(d)** 192 T. Grain-boundary darkening, an indication of a boundary discontinuity, is observed in **(b)**. *(Courtesy of Kim and Laird, Ref 74)*

The factors that control step irreversibility and thus the rate of step growth are not known in detail. The rate clearly does not depend simply on the difference in the degree of slip activity between the two grains.[75] Margolin and coworkers[76] have cited cross-slip as one of the important controlling factors. Taking account of stress distributions in polycrystalline material,[77,78] they conclude that cross-slip resulting from the presence of a stress gradient produces the path discrimination in tension and compression that promotes the irreversibility. In addition, grain-boundary dislocation sources may also contribute to the step formation even after the saturation-cell structure is generally established.[75]

The mechanism described above has a family resemblance to Neu-

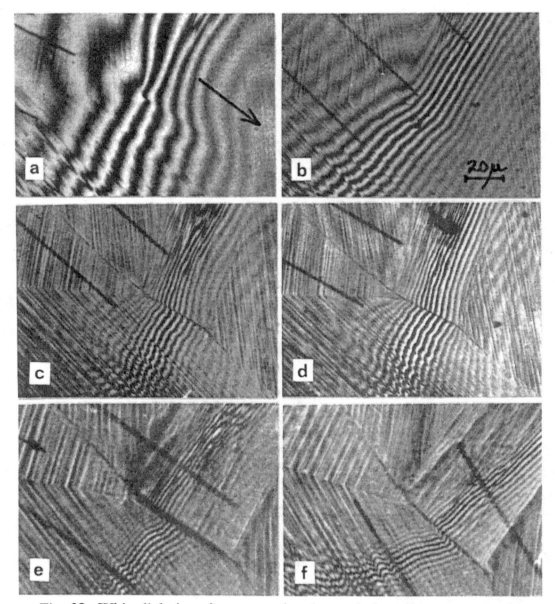

Fig. 28. White-light interferograms showing grain-boundary step growth in copper, cycled at a plastic strain of ±0.0076. Fringes shifted from left to right indicate depressions. **(a)** 10 T, a small step of 0.1 μm is observed. **(b)** Upon load reversal (10C) this step is cancelled. **(c)** 30 T, step height ~0.4 μm. **(d)** 30C, step partly cancelled, 0.15 μm. **(e)** 60 T, step height ~0.9 μm. **(f)** 60C, step partly cancelled. *(Courtesy of Kim and Laird, Ref 74)*

mann's model for the formation of a crack[79] in which duplex slip is involved. This model is shown schematically in Fig. 30 and is designed to explain transgranular nucleation. However, such nucleation has not been observed in polycrystalline material (to this author's knowledge), and it is doubtful whether the duplex-slip requirements would be satisfied

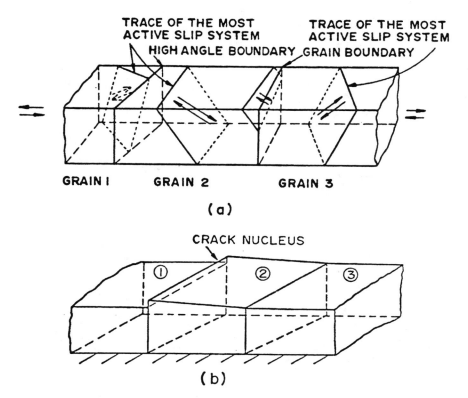

Fig. 29. Schematic representation of the crack-nucleation process in high-strain fatigue. (a) The active slip systems in either grains 1 or 2, or in both of them, are directed at the boundary between grains 1 and 2. (b) After cycling, a step is formed at the boundary between grains 1 and 2. *(Courtesy of Kim and Laird, Ref 75)*

in the absence of a grain boundary or a stress concentration such as a notch.

The grain-boundary crack-nucleation model of interest here and the associated crystallographic requirements are not specific to pure metals, and thus the mechanism may well be a general one.[75] However, in commercial alloys cycled in atmospheres corresponding to those of service conditions, complex environmental effects intervene in the crack-nucleation phenomena.[80,81] Much more work is necessary in exploring these interesting effects.

3. Mechanisms of Fatigue-Crack Propagation

(a) Stage I Growth

Once a crack is nucleated (it is still neither known nor agreed what constitutes a crack[82]) and the applied stress is low, the crack continues to develop along an active slip plane, generally a PSB.[21] Since the PSB's are formed on the most highly stressed slip plane, the orientation

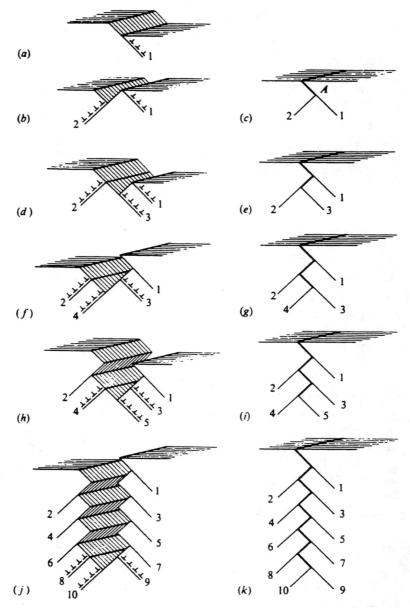

Fig. 30. Model for the formation of a crack at coarse surface steps under symmetrical push-pull loading. **Left:** tensile stages. **Right:** unloading stages. *(Courtesy of Neumann, Ref 79)*

of this crack is usually at 45° to the stress axis. In this configuration, it is known as a Stage I crack. Recent electron-microscope studies of Stage I cracks in PSB's have shown that the cracks can penetrate to depths many times the inter-hedge spacing of the PSB's without altering their dislocation structure significantly.[83] A typical view of such a crack is shown in Fig. 31. The implication of this result is that the PSB deformation processes, which act first to form an extrusion and then

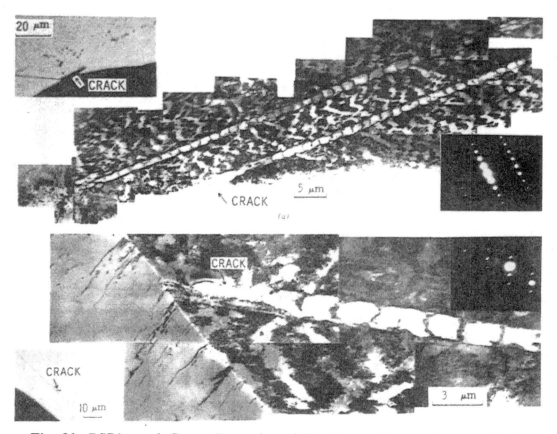

Fig. 31. PSB's and Stage I cracks within them. Insets show optical micrographs of the cracks and electron-diffraction patterns corresponding to the Gaussian images. *(Courtesy of Katagiri et al., Ref 83)*

to form notch-peak topography containing the crack embryo, now can lead to Stage I propagation. The mechanism for this is suggested to develop gradually from that illustrated in Fig. 26 and is shown schematically in Fig. 32.

The main difference between the models shown in Fig. 26 and 32 is connected with the localization of the slip. In Fig. 26, the screw dislocations carrying the PSB strain are distributed with roughly equal probability on all the slip planes within the PSB, and only slight path discrimination in tension and compression can lead to the formation of an extrusion. In the Stage I model, however, a notch, omitted for clarity in Fig. 32(a), now serves to concentrate the slip in tension along a narrow group of slip planes, C-C', at the base of the notch. That is, a stress concentration attracts the screw dislocations into the planes at the tip of the notch, the dislocations cross-slipping from their regular planes to meet the notch. In compression, on the other hand, the notch closes, the stress concentration is not sensed, and the PSB screw dislocations return to the other end of their channel more equally distributed on the planes in the PSB. It will be noted in Fig. 32(a)

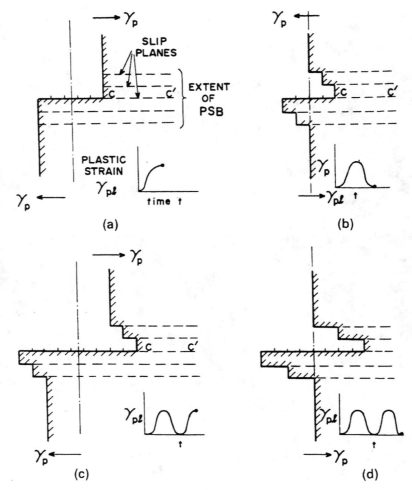

Fig. 32. Schematic representation of Stage I propagation in a PSB, via the deformation produced in a surface normal to the primary Burgers vector. A notch, omitted for clarity from (a), serves to concentrate the strain during tension on slip plane C-C'. The inserted γ_{pl}-time plot indicates the loading sequence corresponding to the slip behavior.

that the stress concentration has caused a slip offset of the height of five Burgers vectors, indicated by small marks on the surface of the step, because all the slip is on plane C-C'. In Fig. 32(b), the compression slip is equally distributed on the five slip planes, so that the C-C' slip offset is cancelled by one Burgers vector. In the next tension stroke, Fig. 32(c), the strain is again concentrated on C-C', and in the following compression, Fig. 32(d), the resultant offset is eight Burgers vectors. However, since the penetration of the crack is associated with an extrusion, the over-all penetration of the crack is only four Burgers vectors, two per cycle. By this model, the volume of the crack is exactly balanced by the volume of an extrusion. Since the crack is very narrow in relation to the width of a PSB, a relatively low, but wide, extrusion will be associated with a deeper crack. The association of extrusions with cracks has been very commonly observed, and a typical example is shown in Fig. 33. This is similar to the reversed-shear model attributed to Wadsworth[21] and in some respects to that advanced by Neumann[85] to explain extrusions and intrusions; but here the cyclic-deformation behavior of the PSB has been taken into account, and the model depends on the generation of a stress concentrator before it can operate.

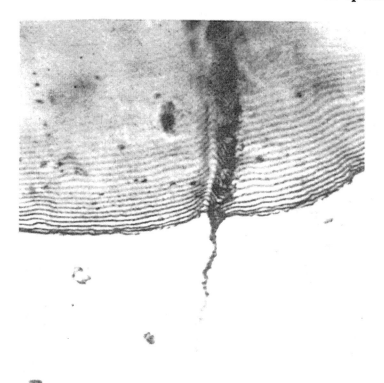

Fig. 33. A Stage I crack associated with an extrusion in Armco iron fatigued at low strain. *(Courtesy of Meleka et al., Ref 84)*

Naturally, an aggressive environment can be expected to play a large role in controlling the kinetics of crack growth by such a model, through "wedging" the crack by corrosion products and through minimizing the cancellation of the critical slip offset during compression strokes.

In principle, this model can be expected to operate in any metal or alloy that shows persistent slip bands. However, until the details of the slip that occurs in complex alloys such as the precipitation-hardened type are finally elucidated, any further discussion about them would be speculative.

(b) Stage II Growth

How deep a Stage I crack grows depends on the applied stress. In low-strain fatigue, a Stage I crack may well penetrate to the depth of several grain diameters before the increasing stress intensity[86,87] of the crack promotes slip on systems other than the primary. A dislocation-cell structure normally forms at the crack tip under these conditions, as frequently observed,[88,89] and the PSB structure is broken down. Since slip is no longer confined to planes at 45° to the stress axis, the crack begins to propagate normal to the stress axis and is known to be in Stage II. In high-strain fatigue, the stress intensity is so large at a crack nucleus that the crack will almost immediately propagate by the Stage II process.[74,75,90,91]

Stage II fracture surfaces resulting from fatigue-crack growth in air or mildly corrosive environments are easily recognizable in the scanning electron microscope or by replication electron microscopy via their characteristic fatigue striations. Not all materials show striations, however. Pure metals and ductile alloys show them clearly, but in many steels they occur infrequently or not at all, and in cold-worked metals, they are so tortured as to be barely recognizable. Zapffe and Worden[92] were the first to recognize fatigue striations on optical photographs, and their observation stimulated many researches. For long it was not clear how the striations were related to the cycling, until Forsyth and Ryder[93] published a fractograph taken from a specimen broken by a variable-loading sequence. Since the spacings and numbers of the observed striations corresponded to the numbers of cycles in the blocks of the loading sequence, it was clearly established that striations represent the propagation distance of a fatigue crack during each stress cycle.

Since the striation now could be connected to the mechanism of crack propagation, attempts were made to interpret its profile, of which a variety of forms has been observed.[94] Initially, it was believed that the profile resulted from a brittle-fracture–ductile-fracture mechanism (for review, see Ref 94), but this was challenged by Laird and Smith,[90] who made sections through cracks in specimens unloaded from different parts of the straining cycle. These sections revealed the profile of the crack tip as it was loaded and unloaded, and they provided direct evidence for a purely ductile model of crack propagation. The striation normally has a profile consisting of a more or less flat region (actually curved when the striation spacing is small) bounded by a trench. Laird and Smith[90] initially believed that the trench was formed at the beginning of the tensile part of the fatigue cycle, but they were misled by the fact that the trench was actually left over from the compression part of the previous cycle.[94] Accumulating evidence caused Laird to present a model of fatigue-crack propagation that he called the "plastic-blunting process."[94] He still believes that model is an accurate representation

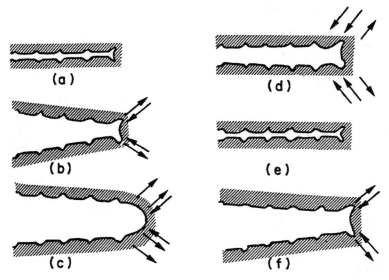

Fig. 34. Schematic representation of the plastic-blunting process of fatigue-crack propagation in Stage II: **(a)** zero load; **(b)** small tensile load; **(c)** maximum tensile load of the cycle; **(d)** closure; **(e)** maximum compressive load of the cycle; **(f)** small tensile load in the succeeding cycle. The double arrowheads in **(c)** and **(d)** signify the widening of slip bands at the crack in these stages of the process. The stress axis is vertical. *(Courtesy of Laird, Ref 94)*

of the ductile fatigue-crack growth mechanism, and it is reproduced in Fig. 34.

The initially unloaded crack tip is shown in Fig. 34(a). As tensile loading is applied, slip is concentrated in sharp bands at the double notch of the crack tip. This slip helps to maintain a roughly square geometry, as shown in Fig. 34(b). Since Laird and Smith[90] frequently observed rounded crack tips at the fully loaded condition, they argued that the slip is spread during loading, and this broadens the slip zones; thus, crack-tip blunting occurs. However, they also observed tips with a leading notch.[90] As pointed out by Cottrell,[95] this process could ideally be represented by the model shown in Fig. 35. Here the material is taken as a perfectly plastic solid, so that the narrow slip bands emerging from the crack tip pass right through the specimen. The hysteresis loop shown in Fig. 35(f) identifies the loading state of the specimen at different parts of the cycle. The slip processes shown in Fig. 35(b) and (c) can be regarded as occurring either simultaneously or in rapid alternation. Note that the crack progresses a distance proportional to the sliding off. This model has been directly supported[96] by in situ observations on single crystals, where the precision of the slip was controlled by

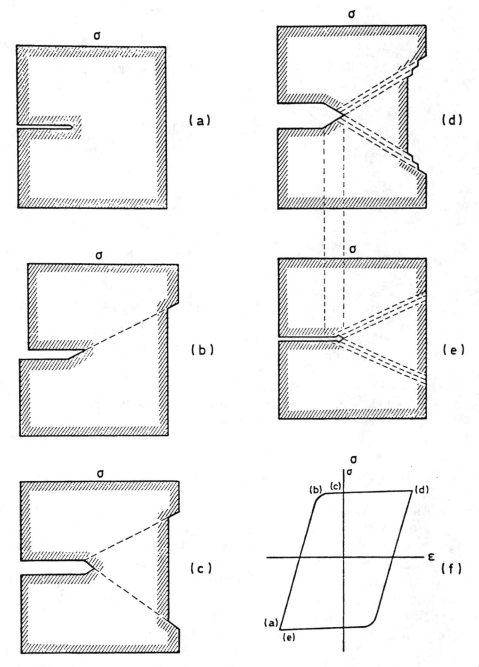

Fig. 35. An idealized representation of the plastic-blunting process. *(Courtesy of Laird and de la Veaux, Ref 103)*

a careful choice of single-crystal orientation. Note that the model shown in Fig. 30 is similar to the plastic-blunting process.

On reversing the loading direction, the slip in the zones is reversed, the crack faces close, and the new crack surface just created is folded [Fig. 34(d)] by buckling of material at the crack tip, producing the double notch and unit of crack advance shown in Fig. 34(e).

It will be noted that there is a basic similarity between the mechanisms of Stage I and Stage II crack propagation, in that both involve crack advance by a "slipping-off" process. The difference between them is that only one slip system need be involved in Stage I, but duplex slip occurs in the Stage II crack plastic zones.

Numerous authors[97-101] have published alternative representations of the plastic-blunting process. Most involve the essential "slipping-off" process of the plastic-blunting mechanism. One of them[100] appears to be in conflict by asserting that not all deformation surrounding the crack tip will contribute to advance of the crack; its authors prefer the term "unzipping process." However, the present author sees no substantive difference between the unzipping model and the plastic-blunting process except in name. Occasionally, real differences with respect to the formation of the striation have been expressed,[101,102] but they have been successfully refuted.[103]

At present, there appear to be three outstanding criticisms of this class of models.[104] They are: (a) The models were developed for pure metals and alloys and may not apply to commercial materials. (b) They were developed from observations on cracks where high rates of growth were necessary to reveal the details of the mechanisms and thus may not be representative of the majority of the crack-propagation life. (c) Bowles finds it difficult to visualize how the double notch will form at a crack tip during the compression stroke and asserts, "If . . . dislocation motion is the mechanism by which crack-tip deformation occurs, it is perhaps more logical to suggest that reversed plastic flow will result in a sharp, singly notched type."[104]

These criticisms can be answered as follows: (a) If a commercial material is ductile enough to prevent interference of the plastic-blunting process by other modes of static failure, there is no reason why it should not apply, because it is a "geometrical" model.[103] (b) Laird has argued from a comparison of the fine details of striations that the model does indeed apply to lower rates of crack propagation where striations are formed.[94] (c) The logic of a given investigator is irrelevant to the issue of the double notch. It has been *observed*.[90,94] Moreover, Laird has made in situ observations by optical microscopy of a rounded crack tip subject to compression[105] and claims to have seen what he has schematically modeled.[94] Unfortunately, depth-of-field problems prevented him from photographing his observations. Now, however, the improved depth of field available in the scanning electron microscope has been applied to this question (unfortunately, only for somewhat brittle, high-strength aluminum alloys) via a plastic casting infiltration technique.[104] The results were complicated by closure problems and brittle mechanisms of propagation; nevertheless, for striations where the trenches are especially well-marked, Bowles concludes that they are formed as shown schematically in Fig. 34 (see p 35 of Ref 104).

(c) Crack-Propagation Theories

The purpose of a crack-propagation theory is to describe accurately the kinetics of crack propagation. Since the stress intensity for crack growth has been worked out for the most usual configurations and conditions of crack growth,[86,87] it is customary to represent those kinetics by a plot of rate versus stress-intensity range, ΔK, of the type shown schematically in Fig. 36. The main features of such a plot are: (a) the existence of a threshold, below which the crack will not propagate; (b) a gradual transition from the threshold to a linear portion of the curve, in which Stage I propagation can occur; (c) the "linear" region proper, associated with Stage II growth; and (d) a region where the growth rate becomes increasingly sensitive to the stress intensity, at high rates. The details of material behavior have been thoroughly documented,[106] and the fracture-mechanics approach has been developed to very sophisticated levels (see Ref 107-109, for example).

A natural consequence of the plastic-blunting process (or the unzipping model) is that crack growth per cycle is a large fraction of the crack-tip opening displacement, as suggested by Laird[105] and McClintock.[110] Liu[111] has pointed out that for such a model, dimensional analysis dictates that the crack growth must vary linearly with crack length, since for a crack in a large part the only significant length parameter is the crack length itself. Consequently, the crack-propagation rate should depend on the stress intensity squared. Kuo and Liu have performed

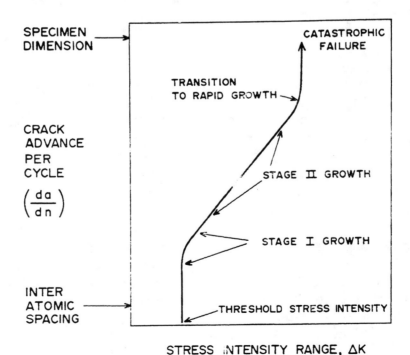

Fig. 36. Schematic representation of crack-growth rate as a function of stress-intensity range.

a finite-element calculation to determine the crack-tip opening displacement and thus to arrive at a description of crack-propagation kinetics,[100] having the appropriate dependence on the stress intensity. They have employed the cyclic stress-strain curve to describe the hardening behavior at the crack tip—a reasonable procedure. Consequently, Liu and his coworkers have produced very good agreement with experimental results of fatigue-crack growth in steels.[100,112,113] A typical comparison of crack-growth measurements obtained by Hahn et al.,[114] Barsom[115] and Bates and Clark[116] with that predicted by the unzipping model is shown in Fig. 37. The predictions of the Dugdale model[117] are also shown and do not emerge well.

It is well-known, however, from careful examinations of data[106] that there are many occasions when the linear portion of the curve shown in Fig. 36 shows a ΔK dependence much greater than that of the power 2. There is, of course, a systematic increase of the dependence near catastrophic failure because static-failure modes ahead of the crack tip increase the rate of crack advance. But even at low crack-propagation rates where regular Stage II growth behavior would be expected, deviations are frequent. They sometimes occur among well-behaved ductile metals but more often are found in high-strength materials where complex embrittlement processes[104,118] may well influence the behavior by mechanisms still to be understood.

McClintock[110,119] has treated a possible mechanism that causes the exponent of the dependence of growth rate on crack length to be greater than unity—namely, one of renucleation of cracks at inclusions ahead of the main crack. He shows that there should be a second-power dependence of crack growth on crack length,[119] if growth is produced only by such damage. Observations of such a dependence have been very frequently observed. However, many other possible mechanisms can contribute to crack growth, and treating them is difficult. Such a model by McClintock is one of a class of models that depend on the ductility of the material ahead of the crack to be "exhausted" by the cyclic straining and thus to produce an increment of crack growth. Obtaining a failure criterion is a major difficulty. Often the Coffin-Manson law is used as the basis for deciding when the ligament ahead of a crack tip is to fail. Since the Coffin-Manson law describes the failure of specimens by crack nucleation and propagation, its use in this connection is circular and unsatisfying. A considerable improvement in our understanding of fracture mechanisms will have to be obtained before this part of the field can be significantly advanced.

The number of different theories dealing with fatigue-crack propagation exceeded 30 when the present author did a count several years ago. The number must now exceed 50, and most of them can be classified within one of the two types described above. For a recent but incomplete tabulation, see Yokobori.[120] In addition, there are theories that are not mechanism-specific, such as that of Weertman.[121] Here, crack

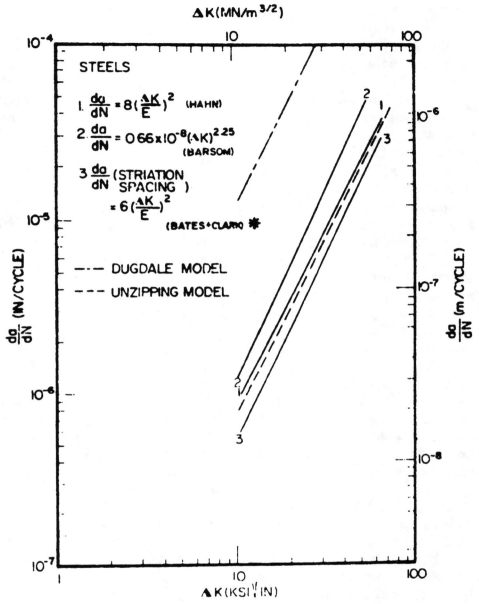

Fig. 37. Comparison of the empirical fatigue-crack-growth rates in steels observed by Hahn et al.,[114] Barsom,[115] and Bates and Clark,[116] with that predicted by the Dugdale model and the unzipping model. *(Courtesy of Liu et al., Ref 113)*

propagation can be considered in terms of the plastic work done in the region of the crack tip. For the situation in which the work is approximately constant with stress, a fourth-power dependence of growth rate on stress intensity is predicted. When the work varies with stress in proportion to ΔK^2, a second-power dependence obtains. Although the plastic work is experimentally measurable,[122] its magnitude will depend sensitively on microstructural parameters and will be difficult to predict. Moreover, the theory predicts a strong environmental effect on crack-growth rate for material obeying the fourth-power equation, but such is often not observed.[123] More space is available for general discussion of this theory in the present volume.[124]

No satisfactory theories are available for Stage I crack growth except that: (a) many Stage II theories can be adapted to Stage I growth without conceptual difficulty,[94] and (b) Purushothaman and Tien have treated Stage I by an interesting discontinuous-growth mechanism analogous to ledge growth in phase transformations.[125,126] Incidentally, this is the only model in which an attempt has been made to predict a threshold for fatigue-crack growth. The details of Stage I growth are extremely complicated,[118,127-130] and the small rates of crack advance make its study difficult. Much more work is required to elucidate its complexities and to explain the threshold for crack propagation.

4. Practical Implications of Fundamental Studies

Although fundamental studies are normally (and adequately) justifiable for the understanding they provide, there are occasions when they provide practical results difficult to obtain by any other approach. Two examples of this are provided here: (a) on the fatigue limit[50] chosen because cyclic deformation is used to obtain a result pertaining to fatigue fracture, and (b) on the fatigue limit of notched bodies,[131] chosen because results of cyclic deformation and fracture mechanics (normally two areas between which there is limited communication) are brought together.

(a) The Fatigue Limit

As mentioned in section 1, recent studies of cyclic deformation in fcc single crystals have identified the critical properties of PSB's, in which cracks form. If the applied strain is below that of the plateau, region A ($\gamma_{pl} \lesssim 6 \times 10^{-5}$), or if the applied stress (worked up gradually) is below that of the plateau (28 MPa for copper), then PSB's will not form, and the material should last indefinitely.[50] That is, there should be a true fatigue limit for nonferrous metals and alloys just as there is for steels, as distinct from the endurance limit (specification of a life at which failure was expected at a fairly low stress), which was traditionally applied to nonferrous metals in ignorance of how they behaved at very low stresses.

Helgeland was the first to observe and claim a fatigue limit for copper[9] (actually the plateau stress, although it was not recognized as such at the time). Unfortunately, his results were apparently contradicted by those of Kettunen,[11] who observed failures at stresses down to 17.7 MPa. This difficulty was resolved by Laird,[50] who showed that Lukáš and Klesnil's long-life Coffin-Manson plots showed failures to occur only down to the plastic-strain fatigue limit;[132] at lower strains, no failures were observed in the testing time available (Fig. 38). However,

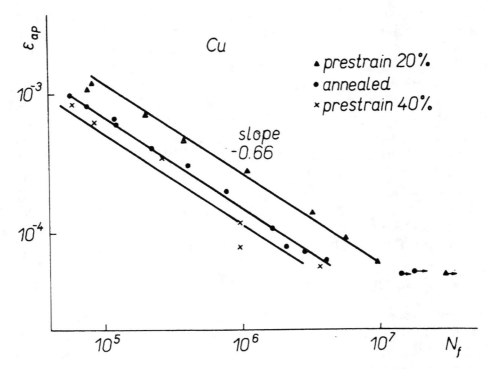

Fig. 38. Applied plastic-strain amplitude versus fatigue-life curves for copper at long life. *(Courtesy of Lukáš and Klesnil, Ref 132)*

Lukáš et al.[133] also carried out stress-cycling tests, in which they monitored the plastic strain. Specimens that had been *stress-cycled* yielded a plot of saturation plastic-strain amplitude versus life, where failures occurred at strains as low as 10^{-5}. The difference between these tests is that in strain cycling, the stress is low in the initial cycles and increases to saturation, whereas in stress cycling, full application of the load in the first cycle causes a large strain in a soft material. This initial large strain creates the PSB cell structures, which would not otherwise form in a constant-strain test. Since Kettunen applied the full load to his specimens, failures were observed at stresses below that of the plateau. Helgeland, on the other hand, although he was stress cycling, imposed a low stress at the start of his tests and increased it gradually to the chosen value.[9]

Laird has discussed the structure sensitivity of the fatigue limit.[50]

(b) Fatigue Limit of Notched Bodies

Since, at least in steels, PSB's and associated microcracks can be observed at stresses lower than the fatigue limit,[134] Lukáš and Klesnil approach the definition of the fatigue limit as the set of conditions under which already-nucleated microcracks can propagate.[131] They assume that this process is controlled by the threshold stress intensity for crack propagation K_{atb} and write:

$$K_{atb} = 1.122\, \sigma_{ac}\, (\pi l_c)^{1/2}$$

where σ_{ac} is the PSB fatigue limit for a smooth specimen and l_c the critical length of the microcrack. Actually, this expression constitutes a definition of the critical length, a material constant.[131]

Lukáš and Klesnil assume that the fatigue limit of notched specimens is given by the same process as in the case of smooth specimens. First the critical crack must be nucleated, and they take it as occurring when the local stress and strain values at the notch root are just equal to the plateau stress (for polycrystals, if need be) and the critical strain required to form a PSB. Using Neuber's rule to relate the notch stress and strain-concentration factors to the nominal stress-concentration factor K_t, and using the empirical relation for the cyclic stress-strain curve to relate the stress and strain, Lukáš and Klesnil derive a relationship between K_t and K_f, the fatigue-notch factor, defined as the ratio of the smooth-specimen fatigue limit and the nominal stress at the fatigue limit of the notched specimen. They conclude that the effect of cyclic plasticity on decreasing the stress concentration at the notch root is quite weak.[131]

Once the above condition is satisfied and a crack of critical length, l_c, is nucleated at the notch root, the condition for its propagation is that the applied stress intensity be greater than K_{atb}.[131] Since the stress intensity is related to the nominal stress, and K_{atb} to the fatigue limit of the smooth body, the following relationship is derived:

$$K_t/K_f = (1 + 4.5\, l_c/\rho)^{1/2}$$

where ρ is the notch radius.

To test this relationship, Lukáš and Klesnil carried out the following experiments: (a) measurement of the fatigue limit on smooth specimens of two steels (Czech designations 12010 and 12060); (b) measurement of the threshold stress for crack propagation in these steels, approximately 4 MPa m$^{1/2}$; and (c) determination of the fatigue limit of sheet specimens having a central hole of diameter ranging from 0.3 mm to 4 mm. Experiments 1 and 2 yield the values of l_c for the steels and experiments 1 and 3 experimental values for K_f. Relevant values of K_t were found in Peterson's charts. Figure 39 shows the comparison of the predicted values of K_t/K_f with those determined experimentally, and it is clear that the agreement between the values is quite good.[131]

Since both the threshold stress and the smooth-specimen fatigue limit are known for many materials (to give l_c), it is now possible to apply Lukáš and Klesnil's relationship for practical purposes. Their contribution represents a most interesting example of where a fracture-mechanics result is used to strengthen a traditional engineering approach to fatigue design, in which crack nucleation is of main interest, and which is still widely used.

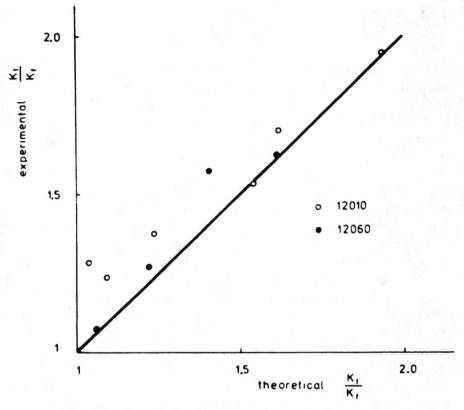

Fig. 39. Comparison of the theoretically predicted and the directly measured values of the K_t/K_f ratio. *(Courtesy of Lukáš and Klesnil, Ref 131)*

Summary

An overview of fatigue mechanisms is presented; fatigue is regarded as taking place first by cyclic deformation (which conditions the material for failure), second by crack nucleation, and finally by propagation. Generally, different fatigue processes take place at high strains and low strains, but they often have points of similarity. Although the mechanisms are discussed mainly with respect to pure metals, the results can frequently be applied to more complex materials, provided due consideration is given to the special properties of those materials.

In low-strain cyclic deformation, an initially annealed metal hardens rapidly by debris accumulation, the debris consisting of dense veins of dislocation dipoles in strong edge orientation. The plastic strain is carried both by screw dislocations in the channels between veins and by the veins themselves, deforming by flip-flop motions of the dipoles. As the strain is increased, the veins are no longer able to carry the strain, and persistent slip bands (PSB's) form when the cyclic stress saturates. The PSB's deprive the matrix veins of strain and adjust their volume fraction to accommodate the strain applied. The structure of the PSB consists of dense dislocation hedges aligned normal to the primary Burgers vector and to the slip plane, and spaced about 1.5 μm apart (at least at ambient temperature in copper). The structure of the hedges consists of even-numbered dipolar walls in which loop debris is densely mixed. The strain in the PSB is carried by screw dislocations, operating in loose coordination across the hedges so as to maintain the balance of the dislocations in the walls. The average strain localized in the PSB is about 1% and is largely reversible in a mechanical sense except at the level of the finest slip. When the applied strain is greater than 1%, single slip gives way to duplex slip, the structure consists of homogeneous dislocation cells, and the strain is carried uniformly in the structure by dislocations shuttling in the cells. It is possible to relate the behavior of monocrystals and polycrystals in detailed fashion through use of Taylor's orientation factors.

Once the saturated condition is reached, crack-nucleation processes begin. Old theories of nucleation are briefly reviewed and, although they are sometimes accurate, they are generally shown not to fit with our improved understanding of cyclic deformation. In low-strain fatigue, cracks form in the PSB by a mechanism of random slip, which induces a notch-peak topography. However, if the paths of the screw dislocations that glide in the PSB are discriminated in tension and compression, extrusions or intrusions can form in a systematic fashion. At high cyclic strains, the mechanism of crack nucleation is different from that at low strain because the slip is no longer localized. The geometry of the slip process remains important in that slip directed at the line of intersection of a grain boundary with the specimen surface leads to step formation in the boundary. Initially, a step formed in a tension cycle is cancelled in compression, but little-understood processes gradually induce step irreversibility, and a step large enough to consist of a crack builds up.

At low strain, a crack nucleated in a PSB continues to propagate along its plane in Stage I growth. This growth proceeds to depths many times the PSB inter-hedge spacing without altering the PSB structure significantly near the crack tip. It is argued, therefore, that an unslipping mechanism causes crack growth and that this must take place by action of the PSB screw dislocations. As the stress intensity is increased, by either growth of the crack or higher applied stress, slip begins to occur near the crack tip on planes other than the primary. This duplex slip induces Stage II crack propagation by "slipping-off" coordinated on slip planes oriented roughly at 45° to the plane of the crack. This mechanism is known as the plastic-blunting process, or equivalently in more recent publications, as the unzipping process. Crack propagation per cycle can then be linked to the crack-opening displacement, and many theories of crack-growth kinetics can be developed from this model if different approaches to calculating that displacement are used. Recent application of the finite-element method appears to give very good agreement between theory and measurement for ductile metals. The plastic-blunting process is capable of explaining many phenomena of crack propagation, such as the formation of striations on the fracture surface. This model predicts a quadratic dependence on stress intensity, which is frequently observed. In many cases, however, a greater sensitivity to stress intensity is observed for reasons that are still not properly understood. Static or brittle modes of failure appear to increase the dependency of crack propagation on the stress intensity. "Exhaustion of ductility" ideas are used to describe such growth kinetics.

Finally, practical applications of results arising from fundamental studies are considered for two examples: (a) it is shown how in nonferrous metals a fatigue limit and its fragile nature can be demonstrated from PSB behavior, and (b) it is shown how the fatigue limits of notched specimens can be predicted from the fatigue limit of smooth specimens and the threshold stress intensity for crack propagation.

> **Note added in proof:** There have been recent developments in understanding the friction stress, back stress and dislocation behavior during rapid hardening at low plastic strains, which modify the understanding briefly reported here. This work has been done by D. Kuhlmann-Wilsdorf (to be published in Mat. Sci. Eng., 1979), and the interested reader is directed to it for details.

Acknowledgments

The helpful discussions of Dr. S. P. Bhat are gratefully acknowledged. During the preparation of this overview, the author received support from the following sources: (a) the materials-failure coherency group of the Laboratory for Research on the Structure of Matter, University of Pennsylvania under Grant No. NSF-MRL-DMR-80994-A01; (b) the National Science Foundation under Grant No. DMR77-13934; and (c) the Army Research Office under Grant No. DAAG 29-78-C-0039. He deeply appreciates this support.

References

1. L. F. Coffin, "Fatigue in Machines and Structures—Power Generation," this volume.
2. T. D. Cooper and C. A. Kelto, "Fatigue in Machines and Structures—Aircraft," this volume.
3. D. H. Breen and E. M. Wene, "Fatigue in Machines and Structures—Ground Vehicles," this volume.
4. O. Buck and G. A. Alers, "New Techniques for Detection and Monitoring of Fatigue Damage," this volume.
5. M. S. Paterson, Acta Met., vol 3, 1955, p 491.
6. D. S. Kemsley and M. S. Paterson, Acta Met., vol 8, 1960, p 453.
7. D. H. Avery and W. A. Backofen, Acta Met., vol 11, 1963, p 352.
8. N. J. Wadsworth, Acta Met., vol 11, 1963, p 663.
9. O. Helgeland, J. Inst. Met., vol 93, 1964-65, p 570.

10. C. E. Feltner, Phil. Mag., vol 12, 1965, p 1229.
11. P. O. Kettunen, Acta Met., vol 15, 1967, p 1275.
12. J. R. Hancock and J. C. Grosskreutz, Acta Met., vol 17, 1969, p 77.
13. S. J. Basinski, Z. S. Basinski, and A. Howie, Phil. Mag., vol 19, 1969, p 899.
14. W. N. Roberts, Phil. Mag., vol 20, 1969, p 675.
15. P. J. Woods, Phil. Mag., vol 28, 1973, p 155.
16. A. T. Winter, Phil. Mag., vol 30, 1974, p 719.
17. J. M. Finney and C. Laird, Phil. Mag., vol 31, 1975, p 339.
18. D. Kuhlmann-Wilsdorf and C. Laird, Mat. Sci. Eng., vol 21, 1977, p 137.
19. H. Mughrabi, Mater. Sci. Eng., vol 33, 1978, p 207.
20. D. Kuhlmann-Wilsdorf and C. Laird, Mater. Sci. Eng., in press.
21. N. Thompson and N. J. Wadsworth, Adv. Phys., vol 7, 1958, p 72.
22. R. L. Segall, Adv. Mater. Res., vol 3, 1968, p 109.
23. J. C. Grosskreutz, Phys. Stat. Solidi., vol 47, 1971, p 11.
24. C. Laird, Treatise on Materials Science and Technology, vol 6, Ed. R. J. Arsenault, Academic Press, N. Y., 1975, p 101.
25. J. C. Grosskreutz and H. Mughrabi, "Constitutive Equations in Plasticity," Ed. A. S. Argon, MIT Press, Cambridge, Mass., 1975, p 251.
26. D. V. Wilson, Acta Met., vol 13, 1965, p 807.
27. J. D. Atkinson, L. M. Brown, and W. M. Stobbs, Phil. Mag., vol 30, 1974, p 1247.
28. D. Kuhlmann-Wilsdorf and C. Laird, to be published.
29. P. Rapps, K-H. Katerbau, H. Mughrabi, K. Urban, and M. Wilkens, Phys. Stat. Sol., vol 47(a), 1978, p 479.
30. P. Neumann, Acta Met., vol 17, 1969, p 1219.
31. A. T. Winter, Phil. Mag., vol 34, 1978, p 457.
32. J. G. Antonopoulos and A. T. Winter, Phil. Mag., vol 33, 1976, p 87.
33. C. E. Feltner and C. Laird, Acta Met., vol 15, 1967, p 1621.
34. C. E. Feltner and C. Laird, Acta Met., vol 15, 1967, p 1633.
35. C. Laird, in "Work Hardening in Tension and Fatigue," Ed. A. W. Thompson, The Metallurgical Society, 1977, p 150.
36. D. Kuhlmann-Wilsdorf, Ibid., p 1.
37. C. E. Feltner and C. Laird, Trans. Am. Inst. Met. Eng., vol 245, 1969, p 1372.
38. C. Laird, J. M. Finney, A. Schwartzman, and R. de la Veaux, J. Testing and Evaluation, vol 3, 1975, p 435.
39. H. D. Nine, Scripta Met., vol 4, 1970, p 887.
40. H. D. Nine, Phil. Mag., vol 26, 1972, p 1409.
41. R. Neumann, Zeit. Metallk., vol 66, 1975, p 26.
42. H. Mughrabi and C. Wüthrich, Phil. Mag., vol 33, 1976, p 963-984.
43. H. Mughrabi, K. Herz, and X. Stark, Acta Met., vol 24, 1976, p 659.
44. H. Mughrabi, F. Ackermann and K. Herz, "PSB's in Fatigued F.C.C. and B.C.C. Metals," Symposium on Fatigue Mechanisms, Kansas City, May 22-24, 1978. To be published in ASTM, STP.
45. P. O. Kettunen, Phil. Mag., vol 16, 1967, p 253.
46. S. P. Bhat and C. Laird, Scripta Met., vol 12, 1978, p 687.
47. A. Saxena and S. Antolovich, Met. Trans., vol 6A, 1975, p 1809.
48. P. Lukáš and M. Klesnil, Mat. Sci. Eng., vol 11, 1973, p 345.
49. P. Lukáš and J. Polák, in "Work Hardening in Tension and Fatigue," Ed. A. W. Thompson, The Met. Soc., AIME, 1977, p 177.

50. C. Laird, Mat. Sci. Eng., vol 22, 1976, p 231.
51. C. Calabrese and C. Laird, Mat. Sci. Eng., vol 13, 1974, p 141.
52. R. E. Stoltz and A. G. Pineau, Mat. Sci. Eng., vol 34, 1978, p 275.
53. S. P. Bhat, Ph.D. Thesis, University of Pennsylvania, 1978.
54. R. J. Asaro, Acta Metallurgica, in press, 1978.
55. H. J. Gough, Proc. ASTM, vol 33, 1933, p 3.
56. C. Laird and D. J. Duquette, "Corrosion Fatigue," NACE-2, 1972, p 88-117.
57. A. J. Kennedy, "Processes of Creep and Fatigue in Metals," John Wiley and Sons, Inc., New York, 1963.
58. L. M. Brown, Metal. Sci., vol 11, 1977, p 315.
59. H. D. Nine, J. Appl. Phys., vol 38, 1967, p 1678.
60. T. H. Alden and W. A. Backofen, Acta Met., vol 9, 1961, p 352.
61. W. A. Wood, "Fatigue in Aircraft Structures," Academic Press, New York, 1956.
62. A. N. May, Nature, vol 185, 1960, p 303.
63. A. N. May, Nature, vol 188, 1960, p 573.
64. P. J. E. Forsyth, Nature, vol 171, 1953, p 172.
65. P. J. E. Forsyth, J. Inst. Met., vol 83, 1955, p 395.
66. P. J. E. Forsyth, Proc. Roy. Soc., vol A242, 1957, p 198.
67. D. S. Kemsley, Phil. Mag., vol 2, 1957, p 131.
68. C. Laird and G. C. Smith, Phil. Mag., vol 8, 1963, p 1945.
69. D. S. Kemsley, J. Inst. Met., vol 85, 1956, p 420.
70. J. Porter and J. C. Levy, J. Inst. Met., vol 89, 1960, p 86.
71. R. C. Boettner, C. Laird, and A. J. McEvily, Trans., A.I.M.E., vol 233, 1965, p 379.
72. W. H. Kim, Ph.D. Thesis, University of Pennsylvania, 1977.
73. W. H. Kim and C. Laird, Mat. Sci. Eng., vol 33, 1978, p 225.
74. W. H. Kim and C. Laird, Acta Met., vol 26, 1978, p 777.
75. W. H. Kim and C. Laird, Acta Met., vol 26, 1978, p 789.
76. H. Margolin, Y. Mahajan, and Y. Saleh, Scripta Met., vol 10, 1976, p 1115.
77. H. Margolin and M. S. Stanescu, Acta Met., vol 23, 1975, p 1411.
78. Y. Chuang and H. Margolin, Metall. Trans., vol 4, 1973, p 1905.
79. P. Neumann, Acta Met., vol 17, 1969, p 1219.
80. B. Hodgson, Met. Sci. J., vol 2, 1968, p 235.
81. L. F. Coffin and C. J. McMahon, Met. Trans., vol 1, 1970, p 3443.
82. J. C. Grosskreutz, "Fundamental Knowledge of Fatigue Fracture," Int. Congress on Fracture, Munich 1973, PLV-212.
83. K. Katagiri, A. Omura, K. Koyamagi, J. Awatani, T. Shiraishi, and H. Kaneshiro, Met. Trans., vol 8A, 1977, p 1769.
84. A. H. Meleka, W. Barr, and A. A. Baker, B.I.S.R.A., Report No. P/16/60, 1960.
85. P. Neumann, Symposium on Fatigue Mechanisms, Kansas City, May 22-24, 1978, to be published in ASTM, STP.
86. H. Tada, P. Paris, and G. Irwin, "The Stress Analysis of Cracks Handbook," Del Research Corp., Hellertown, Pa., 1973.
87. D. P. Rooke and D. J. Cartwright, "Stress Intensity Factors," HMSO, London, 1976.
88. J. C. Grosskreutz, J. App. Phys., vol 33, 1962, p 1787.
89. M. A. Wilkins and G. C. Smith, Acta Met., vol 18, 1970, p 1035.
90. C. Laird and G. C. Smith, Phil. Mag., vol 7, 1962, p 847.

91. C. Laird and G. C. Smith, Phil. Mag., vol 8, 1963, p 1945.
92. C. A. Zapffe and C. O. Worden, Trans. ASM, vol 43, 1951, p 958.
93. P. J. E. Forsyth and D. A. Ryder, Aircraft Eng., vol 32, 1960, p 96.
94. C. Laird, in "Fatigue Crack Propagation," ASTM STP 415, 1966, p 131.
95. A. H. Cottrell, "Introductory Review of the Basic Mechanisms of Crack Propagation," Proc. of Crack Propagation Symposium, Cranfield, U.K. 1961.
96. P. Neumann, H. Vehoff and H. Fuhlrott, "Fracture 1977," vol 2, ICF 4, Canada, 1977, p 1313.
97. J. Schijve, in "Fatigue Crack Propagation," ASTM STP 415, 1966, p 533.
98. R. M. N. Pelloux, Trans. ASM, vol 62, 1969, p 281.
99. R. M. N. Pelloux, Eng. Frac. Mech., vol 1, 1970, p 697.
100. A. S. Kuo and H. W. Liu, Scripta Met., vol 10, 1976, p 723.
101. B. Tompkins and W. D. Biggs, J. Mat. Sci., vol 4, 1969, p 544.
102. R. J. H. Wanhill, Met. Trans., vol 6A, 1975, p 1587.
103. C. Laird and R. de la Veaux, Met. Trans., vol 8A, 1977, p 657.
104. C. Q. Bowles, "The Role of Environment, Frequency and Wave Shape During Fatigue Crack Growth in Aluminum Alloys," Delft Univ. Report LR-270, May 1978.
105. C. Laird, Ph.D. Thesis, University of Cambridge, 1963.
106. N. E. Frost, K. J. Marsh, and L. P. Pook, "Metal Fatigue," Clarendon Press, Oxford, 1974.
107. "Stress Analysis and Growth of Cracks," ASTM STP 513, 1972.
108. "Mechanics of Crack Growth," ASTM STP 590, 1976.
109. "Fatigue Crack Growth under Spectrum Loads," ASTM STP 595, 1976.
110. F. A. McClintock, in "Fatigue Crack Propagation," ASTM STP 415, 1966, p 170.
111. H. W. Liu, Trans. Am. Soc. Mech. Engrs., vol 85D, 1963, p 116.
112. H. W. Liu, "Analysis of Fatigue Crack Propagation," Syracuse University report, No. MTS-HWL-0472, April 1972.
113. H. W. Liu, C. Y. Yang, and A. S. Kuo, "Cyclic Crack Growth Analyses and Modelling of Crack Tip Deformation," Int. Symposium on Fracture Mechanics, Sept. 1978, to be published.
114. G. T. Hahn, R. E. Hoagland, and A. R. Rosenfield, AF 33615-70-C-1630, Battelle Memorial Inst., Columbus, Ohio, Aug. 1971.
115. J. M. Barsom, ASTM STP 486, 1971, p 1.
116. R. C. Bates and W. G. Clark, Trans. ASM, vol 62, 1969, p 380.
117. D. S. Dugdale, J. Mech. Phys. Solids, vol 8, 1960, p 100.
118. R. O. Ritchie, Metal Science, vol 11, 1977, p 368.
119. F. A. McClintock, in "Fracture of Solids," Interscience Publishers, New York, 1963, p 65.
120. T. Yokobori, Symposium on Fatigue Mechanisms, Kansas City, May 22-24, 1978, to be published in ASTM, STP.
121. J. Weertman, "True Stress Intensity Factor Rationalization of a 2nd and 4th Power Paris Fatigue Crack Growth Equation," Int. Sym. on Fracture Mech., Sept. 1978.
122. S. Ikeda, Y. Izumi, and M. E. Fine, Eng. Fracture Mech., vol 9, 1977, p 123.
123. N. E. Ryan in "Aircraft Structural Fatigue," Commonwealth of Australia, 1977, p 393.

124. J. Weertman, this volume.
125. S. Purushothaman and J. K. Tien, Mat. Sci. Eng., vol 34, 1978, p 241.
126. J. K. Tien and S. Purushothaman, Mat. Sci. Eng., vol 34, 1978, p 247.
127. D. A. Ryder, M. Martin, and M. Abdullah, Metal Sci., vol 11, 1977, p 340.
128. G. Clark and J. F. Knott, Ibid., p 345.
129. P. E. Irving and L. N. McCartney, Ibid., p 351.
130. C. J. Beevers, Ibid., p 362.
131. P. Lukáš and M. Klesnil, Mat. Sci. Eng., vol 34, 1978, p 61.
132. P. Lukáš and M. Klesnil, Mat. Sci. Eng., vol 11, 1973, p 345.
133. P. Lukáš, M. Klesnil, and J. Polak, "High Cycle Fatigue Life of Metals," Inst. Phys. Met., Brno, 1972.
134. M. Hempel, Arch. Eisenhuttenw., vol 38, 1967, p 446.
135. J. M. Finney, "Strain Localization in Cyclic Deformation," Ph.D. Thesis, Univ. of Pennsylvania, 1974.
136. P. J. E. Forsyth, Int. Conf. on Fatigue, Inst. of Mech. Engrs. and ASME, Paper 6.5, 1956.

A Rational Analytic Theory of Fatigue

PAUL C. PARIS
Assistant Professor of Civil Engineering

MARIO P. GOMEZ* and WILLIAM E. ANDERSON
Research Engineers, Boeing Airplane Company

A great deal of effort has recently centered around examination of the factors influencing the growth of fatigue cracks. Fatigue has been considered a multi-phase problem: e.g., initiation of a crack and its growth are often considered as separate phenomena. In contrast, the objective of this work is to show that the growth of an initial "crack-like" imperfection to a critical size, which causes static failure of a structure, may be described by a single rational theory.

Two loading parameters, the nature of the stress field near the tip of a crack and the variation of this field, are taken to control the rate of crack extension in a given material. This hypothesis is proven by using it to correlate data from three independent investigators. Since it shows a positive correlation of all available data for crack-extension rates from 10^{-7} to 10^{-2} in. per cycle, the hypothesis may be used to formulate a theory of fatigue that permits computing the structural lives of complicated geometries from simple laboratory tests of material properties.

The Stress Distribution Near the Tip of a Crack

The form of the stress distribution in the vicinity of a crack root was given by Sneddon[1] in 1946 and has recently been expanded by Irwin[2,3] and Williams.[4] The unique character of this form, as Irwin showed,[2] is a controlling factor in attempts to analyze crack extension under static loads. We will show that this same character becomes fundamental in crack extension under cyclic loading upon the addition of new concepts to describe the cyclic nature of the loading.

---

*Mr. Gomez received his M. S. degree in Metallurgical Engineering in 1958 at the University, after which he worked for Boeing. He is now Senior Scientist at the Missile Systems Division of Lockheed Aircraft Corporation.

Restricting this discussion to cracked bodies in which the geometry and loading of the body are symmetric with respect to the plane of the crack results in very little loss of generality. The nature of cracks

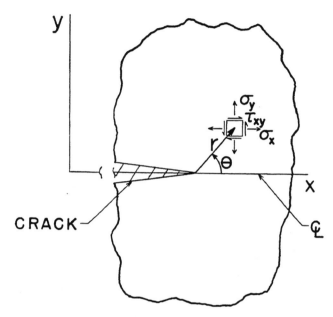

FIG. 1. COORDINATES USED TO DESCRIBE STRESSES NEAR A CRACK TIP (θ; σ_x; τ_y; v)

is to form most often on such planes, i.e., planes perpendicular to maximum-principle tension stresses. Williams[4] and Irwin[5] have given the required forms of stresses for other cases, but these will not be discussed further in this work.

The coordinates of points in a cracked body with

P. C. Paris

M. P. Gomez

W. E. Anderson

SYMBOLS

a	= the half crack length
b	= plate dimension parallel to a crack
F	= a loading force on a body
$f()$ or $F()$	= a function of
K	= the stress singularity-intensity factor at the tip of a crack
K_{cr}	= the critical value of K for a material associated with a crack extension under static load
L	= the length of a plate
N	= the number of load cycles since initial loading
P	= the loading of a body
r, θ	= polar coordinates from the crack tip
x, y	= rectangular coordinates centered with respect to a crack
α	= the correction factor for K in plates of finite width
β	= the ratio of maximum to minimum load on a body during a load cycle
μ	= Poisson's ratio
σ_g	= gross area stress or nominal stress level
$\sigma_x, \sigma_y, \tau_{xy}$	= components of stress near a crack tip
$\sigma_o(x)$	= the normal stress present at a crack location before the crack appeared
$\Delta a/\Delta N$ or da/dN	= the rate of crack extension

crack-plane symmetric geometry may be described as in Fig. 1. If terms of higher order in r are ignored, the elastic solution for stresses in the vicinity of the crack tip for all such problems is

$$\sigma_y = \frac{K}{\sqrt{2r}} \cos\frac{\theta}{2}\left[1 + \sin\frac{\theta}{2}\sin\frac{3\theta}{2}\right],$$

$$\sigma_x = \frac{K}{\sqrt{2r}} \cos\frac{\theta}{2}\left[1 - \sin\frac{\theta}{2}\sin\frac{3\theta}{2}\right],$$

$$\tau_{xy} = \frac{K}{\sqrt{2r}} \sin\frac{\theta}{2}\cos\frac{\theta}{2}\cos\frac{3\theta}{2},$$

$$\tau_{xz} = \tau_{yz} = 0,$$

and

$$\sigma_z = 0 \text{ (plane stress)}$$

or

$$\sigma_z = \mu(\sigma_x + \sigma_y) \text{ (plane strain)}$$

(1)

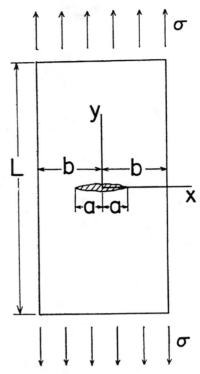

FIG. 2. A TYPICAL FATIGUE-PANEL CONFIGURATION FOR WHICH TEST DATA WERE AVAILABLE

This result implies that the distribution of elastic stress always has the same functional form near the singularity caused by the sharp crack root and that it differs only by a stress singularity-intensity factor, K. This factor is linearly dependent upon the loads on the body and also must contain a geometric factor related to the crack length and other geometric properties of the body. An example is $K = \sigma\sqrt{a}$ for the problem of a crack of length $2a$ in an infinite sheet under a uniform tensile-stress field, σ, perpendicular to the crack.

Now the Griffith-Irwin theory of static strength of bodies containing cracks may be resolved from this discussion in the following fashion: Identical intensity factors of elastic stress will result in identical yield zones near the tips of cracks in the same material if the yield zones are small compared to the region of applicability of the stresses given by Eq. (1). As discussed in previous works,[6] the size of the yield zone may be shown to be small if the nominal stresses in the body are well below the yield point. Therefore, regardless of the appearance of a small yield zone, there will be some critical value of the stress singularity intensity, K_{cr}, near the tip of a crack that will cause static crack extension in a given material. The above hypothesis is the equivalent of the Griffith-Irwin theory, which was originally based on energy considerations.

The Significance of Stress-Intensity Factors in Fatigue

The stress-intensity factor may be considered to be a measure of the effect of the loading and the geometry of a body on the stress intensity near the root of a crack. Therefore, as the loads on a body vary and as the geometry changes by crack extension, the instantaneous values of K reflect the effects of these changes at the crack root.

Let β be the ratio of maximum to minimum load on a cracked body during a cycle of loading. Then, since K is directly proportional to the magnitude of the load, β is also the ratio of K_{max} to K_{min}, regardless of the geometry of the body; that is,

$$\beta = \frac{P_{max}}{P_{min}} = \frac{K_{max}}{K_{min}}. \qquad (2)$$

Therefore the stresses near the root of a crack are completely described by K_{max} and β in a given material, since these two parameters give both the intensity and variation of the effects of loading and geometry.

A theory of fatigue crack extension may now be hypothesized as follows: Since, as has been shown, during a cycle of loading the stresses and strains near the tip of a crack are completely specified by K_{max} and β, we can reasonably assume that any phenomena occurring in this region are controlled by these parameters. The amount of crack extension per cycle of loading is just such a phenomenon, or, in functional form,

$$\frac{\Delta a}{\Delta N} = f(K_{max}, \beta). \qquad (3)$$

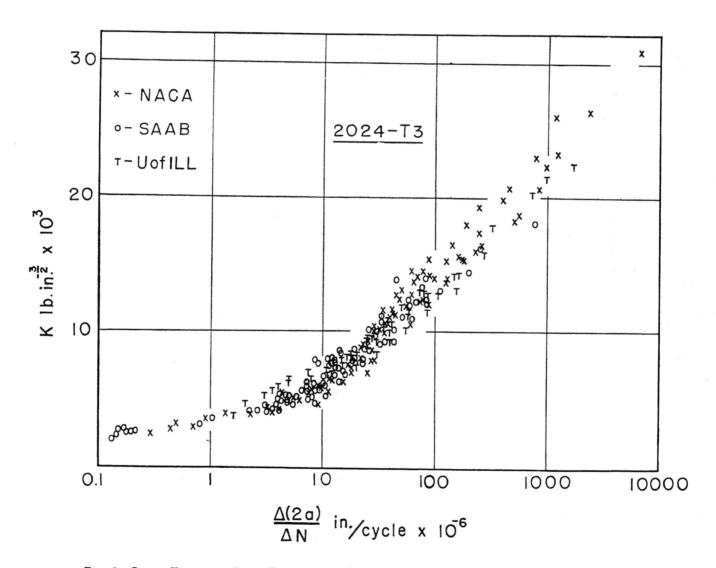

FIG. 3. CRACK EXTENSION-RATE DATA ON 2024-T3 ALUMINUM ALLOY CORRELATED FROM TESTS BY THREE INDEPENDENT INVESTIGATORS

Experimental Evidence

It is pertinent to examine the above hypothesis in the light of available experimental data. Many investigators[7,8,9,10] have measured crack-extension rates due to cyclic loading in aluminum alloys by using the configuration shown in Fig. 2. The majority of these tests have been performed with minimum loads near zero, or $\beta \equiv \infty$; therefore from examination of Eq. (3), these results for a given material should form a single curve on a plot of K_{max} vs $\Delta a/\Delta N$. For such a configuration K_{max} may be computed from[8]

$$K = \alpha \sigma_g \sqrt{a},$$

where

$$\alpha = \frac{\sqrt{4 + 2(a/b)^4}}{2 - (a/b)^2 - (a/b)^4} \qquad (4)$$

if $\qquad a < b.$

The results of this attempt at correlation of crack extension-rate data are shown on Figs. 3 and 4. It is worthy of special note that the data on these curves are from three independent investigators, using many specimen sizes, i.e., widths from 1.8 to 12 in., thicknesses from 0.032 to 0.102 in., and lengths from 5 to 35 in. The testing frequencies varied from 50 to 2000 cpm, and the maximum stresses on the gross area varied from 6 to 30 ksi. On each graph, the materials are both clad metals and bare metals. Therefore the correlation shown is surely more than coincidental.

On the presumption that such curves may be obtained for various values of β for a given material from laboratory tests, this discussion proceeds to formulate the necessary elements of an analytic theory of fatigue, these results being applied in the following sections.

An Analytic Theory of Fatigue

Knowledge of the material-property curves in the form of Figs. 3 and 4, with the addition of curves for other β values, implies the functional form of Eq. (3) as given. Further, given the loading and geometry

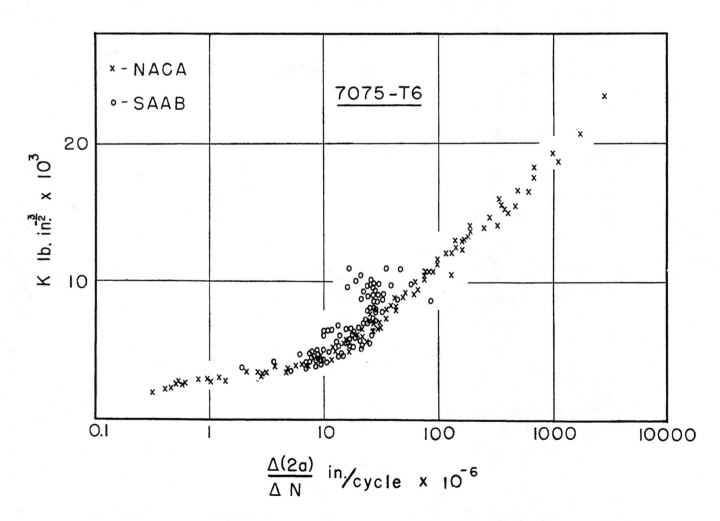

Fig. 4. Crack Extension-Rate Data Correlation for 7075-T6 Aluminum Alloy

of a structure, β is known from the load ratio during any cycle, N, or

$$\beta = \beta(N). \quad (5)$$

Moreover, K may be computed for any crack length, a, and the maximum load as given during the Nth cycle, or

$$K_{\max} = K_{\max}(N, a). \quad (6)$$

Therefore, for a problem with the above specification, Eq. (5) and Eq. (6) may be substituted into Eq. (3) to give

$$da/dN = F(N, a), \quad (7)$$

where the functional form of F is known, point by point, from the data given, and effects of loading history are neglected.

The solution to Eq. (7) may be found provided an initial crack size may be specified, or some equivalent condition may be stated. In practice, maximum imperfection sizes may be stipulated on the basis of material quality, production methods, and inspection technique, for in practice we know that fatigue cracks grow from just such imperfections. Then Eq. (7) may be integrated, at least by numerical procedures, to generate a complete crack history for the structure. The most difficult phase of this analysis is the computation of K for a given load and crack length, as will be commented upon later.

Relationships to Classical Fatigue Theory

Suppose the material-property curves of the form of Figs. 3 and 4 are known for a specimen that has been subjected to an ordinary fatigue test and has developed an easily measurable crack in a given number of cycles. By using these curves and the measured crack length, the crack-extension rates may be integrated backward to determine an effective initial imperfection size.

Using this computed initial imperfection size makes it possible to calculate the number of load cycles required for failure of the specimen at any stress level. Therefore the complete S-N curve for a material may be computed by this process; moreover, the whole S-N curve can be obtained from a single specimen, since curves of the material properties for several β values may be generated during the same test.

An accumulative damage theory is automatically present in the preceding analysis, which replaces Miner's empirical hypothesis of damage. The Soderberg or Goodman diagrams being likewise empirical, this paper presents a rational analytic theory. As is evident, the form of the analysis lends itself well to statistical analyses.

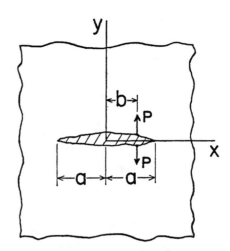

FIG. 5. A Crack in an Infinite Sheet with Concentrated Forces Applied to Crack Surface

Computation of K, the Stress-Intensity Factor

By use of data on materials, in the form of Fig. 3, the inherently nonlinear problem of structural life has been resolved to require only the computation of instantaneous values of K as the crack propagates through the structure. Since K is the elastic-stress intensity factor for stresses near a crack tip, the problem has been essentially linearized from the point of view of mechanics, a method that permits the use of well-known techniques in attacking the problem.

Irwin[3] has given the solution to the problem shown in Fig. 5, which may be applied by superposition technique (as a Green's Function) to solve all problems of sheet-skins containing internal cracks. This method, which is described in detail in Reference 6, will be outlined here.

Consider the general and arbitrary x-axis symmetric plane-stress problem containing a crack, illustrated by Fig. 6(a). This may be considered to be a free body removed from a gross structure, and may be chosen large enough in size that the crack itself has very little influence on the magnitude and distribution of the boundary forces and stresses shown. Thus we may proceed to solve the plane-stress problem for this sheet, but we find considerable difficulty in problems of doubly connected regions. As an alternative, this problem may be considered to be the sum of two problems, i.e., those in Fig. 6 (b) and (c). The first, (b), is the solution to the same problem with no crack present; the second, (c), is the solution with the same geometry as the original problem, (a), but with loads on the crack surfaces only, of equal and opposite intensity to the stresses that occur at the crack location in (b). This approach ensures that the sum of the boundary forces in (b) and (c) are identical to those in (a), and thus their sum is the solution to (a).

Now we desire only to determine the stress inten-

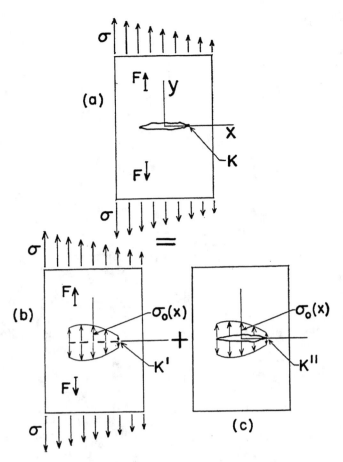

FIG. 6. THE REDUCTION OF A PROBLEM, (a), INTO TWO SIMPLER PROBLEMS, (b) and (c), FOR COMPUTATIONS OF STRESS SINGULARITY - INTENSITY FACTORS

sity factor, K, for the problem, which is the sum of the stress intensity factors, K' and K'', for problems (b) and (c). Since K is the intensity of the singularity of stresses at the crack tip—note $1/\sqrt{r}$ in Eq. (1)—and this singularity is not present, in (b), then

$$K' = 0. \quad (8)$$

The task remaining is to determine K'' for problem (c), since, observing Eq. (2),

$$K = K''. \quad (9)$$

Thus the solution to all problems of the type of (a) is reduced to finding the stress, $\sigma_o(x)$, on the crack axis with no crack present, as in (b), and using this stress as the loading in (c).

To find the value of the stress singularity, K'', for (c), Irwin's solution[3] for the stress singularity in Fig. 5 gives

$$K_P = \frac{P(a+b)^{\frac{1}{2}}}{\pi(a)^{\frac{1}{2}}(a-b)^{\frac{1}{2}}}. \quad (10)$$

Taking (10) as a Green's Function, the stress singularity, K'', is given by

$$K = K'' = \frac{1}{\pi\sqrt{a}} \int_{-a}^{a} \frac{\sigma_o(x)(a+x)^{\frac{1}{2}}}{(a-x)^{\frac{1}{2}}} dx, \quad (11)$$

which may be further simplified by y-axis symmetry,

$$K = \frac{2\sqrt{a}}{\pi} \int_{o}^{a} \frac{\sigma_o(x)}{(a^2-x^2)^{\frac{1}{2}}} dx. \quad (12)$$

Hence all the general problems of the form of Fig. 6(a) may be attacked by solving the problem for the stresses, $\sigma_o(x)$, along the crack axis with the crack absent, and integrating the result by Eq. (11) or Eq. 12.

Equally powerful techniques may be presented for other general classes of problems. Therefore we conclude that, in general, K may be computed easily in most of the problems of interest for engineering purposes.

Conclusion

On the basis of the experimental data given, it is evident that rates of crack growth—for example, those in 2024-T3 and 7075-T6 skins of aircraft structure—may be computed by the theory presented over a wide range of nominal stress levels and crack sizes. The ramifications of such broad correlation imply an analytic theory of fatigue based on a concept of growth from initial imperfections through which structural life may be predicted.

REFERENCES

1. I. N. SNEDDON, "The Stress Distribution in the Neighborhood of a Crack in an Elastic Solid," *Proc. Royal Soc. of London*, Vol. A-187, 1946, pp. 229-260.
2. G. R. IRWIN, "Relation of Stresses Near a Crack to the Crack Extension Force," N.R.L., 1956. (Also *Proc. IXth International Congress of Applied Mech.*, Paper No. 101).
3. G. R. IRWIN, "Analysis of Stresses and Strains Near the End of a Crack Traversing a Plate," *Jour. of Appl. Mech.*, June, 1957.
4. M. L. WILLIAMS, "On the Stress Distribution at the Base of a Stationary Crack," *Jour. of Appl. Mech.*, March, 1957, Vol. 24, No. 1, p. 109.
5. G. R. IRWIN, "Fracture Mechanics," *Proc. Symposium on Naval Structural Mechanics*, Stanford University, Aug. 11, 1958. (See also "Fracture," *Encyclopedia of Physics*, Vol. VI, Springer, 1958.)
6. P. C. PARIS, "The Mechanics of Fracture Propagation and Solutions to Fracture Arrestor Problems," Report D2-2195, The Boeing Airplane Company, Sept., 1957.
7. D. E. MARTIN and G. M. SINCLAIR, "Crack Propagation Under Repeated Loading," *Proc. Third U. S. Cong. Appl. Mech.*, June, 1958, pp. 595-604.
8. A. J. McEVILY, JR., and WALTER ILLG, "The Rate of Fatigue-Crack Propagation in Two Aluminum Alloys," NACA TN 4394, Sept., 1958.
9. WALODDI WEIBULL, "The Propagation of Fatigue Cracks in Light Alloy Plates," SAAB TN 25, 1954.
10. WALODDI WEIBULL, "Effect of Crack Length and Stress Amplitude on Growth of Fatigue Cracks," Mem. 65, Aero. Res. Institute of Sweden, 1956.

Reprinted from *Journal of Basic Engineering*, Vol. 85, pp. 528-534 (December 1963).

A Critical Analysis of Crack Propagation Laws

P. PARIS[1]
Associate Professor of Mechanics and Assistant Director of the Institute of Research, Lehigh University, Bethlehem, Pa.

F. ERDOGAN
Professor of Mechanical Engineering, Lehigh University, Bethlehem, Pa. Mem. ASME

The practice of attempting validation of crack-propagation laws (i.e., the laws of Head, Frost and Dugdale, McEvily and Illg, Liu, and Paris) with a small amount of data, such as a few single specimen test results, is questioned. It is shown that all the laws, though they are mutually contradictory, can be in agreement with the same small sample of data. It is suggested that agreement with a wide selection of data from many specimens and over many orders of magnitudes of crack-extension rates may be necessary to validate crack-propagation laws. For such a wide comparison of data a new simple empirical law is given which fits the broad trend of the data.

Introduction

SEVERAL crack-propagation laws have been presented in the past few years, which claim to be verified by the experimental data analyzed in their respective papers. They are specifically the work of Head [1],[2] Frost and Dugdale [2], McEvily and Illg [3], Liu [4, 5], and Paris, Gomez, and Anderson [6, 7, 8].

This paper will attempt to show that basing validation of a crack-propagation theory on a limited amount of data is a poor test of any theory. Moreover, a wide range of test data is now available in a convenient form [9] which may be employed to analyze critically these several crack propagation laws.

Therefore the paper will consist of three parts which will include:

1 A review and comparison of existing crack-propagation laws.
2 The erroneous results obtained by comparing each law with a limited range of test data.
3 The results of comparing crack-propagation laws with a wide range of data.

A Review of Existing Crack-Propagation Laws

Crack-propagation laws given in the literature take many forms. In general they treat cracks in infinite sheets subjected to a uniform stress perpendicular to the crack (or can be applied to that configuration) and they relate the crack length, $2a$, to the number of cycles of load applied, N, with the stress range σ, and material constants C_i.[3] The single form in which all crack-propagation laws may be written is

$$\frac{da}{dN} = f(\sigma, a, C_i) \quad (1)$$

Therefore this format will be employed in the discussion to follow.

Chronologically the first crack-propagation law which drew wide attention was that of Head [1] in 1953. He employed a mechanical model which considered rigid-plastic work-hardening elements ahead of a crack tip and elastic elements over the remainder of the infinite sheet. The model required extensive calculations and deductions to obtain a law which may be written as

$$\frac{da}{dN} = \frac{C_1 \sigma^3 a^{3/2}}{(C_2 - \sigma)\omega_0^{1/2}} \quad \text{(Head's law)} \quad (2)$$

where C_1 depends upon the strain-hardening modulus, the modulus of elasticity, the yield stress and the fracture stress of the material, and C_2 is the yield strength of the material. Head defined ω_0 as the size of the plastic zone near the crack tip and presumed it was constant during crack propagation. However, Frost [2] noticed that the plastic-zone size increased in direct proportion to the crack length in his tests. Moreover, Irwin [10] has recently pointed out from analytical considerations that

$$\omega_0 \approx \sigma^2 a \quad (3)$$

for the configuration treated here which is in agreement with Frost's conclusions. Therefore, though Head adopted equation (2) with ω_0 considered constant as his crack-propagation law, we are forced here to introduce equation (3) into equation (2) to obtain a modified or corrected form of Head's crack-propagation law; i.e.,

$$\frac{da}{dN} = \frac{C_3 \sigma^2 a}{(C_2 - \sigma)} \quad \text{(Head's corrected law)} \quad (4)$$

Frost and Dugdale [2] in 1958 presented a new approach to crack-propagation laws. They observed as was introduced in equation (4), that Head's law should be corrected for the variation of plastic-zone size with crack length. They deduced that the corrected result, equation (4), depends linearly on the crack length a. However, they also argued by dimensional analysis

[1] Also Consultant to the Boeing Company, Transport Division, Renton, Wash.
[2] Numbers in brackets designate References at end of paper.
[3] In this discussion σ will imply the stress-fluctuation range and C_i will vary slightly with mean stress in a general interpretation.

Contributed by the Metals Engineering Division and presented at the Winter Annual Meeting, New York, N. Y., November 25-30, 1962, of THE AMERICAN SOCIETY OF MECHANICAL ENGINEERS. Manuscript received at ASME Headquarters, April 11, 1962. Paper No. 62-WA-234.

——Nomenclature——

- a = half crack length
- a_0 = initial half crack length
- B = a function of stress range (and mean stress)
- C = a numerical constant
- C_i = constants (which vary slightly with mean stress)
- D_i = constants which depend on stress level
- $f(\ )$ = a function of
- $F\{\ \}$ = a function of
- $G\{\ \}$ = a function of
- k = stress-intensity factor (range)
- K_N = stress-concentration factor
- m = a numerical exponent
- M = a material constant
- n = a numerical exponent
- N = cycle number
- N_0, N_F = initial and final cycle number, respectively
- $\frac{da}{dN}$ = crack extension per cycle of load
- Δk = stress-intensity-factor range
- ρ, ρ_1 = crack-tip radius
- σ = applied stress (range)
- σ_0 = stress at a crack tip (with finite radius ρ)
- σ_{net} = net section stress
- ω_0 = plastic-zone size

that the incremental increase in crack length da, for an incremental number of cycles dN, should be directly proportional to the crack length a. Hence they concluded that (independent of Head's model)

$$\frac{da}{dN} = Ba \qquad (5)$$

where B is a function of the applied stresses. Then they observed that in order to fit their experimental data:

$$B = \frac{\sigma^3}{C_4} \qquad (6)$$

Combining equations (5) and (6) they obtained the law:

$$\frac{da}{dN} = \frac{\sigma^3 a}{C_4} \quad \text{(Frost and Dugdale's law)} \qquad (7)$$

About the same time McEvily and Illg [3] modified a method of analysis of static strength of plates with cracks used at NASA to obtain a theory of crack propagation. Their arguments were as follows: Presuming that a crack tip in a material has a characteristic (fictitious) radius ρ_1, which allows computation of the stress σ_0, in the element at the crack tip using elastic stress-concentration factor concepts, the stress σ_0 is

$$\sigma_0 = K_N \sigma_{\text{net}} \qquad (8)$$

where K_N is the stress-concentration factor and σ_{net} is the net area stress at the cracked section. For the configuration used here, i.e., an infinite plate with uniform stress σ

$$K_N = 1 + 2(a/\rho_1)^{1/2} \qquad (9)$$

which is based on the elastic solution for an elliptical hole of semimajor axis a and end radius ρ_1.

$$\sigma_{\text{net}} = \sigma \qquad (10)$$

and substituting equations (9) and (10) into (8) gives

$$\sigma_0 = \sigma[1 + 2(a/\rho_1)^{1/2}] \qquad (11)$$

Based on considerations that under cyclic loading work-hardening at the crack tip will raise the local stress to a fracture stress, they concluded that the crack-extension rate will be a function of σ_0 or

$$\frac{da}{dN} = F\{\sigma_0\} = F\{K_N \sigma_{\text{net}}\} \qquad (12)$$

(McEvily and Illg's law)

Therefore for the special configuration of interest here, i.e., introducing equations (8), (9), (10), and (11) into (12), we have

$$\frac{da}{dN} = F\left\{\sigma\left[1 + 2\left(\frac{a}{\rho_1}\right)^{1/2}\right]\right\} \qquad (13)$$

which is the desired form in this discussion upon considering ρ_1 to be a material constant, in likeness to the C_i.

McEvily and Illg go on in an empirical manner to obtain the form of the function $F\{\ \}$, and suggest

$$\log_{10}\left(\frac{da}{dN}\right) = 0.00509 K_N \sigma_{\text{net}} - 5.472 - \frac{34}{K_N \sigma_{\text{net}} - 34} \qquad (14)$$

(McEvily and Illg's law empirically extended)

It should be noted by the reader that McEvily and Illg's laws, equations (12) and (14), are not restricted to the special configuration here, as was the case for the laws of Head, and Frost and Dugdale. The applicability to other configurations and an additional similarity to Paris' work [6, 7, 8] will warrant later comments.

Independent of McEvily and Illg, Paris [11] proposed a crack-propagation theory at about the same time. It is based on the following arguments: Irwin's [12] stress-intensity-factor k reflects the effect of external load and configuration on the intensity of the whole stress field around a crack tip. Moreover, for various configurations the crack-tip stress field always has the same form (i.e., distribution). Therefore it was reasoned that the intensity of the crack-tip stress field as represented by k should control the rate of crack extension.[4] That is to say:

$$\frac{da}{dN} = G\{k\} \qquad (15)$$

In treating the special configuration of interest in this discussion, it should first be observed that (11)

$$k = \sigma a^{1/2}, \qquad (16)$$

whereupon equation (15) may be specialized to read:

$$\frac{da}{dN} = G\{\sigma a^{1/2}\}. \qquad (17)$$

Somewhat later, Liu [4] restated Frost and Dugdale's [2] dimensional analysis in a much more elegant form and argued that the crack-growth rate should depend linearly on the crack length; i.e.,

$$\frac{da}{dN} = Ba \quad \text{(Liu's law)} \qquad (18)$$

which is the same result as equation (5). Liu then presumed that B was in general a function of stress range σ (and mean stress); i.e.,

$$B = B(\sigma) \qquad (19)$$

In a subsequent work, Liu [5] notes that mean stress is of secondary influence and, using a model of crack extension employing an idealized elastic-plastic stress-strain diagram and a concept of total hysteresis energy absorption to failure, reasons that

$$B(\sigma) = C_5 \sigma^2 \qquad (20)$$

which combined with equation (18) gives

$$\frac{da}{dN} = C_5 \sigma^2 a \quad \text{(Liu's modified law)} \qquad (21)$$

The Equivalence of $K_N \sigma_{\text{net}}$ to k

Hardrath [14] observed that $K_N \sigma_{\text{net}}$ for the special configuration employed here, from equations (8) and (10),

$$K_N \sigma_{\text{net}} = \sigma[1 + 2(a/\rho_1)^{1/2}] \qquad (22)$$

is similar to the stress-intensity factor, equation (16),

$$k = \sigma a^{1/2} \qquad (23)$$

if ρ_1 is small compared to a. For aluminum alloys (2024T3 and 7075T6) his colleagues McEvily and Illg [3] had already observed that ρ_1 is less than 0.005 in. so that the condition $\rho_1 \ll a$ is in fact present for cracks of a readily observable length in crack-propagation tests. Thus $K_N \sigma_{\text{net}}$ and k are known to concur for this special configuration and in addition a proof that they are equivalent in general will be offered here.

Irwin [12] observed that the equations for the stress field surrounding the tip of a sharp crack contains the factor $k/(2r)^{1/2}$ which implies a singularity of stress at the crack tip. Now, if a hole of radius ρ is drilled at the crack tip, the maximum stress on the periphery of the hole will be proportional to $k/(2\rho)^{1/2}$ or

$$\sigma_0 = Ck/(2\rho)^{1/2} \qquad (24)$$

which applies for any ρ which is small compared to other planar

---

[4] At about this same time Martin and Sinclair [13] attempted unsuccessfully to correlate crack-extension rates using a similar parameter but did not observe a correlation which does in fact occur.

dimensions of any configuration considered, and C is a constant independent of the configuration. Finally, σ_0 in equation (24) may be interpreted to be the same as $K_N \sigma_{\text{net}}$, from equation (8) and its accompanying discussion, if ρ is taken equal to ρ_1. Then

$$K_N \sigma_{\text{net}} = \sigma_0 \frac{Ck}{(2\rho_1)^{1/2}} \qquad (25)$$

In order to evaluate C, the results for the special configuration considered herein may be employed, i.e., substituting equations (22) and (23) into (25) and noting equivalence for small ρ_1 gives

$$C = \lim_{\rho_1 \to 0} \frac{K_N \sigma_{\text{net}} (2\rho_1)^{1/2}}{k}$$

$$= \lim_{\rho_1 \to 0} \frac{\sigma \left[1 + 2\left(\frac{a}{\rho_1}\right)^{1/2}\right](2\rho_1)^{1/2}}{a} = 2\sqrt{2} \qquad (26)$$

Putting this result into equation (25) and rearranging,

$$k = \lim_{\rho_1 \to 0} \frac{K_N \sigma_{\text{net}} \rho_1^{1/2}}{2} \qquad (27)$$

which implies the general equivalence of $K_N \sigma_{\text{net}}$ and k.

Similarities Between Crack-Propagation Laws

The result of the foregoing discussion, equation (27), implies the direct similarity of McEvily and Illg's law, equation (12), and Paris' result, equation (15). A choice between the two is strictly dependent on a matter of convenience and clarity of accompanying concepts in employing one or the other.[5]

The laws of Head, equation (4), Frost and Dugdale, equation (7), and Liu, equations (18) and (21), can all be approximated by the form

$$\frac{da}{dN} = \frac{\sigma^n a^m}{C_0} \qquad (28)$$

for the special configuration which is treated. Now, it is evident that Paris' result for this configuration, equation (17), implies:

$$m = \frac{n}{2} \qquad (29)$$

which can also be derived from McEvily and Illg's result, equation (13), for ρ_1 small compared to a.

The laws of Head, and Frost and Dugdale do not quite agree with applying equation (29) to equation (28), however, Liu's law does concur with the specified form.

It is pertinent to now show that determining m and n from a limited quantity of data is a doubtful practice. That is to say that plotting data from single test specimens on a logarithmic or semilogarithmic graph on which laws such as Head's, Frost's, and Liu's predict straight line relationships is not a reasonable test of the validity of a crack propagation law.

Erroneous Evaluation of Data From Single Tests

A typical crack-propagation test consists of a wide plate with a central crack of length $2a$, subjected to uniform tension σ, repeatedly applied. During a single test then σ is a constant and data consisting of crack lengths and corresponding cycle numbers are obtained. Let the problem of examining each of the previous crack-propagation laws for a particular test be formulated.

In a constant maximum repeated stress-level test Head's law, equation (2), is reduced to

$$\frac{da}{dN} = D_1 a^{3/2} \qquad (30)$$

---
[5] For very small cracks there is a difference between the two theories which is left unresolved.

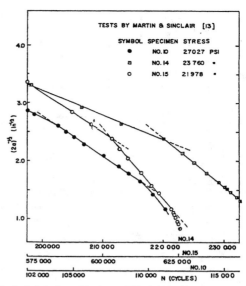

Fig. 1 Typical data from Martin and Sinclair on Head's suggested plot

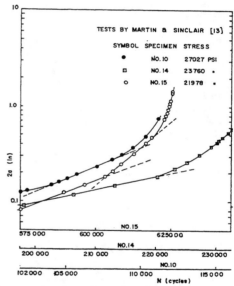

Fig. 2 The same data from Martin and Sinclair on Frost's or Liu's suggested plot

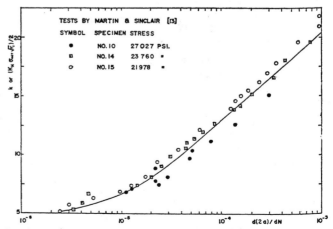

Fig. 3 Again the same data from Martin and Sinclair on McEvily's or Paris' suggested plot

where D_1 is a constant for a given stress range σ. Integrating this expression gives

$$-\frac{2}{a^{1/2}} = D_1 N + \text{const} \qquad (31)$$

Therefore Head suggested plotting $1/a^{1/2}$ versus N and, observing equation (31), implied that obtaining a straight line for each specimen whose data are plotted this way indicates verification of his law.

The data of Martin and Sinclair [13] are employed here as an unbiased source. From their data specimens nos. 10, 14, and 15 were chosen since those specimens were run at medium stress levels and the greatest number of data points per specimen was recorded for them. Fig. 1 shows the type of plot suggested by Head and it is noticed that portions of the data do form straight lines. Does this mean that Head's theory is verified? The reader is warned that this might be a hasty conclusion.

First, consider the corrected law of Head, Frost and Dugdale's law, and Liu's law, equations (4), (7), (18), and (19), respectively. For a constant stress-range test all these laws reduce to the form

$$\frac{da}{dN} = D_2 a \qquad (32)$$

Integrating this result gives:

$$\log a = D_2 N + \text{const} \qquad (33)$$

Therefore plotting $\log a$ versus N and obtaining straight lines has been accepted as verifying these theories. Fig. 2 is this type of plot employing the very same test data point by point of Martin and Sinclair shown in Fig. 1. Again, the test data have straight-line portions. Therefore it now seems to verify both Head's law equation (30) and the other laws with the form of equation (32). But the theories represented by equations (30) and (32) are not the same!

Now let the laws of McEvily and Illg, and Paris, equations (12) and (15), be examined in the light of these same data. Recalling equation (27) these two laws are equivalently expressed by

$$\frac{da}{dN} = G\{k\} \qquad (34)$$

for a given material. To test these theories, data from several specimens should be plotted on a k versus da/dN (or $\log da/dN$) graph and these laws predict that the data of several specimens will form a single curve. Again the same data of Martin and Sinclair are plotted in Fig. 3. Notice that the points which do not fall on the straight-line portions of Figs. 1 and 2 seem to be perfectly acceptable here. Since the data were differentiated to obtain da/dN, an additional amount of scatter is introduced. Therefore the data also imply verification of these two laws as well.

Hence, all of the laws agree with the data and, since these laws are not identical, the method of verification of crack-propagation theories from a limited amount of test data is evidently in error. An alternative approach employing a wider range of test data must be employed.

Comparison of Crack-Propagation Laws With a Wide Range of Test Data

In previous work [6, 8] it was observed that data from several sources [3, 9, 13] may be plotted in the form of Fig. 3 (semilogarithmic) to obtain a single curve on a range of stress-intensity factor, Δk (corresponding to stress range) versus $\log(da/dN)$ diagram, where da/dN covers as many as 6 log cycles. However, replotting these data on a $\log \Delta k$ versus $\log(da/dN)$ graph reveals some pertinent results. The data are replotted in Fig. 4 and the three specimens from Martin and Sinclair's work used in the foregoing for the purpose of illustration are shown to indicate their concurrence with the general trend.

The authors are hesitant but cannot resist the temptation to draw the straight line of slope $1/4$ through the data in Fig. 4. The equation of this line is observed to be

$$\frac{da}{dN} = \frac{(\Delta k)^4}{M} \qquad (35)$$

Equation (35) fits the data almost as well as McEvily and Illg's extended law, equation (14), and is considerably simpler in form.

The advantage of this form is clarified by considering its application to the configuration employed earlier, in which case it becomes

$$\frac{da}{dN} = \frac{\sigma^4 a^2}{M} \qquad (36)$$

The laws of Head, Frost and Dugdale, and Liu depend on a in a manner other than a^2. Clearly, their laws are at variance with the data trend in Fig. 4; i.e., Head predicts the slope of $1/3$ and Frost and Dugdale as well as Liu would predict $1/2$ as indicated in the figure.

Moreover, if one might be so bold as to attempt to integrate equation (36) it becomes

$$\frac{1}{a_0} - \frac{1}{a} = \frac{\sigma^4}{M}(N - N_0) \qquad (37)$$

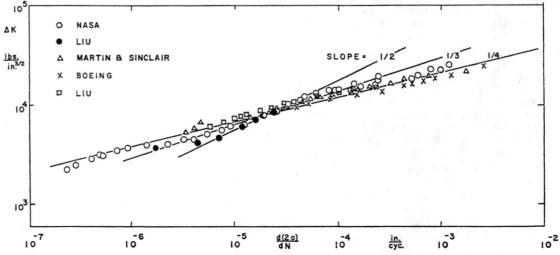

Fig. 4 Broad trend of crack-growth data on 2024-T3 aluminum alloy

where a_0 and N_0 are the initial crack size and cycle number, respectively. For final failure of a specimen, the number of cycles may be equated to N_F, as a approaches infinity whereupon equation (37) reduces to

$$N_F = \frac{M}{\sigma^4 a_0} \qquad (38)$$

which does in fact resemble an *S-N* diagram. The authors do not wish to imply that equation (38) is directly useful, but derive it here to show that such considerations do not immediately lead to the conclusion that equations (35) or (36) are in error.

Data From Wedge-Force Tests

The results of wedge-force tests of the configuration shown in Fig. 5 are useful in critically examining the dimensional analyses of Frost and Dugdale, and Liu in deriving their crack-propagation laws. If the sheet is infinite, the only characteristic dimension of the problem is a. As the crack grows, strictly dimensional arguments imply that for different crack lengths a_1 and a_2 the incremental rates of crack extension are

$$\frac{da_1}{dN} \cdot \frac{1}{a_1} = \frac{da_2}{dN} \cdot \frac{1}{a_2} = B \qquad (39)$$

which might be thought valid for all crack lengths or

$$\frac{da}{dN} = Ba \qquad (40)$$

This result is the basis of equations (5) and (18).

In wedge-force tests Donaldson and Anderson [9] observe that the crack grows slower as it gets larger. This is contrary to equation (40) and hence weakens any arguments based on dimensional analyses which do not include further considerations. Therefore the crack propagation laws of Head, Frost and Dugdale, and Liu must be considered as unclarified, if not totally in error.

Moreover, Fig. 6 is a graph, similar in form to Fig. 4, but for a different material; i.e., 7075T6 aluminum alloy. The data shown are from two independent sources [3 and 9] and in addition wedge-force test data from tests employing configuration of Fig. 5 are also plotted. No further comment seems necessary.

Conclusion

The results here indicate that the practice of using data from single test specimens is not a sensitive evaluation of a crack-propagation law's validity. Randomly chosen data from single specimens analyzed here leads to an apparent agreement of several contradictory laws to the same test data.

For that reason the authors suggest that laws which correlate a wide range of test data from many specimens are perhaps the "correct" laws. The results at least indicate that hasty conclusions have been drawn in many earlier works which should be reexamined before any given crack-propagation law is accepted as valid.

Acknowledgment

This work was supported by the Boeing Company, Transport Division, Renton, Wash. Their encouragement, as well as financial aid, is gratefully acknowledged.

References

1 A. K. Head, "The Growth of Fatigue Cracks," *The Philosophical Magazine*, vol. 44, series 7, 1953, p. 925.

2 N. E. Frost and D. S. Dugdale, "The Propagation of Fatigue Cracks in Sheet Specimens," *Journal of the Mechanics and Physics of Solids*, vol. 6, no. 2, 1958, p. 92.

3 A. J. McEvily and W. Illg, "The Rate of Crack Propagation in Two Aluminum Alloys," NACA Technical Note 4394, September, 1958.

4 H. W. Liu, "Crack Propagation in Thin Metal Sheets Under Repeated Loading," THE JOURNAL OF BASIC ENGINEERING, TRANS. ASME, Series D, vol. 83, 1961, p. 23.

5 H. W. Liu, "Fatigue Crack Propagation and Applied Stress Range—An Energy Approach," JOURNAL OF BASIC ENGINEERING, TRANS. ASME, Series D, vol. 85, 1963, p. 116.

6 P. C. Paris, M. P. Gomez, and W. E. Anderson, "A Rational Analytic Theory of Fatigue," *The Trend in Engineering*, vol. 13, no. 1, January, 1961, p. 9.

7 W. E. Anderson and P. C. Paris, "Evaluation of Aircraft Material by Fracture," *Metals Engineering Quarterly*, vol. 1, no. 2, May, 1961, p. 33.

8 P. C. Paris, "Crack Propagation Caused by Fluctuating Loads," ASME Paper No. 62—Met-3.

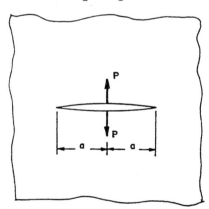

Fig. 5 Wedge-force test configuration

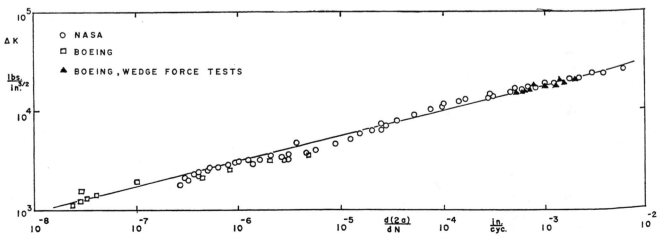

Fig. 6 Broad trend of crack-growth data on 7075-T6 aluminum alloy including wedge-force tests

9 D. R. Donaldson and W. E. Anderson, "Crack Propagation Behavior of Some Airframe Materials," presented at the Crack Propagation Symposium, Cranfield, England, September, 1961; proceedings to be published. Also available on request to The Boeing Company, Transport Division, Document No. D6-7888.

10 G. R. Irwin, "Fracture Mode Transition for a Crack Traversing a Plate," JOURNAL OF BASIC ENGINEERING, TRANS. ASME, Series D, vol. 82, no. 2, June, 1960, p. 417.

11 P. C. Paris, "A Note on the Variables Effecting the Rate of Crack Growth Due to Cyclic Loading," The Boeing Company, Document No. D-17867, Addendum N, September 12, 1957.

12 G. R. Irwin, "Analysis of Stresses and Strains Near the End of a Crack Traversing a Plate," *Journal of Applied Mechanics*, vol. 24, TRANS. ASME, vol. 79, 1957, p. 361.

13 D. E. Martin and G. M. Sinclair, "Crack Propagation Under Repeated Loading," Proceedings of the Third U.S. National Congress of Applied Mechanics, June, 1958, p. 595.

14 H. F. Hardrath and A. J. McEvily, "Engineering Aspects of Fatigue Crack Propagation," presented at the Crack Propagation Symposium, Cranfield, England, September, 1961.

DISCUSSION

W. E. Anderson[6]

We would like to thank the authors for reminding us of the dangers involved when attempting to generalize from limited amounts of data. We think that most experimenters really recognize this point while at the same time they are dangling their legs over the pit, so to speak. Speculative theories about "laws" of behavior are frequently published in several American technologies; and this may be more useful than damaging.

In this light we wish to draw attention to the data in Figs. 4 and 6. Another interpretation of these results is that the slope trend is $1/5$, composed of two segments joined by a transition slope of about $1/2$. The midtransition occurs at growth rates of about 10^{-5} in. per cycle but the 2024-T3 transition occurs at somewhat lower rates than the transition in 7075-T6.

We shall now speculate that there is a general cracking mechanism, causing the 1:5 slope, supervened, in one region, by another mechanism having superior capacity for cyclic energy dissipation, causing the transition.

It is actually observed that the intermetallic constituent optically visible in these alloys trend downward in size from dimensions on the order of 5×10^{-5} in.; the sizes in alloy 2024-T3 consistently smaller than in alloy 7075-T6.

We can superimpose the two jogged curves we have drawn in Figs. 4 and 6 and note that the 2024 curve is regularly parallel to but shifted to lower rates than the 7075 curve. We recollect that (a) the solution treatment temperatures for 2024 are consistently higher than for 7075; and (b) there has occassionally appeared, in the literature, suggestions that something equivalent to local annealing occurs at the tip of fatigue cracks in aluminum alloys. The discusser is hesitant, but cannot resist the temptation to continue. The summary thesis being made (with considerable reluctance) that:

1 Somehow, one or another, or perhaps all the intermetallic constituents in commercial alloys 2024 and 7075 act as to inhibit crack growth over regions where the growth rate per cycle is of order of the intermetallic particle sizes.

2 The general trend (1:5 slope) of crack growth due to fluctuating loads in these alloys is perhaps due to the overbearing characteristics of mechanical stress (or strain) state experienced at crack tips and as measured by the intensity factor, but the growth rate differences are directly related to the annealing resistance of the alloys; the more the annealing resistance the slower the growth rate.

The need for much additional and relevant data is at once clear. It must be wide ranging and from tests where loadings are well controlled, as the authors have so clearly demonstrated.

We shall conclude, however, by suggesting that the authors pursue their quest for power laws no further. The course of events has proceeded from Head's linear power law to, now, a fifth power law. Further efforts will surely lead only to flatter slopes; and this will mean no relation at all!

A. J. McEvily, Jr.[7]

It may be of interest to point out that a recent evaluation [15][8] of fatigue crack propagation theories has led to conclusions which are quite similar to those which have been reached by Paris and Erdogan. Power laws of the type

$$\frac{da}{dN} = Ba^n \quad (41)$$

were assessed in the following manner.

Consider two specimens in which fatigue cracks are growing at equal rates; then

$$\frac{da_1}{dN} = \frac{da_2}{dN} \quad (42)$$

and

$$B_1 a_1^n = B_2 a_2^n, \quad (43)$$

where the subscripts refer to the two specimens. It has been established [16] that the rate is uniquely defined by the parameter $K_N S_{NET}$, and it has been further shown [17] that even for specimens of finite width

$$K_N S_{NET} \approx \sigma_g \sqrt{a}. \quad (44)$$

Here σ_g refers to the average stress based on the gross section area. Since the rates in the two specimens under consideration are equal, it follows that

$$\sigma_{g1}\sqrt{a_1} = \sigma_{g2}\sqrt{a_2}. \quad (45)$$

Equation (44) can be written as

$$A\sigma_{g1}^{2n}a_1^n = A\sigma_{g2}^{2n}a_2^n \quad (46)$$

where A is a material constant. Equating the coefficients of the a^n terms in (43) and (46) we find that

$$B = A\sigma_g^{2n}, \quad (47)$$

which leads to

$$\frac{da}{dN} = A\sigma_g^{2n}a^n \quad (48)$$

which can be rewritten as

$$\log \frac{da}{dN} = \log A + 2n \log \sigma_g \sqrt{a}. \quad (49)$$

Equation (49) was compared with experimental results, and it was found that a value for n of 2 best suited the results for high strength aluminum alloys, in agreement with the authors' finding. On the other hand, the best value for n for a series of copper alloys was closer to 3. However, the net section stresses for some of these alloys were in the plastic range, which, as will now be shown, results in values for n greater than 2.

When one reflects upon the factors responsible for the rate of crack growth in a given material it appears two factors are of primary importance. One of these is the peak stress at the tip of the crack, and the other is the condition of the material immediately ahead of the advancing crack. The way in which both of these factors are coupled is through the strain energy stored in the portion of the plastic zone at the crack tip through which the crack must grow. In a strain-hardening material, we can approximate the energy per unit volume as being proportional to the square of the peak stress. The volume of affected material can

---

[6] Research Specialist, Boeing Co., Renton, Wash.

[7] Scientific Laboratory, Ford Motor Co., Dearborn, Mich.

[8] Numbers in brackets designate Additional References at end of this discussion.

be related to the radius of the plastic zone at the crack tip. However, the entire zone need not be considered, but only the rectangular slice immediately ahead of the advancing crack. This region is given by eR, where e is a constant and R is the radius of the zone, which according to Irwin [18] and McClintock [19] can be expressed as

$$R \approx \sigma_o^2 a. \qquad (50)$$

Upon making the reasonable assumption that the rate of crack growth is proportional to the energy stored in the rectangular slice at the crack tip, one obtains

$$\frac{da}{dN} \approx (\sigma_o^2 a)(e\sigma_o^2 a), \qquad (51)$$

which is the relationship that has been found to be in agreement with the data for the high strength aluminum alloys. However, as the applied stress approaches the yield strength the size of the plastic zone increases more rapidly than predicted by equation (50) [19], so that an even stronger dependence of the rate on the stress and crack length results. This last situation applies to the copper alloys investigated [15] for which a value for n of 3 was found to be appropriate.

With respect to the authors' paper, it is not clear how their treatment deals with sheets of finite width and with the effect of a variation in mean stress. Finally, it is pointed out that at low values of the applied stress the simple form of any power law would have to be modified since a minimum stress is required before a crack will propagate.

Additional References

15 A. J. McEvily, Jr., and R. C. Boettner, "On Fatigue Crack Propagation in FCC Metals," International Conference on Mechanisms of Fatigue, Orlando, Fla., November, 1962.

16 A. J. McEvily, Jr., and W. Illg, "The Rate of Fatigue Crack Propagation in Two Aluminum Alloys," NACA TN 4394, September, 1958.

17 H. F. Hardrath and A. J. McEvily, "Engineering Aspects of Fatigue Crack Propagation," Crack Propagation Symposium, Cranfield, England, September, 1961.

18 G. R. Irwin, "Fracture Mode Transition for a Crack Transversing a Plate," JOURNAL OF BASIC ENGINEERING, TRANS. ASME, Series D, vol. 82, June, 1960, p. 417.

19 F. A. McClintock, JOURNAL OF BASIC ENGINEERING, TRANS. ASME, Series D, vol. 82, June, 1960, p. 423.

Authors' Closure

The points made by Mr. Anderson and Dr. McEvily in the discussion are most appropriate. They contribute in a constructive way toward emphasizing the relative importance of the views taken in the paper, as well as adding some new ideas.

Mr. Anderson's observation and plausible explanation of the "Jogg" in the data in both Figs. 4 and 6 provides some motivation for further study of the Micro-Mechanisms involved in crack growth. Though such studies were not necessary to arrive at the conclusions drawn in the paper, they would add immeasurably to a better understanding of many detailed phenomena of crack growth. Nevertheless, empirically, the broad trends shown in the Figs. 4 and 6 remain valid.

Dr. McEvily's alternate means of arrival at an acceptable power law is most enlightening. Moreover, it has come to the authors' attention that laws similar to equations (36) and (48) have been expounded by McClintock [20] and Schijve [21] as the results of still different approaches. In all of these, the continuum parameter, k or $K_n \sigma_{net}$, was found to be appropriate to employ.

In this connection one may also mention the conclusion reached by the staff of Battelle Memorial Institute in their analysis of the crack propagation data from various sources [22]. They observed that, under constant load level, the crack growth rate was approximately proportional to the crack length.

Further, McEvily's observation that the exponent of the power law apparently changes when the net section stress exceeds the yield point is a result which is expected. Since the parameter k or $K_n \sigma_{net}$, is based on elastic stress analysis, it is evident that its role must change when the behavior of a specimen changes from predominantly elastic to predominantly plastic. Thus, as McEvily notes, an assumption of the power law expressed in the paper is that nominal stresses shall remain below the yield point. The full assessment of this restriction and the influence of the mean stress requires further studies.

Additional References

20 F. A. McClintock, "On the Plasticity of Growth of Fatigue Cracks," presented at the International Conference on Fracture, Maple Valley, Washington, August, 1962.

21 J. Schijve, "Fatigue Crack Propagation in Light Alloy Sheet Material and Structures," Advances in Aeronautical Sciences, vols. 3–4, Pergamon Press, 1961, pp. 387–407. See also, National Aeronautical Research Institute, Amsterdam, Reports NLR-TN M2092 and NLR-TN M2094, 1961, by J. Schijve, D. Broek, and P. de Rijk.

22 "Prevention of Failure of Metals Under Repeated Stress," a Handbook prepared at the Battelle Memorial Institute, John Wiley & Sons, Inc., New York, N. Y., 1949, p. 228.

Numerical Analysis of Crack Propagation in Cyclic-Loaded Structures[1]

R. G. FORMAN
Air Force Flight Dynamics Laboratory (FDTR),
Wright-Patterson Air Force Base, Ohio.
Mem. ASME

V. E. KEARNEY
R. M. ENGLE
Air Force Flight Dynamics Laboratory, (FDTR),
Wright-Patterson Air Force Base, Ohio

An improved theory is proposed for the crack-growth analysis of cyclic-loaded structures. The theory assumes that the crack tip stress-intensity-factor range, ΔK, is the controlling variable for analyzing crack-extension rates. The new theory, however, takes into account the load ratio, R, and the instability when the stress-intensity factor approaches the fracture toughness of the material, K_c. Excellent correlation is found between the theory and extensive experimental data. A computer program has been developed using the new theory to analyze the crack propagation and time to failure for cyclic-loaded structures.

Introduction

RECENT work by P. C. Paris [1][2] indicates that the crack-propagation rate in a structure caused by cyclic loading is primarily a function of the stress-intensity-factor range ΔK. Based on data which were presented, Paris proposed an exponential equation relating the crack-propagation rate, da/dN, to the stress-intensity-factor range, ΔK. He then solved the equation for an infinite plate subjected to uniaxial tension.

Unfortunately, if a large variation in data is studied, several discrepancies are discovered in the proposed crack-propagation theory. Essentially, two effects occur which must be taken into account. One is the layering of the data owing to the load ratio, R. The other is the instability of the crack growth when the stress-intensity factor approaches the K_c-value for the material.

The object of this paper is to modify the exponential crack-propagation equation to cover the effects owing to the load ratio and the maximum stress-intensity factor. With the modified equation, an initial-value problem is formulated, and a solution is obtained by direct numerical integration using the Runge-Kutta method. Finally, some suggestions are made concerning further improvements which can be made in crack-propagation analysis.

---

[1] This research was conducted as part of an inhouse fatigue and fracture mechanics effort in the Air Force Flight Dynamics Laboratory. The Government reserves the right to reproduce this article in such numbers and in any manner that is required for any subsequent government use.

[2] Numbers in brackets designate References at end of paper.
Contributed by the Metals Engineering Division and presented at the Winter Annual Meeting and Energy Systems Exposition, New York, N. Y., November 27–December 1, 1966, of THE AMERICAN SOCIETY OF MECHANICAL ENGINEERS. Manuscript received at ASME Headquarters, August 4, 1966. Paper No. 66—WA/Met-4.

Governing Differential Equation

Based on data which were presented for 7075-T6 and 2024-T3 aluminum-alloy plates, Paris suggested the following equation for the correct crack-propagation law:

$$\frac{da}{dN} = C(\Delta K)^4 \quad (1)$$

This equation appeared to have very good correlation with the test data that were presented. However, if comparisons are made with a larger range of data, such as higher load ratios and crack-growth rates, the correlation is not good. In fact, equation (1) does not appear to be complete. Two effects occur which are not taken into account. One is the variation in the crack-growth rate owing to the load ratio, R. The other is the instability of the crack growth when the value of the maximum stress-intensity factor approaches the fracture toughness of the material, K_c.

Broek and Schijve [2] studied the layering effect owing to the load ratio, R, and proposed an exponential equation incorporating R for the case of a centrally cracked aluminum plate. Their crack-growth equation can be written in the form

$$\frac{da}{dN} = C_1 e^{-C_2 R}(T\sqrt{a})^3[1 + 10(a/b)^2] \quad (2)$$

This equation is similar to equation (1), though, because

$$K = T\sqrt{\pi a}\,[1 + 10(a/b)^2]^{1/2} \quad (3)$$

is an approximate stress-intensity factor for a central crack in a finite-width plate.

Neither equation (1) nor equation (2) properly predicts the crack-propagation rate approaching fast fracture, or the instability of the crack growth. Assuming that current fracture-mechanics theory is valid, a correct crack-growth law should have the criteria

$$\lim_{\max K \to K_c} \frac{da}{dN} = \infty \quad (4)$$

Furthermore, making the substitution

Nomenclature

- a = half-crack length
- a_o, a_f = initial and final half-crack length, respectively
- b = half-plate width
- $f(\ )$ = a function of
- k = ratio of crack length to plate width
- n = a numerical exponent
- t = plate thickness
- C, C_1, C_2 = material constants
- K = fracture mechanics stress-intensity factor
- ΔK = stress-intensity-factor range (maximum K − minimum K)
- K_c = critical stress-intensity factor for fracture
- N = cycle number
- N_o, N_f = initial and final cycle number, respectively
- N_c = cycle number at crack instability
- R = ratio of minimum K to maximum K in a given cycle
- T = applied tensile stress normal to crack
- ΔT = applied tensile-stress range
- da/dN = crack extension per cycle of load
- β = plate-finite-width correction factor

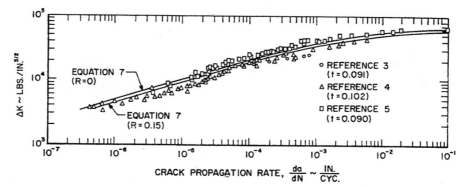

Fig. 1 Comparison of experimental and theoretical crack-propagation rates in 7075-T6 aluminum plate for $R = 0$ to $R = 0.15$

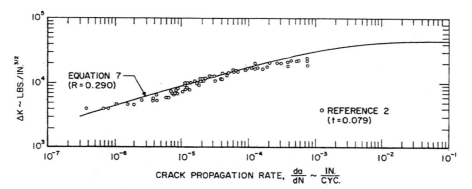

Fig. 2 Comparison of experimental and theoretical crack-propagation rates in 7075-T6 aluminum plate for $R = 0.285$ to $R = 0.297$

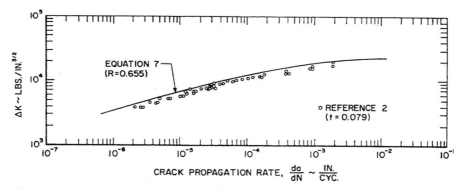

Fig. 3 Comparison of experimental and theoretical crack-propagation rates in 7075-T6 aluminum plate for $R = 0.655$

$$\max K = \frac{\Delta K}{1 - R} \quad (5)$$

where R is the ratio of the minimum stress-intensity factor to the maximum stress-intensity factor, an even more general requirement for crack-growth rate is

$$\lim_{\Delta K \to (1-R)K_c} \frac{da}{dN} = \infty \quad (6)$$

Assuming, then, that a correct crack-growth equation has an exponential form, but in addition has a singularity at $(1 - R)K_c - \Delta K$, an improved equation can be shown to be

$$\frac{da}{dN} = \frac{C(\Delta K)^n}{(1 - R)K_c - \Delta K} \quad (7)$$

Equation (7) is plotted in Figs. 1 through 6 and compared with test results for 7075-T6 and 2024-T3 aluminum alloys. The wide range of test data in the figures particularly show the correlation for the layering effects at different R-values and for the predicted theoretical asymptote. In addition, there is the interesting agreement between theory and test results shown in Fig. 4 for the load ratio of $R = -1$, or the case of completely reversed tension-compression cycling.

The material constants used for calculating the theoretical curves in Figs. 1 through 6 are listed as follows:

(a) For 7075-T6 aluminum alloy:

$$K_c = 68{,}000 \text{ lb/in.}^{3/2}$$
$$C = 5 \times 10^{-13}$$
$$n = 3$$

(b) For 2024-T3 aluminum alloy:

$$K_c = 83{,}000 \text{ lb/in.}^{3/2}$$
$$C = 3 \times 10^{-13}$$
$$n = 3$$

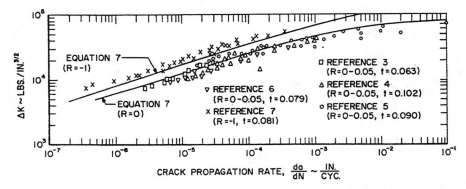

Fig. 4 Comparison of experimental and theoretical crack-propagation rates in 2024-T3 aluminum plate for $R = 1$ and $R = 0$

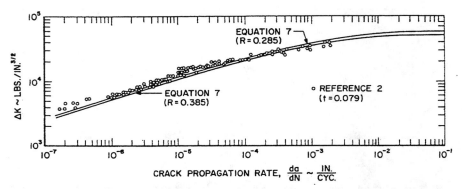

Fig. 5 Comparison of experimental and theoretical crack-propagation rates in 2024-T3 aluminum plate for $R = 0.285$ to $R = 0.385$

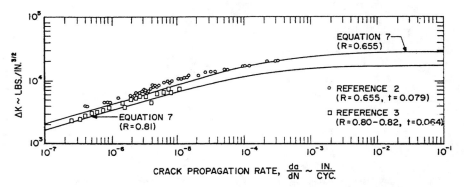

Fig. 6 Comparison of experimental and theoretical crack-propagation rates in 2024-T3 aluminum plate for $R = 0.60$ to $R = 0.82$

The listed values for K_c, C, and n were selected to give the best fit for the test data shown in the figures. The values for K_c were compared with published values, such as in reference [8], and were found to be reasonably accurate.

Formulation and Solution of the Initial-Value Problem

For problems of varying load of uniform character, e.g., ΔT and R constant, equation (7) has the form

$$\frac{da}{dN} = f(a) \tag{8}$$

Equation (8) is a simple first-order linear differential equation. For most practical problems, an initial crack size is known at an initial value of N, such as $N = 0$. The problem is to determine the crack length (or additionally the stress-intensity factor) after a given number of cycles. Mathematically, this is an initial-value problem and can be solved by various direct-numerical-integration methods.

Based on equation (8) and the use of the Runge-Kutta numerical-integration method, the authors have written a computer program[3] which has been applied to several problems and compared with experimental results. In theory, any problem can be solved for which there exists an analytic equation for the stress-intensity factor. The most useful problem which has been solved is the central crack in a finite-width plate subjected to uniaxial tension. For this problem, the equation for ΔK was obtained from reference [9] and is written as follows:

$$\Delta K = \Delta T \sqrt{\pi a \beta^2} \tag{9}$$

---

[3] The program was written and all calculations carried out by the authors on the IBM 7094 computer at Wright-Patterson AFB. A variable step-size method was used in the direct integration to select an optimum step size at every step according to a prescribed accuracy.

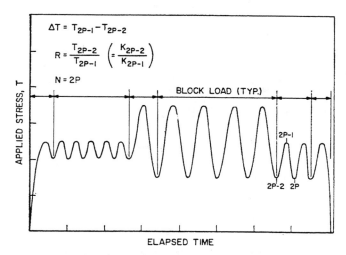

Fig. 7 Description of cyclic stress and stress ratio for block-loading problem

where

$$\beta^2 = 1 + 1.18968k^2 + 1.30162k^4 + 1.36502k^6 \\ + 1.37394k^8 + 1.47638k^{10}$$

and

$$k = a/b$$

The simplest crack-propagation problems that can be solved are for the cases when the loading parameters, ΔT and R, are constant. However, the computer program is capable of determining the crack propagation for nonuniform loading of the type shown in Fig. 7. For data input, the loading must be described in blocks for which ΔT and R are constant in each block. The alternating load and load ratio are defined by the equations

$$\Delta T = T_{2p-1} - T_{2p-2} \quad (10)$$

$$R = \frac{K_{2p-2}}{K_{2p-1}} \approx \frac{T_{2p-2}}{T_{2p-1}} \quad (11)$$

where $2p$ equals N-cumulative cycles. These equations could be written in several different ways. They were selected as shown, though, because of the importance of expressing the maximum stress exactly for calculating the stress-intensity factor. The use of the approximate sign in equation (11) is valid if the crack extension between $2p-2$ and $2p-1$ is relatively small.

Special Case of an Infinite Plate

For an infinite plate with a crack, β in equation (9) is unity, and the expression for ΔK simplifies to

$$\Delta K = \Delta T \sqrt{\pi a} \quad (12)$$

If the loading is assumed to be uniform, then it is possible to substitute equation (12) into equation (7) and integrate equation (7) exactly. For the case of $n = 3$, the result is

$$N_f - N_o = \frac{2(1-R)K_c}{C(\Delta T \sqrt{\pi})^3} \left[\frac{1}{\sqrt{a_o}} - \frac{1}{\sqrt{a_f}} \right] \\ + \frac{1}{C(\Delta T \sqrt{\pi})^2} [\ln a_o - \ln a_f] \quad (13)$$

where the terms N_o and a_o are initial values and N_f and a_f are final values. Making the substitution $\Delta T = (1 - R)T$, equation (13) can be simplified to

$$N_f - N_o = \frac{2}{\pi C (1-R)^2 T^2} \left[\frac{K_c}{K_o} - \frac{K_c}{K_f} - \ln \frac{K_f}{K_o} \right] \quad (14)$$

Table 1 Comparison of theoretical and experimental values of $N_c - N_o$

Material	Maximum T, psi	R	Theoretical $N_c - N_o$	Experimental[a] $N_c - N_o$
7075-T6 Al	12087	0.059	77000	61000
	10950	0.168	137200	89660
	9960	0.285	254000	179000
	22050	0.161	12550	8850
	18500	0.385	43300	20100
	16350	0.565	132000	70530
	26300	0.297	9470	6880
	22750	0.500	32300	16330
	20600	0.655	94500	52510
2024-T3 Al	12100	0.059	166200	215890
	10950	0.168	294000	398200
	9960	0.285	542000	743000
	22050	0.161	28200	36280
	18500	0.385	112000	116900
	16350	0.565	290000	432000
	26300	0.297	21700	24720
	22750	0.500	71200	80990
	20600	0.655	210000	301000

[a] Broek and Schijve [2] data. Each listed value is an average for three tests. The initial half-crack length, a_o, was 0.788 in. for all tests.

The importance of equation (14) is that it can be used to determine the number of cycles required to cause crack instability. This occurs when K_f equals K_c, or when

$$N_c - N_o = \frac{2}{\pi C (1-R)^2 T^2} \left[\frac{K_c}{K_o} - 1 - \ln \frac{K_c}{K_o} \right] \quad (15)$$

where N_c is the value of N at crack-growth instability.

Using equation (15), calculations were made for a number of problems and the results compared with reference [2] test data. The results of this comparison are shown in Table 1.

Additional Refinements

In actual cyclic-loaded structures, a large number of variables have an effect upon the crack-propagation rate. Even the changes made in going from equations (1) and (7) still leaves a need for further improvements. A number of factors which should be studied and possibly incorporated into the computer program are listed as follows:

1 Effect of crack-surface geometry.
2 Effect of loading sequence for combined high and low-level loads.
3 Out-of-plane crack buckling.
4 Stress corrosion at the crack tip.
5 Thickness of the material.
6 Creep at the crack tip.
7 Load cycles required for fatigue-crack initiation from flaws.
8 Variations in the material parameter, K_c.
9 Effects owing to temperature.

The authors are studying many of the aforementioned problems and feel that most of the problems can be solved and incorporated into the computer analysis.

First, some insight has been obtained on the change in crack-surface geometry. Test data in reference [10] indicate that the transition from a 90-deg tensile to a 45-deg shear-type crack face always occurs at a specific crack-growth rate. For 7075-T6 and 2024-T3 aluminum alloys, the transition is at approximately 5×10^{-6} to 7×10^{-6} in/cycle. Therefore it appears that two crack-propagation curves should be used (such as shown in Fig. 8). The curve that applies would depend upon the crack-propagation rate, or alternately the values of ΔK and R. This introduces no problems in the computer analysis because both curves can be described by equation (7).

For the problem of loading sequence for combined low and high-level loads, the Dugdale [11] yield-zone model may be useful.

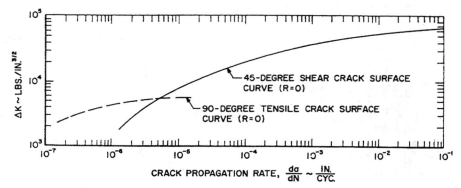

Fig. 8 Crack-propagation rates in 7075-T6 aluminum alloy for both 45-deg shear and 90-deg tensile-type crack surfaces

When changing from a high-load cycle to a low-load cycle, there is often a temporary arrest in the crack growth. The yield-zone model can be used to calculate the increase in the crack-tip radius owing to the high-load cycle. A fatigue crack is then assumed to reinitiate from the rounded crack tip.

The solution to item 7, the load cycles required to initiate a fatigue crack from a flaw, also appears possible. Since most flaws are relatively sharp, such as mechanical scratches and weld flaws, the stress-intensity factor can still be used to approximately define the stress field surrounding the flaw. Preliminary test data indicate that the ΔK concept can be used to predict cycles to fatigue-crack initiation with as much accuracy as exists for predicting crack-growth rates.

Conclusions

A new theory is proposed for predicting the crack-growth rate in cyclic-loaded structures. The theory is an improvement over other theories because the effect owing to the load ratio, R, and the crack instability at onset of fast fracture are taken into account. The theory shows excellent agreement with a wide range of test data. By use of the new theory and with numerical-integration techniques using a digital computer, a more accurate analysis is now possible for the crack-growth behavior of cyclic-loaded structures.

References

1 P. C. Paris and F. Erdogen, "A Critical Analysis of Crack Propagation Laws," JOURNAL OF BASIC ENGINEERING, TRANS. ASME. Series D, vol. 85, 1963, pp. 528–534.

2 D. Broek and J. Schijve, "The Influence of the Mean Stress on the Propagation of Fatigue Cracks in Aluminum Alloy Sheet," National Aero- Research Institute, Amsterdam, The Netherlands, Report NLR-TN M.2111, January, 1963.

3 D R. Donaldson and W. E. Anderson, "Crack Propagation Behavior of Some Airframe Materials," *Proceedings of Crack Propagation Symposium*, College of Aeronautics, Cranfield, England, vol. 2, September, 1961, pp. 375–441.

4 A. J. McEvily and W. Illg, "The Rate of Crack Propagation in Two Aluminum Alloys," NACA TN 4394, September, 1958.

5 C. M. Hudson and H. F. Hardrath, "Effects of Changing Stress Amplitude on the Rate of Fatigue Crack Propagation in Two Aluminum Alloys," NASA TN D-960, September, 1961.

6 W. Weibull, "The Propagation of Fatigue Cracks in Light-Alloy Plates," Svenska Aeroplan Akliebalaget, Linkaping, SAAB TN 25, January, 1954.

7 W. Illg and A. J. McEvily, Jr., "The Rate of Fatigue-Crack Propagation for Two Aluminum Alloys Under Completely Reversed Loading," NASA TN D-52, October, 1959.

8 R. G. Forman, "Experimental Program to Determine Effect of Crack Buckling and Specimen Dimensions on Fracture Toughness of Thin Sheet Materials," AFFDL-TR-65-146, September, 1965.

9 R. G. Forman and A. S. Kobayashi, "On the Axial Rigidity of a Perforated Strip and the Strain Energy Release Rate in a Centrally Notched Strip Subjected to Uniaxial Tension," JOURNAL OF BASIC ENGINEERING, TRANS. ASME, Series D, vol. 86, 1964, pp. 693–699.

10 D. Broek, P. DeRijk, and P. J. Sevenhuysen, "The Transition of Fatigue Cracks in Alclad Sheet," National Aero- Research Institute, Amsterdam, The Netherlands, Report NLR-TR M.2100, November, 1962.

11 D. S. Dugdale, "Yielding of Steel Sheets Containing Slits," *Journal of Mechanics and Physics of Solids*, vol. 8, 1960, pp. 100–104.

DISCUSSION

D. R. Miller[4]

The authors deserve commendation for development of a crack-propagation equation which accounts for the effects of mean stress and the effect of the approach to the critical flaw size for unstable fracture propagation. The writer will be interested in seeing checks of the applicability of this equation to materials which are used in pressure vessels in the thermal-power industry. This equation may provide a more rational basis than now exists for evaluation of the low-cycle fatigue resistance of bolts, in which cracks may initiate early in life, and in which the total life is determined primarily by crack propagation and terminal fracture considerations.

Two interesting questions arise in regard to the authors' equation and its application. Should the equation have the form

$$\frac{da}{dN} = \frac{C(\Delta K)^n}{K_c - \dfrac{\Delta K}{1-R}} \qquad (16)$$

for materials in which the crack-propagation rate is insensitive to mean stress as was observed by Frost [12][5] for mild steel and copper? Should the value of K_c in the equation be lower than the value for the virgin material because of embrittlement which is produced by cyclic strains in the material ahead of the advancing crack?

The writer has used equation (7) as a basis for deriving an equation which can be used to determine, for a particular number of cycles to complete fracture, the relation between the amplitude of alternating stress and the maximum tensile stress during the cycle. This relation has a form which is somewhat similar to that which has been observed by Trapp [13] in room-temperature fatigue tests under axial loading on specimens of SAE 4340 steel. The yield and ultimate strength of unnotched specimens was 147,000 and 159,000 psi, respectively. The specimen diameter in the unnotched region was 0.45 in., the circumferential notch was 0.025 in. deep, the notch root radius was 0.010 in., and the notch had an included angle of 60 deg.

---

[4] Consulting Engineer—Mechanics, Knolls Atomic Power Laboratory (Operated for the United States Atomic Energy Commission under Contract No. W-31-109-Eng-52), General Electric Co., Schenectady, N. Y. Mem. ASME.

[5] Numbers in brackets designate Additional References at end of discussion.

If we assume a particular shape of crack and assume that the crack depth is small in comparison with the section thickness, we can write

$$\Delta K = 2\sigma_a A \sqrt{a} \quad (17)$$

where σ_a is the amplitude of the alternations of nominal stress, A is a constant which depends upon crack shape ($A = \sqrt{\pi}$ in equation (3)) and a is the maximum depth of the crack.

Substitution of equation (17) in equation (7), integrating and evaluating the constant of integration B, gives

$$\frac{2(1-R)K_c a^{\frac{2-n}{2}}}{2-n} - \frac{4\sigma_a A a^{\frac{3-n}{2}}}{3-n} + B = 2^n C \sigma_a^n A^n N \quad (18)$$

and

$$B = \frac{4\sigma_a A a_0^{\frac{3-n}{2}}}{3-n} - \frac{2(1-R)K_c a_0^{\frac{2-n}{2}}}{2-n} \quad (19)$$

If we denote the crack depth for unstable crack growth by a_c, we can write

$$K_c = \frac{2\sigma_a A \sqrt{a_c}}{1-R}$$

whence

$$a_c = \left[\frac{K_c(1-R)}{2\sigma_a A}\right]^2 \quad (20)$$

Substitution of equation (20) in equation (18) and using N_c to denote the number of cycles to complete fracture gives

$$\frac{2^{n-1}(1-R)^{3-n}(A\sigma_a)^{n-2}K_c^{3-n}}{(2-n)(3-n)} + B = 2^n C \sigma_a^n A^n N_c \quad (21)$$

We now let σ_m denote the maximum nominal tensile stress, and substitute $2\sigma_a/\sigma_m$ for $(1-R)$ in equations (19) and (21). We then combine the two new equations to eliminate B and obtain

$$\frac{4A^{n-2}K_c^{3-n}}{(2-n)(3-n)\sigma_m^{3-n}} + \frac{4Aa_0^{\frac{3-n}{2}}}{3-n} - \frac{4K_c a_0^{\frac{2-n}{2}}}{(2-n)\sigma_m}$$
$$= 2^n C \sigma_a^{n-1} A^n N_c \quad (22)$$

We now use $n = 4$ as suggested by Paris, and equation (22) becomes

$$\frac{2A^2 \sigma_m}{K_c} - \frac{4A}{a_0^{1/2}} + \frac{2K_c}{\sigma_m a_0} = 16 C \sigma_a^3 A^4 N_c \quad (23)$$

We now introduce the symbol σ_0 denoting the maximum nominal tensile stress which would have caused unstable crack propagation before the initial crack was extended by cyclic loading. This stress is

$$\sigma_0 = \frac{K_c}{A\sqrt{a_0}} \quad (24)$$

Substitution of equation (24) in equation (23) gives

$$\frac{\sigma_m}{\sigma_0} + \frac{\sigma_0}{\sigma_m} - 2 = 8C\sigma_a^3 a_0^{1/2} A^3 N_c \quad (25)$$

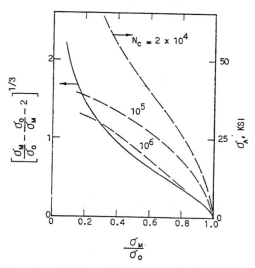

Fig. 9 Comparison of theoretical shape of stress amplitude-vs-maximum stress relation (solid line) with experimental data

This equation indicates that the value of σ_a would decrease monotonically from positive infinity when $\sigma_m/\sigma_0 = 0$ to zero when $\sigma_m/\sigma_0 = 1$. However, if we restrict use of equation (25) to situations where the mean nominal stress is tensile, the values of σ_a remain finite. Values of σ_a are proportional to the cube root of the left-hand side of equation (25).

The solid line in Fig. 9 shows how this quantity varies with the ratio of σ_m/σ_0. The broken lines represent Trapp's data. Their similarity to the theoretical curve is most notable for the shortest cyclic life, in which crack propagation occupies the greatest fraction of life. The ratio of alternating stress amplitude for a life of 2×10^4 cycles to that for a life of 10^5 cycles is about 1.53 instead of the value of 1.71 which is indicated by equation (25). A better correspondence between theory and experiment might have been obtained in larger specimens in which the crack depth at instability is a smaller fraction of the diameter.

Additional References

12 N. E. Frost, "Effect of Mean Stress on the Rate of Growth of Fatigue Cracks in Sheet Materials," *The Journal of Mechanical Engineering Science*, vol. 4, no. 1, 1962, pp. 22–35.

13 W. J. Trapp, "Elevated Temperature Fatigue Properties of SAE 4340 Steel," Wright Air Development Center, Report no. WADC-TR-52-325, part 1, 1952.

Authors' Closure

The authors wish to thank Mr. Miller for his valuable discussion which offered further verification of the improved crack propagation theory. In reply to Mr. Miller's questions, the authors can only state that equation (7) still requires further study and comparison with experimental results. Mainly, more has to be known about the behavior of the material constants K_c, C, and n. For instance, does material embrittlement due to strain cycling or temperature effects cause changes in all three material constants or in only the fracture toughness, K_c?

O. E. WHEELER
Convair Aerospace Division,
General Dynamics Corp.,
Fort Worth, Texas

Spectrum Loading and Crack Growth[1]

An analytical device for improving the accuracy of crack growth predictions in metal subjected to variable amplitude cyclic loading is presented. A modification to the linear cumulative growth idea is proposed which incorporates a consideration of prior load history by taking into account the yield zone ahead of the crack tip. Correlation between analysis and experimental results for six different cases shows that the scheme, even though only a first order improvement on the Miner idea, is sound and can be used with confidence for design and analysis.

Introduction

Leaving aside all questions of crack nucleation or flaw occurrence, this paper addresses the problem of predicting the growth of existing cracks under a sequence of varying loads. Specifically it is concerned with a computational technique for including the effects of load order in those predictions. The problem arises because the empirical data used in crack growth predictions is usually generated using constant amplitude load cycles. These data are most often used in a growth calculation based on the Miner idea which computes a linear accumulation of growth under individual loads. That approach works well if the load is of constant cyclic amplitude. If however, the load is of arbitrarily varying amplitude, the linear cumulative growth scheme is often found to be ultra-conservative. The difference between predicted and observed growth under such loading conditions is due, at least in part, to what happens at the tip of the crack when load is applied. The prediction technique described herein is an attempt to reduce this ultra-conservatism by making some allowance in the analysis for the strain history ahead of the crack tip.

The retardation of crack growth due to varying load levels has been described in the literature [1-4].[2] The preponderance of the published data on this subject is concerned with crack growth in thin skin materials. Hudson and Hardrath [3] present some data on fairly thick sections but the primary interest has been on structure that can sustain a reasonably long crack. The objective has been to determine a safe inspection interval which will allow detection of cracks without catastrophy. It has been recognized that the Miner approach to computing crack growth was in error but tests and service usage both indicated that it was conservative and, for the stated purpose, acceptable.

The analytical attempts to predict crack growth with greater accuracy than the Miner approach have concentrated on statistical techniques [5-6]. These efforts have not been generally adopted for computational purposes even through they show some promise of correlation with test data [7]. In all, there have been in excess of 70 papers published which touch on growth retardation in some way, but there has not been a computational technique presented which is sufficiently simple and sufficiently ac-

---

[1] This work was funded under Air Force Contract AF33(657)-13403.
[2] Numbers in brackets designate References at end of paper.
Contributed by the Metals Engineering Division for publication (without presentation) in the Journal of Basic Engineering. Manuscript received at ASME Headquarters, May 10, 1971. Paper No. 71-Met-X.

Nomenclature

- a = crack length or half length
- a_r = crack length after r load applications
- a_0 = initial crack length
- a_p = maximum excursion of elastic-plastic interface
- B = thickness of compact tension test specimen
- b = plate thickness
- C = constant in exponential crack growth relationship
- C_p = retardation parameter
- C_{p_i} = retardation parameter at ith load
- da/dN = crack extension per cycle of load
- K_I = mode one stress intensity factor
- ΔK = change in stress intensity factor
- M = back face correction factor
- m = retardation shaping exponent
- n = exponent in exponential crack growth relationship
- P = applied load on compact tension test specimen
- R_y = extent of current yield zone
- w = distance from load application to back of compact tension test specimen
- ϕ = complete elliptic integral
- σ = gross section stress
- σ_{ys} = uniaxial yield stress

curate to gain widespread use. The following paragraphs describe a technique which appears to yield good growth predictions.

Analysis

The customary cumulative growth approach is based on the premise that some functional relationship can be established between the rate of crack growth under load and the change in stress intensity factor. (A thorough review of the ideas and terminology in fracture mechanics has been presented by Irwin [8].) This relationship may well include variables other than change in stress intensity factor but it is the variable of primary importance. There are different forms of this function available [5, 9] so only the functional dependence will be indicated here as follows.

$$\frac{da}{dN} = f(\Delta K) \quad (1)$$

where

a = crack length

ΔK = change in stress intensity factor

da/dN = crack extension per cycle of load.

Following Miner [10], the idea is to sum the damage incurred during each load application to get the total damage after r load applications.

$$a_r = a_0 + \sum_{i=1}^{r} f(\Delta K_i) \quad (2)$$

where

a_0 = initial crack length.

This relationship is linear in that the order of load application does not influence the total growth of the crack. This is due to the way equation (1) is formulated. There is a hidden assumption embedded in it that the functional relationship is independent of prior load history. This assumption is perpetuated in the way experimental data on crack growth is often collected and presented, i.e., as da/dN versus ΔK plots.

Experimental data suggests that this prediction scheme could be improved by introducing some factor to delay the crack growth after a high load application [1]. The plot in Fig. 1, which is representative of this behavior, suggests the following form.

$$a_r = a_0 + \sum_{i=1}^{r} C_{pi} f(\Delta K_i) \quad (3)$$

where

C_{pi} = retardation parameter.

The experimental data implies that this factor should have certain properties. It should be bounded between zero and one, and it should increase from its smallest value immediately after a high load to its maximum value of one at some later time. The other principal requirement for C_p is that it be representative of the strain state of the material at the crack tip.

The requirements stated in the last paragraph are satisfied if the retardation parameter is taken in the following form.

$$C_p = \left(\frac{R_y}{a_p - a}\right)^m; \quad a + R_y < a_p \quad (4)$$

or

$$C_p = 1; \quad a + R_y \geq a_p$$

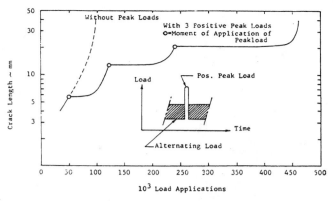

Fig. 1 Typical crack growth retardation

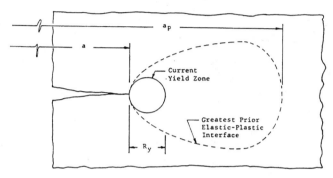

Fig. 2 Crack tip yield zones

where

R_y = extent of current yield zone

$a_p - a$ = distance from crack tip to elastic-plastic interface

m = shaping exponent.

The picture is as in Fig. 2. The dotted line represents the extent of yielding due to some prior high load application and the solid line the current yield zone. This ratio is always bounded between zero (but obviously never equal to zero) and one until the current yield zone moves out of the previously yielded area. After that, C_p is set equal to one. The exponent, m, provides the flexibility to shape this scaling parameter to correlate with test data. The parameter C_p is thus made to depend on the current load application, through R_y, and on prior load history, through the location of the plastic front, a_p, and the crack tip, a.

The computational scheme for incorporating this retardation parameter in crack growth predictions requires that the crack be grown one load application at a time. This amounts to a piecewise linearization of a highly nonlinear process. The algorithm which is used must keep track not only of loads, stresses, and crack tip location, but also of yield zone size and the elastic-plastic interface location. Such a procedure has been written and is now in use. The core of this algorithm is a numerical integration of crack length as a function of load applications. For each load application, the various quantities are computed, e.g., stress intensity factor, change in stress intensity factor, yield zone size, etc. The retardation parameter is computed according to equation (4) and then crack length and elastic-plastic interface location are up-dated. This simple summing is repeated until all the loads are exhausted (arbitrary cutoff), the crack reaches a specified critical length, or the stress intensity factor reaches a critical value.

Experimental Data

The exponent, m, in equation (4) must be established experimentally. Intuition leads one to the value of 2 since this is an

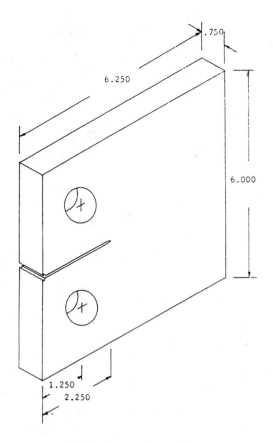

Fig. 3 Fracture test specimen

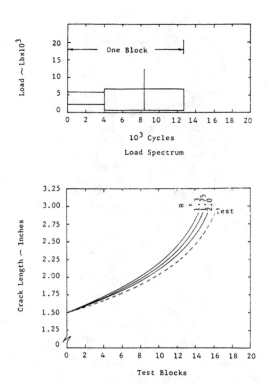

Fig. 4 Predicted and actual crack growth—12.4 KIP spike

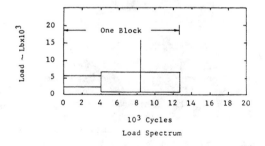

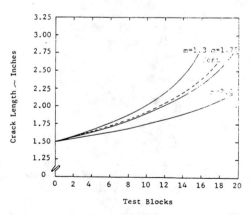

Fig. 5 Predicted and actual crack growth—15.5 KIP spike

energy related phenomenon and the quadratic form is usually appropriate in such cases. Test data indicates different values however, for the different materials investigated. (For the steel described below, $m = 1.3$ and for the titanium, $m = 3.4$.) The extent of the retardation effect in D6ac steel was explored using specimens of the type shown in Fig. 3. The material is described in Table 1. The first tests that were conducted utilized a simple loading pattern containing two levels of repeated alternating load and a single load spike within each block. Fig. 4 shows a comparison of the predicted growth versus test data for several values of m. In these computations the yield zone size, stress intensity factor, and crack growth were computed according to the following formulas.

$$R_y = \frac{1}{4\sqrt{2}\,\pi}\left(\frac{K_I}{\sigma_{ys}}\right)^2 \quad (5)$$

$$K_I = \frac{P}{B\sqrt{w}}\left(29.6\left(\frac{a}{w}\right)^{1/2} - 185.5\left(\frac{a}{w}\right)^{3/2}\right.$$
$$\left. + 655.7\left(\frac{a}{w}\right)^{5/2} - 1017\left(\frac{a}{w}\right)^{7/2} + 638.9\left(\frac{a}{w}\right)^{9/2}\right) \quad (6)$$

$$a_r = a_0 + \sum_{i=1}^{r} C_{p_i}(C(\Delta K_i)^n) \quad (7)$$

The expression for plane strain yield zone size, equation (5), can be developed from Williams' work [11], it is also given by Irwin [12]. The maximum stress intensity factor is used to compute R_y. The expression for K_I was taken from [13]. For this steel, $C = 0.0022 \times 10^{-6}$ and $n = 2.55$ for ΔK in ksi$\sqrt{\text{in}}$. Fig. 4 shows the result when the load spike was 12,400 lb. There is not much difference between various values of the exponent m. Fig. 5 shows that something between 1.3 and 1.75 is appropriate if the spike is 15,500 lb. (The peak loads were far enough apart, in both cases, so that there was no interaction between them.)

Additional tests, using the same type of specimen, had been run with an eight level load block. This was a spectrum intended to show the influence of load order on the crack growth. Fig. 6 is a plot of calculated growth curves compared with test data. This particular test was used to determine m more precisely than

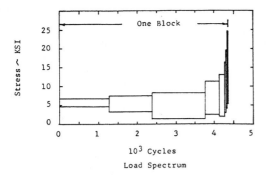

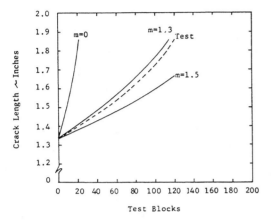

Fig. 6 Crack growth—ascending load spectrum

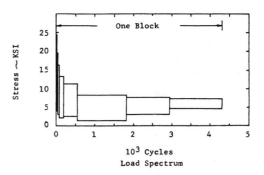

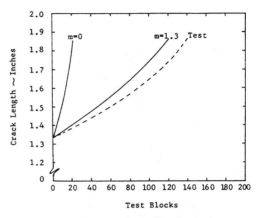

Fig. 7 Crack growth—descending load spectrum

had been done with the three level test data. The figure shows that 1.3 is about right. The $m = 0$ plot is the unretarded (Miner) growth curve for this test. This test was run with the loads in each block arranged in an ascending order. As a check on the value of 1.3 for m, the growth of a crack under loading with the sequence reversed (descending) was computed. The computation and experimental result are shown in Fig. 7. The prediction is reasonably accurate (17 percent in error) and conservative. The Miner prediction ($m = 0$) is the same for both ascending and descending load application.

The tests and analyses described above were all concerned with the same configuration, that shown in Fig. 3. Another configuration, a semicircular part through crack in a wide plate, had been tested using still another load spectrum. This test presented a reasonably independent check on the adequacy of the value of 1.3 for m since both spectrum (13 level) and specimen were different from those used in arriving at that value. Equations (5) and (7) were used in the computation with the following expression for the stress intensity factor [12].

$$K_I = M\sigma \left(\frac{1.21\pi a}{\phi^2 - 0.212\left(\frac{\sigma}{\sigma_{ys}}\right)^2} \right)^{1/2} \quad (8)$$

where

$\phi^2 = 2.46$ for a semicircular crack

and

$$M = 0.899 + 0.184\frac{a}{b} + 0.279\left(\frac{a}{b}\right)^2 + 0.108\left(\frac{a}{b}\right)^3 \quad (9)$$

The factor M is the so-called back face correction factor [14], b is the plate thickness, and a is the crack depth. The correlation is shown in Fig. 8 and is considered acceptable.

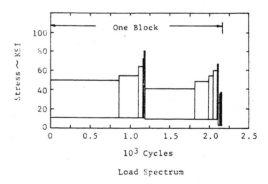

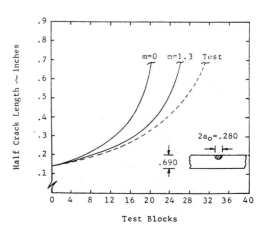

Fig. 8 Part through crack correlation

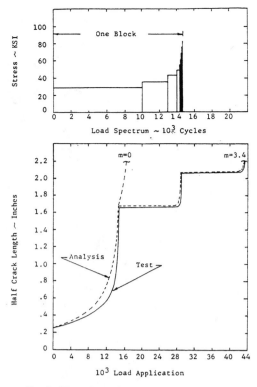

Fig. 9 Through crack growth in titanium panel

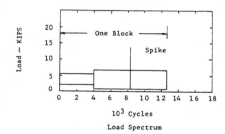

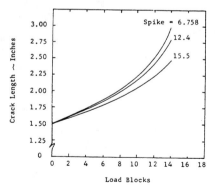

Fig. 10 Influence of spike magnitude on crack-growth

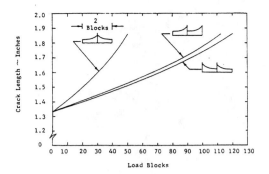

Fig. 11 Influence of load order on crack growth

Table 1 Material description

D6ac—General Dynamics Specification FMS-1011
  Ultimate Tensile Strength 220/240 KSI
  Chemistry (Average Values)
    C— .45%
    Mn— .75%
    Si— .225%
    Cr—1.065%
    Mo—1.025%
    Ni— .55%
    V— .075%
    Fe—Balance
  Heat Treat
    1700° ± 25 deg F for 1 hour.
    Quench to 975 deg F and stabilize.
    Quench into 140 deg F oil.
    Double temper at 1025 deg F for 2 hours, each tempering.
Ti-6Al-4V—Specification Mil-T-9046
  Ultimate Tensile Strength 130 KSI
  Heat Treat
    Annealed per Mil-T-9046. Slightly modified by diffusion bonding process.

A comparison of predicted and observed crack growth in titanium is presented in Fig. 9. The material was Ti-6Al-4V (see Table 1) and the specimen was an integrally stiffened panel with a through crack located midway between a pair of stiffeners [15]. Equations (5) and (7) were used with the following expression for stress intensity factor.

$$K_I = \sigma\sqrt{\pi a} \qquad (10)$$

where

a = half crack length.

The influence of the stiffeners was ignored since the spacing was large (4.75 in.) compared to the plate thickness (0.10 in.). Crack growth was terminated in the analysis at a half length of 2.2 in. For this material, $C = 0.00181 \times 10^{-6}$ and $n = 2.8$ with ΔK expressed in ksi$\sqrt{\text{in}}$. Trial and error runs resulted in a value of 3.4 for m and produced an analytical result which matches the test data very well.

Discussion

The business of predicting crack growth is of necessity something of a black art. The phenomenon occurs on a scale where consideration of dislocations is proper but the analytical tools are grounded firmly in continuum mechanics. Even in its simplest form [5]

$$\frac{da}{dN} = C(\Delta K)^n \qquad (11)$$

the growth relationship contains two constants which must be empirically determined, C and n. It is felt, for this reason, that a more sophisticated analysis than that presented here is probably not justified at this time. The retardation parameter presented in equation (4) does reflect the loading, the geometrical configuration, and the material properties of the problem. It also produces retardation of crack growth consonant with what intuition and experience would suggest. For example, Fig. 10 shows the result of varying the spike load in the three level spectrum described earlier. The higher the spike load, the greater the retardation, all other things being equal. Fig. 11 shows the result of altering load order within the eight level block without changing the number or magnitude of load applications. The back to back arrangement is the worst possible without changing block size since the high loads after the peak tend to wipe out the benefit of those before it. (The material for these analyses was D6ac steel.)

The degree of correlation obtained for D6ac with the five test spectra (2.5 percent to 20 percent conservative error) could be improved by altering the exponent, m, in equation (4) for each separate case. Considering the fact that this is still a third empirical parameter in the computation, making it a variable does not seem advisable. This exponent will undoubtedly depend on which expression of the several available is used to compute R_y, and even which of the available growth relationships is used. It happens that for D6ac, in this heat treat, using equations (5) and (9), a value of 1.3 yields fairly accurate conservative results. The value will be different for other materials, as indicated for titanium. (The value of 3.4 shown here for titanium should be confirmed by other specimen and spectra since only one test is represented here.) The proper value should be determined by testing for each application.

The use of equation (3), in which C_p has a different value for each load application, requires that the crack growth be computed one load cycle at a time. Until recently, this would have been a prohibitive feature and even now, it is a discouraging one. The use of a high speed digital computer is obviously required to perform any realistic analysis. An algorithm has been developed, and should be easy for anyone else to develop, which applies about 50,000 load cycles per minute [16]. The use of double precision arithmetic is mandatory for most problems because sums of very small numbers are involved. Several runs have been made involving 800,000 load applications and one involving about 4,000,000 load cycles.

Conclusion

The retardation of crack growth due to variations in applied loading can be computed conservatively and with reasonable accuracy. The use of a retardation parameter of the form of equation (4) appears to be consistent with the other expressions used in such computations. This parameter has been used successfully to predict the growth of cracks in specimens subjected to six different spectra, having three different physical configurations, and made of two materials. It is felt that this approach represents a useful improvement on the idea of linear cumulative crack growth which can be used with confidence in design and analysis.

References

1 Schijve, J., "Fatigue Crack Propagation in Light Alloy Sheet Material and Structures," *Advances in Aeronautical Sciences*, Pergamon Press, New York, N. Y., 1961, pp. 387–408.

2 Hudson, C. M., and Hardrath, H. F., "Effects of Changing Stress Amplitude on the Rate of Fatigue Crack Propagation in Two Aluminum Alloys," NASA TN-D 960, Sept. 1961.

3 Hudson, C. M., and Hardrath, H. F., "Effects of Variable-Amplitude Loadings on Fatigue Crack Propagation Patterns," NASA TN-D 1803, Aug. 1963.

4 Schijve, J., Jacobs, F. A., and Tromp, P. J., "Crack Propagation in Aluminum Alloy Sheet Materials Under Flight-Simulation Loading," NLR TR 68117 U (AD 692467), Dec. 1968.

5 Paris, P. C., "The Growth of Cracks Due to Variations in Load," PhD dissertation, Lehigh University, 1962.

6 Smith, S. H., "Random-Loading Fatigue Crack Growth Behavior of Some Aluminum and Titanium Alloys," *Structural Fatigue in Aircraft*, ASTM Special Technical Publication No. 404, Philadelphia, Pa., 1966, pp. 74–100.

7 Paris, P. C., "The Fracture Mechanics Approach to Fatigue," *Fatigue—An Interdisciplinary Approach*, Syracuse University Press, Syracuse, N. Y., 1964, pp. 107–132.

8 Irwin, G. R., "Fracture Mechanics," *Structural Mechanics—Proceedings of the First Symposium on Naval Structural Mechanics*, Pergamon Press, New York, N. Y., 1960, pp. 557–592.

9 Forman, R. G., Kearney, V. E., and Engle, R. M., "Numerical Analysis of Crack Propagation in Cyclic-Loaded Structures," JOURNAL OF BASIC ENGINEERING, TRANS. ASME, Series D, Vol. 89, 1967, pp. 459–464.

10 Miner, M. A., "Cumulative Damage in Fatigue," *Journal of Applied Mechanics*, Vol. 12, TRANS. ASME, Vol. 67, 1945, pp. A-159–A-164.

11 Williams, M. L., "On the Stress Distribution at the Base of a Stationary Crack," *Journal of Applied Mechanics*, Vol. 24, TRANS. ASME, Vol. 79, Series E, No. 1, Mar. 1957, pp. 109–114.

12 Irwin, G. R., "The Crack Extension Force for a Part Through Crack in a Plate," *Journal of Applied Mechanics*, Vol. 29, TRANS. ASME, Vol. 84, Series E, No. 3, Sept. 1962, pp. 651–654.

13 "Proposed Method of Test for Plane Strain Fracture Toughness of Metallic Materials," ASTM Standards Test Method E339, 1970.

14 Masters, J. N., Haese, W. P., and Finger, R. W., "Investigation of Deep Flaws in Thin Walled Tanks," NASA CR-72606, Dec., 1969.

15 "F-111 Wing Carry-Through Box Assembly Feasibility Study," General Dynamics Corporation Report FZM-12-10732-II, Fort Worth, Texas, 27 June 1969, pp. 32–37.

16 Simodynes, E. E., "Cycle by Cycle Simulation of Fatigue Crack Growth for Spectrum Loading," General Dynamics Corporation Report MRL-127, Fort Worth, Texas, 15 June 1970.

FATIGUE CRACK CLOSURE UNDER CYCLIC TENSION

WOLF ELBER

Institut für Festigkeit, Mülheim, Germany

Abstract — Results of an investigation are presented, which indicate that a fatigue crack, propagating under zero-to-tension loading may be partially or completely closed at zero load. An analysis of the stress distribution acting on the fracture surfaces shows that the local compressive stress maxima may exceed the yield stress of the material. Fractographic evidence is presented to show that crack closure may influence the shape of the striation pattern on the fracture surfaces.

NOTATION

P, T_1, T_2, C testing forces or residual resultant forces
ϵ, e_1, e_2, e testing eccentricities or residual eccentricities
$\delta, \delta_0, \delta_1$ displacements between sections
$F, \dfrac{d\delta}{dP}$ flexibility of the structure for any set of displacements and forces
σ_{max} the local peak stresses on the fracture surface over areas smaller than the width of one striation
σ_{nom} local nominal stresses averaged over the area of at most several striations
C_0 compatibility factor on the micro-scale
$\Delta\sigma$ gross stress increment in one loading cycle
ΔK appropriate rise in the stress intensity factor
α constant associated with crack closure

1. INTRODUCTION

IN THE study of fatigue crack propagation, crack closure has usually only been considered to occur under compressive loads. Rice[1] has excluded the occurrence of crack closure under cyclic tension loading in the stress analysis of the crack tip. This, however, applies only to an idealized crack which is not propagating. As a consequence of the permanent tensile plastic deformation left in the wake of a fatigue crack, one should expect partial crack closure after unloading the specimen[2]. This question is studied in the present paper.

The experiments described here, show that a fatigue crack propagating under zero-to-tension loading was fully closed at zero load due to internal forces existing in the specimen. The experiments were based on the following two principles:

(1) The net internal force acting across a section in a statically indeterminate structure can be obtained by cutting the section and measuring the force system required to reverse the displacement system produced by the cut.

(2) A crack in a structural member is closed, when the stiffness of the member is the same as the stiffness of an identical uncracked member under the same load system.

The crack closure phenomenon was first observed when a partially cracked sheet specimen of the type shown in Fig. 1, was cut open in order to allow fractographic investigations of the fracture surfaces. This cutting was accompanied by deformations large enough to be observed by the naked eye. The specimen was then equipped with strain gauge targets (Fig. 1), and a force system (P, ϵ) was applied. The relationship between the force P, and the displacement at the edge of the specimen is shown in Fig. 2. This relationship is elastic but non-linear with a stiffness increasing with the

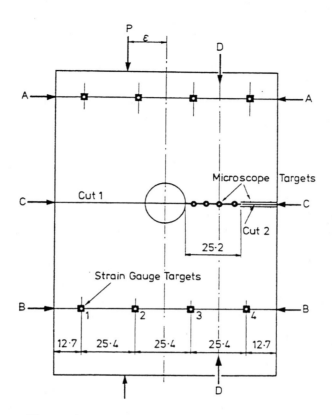

Fig. 1. Specimen configuration and instrumentation. (Clamping area removed.)

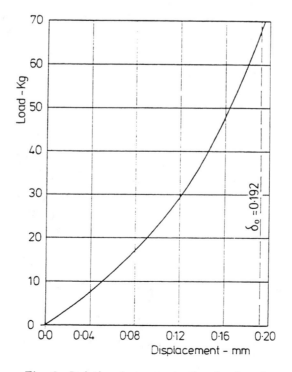

Fig. 2. Relation between testing load and edge displacement.

applied force. This immediately suggests that the system has a varying geometry, which in turn can only mean that the crack is closing. Following this preliminary experiment, a more complete analysis of the internal force system was made.

2. ANALYSIS OF THE INTERNAL FORCE SYSTEMS

From the preliminary results, it was concluded that the stress distribution across the critical section of the specimen must be of the form shown in Fig. 4(a). There are three regions in which compressive stresses act.

(1) *A–B* The cracked section.
(2) *B–C* Part of the plastic zone ahead of the crack tip.
(3) *D–E* Part of the plastic zone on the uncracked side of the notch.

The remaining sections would be under tension.

Instead of carrying out a complete analysis of this statically indeterminate system, it is possible to treat the specimen as if it consisted of three main members each carrying a two parameter force system (P, ϵ), representing the net force 'P' on the section with an eccentricity 'ϵ' from the centre of the sheet.

The force system on the uncracked side of the specimen will be labelled (T_1, e_1), the force system acting on the crack surfaces by (C, e) and the force system acting on the remaining portion of the cracked side of the specimen will be labelled (T_2, e_2), see Fig. 4(a). In order to determine the tensile force systems, the section over which it acts is cut and the displacement system resulting from this cut has to be obtained as a two parameter system. It was found that the displacement system between sections *A–A* and *B–B* in Fig. 1 was linear across the specimen width when a 100 mm base length was used.

The force system (T_1, e_1) and (T_2, e_2) cannot both be determined from the same specimen. Therefore a series of identically cracked specimens was required.

The force system (C, e) can be obtained from the two tensile force systems using the following equations.

$$C = T_1 + T_2$$
$$e = (T_2 \cdot e_2 - T_1 \cdot e_1)/C.$$

Preparation of specimens

The material of the specimen was an aluminum alloy 65 S-T6 (composition Mg: 1·0, Si: 0·6, Cu: 0·28, Cr: 0·25) with the following static properties: $S_u = 38$ kg/mm^2, $S_{0·2} = 37$ kg/mm^2 and elongation 10 per cent. Dimensions of the specimens were: width 100 mm, thickness 6·35 mm, hole diameter 19 mm and crack length 25 mm, see also Fig. 1.

Two specimens were required for the determination of the force systems (T_1, e_1), and (T_2, e_2). They had to satisfy the following requirements:

(1) The crack length had to be equal and approx. 25 mm in length.
(2) The modes of failure had to be similar, being single cracks and preferably of the double shear mode type to prevent distortion during sectioning.
(3) The load spectrum producing the cracks was required to have a minimum tension load as close as possible to zero, so that the examination in the unloaded

condition would reveal the residual stress distribution at the minimum load of the applied load spectrum.

The fatigue load adopted induced the following gross stresses in the specimen: $S_{max} = 10$ kg/mm² $S_{min} = 0.3$ kg/mm².

Displacements between sections A-A and B-B were measured by means of Huggenberger Tensotast equipment using the ball-point targets shown in Fig. 1.

Force systems (T_1, e_1) *and* (T_2, e_2)

The first specimen was sectioned along the line marked 'Cut 1' and the displacement system (δ_0, δ_1) was measured. The force system (P, ϵ) was then applied, and the eccentricity ϵ was increased until a displacement system $(-\delta_0, -\delta_1)$ was obtained. (δ_0 and δ_1 are the displacements at the two edges of the specimen).

This experiment was repeated using the second specimen with a cut labelled 'Cut 2', and the force system was found which would reverse the displacement system caused by Cut 2.

The resulting two force systems (T_1, e_1), (T_2, e_2) and the calculated force system (C, e) are shown in Fig. 4(b).

It must be noted that these force systems are only estimates of the real force systems acting in the uncut specimen, because the point of application of the force system (P, ϵ) is not identical with the centroid of the force system (T_1, e_1) or (T_2, e_2). However, the resulting errors are believed to be relatively small.

Flexibility studies

Since a total compressive force of 395 kg was acting across the fracture faces, the crack in the specimen is at least partially closed in the unloaded state. Further experimentation was required to ascertain whether the crack was completely or only partially closed. Microscopic observation at the specimen surface appeared to be inadequate, since at the surface the crack always appears to be open. Repolishing of the specimen could not be attempted because of the instrumentation on the surfaces.

It was therefore decided to ascertain the degree of crack closure by studying the flexibility of the specimen under the force system (T_1, e_1), and comparing this with the flexibility of a similar specimen also sectioned along Cut 1, but containing an artificial sawcut crack instead of the fatigue crack.

For this purpose a third specimen was sectioned along Cut 1, and subjected to the force system (T_1, e_1). The displacement between sections AA and BB was measured at the edge of the specimen in order to obtain the greatest sensitivity. This displacement at the free edge of the specimen has been labelled 'δ'.

The relation between the force T_1 and the displacement δ was found to be linear and can be expressed by the equation

$$\delta = F \cdot T_1$$

where F is a flexibility constant. This constant F was obtained for various lengths of the artificial sawcut crack in the specimen.

As a next step, specimen 1 was again subjected to the force system (T_1, e_1), and the relation between the force T_1 and the displacement δ was established. This relation is nonlinear, although the specimen appears to behave elastically. This indicates that there is a progressive change in the geometry of the specimen, which must be caused by the closing of the crack under the force T_1. The relation is shown graphically in Fig. 2. The dotted line indicates the displacement δ_0 required to reverse the displacement caused by the sectioning.

The cotangent of the load-displacement curve represents the flexibility of the specimen. This is given by

$$\frac{d\delta}{dT_1} = F.$$

The relation between the flexibility F and the displacement δ was obtained from Fig. 2.

If it is assumed that the flexibility of the specimen containing the artificial crack is the same as the flexibility of Specimen 1 with the crack being open to the same length, then a relation can be established between the load T_1 and the open crack is shown in Fig. 3.

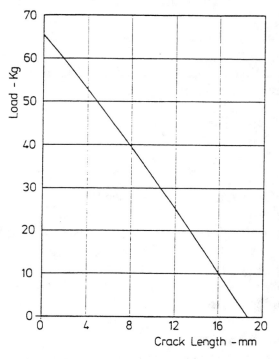

Fig. 3. Relation between testing load and length of open crack.

From this figure it can be seen that at a load of 66 kg, which is just below the load required to reverse the displacement system caused by sectioning specimen 1, the open crack length is zero. This means that in the unloaded state the crack is completely closed.

When the load T_1 is zero, the open crack length is 18·5 mm, whereas the total crack length is 25·2 mm. This means that after sectioning along Cut 1 the crack still appears to be partially closed.

The free shape of the fracture surfaces

From the experiments just described, it can be concluded that the fracture surfaces are not plane, and that they are incompatible in the free state. It was possible to obtain an estimate of the free shape of the fracture surfaces by taking further simultaneous measurements of the opening of the crack and the displacement of sections AA and BB.

Since the crack of Specimen 1 was still partially shut after sectioning along Cut 1, the remainder of the section was progressively sectioned along Cut 2 and measurements were taken of the crack opening, the displacements across the cut and the displacements between the sections AA and BB.

Using the fact that the body deformations between sections AA and CC are equal to half the difference between the relative displacements of sections AA and BB at any section DD (see Fig. 1) and the crack opening at section DD, it is possible to calculate the free shape of the fracture surfaces. Using data from a total of seven observation points along section CC, the free shape of the fracture surfaces can be constructed. This free shape is shown on a distorted scale in Fig. 4(c).

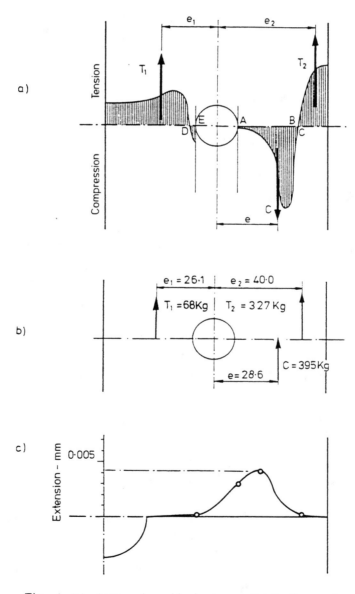

Fig. 4. (a) Assumed residual stress distribution and resultant force systems. (b) Experimentally obtained force systems. (c) Shape of the free fracture surface.

Estimation of crack closure stresses

Given the free shape of the two incompatible fracture surfaces, it is possible to establish by methods of elastic analysis a stress distribution which would satisfy compatibility along the section CC at every point. It is felt, however, that the experimental techniques used in obtaining the free shape of the fracture surfaces only justify an estimate of the maximum compressive stresses acting across the fracture surfaces.

Such an estimate can be based on the force system (C, e), using the assumption that the crack closure stresses are distributed such that they increase monotomically from zero at the original notch root to a maximum at the crack tip.

The centroid of the force C is approximately at one quarter of the crack length from the crack tip. A parabolic stress distribution would also have its centroid at that point. The maximum stress at the crack tip would be $7 \cdot 5 \text{ kg/mm}^2$, if a parabolic distribution is assumed.

In view of the deformation of the fracture as illustrated by Fig. 4(c), it is expected that the stress will rise sharper than parabolically. Hence the value of $7 \cdot 5 \text{ kg/mm}^2$ is probably too low and a value in the order of 10 kg/mm^2 seems to be more appropriate for the discussion.

Micro compatibility of the fracture surfaces

A fractographic investigation of the fracture surfaces has shown that only a fraction of the surface area is capable of making contact with the opposite surface. This means that the average stresses calculated are transmitted over a reduced area. If we define a local compatibility factor 'C_o' as the ratio of load bearing area to total area, then the local peak stress σ_{max} is given by

$$\sigma_{max} = C_0 \times \sigma_{nom}.$$

If we suppose that the shape of the fracture surface at the moment of formation has a shape of either the non-mating sawtooth model or the more general non-mating continually undulated surface, then during the first instant of unloading the compatibility factor would be zero. During the process of crack closure the peak local stresses are limited to a yield level, while the nominal stress σ_{nom} increases to the measured and calculated value. As a result, the compatibility factor must vary by either elastic or plastic deformations. Hence, the shape of the striations as formed during the crack propagation process will be changed during crack closure until sufficient compatible surface is available to carry the compressive forces.

Using an approximate yield stress of 30 kg/mm^2 for the aluminum alloy under repeated hammering, and the calculated maximum nominal stress of 10 kg/mm^2, a surface compatibility of approximately one third is required.

Fractographic investigations have been carried out on the fracture surfaces. The surface shown in Fig. 5 can be interpreted to consist of a generally flat surface in which the striations represent grooves.

The area encircled and labelled (a) shows striation grooves wider than the areas (b) and (c). The degree of surface compatibility can be estimated at approximately 50 per cent, whereas in area (c) the surface compatibility approaches 100 per cent.

On the assumptions that the maximum nominal stress is 10 kg/mm^2, the yield stress at which the surface can be deformed is 30 kg/mm^2, we obtain a surface compatibility of 33 per cent, below which the surfaces must be deformed in order to carry the crack closure stresses.

Conversely, since the observations are made after the deformation process, the lowest surface compatibility found should be of the order of 33 per cent.

Systematic investigation of the fracture surfaces has shown that areas of a surface compatibility of less than 20 per cent were extremely rare and never extended over large areas.

The fractographic evidence can be used to reverse the argument in the following way. It is assumed that the flat surface features were not formed as such during the crack propagation, because this would require cleavage fracture. Hence deformations during crack closure are responsible for the flattening of the striation peaks.

3. CONCLUSIONS

From the results presented above we can draw the following conclusions:

(1) The crack closure phenomenon is a direct consequence of the permanent tensile plastic deformations left in the wake of the propagating crack. These deformations make the mating fracture surfaces incompatible.

(2) A crack in a fatigue specimen is fully open for only a part of the load cycle, even when the loading cycle is fully in tension.

(3) The striation shape on the fracture surfaces is determined not only by the crack propagation process, but also by the crack closure stresses, when these reach load maxima above the yield stress in compression.

(4) Under constant amplitude loading, the loading conditions at the crack tip cannot be determined solely by the stress intensity factor (ΔK) resulting from the whole stress increment $\Delta \sigma$. If the crack is fully open for a fraction '$\alpha \times \Delta \sigma$' of the load cycle, then it may be possible to correlate the rate of crack propagation with $\alpha \times \Delta K$. If the fraction α is constant with crack length and mean stress level, the fourth power relationship between ΔK and the rate of crack propagation would not be affected.

(5) Under variable amplitude loading, crack closure may be responsible for at least part of the interaction effects between stress levels.

Further work, both theoretical and experimental, is required to establish the influence of crack closure on the crack propagation process. The most important aspects of this are:

(1) The determination of the 'crack opening load', that load at which contact between the fracture surfaces is broken, and at which the stress distribution at the crack tip first experiences the singularity of the theoretically sharp crack.

(2) The determination of the 'crack closure load', that load at which contact between the fracture surfaces is established on unloading. This load should always be higher than the 'crack opening load'.

(3) The determination of the difference between the 'crack closure load' and the subsequent 'crack opening load' as a function of the intermediate load minimum.

(4) The determination of the rate at which the 'crack opening load' decreases, if the load cycling has been dropped below this opening level. Such low level cycling (for instance ground-to-air cycle in aeronautical fatigue) would obviously increase the crack closure stresses and would cause compressive deformations on the fracture surfaces, thus decreasing the load at which the crack will again reopen.

At the present stage, it appears that theoretical calculation of the crack closure phenomenon presents great difficulties. Hence, the data required would have to be obtained by experimental means.

Fig. 5. Electromicrograph of a typical fracture surface from one of the specimens (×6000).

REFERENCES

[1] J. R. Rice, Mechanics of crack tip deformation and extension by fatigue. *Fatigue Crack Propagation, ASTM Spec. Tech. Publ. No.* 415, p. 247. Am. Soc. Testing Mater (1967).
[2] W. Elber, Fatigue crack propagation. Ph.D. Thesis, Univ. of N.S.W., Australia (1968).

(*Received* 17 *June* 1969)

Résumé—On présente les résultats d'une recherche indiquant qu'une fissure due à la fatigue, se propageant sous une charge allant de zéro à la tension qui peut être partiellement ou complètement refermée à la charge zéro. Une analyse de la distribution de la tension agissant sur les surfaces de fracture, montre que les maxima de la force locale de compression, peuvent dépasser la tension limite du matériau. La preuve fractographique est présentée dans le but de montrer que la fermeture de la fissure peut influencer la forme du modèle de striation sur les surfaces de fracture.

Zusammenfassung—Es werden die Ergebnisse einer Untersuchung dargelegt, die daraufhindeuten, dass ein unter Ursprungsbelastung wachsender Ermündungsbruch im unbelasteten Zustand teilweise oder ganz gescholssen sein kann. Eine Analyse der auf der Bruchfläche vorhandenen Spannungsverteilung zeigt, dass die örtlichen Druckspannungshöchstwerte die Fliessgrenze des Werkstoffes übertreffen können. Es wird fraktographisches Beweismaterial geliefert um zu zeigen, dass ein Schliessen des Risses die Form des Markierungsmusters auf den Bruchflächen beeinflussen kann.

The Significance of Fatigue Crack Closure

Wolf Elber[1]

REFERENCE: Elber, Wolf, "The Significance of Fatigue Crack Closure," *Damage Tolerance in Aircraft Structures, ASTM STP 486*, American Society for Testing and Materials, 1971, pp. 230–242.

ABSTRACT: Experiments on 2024-T3 aluminum alloy sheet are described which confirm the occurrence of fatigue crack closure under cyclic tensile loading. The results show that a fatigue crack can be closed at the crack tip for up to half of the loading amplitude, leaving this portion of the cycle ineffective in propagating the crack. An expression for the crack propagation rate in terms of effective stress amplitude is proposed. This expression is fitted to existing constant amplitude crack propagation data for 2024-T3 aluminum alloy. The parameters evaluated provide a better fit to the data than other empirical expressions available. Analysis of qualitative experiments on variable amplitude loading shows that the crack closure phenomenon could account for acceleration and retardation effects in crack propagation.

KEY WORDS: aircraft, cracking (fracturing), crack propagation, fatigue (materials), fatigue tests, closure, fracture properties, loads (forces), cyclic loads, stresses, plastic deformation, aluminum alloys, correlation

Nomenclature

- a Distance from the center of a sheet to the crack tip
- da/dN Crack propagation rate
- K_{max} Maximum applied stress intensity
- l Crack length
- R Stress ratio, S_{min}/S_{max}, in a cycle
- S Applied stress
- S_{max} Maximum applied stress in a cycle
- S_{min} Minimum applied stress in a cycle
- S_{op} Stress level at which the crack is just fully open
- U Effective stress intensity range ratio
- ΔK Stress intensity range in a cycle
- ΔK_{eff} Effective stress intensity range
- δ Gage displacement
- δ_{fc} Crack opening displacement of a fatigue crack
- δ_0 Residual displacement

[1] Resident research associate, National Academy of Sciences, NASA Langley Research Center, National Aeronautics and Space Administration, Hampton, Va. 23365.

δ_{sc} Crack opening displacement of a saw cut crack
ϵ_{0y} Residual tensile strain perpendicular to a crack

NOTE—All stresses are gross section stresses.

Recent work [1, 2][2] has shown that fatigue cracks in sheets of aluminum alloy close before all tensile load is removed. Significant compressive stresses are transmitted across the crack at zero load. In previous work, usually the assumption has been made implicitly that a crack is closed under compressive loads and open under tensile loads. This assumption is based on the behavior of a saw cut crack of zero width. However, a fatigue crack differs from a saw cut crack primarily because during crack propagation a zone of residual tensile deformation is left in the wake of the moving crack tip. These deformations effectively decrease the amount of crack opening displacement from that of the saw cut crack. On unloading, this can cause crack closure above zero load. The determination of the crack closure stress must, therefore, be a necessary step in the stress analysis of a cracked structure.

The threefold purpose of the work described here was to show that the loads at which a crack closes can be determined by continuously monitoring the crack opening displacement in the vicinity of the crack tip, to develop an equation for the rate of crack propagation based on crack closure for constant amplitude loading, and to demonstrate the applicability of this concept to variable amplitude loading. In order to achieve these results, constant and variable amplitude loading tests were carried out on sheets of 2024-T3 aluminum alloy. An empirical relation was obtained for the crack opening stress level and this relation was used as a basis for a crack propagation equation. The empirical parameters for this equation were obtained by carrying out a statistical correlation to experimental data from the literature, and the fit of this equation to the data was compared to those of the equations of Forman [3] and Erdogan [4]. From the results of variable amplitude tests, comparisons were made between the trends of the crack propagation rates and the trends of the predictions based on the crack closure phenomenon.

The Crack Closure Phenomenon

Concepts

The application of the fracture mechanics concept to fatigue crack propagation is based on the assumption that a fatigue crack can be represented by a zero-width saw cut. An analysis of elastoplastic behavior at the tip of such an ideal crack was preformed by Rice [5]. The results showed that under cyclic tensile loading the crack would be fully open above zero load. Previous experimental work [1, 2] has shown, however, that a fatigue crack produced under zero-to-tension loading closes during unloading and that large, residual compressive stresses exist normal to the fracture surfaces at zero load.

[2] Italic numbers in brackets refer to the list of references at the end of this paper.

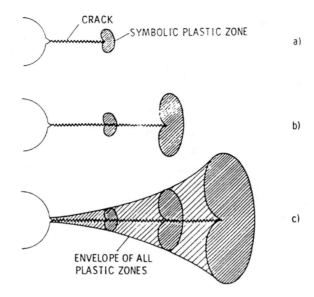

FIG. 1—*Development of a plastic zone envelope around a fatigue crack.*

The load at which the crack closes is therefore tensile rather than zero or compressive. As the crack cannot propagate while it is closed at the crack tip, a knowledge of the crack opening load is essential to refine the prediction of the crack propagation rate.

The difference between the behavior of a fatigue crack and that of the ideal zero-width cut can be explained by the existence of a zone of material behind the crack tip having residual tensile strains. In Fig. 1 a fatigue crack produced under constant amplitude loading is shown at three crack lengths. Figure 1a shows the crack tip, surrounded by a plastic zone as it is represented normally. Figure 1b shows the crack at a greater crack length surrounded by a larger plastic zone, because the stress intensity is higher. The plastic zone of Fig. 1a has been retained to show that the material had been subjected previously to plastic deformations. Figure 1c represents the crack surrounded by the envelope of all zones which during crack growth had been subjected to plastic deformations. During a single cycle of crack growth, residual tensile deformations are left in the material behind the moving crack front, as only elastic recovery occurs after separation of the surfaces. Just behind the crack tip, these deformations are about the same as the plastic deformation at the crack tip.

In Fig. 2 a comparison is made between a saw cut crack and a fatigue crack to show the significance of these residual tensile deformations. At an arbitrary section Y-Y behind the crack tip, the residual strains ϵ_{0y} existing inside the envelope of all previous plastic zones are shown.

The residual deformation δ_0 of the fatigue specimen at section Y-Y can be obtained from the equation

$$\delta_0 = \int \epsilon_{0y} dy$$

At the same section the saw cut crack has no residual strains. The crack opening displacement δ_{fc} of the fatigue crack at section Y-Y is therefore less than δ_{sc} by an amount δ_0.

On unloading, the crack opening displacements of both cracks will decrease at the same rate. Because of the smaller maximum value of δ_{sc}, the fatigue crack will close, $\delta_{fc} = 0$, before δ_{sc} will reach zero.

Experimental Work

Specimens—Sheets of 2024-T3 aluminum alloy were tested under tensile cyclic loads. The specimens were 5 mm thick and 130 mm wide. Cracks were initiated from jeweler's saw cuts in a 30-mm-diameter central hole. The details of the specimen are shown in Fig. 3. The mechanical properties of the material tested are listed in Ref 6.

Equipment—A displacement pickup (Fig. 4) was developed with a gage length of 1.5 mm between contact points. This gage was mounted on the surface of the specimen straddling the crack. The electrical signal from a foil strain gage bridge in the pickup was displayed on a cathode ray oscilloscope screen. A photo attachment was used to record the trace during load cycling. Depending on the type of experiment carried out, the gage was mounted at the crack tip for the duration of several cycles, or was mounted ahead of the crack tip to record deformations as the crack grew through the gage line. The latter method must be used if residual deformations are required.

A 0.3-MN-capacity fatigue machine with a horizontal test bed was used for all tests.

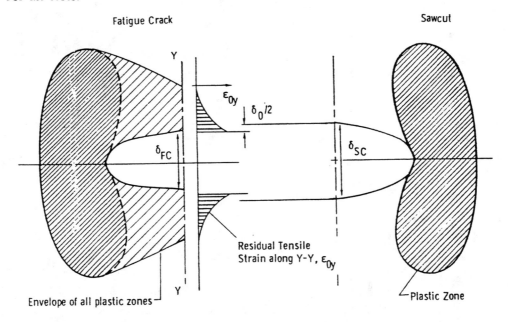

FIG. 2—*Comparison of deformations near the crack tip for a fatigue crack and a saw cut crack.*

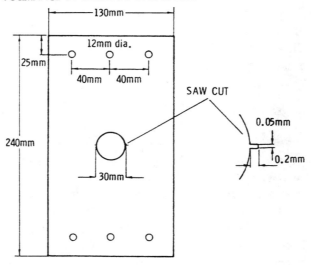

FIG. 3—*Crack propagation specimen.*

Loading Conditions—Constant amplitude tests at 1 Hz were performed to obtain the basic characteristics of the relationship between applied load and crack opening displacement. These tests were performed with a stress ratio $R = 0$ and a maximum applied stress of 150 MN/m^2.

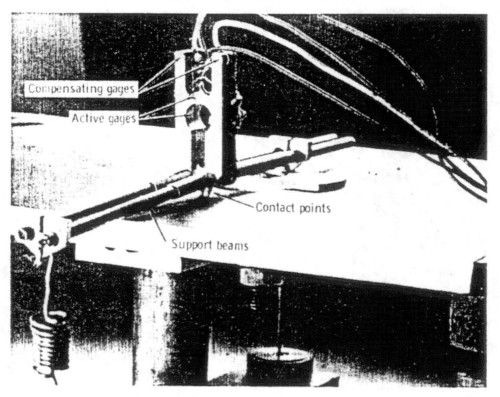

FIG. 4—*Crack opening displacement gage.*

Results—Determination of Crack Opening Stress

Figure 5a shows the crack configuration and gage location for a typical test. Figure 5b shows the relation between the applied stress and the displacement measured by the gage. The main characteristic of this relationship is its nonlinearity. Nonlinear behavior of the structure can have only two causes, material nonlinearity (plasticity) and a change of configuration. In a particular situation, the cause of the nonlinear behavior can be identified by analyzing the curvature of the stress displacement relation. Between points A and B the relation is linear and the measured stiffness $dS/d\delta$ is equal to the stiffness of the uncracked sheet. The stress displacement plot for the uncracked sheet is shown for comparison. Between points C and D the relation also is linear and the measured stiffness is equal to the measured stiffness of an identical sheet containing a saw cut of the same length as the fatigue crack. Between points B and C the curvature $d^2S/d\delta^2$ is negative. Because plastic behavior of the material would produce a positive curvature on unloading, the only possible cause for a negative curvature is a change in configuration which increases the stiffness for decreasing loads. This change of configuration can be explained by crack closure only. From these considerations, the conclusion is reached that the crack is fully open between points D and C during unloading. The crack closes gradually between points C and B and is closed between points B and A.

Figure 6 shows a relationship between stress and displacement recorded at the crack tip of the same specimen under the same loading conditions. The behavior measured at that point is not fully elastic but has similarities with the behavior shown in Fig. 5. In Fig. 6 the relation has a plastic deformation effect superimposed on the configuration change effect. Between points A and

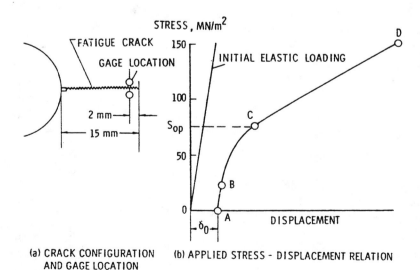

FIG. 5—*Crack configuration and applied stress-displacement relationship.*

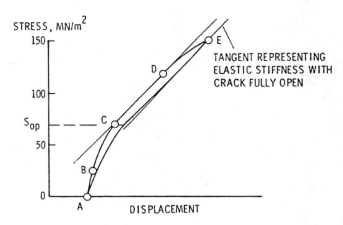

FIG. 6—*Relationship between applied stress and gage displacement at the crack tip.*

B the relation is linear, showing that the crack is closed; between points B and C the negative curvature indicates that the crack is opening; between points C and D the curve is linear; and between points D and E the curve again has a negative curvature. Since plastic tensile deformations can occur only after the crack is fully open, the curvature between B and C is due entirely to crack opening and the curvature between D and E is due entirely to plastic deformations in the plastic zone. In some cases the length between C and D is relatively short compared to the total length of the curve, and a straight line portion cannot be identified. Nevertheless, there must be a stress level at which the crack is fully open and at which the stiffness corresponds to the elastic stiffness of the fully open crack. This stiffness is obtained as the slope of the unloading branch at the maximum load of the previous cycle, where the crack is fully open and the material behaves elastically. A line with this slope can be used as a tangent to the loading branch to determine the crack opening load. When the tangency condition exists for some length of curve, the crack opening load is given by the lowest stress level satisfying the tangency condition. This is shown in Fig. 6, where C represents the lowest tangency point.

As further elucidation of the curvatures of the relation between applied stress and gage displacement, Fig. 7 contains the interpretations of curvatures for loading and unloading conditions.

Effects of Crack Closure

Constant Amplitude Loading

Concepts—Crack propagation can occur only during that portion of the loading cycle in which the crack is fully open at the crack tip; therefore, in attempting to predict analytically crack propagation rates it seems reasonable that the crack opening stress level should be used as a reference stress level from which an effective stress range is obtained. The effective stress range is defined therefore as

$$\Delta S_{eff} = S_{max} - S_{op}$$

where S_{op} is the crack opening stress. An effective stress range ratio is then defined as

$$U = \frac{(S_{max} - S_{op})}{(S_{max} - S_{min})} = \frac{\Delta S_{eff}}{\Delta S}$$

It generally has been accepted that crack propagation rate is a function of ΔK, the stress intensity range during a cycle. Based on the above, it seems reasonable to expect that a better analysis of crack propagation rates might utilize the effective stress intensity range concept. The following functional form of the crack propagation equation will be tested:

$$da/dN = C(\Delta K_{eff})^n = C(U\Delta K)^n \dots \dots \dots \dots (1)$$

Experimental Work—The materials, specimens, and equipment were the same as those used in the experiments described in the previous section of this paper. Two loading frequencies were used, a high frequency of 30 Hz for crack propagation and a low frequency of 1 Hz for observation of the crack opening displacement.

A series of constant amplitude tests were performed to establish the crack opening load under various conditions of stress intensity range, load ratio, and crack length. The stress intensity was varied over the range $13 < \Delta K < 40$ MN/m$^{3/2}$ and the stress ratio was varied over the range $-0.1 < R < 0.7$. Antibuckling guides were used when compressive loads were applied.

	$\frac{d^2P}{d\delta^2}$	Shape	Loading	Unloading
I	>0	 ⌣	Generally impossible for metals	1) Plastic behavior 2) Plastic behavior > configuration change
II	<0	 ⌢	1) Configuration change 2) Plastic behavior	1) Configuration change 2) Configuration change > plastic behavior
III	=0	 ╱	Elastic behavior at constant configuration	1) Elastic behavior at constant configuration 2) Plastic behavior - configuration change
IV	→0	 ╱	Transition from changing configuration to constant configuration	Transition to configuration change > plastic behavior

FIG. 7—*Curvature effects in the load-displacement relationship.*

Results and Discussion—Constant amplitude loading tests were conducted to establish the relationship between U and three variables which were anticipated to have a significant effect on U, namely, stress intensity range, crack length, and stress ratio.

The results are shown in Fig. 8. For the given range of testing conditions only the stress ratio R is a significant variable. The relation between U and R is linear and can be expressed as

$$U = 0.5 + 0.4R, \text{ where } -0.1 < R < 0.7 \quad\quad\quad\quad (2)$$

for 2024-T3 aluminum alloy.

With the substitution of Eq 2 for U in terms of the known quantity R, Eq 1 becomes

$$da/dN = C[(0.5 + 0.4R)\Delta K]^n$$

for this material. A least squares fit of this equation was performed to the data reported by Hudson [7]. The parameters were found to be

$$C = 1.21 \times 10^{-9}$$

$$n = 3.62$$

when ΔK has units of MN/m$^{3/2}$ and da/dN has units of m/cycle. This relation is shown in Fig. 9. For comparison of the least squares fit of this equation and the equations of Forman [3] and Erdogan [4], the sums of squares of residuals have been computed. These are listed in the accompanying table

Equation	Sum of Squares of Residuals
Forman	28
Erdogan	27
Crack closure	21

The correlation of the data was found to be best for the crack closure equation of this work.

Variable Amplitude Loading

Concepts—One of the most important problems in aircraft structures is the inablity to predict accurately the rate of fatigue crack propagation under variable amplitude loading. Attempts to calculate these crack rates on the basis of constant amplitude data usually ignore interaction effects and lead to errors of significant magnitude.

Crack closure may be a significant factor in causing these interaction effects. This can be shown by the following example: Assume a crack in 2024-T3 aluminum alloy is propagating under the conditions $R = 0$ and $K_{max} = 20$ MN/m$^{3/2}$. Under these conditions the crack opening level is at

$K_{op} = 10$ MN/m$^{3/2}$. If the stress intensity range suddenly is halved, the new conditions are $K_{max} = 10$ MN/m$^{3/2}$ and $R = 0$. The crack opening level, however, is still at $K_{op} = 10$ MN/m$^{3/2}$, equal to the new peak stress intensity, so the crack does not open. Therefore, the crack does not propagate until the crack opening level changes. The behavior of the crack opening stress level under variable amplitude loading must therefore be investigated.

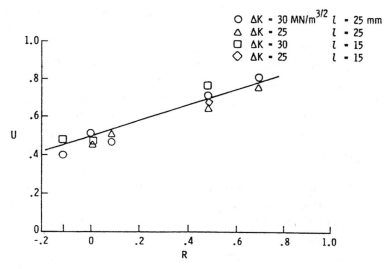

FIG. 8—*Relationship between effective stress range ratio and stress ratio for 2023-T3 aluminum alloy.*

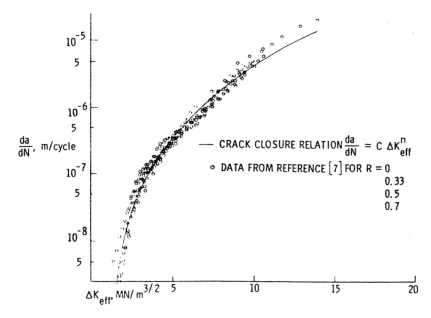

FIG. 9—*Relationship between crack propagation rate and effective stress intensity range.*

For these experiments, the same equipment and specimens were used as in the previous experiments. A crack was grown to a length of 4 mm at a stress level of 100 MN/m². The stress sequence shown in Fig. 10a, containing a single high load, was then applied to the specimen. The gage was located 0.1 mm ahead of the crack tip, and crack opening displacement records were taken during the last cycle before the high load, during the high-load cycle, and at intervals after the high-load cycle.

A second experiment was performed with a stepped program load. As in the previous experiment, a 4-mm-long crack was grown at a stress level of 100 MN/m² and the stress level was then changed to 180 MN/m²; during and after the stress level step, records of the crack opening displacement were taken. The stress program is shown in Fig. 11a.

Results and Discussion—Figure 10b shows the stress-displacement record for the single high-load sequence. The stress-time record shows the high-load cycle and the two load cycles before and after the high-load cycle. In addition, it shows the 1000th load cycle after the high-load cycle. The stress-displacement record shows the gage displacements during these cycles. In the first cycle, no hysteresis is registered, so loading and unloading curves are colinear; hence, the crack opening stress can be obtained as the stress at which the curve deviates from the linear portion of the curve through the point $\bar{1}$. This shows that U is approximately 0.5. For the loading $(2-\bar{2})$ the crack opening stress is the same, so $U = 0.25$. This high load produces a residual displacement which is larger than the displacement at which the crack previously opened; hence, at point 3 the crack cannot be closed over the previous fracture surfaces.

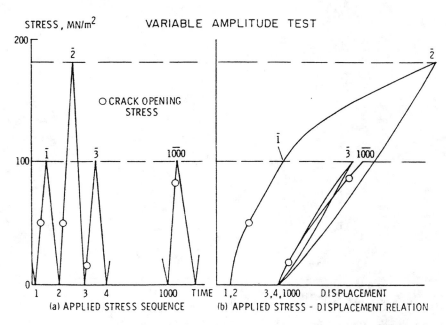

FIG. 10—*Variation of crack opening stress caused by a single high load.*

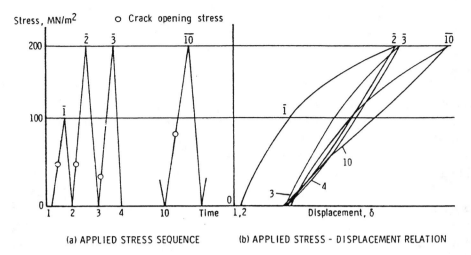

FIG. 11—*Variation of crack opening stress caused by a program step.*

The loading curve (*3–3̄*) is nonlinear and, if the tangent from the unloading curve at *2̄* is drawn through *3–3̄*, the lowest stress satisfying tangency is approximately 30 MN/m², making $U = 0.7$. This implies that the crack has closed only over the single striation produced by the high load. After 1000 load cycles the crack has propagated approximately 0.3 mm, and the loading branch (*1000–1̄0̄0̄0̄*) shows that the crack is almost continuously closed with U at approximately 0.1.

Observation of the crack propagation rate showed that the crack propagated by one large step during the high-load cycle where the effective stress intensity range was largest. The crack continued to grow at a decreasing rate during the next 150 cycles and the effective stress intensity range dropped to zero. The crack then became stationary and no further change in the stress-displacement relation could be observed. The test was discontinued at that point.

The fact that a crack will continue to grow for some time after some high loads had been observed previously by Schijve [*8*] and had been termed "delayed retardation." The delayed retardation of crack growth after a single high load can be explained by examining the behavior of the large plastic zone left by the high-load cycle ahead of the crack tip. The elastic material surrounding this plastic zone acts like a clamp on this zone, causing the compressive residual stresses. As long as this plastic zone is ahead of the crack tip, this clamping action does not influence the crack opening. As the crack propagates into the plastic zone, the clamping action will act on the new fracture surfaces. This clamping action, which builds up as the crack propagates into the plastic zone, requires a larger, externally applied stress to open the crack; hence, the crack will propagate at a decreasing rate into this zone and may come to a standstill.

Schijve [8] and others have presumed that the retardation was caused by the action of residual stresses ahead of the crack tip. The results of the experiments here suggest that the retardation is caused by the residual deformations appearing behind the crack tip, as the crack propagates into the plastic zone.

In the second experiment a crack was propagated under the same initial conditions. Instead of dropping back to the low stress level after one cycle, the high-stress cycles were continued (Fig. 11a). The crack opening stresses in cycles $(1-\bar{1})$, $(2-\bar{2})$, and $(3-\bar{3})$ are determined from Fig. 11b. These stresses are identical to the crack opening stresses in Fig. 10b. Under the continued high-load cycling, the crack opening stress level rises and reaches the new equilibrium level for the high-load cycle after 10 cycles. The first cycles have a greater effective stress intensity range than the tenth cycle, causing an initial crack propagation rate larger than the final equilibrium rate. This phenomenon has been observed previously by fractographic methods. A typical fractograph representing this phenomenon can be found in Ref 9.

Conclusions

The results of a study to determine the significance of fatigue crack closure on crack propagation in 2024-T3 aluminum alloy sheet have been presented. From this study, the following were concluded:

1. Fatigue cracks are closed for a significant portion of the tensile load cycle.

2. Under constant amplitude loading, an expression was derived for the rate of crack propagation in terms of an effective stress intensity range. This expression provides a good fit to existing data.

3. Under variable amplitude loading the crack closure phenomenon accounts for acceleration and retardation effects in crack propagation.

References

[1] Elber, W., "Fatigue Crack Propagation," Ph.D. thesis, University of New South Wales, Australia, 1968.
[2] Elber, W. in *Engineering Fracture Mechanics*, Vol. II, No. 1, Pergamon Press, July 1970.
[3] Forman, R. G., Kearney, V. E., and Engle, R. M., *Journal of Basic Engineering*, ASME Transactions, Series D, JBAEA, Vol. 89, No. 3, Sept. 1967, pp. 459–464.
[4] Erdogan, Fazil, "Crack Propagation Theories," NASA CR 901, National Aeronautics and Space Administration, 1967.
[5] Rice, J. R. in *Fatigue Crack Propagation*, ASTM STP 415, American Society for Testing and Materials, 1967, p. 247.
[6] Jacoby, G., *Fortschritt-berichte VDI-2*, FBVBA, Reihe 5, Nr. 7, April 1969.
[7] Hudson, C. M., "Effect of Stress Ratio on Fatigue-Crack Growth in 7075-T6 and 2024-T3 Aluminum-Alloy Specimens," NASA TN D-5390, National Aeronautics and Space Administration, 1969.
[8] Schijve, J., "Fatigue Crack Propagation in Light Alloy Sheet Material and Structures," Rep. MP 195, National Luchtvaart Laboratorium, Amsterdam, Aug. 1960.
[9] McMillan, J. C. and Pelloux, R. M. N. in *Fatigue Crack Propagation*, ASTM STP 415, American Society for Testing and Materials, 1967, p. 516.

FOUR LECTURES ON FATIGUE CRACK GROWTH

J. SCHIJVE

Delft University of Technology, Department of Aerospace Engineering, 1 Kluyverweg, Delft 9, The Netherlands

Abstract—During the course year 1976–1977 the author presented a series of eight lectures on fatigue crack growth at the Department of Aerospace Engineering of the Delft University of Technology. Four of these lectures in a slightly modified version were presented in August 1977 as part of a Seminar on Fatigue, Fundamental and Applied Aspects, organized by the Linköping Institute of Technology (Prof. T. Ericsson). These lectures are reproduced here. Titles and summaries are given below.

I. FATIGUE CRACK GROWTH AND FRACTURE MECHANICS

Summary

Aspects of the technical meaning of fatigue considerations in practice are indicated. The fatigue life is subdivided into a crack nucleation period and a crack propagation period. The significance of recognizing these periods for practical problems is illustrated by several examples. The similarity approach for correlating fatigue data is introduced. The meaning of the stress intensity factor for fatigue crack growth is discussed.

II. FATIGUE CRACKS, PLASTICITY EFFECTS AND CRACK CLOSURE

Summary

Concepts introduced are residual plastic deformation, residual stress, reversed plastic deformation, plastic deformation in the wake of the crack and crack closure under tensile load. COD measurements as a method to determine the crack closure level are discussed. The significance of crack closure for fatigue crack growth is analysed and illustrated by several examples, including effects of yield stress, stress ratio and delayed crack growth after a peak load. Finally some attention is paid to three dimensional aspect following from thickness effects, shear lips and curved crack fronts.

III. FATIGUE CRACK PROPAGATION, PREDICTION AND CORRELATION

Summary

Two prediction techniques are introduced, (1) cycle-by-cycle prediction and (2) prediction by correlation. Attention is paid to the problem of describing variable-amplitude loading in terms of load cycles. Aspects of fatigue damage are reviewed with reference to interaction effects and weaknesses in cycle-by-cycle prediction methods. The discussion on prediction by correlation is restricted to constant-amplitude loading. The validity of the similarity concept based on K-factors is reconsidered. Application of simple specimen data to complex structures is shown. Finally a variety of crack growth equations is reviewed, including aspects of curve fitting, a comparison between formulas of Walker and Elber and asymptotic values in the $da/dn - \Delta K$ relation.

IV. FATIGUE CRACK GROWTH UNDER VARIABLE-AMPLITUDE LOADING

Summary

Stationary and non-stationary types of variable-amplitude loading are specified. Some attention is paid to the description of stationary random load. The stress intensity factor is then applied to crack growth under stationary variable-amplitude loading by defining first characteristic stresses and characteristic stress intensity factors. This is done for random loading, non-random loading and flight-simulation loading. It is discussed how and why this concept may break down if the stationarity is lost. Attention is paid to truncation of high loads in a flight-simulation test and the analogous problem of the crest factor under random loading. The significance of crack closure for understanding crack growth under variable-amplitude loading is emphasized.

CONTENTS

I. Fatigue crack growth and fracture mechanics
   1. Introduction
   2. Fatigue crack initiation
   3. Fatigue crack growth
   4. The stress intensity factor K
   5. K and cyclic loading
   6. References

II. Fatigue cracks, plasticity effects and crack closure
   1. Introduction
   2. Residual stress after local plastic deformation
   3. Crack closure
   4. How to measure crack closure
   5. Crack closure and fatigue crack growth
   6. Some consequences of crack closure
   7. Crack closure and the effect of thickness
   8. Crack closure and shear lips
   9. Crack closure and crack front curvature
   10. References

III. Fatigue crack propagation, prediction and correlation
   1. Introduction
   2. Cycle-by-cycle prediction for variable-amplitude loading
   3. Crack growth prediction by correlation
   4. Crack growth equations for constant-amplitude loading
   5. References

IV. Fatigue crack growth under variable-amplitude loading
   1. Introduction
   2. Different types of variable-amplitude loading
   3. How to describe random load
   4. K-concept and random load
   5. K-concept and stationary non-random load
   6. Crack growth under flight-simulation loading
   7. Truncation of high loads
   8. Crack closure and non-stationary variable-amplitude loading
   9. References

I. FATIGUE CRACK GROWTH AND FRACTURE MECHANICS

Abstract—Aspects of the technical meaning of fatigue considerations in practice are indicated. The fatigue life is subdivided into a crack nucleation period and a crack propagation period. The significance of recognizing these periods for practical problems is illustrated by several examples. The similarity approach for correlating fatigue data is introduced. The meaning of the stress intensity factor for fatigue crack growth is discussed.

1. INTRODUCTION

AS AN INTRODUCTION we will consider a kind of an overall picture about the relevance of fatigue. Fatigue is not an important problem because it is highly interesting to see a tip of a fatigue crack growing under a microscope at high magnification. That is why fatigue is an interesting phenomenon, but the essential arguments why we have to consider fatigue as a relevant problem are based on:

—economy ($),
—safety (men).

Who should bother about fatigue? Probably the designer of a structure, but that answer is too specific and certainly not complete. At least two groups of people should be recognized, which are to be found in:

—the industry, the manufacturer of the structure,
—the operator, people using the structure in service.

Note that the first group involves more than the designer alone. For instance the production shop is adding to the fatigue quality of a structure.

Now let us consider for a while why those two groups will have a different approach to the fatigue problem. The manufacturer will try to produce a structure, that will be free from fatigue as far as possible. In other words he will design and produce as to obtain a high fatigue resistance. Aspects of prime importance are:

—cyclic stresses,
—material properties,
—surface quality etc.

The operator on the other hand should consider reasons why fatigue still might occur, the consequences it could have and how to prevent it. There are several reasons why fatigue still may occur, such as:

—a more severe utilization of the structure,
—fatigue damaging aspects overlooked by the designer,
—poor production standards,
—incidental damage,
—aggressive chemical environments, etc.

As a result the operator's fatigue problems are more associated with maintenance, inspections and non-destructive testing techniques. In terms of fatigue some questions of the operator are: at which locations will cracks start and how fast will they be growing?

At this point it should be said that the appreciation of the fatigue problem will be different depending on the type of structure to be considered. Let us consider three typical cases:

1. Motor car engine: Cracks should not occur, "infinite" life is required, crack growth is not of interest. Design and production against crack nucleation is of prime importance.

2. Nuclear pressure vessel: Initial flaws and defects in a welded steel structure have to be expected. Consequently crack nucleation is of little interest, but (very slow) crack growth should be considered.

3. Aircraft structure: A finite life has to be accepted. Hence both crack nucleation and propagation are significant.

In terms of crack nucleation and crack growth different situations are illustrated in Fig. 1, which shows schematic crack growth curves for both finite and infinite life. In the absence of

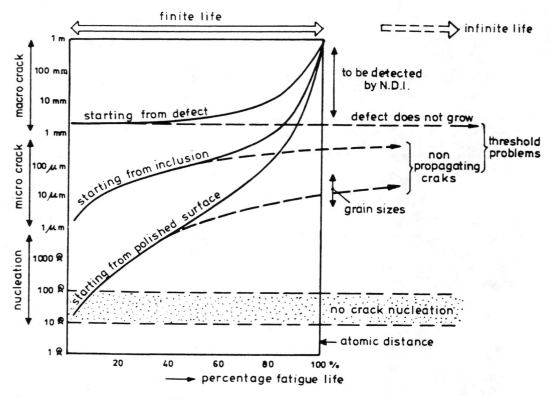

Fig. 1. Survey of different crack growth cases.

initial defects some 90% of the fatigue life is spent in the micro-range. If we then ignore the remaining 10% the error made is small. In other words in such cases the fatigue damaging process is largely occurring in a very small volume of the material and it will be highly depending on local conditions. This is significant for prediction problems. In other cases, where macrocrack growth has to be considered bulk properties of the material will also be involved.

Let us now limit the scope of the discussion by considering the prediction of fatigue properties of a technical structure. Predictions in the sense of Applied Mechanics imply prediction of the fatigue properties, usually starting from data of a more general and a more simple nature. Classical examples are:

—prediction of fatigue properties of a notched element from available data for the unnotched material (notch effect, size effect, stress distribution, K_t, stress gradient).

—prediction of fatigue life under complex load-time histories from available $S-N$ data (cumulative fatigue damage problem).

—prediction of fatigue behavior under biaxial loading from the behavior under uni-axial loading (fracture or yield criteria).

Solutions for these problems have the character of extrapolating available data to other conditions, assuming that some correlation exists between fatigue in the two cases, (1) one case for which data are known and (2) the second one for which predictions have to be made. Now a correlation between the two cases seems to be justified only if the same fatigue mechanism can and will occur in both cases. This question can be answered only if we know what is going on during fatigue of metals and which factors are having an influence on this phenomenon. In other words: application of *fracture mechanics* (extrapolation) requires some knowledge of the *fracture mechanism* (physical understanding). Note the almost similar writing, but the highly different meaning.

This lecture starts with some observations on fatigue of metals including crack nucleation and crack growth. At a later stage the stress intensity factor is defined in order to describe the "stress system in a crack tip area", with some comments on its significance.

2. FATIGUE CRACK INITIATION

The fatigue life can be subdivided in some periods as shown in the figure below.

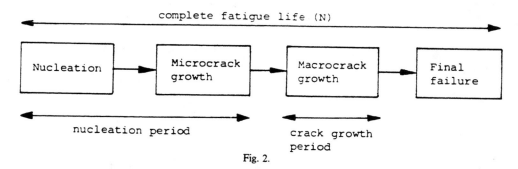

Fig. 2.

The beginning and the end of each period is not easily defined, except for the last one. Final failure occurs in the very last cycle of the life and usually this part of the failure is supposed to be quasi-static rather than fatigue.

Microscopical studies have shown that crack nucleation starts early in the fatigue life. We may ask when incipient cracking should be considered to be a microcrack, however, this question will be ignored since an answer is not of great interest for practical problems. The simplified picture than becomes:

fatigue life = (nucleation period) + (crack growth period).

An obvious question now is: when does a microcrack become a macrocrack? The classical definition of a macrocrack is that the crack is large enough to be seen by the naked eye. This is not a very exact definition. An alternative definition is to consider a crack to be a macrocrack if it had sufficient depth (or length) to be sure that local conditions, responsible for crack nucleation do no longer affect crack growth. In other words crack growth will then depend on bulk properties. Physically this definition appears to be more reasonable, although not yet very precise. A third definition with some appeal to the present course might be: A crack is a macrocrack as soon as Fracture Mechanics are applicable. Note that this definition is depending on the definition of Fracture Mechanics. Perhaps the definition should be rephrased as: A crack is a macrocrack as soon as the stress intensity factor K has a real meaning for describing its growth.

Some observations will now be reviewed to illustrate the usefulness of considering crack nucleation and crack propagation as different phases. Fatigue cracks generally start at the surface of the material (macroscopic observation). A number of reasons may contribute:

—high stress level (K_t always >1),
—surface roughness (inhomogeneous stress distribution on a small scale),
—environmental effects,
—lower restraint on plasticity.

These aspects will promote crack nucleation at the surface. Material structure at the surface (decarburizing) and residual stresses may also contribute, but they are not necessarily unfavourable for fatigue (e.g. shot peening).

The number of fatigue cracks occurring in one specimen has been observed in some laboratory studies.

They all show the same trend. The number is larger at a higher stress level (Fig. 3). However, at stress levels near the fatigue limit S_f it may well happen that only one crack has been nucleated. Apparently there is a typically weakest link in a material which may have as many as 1000 grains in 1 mm^2. The weak link will be a compilation of unfavourable features, which will not be discussed here. Anyhow, it is a highly local phenomenon in the nucleation period, while at higher stress levels several weak links are ready to produce a crack. Even then the fatigue phenomenon is still a local process for a long time, i.e. microcracks are present in a rather small volume of the material only. It is highly worthwhile to have this observation in

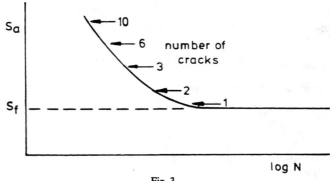

Fig. 3.

mind if the similarity between fatigue in different components or specimens of the same material is analysed. It has an essential meaning for the validity and the limitations of similarity concepts mentioned before.

The significance of surface roughness is shown in Fig. 4. Apparently the quality of the surface finish has a large effect on the nucleation period, whereas the effect on the crack growth period was negligible. Macrocrack growth is depending on bulk properties of the material, since it is no longer a localized phenomenon.

Another consequence is related to scatter. Crack nucleation being a highly localized phenomenon can be subject to considerable scatter, depending on local conditions. However, crack propagation, being dependent on bulk properties only, usually shows much less scatter.

Figure 5 shows scatter bands of S-N data for an aluminium alloy (unnotched specimen). The bare material is highly depending on surface quality, especially at low stress levels, and

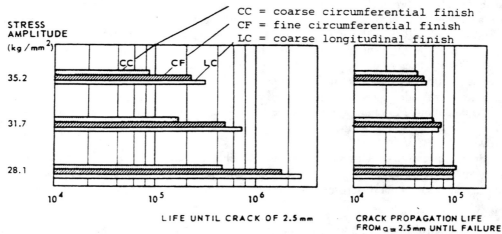

Fig. 4. Effect of surface finish on the pre-crack life and the crack propagation life of unnotched rotating beam specimens of 0.2% C steel (SAE 1020) (results of DeForest).

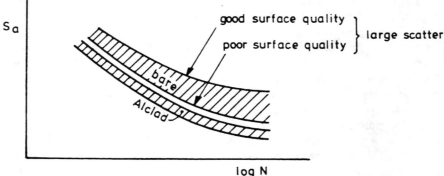

Fig. 5. Scatter bands for a strong Al-alloy.

scatter is large. At high stress levels more cracks will be nucleated which implies less scatter. In the same alloy with soft cladding layers of pure aluminium (corrosion protection) crack nucleation in the cladding is very easy at any stress level. This implies a well reproducible surface with a low fatigue resistance. Scatter is much lower then.

It is a good question whether crack nucleation is always followed by crack propagation. Under certain conditions it does not and then non-propagating cracks are present. Such cracks are found under compressive mean stresses and this explains the asymmetric shape of fatigue diagrams. In area A (Fig. 6) microcracks are initiated early in the life, but these cracks do not grow, due to compressive stresses. In area B microcracks are just not initiated, while above the S_f-line cracks are initiated, which then will grow further.

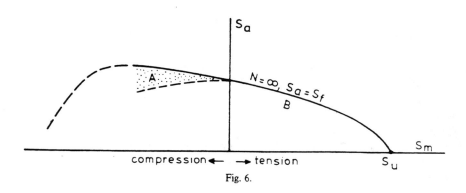

Fig. 6.

Non-propagating cracks have also been observed at positive mean stresses, especially for specimens with sharp notches. Frost[2] observed the trend as shown in Fig. 7. Apparently for moderate K_t-values the fatigue limit is defined as the minimum stress for crack initiation. However, for high K_t-values the fatigue limit is the minimum stress for crack growth. Apparently there are two essentially different definitions of the fatigue limit.

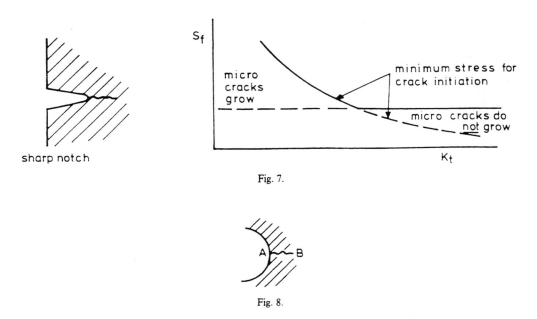

Fig. 7.

Fig. 8.

It still requires an explanation why a microcrack, once being initiated, should not grow any further. For a homogeneous material the more obvious argument is the increased restraint on (cyclic) plasticity going from A (free surface) to B (triaxial state of stress).

Difficulties in the growth of a small microcrack may also affect the shape of the S–N curve. Two typical shapes are shown below.

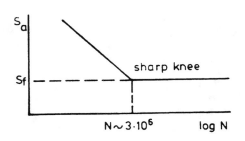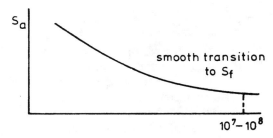

Fig. 9.

In the first case (sharp knee) nucleation of a crack is the problem to obtain a finite life. Crack growth will follow anyhow. In the second case crack initiation at a very low stress amplitude is possible, for instance from an inclusion (in high strength steel), or by fretting corrosion, or from a cladding layer. However, in view of the low stress level crack growth initially occurs extremely slowly and complete failure may require 10^7–10^8 cycles.

3. FATIGUE CRACK GROWTH

For technical materials fatigue crack propagation has been shown to be crack extension in every load cycle (striations, fractographic observations by the electron microscope). Since crack extension implies decohesion in the material a fracture mechanism is operating. Details of the mechanism on an atomic level actually are unknown, but on a larger scale several observations have been made, e.g.

Microscopic level:
—Crack propagation in several metals follows a transcrystalline path.
—Crack propagation occurs along slipplanes in some materials only (Forsyth, Stage I), while in general it will not do so (Stage II). Slip on more than one crystallographic plane is operative then (in the latter case Stage I may still occur in the nucleation period and also later on at the free surface, due to low restraint on slip deformations and crack path).

Macroscopic level:
—Cracks usually grow in a macroscopic plane perpendicular to the main principal stress, at least as long as the crack rate is low (tensile mode fracture). At faster crack rates the growth direction remains perpendicular to the maximum principal stress, but the plane of the fatigue fracture (for several materials) will be under an angle of 45° with that stress (shear mode fracture). In sheet materials this leads to shear lips as shown in Fig. 10.
—Crack propagation is strongly affected by the environment, the crack rate being slower in inert environments and faster in aggressive environments.

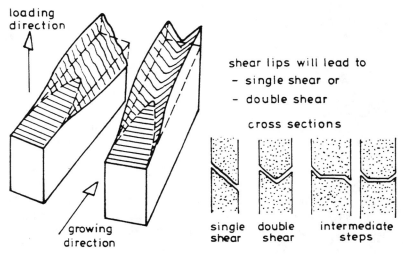

Fig. 10.

Humid air for several materials should be considered to be aggressive as compared to inert environments. Obviously salt water is still more aggressive. The environment may also affect the fracture mode. For Al-alloys an aggressive environment is postponing the transition from the tensile mode to the shear mode.

The crack growth rate, usually denoted as da/dn (= slope of crack growth curve), should be considered to be the crack extension Δa of a crack (length a) occurring in one cycle.

$$da/dn = \Delta a.$$

From the previous arguments it will be clear that Δa will depend on:
—the cyclic stress on the crack tip area,
—the (cyclic) elasto–plastic response of the material in the same area,
—the environment,
—some fracture criterion.

The plastic deformations around a tip of a fatigue crack are depending on the cyclic) strain-hardening behavior of the material. The distribution of stress and strain in the crack tip area will be highly inhomogeneous (large gradients) and this makes exact calculations for a cyclic loading difficult and expensive. Even if this information would be available a fracture criterion, including environmental effects, is not available, although some speculations have been published. It should be concluded that predicting Δa from first principles is a tremendous problem as yet unsolved.

To overcome this problem a practical approach is to correlate crack growth rates under similar conditions. The *similarity approach* implies:

similar conditions, applied to the same system, will cause similar consequences
↓ ↓ ↓
$\begin{pmatrix} \text{same } K\text{-values} \\ \text{same environment} \end{pmatrix}$ + $\begin{pmatrix} \text{same material} \\ \text{in crack tip area} \end{pmatrix}$ ⟶ (same crack rates).

Similar conditions imply the same loading on the crack tip area and the same environment surrounding the crack tip. The loading on the crack tip area can be described by the *stress intensity factor K*, to be discussed later.

The similarity approach is physically sound, but it should be carefully examined whether the required similarity is satisfied. It may be recalled that the similarity approach is also adopted for predicting the fatigue limit of a notched component by employing $K_f = K_t$. It is based on the fact that similar cyclic peak stresses in a notched and an unnotched component should cause (or should not cause) crack nucleation in *both* components. Actually the similarity approach is the basis for many predictions in Applied Mechanics. A fundamental understanding of the mechanism going on in the material is not required for the applicability of the similarity approach. However, understanding of the phenomenon is essential to see the limitations of the validity.

4. THE STRESS INTENSITY FACTOR "K"

With the theory of elasticity, assuming linearly elastic behavior (Hooke's law), the stress distribution in a cracked element can be calculated. The most simple problem is an infinite sheet loaded by a tensile stress S. The solutions for the stresses in the sheet are still fairly complex, and cannot be given as explicit functions.

However, if we consider stresses in the neighbourhood of the crack tip only (i.e. $r \ll a$) the formulas for the stress components become relatively simple[3]:

$$\sigma_x = \frac{K}{\sqrt{(2\pi r)}} \cos\frac{\theta}{2} \left(1 - \sin\frac{\theta}{2} \sin\frac{3\theta}{2}\right) \qquad (1)$$

$$\sigma_y = \frac{K}{\sqrt{(2\pi r)}} \cos\frac{\theta}{2} \left(1 + \sin\frac{\theta}{2} \sin\frac{3\theta}{2}\right) \qquad (2)$$

$$\tau_{xy} = \frac{K}{\sqrt{(2\pi r)}} \cos\frac{\theta}{2} \sin\frac{\theta}{2} \cos\frac{3\theta}{2} \qquad (3)$$

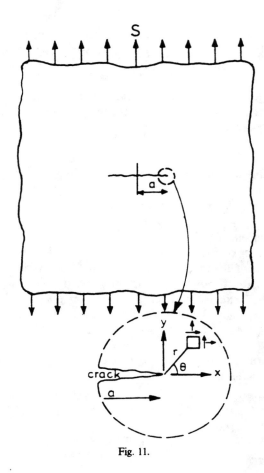

Fig. 11.

or

$$\sigma_{ij} = \frac{K}{\sqrt{(2\pi r)}} f_{ij}(\theta) \qquad (4)$$

with the stress intensity factor

$$K = S\sqrt{(\pi a)}. \qquad (5)$$

If the sheet has finite dimensions, but the load and the geometry are symmetric with respect to the X-axis (crack opening Mode I) eqns (1)–(4) still apply, while a geometry correction factor C (dimensionless) has to be added to eqn (5). This factor accounts for the shape of the component.

$$K = CS\sqrt{(\pi a)}. \qquad (6)$$

Equations (1)–(6) imply that stress distributions around crack tips are all geometrically similar, while the stress intensity factor fully determines all stress components. The similarity of stress distributions around different cracks is illustrated in Fig. 12.

The equations show,

$$\sigma_{ij} \div S \qquad \text{(Hooke's law)}$$

$$\sigma_{ij} \div r^{-1/2} \qquad \text{(square root stress singularity)}.$$

The singularity implies that $\sigma_{ij} \to \infty$ if $r \to 0$. Infinite elastic stresses are impossible and ductile materials will always show some plasticity at the tip of a crack. Two limitations of K to describe the stress field around a crack tip are apparent now:

(1) Equations (1)–(4) are valid only if $r \ll a$ (asymptotic solutions for the near field stresses).
(2) Equations (1)–(4) cannot be valid if $r \to 0$ (plasticity).

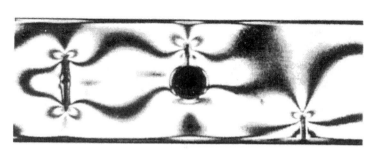
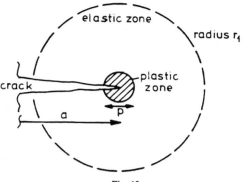

Fig. 12. Photo-elastic picture of a beam with three different cracks[4].

Fig. 13.

The first limitation need not bother us too much because we are interested in the crack tip area only. However, the second one should be a matter of concern.

In Fig. 13 an elastic zone (radius r_1) is considered, which is sufficiently small ($r_1 \ll a$) for eqns (1)–(6) to be valid. If no plasticity occurs the stresses acting on the periphery of this zone are determined by K. If a small plastic zone is formed some redistribution of stress and strain in the elastic zone (r_1) will occur, but as long as $p \ll r_1$ the redistribution will have largely faded at the periphery of the elastic zone (r_1). Consequently it may well be expected that the same K-value will always produce the same plastic zone as long as the plastic zone is small (small-scale yielding). In other words the similarity approach predicts: the same K-values will produce (in the same material) the same elastic–plastic deformation in the crack tip zone. As a result the same amount of crack extension Δa has to be expected.

5. K AND CYCLIC LOADING

From the definition: $K = CS\sqrt{(\pi a)}$ it follows that a cyclic variation of S will cause a similarly cyclic variation of K. The stress intensity in the crack tip area will thus be characterized by K_{max} and K_{min}.

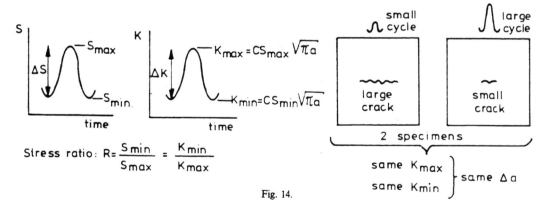

Fig. 14.

The similarity approach thus predicts

$$\left.\begin{array}{l}\dfrac{da}{dn} = f(K_{min}, K_{max}) \\[2mm] \text{where} \\[2mm] K_{max} = \dfrac{\Delta K}{1-R} \\[2mm] \text{and} \\[2mm] K_{min} = \dfrac{R\Delta K}{1-R}.\end{array}\right\} \quad \text{or:} \quad \begin{array}{l}\dfrac{da}{dn} = f(\Delta K, R) \quad (7) \\[3mm] \dfrac{da}{dn} = f_R(\Delta K). \quad (8)\end{array}$$

A formula proposed by Forman[5] is

$$\frac{da}{dn} = \frac{C\Delta K^m}{(1-R)(K_c - K_{max})}. \tag{9}$$

Figure 15 gives an illustration of the R-effect for an Al-alloy. Different S_a and S_m values produced the same $da/dn - \Delta K$ relation, provided the same R value applied.

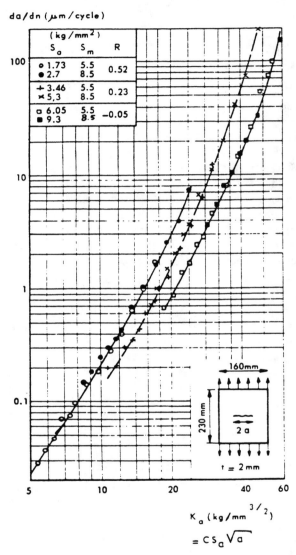

Fig. 15. Correlation between da/dn and K_a. Effect of stress ratio. Sheet specimens of an Al-alloy (2024–T3)[4].

Another illustrative example of the applicability of K is the comparison between crack growth results from specimens with end loading and specimens with crack edge loading, see Fig. 16, [6]. For crack edge loading the K-factor is decreasing for increasing crack length! There is a full agreement between the crack growth rates for the two types of loading if plotted as a function of ΔK. In other words the results of one specimen can be predicted if the results for the other one are known. For this prediction the functional relationship in eqns (7) and (8) need not be known analytically. The similarity approach leads to a correlation between the two cases, but not to analytical relations for crack growth. Such relations can be assumed by employing empirical crack growth data (curve fitting), but unfortunately they cannot be derived from physical arguments so far.

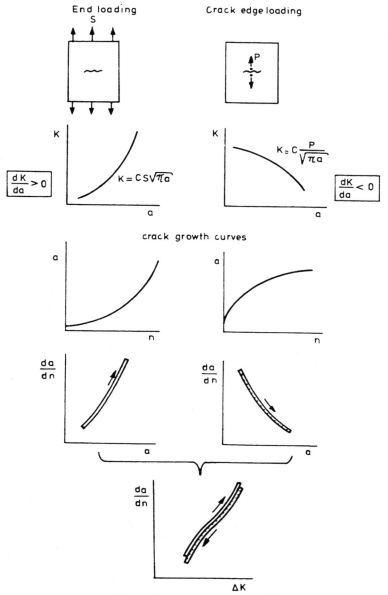

Fig. 16. Results in one scatter band[6].

6. REFERENCES

[1] A. V. de Forest, The rate of growth of fatigue cracks. *J. Appl. Mech.* **3**, A-23 (1936).
[2] N. E. Frost, K. J. Marsh and L. P. Pook, *Metal Fatigue*. Clarendon Press, Oxford (1974).
[3] P. C. Paris and G. C. Sih, Stress analysis of cracks. *ASTM STP* **381**, 30 (1965).
[4] J. Schijve, Fatigue crack propagation and the stress intensity factor. Delft University of Technology, Dept. of Aerospace Engng, *Rep.* M-191 (1973).
[5] R. G. Forman, V. E. Kearney and R. M. Engle, Numerical analysis of crack propagation in cyclic-loaded structures. *J. Basic Engng, Trans. ASME D* **89**, 459 (1967).
[6] I. G. Figge and J. C. Newman, Fatigue crack propagation in structures with simulated rivet forces. *ASTM STP* **415**, 71 (1967).

(*Received* 18 *April* 1978; *received for publication* 15 *June* 1978)

II. FATIGUE CRACKS, PLASTICITY EFFECTS AND CRACK CLOSURE

Abstract—Concepts introduced are residual plastic deformation, residual stress, reversed plastic deformation, plastic deformation in the wake of the crack and crack closure under tensile load. COD measurements as a method to determine the crack closure level are discussed. The significance of crack closure for fatigue crack growth is analysed and illustrated by several examples, including effects of yield stress, stress ratio and delayed crack growth after a peak load. Finally some attention is paid to three dimensional aspects following from thickness effects, shear lips and curved crack fronts.

1. INTRODUCTION

PLASTIC deformation occurring around the tip of a fatigue crack is usually considered to be "small scale yielding" in technical materials with a limited ductility, having a relatively high yield stress. However, small plastic zones are still sufficient to leave residual deformations and stresses and to cause closure. This is the main topic of the present lecture.

2. RESIDUAL STRESS AFTER LOCAL PLASTIC DEFORMATION

Let us consider a strip with a central hole loaded in tension (Fig. 1). If the load is sufficiently high plastic deformation will occur in small zones at the edge of the hole and this implies permanent deformation. If the strip is unloaded ($\Delta P \to 0$) the plastic zones do no longer fit in the elastic surroundings as a result of the plastic deformation. This misfit causes residual stresses. Assuming that unloading will cause elastic spring back only, a residual stress distribution as shown in Fig. 2 will occur along the X-axis.

It should be noted that there is a two-dimensional residual stress field which obviously will extend outside the plastic zones.

Let us now consider a strip with a much sharper notch, say a narrow elliptical hole, which is loaded and unloaded (Fig. 3). During uploading a plastic zone will be formed (along the X-axis approximately between A and C). Full elastic unloading will cause residual compressive stresses at A exceeding the compressive yield limit. As a result reversed plastic deformation will occur between A and B. Clearly enough AB will be considerably smaller than AC.

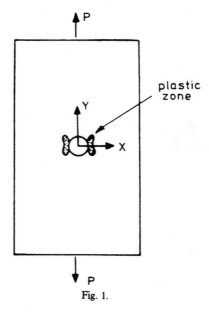

Fig. 1.

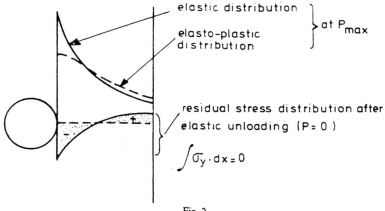

Fig. 2.

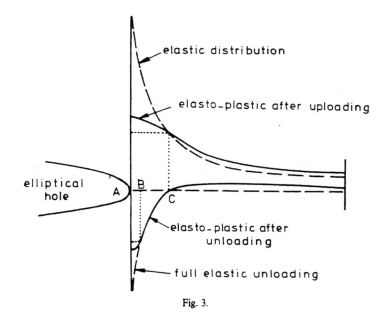

Fig. 3.

A similar picture applies if we consider a crack during uploading followed by unloading (Fig. 4). The monotonic plastic zone is significantly larger than the reversed plastic zone. As a first approximation the size of a plastic zone (r_p) is inversely proportional to the square of the yield stress. During unloading the stress increment to cause yielding in the reversed direction may be assumed to be twice the yield stress during uploading (Fig. 5, Bauschinger effect) and as a consequence [1]:

$$(r_p)_{uploading} \div \frac{1}{(\sigma_{yield})^2}$$

$$(r_p)_{unloading} \div \frac{1}{(2\sigma_{yield})^2}$$

and

$$\begin{pmatrix} \text{reversed} \\ \text{plastic} \\ \text{zone size} \end{pmatrix} = \frac{1}{4} \begin{pmatrix} \text{monotonic} \\ \text{plastic} \\ \text{zone size} \end{pmatrix}$$

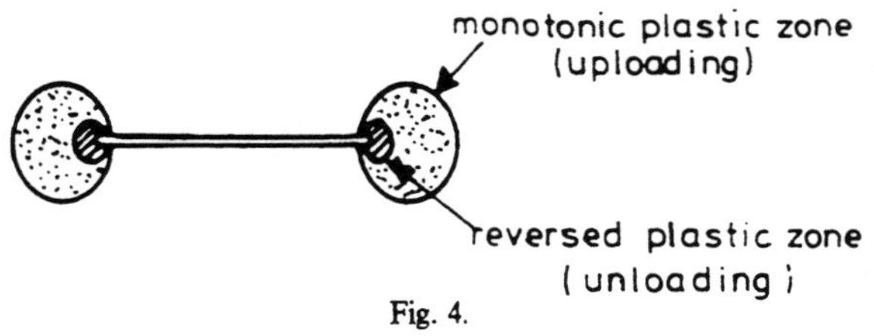

Fig. 4.

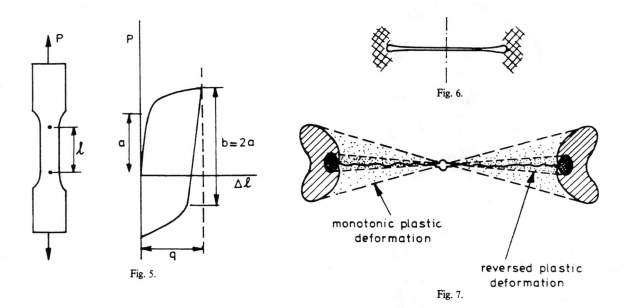

Fig. 5.

Fig. 6.

Fig. 7.

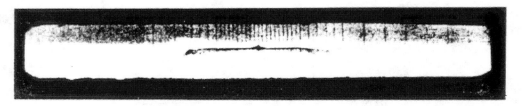

Fig. 8.

Question

Why will reversed plasticity start during unloading? Although uploading caused crack tip blunting, the crack is still an extremely sharp notch (very high K_t). Consequently unloading will cause reversed plasticity almost immediately upon reversion of the loading direction. A second question then is whether this will cause closing of the crack tip. To answer this question it should be realised that there is a difference between a stationary slit and a growing fatigue crack and this should be discussed first.

In a plate with a slit (in a test a very fine saw cut) plasticity during uploading will mainly occur ahead of the tip(s) of the slit (crack) with tip blunting as a result. Despite reversed plasticity during unloading the tip and the full slit will be open after complete unloading ($P = 0$). To close the crack it is necessary to apply a compressive load. As shown by elasto–plastic finite-element calculations (Ref. [2]) closing will start at the center line of the crack (Fig. 6), followed by more closure spreading outwards. De Koning[2] also analysed a growing crack, i.e. a crack extending in length during uploading and thus growing into the plastic zone created during that uploading. In this case the finite-element calculations showed that reversed plasticity caused crack closure at the crack tip during unloading (i.e. under tensile load).

3. CRACK CLOSURE

Let us now consider a growing fatigue crack (Fig. 7). During its growth the plastic zone is moving with the tip of the crack. It is also increasing in size and as a first approximation (central crack)

$$r_p \div K^2 \div a.$$

That means that the plastic zone size is proportional to the crack length a. The same will be true for the reversed plastic zone. Since the monotonic plastic zone is considerably larger than the reversed plastic zone, the consequence of the growing fatigue crack is that monotonic plastic deformation has been left in the wake of the crack. This deformation involves elongation in the y-direction. As a result of this elongation the crack will close (at least partly) during unloading, and after full unloading ($P = 0$) compressive residual stresses will be present in the wake of the crack. It means that residual compressive stresses are transmitted through the crack, because the fracture surfaces are pressed together by the plastic deformation left in the wake of the crack.

The phenomenon that the upper and lower fracture surface of a fatigue crack come together before complete unloading (i.e. at $P > 0$, tensile load) implies that the crack is no longer fully open. This phenomenon in the literature is referred to as "crack closure". It was first observed by Elber[3] and it is sometimes referred to as the Elber mechanism. The proof of the occurrence of crack closure can be obtained from direct evidence, such as stiffness measurements (COD measurements) to be discussed later. It also may come from indirect evidence, which is the effect on fatigue crack growth.

The picture in Fig. 8 shows two fatigue cracks started from a small notch (hole with two saw cuts) as viewed through a rectangular window. The plastic zone in the wake of the crack is visible due to a slight lateral contraction of the material and a simple illumination technique.

From the picture the wedge opening angle α can be measured: $tg\alpha \sim 0.11$. As a first approximation:

$$r_p = \frac{1}{\beta\pi}\left(\frac{K_{max}}{\sigma_{yield}}\right)^2 = \frac{1}{\beta}\left(\frac{CS_{max}}{\sigma_{yield}}\right)^2 \cdot a \rightarrow tg\alpha = \frac{r_p}{a} = \frac{1}{\beta}\left(\frac{CS_{max}}{\sigma_{yield}}\right)^2.$$

With $a \sim 21$ mm, $2w = 160$ mm, $S_{max} = 161$ N/mm^2 and $\sigma_{yield} \sim 450$ N/mm^2 the result is $\beta = 1.39$. This order of magnitude agrees with suggested values for plane stress.

4. HOW TO MEASURE CRACK CLOSURE

The most well known method to indicate crack closure is by COD measurements (COD = crack opening displacement). For crack opening between two points A and B, close to the edges of the crack and in the center of the panel, the relation for an infinite sheet is

$$\text{COD} = \frac{4a\sigma}{E}.$$

It means that COD is linearly proportional to σ (Hooke's law) and to the crack length a. For a finite sheet a geometry correction factor has to be added. Measurements on panels with fine saw cuts (see Fig. 10) have confirmed the linear relationship with excellent agreement between measured slopes (σ/COD) and theoretical values. If a similar test is carried out on a panel with a fatigue crack the σ-COD record shows a non-linear part. Above point A the record is linear and the slope is in agreement with the crack length, which implies that the crack is fully open.

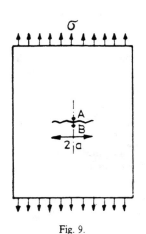

Fig. 9.

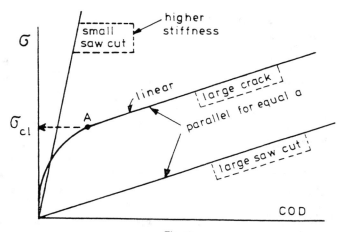

Fig. 10.

Below point A the slope (tangent to σ-COD record) is larger, which implies that the panel behaves as a panel with a shorter crack. This indicates that the crack is partly closed. The stress corresponding to point A is called the crack closure stress (σ_{cl}) or the crack opening stress (σ_{op}). A full loop of a COD record (Fig. 11) usually shows a slight hysteresis, but there is no doubt about the occurrence of full crack opening at A' the onset of crack closure at A''. The existence of a non-linear part followed by a linear part is easily observed, but the problem is to determine accurately the point A where the transition occurs. Measurements suggest that A' and A'' coincide, however, experience shows that the unloading branch (A'') gives a slightly better reproduction and a more unambiguous determination of the closure stress. Nevertheless it cannot be denied that it is difficult to achieve a high accuracy.

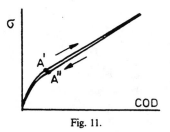

Fig. 11.

In order to improve the accuracy Paris[4] suggested a compensation method, illustrated by Fig. 12. Instead of recording the COD signal it is compensated by a signal that would have been obtained under full linear behaviour (i.e. no crack closure). This leads to a vertical line as long as the crack is fully open. The compensated COD signal now allows a much larger amplification which will bring out the transition point A more clearly. If the amplification is selected too high the linear part (vertical line) may become erratic.

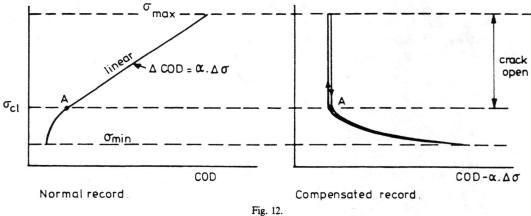

Fig. 12.

Another possibility to improve the sensitivity of the crack closure measurement is to locate the COD meter more closely to the crack tip. The effect on the σ-COD record is shown in Fig. 13.

The non-linear part ($A'' - A'$) of the record is small and the transition is more easily observed. This method can be used for a particular crack length, but if the crack is growing the COD meter has to be moved also. If the COD-meter is too close to the crack tip plasticity effects may obscure the measurements.

Several other methods to measure crack closure have been proposed in the literature (for a survey see [5]). Some of them may be mentioned here.

(1) *Electrical potential method.* This method was developed for automatic crack length measurements. It is possible to use this method, but difficulties may arise due to electrical short circuits over the crack, e.g. at the shear lips or some other minor asperities. A mechanically open crack is not the same as an electrically open crack. The risk of problems is smaller (but not absent) if the material forms an oxide layer, which acts as an insulator on the fracture surface (which will not happen in an inert environment, e.g. in vacuum).

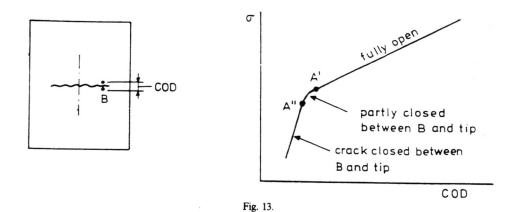

Fig. 13.

(2) Strain gauges on the surface near the crack. Indications of non-linear behaviour are obtained, but interpretation problems may occur.

(3) Ultrasonic transmission of surface waves through the crack line. Transmission is possible if the crack is closed. However, acoustic short circuiting is possible. A mechanically open crack is not the same as an "acoustically" open crack.

5. CRACK CLOSURE AND FATIGUE CRACK GROWTH

In 1970 W. Błazewicz carried out fatigue tests in Delft (unpublished results). He made ball impressions on 2024–T3 sheet specimens before the crack growth test was started (see Fig. 14).

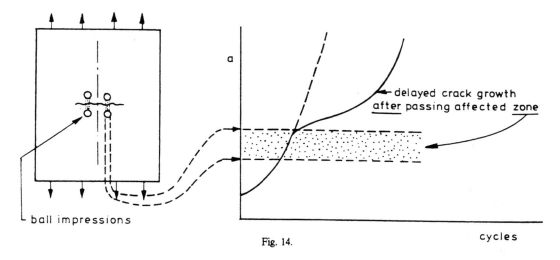

Fig. 14.

As a result there is a zone between the impressions with residual compressive stresses. This caused a delay in the crack growth, but (surprisingly enough at that time) the delay was small during the growth through the zone between the impressions, whereas it was significant at a later stage. The explanation is that the deformations of the ball impressions were the cause of crack closure after the crack had grown through the affected zone.

Elber[3] suggested the following relation between crack closure and crack growth. During a stress cycle a fatigue crack will be partly or fully closed as long as $\sigma < \sigma_{cl.}$ (Fig. 15). He then suggested that the stress variation will contribute to crack extension only if

$$\sigma < \sigma_{cl.}$$

which leads to the definition of an effective stress range:

$$\Delta\sigma_{\text{eff.}} = \sigma_{\max} - \sigma_{cl.}$$

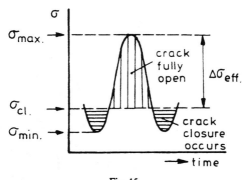

Fig. 15.

and an effective stress intensity factor:

$$\Delta K_{\text{eff.}} = C \Delta \sigma_{\text{eff.}} \sqrt{(\pi a)}.$$

The crack rate was supposed to be dependent on $\Delta K_{\text{eff.}}$ only

$$\frac{da}{dn} = f(\Delta K_{\text{eff.}}).$$

This relation includes the effect of the stress ratio R because crack closure (and thus $\Delta K_{\text{eff.}}$) will depend on R. For 2024–T3 material Elber found that $\sigma_{\text{cl.}}$ was approximately constant during a fatigue test, implying that $\sigma_{\text{cl.}}$ was independent of the crack length a. This is an empirical result. He defined the ratio:

$$U = \frac{\Delta K_{\text{eff.}}}{\Delta K} \left(= \frac{\Delta \sigma_{\text{eff.}}}{\Delta \sigma} \right).$$

and the test results indicated the relation:

$$U = 0.5 + 0.4R.$$

This is again an empirical result. Combining the above equations leads to

$$\log \Delta K_{\text{eff.}} = \log \Delta K + \log (0.5 + 0.4R).$$

The difference between $\log \Delta K_{\text{eff.}}$ and $\log \Delta K$ is constant for constant R. The assumption $da/dn = f(\Delta K_{\text{eff.}})$ then implies that plots of da/dn as a function of $\log \Delta K$ should give parallel curves for different R-values. Results from Ref. [6] in the left graph of Fig. 16 confirm this trend. In these tests crack closure measurements were not made, however, adopting Elber's formula for $\Delta K_{\text{eff.}}$ the right hand graph of Fig. 16 shows that all three curves practically coincide.

Some comments on crack closure and Elber's proposals are appropriate now.

(1) Although it is not easy to measure accurately the stress level at which crack closure does occur, there is abundant evidence that it occurs in several materials under tensile load.

(2) The relation between $\Delta K_{\text{eff.}}$ and R will depend on the type of material. The above relation ($U = 0.5 + 0.4R$) was found for an Al-alloy (2024–T3) as an empirical result. It may be expected that such relations can be calculated by elasto–plastic finite element techniques. The same applies to the effect of crack length and type of specimen. Elber found $\Delta K_{\text{eff.}}$ to be independent of crack length for a center cracked specimen.

(3) The relation $da/dn = f(\Delta K_{\text{eff.}})$ assumes that cyclic stress is significant only if the crack is fully open, whereas it is insignificant as long as the crack is partly closed. The basic idea is that there is no stress singularity at the real tip of the crack unless it is fully open. However, if the crack tip is almost open it should be expected that some preliminary plastic deformation will occur. As a first approximation it is ignored in Elber's proposals.

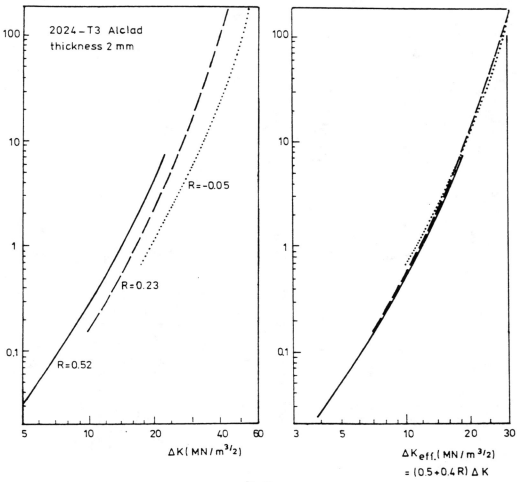

Fig. 16.

6. SOME CONSEQUENCES OF CRACK CLOSURE

Crack closure is caused by plastic deformations left in the wake of the crack. It should be expected that material with a lower ductility will form smaller plastic zones and thus show less crack closure. This was recently explored by comparing 2024–$T3$ (as received) with the same material after 3% plastic prestraining[7]. The prestraining raised the yield stress from 428 to 480 MN/m^2. The material was then cycled between 13.8 and 138 MN/m^2. As a result of the prestraining the crack closure level was systematically lower (Fig. 17) in agreement with the higher yield stress. The crack growth rate was about twice as high as for the unprestrained material as shown in the left hand graph of Fig. 18. The same results plotted as a function of ΔK_{eff} (calculated from the measured closure stresses) are shown in the right hand graph. The two curves come closer together, but they do not coincide. Obviously, crack closure can explain half of the effect of the higher yield stress, but it cannot explain the full difference. It should be said that plastic prestraining also implies that the material itself is changed by strain-hardening. However, without knowing $\Delta\sigma_{\text{eff}}$ an explanation anyhow will be incomplete.

An intermediate peak load applied in a constant-amplitude test causes the well known crack growth delay. An example from recent tests is shown in Fig. 19[8]. When a crack length a of 15 mm was reached a peak load was applied causing a delay of 33,000 cycles as compared to the original crack propagation curve for constant-amplitude loading only. Crack closure measurements until the peak load occurred were measured on other specimens. The closure level was lower than the minimum stress of the cycle ($\sigma_{\text{cl}} < \sigma_{\text{min}}$). As a result of the peak load it decreased even further, because the plasticity ahead of the crack induced by the peak load

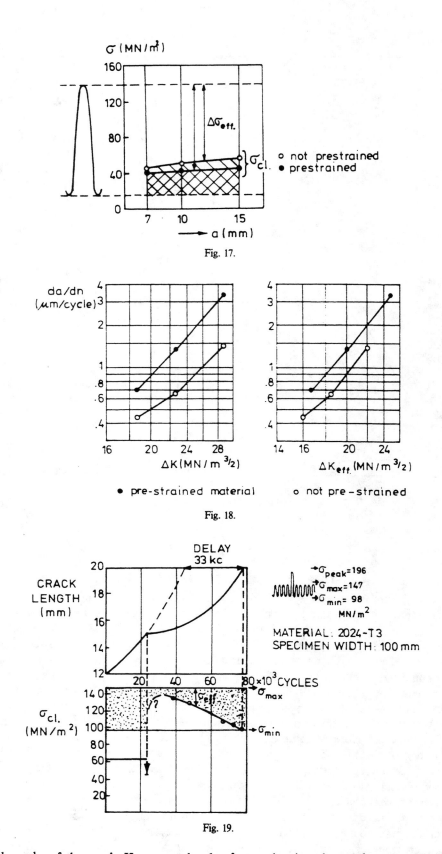

Fig. 17.

Fig. 18.

Fig. 19.

opens the wake of the crack. However, shortly afterwards when the crack penetrated into the plastic zone of the peak load the closure stress raised significantly up to a level beyond σ_{min}. Then $\Delta\sigma_{eff}$ is reduced and a delaying effect should be expected. Later on σ_{cl} decreased and when it dropped below σ_{min} no further delay was observed. As Fig. 19 shows the delay occurred during a crack length increment $\Delta a = 5$ mm which is significantly larger than the plastic zone size associated with the peak load. It confirms that crack closure can still occur after the crack has fully grown through the plastic zone of the peak load. A similar indication was obtained in the tests of Błazewicz as mentioned before.

Crack closure was also observed in more complex variable-amplitude tests, such as a flight-simulation tests. This will be discussed in another lecture.

Crack closure measurements were also made on two Al-alloys (2024–$T3$ and 7075–$T6$) tested in three environments, viz. vacuum, humid air and salt water[9]. Although the crack growth rates in the three environments were highly different, the crack closure stress was the same in all environments. The size of plastic zones should not be expected to be dependent on the environment and consequently the same amount of plasticity in the wake of the crack will occur, which implies similar σ_{cl} values. Environmental aspects should therefore be explained on the basis of other arguments[10].

7. CRACK CLOSURE AND THE EFFECT OF THICKNESS

In the preceding sections fatigue crack growth in a sheet or plate has been considered as a two-dimensional problem with crack growth in X-direction and loading in the Y-direction. The thickness of the material is adding the third dimension (Z-direction) and it should be expected that this will complicate the problem. The most simple case is a crack with a straight crack front perpendicular to the surface of the material, as depicted in Fig. 20. The reason why the thickness may be important is the difference between the state of stress at the free surface (plane stress, $\sigma_z = 0$) and the state of stress along the crack front away from the free surface (approximately plane strain, $\epsilon_z = 0$). Frequently quoted formula's for the plastic zone size for the two states of stress are

$$\text{Plane stress:} \quad r_p = \frac{1}{2\pi} \left(\frac{K}{\sigma_{\text{yield}}} \right)^2,$$

$$\text{Plane strain:} \quad r_p = \frac{1}{6\pi} \left(\frac{K}{\sigma_{\text{yield}}} \right)^2.$$

The formulas suggest that the plastic zone size is three times larger in the plane stress condition. Larger plastic zones imply more deformation left in the wake of the growing crack and hence more crack closure. Increasing the thickness will lead to a relatively smaller part of the crack front in plane stress (see Fig. 21) and as a consequence less effective crack closure. This was confirmed in some recent tests[8] on an Al-alloy as shown in Fig. 22. Whether this should fully account for faster crack growth in thicker material is doubtful, because a thickness effect on crack rate is also observed if $\sigma_{cl} < \sigma_{min}$.

The above picture about the relation between thickness and crack closure suggests that crack closure should predominantly occur near the free surface and to a much lesser extent at the interior of the material. This was checked in fatigue tests[11] on material with an original thickness of 10 mm which gave a crack closure level of 33 MN/mm^2 (see Fig. 23). The thickness of the specimen was then reduced from both sides by removing the surface material. After

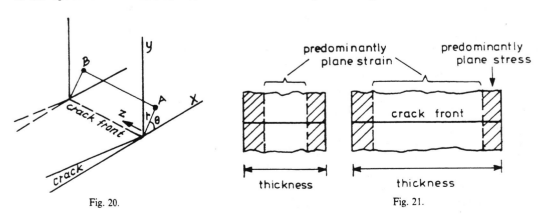

Fig. 20.

Fig. 21.

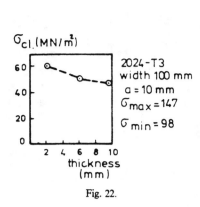

Fig. 22.

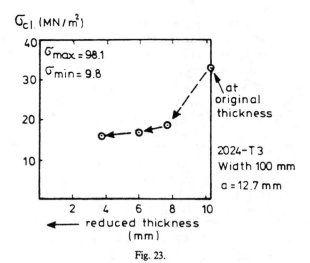

Fig. 23.

some steps σ_{cl} was measured again and the results indicate that the major part of crack closure was associated with larger plastic zones at the free surface.

Another elegant confirmation that crack closure predominantly occurs at the free surface was presented by McEvily[12]. He found in tests on the Al-alloy 6061 (Al Si Mg) that a peak load introduced a significant crack growth delay. He then reduced the thickness of the specimen immediately after the peak load and a much smaller delay occurred.

The real elasto–plastic three-dimensional problem is more complicated than discussed before. Along the crack front there is a transition from plane stress at the surface to approximately plane strain at the interior. However, the r_p-formulas given before, apply to either pure plane stress (extremely thin material) or pure plane strain (very thick material). The formulas then suggest that the plastic zone is three times larger in plane stress. Probably the factor will be less for intermediate thicknesses. Consider a line AB in Fig. 20 from one free surface to the other one (r = constant, θ = constant). A plane stress zone at the surface and a plane strain zone at the interior cannot deform independently. Both zones have to be coherent and as a first approximation it may be assumed that the line AB will remain a straight line after deformation. This is equivalent to assuming that ϵ_x and ϵ_y are independent of z. Different zone sizes will still be obtained because at the free surface due to the plane stress situation a lower effective yield limit will apply (von Mises criterion, $\sigma_z = 0$). As a consequence the plastic zone will still be larger at the free surface.

In the literature it is sometimes suggested that a plane stress situation through the full thickness will be obtained, if the estimated plastic zone size (r_p) calculated with the plane stress formula is equal or larger than half the material thickness, i.e.

$$r_p \geq \frac{1}{2} t.$$

The paradox may be that adopting the plane strain formula will indicate $r_p < (1/2)t$. Furthermore considering cyclic plasticity at the crack tip, which is causing crack growth, the reversed plastic zone size estimated with the plane strain formula is:

$$r_p = \frac{1}{6\pi} \left(\frac{\Delta K}{2\sigma_{\text{yield}}} \right)^2.$$

Considering $\Delta \sigma$ values in the order of $0.5\sigma_{\text{yield}}$ or smaller we obtain:

$$r_p \leq \frac{1}{6\pi} \left(\frac{0.5\sqrt{(\pi a)}}{2} \right)^2$$

or

$$\frac{r_p}{a} \leq \sim 0.01.$$

It then may be expected that many fatigue cracks will be predominantly in plane strain during the larger part of crack growth. However, further study of this issue is necessary.

8. CRACK CLOSURE AND SHEAR LIPS

In general the fracture surface of a fatigue crack is perpendicular to the main principal stress. However, at a free surface the growing fatigue crack is forming shear lips (see previous lecture) which become wider as the crack is growing faster. In the past shear lips were associated with plane stress conditions at the free surface. If this were true shear lips might be important for the occurrence of crack closure. However, since we know more about environmental effects on fatigue fracture behaviour, the correlation between shear lips and plane stress conditions seems to be less positive. In aggressive environments shear lips are much smaller than in inert environments under the same cyclic loading[10]. In other words shear lip formation is also depending on the fatigue fracture mechanism and since the environment will not affect the state of stress the width of shear lips cannot be an unambiguous indication of the plastic zone size.

In spite of the above complexity it still should be expected that the free surface edges of shear lips were part of larger plane stress plastic zones. As a consequence crack closure during unloading will start at the shear lip edges[13]. If contact between the upper and lower fracture surface start in the slant shear lips some rubbing should be expected. As a matter of fact black debris is observed on the shear lips of fatigue fractures in aluminium alloy material, whereas it is absent on the tensile mode part of the fracture.

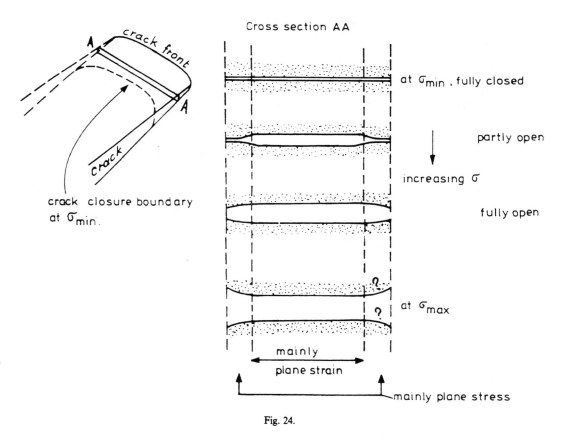

Fig. 24.

9. CRACK CLOSURE AND CRACK FRONT CURVATURE

In the previous sections it was assumed that the crack front in a plate was a straight line perpendicular to the material surface. In reality a tendency to slightly curved crack fronts is frequently observed. Under certain conditions large curvatures may occur (tongues) but they will not be discussed here.

It is not fully correct that a curved crack front implies a slower crack growth at the free surface. If the amount of curvature of a growing crack does not change the crack rate is the same all along the crack front. However, the curvature anyhow implies that crack growth is lagging behind at the surface. It similarly could be said that crack growth at the interior remains ahead of the growth at the surface. It is a good question to ask why this can occur. In terms of crack closure a straightforward explanation seems to be possible (see Fig. 24). More crack closure at the surface and less crack closure at the interior implies a higher ΔK_{eff} at the interior. Figure 24 suggests that the crack tip will be fully closed at σ_{min}, but it is possible that it will not fully close at the interior. Plastic infiltrations of cracks, now carried out by Bowles[14] may answer this question.

In addition to the above arguments it should not be overlooked that the crack can lag behind at the surface because plastic deformation is more easy there (lower restraint) which may delay the cracking mechanism.

Elastic deformations are also more easy at the surface. In a recent publication[15] Newman analysed the elastic stress intensity along a crack with a straight crack front. Three-dimensional FEM calculations showed a drop of the K-value near the free surfaces of the plate.

10. REFERENCES

[1] J. R. Rice, The mechanics of crack tip deformation and extension by fatigue. *ASTM STP* **415**, 247 (1967).
[2] A. V. de Koning, Nat. Aerospace Lab. NLR, Amsterdam (to be published).
[3] W. Elber, Fatigue crack closure under cyclic tension. *Engng Fracture Mech.* **2**, 37 (1970); The significance of fatigue crack closure. *ASTM STP* **486**, 230 (1971).
[4] P. C. Paris. Private communication.
[5] V. Bachmann and D. Munz, Crack closure in fatigue crack propagation. *Proc. S.E.E. Conf. Fatigue Testing and Design*, London (April 1976).
[6] J. Schijve, Fatigue crack propagation and the stress intensity factor. Delft University of Technology, Dept. of Aerospace Engng, *Memo.* M-191 (Feb. 1973).
[7] J. Schijve, The effect of pre-strain on fatigue crack growth and crack closure. *Engng Fracture Mech.* **8**, 575 (1976).
[8] W. J. Arkema, results quoted in J. Schijve, Observations on the prediction of fatigue crack growth propagation under variable-amplitude loading. *ASTM STP* **595**, 3 (1976).
[9] J. Schijve and W. Arkema, Crack closure and the environmental effect on fatigue crack growth. Delft University of Technology, Dept. of Aerospace Engng, *Rep.* VTH-217 (1976).
[10] L. B. Vogelesang, Some factors influencing the transition from tensile mode to shear mode under cyclic loading. Delft University of Technology, Dept. of Aerospace Engng, *Rep.* CR-222 (1976).
[11] R. Th. Furnee, Thesis (in Dutch), Department of Metallurgy, Delft University of Technology (Mar. 1977).
[12] A. J. McEvily, Current aspects of fatigue. Appendix: Overload experiments. *Fatigue 1977 Conf.*, University of Cambridge (28–30 March 1977).
[13] T. C. Lindley and C. E. Richards, The relevance of crack closure to fatigue crack propagation. *Mat. Sci. Engng* **14**, 281 (1974).
[14] C. Q. Bowles, An experimental technique for vacuum infiltration of cracks with plastic and subsequent study in the scanning electron microscope. Delft University of Technology, Dept. of Aerospace Engng, *Rep.* LR-249, (June 1977).
[15] I. S. Raju and J. C. Newman, Three-dimensional finite-element analysis of finite-thickness fracture specimens. *NASA TN* D-8414 (1977).

(*Received* 18 *April* 1978; *received for publication* 15 *June* 1978)

III. FATIGUE CRACK PROPAGATION, PREDICTION AND CORRELATION

Abstract—Two prediction techniques are introduced, (1) cycle-by-cycle prediction and (2) prediction by correlation. Attention is paid to the problem of describing variable-amplitude loading in terms of load cycles. Aspects of fatigue damage are reviewed with reference to interaction effects and weaknesses in cycle-by-cycle prediction methods. The discussion on prediction by correlation is restricted to constant-amplitude loading. The validity of the similarity concept based on K-factors is reconsidered. Application of simple specimen data to complex structures is shown. Finally a variety of crack growth equations is reviewed, including aspects of curve fitting, a comparison between formulas of Walker and Elber and asymptotic values in the $da/dn - \Delta K$ relation.

1. INTRODUCTION

BASICALLY there are two different prediction techniques:
1. Cycle-by-cycle prediction.
2. Prediction by correlation, employing characteristic K-values.

In the literature the first technique has been adopted by several authors for predicting crack growth under variable-amplitude loading (*VA-loading*). All methods of this type include some correlation with crack growth data obtained under constant-amplitude loading (*CA-loading*). The second technique is well-known for *CA*-loading, but application to *VA*-loading received limited attention.

In this lecture problems involved in the first technique are summarized, while the discussion on the second one is restricted to *CA*-loading. In a fourth lecture more attention is paid to crack growth under *VA*-loading.

2. CYCLE-BY-CYCLE PREDICTION FOR VARIABLE-AMPLITUDE LOADING

Before we can discuss the cycle-by-cycle prediction methods it is necessary to specify in some detail variable-amplitude loading (*VA-loading*) as a load-time history. Two samples of *VA*-loading, as they may occur on a structure in service, are shown in Fig. 1. Although many more types of *VA*-loading are possible (and will be defined in the fourth lecture) these samples are shown here, because in the first one a definition of load cycles seems relatively easy, whereas in the second one it is more complex.

Whatever the type of a cyclic load is, the significant events for the fatigue behaviour of the material are the maxima and the minima, i.e. the moments at which the loading direction is reversed. That is what the material feels. After passing a maximum or a minimum of the load

a. Narrow-band random load.

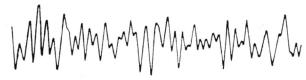

b. Broad-load random load.

Fig. 1. (a) Narrow-band random load; (b) Broad-load random load.

the reversion of stress and strain will start immediately. For a crack tip it implies in view of the stress and strain singularity, that reversed plasticity will also start immediately as soon as the loading direction is reversed.

In addition to the minima and the maxima the load-time history is characterized by a loading rate, which can vary in some systematic way (e.g. sinusoidal loading) or in a random way. If time dependent effects on fatigue are involved (corrosion fatigue e.g.) the loading rate has to be considered (effects of wave shape, and cyclic frequency). If time dependent effects are not involved the load-time history is described in sufficient detail by a series of successive minima and maxima.

The load on a structure in service can be measured for a long period in order to be representative. As a result the series of maxima and minima will become a very large one and for practical reasons it is then condensed to statistical distribution functions of peak values (maxima and minima) or ranges (differences between successive maxima and minima) or both. This can be done by counting how many times events of a certain magnitude have occurred in a certain period, preferably a long period. A variety of counting methods have been developed [1–3] but a discussion here is outside the scope. It should be pointed out, however, that a distribution function of counted events does not give information on the sequence in which the events have occurred. The counted results will be informative only if information on the sequence is available also. For a stationary random load (e.g. gust loads on an aircraft wing during constant weather conditions) and for systematically planned service loads (e.g. pressure cycles on a pressure vessel) this information can be provided. For unstationary randon loading and unscheduled manoeuvre loadings assumptions on the sequence can sometimes be made.

For simplicity let us assume that load cycles occurring under VA-loading are fully described by a minimum and an associated maximum, from which related stress values (σ_{min} and σ_{max}) and K-values (K_{max} and K_{min}, or ΔK and R) can be derived. The crack length after N cycles according to the non-interaction cycle-by-cycle prediction technique can be obtained by calculating:

$$a_N = a_o + \sum_{i=1}^{N} \Delta a_i \tag{1}$$

with

$$\Delta a_i = \frac{da}{dn} \quad \text{for } \Delta K \text{ and } R \text{ in cycle } i \text{ as obtained under } CA\text{-loading.} \tag{2}$$

In this relation a_o is the initial value (a for $N = 0$). Each cycle will contribute a crack length increment (Δa_i) which is assumed to be equal to the increment obtained at the same ΔK and R value in a constant-amplitude test. This increment is equal to the crack rate observed under CA-loading. The approach assumes that Δa_i in cycle i is independent of the history of the preceding crack growth. Unfortunately in view of so-called interaction effects this is not correct. A well-known example of interaction effects is delayed crack growth after a high peak load (Ref. [4]). In general terms the crack length increment Δa in a cycle will be a function of:

1. the crack geometry being present before the cycle started,
2. the condition of the crack tip material, and
3. the magnitude of the load cycle.

Only the magnitude of the cycle and one dimension of the crack (a) is accounted for by ΔK and R. However, significant aspects resulting from the preceding fatigue cycles are disregarded by ΔK and R. This is illustrated in Fig. 2.

Various interaction effects between the "fatigue damage" as left from preceding cycles and the increments of damage in future cycles are possible, depending on changes of the cyclic load (i.e. in VA-loading). Several interaction effects, surveyed in Ref. [4], can be understood qualitatively by considering crack closure (see fourth lecture). However, such interaction effects are ignored in eqns (1) and (2) and as a result predictions obtained are generally conservative and sometimes highly conservative (see e.g. [5]).

The conservatism of the non-interaction prediction has prompted some models for crack growth under VA-loading with simple assumptions for the effect of high load cycles on delayed crack growth in subsequent cycles of a lower magnitude. Two models [6, 7] receiving more

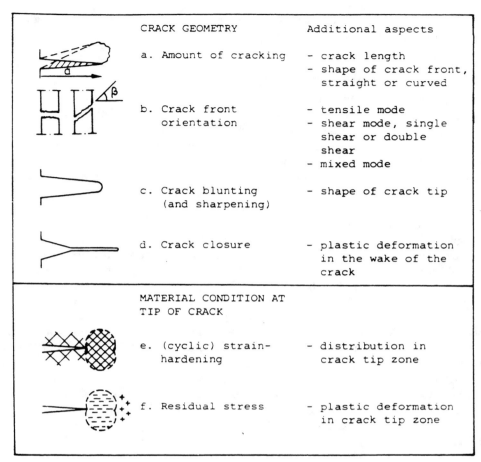

Fig. 2. Aspects of fatigue damage.

attention in the literature are referred to as:
—Willenborg-model and the
—Wheeler-model.

The main aim of the models was to account for delayed growth and the assumptions made by Wheeler[7] are certainly more reasonable than for the other model.

He replaced eqn (2) by

$$\Delta a_i = \left(\frac{da}{dn}\right)_{VA, a=a_i} = \beta \left(\frac{da}{dn}\right)_{CA, a=a_i} \qquad (3)$$

with a delay factor β (≤ 1) depending on plastic zone sizes. Assumptions are made about this dependence including the introduction of one empirical material constant. In view of this constant there is a possibility to adjust its value in order to fit empirical crack growth dates obtained under VA-loading. This may well offer a possibility to correlate crack growth data obtained under specific types of VA-loading occurring in service[8] and thus offer some promise for crack growth predictions in practical problems. At the same time we should realize that such a model has to be physically incorrect because it is fundamentally uncapable to account for physical observations, such as: (1) accelerated crack growth under certain load variations ($\beta > 1$), (2) delayed retardation after a high peak load[9] and (3) a sustained delay after a high peak load even when the crack has already fully penetrated through the plastic zone created by the high peak load (see previous lecture). Qualitatively these observations can be explained by crack closure as will be shown in the fourth lecture.

In closing it should be noted that eqn (3) implies that it is still tried to correlate crack extension under VA-loading with crack extension under CA-loading! In view of the complexity of fatigue crack extension and possible interaction mechanisms, one cannot be very optimistic about deriving a reliable prediction technique with some general validity based on equations like eqn (3).

3. CRACK GROWTH PREDICTION BY CORRELATION

Predictions made by correlation are based on the similarity approach discussed in the first lecture. Briefly, similar conditions on the same system will cause similar consequences. For crack growth it implied: the same ΔK and R in a load cycle will produce the same crack growth increment $\Delta a = da/dn$. This approach seems reasonable if applied to CA-loading. However, it will be clear that the similarity is no longer satisfied if a single cycle of VA-loading is compared to a single cycle of CA-loading, in spite of the same ΔK and R being applicable. The reason is that we do not consider the same system, because that includes,

—the same material in the plastic zone
—the same geometry at the crack tip.

Under VA-loading this cannot be guaranteed since all aspects mentioned in Fig. 2 will depend on previous history. In the fourth lecture it will be shown that the stress intensity factor can still be used to correlate crack growth rates as obtained under similar VA-loadings, but it has to be concluded here that the K-concept is unsuitable to correlate crack growth between dissimilar types of cyclic loading.

Restricting ourselves to CA-loading the requirement of having the same material in the plastic zone should be given some more attention. This material has seen monotonic plasticity and considerable cyclic plasticity. Similar material implies similar (cyclic) strain-hardening. In Fig. 3 two cracks are compared, obtained under a low and a high CA-loading respectively. For the crack length shown in the figure they have the same ΔK (and R) and hence the same plastic zone size. However, the cyclic strain-hardening in these zones (and also other damage aspects) will be similar only if the crack growth in the preceding cycles occurred with the same growth rate in both cases. That requires equal da/dn values in previous cycles, implying that the derivative

$$\frac{d\left(\frac{da}{dn}\right)}{dn}$$

should be equal. In the first lecture we arrived at

$$\frac{da}{dn} = f_R(\Delta K) \qquad (4)$$

and thus

$$\frac{d\left(\frac{da}{dn}\right)}{dn} = f'(\Delta K) \cdot \frac{d\Delta K}{da}. \qquad (5)$$

Consequently similar cyclic strain-hardening requires that dK/da should be similar also (in addition to similar ΔK and R values). However, the requirements of both equal K-values and equal dK/da-values are not compatible in general, as illustrated in Fig. 4.

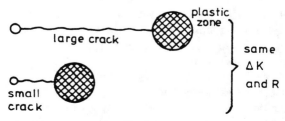

Fig. 3.

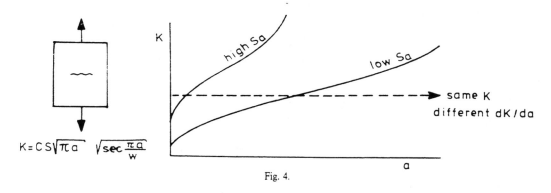

Fig. 4.

As a result we should rewrite eqn (4) more generally as:

$$\frac{da}{dn} = f_R(\Delta K, dK/da). \qquad (6)$$

Results obtained in tests with CA-loading do not reveal any noticeable effect of dK/da. It should be realized that this is an empirical observation, although it is a rather fortunate finding. One extreme example of a negligible dK/Da influence was given in the first lecture (Fig. 16) by comparing the $da/dn - \Delta K$ results of a specimen with end loading and a specimen with crack edge loading. For the first specimen dK/da is positive and for the second one it is negative. Nevertheless $da/dn = f_R(\Delta K)$ was apparently applicable.

If dK/da has little effect on crack growth we may feel more confident that crack growth results of CA-tests on simple specimens will be applicable to CA-loading on structures with a more complex relation between K and a. This will be illustrated by considering crack growth in a stiffened panel (see Fig. 5). The crack growth data obtained in tests on a simple sheet specimen are used as a calibration curve, i.e. the relation $da/dn = f_R(\Delta K)$. For the stiffened panel (a skin with a number of stringers) the relation between ΔK and the crack length has to be calculated. For simple stiffened panels analytical solutions are available, but otherwise finite-element methods have to be adopted. The prediction of crack growth in the stiffened panel is then a simple conversion of the $\Delta K - a$ relation of the panel into a $da/dn - a$ relation by application of the calibration curve (Fig. 5). An example of predicted an observed growth rates in the skin of a stiffened panel is shown in Fig. 6. The prediction is that da/dn will initially increase, but when approaching a stringer the stress intensity factor of the crack in the skin is reduced by the stringer. As a result the growth rate is decreasing. The agreement between prediction and observation is considered to be promising.

In the literature K-values for a large variety of geometries have been compiled in handbooks[11–13] and the number of available solutions is steadily increasing. Nevertheless for several practical cases K-values cannot be found in the literature, but sometimes clever estimates can be derived from known solutions. If this is not sufficiently accurate, calculations (FEM) have to be made, while experimental techniques have also been developed.

4. CRACK GROWTH EQUATIONS FOR CONSTANT-AMPLITUDE LOADING

In the literature many crack growth equations can be found. Older equations recognizing the effects of crack length, σ_a and σ_m were of the type,

$$\frac{da}{dn} = C\sigma_a^\alpha \sigma_m^\beta a^\gamma \qquad (7)$$

where α, β and γ had to be determined from empirical data. A significant step was the recognition that K is a good parameter to indicate the severity of the stress field at the tip of a crack (Irwin) and to relate da/dn to ΔK (Paris)

$$\frac{da}{dn} = f(\Delta K). \qquad (8)$$

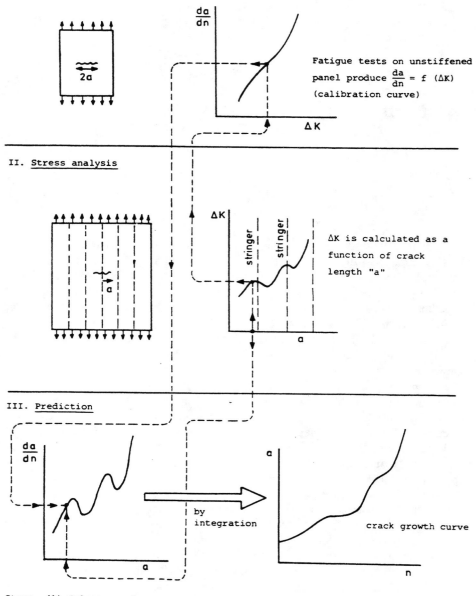

Fig. 5. Prediction of crack growth in a stiffened panel.

An example frequently referred to as the Paris relation is,

$$\frac{da}{dn} = C\Delta K^m \tag{9}$$

with C and m being material constants[14]. The function implies a linear relation in a double-log plot. When more data became available it was obvious that the linearity was not generally confirmed, see Fig. 15 of the first lecture. Moreover eqn (9) does not include an effect of mean stress or stress ratio. It was suggested by Walker[15] that this effect could be accounted for by assuming,

$$\frac{da}{dn} = f(\Delta K^\alpha \cdot K_{max}^\beta) \tag{10}$$

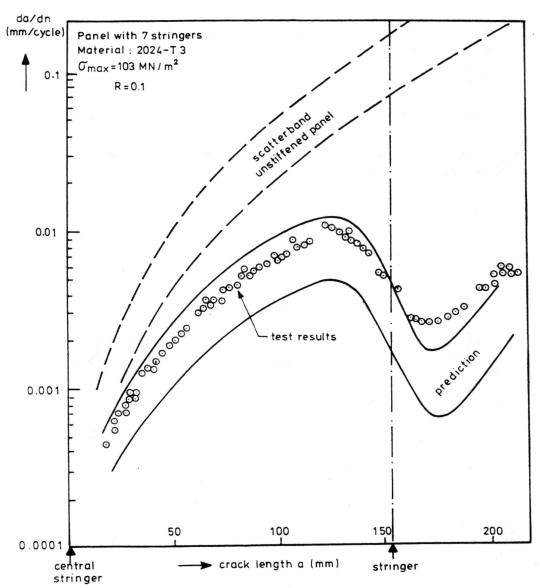

Fig. 6. Crack growth rate in stiffened panel, comparison between prediction and test results[10].

which he also wrote as,

$$\frac{da}{dn} = f(\overline{\Delta K}) \tag{11}$$

with

$$\overline{\Delta K} = \Delta K^{\alpha/(\alpha+\beta)} \cdot K_{max}^{\beta/(\alpha+\beta)} = \Delta K^m \cdot K_{max}^{1-m} \tag{12}$$

or

$$\overline{\Delta K} = C\Delta\sigma^m \cdot \sigma_{max}^{1-m} \sqrt{(\pi a)} = C\overline{\Delta\sigma}\sqrt{(\pi a)} \tag{13}$$

with

$$\overline{\Delta\sigma} = \Delta\sigma^m \cdot \sigma_{max}^{1-m} = \frac{\Delta\sigma^m}{(1-R)^{1-m}}. \tag{14}$$

Walker referred to $\overline{\Delta\sigma}$ as an "effective stress".

A relation proposed in [16] for 2024–T3 material was

$$\frac{da}{dn} = C\Delta K \cdot K_{\max}^2 \qquad (15)$$

or in terms of Walker's formula,

$$\frac{da}{dn} = C(\overline{\Delta K})^3 \quad \text{with} \quad \overline{\Delta K} = \Delta K^{1/3} \cdot K_{\max}^{2/3} \quad \text{or} \quad m = 1/3.$$

However, Walker found a better fit for this alloy with $m = 0.5$, while he found $m = 0.425$ for the 7056–T6 alloy.

With $K_{\max} = \Delta K/(1 - R)$ eqn (10) becomes

$$\frac{da}{dn} = f\left[\frac{\Delta K^m}{(1-R)^{1-m}}\right]. \qquad (16)$$

Some similarity with Forman's equation[17] is apparent,

$$\frac{da}{dn} = \frac{C\Delta K^m}{(1-R)K_c - \Delta K}. \qquad (17)$$

Both formulas account for effects of ΔK and R and both fit test data quite well, but it should be realized that the formulas are purely empirical, i.e. no physical arguments were involved in the derivation.

Comparison between the formulas of Elber and Walker

It should be noted from eqn (14) that Walker's effective stress $\overline{\Delta\sigma} \geq \Delta\sigma$ for $R \geq 0$, contrary to Elbers effective stress range $\Delta\sigma_{\text{eff.}} \leq \Delta\sigma$ since crack closure can only reduce $\Delta\sigma$.

Elber started from physical arguments (crack closure → $\Delta K_{\text{eff.}}$, see previous lecture) to account for the R-effect. For 2024–T3 material he then arrived at the empirical relation:

$$U = \frac{\Delta S_{\text{eff.}}}{\Delta S} = \frac{\Delta K_{\text{eff.}}}{\Delta K} = 0.5 + 0.4R.$$

Walker's result for the same material is ($m = 0.5$): $\overline{\Delta K} = \Delta K^{1/2} \cdot K_{\max}^{1/2}$ and with $K_{\max} = \Delta K/(1 - R)$ we obtain:

$$\frac{\overline{\Delta K}}{\Delta K} = \frac{1}{\sqrt{(1-R)}}. \qquad (19)$$

The empirical relations from Elber and Walker are evidently different, but it is interesting to see the quantitative differences for R-ratios between 0 and 0.6 covering most available test data.

	R:	0	0.1	0.2	0.3	0.4	0.5	0.6
Elber:	$\Delta K_{\text{eff}}/\Delta K$	0.50	0.54	0.58	0.62	0.66	0.70	0.74
Walker:	$\overline{\Delta K}/\Delta K$	1	1.05	1.12	1.20	1.29	1.41	1.58
Ratio:	$\Delta K_{\text{eff}}/\overline{\Delta K}$	0.50	0.51	0.52	0.52	0.51	0.49	0.47

The last line indicates that there is an almost constant ratio between $\Delta K_{\text{eff.}}$ and $\overline{\Delta K}$. Consequently a satisfactory correlation between data for different R-values obtained with ΔK_{eff} will automatically imply that a satisfactory correlation will also be obtained with $\overline{\Delta K}$. Different relations for the R-effect are apparently fitting the data equally well and hence the data are not a good instrument to discriminate between different formulas. It may be stressed once again that both R-effect formulas are the result of curve fitting (regression analysis). Some people prefer $\overline{\Delta K}$ and $\overline{\Delta S}$ because it does not have the pretension or the necessity to know something about crack closure, while the formulas are relatively simple.

Asymptotic K-values

Considering the relation between da/dn and ΔK there are two obvious limitations to ΔK. If ΔK becomes too large a static failure will follow immediately, which implies that K_{max} exceeded the fracture toughness K_c, which is equivalent to ΔK exceeding $(1-R)K_c$. The other limitation comes from the fact that cracks will not grow if ΔK is too low. There is a threshold value ΔK_o implying that crack growth requires that $\Delta K > \Delta K_o$. As a result of the two limitations the trend is a sigmoidal relation on a double log plot with two vertical asymptotes (see Fig. 7).

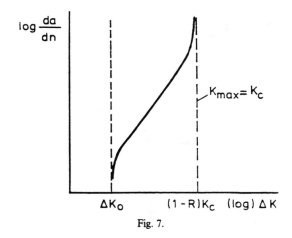

Fig. 7.

For an analytical relation the two limitations imply,

$$\frac{da}{dn} \to 0 \quad \text{if } \Delta K - \Delta K_o \to 0 \tag{20}$$

$$\frac{da}{dn} \to \infty \quad \text{if } K_c - K_{max} \to 0. \tag{21}$$

In analytical relations these limitations are easily satisfied by having $\Delta K - \Delta K_o$ in the numerator and $K_c - K_{max}$ in the denominator. Equation (17) of Forman can also be written as

$$\frac{da}{dn} = \frac{C\Delta K^m}{(1-R)(K_c - K_{max})}. \tag{22}$$

The second limitation has been included although it could be done in a variety of ways, e.g.

$$\frac{da}{dn} = \frac{C\Delta K^m}{(K_c - K_{max})^\beta} \cdot f(R). \tag{23}$$

Apparently Forman selected $\beta = 1$ and $f(R) = 1/(1-R)$, which fitted the data reasonably well. In the Forman equation a lower asymptote is missing. Such an asymptote is not clearly evident from data for Al-alloys, but it seems applicable to carbon steels. The variety of formulas suggested in the literature, including both asymptotes, will be illustrated by two examples. The first one is proposed by Priddle[8]

$$\frac{da}{dn} = C\left[\frac{\Delta K - \Delta K_o}{K_c - K_{max}}\right]^m \tag{24}$$

and a somewhat more complex one by Branco et al.[19]

$$\frac{da}{dn} = C\left[\frac{K_m \cdot \Delta K(\Delta K - \Delta K_o)}{K_c^2 - K_{max}^2}\right]^m. \tag{25}$$

Both formulas clearly satisfy the asymptotic requirements of eqns (20) and (21). The formulas were mainly checked for various types of steel and it was shown that the threshold ΔK_o was depending on R (i.e. on mean stress). A relation proposed is

$$\Delta K_o = A(1 - R)^\gamma \tag{26}$$

with $\gamma = 0.5 - 1.0$ [18–20] depending on the type of steel.

None of the formulas disussed before can claim a physical background. The formulas are adjusted to agree with observed trends, while constants in the formulas are determined by regression analysis. It sometimes require K_c-values different from measured fracture toughness values. The usefullness of the formulas is that they show the effect of the variables involved as they are observed under CA-loading. Secondly they represent a large amount of data in a condensed analytical form, which can be useful in certain computerized life calculation techniques.

5. REFERENCES

[1] J. Schijve, The analysis of random load-time histories with relation to fatigue tests and life calculations. In *Fatigue of Aircraft Structures*, p. 115. Pergamon Press, Oxford (1963).
[2] O. Buxbaum, Statistische Zählverfahren also Bindeglied zwischen Beanspruchungsmessungen und Betriebsfestigkeitversuch. Lab. für Betriebsfestigkeit TR No. TB-64 (Darmstadt 1966).
[3] G. M. van Dijk, Statistical load data processing. *Advanced Approaches to Fatigue Evaluation. NASA SP*-309, 565 (1972).
[4] J. Schijve, Observations on the prediction of fatigue crack propagation under variable-amplitude loading. *Fatigue Crack Growth Under Spectrum Loads, ASTM-STP* **595**, 3 (1976).
[5] J. Schijve, F. A. Jacobs and P. J. Tromp, Fatigue crack growth in aluminium alloy sheet material under flight-simulation loading. Effects of design stress level and loading frequency. *NLR-TR* 72018 (1972).
[6] J. D. Willenborg, R. M. Engle and H. A. Wood, A crack growth retardation model using an effective stress concept. *AFFDL-TM-FBR*-71-1, Air Force Flight Dynamics Lab. (1971).
[7] O. E. Wheeler, Spectrum loading and crack growth. *J. Basic Engng, Trans. ASME*, 181 (1972).
[8] D. Broek and S. H. Smith, *Spectrum Loading Fatigue-Crack-Growth Predictions and Safety-Factor Analysis*. Naval Air Dev. Center NADC-7683-30 (Sep. 1976).
[9] V. W. Trebules, R. Roberts and R. W. Hertzberg, Effect of multiple overloads on fatigue crack propagation in 2024–T3 aluminium alloy. *Progress in Flaw Growth and Fracture Toughness Testing, ASTM-STP* **536**, 115 (1973).
[10] C. C. Poe, Fatigue crack propagation in stiffened panels. *Damage Tolerance in Aircraft Structures, ASTM-STP* **486**, 79 (1971).
[11] H. Tada, P. Paris and G. Irwin, The stress analysis of cracks handbook. Del Research Corp., Hellertown, Pa. (1973).
[12] G. C. Sih, Handbook of stress intensity factors. Lehigh University (1973).
[13] D. P. Rooke and D. J. Cartwright, Stress intensity factors. London, Her Majesty's Stationary Office (1976).
[14] P. Paris and F. Erdogan, A critical analysis of crack propagation laws. *Trans ASME, Series D* **85**, 528 (1963).
[15] E. K. Walker, The effect of stress ratio during crack propagation and fatigue for 2024–T3 and 7075–T6 aluminium. *Effects of Environments and Complex Load History on Fatigue Life. ASTM-STP* **462**, 1 (1970).
[16] D. Broek and J. Schijve, The influence of the mean stress on the propagation of fatigue cracks in aluminium alloy sheets. *NLR TR M*2111, (1963).
[17] R. G. Forman, V. E. Kearney and R. M. Engle, Numerical analysis of crack propagation in cyclic-loaded structures. *J. Basic Engng, Trans. ASME-D* **89**, 459 (1967).
[18] E. K. Priddle, High cycle fatigue crack propagation under random and constant amplitude loadings. *Int. J. Pres. Ves. & Piping*. **4**, 89 (1976).
[19] C. M. Branco, J. C. Radon and L. E. Culver, Growth of fatigue cracks in steel. *Metal Science* **10**, 149 (1976).
[20] A. Ohta and E. Sasaki, Influence of stress ratio on the threshold load for fatigue crack propagation in high strength steels. *Engng Fracture Mech.* **9**, 307 (1977).

(*Received* 18 *April* 1978; *received for publication* 15 *June* 1978)

IV. FATIGUE CRACK GROWTH UNDER VARIABLE-AMPLITUDE LOADING

Abstract—Stationary and non-stationary types of variable-amplitude loading are specified. Some attention is paid to the description of stationary random load. The stress intensity factor is then applied to crack growth under stationary variable-amplitude loading by defining first characteristic stresses and characteristic stress intensity factors. This is done for random loading, non-random loading and flight-simulation loading. It is discussed how and why this concept may break down if the stationarity is lost. Attention is paid to truncation of high loads in a flight-simulation test and the analogous problem of the crest factor under random loading. The significance of crack closure for understanding crack growth under variable-amplitude loading is emphasized.

1. INTRODUCTION

IN THE PREVIOUS lecture we have seen that the K-concept was quite successful in correlating crack growth data obtained under CA-loading (constant-amplitude loading). In the present lecture it will be explored if the K-concept can be useful for VA-loading (variable-amplitude loading) if this type of loading has a stationary character. For this purpose several aspects of VA-loading have to be specified first. Characteristic K-values for both random loading and flight-simulation loading will then be defined and their usefulness will be discussed. If the meaning of characteristic K-values breaks down the significance of interaction effects has to be reconsidered. This leads to a discussion on load spectrum truncation effects and the significance of crack closure during VA-loading.

2. DIFFERENT TYPES OF VARIABLE-AMPLITUDE LOADING

For the purpose of the present discussion VA-loading should be defined in two categories:
1. Stationary VA-loading.
2. Non-stationary VA-loading.

Simple examples of both categories are shown in Fig. 1.

From these illustrations it appears that stationary VA-loading is defined by the following requirement: After a sufficiently small period (return period, recurrence period) the same sequence of load cycles is repeated.

Mathematically in terms of Fourier series the definition is that stationary VA-loading is a periodic type of loading, which can be developed into a Fourier-series with a finite number of terms.

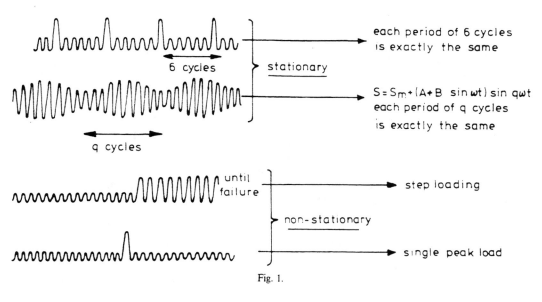

Fig. 1.

However, a stationary VA-loading has a broader meaning if we consider random loading caused by some random process. Examples from the field of aeronautics are: air turbulence causing gusts loads on a wing, and random noise from jet engines, causing sheet bending in stiffened skin panels. From other fields we can refer to loads on ships and off-shore structures, caused by sea waves, and loads on transport vehicles riding on a variety of roads with random roughness. A random process is defined as a stationary random process if its statistical properties are constant, i.e. independent of time.

In various practical cases service loads are a mixture of deterministic loads and random loads. As an example a transport aircraft will see deterministic ground-air-ground transitions (one cycle per flight) and random gust loads (many cycles per flight) (see Fig. 12 later on). As a more general definition of stationary VA-loading we now require that the description of the loads, including both deterministic and random loads, should not vary as a function of time.

3. HOW TO DESCRIBE RANDOM LOAD

Several random processes are supposed to behave as stationary Gaussian random noise, and if it is "Gaussian" interesting input–output calculations for linear systems can be made (e.g. [1]), but this is outside the scope of this lecture. However, some comments on the power spectral density function (PSD) should be made. For Gaussian random noise this function fully describes its statistical properties. It should be considered as a density distribution function of the energy in the frequency domain (Fig. 2). To evaluate this concept a bit further, we will consider a Fourier series with an infinite number of terms with infinitesimally small differences between frequencies of successive terms, and coefficients A being a function of ω, see Fig. 3. In an oscillation the energy is proportional to the square of the amplitude, which implies

$$\Phi(\omega) \div [A(\omega)]^2.$$

The sum of the infinite Fourier series is random noise. Some examples of random noise and corresponding power spectral density functions are shown in Fig. 4, taken from [2].

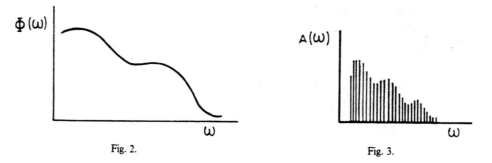

Fig. 2.

Fig. 3.

In Fig. 4(a) the energy is concentrated in a narrow frequency band and as a result the load-time history is somewhat similar to an amplitude modulated signal, in this case with a random modulation. This *narrow band random loading* is typical for resonance systems, which predominantly respond at one single resonance frequency if activated by some external random process over a broader frequency range.

In Fig. 4(c) the spectral density function of the random signal covers a much wider frequency band and the corresponding broad band random loading shows a higher degree of irregularity. White noise is referring to $\Phi(\omega)$ = constant, i.e. the same energy at all frequencies.

Since $\Phi(\omega)$ fully characterizes random load some characteristic properties can be derived from $\Phi(\omega)$[1]. If we consider a stress $\sigma(t)$ varying randomly around a (constant) mean stress σ_m, the r.m.s. of $(\sigma - \sigma_m)$ follows from

$$\sigma_{r.m.s.} \equiv \left[\lim_{T \to \infty} \frac{1}{T} \int_0^T (\sigma - \sigma_m)^2 \, dt \right]^{1/2} = \left[\int_0^\infty \Phi(\omega) \, d\omega \right]^{1/2}. \tag{1}$$

Characteristic events of a random signal are the peaks and the mean-crossings (see Fig. 5). It

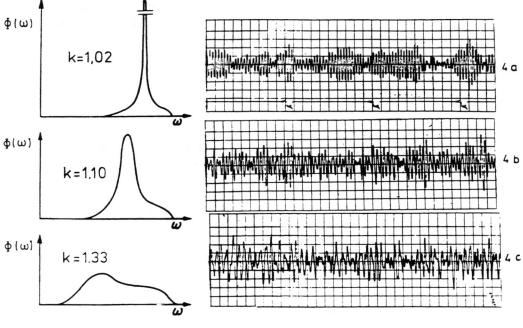

Fig. 4.

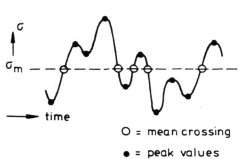

Fig. 5.

will be clear that the number of peaks of a random signal will always be larger than the number of mean-crossings, and this will be more so for more irregular signals. This has led to defining an irregularity factor k as the ratio between the number of peaks and the number of mean-crossings with $k \geq 1$. The relations to $\Phi(\omega)$ are given below.

$$\text{Number of mean-crossings per sec} = N_o = \frac{1}{\pi} \left[\frac{\int_0^\infty \omega^2 \Phi(\omega)\, d\omega}{\int_0^\infty \Phi(\omega)\, d\omega} \right]^{1/2}. \qquad (2)$$
($\nearrow$ and $\searrow$)

$$\text{Number of peaks per sec} = N_1 = \frac{1}{\pi} \left[\frac{\int_0^\infty \omega^4 \Phi(\omega)\, d\omega}{\int_0^\infty \omega^2 \Phi(\omega)\, d\omega} \right]^{1/2}. \qquad (3)$$
(max and min)

$$\text{Irregularity factor } k = \frac{\text{number of peaks}}{\text{number of mean-crossings}} = \left[\frac{\int_0^\infty \Phi(\psi)\, d\omega \int_0^\infty \omega^4 \Phi(\omega)\, d\omega}{\left\{ \int_0^\infty \omega^2 \Phi(\omega)\, d\omega \right\}^2} \right]^{1/2}. \qquad (4)$$

The distribution density function of the peak values can also be derived from $\Phi(\omega)$ but it is a fairly complex equation. For narrow band random loading ($k \sim 1$) it reduces to a simple relation, which is known as the Rayleigh distribution.

With a "normalized" stress

$$\alpha = \frac{\sigma - \sigma_m}{\sigma_{\text{r.m.s.}}} \qquad f(\alpha) = \alpha \cdot e^{-\alpha^2/2}. \tag{5}$$

Figure 6 shows the distribution density function (for maxima only). It confirms for narrow band random loading ($k \sim 1$) that all maxima occur above the mean ($\sigma_{\max}$ always $> \sigma_m$), whereas for more irregularly random loading ($k = 1.4$) $\sigma_{\max}$-values below σ_m do occur.

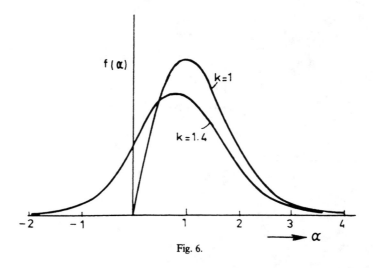

Fig. 6.

Perhaps it should be pointed out that the irregularity factor k gives a good indication of the irregularity of random loading, but it does not fully describe random loading. This can only be done by $\Phi(\omega)$ and theoretically different $\Phi(\omega)$ functions can still give the same k-value. (Note: In the literature some authors define the irregularity factor as $1/k$.)

4. K-CONCEPT AND RANDOM LOAD

If a fatigue crack is growing under a random load, $\sigma(t)$, the stress intensity in the crack tip area will vary in the same random way. We when can define a r.m.s. value of the variation of the stress intensity factor K:

$$K_{\text{r.m.s.}} = C\sigma_{\text{r.m.s.}}\sqrt{(\pi a)}$$

with $\sigma_{\text{r.m.s.}}$ as defined in eqn (1) and C and a as defined in the first lecture. For a certain spectral density function we may then expect that $K_{\text{r.m.s.}}$ will fully describe the random variation of the stress intensity at the tip of the crack.

It should then be expected that the crack rate will be a function of $K_{\text{r.m.s.}}$, or

$$\frac{da}{dn} = f(K_{\text{r.m.s.}}). \tag{7}$$

The relation will depend on the mean stress σ_m. A stress ratio R as defined for CA-loading ($R = \sigma_{\min}/\sigma_{\max}$) cannot be defined in the same way for random loading, but the relative severity of the mean stress is given by

$$\gamma = \frac{\sigma_m}{\sigma_{\text{r.m.s.}}}. \tag{8}$$

Similarity then implies that random stress variations will be similar if both $K_{\text{r.m.s.}}$ and γ are equal, and consequently eqn (7) should read,

$$\frac{da}{dn} = f(K_{r.m.s.}, \gamma) \quad \text{or} \quad \frac{da}{dn} = f_\gamma(K_{r.m.s.}). \tag{9}$$

Limited evidence in the literature has confirmed the applicability of this approach[3–5]. The function f_γ might depend on the PSD function, although some experiments have suggested that this effect will be small[3, 4].

5. K-CONCEPT AND STATIONARY NON-RANDOM LOADING

A simple example as shown in Fig. 7 can be considered as a periodic phenomenon, for which a period and a characteristic stress are easily defined. Similarly to eqns (6)–(9) we now define a characteristic stress intensity factor:

$$K_{char} = C\sigma_{char}\sqrt{(\pi a)} \tag{10}$$

and it may then be expected that,

$$\Delta a/\text{period} = f(K_{char}). \tag{11}$$

Since there is always the same number of cycles in one period eqn (11) can be written as

$$\frac{da}{dn} = f(K_{char}). \tag{12}$$

The larger cycles will delay crack growth during the smaller cycles, and this will affect the function f. However, if crack growth, including delay effects, can still be considered to be a stationary process the similarity approach should still be expected to be valid. Of course the two sequences shown in Fig. 8 will not have the same $f(K_{char})$ since the "wave shapes" are different. It implies that eqn (12) should be more generally written as

$$\frac{da}{dn} = f(K_{char}, \text{wave shape}). \tag{13}$$

Let us now consider an example with a much longer period as shown below (Fig. 9). This sequence is almost CA-loading with periodically inserted peak loads, which subsequently cause noticeable crack growth delay (Fig. 10).

This period is supposed to be sufficiently long (i.e. to continue a sufficiently large number of cycles) in such a way that the delay is over before the next peak occurs. That implies that crack growth on a macro-scale apparently is no longer a stationary process. The size of the discontinuity in the crack growth curve is of a similar magnitude as the plastic zone (r_p) caused by the peak load. It should be expected that eqn (13) is no longer applicable.

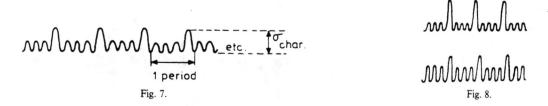

Fig. 7.

Fig. 8.

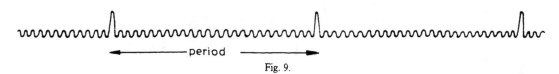

Fig. 9.

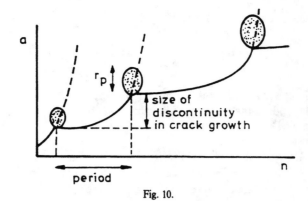

Fig. 10.

If the period is shorter the delays still occur, but the delays are less obvious from the crack growth curve (see Fig. 11) (except perhaps for the first one). The crack growth curve suggests an almost quasi-stationary process, while in reality a continuous superposition of delay effects is causing a most effective growth retardation. The applicability of eqn (13) might be questionable.

Since high peak loads have a large effect on crack growth under VA-loading, the return period of the high loads has a significant meaning for the stationarity of VA-loading. Large return periods will upset the stationarity of VA-loading. This can apply to flight-simulation loading and random loading, as discussed later (truncation of high loads, crest factor).

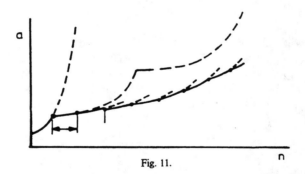

Fig. 11.

6. CRACK GROWTH UNDER FLIGHT-SIMULATION LOADING

A flight-simulation test as applied nowadays to full-scale aircraft structures, components and specimens, consists of a sequence of loads as they occur in service. The ground-air-ground transition is a deterministic load cycle, occurring once per flight. Statistically variable loads will be added in flight (gusts, manoeuvres) or in the ground phase (taxiing loads). The example shown in Fig. 12[6] applies to a civil transport aircraft for which gust loads in flight are predominant, while taxiing loads were omitted since they occurred in compression. Ten different types of flight (A–K) were applied in a random sequence and the gusts in each flight were also applied in a random sequence.

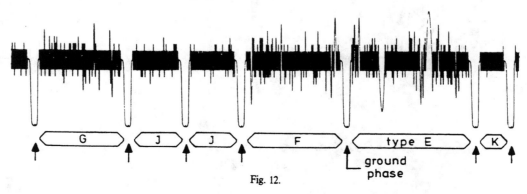

Fig. 12.

The question now is whether such a flight-simulation loading could still be a stationary VA-loading. Let us consider four possibilities of aircraft operation in service.

(a) Always the same flight under the same weather conditions (apparently not in Fig. 12).

(b) Always the same flight under various weather conditions, but the variation of the weather being a stationary random process (Fig. 12).

(c) As (b) but different flights.

(d) Good weather conditions during summer and poor weather conditions during winter time.

In spite of the mixture of deterministic and random loading as it occurs in flight-simulation loading, it still may be expected to be stationary VA-loading, if the statistical properties of all types of loads are independent of time. We then should expect that cases (a) and (b) are examples of stationary VA-loading, while the same could apply to case (c), provided that the variation of the flights has a stationary character. However, case (d) appears to have a systematic return period of 1 yr, which could be too long for crack growth and thus upset the stationarity. Another restriction may be caused by rarely occurring very high loads, as will be discussed later.

We now have to define a characteristic stress for flight-simulation loading. The most obvious choice from a design point of view is the ultimate design stress level. However, any other typical stress related to this level is equally suitable. It is common by now to adopt the mean stress in flight (1 g-condition) as a characteristic stress for flight-simulation loadings. Denoting this stress as σ_{mf} a characteristic stress intensity factor for flight-simulation loading can be defined as:

$$K_{FS} = C\sigma_{mf}\sqrt{(\pi a)}. \tag{14}$$

Both σ_{mf} and K_{FS} can be meaningful only if all other stresses are linearly related to σ_{mf}, and we may ask whether that is a practical situation. Fortunately it is, because:

(1) A change of a local stress level in a structure by a structural modification (effect on nominal stress level, effect on stress concentration) will affect all induced stresses by gusts, manoeuvres, ground-air-ground transitions, etc. in a proportional way.

All stresses will change with the same percentage as σ_{mf}.

(2) If we compare stress levels at two different locations in the same structure all induced stresses will occur in the same ratio. Consequently the ratio of σ_{mf} values will be characteristic.

The effect of changing σ_{mf} on the load spectrum is illustrated by Fig. 13.

It now might be hoped that the crack growth rates for different σ_{mf}-values can be correlated by a single function:

$$\frac{da}{dn} = f(K_{FS}). \tag{15}$$

This was checked in crack growth tests on 2024–T3 and 7075–T6 sheet material[6]. The results of the former alloy are plotted in Fig. 14. If eqn (15) were applicable the five curves in the right

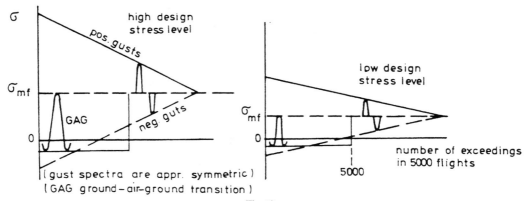

Fig. 13.

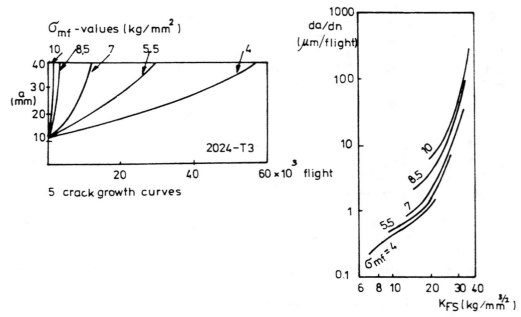

Fig. 14.

hand graph should have merged into a single scatter band. Unfortunately they do not, and this is unfortunate indeed, because otherwise eqn (15) would allow to extrapolate results obtained for a certain design stress level to another stress level. This would be most helpful in a design office in studies on the effect of design stress levels.

If $da/dn = f(K_{r.m.s.})$ (eqn 7) is still valid to random load, why does eqn (15) not apply to flight-simulation loading? It should be expected that an insufficient stationary character of flight-simulation loading will be significant here. This will be discussed in the following section.

7. TRUNCATION OF HIGH LOADS

A "load spectrum" as it occurs in service is frequently presented as an exceedance curve, i.e. the number of times that load levels are exceeded in a specified time interval. As a qualitative example Fig. 15 shows how many times stress amplitude levels are exceeded in a wing structure during 5000 flights. Such a curve is closely related to a distribution function. It requires a good deal of effort to arrive as such a curve. First the anticipated utilisation of the aircraft has to be analysed. Second, gust load statistics for all relevant flying conditions have to be compiled. Finally the gust loads have to be converted into stress levels in the structure.

A weak point in the gust load statistics is the upper end of the exceedance curve, associated with the very severe gust loads (stormy weather), having a very low propability of occurrence. In this area gust load statistics are less reliable. Secondly, we cannot be sure whether all aircraft

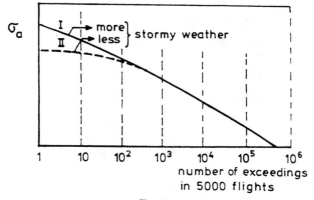

Fig. 15.

will meet the same severe storms. It is even more likely that they will not, which implies that the upper end of the load spectrum will vary from aircraft to aircraft. Unfortunately the very high loads can have a predominant effect on fatigue life and crack growth. Fatigue critical components of an aircraft are carrying a positive mean stress in flight. It implies that the extreme stress levels induced by a symmetric gust load spectrum will be larger in tension than in compression (see Fig. 16), or

$$(\sigma_{\text{max, upwards}}) > |\sigma_{\text{max, downwards}}|. \tag{16}$$

Very high loads will cause local yielding at notches and in view of eqn (16) the probability of having compressive residual stresses (which are favourable) are much better than for tensile residual stresses (which are unfavourable). It implies that load Spectrum I in Fig. 15 (higher maximum loads) will give a larger life than load Spectrum II (lower maximum load). The difference can even be larger for crack growth, since negative loads will close the crack and the crack will not be a stress raiser under high downward loads (see also the 2nd lecture).

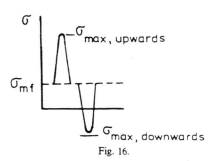

Fig. 16.

As a consequence flight-simulation tests will produce more conservative results (i.e. lower fatigue lives, faster crack growth) if Spectrum II is adopted, although Spectrum I may be the best estimate of the "average" load spectrum. Spectrum II implies that the rarely occurring high load amplitudes of Spectrum I are truncated (not omitted!) to a lower level. Comparative fatigue tests with flight-simulation loading have amply confirmed the predominant effect of truncating high loads on fatigue life of notched specimens (e.g. [7–9]) and on crack propagation (e.g. [7, 10]). As an illustration Fig. 17 shows results for three truncation levels. It will be noted that a higher truncation level implies longer crack initiation periods and longer crack growth periods.

In another test program [11] with a somewhat more severe gust spectrum the truncation level was selected at a level occurring in 5 out of 4000 flights only (flights of type A, B and C). An example of average crack growth curves obtained is presented in Fig. 18. All three curves show growth delays caused by the severe flights. Obviously crack growth was not a stationary process in all three environments, although it almost appears to be so in dry air at a later stage (compare Fig. 18 with Figs. 10 and 11). The more severe flights upset the stationary character of crack growth and this may well explain why eqn (15) was unable to correlate data obtained at different σ_{mf}-values (see Fig. 14). Actually the invalidity of eqn (15) can also be interpreted to be a result of a dK/da-effect discussed in the 3rd lecture. This effect was expected as a result of the preceding K-history. Fortunately it was absent under CA-loading. However, crack growth delays as observed under flight simulation loading are clear manifestations of history effects.

8. CRACK CLOSURE AND NON-STATIONARY VARIABLE-AMPLITUDE LOADING

From the previous section we have learned that apparently stationary flight-simulation loading was insufficiently stationary due to rarely occurring high loads. A more simple example of unstationary VA-loading was discussed before with reference to Figs. 10 and 11. Under such conditions a prediction of crack growth by correlation, employing a characteristic K-value

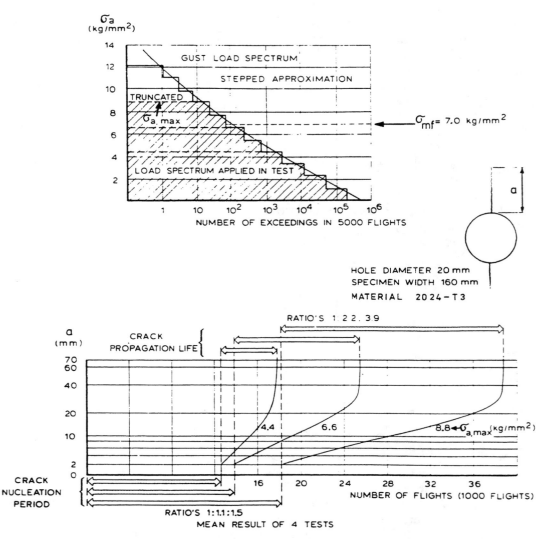

Fig. 17. (Ref. [7]).

breaks down. We then have to return to the equation discussed in the third lecture:

$$a_N = a_o + \sum_{i=1}^{N} \Delta a_i. \qquad (17)$$

It was pointed out in that lecture that crack growth could not be such a simple additive process as assumed in eqn (17), because Δa_i will depend on the fatigue damage caused by the preceding load history (size of crack, mode of crack, crack closure, tip blunting, etc.). It implies that Δa_i will differ from the expected value (da/dn) based on the magnitude of the cycle and test data obtained under CA-loading. Effects causing such deviations are referred to as *interaction effects*, which can be either positive or negative. For a positive interaction effect Δa is smaller than expected (retardation) and for a negative interaction effect Δa is larger than expected (accelerated growth). For certain load sequences retardations are easily observed as crack growth delays. Acceleration effects are not easily observed, but they have been indicated from striation observations.

Various mechanisms can contribute to interaction effects and they probably will do so. However, it seems that crack closure could be a predominant one. This was already proposed by Elber[12] for a test with step loading. The principle how it could introduce an interaction effect is illustrated qualitatively in Fig. 19 by indicating how the crack closure level (σ_{cl}) changes from one stable level to another one. During the transition period ΔK_{eff} will differ from

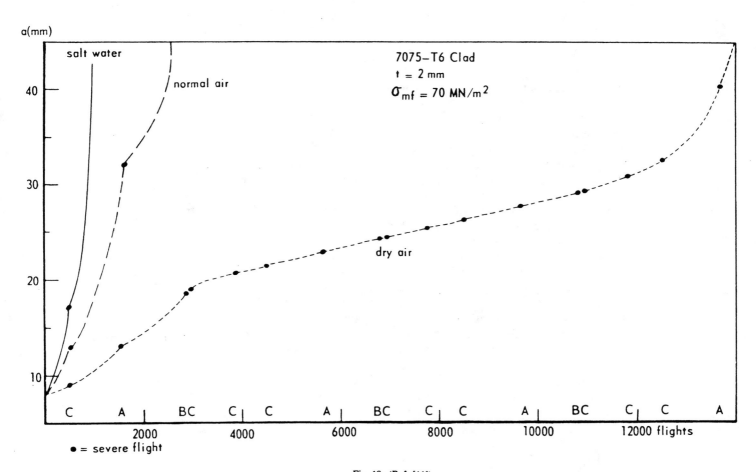

Fig. 18. (Ref. [11]).

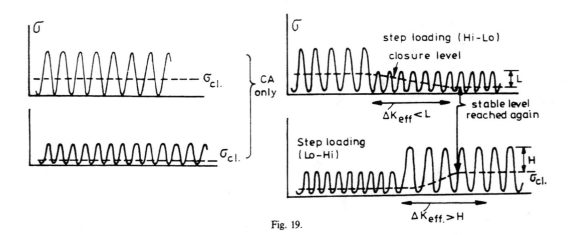

Fig. 19.

the expected value observed under *CA*-loading, and as a result crack growth will be slower (Hi–Lo case) or faster (Lo–Hi case).

A similar transient behavior can occur after a peak load, see Fig. 20. The peak load itself will reduce σ_{cl} but when the crack grows into the plastic zone created by the peak load it will increase beyond the original level, while it will return to the original stable level at a later time. As a consequence the peak load will be followed by an acceleration first, but later on the retardation will take over. This was already discussed in the second lecture with reference to Fig. 19 of that lecture. If the peak load is relatively high σ_{cl} can increase beyond σ_{max} and crack growth will stop (lower part of Fig. 20). It was shown that a Limit Load applied to a wing structure successfully stopped the growth of fatigue cracks[7].

Some empirical evidence of the transitional behavior of σ_{cl} and the associated effects on crack growth was discussed in the 2nd lecture. Further, crack closure and the transitional behavior during step loading (Fig. 19) also follow from elasto–plastic calculations employing finite-element techniques[14, 15]. Such calculations are fairly expensive, but it is highly instructive to know that crack closure considerations can be checked by calculations also. Beyond any doubt a physically sound theory on crack growth under *VA*-loading should include crack closure. However, it should be kept in mind that interaction effects may stem from other sources as well. This should be clear from the discussion in the two previous lectures. The significance of other sources can be explored in experiments, provided that crack closure is measured also. It certainly will remain a difficult issue to separate the contribution of all possible mechanisms. Fortunately crack closure is a mechanism that can be observed experi-

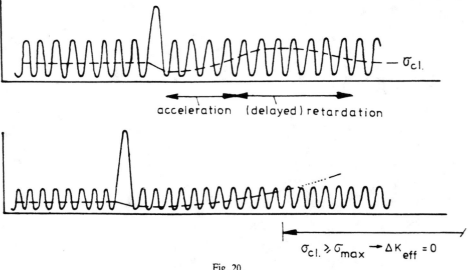

Fig. 20.

mentally, and it would be stimulating to see that crack closure has a predominant effect during VA-loading with service load-time histories, which might justify to ignore other interaction mechanisms.

Some exploratory COD-measurements during flight-simulation loading have clearly confirmed the occurrence of crack closure[8]. An example of measurements during a severe flight is shown in Fig. 21. After each severe upward gust cycle the position of the recorder pen was given a small horizontal shift to avoid coinciding σ-COD records. The non-linear behavior can be observed easily and the arrows indicate the crack closure stress.

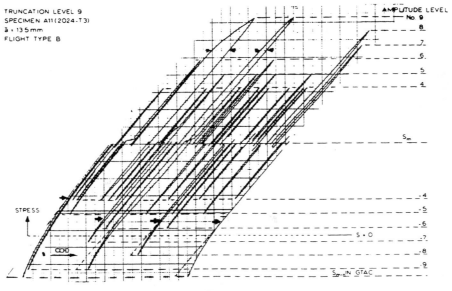

Fig. 21.

After the first maximum load $\sigma_{cl.}$ was reduced to a lower level, in agreement with the results of more simple tests (e.g. Fig. 19 of the 2nd lecture). After some crack growth in subsequent flights the crack closure level increased again as should be expected. More systematic measurements are now carried out at the University in Delft, including the effect of truncation level.

Crack closure measurements during random load test were recently carried out by Elber[16]. The variation of $\sigma_{cl.}$ during his tests on 7075–T6 sheet specimens was fairly small (see Fig. 22) and he therefore proposed to adopt the average value of a test to define $\Delta K_{eff.}$-values for all individual cycles. Employing these values and a linear damage summation with $a = a_o + \Sigma \Delta a$ (eqn 17) and $\Delta a = da/dn = f(\Delta K_{eff.})$ he could make a comparison between prediction and test results. A reasonable agreement was found, but the number of tests was still fairly low. Moreover the return period (~ 1500 cycles) was small. Anyhow, it seems worthwhile to study this average $\Delta K_{eff.}$-approach in more detail in the future.

One aspect that should be studied is the significance of crack closure in explaining the truncation effect mentioned in the previous section. Higher truncation levels will give larger plastic zones and possibly more crack closure in flight-simulation tests. The same problem can arise under pure random loading. For the larger load amplitudes the distribution density function is approximately given by [17],

$$f(\alpha) = \frac{\alpha}{k} e^{-\alpha^2/2}$$

(compare to eqn 5) and the distribution function by

$$F(\alpha) = \int_\alpha^\infty f(\alpha)\, d\alpha = \frac{1}{k} e^{-\alpha^2/2} \qquad (19)$$

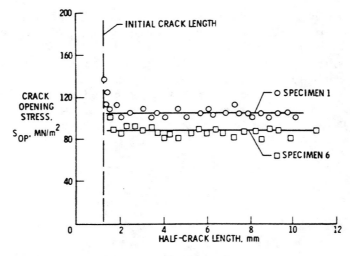

Fig. 22.

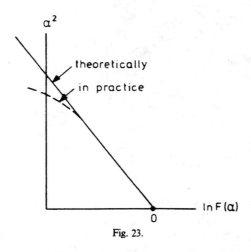

Fig. 23.

or

$$\ln F(\alpha) = -\ln k - \frac{\alpha^2}{2}. \tag{20}$$

Plotting α^2 as a function of $\ln F(\alpha)$ a linear relationship should be obtained. In practice deviations are found for larger α-values, which implies that the load spectrum is truncated (see Fig. 23). The deviations are due to physical restraints, such as damping and non-linear behavior at high α-values. Equation (19) would theoretically allow an in finite amplitude, which is obviously impossible. It is usual to specify the physical restraint by the so-called crest factor (or clipping ratio), which is defined as:

$$\text{crest factor} = \frac{\text{maximum amplitude}}{\text{r.m.s.-value}}.$$

For the normalized stress amplitude α the crest factor would simply be $\alpha_{\max}$. The magnitude of the crest factor will depend on the dynamic response of the vibrating system. Since the crest factor is representative for the truncation of random load, more should be known about its effect on fatigue. However, this problem hardly received any attention in the literature. Several investigators reporting results of random load fatigue tests even do not specify the value of the crest factor in their tests. An exploratory test program seems to be worthwhile.

9. REFERENCES

[1] J. S. Bendat, *Principles and Applications of Random Noise Theory*. Wiley, New York (1958).
[2] J. Kowalewski, On the relation between fatigue lives under random loading and under corresponding program loading. Full-Scale Fatigue Testing of Aircraft Structures, p. 60. Pergamon Press, Oxford (1961).
[3] P. C. Paris, The fracture mechanics approach to fatigue. Fatigue, an Interdisciplinary Approach, Syracuse University Press, p. 107 (1964).
[4] S. H. Smith, Fatigue crack growth under axial narrow and broad band random loading. Acoustical Fatigue in Aerospace Structures, Syracuse University Press, p. 331 (1965).
[5] S. R. Swanson, F. Cicci and W. Hoppe, Crack propagation in Clad 7079–$T6$ aluminium alloy sheet under constant and random amplitude fatigue loading. *ASM STP* **415**, 312 (1967).
[6] J. Schijve, F. A. Jacobs and P. J. Tromp, Fatigue crack growth in aluminium alloy sheet material under flight-simulation loading. Effects of design stress level and loading frequency. NLR TR 72018, Amsterdam (1972).
[7] J. Schijve, The accumulation of fatigue damage in aircraft materials and structures. *AGARDograph No.* 157 (1972).
[8] J. Schijve, F. A. Jacobs and P. J. Tromp, Flight simulation tests on notched elements. *NLR TR* 74033, Amsterdam (1974).
[9] D. Schütz and H. Lowak, Application of design rules for fatigue life estimates based on service simulation test results (in German). LBF Bericht FB-109, Darmstadt (1976).
[10] W. Schütz and H. Zenner, Crack growth in titanium alloy sheets under flight-simulation loading (in German). IABG *Rep.* TF-224, Ottobrunn (1972).
[11] J. Schijve, F. A. Jacobs and P. J. Tromp, Environmental effects on crack growth in flight-simulation tests on 2024–$T3$ and 7075–$T6$ material. *NLR TR* 76104, Amsterdam (1976).
[12] W. Elber, The significance of crack closure. *ASTM STP* **486**, 230 (1971).
[13] J. Schijve, Observations on the prediction of fatigue crack propagation under variable-amplitude loading. *ASTM STP* **595**, 3 (1976).
[14] K. Ohji, K. Ogura and Y. Ohkuba, Cyclic analysis of a propagating crack and its correlation with fatigue crack growth. *Engng Fracture Mech.* **7**, 457 (1975).
[15] J. C. Newman, A finite-element analysis of fatigue crack closure. *ASTM STP* **590**, 281 (1976).
[16] W. Elber, Equivalent constant-amplitude concept for crack growth under spectrum loading. *ASTM STP* **595**, 236 (1976).
[17] J. Schijve, Some notes on random load and fatigue (in Dutch) *NLR TM M*2071 (1960).

(*Received* 18 *April* 1978; *received for publication* 15 *June* 1978)

Author Index

Anderson, W.E., *539*
Barenblatt, G.I., *285*
Bueckner, H.F., *152, 424*
Carlsson, A.J., *379*
Dugdale, D.S., *280*
Eftis, J., *363, 403*
Elber, Wolf, *564, 574*
Engle, R.M., *552*
Erdogan, F., *545*
Felbeck, D.K., *146*
Forman, R.G., *552*
Forsyth, P.J.E., *465*
Gilman, John J., *105*
Gomez, M.P., *539*
Green, A.E., *55*
Griffith, A.A., *60, 96*
Inglis, C.E., *3*
Irwin, G.R., *116, 136, 167, 171, 175, 215, 267*

Kearney, V.E., *552*
Kies, J.A., *136*
Laird, Campbell, *484*
Larsson, S.G., *379*
Lee, J.D., *403*
Liebowitz, H., *363, 403*
Orowan, E., *142, 146*
Paris, P.C., *539, 545*
Rice, James R., *393, 442*
Sanford, R.J., *373*
Schijve, J., *587*
Sih, G.C., *360*
Sneddon, I.N., *23, 55*
Wells, A.A., *215*
Westergaard, H.M., *18*
Wheeler, O.E., *558*
Williams, M.L., *161*
Wood, W.A., *453*

Subject Index

Bauschinger effect, *587*
bi-axial, *403*
branching, *161, 175, 215*
brittle fracture, *3, 60, 96, 167, 175, 215, 285*

cleavage, *96, 105, 116, 136, 142, 146, 175, 215, 393, 465*
crack opening displacement (COD), *215, 574, 587*
cohesion, *285*
compliance, *175, 442, 564*
crack
    co-linear, *175, 285*
    edge, *3, 161, 285, 424*
    elliptical, *171, 215*
    internal, *3, 18, 60, 167, 175, 285, 403*
    nucleation, *484, 587*
    periodic, *18, 167, 175, 363*
    point loads, *18, 167, 175*
    pressurized, *18, 23, 55, 175, 285, 442*
    profile, *23, 55, 116, 161, 167, 171, 175, 285, 379, 403*
    surface, *60, 171, 215, 558*
    three-dimensional, *23, 55, 171, 175, 285, 442*
crack arrest, *215*
crack extension force, *167, 175, 215*
crystals, *60, 96, 105, 175*

dislocation, *453, 465, 484*
distortional strain energy (Von Mises), *161*
ductile fracture, *116, 136, 142, 146, 215*
dynamic fracture, *116, 136, 175, 215, 285*

electrical potential, *587*

environment, *60, 465, 484, 587*
extrusion, *453, 465, 484*

failure criteria
    equilibrium, *280, 285*
    Griffith, *23, 60, 96, 116, 142, 146, 152, 175*
    Irwin-Orowan, *23, 116, 142, 146, 152, 175, 215, 539*
    plastic strain, *215*
    strain energy, *116, 136, 142, 215*
    strain energy density, *403*
fatigue
    closure, *564, 574, 587*
    corrosion, *465*
    crack growth, *465, 539, 545, 558, 587*
    damage, *484, 587*
    high strain, *484*
    initiation, *175, 453, 465, 484*
    life, *539, 552, 558, 587*
    mechanisms, *136, 215, 453, 465, 484, 564, 587*
    models, *215, 539, 545, 552, 558, 587*
    retardation, *465, 558, 587*
    stage I, *453, 465, 484*
    stage II, *465, 484, 539, 545, 558*
finite element analysis, *379, 403*
fractography, *175, 215, 453, 465, 484, 564*
fracture process zone, *215*
fracture testing, *105, 116, 136, 146, 175, 215, 280, 379*
free edge correction factor, *171*

glass, *60, 96, 175, 215, 285*
grain boundary, *465, 484*

holes
    cracked, *3, 285*
    elliptical, *3, 60, 96*

intrusion, *453, 465, 484*

J-integral, *393, 403*

loading
    constant amplitude, *539, 545, 552, 587*
    spectrum, *558, 587*
    random, *587*
    remote, *18, 55*
    variable amplitude, *558, 564, 574, 587*

metals, *60, 105, 116, 136, 142, 175, 215, 465, 484, 552, 574*
microcrack, *60, 96, 587*
mode
    anti-plane shear, *175, 215*
    forward shear, *161, 175, 215, 360*
    opening, *161, 175, 215, 360*

near-field equations, *161, 167, 175, 215, 363, 393, 403, 539*
Neuber, *175*
notch bar testing, *175, 215*

photoelasticity, *23, 161, 175, 373, 403, 587*
plane strain constraint, *161, 167, 171, 175, 215, 267*
plastic strain, *453, 465, 484, 564, 574*
plastic zone, *3, 142, 161, 171, 175, 215, 267, 280, 379, 393, 465, 545, 558, 587*
potential energy, *23, 60, 96, 285*
probability of failure, *175, 215*

shear lip, *587*
size effect, *175, 215, 267, 587*

slip, *453, 465, 484, 587*
slipbands, *453, 465, 484*
small scale yielding, *142, 167, 215, 267, 285, 379, 393, 539, 587*
strain energy, *60, 96, 116, 136, 142, 152, 215, 442*
strain energy density, *424*
strain energy release rate G, *167, 171, 175*
strain gage, *167, 587*
strain hardening, *465, 484, 545*
stress
    crack closure, *564, 574*
    effective, *574, 587*
    mean, *587*
    nonsingular, *161, 167, 175, 360, 363, 373, 379, 393, 403*
    residual, *465, 564, 574, 587*
stress function
    complex variables, *55, 285, 360, 363, 373, 424*
    real, *3, 23, 161, 403*
    Westergaard, *18, 23, 167, 175, 360, 363, 373*
stress intensity factor K, *167, 171, 215, 379, 424, 442, 539, 545, 587*
stress ratio, *539, 552, 587*
striation, *453, 465, 484, 564*
strip zone, *215, 280*
subcritical crack growth, *285, 587*
surface roughness, *587*
surface tension (energy), *23, 60, 96, 105, 116, 142, 146, 175, 285*

temperature, *424, 552*

ultrasonic, *587*

void coalescence, *175*

wedge, *285*
weight function, *424, 442*